BONUS S .... OR YOU!

Wolters Kluwer offers motion analysis software specially formulated for this textbook!

WITHDRAWN

Visit **http://thepoint.lww.com/Hamill4e** to access Innovision Systems' MaxTRAQ software specially formulated for *Biomechanical Basis of Human Movement, Fourth Edition.* This downloadable motion analysis software offers you an easy-to-use tool to track data and analyze various motions selected by the authors.

WITHDRAWN

The MaxTRAQ software includes:

- Video clips of a variety of motions such as golf swing and gait
- 2D Manual Tracking
- Coverage of distance and angles

Student Login Instructions:

- Go to **http://thepoint.lww.com/Hamill4e** and click on Student Resources
- Click on MaxTRAQ software link
- Download and install the MaxTRAQ software
- Run the MaxTRAQ software
- Enter your MaxTRAQ access code (to the right) when prompted

 Wolters Kluwer

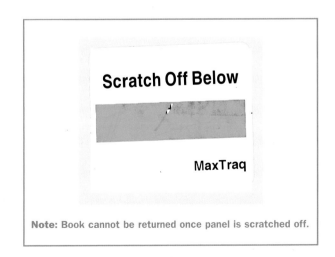

Scratch Off Below

MaxTraq

Note: Book cannot be returned once panel is scratched off.

Biomechanical Basis
of Human
Movement

4th
EDITION

Biomechanical Basis
of Human
Movement

4th
EDITION

LIS - LIBRARY

Date	Fund
19.2.15	S-che

Order No.

2576417

University of Chester

Joseph Hamill, PhD

Professor, Department of Kinesiology
University of Massachusetts at Amherst
Amherst, Massachusetts

Kathleen M. Knutzen, PhD

Professor, Department of Physical Education and Kinesiology
Dean, School of Social Sciences and Education
California State University
Bakersfield, California

Timothy R. Derrick, PhD

Professor, Department of Kinesiology
Iowa State University
Ames, Iowa

Wolters Kluwer

Health

Philadelphia • Baltimore • New York • London
Buenos Aires • Hong Kong • Sydney • Tokyo

Acquisitions Editor: Emily Lupash
Managing Editor: Kristin Royer, Staci Wolfson
Marketing Manager: Shauna Kelley
Production Project Manager: Marian Bellus
Designer: Stephen Druding
Compositor: Integra

Fourth Edition

Copyright © 2015, 2009, 2003, 1995 Lippincott Williams & Wilkins, a Wolters Kluwer business.

Two Commerce Square	351 West Camden Street
2001 Market Street	Baltimore, MD 21201
Philadelphia, PA 19103	

Printed in China

Not authorised for sale in United States, Canada, Australia, New Zealand, Puerto Rico, and U.S. Virgin Islands.

All rights reserved. This book is protected by copyright. No part of this book may be reproduced or transmitted in any form or by any means, including as photocopies or scanned- in or other electronic copies, or utilized by any information storage and retrieval system without written permission from the copyright owner, except for brief quotations embodied in critical articles and reviews. Materials appearing in this book prepared by individuals as part of their official duties as U.S. government employees are not covered by the above- mentioned copyright. To request permission, please contact Lippincott Williams & Wilkins at 530 Walnut Street, Philadelphia, PA 19106, via email at permissions@lww.com, or via website at lww.com (products and services).
9 8 7 6 5 4 3 2 1

Library of Congress Cataloging-in-Publication Data

Hamill, Joseph, 1946-author.
  Biomechanical basis of human movement/Joseph Hamill, Kathleen M. Knutzen, Timothy R. Derrick.—Fourth edition.
    p. ; cm.
  Human movement
  Includes bibliographical references and index.
  ISBN 978-1-4511-7730-5
  I. Knutzen, Kathleen, author.   II. Derrick, Timothy R., author.   III. Title.   IV. Title: Human movement.
  [DNLM: 1. Movement.   2. Biomechanics. WE 103]
  QP303
  612.7'6—dc23
                                                                    2013044474

Care has been taken to confirm the accuracy of the information present and to describe generally accepted practices. However, the authors, editors, and publisher are not responsible for errors or omissions or for any consequences from application of the information in this book and make no warranty, expressed or implied, with respect to the currency, completeness, or accuracy of the contents of the publication. Application of this information in a particular situation remains the professional responsibility of the practitioner; the clinical treatments described and recommended may not be considered absolute and universal recommendations.

The authors, editors, and publisher have exerted every effort to ensure that drug selection and dosage set forth in this text are in accordance with the current recommendations and practice at the time of publication. However, in view of ongoing research, changes in government regulations, and the constant flow of information relating to drug therapy and drug reactions, the reader is urged to check the package insert for each drug for any change in indications and dosage and for added warnings and precautions. This is particularly important when the recommended agent is a new or infrequently employed drug.

Some drugs and medical devices presented in this publication have Food and Drug Administration (FDA) clearance for limited use in restricted research settings. It is the responsibility of the health care provider to ascertain the FDA status of each drug or device planned for use in their clinical practice.

To purchase additional copies of this book, call our customer service department at (800) 638-3030 or fax orders to (301) 223-2320. International customers should call (301) 223- 2300.

  Visit Lippincott Williams & Wilkins on the Internet: http://www.lww.com. Lippincott Williams & Wilkins customer service representatives are available from 8:30 am to 6:00 pm, EST.

CCS0714

TO OUR FRIEND AND MENTOR B.T. BATES,
AND TO OUR FAMILIES.

Preface

Biomechanics is a quantitative field of study within the discipline of exercise science. This book is intended as an introductory textbook that stresses this quantitative (rather than qualitative) nature of biomechanics. It is hoped that, while stressing the quantification of human movement, this fourth edition of *Biomechanical Basis of Human Movement* will also acknowledge those with a limited background in mathematics. The quantitative examples are presented in a detailed, logical manner that highlight topics of interest. The goal of this book, therefore, is to provide an introductory text in biomechanics that integrates basic anatomy, physics, calculus, and physiology for the study of human movement. We decided to use this approach because numerical examples are meaningful and easily clear up misconceptions concerning the mechanics of human movement.

Organization

This book is organized into three major sections: Part I: Foundations of Human Movement; Part II: Functional Anatomy; and Part III: Mechanical Analysis of Human Motion. The chapters are ordered to provide a logical progression of material essential toward the understanding of biomechanics and the study of human movement.

Part I, Foundations of Human Movement, includes Chapters 1 through 4. Chapter 1, "Basic Terminology," presents the terminology and nomenclature generally used in biomechanics. Chapter 2, "Skeletal Considerations for Movement," covers the skeletal system with particular emphasis on joint articulation. Chapter 3, "Muscular Considerations for Movement," discusses the organization of the muscular system. Finally, in Chapter 4, "Neurologic Considerations for Movement," the control and activation systems for human movement are presented. In this edition, some of the foundation material was reorganized and new material was added in areas such as physical activity and bone formation, osteoarthritis, osteoporosis, factors influencing force and velocity development in the muscles, and the effect of training on muscle activation.

Part II, Functional Anatomy, includes Chapters 5 through 7 and discusses specific regions of the body: the upper extremity, lower extremity, and trunk, respectively. Each chapter integrates the general information presented in Part I relative to each region. In this edition, the information on muscles and ligaments was moved from the appendix into the chapter text to facilitate review of muscle and ligament locations and actions. The exercise section was reorganized to provide samples of common exercises used for each region. Finally, the analysis of selected activities at the end of each chapter includes a more comprehensive muscular analysis based on the results of electromyographic studies.

Part III, Mechanical Analysis of Human Motion, includes Chapters 8 through 11, in which quantitative mechanical techniques for the analyses of human movement are presented. Chapters 8 and 9 present the concepts of linear and angular kinematics. Conventions for the study of linear and angular motion in the analysis of human movement are also detailed in these two chapters. A portion of each chapter is devoted to a review of the research literature on human locomotion, wheelchair locomotion, and golf. These activities are used throughout Part III to illustrate the quantitative techniques presented. Chapters 10 and 11 present the concepts of linear and angular kinetics, including discussions on the forces and torques that act on the human body during daily activities. The laws of motion are provided and explained. Included here is a discussion of the forces and torques applied to the segments of the body during motion.

Although the book follows a progressive order, the major sections are generally self-contained. Therefore, instructors may delete or deemphasize certain sections. Parts I and II, for example, could be used in a traditional kinesiology course, and Part III could be used for a biomechanics course.

Features

Each chapter contains a list of **Chapter Objectives** to enable the student to focus on key points in the material, and **Chapter Outlines** provide a guide to the content discussed. **Boxes** are included throughout to highlight important material, and relevant **Questions** are pulled out to help the student briefly review a concept. **Chapter Summaries** at the end of each chapter recap the major concepts presented. Each chapter contains **Review Questions**, both true/false and multiple choice, to challenge students and help them digest and integrate the material presented. A **Glossary** is presented at the end of this book, defining terms found in each chapter and to be used as a source of reinforcement and reference. Finally, four appendices present information on units of measurement, trigonometric functions, and hands-on data.

Illustrations of the principles of human movement are easily seen in most sports examples, but in this edition of *Biomechanical Basis of Human Movement*, new and updated illustrations include applications from ergonomics, orthopedics, and exercise. These are supplemented with references from the current biomechanics literature. With these and the content and features mentioned above, the full continuum of human movement potential is considered.

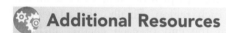 **Additional Resources**

Biomechanical Basis of Human Movement, Fourth Edition, includes additional resources for both instructors and students that are available on the book's companion web site at http://thePoint.lww.com/hamill4e.

INSTRUCTORS

Approved adopting instructors will be given access to the following additional resources:

- Brownstone test generator
- PowerPoint presentations
- Image bank
- WebCT and Blackboard Ready Cartridges

STUDENTS

Students who have purchased the text have access to the following additional resources:

- Answers to the review questions in the text
- Student practice quizzes
- MaxTRAQ motion analysis software, accessible via hyperlink from thePoint

See the inside front cover of this text for more details, including the passcode you will need to gain access to the website.

Acknowledgments

To those who reviewed this edition of the book and who made a substantial contribution to its development, we express our sincere appreciation. We also thank Kristin Royer (product manager), Emily Lupash (acquisitions editor), and Shauna Kelley (marketing manager) of Wolters Kluwer Health/Lippincott Williams & Wilkins for their expertise throughout the publishing process. A special thanks to Nic Castona and Nike, Inc., for the photography used throughout.

Contents

Preface vii

Acknowledgements ix

SECTION I	Foundations of Human Movement	1
1	Basic Terminology	3
2	Skeletal Considerations for Movement	25
3	Muscular Considerations for Movement	59
4	Neurologic Considerations for Movement	99

SECTION II	Functional Anatomy	129
5	Functional Anatomy of the Upper Extremity	131
6	Functional Anatomy of the Lower Extremity	172
7	Functional Anatomy of the Trunk	241

SECTION III	Mechanical Analysis of Human Motion	281
8	Linear Kinematics	283
9	Angular Kinematics	318
10	Linear Kinetics	346
11	Angular Kinetics	391

APPENDIX A	The Metric System and SI Units	441
APPENDIX B	Trigonometric Functions	445
APPENDIX C	Sample Kinematic and Kinetic Data	449
APPENDIX D	Numerical Example for Calculating Projectile	457

Glossary 459

Index 473

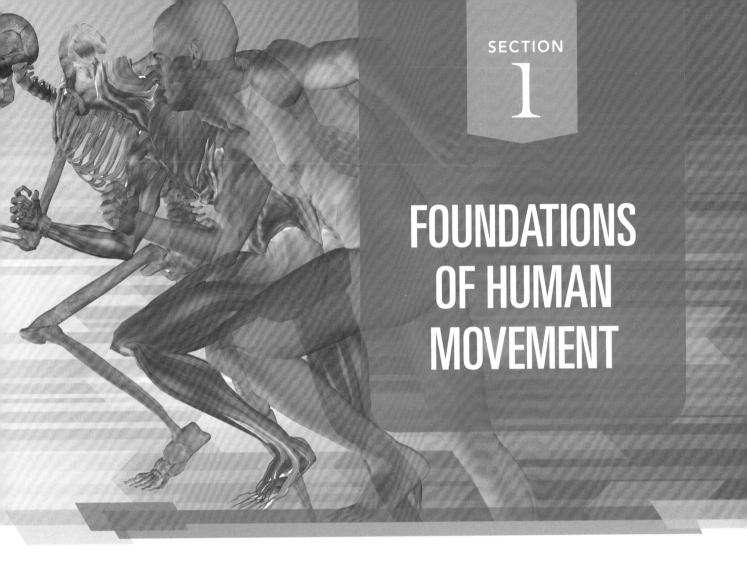

FOUNDATIONS OF HUMAN MOVEMENT

CHAPTER 1
Basic Terminology

CHAPTER 2
Skeletal Considerations for Movement

CHAPTER 3
Muscular Considerations for Movement

CHAPTER 4
Neurologic Considerations for Movement

BASIC TERMINOLOGY

OBJECTIVES

After reading this chapter, the student will be able to:

1. Define mechanics, biomechanics, and kinesiology, and differentiate among their uses in the analysis of human movement.

2. Define and provide examples of linear and angular motion.

3. Define kinematics and kinetics.

4. Describe the location of segments or landmarks using correct anatomical terms, such as medial, lateral, proximal, and distal.

5. Identify segments by their correct names, define all segmental movement descriptors, and provide specific examples in the body.

6. Explain the difference between relative and absolute reference systems.

7. Define sagittal, frontal, and transverse planes along with corresponding frontal, sagittal, and longitudinal axes. Provide examples of human movements that occur in each plane.

8. Explain degree of freedom, and provide examples of degrees of freedom associated with numerous joints in the body.

OUTLINE

Core Areas of Study
    Biomechanics versus Kinesiology
    Anatomy versus Functional Anatomy
    Linear versus Angular Motion
    Kinematics versus Kinetics
    Statics versus Dynamics
Anatomical Movement Descriptors
    Segment Names
    Anatomical Terms
    Movement Description

Reference Systems
    Relative versus Absolute
    Planes and Axes
Summary
Review Questions

To study kinesiology and biomechanics using this textbook requires a fresh mind. Remember that human movement is the theme and the focus of study in both disciplines. A thorough understanding of various aspects of human movement may facilitate better teaching, successful coaching, more observant therapy, knowledgeable exercise prescription, and new research ideas. Movement is the means by which we interact with our environment, whether we are simply taking a walk in a park, strengthening muscles in a bench press, competing in the high jump at a collegiate track meet, or stretching or rehabilitating an injured joint. Movement, or motion, involves a change in place, position, or posture relative to some point in the environment.

This textbook focuses on developing knowledge in the area of human movement in such a manner that you will feel comfortable observing human movement and solving movement problems. Many approaches can be taken to the study of movement, such as observing movement using only the human eye or collecting data on movement parameters using laboratory equipment. Observers of activities also have different concerns: A coach may be interested in the outcome of a tennis serve, but a therapist may be interested in identifying where in the serve an athlete with tendinitis is placing the stress on the elbow. Some applications of biomechanics and kinesiology require only a cursory view of a movement such as visual inspection of the forearm position in the jump shot. Other applications, such as evaluating the forces applied by a hand on a basketball during a shot, require some advanced knowledge and the use of sophisticated equipment and techniques.

Elaborate equipment is not needed to apply the material in this text but is necessary to understand and interpret numerical examples from data collected using such intricate instruments. Qualitative examples in this text describe the characteristics of movement. A **qualitative analysis** is a nonnumeric evaluation of motion based on direct observation. These examples can be applied directly to a particular movement situation using visual observation or video.

This text also presents quantitative information. A **quantitative analysis** is a numeric evaluation of the motion based on data collected during the performance. For example, movement characteristics can be presented to describe the forces or the temporal and spatial components of the activity. The application of this material to a practical setting, such as teaching a sport skill, is more difficult because it is more abstract and often cannot be visually observed. Quantitative information can be important, however, because it often substantiates what is seen visually in a qualitative analysis. It also directs the instructional technique because a quantitative analysis identifies the source of a movement. For example, a front handspring can be qualitatively evaluated through visual observation by focusing on such things as whether the legs are together and straight, the back is arched, and the landing is stable and whether the handspring was too fast or slow.

But it is through the quantitative analysis that the source of the movement, the magnitude of the forces generated, can be identified. A force cannot be observed qualitatively, but knowing it is the source of the movement helps with qualitative assessment of its effects, that is, the success of the handspring.

This chapter introduces terminology that will be used throughout the remainder of the text. The chapter begins by defining and introducing the various areas of study for movement analysis. This will be the first exposure to the areas presented in much greater depth later in the text. Then the chapter discusses methods and terminology describing how we arrive at the basic mechanical properties of various structures. Finally, the chapter establishes a working vocabulary for movement description at both structural and whole-body levels.

 ## Core Areas of Study

BIOMECHANICS VERSUS KINESIOLOGY

Those who study human movement often disagree over the use of the terms *kinesiology* and *biomechanics*. Kinesiology can be used in one of two ways. First, **kinesiology** as the scientific study of human movement can be an umbrella term used to describe any form of anatomical, physiologic, psychological, or mechanical human movement evaluation. Consequently, kinesiology has been used by several disciplines to describe many different content areas. Some departments of physical education and movement science have gone so far as to adopt *kinesiology* as their department name. Second, kinesiology describes the content of a class in which human movement is evaluated by examination of its source and characteristics. However, a class in kinesiology may consist primarily of functional anatomy at one university and strictly biomechanics at another.

Historically, a kinesiology course has been part of college curricula as long as there have been physical education and movement science programs. The course originally focused on the musculoskeletal system, movement efficiency from the anatomical standpoint, and joint and muscular actions during simple and complex movements. A typical student activity in the kinesiology course was to identify discrete phases in an activity, describe the segmental movements occurring in each phase, and identify the major muscular contributors to each joint movement. Thus, if one were completing a kinesiological analysis of the act of rising from a chair, the movements would be hip extension, knee extension, and plantarflexion via the hamstrings, quadriceps femoris, and triceps surae muscle groups, respectively. Most kinesiological analyses are considered qualitative because they involve observing a movement and providing a breakdown of the skills and identification of the muscular contributions to the movement.

The content of the study of kinesiology is incorporated into many biomechanics courses and is used as a precursor to the introduction of the more quantitative biomechanical content. In this text, *biomechanics* will be used as an umbrella term to describe content previously covered in courses in kinesiology as well as content developed as a result of growth of the area of biomechanics.

In the 1960s and 1970s, biomechanics was developed as an area of study in undergraduate and graduate curricula across North America. The content of biomechanics was extracted from mechanics, an area of physics that consists of the study of motion and the effect of forces on an object. Mechanics is used by engineers to design and build structures and machines because it provides the tools for analyzing the strength of structures and ways of predicting and measuring the movement of a machine. It was a natural transition to take the tools of mechanics and apply them to living organisms. Biomechanics is the study of the structure and function of biological systems by means of the methods of mechanics (1). Another definition proposed by the European Society of Biomechanics (2) is "the study of forces acting on and generated within a body and the effects of these forces on the tissues, fluid, or materials used for the diagnosis, treatment, or research purposes."

A biomechanical analysis evaluates the motion of a living organism and the effect of forces on the living organism. The biomechanical approach to movement analysis can be qualitative, with movement observed and described, or quantitative, meaning that some aspect of the movement will be measured. The use of the term *biomechanics* in this text incorporates qualitative components with a more specific quantitative approach. In such an approach, the motion characteristics of a human or an object are described using parameters such as speed and direction; how the motion is created through application of forces, both inside and outside the body; and the optimal body positions and actions for efficient and effective motion. For example, to biomechanically evaluate the motion of rising from a chair, one attempts to measure and identify joint forces acting at the hip, knee, and ankle along with the force between the foot and the floor, all of which act together to produce the movement up and out of the chair. The components of a biomechanical and kinesiologic movement analysis are presented in Figure 1-1. We now examine some of these components individually.

ANATOMY VERSUS FUNCTIONAL ANATOMY

Anatomy, the science of the structure of the body, is the base of the pyramid from which expertise about human movement is developed. It is helpful to develop a strong understanding of regional anatomy so that for a specific region such as the shoulder, the bones, arrangement of muscles, nerve innervation of those muscles, and blood supply to those muscles and other significant structures (e.g., ligaments) can be identified. A knowledge of anatomy can be put to good use if, for example, one is trying to assess an injury. Assume a patient has a pain on the inside of the elbow. Knowledge of anatomy allows one to recognize the medial epicondyle of the humerus as the prominent bony structure of the medial elbow. It also indicates that the muscles that pull the hand and fingers toward the forearm in a flexion motion attach to the epicondyle. Thus, familiarity with anatomy may lead to a diagnosis of medial epicondylitis, possibly caused by overuse of the hand flexor muscles.

Functional anatomy is the study of the body components needed to achieve or perform a human movement or function. Using a functional anatomy approach to analyze a lateral arm raise with a dumbbell, one should identify the deltoid, trapezius, levator scapulae, rhomboid, and supraspinatus muscles as contributors to upward rotation and elevation of the shoulder girdle and abduction of the arm. Knowledge of functional anatomy is useful in a variety of situations, for example, to set up an exercise or weight training program and to assess the injury potential in a movement or sport or when establishing training techniques and drills for athletes. The prime consideration

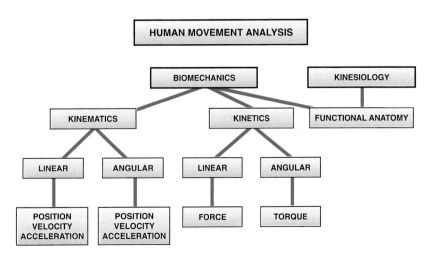

FIGURE 1-1 Types of movement analysis. Movement can be analyzed by assessing the anatomical contributions to the movement (functional anatomy), describing the motion characteristics (kinematics), or determining the cause of the motion (kinetics).

of functional anatomy is not the muscle's location but the movement produced by the muscle or muscle group.

LINEAR VERSUS ANGULAR MOTION

Movement or motion is a change in place, position, or posture occurring over time and relative to some point in the environment. Two types of motion are present in a human movement or an object propelled by a human. First is **linear motion**, often termed *translation* or *translational motion*. Linear motion is movement along a straight or curved pathway in which all points on a body or an object move the same distance in the same amount of time. Examples are the path of a sprinter, the trajectory of a baseball, the bar movement in a bench press, and the movement of the foot during a football punt. The focus in these activities is on the direction, path, and speed of the movement of the body or object. Figure 1-2 illustrates two focal points for linear movement analysis.

The center of mass of the body, of a segment, or of an object is usually the point monitored in a linear analysis (Fig. 1-2). The center of mass is the point at which the mass of the object appears to be concentrated, and it represents the point at which the total effect of gravity acts on the object. However, any point can be selected and evaluated for linear motion. In skill analysis, for example, it is often helpful to monitor the motion of the top of the head to gain an indication of certain trunk motions. An examination of the head in running is a prime example. Does the head move up and down? Side to side? If so, it is an indication that the central mass of the body is also moving in those directions. The path of the hand or racquet is important in throwing and racquet sports, so visually monitoring the linear movement of the hand or racquet throughout the execution of the motion is beneficial. In an activity such as sprinting, the linear movement of the whole body is the most important component to analyze because the object of the sprint is to move the body quickly from one point to another.

The second type of motion is **angular motion**, which is motion around some point so that different regions of the same body segment or object do not move through the same distance in a given amount of time. As illustrated in Figure 1-3, swinging around a high bar represents angular motion because the whole body rotates around the contact point with the bar. To make one full revolution around the bar, the feet travel through a much greater distance than the arms because they are farther from the point of turning. It is typical in biomechanics to examine the linear motion characteristics of an activity and then follow up with a closer look at the angular motions that create and contribute to the linear motion.

All linear movements of the human body and objects propelled by humans occur as a consequence of angular contributions. There are exceptions to this rule such as skydiving or free falling, in which the body is held in a position to let gravity create the linear movement

FIGURE 1-2   Examples of linear motion. Ways to apply linear motion analysis include examination of the motion of the center of gravity or the path of a projected object.

downward, and when an external pull or push moves the body or an object. It is important to identify the angular motions and their sequence that make up a skill or human movement because the angular motions determine the success or failure of the linear movement.

Angular motions occur about an imaginary line called the axis of rotation. Angular motion of a segment, such as the arm, occurs about an axis running through the joint. For example, lowering the body into a deep squat entails angular motion of the thigh about the hip joint, angular motion of the leg about the knee joint, and angular motion of the foot about the ankle joint. Angular motion can also occur about an axis through the center of mass.

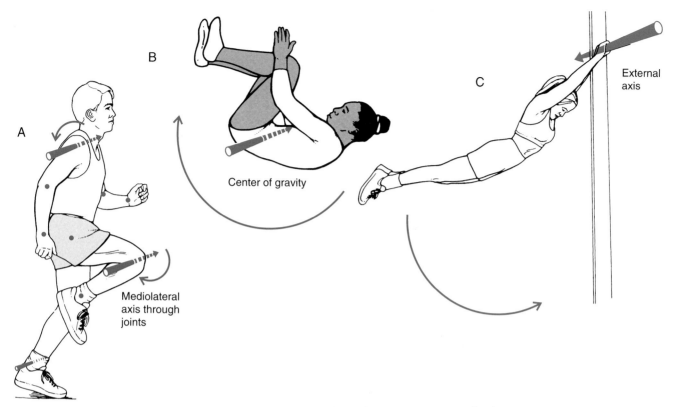

FIGURE 1-3 Examples of angular motion. Angular motion of the body, an object, or segment can take place around an axis running through a joint (**A**), through the center of gravity (**B**), or about an external axis (**C**).

Examples of this type of angular motion are a somersault in the air and a figure skater's vertical spin. Finally, angular motion can occur about a fixed external axis. For example, the body follows an angular motion path when swinging around a high bar with the high bar acting as the axis of rotation.

For proficiency in human movement analysis, it is necessary to identify the angular motion contributions to the linear motion of the body or an object. This is apparent in a simple activity such as kicking a ball for maximum distance. The intent of the kick is to make contact between a foot traveling at a high linear speed and moving in the proper direction to send the ball in the desired direction. The linear motion of interest is the path and velocity of the ball after it leaves the foot. To create high speeds and the correct path, the angular motions of the segments of the kicking leg are sequential, drawing speed from each other so that the velocity of the foot is determined by the summation of the individual velocities of the connecting segments. The kicking leg moves into a preparatory phase, drawing back through angular motions of the thigh, leg, and foot. The leg whips back underneath the thigh very quickly as the thigh starts to move forward to initiate the kick. In the power phase of the kick, the thigh moves vigorously forward and rapidly extends the leg and foot forward at very fast angular speeds. As contact is made with the ball, the foot is moving very fast because the velocities of the thigh and leg have been transferred to the foot.

Skilled observation of human movement allows the relationship between angular and linear motion shown in this kicking example to serve as a foundation for techniques used to correct or facilitate a movement pattern or skill.

KINEMATICS VERSUS KINETICS

A biomechanical analysis can be conducted from either of two perspectives. The first, **kinematics**, is concerned with the characteristics of motion from a spatial and temporal perspective without reference to the forces causing the motion. A kinematic analysis involves the description of movement to determine how fast an object is moving, how high it goes, or how far it travels. Thus, position, velocity, and acceleration are the components of interest in a kinematic analysis. Examples of linear kinematic analysis are the examination of the projectile characteristics of a high jumper or a study of the performance of elite swimmers. Examples of angular kinematic analysis are an observation of the joint movement sequence for a tennis serve or an examination of the segmental velocities and accelerations in a vertical jump. Figure 1-4 presents both an angular (*top*) and a linear (*bottom*) example of the kinematics of the golf swing. By examining an angular or linear movement kinematically, we can identify the segments involved in that movement that require improvement or obtain ideas and technique enhancements from elite performers or break a skill down into its component

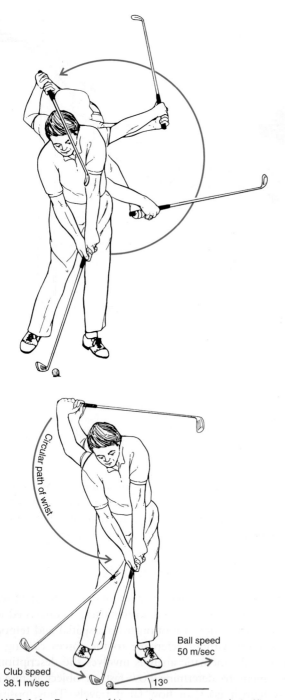

FIGURE 1-4 Examples of kinematic movement analysis. Kinematic analysis focuses on the amount and type of movement, the direction of the movement, and the speed or change in speed of the body or an object. The golf shot is presented from two of these perspectives: The angular components of the golf swing (*top*) and the direction and speed of the club and ball (*bottom*).

parts. By each of these, we can further our understanding of human movement.

Pushing on a table may or may not move the table, depending on the direction and strength of the push. A push or pull between two objects that may or may not result in motion is termed a *force*. **Kinetics** is the area of study that examines the forces acting on a system,

such as the human body, or any object. A kinetic movement analysis examines the forces causing a movement. A kinetic movement analysis is more difficult than a kinematic analysis to both comprehend and evaluate because forces cannot be seen (Fig. 1-5). Only the effects of forces can be observed. Watch someone lift a 200-lb barbell in a squat. How much force has been applied?

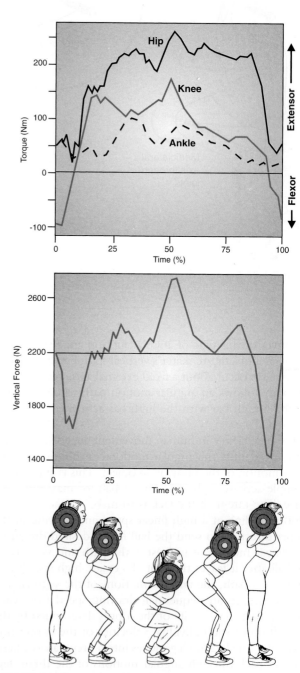

FIGURE 1-5 Examples of kinetic movement analysis. Kinetic analysis focuses on the cause of movement. The weight lifter demonstrates how lifting can be analyzed by looking at the vertical forces on the ground that produce the lift (linear) and the torques produced at the three lower extremity joints that generate the muscular force required for the lift. (Redrawn from Lander, J. et al. [1986]. Biomechanics of the squat exercise using a modified center of mass bar. *Medicine & Science in Sports & Exercise*, 18:469–478.)

Because the force cannot be seen, there is no way of accurately evaluating the force unless it can be measured with recording instruments. A likely estimate of the force is at least 200 lb because that is the weight of the bar. The estimate may be inaccurate by a significant amount if the weight of the body lifted and the speed of the bar were not considered.

The forces produced during human movement are important because they are responsible for creating all of our movements and for maintaining positions or postures having no movement. The assessment of these forces represents the greatest technical challenge in biomechanics because it requires sophisticated equipment and considerable expertise. Thus, for the novice movement analyst, concepts relating to maximizing or minimizing force production in the body will be more important than evaluating the actual forces themselves.

A kinetic analysis can provide the teacher, therapist, coach, or researcher with valuable information about how the movement is produced or how a position is maintained. This information can direct conditioning and training for a sport or movement. For example, kinetic analyses performed by researchers have identified weak and strong positions in various joint positions and movements. Thus, we know that the weakest position for starting an arm curl is with the weights hanging down and the forearm straight. If the same exercise is started with the elbow slightly bent, more weight can be lifted.

Kinetic analyses also identify the important parts of a skill in terms of movement production. For example, what is the best technique for maximizing a vertical jump? After measuring the forces produced against the ground that are used to propel the body upward, researchers have concluded that the vertical jump incorporating a very quick drop downward and stop-and-pop upward action (often called a countermovement action) produces more effective forces at the ground than a slow, deep gather jump.

Lastly, kinetics has played a crucial role in identifying aspects of a skill or movement that make the performer prone to injury. Why do 43% of participants and 76% of instructors of high-impact aerobics incur an injury (3)? The answer was clearly identified through a kinetic analysis that found forces in typical high-impact aerobic exercises to be in the magnitude of four to five times body weight (4). For an individual weighing 667.5 N (newtons) or 150 lb, repeated exposure to forces in the range of 2670 to 3337.5 N (600 to 750 lb) partially contributes to injury of the musculoskeletal system.

Examination of both the kinematic and kinetic components is essential to full understanding of all aspects of a movement. It is also important to study the kinematic and kinetic relationships because any acceleration of a limb, of an object, or of the human body is a result of a force applied at some point, at a particular time, of a given magnitude, and for a particular duration. Although it is of some use to merely describe the motion characteristics kinematically, one must also explore the kinetic sources before a thorough comprehension of a movement or skill is possible.

STATICS VERSUS DYNAMICS

Examine the posture used to sit at a desk and work at a computer. Are forces being exerted? Yes. Even though there is no movement, there are forces between the back and the chair and the foot and the ground. In addition, muscular forces are acting throughout the body to counteract gravity and keep the head and trunk erect. Forces are present without motion and are produced continuously to maintain positions and postures that do not involve movement. Principles of statics are used to evaluate the sitting posture. **Statics** is a branch of mechanics that examines systems that are not moving or are moving at a constant speed. Static systems are considered to be in equilibrium. Equilibrium is a balanced state in which there is no acceleration because the forces causing a person or object to begin moving, to speed up, or to slow down are neutralized by opposite forces that cancel them out.

Statics is also useful for determining stresses on anatomical structures in the body, identifying the magnitude of muscular forces, and identifying the magnitude of force that would result in the loss of equilibrium. How much force generated by the deltoid muscle is required to hold the arm out to the side? Why is it easier to hold an arm at the side if you lower the arm so that it is no longer perpendicular to the body? What is the effect of a lordosis (increased curvature of the back, or swayback) on forces coming through the lumbar vertebrae? These are the types of questions static analysis may answer. Because the static case involves no change in the kinematics of the system, a static analysis is usually performed using kinetic techniques to identify the forces and the site of the force applications responsible for maintaining a posture, position, or constant speed. Kinematic analyses, however, can be applied in statics to substantiate whether there is equilibrium through the absence of acceleration.

To leave the computer workstation and get up out of the chair, it is necessary to produce forces in the lower extremity and on the ground. **Dynamics** is the branch of mechanics used to evaluate this type of movement because it examines systems that are being accelerated. Dynamics uses a kinematic or kinetic approach or both to analyze movement. An analysis of the dynamics of an activity such as running may incorporate a kinematic analysis in which the linear motion of the total body and the angular motion of the segments are described. The kinematic analysis may be related to a kinetic analysis that describes forces applied to the ground and across the joints as the person runs. Because this textbook deals with numerous examples involving motion of the human or a human-propelled object, dynamics is addressed in detail in specific chapters on linear and angular kinematics and kinetics.

Anatomical Movement Descriptors

SEGMENT NAMES

It is important to identify segment names correctly and use them consistently when analyzing movement. To flex the shoulder, does one lift the arm with weights in the hand or raise the whole arm in front? Whatever interpretation is placed on the segment name, the term *arm* will determine the type of movement performed. The correct interpretation of flexing at the shoulder is to raise the whole arm because the arm is the segment between the shoulder and the elbow, not the segment between the elbow and the wrist or the hand segment. A review of segment names is worthwhile preparation for more extensive use of them in the study of biomechanics.

The head, neck, and trunk are segments comprising the main part of the body, or the **axial** portion of the skeleton. This portion of the body accounts for more than 50% of a person's weight, and it usually moves much more slowly than the other parts of the body. Because of its large size and slow speed, the trunk is a good segment to observe visually when one is learning to analyze movement or following the total body activity.

The upper and lower extremities are termed the **appendicular** portion of the skeleton. Generally speaking, as one moves away from or distal to the trunk, the segments become smaller, move faster, and are more difficult to observe because of their size and speed. Thus, whereas shoulder flexion is raising the upper extremity in front, forearm flexion describes a movement at the elbow. The movements of the arm are typically described as they occur in the shoulder joint, forearm movements are described in relation to elbow joint activity, and hand movements are described relative to wrist joint activity. Figure 1-6 illustrates the axial and appendicular regions of the body with the correct segment names.

In the lower extremity, the thigh is the region between the hip and knee joints, the leg is the region between the knee and ankle joints, and the foot is the region distal to the ankle joint. The movement of the thigh is typically described as it occurs at the hip joint, leg movement is described by actions at the knee joint, and foot movements are determined by ankle joint activity.

ANATOMICAL TERMS

The description of a segmental position or joint movement is typically expressed relative to a designated starting position. This reference position, or the **anatomical position**, has been a standard reference point used for many years by anatomists, biomechanists, and the medical profession. In this position, the body is in an erect stance with the head facing forward, arms at the side of the trunk with palms facing forward, and the legs together with the feet

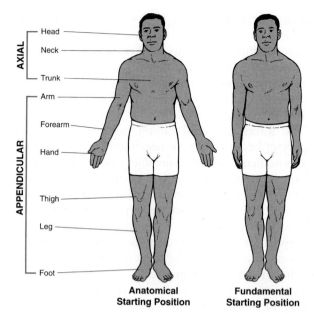

FIGURE 1-6   Anatomical versus fundamental starting position. The anatomical and fundamental starting positions serve as a reference point for the description of joint movements.

pointing forward. Some biomechanists prefer to use what is called the **fundamental position** as the reference position. This reference position is similar to the anatomical position except that the arms are in a more relaxed posture at the sides with the palms facing in toward the trunk. Whatever starting position is used, all segmental movement descriptions are made relative to some reference position. Both of these reference positions are illustrated in Figure 1-6.

To discuss joint position, we must define the **joint angle**, or more correctly, the relative angle between two segments. A **relative angle** is the included angle between the two segments (Fig. 1-7). The calculation of the relative angle is illustrated in Chapter 9 in this book.

The starting position is also called the zero position for description of most joint movements. For example, when a person is standing, there is zero movement at the hip joint. If the thigh is flexed or rotated internally or externally (in or out), the amount of movement is described relative to the fundamental or anatomical starting position. Most zero positions appear to be quite obvious because there is usually a straight line between the two segments so that no relative angle is formed between them. Zero position in the trunk occurs when the trunk is vertical and in line with the lower extremity. The zero position at the knee is found in the standing posture when there is no angle between the thigh and the leg. One not so obvious zero position is at the ankle joint. For this joint, the zero position is assumed in stance with the sole of the foot perpendicular to the leg.

Movement description or anatomical location can best be presented using terminology that is universally accepted and understood. Movement terms should become a part of a working vocabulary, regardless of the level of application

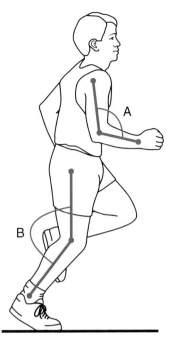

FIGURE 1-7   Relative angles of the elbow (**A**) and knee (**B**).

of terms to instruct students or athletes. A standardized set of terms is most helpful in this situation.

The anatomical terms describing the relative position or direction are illustrated in Figure 1-8. The term **medial** refers to a position relatively close to the midline of the body or object or a movement that moves toward the midline. In the anatomical position, the little finger and the big toe are on the medial side of the extremity because they are on the side of the limb closest to the midline of the body. Also, pointing the toes toward the midline of the body is considered a movement in a medial direction. The opposite of medial is **lateral**, that is, a position relatively far from the midline or a movement away from the midline. In the anatomical position, the thumb and the little toe are on the lateral side of the hand and foot, respectively, because they are farther from the midline. Likewise, pointing the toes out is a lateral movement. Landmarks are also commonly designated as medial or lateral based on their relative position to the midline, such as medial and lateral condyles, epicondyles, and malleoli.

Proximal and **distal** are used to describe the relative position with respect to a designated reference point, with proximal representing a position closer to the reference point and distal being a point farther from the reference. The elbow joint is proximal, and the wrist joint is distal relative to the shoulder joint. The ankle joint is proximal, and the knee joint is distal relative to the point of heel contact with the ground. Both proximal and distal must be expressed relative to some reference point.

A segment or anatomical landmark may lie on the **superior** aspect of the body, placing it above a particular reference point or closer to the top of the head. It may lie on an **inferior** aspect, that is, lower than a reference

of kinesiology required. Development of solid knowledge of the movement characteristics of the various phases of a human movement or sport skill can improve the effectiveness of teaching a skill, assist in correcting flaws in a performance, identify important movements and segments for emphasis in conditioning, and identify aspects of the skill that may be associated with injury. The experienced researcher, coach, or teacher can determine the most relevant movements in a skill and will use a specific vocabulary

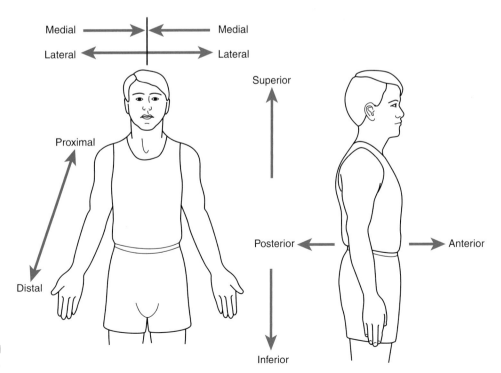

FIGURE 1-8   Anatomical terms used to describe relative position or direction.

segment or landmark. For example, the head is positioned superior to the trunk, the trunk is superior to the thigh, and so on. The greater trochanter is located on the superior aspect of the femur, and the medial epicondyle of the humerus is located on the inferior end of the humerus.

The location of an object or a movement relative to the front or back is **anterior** or **posterior**, respectively. Thus, whereas the quadriceps muscle group is located on the anterior region of the thigh, the hamstrings muscle group is located on the posterior region of the thigh. *Anterior* is also synonymous with **ventral** for a location on the human body, and *posterior* refers to the **dorsal** surface or position on the human body.

The term **ipsilateral** describes activity or location of a segment or landmark positioned on the same side as a particular reference point. Actions, positions, and landmark locations on the opposite side can be designated as **contralateral**. Thus, when a person lifts the right leg forward, there is extensive muscular activity in the iliopsoas muscle of that leg, the ipsilateral leg, and extensive activity in the gluteus medius of the contralateral leg to maintain balance and support. In walking, as the ipsilateral lower limb swings forward, the other limb, the contralateral limb, pushes on the ground to propel the walker forward.

MOVEMENT DESCRIPTION

Basic Movements

Six basic movements occur in varying combinations in the joints of the body. The first two movements, flexion and extension, are movements found in almost all of the freely movable joints in the body, including the toe, ankle, knee, hip, trunk, shoulder, elbow, wrist, and finger. **Flexion** is a bending movement in which the relative angle of the joint between two adjacent segments decreases. **Extension** is a straightening movement in which the relative angle of the joint between two adjacent segments increases as the joint returns to the zero or reference position. Numerous examples of both flexion and extension are provided in Figure 1-9. A person can also perform **hyperflexion** if the flexion movement goes beyond the normal range of flexion. For example, this can happen at the shoulder only when the arm moves forward and up in flexion through 180° until it is at the side of the head, and then hyperflexes as it continues to move past the head toward the back. **Hyperextension** can occur in many joints as the extension movement continues past the original zero position. It is common to see hyperextension movements in the trunk, arm, thigh, and hand.

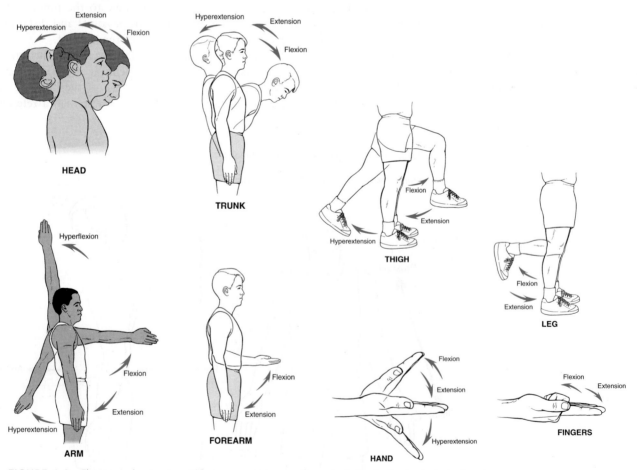

FIGURE 1-9 Flexion and extension. These movements occur in many joints in the body, including the vertebra, shoulder, elbow, wrist, metacarpophalanx, interphalanx, hip, knee, and metatarsophalanx.

A toe-touch movement entails flexion at the vertebral, shoulder, and hip joints. The return to the standing position involves the opposite movements of vertebral extension, hip extension, and shoulder extension. The power phase of the jump shot is produced via smooth timing of lower extremity hip extension, knee extension, and ankle extension coordinated with shoulder flexion, elbow extension, and wrist flexion in the shooting limb. This example illustrates the importance of the lower extremity extension movements to the production of power. Lower extremity extension often serves to produce upward propulsion that works against the pull of gravity. It is opposite in the shoulder joint, where flexion movements are primarily used to develop propulsion upward against gravity to raise the limb.

Abduction and adduction is another pair of movements that is not as commonly known as flexion and extension, occurring only in particular joints, such as the metatarsophalangeal (foot), hip, shoulder, wrist, and metacarpophalangeal (hand) joints. Many of these movements are presented in Figure 1-10. **Abduction** is a movement away from the midline of the body or the segment. Raising an arm or leg out to the side or the spreading of the fingers or toes is an example of abduction. **Hyperabduction** can occur in the shoulder joint as the arm moves more than 180° from the side all the way up past the head. **Adduction** is the return movement of the segment back toward the midline of the body or segment. Bringing the arms back to the trunk, bringing the legs together, and closing the toes or fingers are examples of adduction. **Hyperadduction** occurs frequently in the arm and thigh as the adduction continues past the zero position so that the limb crosses the body. These side-to-side movements are commonly used to maintain balance and stability during the performance of both upper and lower extremity sport skills. Controlling or preventing abduction and adduction movements of the thigh is especially crucial to the maintenance of pelvic and limb stability during walking and running.

The last two basic movements involve rotations, illustrated in Figure 1-11. A **rotation** can be either medial (also known as internal) or lateral (also known as external). Rotations are designated as right and left for the head and trunk only. When in the fundamental

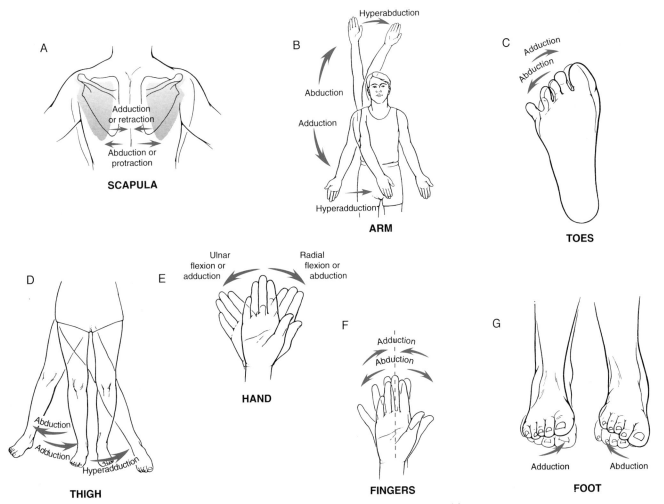

FIGURE 1-10 Abduction and adduction. These movements occur in the sternoclavicular, shoulder, wrist, metacarpophalangeal, hip, intertarsal, and metatarsophalangeal joints.

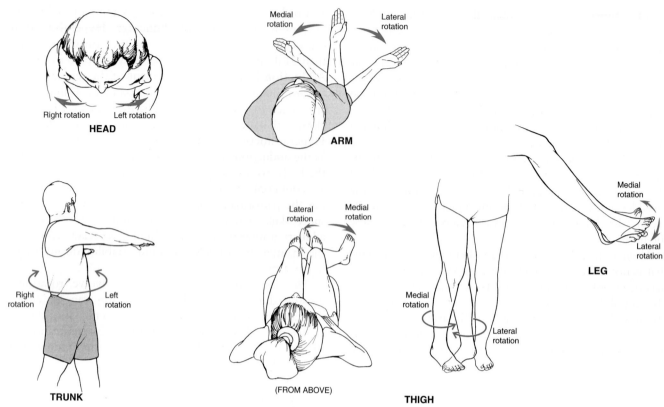

FIGURE 1-11   Rotation. This occurs in the vertebral, shoulder, hip, and knee joints.

starting position, medial or internal rotation refers to the movement of a segment about a vertical axis running through the segment so that the anterior surface of the segment moves toward the midline of the body while the posterior surface moves away from the midline. Lateral or external rotation is the opposite movement in which the anterior surface moves away from the midline and the posterior surface of the segment moves toward the midline. Because the midline runs through the trunk and head segments, the rotations in these segments are described as left or right from the perspective of the performer. Right rotation is the movement of the anterior surface of the trunk so that it faces right while the posterior surface faces left, and left rotation is the opposite movement so that the anterior trunk faces left and the posterior trunk faces right. Rotations occur in the vertebrae, shoulder, hip, and knee joints. Rotation movements are important in the power phase of sport skills involving the trunk, arm, or thigh. For throwing, the throwing arm laterally rotates in the preparatory phase and medially rotates in the power and follow-through phases. The trunk complements the arm action with right rotation in the preparatory phase (right-handed thrower) and left rotation in the power and follow-through phase. Likewise, the right thigh laterally rotates in the preparatory phase and medially rotates until the lower extremity comes off the ground in the power phase.

Specialized Movement Descriptors

Specialized movement names are assigned to a variety of segmental movements (Fig. 1-12). Although most of these segmental movements are technically among the six basic movements, the specialized movement name is the terminology commonly used by movement professionals. Right and left **lateral flexion** applies only to movement of the head or trunk. When the trunk or head is tilted sideways, the movement is termed *lateral flexion*. If the right side of the trunk or head moves so that it faces down, the movement is termed *right lateral flexion* and vice versa.

The shoulder girdle has specialized movement names that can best be described by observing the movements of the scapula. Whereas raising the scapula, as in a shoulder shrug, is termed **elevation**, the opposite lowering movement is called **depression**. If the two scapulae move apart, as in rounding the shoulders, the movement is termed **protraction**. The return movement, in which the scapulae move toward each other with the shoulders back, is called **retraction**. Finally, the scapulae can swing out such that the bottom of the scapula moves away from the trunk and the top of the scapula moves toward the trunk. This movement is termed **upward rotation**, and the opposite movement, when the scapula swings back down into the resting position, is **downward rotation**.

In the arm and the thigh segments, a combination of flexion and adduction is termed **horizontal adduction**, and a combination of extension and abduction is called

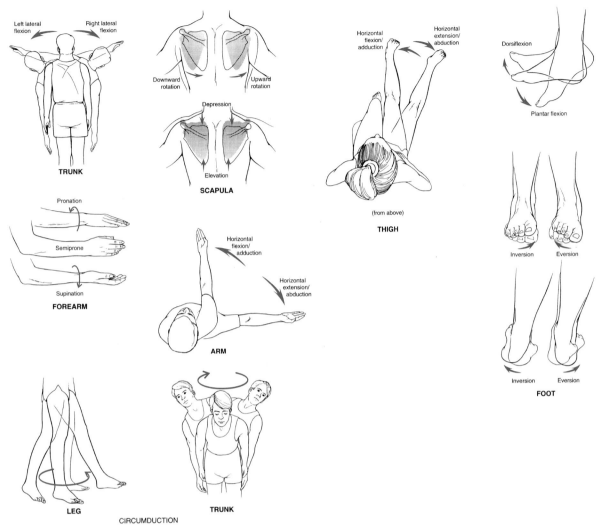

FIGURE 1-12 Examples of specialized movements. Some joint movements are designated with specialized names, even though they may technically be one of the six basic movements.

horizontal abduction. Horizontal adduction, sometimes called *horizontal flexion*, is the movement of the arm or thigh across the body toward the midline using a movement horizontal to the ground. Horizontal abduction, or horizontal extension, is a horizontal movement of the arm or thigh away from the midline of the body. These movements are used in a wide variety of sport skills. The arm action of the discus throw is a good example of the use of horizontal abduction in the preparatory phase and horizontal adduction in the power and follow-through phase. Many soccer skills use horizontal adduction of the thigh to bring the leg up and across the body for a shot or pass.

In the forearm, pronation and supination occur as the distal end of the radius rotates over and back on the ulna at the radioulnar joints. **Supination** is the movement of the forearm in which the palm rotates to face forward from the fundamental starting position. **Pronation** is the movement in which the palms face backward. Supination and pronation joint movements have also been referred to as *external* and *internal rotation*, respectively. As the forearm moves from a supinated position to a pronated position, the forearm passes through the semiprone position, in which the palms face the midline of the body with the thumbs forward. The actions of forearm pronation and supination are used with arm rotation movements to increase the range of motion, add spin, enhance power, and change direction during the force application phases in racquet sports, volleyball, and throwing.

At the wrist joint, whereas the movement of the hand toward the thumb is called **radial flexion**, the opposite movement of the hand toward the little finger is called **ulnar flexion**. These specialized movement names are easier to remember because they do not depend on forearm or arm position, as do the interpretation of abduction and adduction, and they can easily be interpreted if the locations of the radius (thumb side) and the ulna (little finger side) are known. Ulnar flexion and radial flexion are important in racquet sports for control and stabilization of the racquet. Also, in volleyball, ulnar flexion is a valuable component of the forearm pass because it helps maintain the extended arm position and increases the contact area of the forearms.

In the foot, *plantarflexion* and *dorsiflexion* are specialized names for foot extension and flexion, respectively. **Plantarflexion** is the movement in which the bottom of the foot moves down and the angle formed between the foot and the leg increases. This movement can be created by raising the heel so the weight is shifted up on the toes or by placing the foot flat on the ground in front and moving the leg backward so that the body weight is behind the foot. **Dorsiflexion** is the movement of the foot up toward the leg that decreases the relative angle between the leg and the foot. This movement may be created by putting weight on the heels and raising the toes or by keeping the feet flat on the floor and lowering with weight centered over the foot. Any foot–leg angle greater than 90° is termed a plantarflexed position, and any foot–leg angle less than 90° is termed dorsiflexion.

The foot has another set of specialized movements, called *inversion* and *eversion*, that occur in the intertarsal and metatarsal articulations. **Inversion** of the foot takes place when the medial border of the foot lifts so that the sole of the foot faces medially toward the other foot. **Eversion** is the opposite movement of the foot: The lateral aspect of the foot lifts so that the sole of the foot faces away from the other foot.

Often confusion exists over the use of the terms *inversion* and *eversion* and the popularized use of *pronation* and *supination* as descriptors of foot motion. Inversion and eversion are not the same as pronation and supination; in fact, they are only a part of pronation and supination. Pronation of the foot is actually a set of movements consisting of dorsiflexion at the ankle joint, eversion, and abduction of the forefoot. Supination is created through ankle plantarflexion, inversion, and forefoot adduction. Pronation and supination are dynamic movements of the foot and ankle that particularly occur when the foot is on the ground during a run or walk. These two movements are determined by the structure and laxity of the foot, body weight, playing surfaces, and footwear.

The final specialized movement, **circumduction**, can be created in any joint or segment that has the potential to move in two directions, such that the segment can be moved in a conic fashion as the end of the segment moves in a circular path. An example of circumduction is placing the arm out in front and drawing an imaginary circle in the air. Circumduction is not a simple rotation; rather, it is four movements in sequence. The movement of the arm in the creation of the imaginary O is actually a combination of flexion, adduction, extension, and abduction. Circumduction movements are also possible in the foot, thigh, trunk, head, and hand. The movements of all of the major segments are reviewed in Table 1-1.

TABLE 1-1	Movement Review		
Segment	**Joint**	**df**	**Movements**
Head	Intervertebral	3	Flexion, extension, hyperextension, R/L lateral flexion, R/L rotation, circumduction
	Atlantoaxial (3 joints)	1 each	R/L rotation
Trunk	Intervertebral	3	Flexion, extension, hyperextension, R/L rotation, R/L lateral flexion, circumduction
Arm	Shoulder	3	Flexion, extension, hyperextension, abduction, adduction, hyperabduction, hyperadduction, horizontal abduction, horizontal adduction, med/lat rotation, circumduction
Arm/shoulder Girdle	Sternoclavicular	3	Elevation, depression, abduction, adduction (protraction, retraction), rotation
	Acromioclavicular	3	Abduction, adduction (protraction, retraction), upward/downward rotation
Forearm	Elbow	1	Flexion, extension, hyperextension
	Radioulnar	1	Pronation, supination
Hand	Wrist	2	Flexion, extension, hyperextension, radial flexion, ulnar flexion, circumduction
Fingers	Metacarpophalangeal	2	Flexion, extension, hyperextension, abduction, adduction, circumduction
	Interphalangeal	1	Flexion, extension, hyperextension
Thumb	Carpometacarpal	2	Flexion, extension, abduction, adduction, opposition, circumduction
	Metacarpophalangeal	1	Flexion, extension
	Interphalangeal	1	
Thigh	Hip	3	Flexion, extension, hyperextension, abduction, adduction, hyperadduction, horizontal adduction, horizontal abduction, med/lat rotation, circumduction
Leg	Knee	2	Flexion, extension, hyperextension, med/lat rotation
Foot	Ankle	1	Plantarflexion, dorsiflexion
	Intertarsal	3	Inversion, eversion
Toes	Metatarsophalangeal	2	Flexion, extension, abduction, adduction, circumduction
	Interphalangeal	1	Flexion, extension

R/L, right–left; med/lat, medial–lateral.

Reference Systems

RELATIVE VERSUS ABSOLUTE

A reference system is essential for accurate observation and description of any type of motion. The use of joint movements relative to the fundamental or anatomical starting position is an example of a simple reference system. This system was previously used in this chapter to describe movement of the segments. To improve on the precision of a movement analysis, a movement can be evaluated with respect to a different starting point or position.

A **reference system** is necessary to specify position of the body, segment, or object so as to describe motion or identify whether any motion has occurred. The reference frame or system is arbitrary and may be within or outside of the body. The reference frame consists of imaginary lines called **axes** that intersect at right angles at a common point termed the **origin**. The origin of the reference frame is placed at a designated location such as a joint center. The axes are generally given letter representations to differentiate the direction in which they are pointing. Any position can be described by identifying the distance of the object from each of the axes. In two-dimensional or planar movement, there are two axes, a horizontal axis and a vertical axis. In a three-dimensional movement, there are three axes, two horizontal axes that form a plane and a vertical axis. It is important to identify the frame of reference used in the description of motion.

An example of a reference system placed outside the body is the starting line in a race. The center of an anatomical joint, such as the shoulder, can be used as a reference system within the body. The arm can be described as moving through a 90° angle if abducted until perpendicular to the trunk. If the ground is used as a frame of reference, the same arm abduction movement can be described with respect to the ground, such as movement to a of 1.6 m from the ground.

When angular motion is described, the joint positions, velocities, and accelerations can be described using either an absolute or a relative frame of reference. An **absolute reference frame** is one in which the axes intersect at the center of the joint and movement of a segment is described with respect to that joint. The axes are generally oriented horizontally and vertically. The horizontal axis is generally called the x-axis and the vertical axis the y-axis, although these axes may be called by any name as long as they are defined and consistent. A **segment angle** is measured from the right horizontal axes (Fig. 1-13A) and defines the orientation of the segment in space. The absolute positioning of an abducted arm perpendicular to the trunk is 0° or 360° when described relative to the axes running through the shoulder joint. A **relative reference frame** is one in which the movement of a segment is described relative to the adjacent segment. This type of reference frame is often used to describe a joint angle. The axes in this reference frame are not horizontal and vertical. Figure 1-13B shows the y-axis placed along one segment, the leg, and the x-axis perpendicular to the y-axis. The knee angle can then be determined from the lower portion of the y-axis to the dotted line describing the thigh segment.

In the previously described example of the arm, with abduction perpendicular to the trunk, the relative positioning of the arm with respect to the trunk is 90°. The reference frame should be clearly identified so that the results can be interpreted accordingly and, because reference systems vary among researchers, the reference system and reference point must be identified before comparing and contrasting results between studies. For example, some researchers label a fully extended forearm as a 180° position, and others label the position 0°. After 30° of

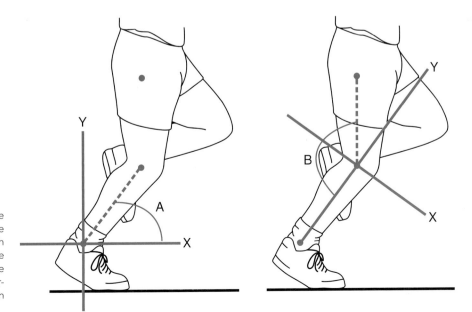

FIGURE 1-13 Absolute versus relative reference frame. *Left,* An absolute reference frame measures the segment angle **(A)** with respect to the distal joint. *Right,* A relative reference frame measures the relative angle **(B)** formed by the two segments. It is important to designate the reference frame in movement description.

flexion at the elbow joint, the final position is 150° or 30°, respectively, for the two systems described above. Considerable confusion can arise when trying to interpret an article using a different reference system from that of the authors.

PLANES AND AXES

The universally used method of describing human movements is based on a system of planes and axes. A **plane** is a flat, two-dimensional surface. Three imaginary planes are positioned through the body at right angles to each other so they intersect at the center of mass of the body. These are the **cardinal planes** of the body. Movement is said to occur in a specific plane if it is actually along that plane or parallel to it. Movement in a plane always occurs about an **axis of rotation** perpendicular to the plane (Fig. 1-14). If you stick a pin through a piece of cardboard and spin the cardboard around the pin, the movement of the cardboard takes place in the plane, and the pin represents the axis of rotation. The cardboard can spin around the pin while the pin is front to back horizontal, vertical, or sideways, for movement of the cardboard in all three of the planes. This example can be applied to describe imaginary lines running through the total body center of mass in the same three pin directions. These planes allow full description of a motion and contrast of an arm movement straight out in front of the body with one straight out to the side of the body. The planes and axes of the human body for motion description are presented in Figure 1-15.

The **sagittal plane** bisects the body into right and left halves. Movements in the sagittal plane occur about a **mediolateral axis** running side to side through the center of mass of the body. Sagittal plane movements involving the whole body rotating around the center of mass include somersaults, backward and forward handsprings, and flexing to a pike position in a dive. The **frontal** or **coronal plane** bisects the body to create front and back halves.

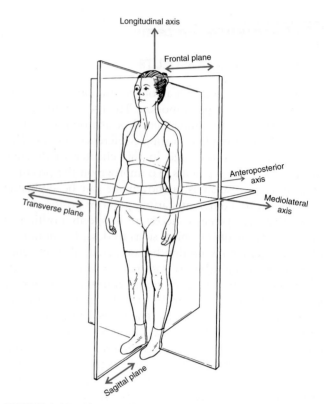

FIGURE 1-15 Planes and axes on the human body. The three cardinal planes that originate at the center of gravity are the sagittal plane, which divides the body into right and left; the frontal plane, dividing the body into front and back; and the transverse plane, dividing the body into top and bottom. Movement takes place in or parallel to the planes about a mediolateral axis (sagittal plane), an anteroposterior axis (frontal plane), or a longitudinal axis (transverse plane).

The axis about which frontal plane movements occur is the **anteroposterior axis** that runs anterior and posterior from the plane. Frontal plane motions of the whole body about the center of mass are not as common as movements in the other planes. The **transverse** or **horizontal plane** bisects the body to create upper and lower halves. Movements occurring in this plane are primarily rotations about a **longitudinal axis**. Spinning vertically around the body, as in a figure skating spin, is an example of transverse plane movement about the body's center of mass.

Although we have described the sagittal, transverse, and frontal cardinal planes, actually any number of other planes can pass through the body. For example, we can define many sagittal planes that do not pass through the center of mass of the body. The only requirement for defining such a plane is that it is parallel to the cardinal sagittal plane. Likewise, we can have multiple transverse or frontal planes. Defining these **noncardinal planes** is useful for describing joint or limb movements. The intersection of the three planes is placed at the joint center so that joint actions can be described in a sagittal, transverse, or frontal plane (Fig. 1-16). Noncardinal planes can also be used in examining movements that take place about an external axis.

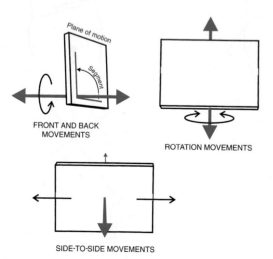

FIGURE 1-14 The plane and axis. Movement takes place in a plane about an axis perpendicular to the plane.

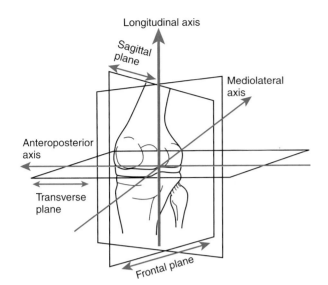

FIGURE 1-16 Planes and axes for the knee.

Most planar or two-dimensional analyses in biomechanics are concerned with motion in the sagittal plane through a joint center. Examples of sagittal plane movements at a joint can be demonstrated by performing flexion and extension movements, such as raising the arm in front, bending the trunk forward and back, lifting and lowering the leg in front, and rising on the toes. Examples of sagittal plane movements of the body about an external support point include rotating the body over the planted foot and running and rotating the body over the hands in a vault. The most accurate view of any motion in a plane is obtained from a position perpendicular to the plane of movement to allow viewing along the axis of rotation. Therefore, sagittal plane movements are best viewed from the side of the body to allow focus on a frontal axis of rotation (Fig. 1-17).

Similar to sagittal plane movements, frontal plane movements can occur about a joint. Characteristic joint movements in the frontal plane include thigh abduction and adduction, finger and hand abduction and adduction, lateral flexion of the head and trunk, and inversion and eversion of the foot. Frontal plane motion about an external point of contact can especially be seen often in dance and ballet as the dancers move laterally from a pivot point and in gymnastics with the body rotating sideways over the hands, such as when doing a cartwheel. The best position to view frontal plane movements is in front or in back of the body to focus on the joint or the point about which the whole body rotates (Fig. 1-18).

Examples of movements in the transverse plane about longitudinal joint axes are rotations at the vertebrae, shoulder, and hip joints. Pronation and supination of the forearm at the radioulnar joints is also a transverse

FIGURE 1-17 Movements in the sagittal plane. Sagittal plane movements are typically flexions and extensions or some forward or backward turning exercise. The movements can take place about a joint axis, the center of gravity, or an external axis.

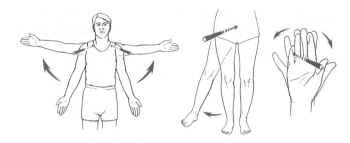

FRONTAL PLANE MOVEMENTS ABOUT JOINT AXES

FRONTAL PLANE MOVEMENT ABOUT THE CENTER OF GRAVITY

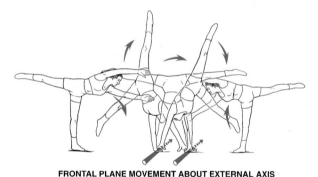

FRONTAL PLANE MOVEMENT ABOUT EXTERNAL AXIS

FIGURE 1-18 Movements in the frontal plane. Segmental movements in the frontal plane about anteroposterior joint axes are abduction and adduction or some specialized side-to-side movement. Frontal plane movements about the center of gravity or an external point involve sideways movement of the body, which is more difficult than movement to the front or back.

plane movement. The axis for all of these movements is an imaginary line running longitudinally through the vertebrae, shoulder, radioulnar, or the hip joints. This is a common movement in gymnastics, dance, and ice skating. Additionally, numerous examples can be found in dance, skating, and gymnastics in which the athlete performs transverse plane movements about an external axis running through a pivot point between the foot and the ground. All spinning movements that have the whole body turning about the ground or the ice are examples. Although transverse plane motions are vital aspects of most successful sport skills, these movements are difficult to follow visually because the best viewing position is either above or below the movement, perpendicular to the plane of motion. Consequently, rotation motions are evaluated by following the linear movement of some point on the body if vertical positioning cannot be achieved. Examples of movements in the transverse plane are presented in Figure 1-19.

Most human movements take place in multiple planes at the various joints. In running, for example, the lower extremity appears to move predominantly in the sagittal plane as the limbs swing forward and backward throughout the gait cycle. Upon closer examination of the limbs and joints, one finds movements in all of the planes. At the hip joint, for example, the thigh performs flexion and extension in the sagittal plane, abduction and adduction in the frontal plane, and internal and external rotation in the transverse plane. If human movements were confined to single-plane motion, we would look like robots as we performed our skills or joint motions. Examine the three-dimensional motion for an overhand throw presented in Figure 1-20. Note the positioning for viewing motion in each of the three planes.

The movement in a plane can also be described as a single **degree of freedom** (df). This terminology is commonly used to describe the type and amount of motion structurally allowed by the anatomical joints. A joint with 1 df indicates that the joint allows the segment to move through one plane of motion. A joint with 1 df is also termed uniaxial because one axis is perpendicular to the plane of motion about which movement occurs. A 1-df joint, the elbow, allows only flexion and extension in the sagittal plane.

Conventionally, most joints are considered to have 1, 2, or 3 df, offering movement potential that is uniaxial, biaxial, or triaxial, respectively. The shoulder is an example of a 3-df, or triaxial, joint because it allows the arm to move in the frontal plane via abduction and adduction, in the sagittal plane via flexion and extension, and in the transverse plane via rotation.

Joints with 3 df include the vertebrae, shoulder, and hip; 2-df joints include the knee, metacarpophalangeal (hand), wrist, and thumb carpometacarpal joints; and 1-df joints include the atlantoaxial (neck), interphalangeal (hand and foot), radioulnar (elbow), and ankle joints. Three degrees of freedom does not always imply great mobility, but it does indicate that the joint allows movement in all three planes of motion. The shoulder is much more mobile than the hip, even though they both are triaxial joints and are capable of performing the same movements. The trunk movements, although classified as having 3 df, are quite restricted if one evaluates movement at a single vertebral level. For example, the lumbar and cervical areas of the vertebrae allow the trunk to flex and extend, but this plane of movement is limited in the middle thoracic portion of the vertebrae. Likewise, the rotation actions of the trunk occur primarily in the thoracic and cervical regions because the lumbar region has limited movement potential in the horizontal plane. It is only the combination of all of the vertebral segments that allows the 3-df motion produced by the spine.

Also, gliding movements occur across the joint surfaces. Gliding movements may be interpreted as adding more degrees of freedom to those defined in the literature. For example, the knee joint is considered to have 2 df for flexion and extension in the sagittal plane and

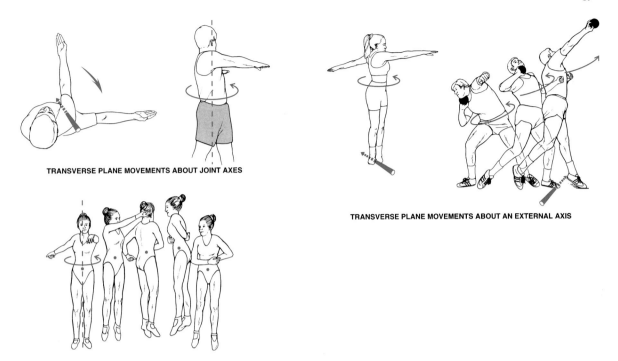

TRANSVERSE PLANE MOVEMENTS ABOUT JOINT AXES

TRANSVERSE PLANE MOVEMENTS ABOUT AN EXTERNAL AXIS

TRANSVERSE PLANE MOVEMENTS ABOUT THE CENTER OF GRAVITY

FIGURE 1-19   Movements in the transverse plane. Most transverse plane movements are rotations about a longitudinal axis running through a joint, the center of gravity, or an external contact point.

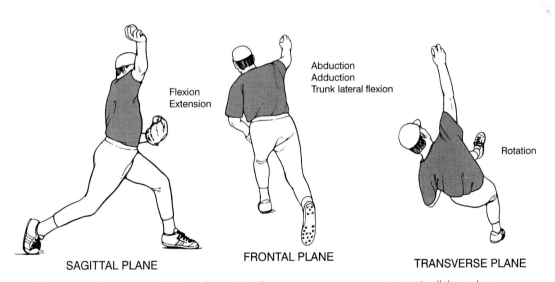

Flexion
Extension

Abduction
Adduction
Trunk lateral flexion

Rotation

SAGITTAL PLANE   **FRONTAL PLANE**   **TRANSVERSE PLANE**

FIGURE 1-20   Movements in all three planes. Most human movements use movement in all three planes. The release phase of the overhand throw illustrates movements in all three planes. The sagittal plane movements are viewed from the side; the frontal plane movements, from the rear; and the transverse plane movements, from above.

rotation in the transverse plane. The knee joint also demonstrates linear translation, however, and it is well known that there is movement in the joint in the frontal plane as the joint surfaces glide over one another to create side-to-side translation movements. Although these movements have been measured and are relatively significant, they have not been established as an additional degree of freedom for the joint. The degrees of freedom for most of the joints in the body are shown in Table 1-1.

A kinematic chain is derived from combining degrees of freedom at various joints to produce a skill or movement. The chain is the summation of the degrees of freedom in adjacent joints that identifies the total degrees of freedom available or necessary for the performance of a movement. For example, kicking a ball might involve an 11-df system relative to the trunk. This would include perhaps 3 df at the hip, 2 df at the knee, 1 df at the ankle, 3 df in the tarsals (foot), and 2 df in the toes.

Summary

Biomechanics, the application of the laws of physics to the study of motion, is an essential discipline for studying human movement. From a biomechanical point of view, human motion can be qualitatively or quantitatively assessed. A qualitative analysis is a nonnumeric assessment of the movement. A quantitative analysis uses kinematic or kinetic applications that analyze a skill or movement by identifying its components or by assessing the forces creating the motion, respectively.

To provide a specific description of a movement, it is helpful to define movements with respect to a starting point or to one of the three planes of motion: sagittal, frontal, or transverse.

Anatomical movement descriptors should be used to describe segmental movements. This requires acknowledgment of the starting position (fundamental or anatomical), standardized use of segment names (arm, forearm, hand, thigh, leg, and foot), and the correct use of movement descriptors (flexion, extension, abduction, adduction, and rotation).

REVIEW QUESTIONS

True or False

1. _____ Functional anatomy is the science of the structure of the body.

2. _____ Movement is a change in place, position, or posture occurring over time.

3. _____ Kinesiology is the study of human motion.

4. _____ The axial skeleton includes the upper and lower extremities.

5. _____ Any angular motion has an axis of rotation.

6. _____ A relative angle is the same as a segment angle.

7. _____ When the joint angle between two segments increases, the action that occurs is flexion.

8. _____ Sagittal plane motion of the ankle is called pronation and supination.

9. _____ The right arm is ipsilateral to the right leg.

10. _____ The axial skeleton is the same as the appendicular skeleton.

11. _____ Medial and lateral are the same as left and right.

12. _____ The anatomical position is the only position used by biomechanists.

13. _____ Lateral flexion applies only to the movement of the head or trunk.

14. _____ Lower extremity motion in running occurs mostly in the sagittal plane.

15. _____ There is only one cardinal plane in the human body.

16. _____ Plantarflexion occurs at the knee joint.

17. _____ A mediolateral axis runs from front to back.

18. _____ The transverse plane has a longitudinal axis.

19. _____ Statics is a branch of mechanics that studies systems that are not moving.

20. _____ The axis of rotation is always perpendicular to the plane of motion.

21. _____ The axes of a reference frame intersect at the origin.

22. _____ The elbow joint has primarily 3 df.

23. _____ A joint that has 2 df can also be called a biaxial joint.

24. _____ All human joints have at least 3 df.

25. _____ Medial rotations are also known as external rotations.

Multiple Choice

1. In which time period was biomechanics developed as an area of study?
   a. 1920s and 1930s
   b. 1940s and 1950s
   c. 1960s and 1970s
   d. 1980s and 1990s

2. Which of the following is an example of quantitative analysis?
   a. A coach correcting a free throw
   b. Determination of the force acting on the femur in a long jumper
   c. A physical therapist watching a patient exercise
   d. All of the above

3. Which of the following is an example of angular motion?
   a. The arm of a pitcher throwing a ball
   b. A parachutist in free fall
   c. The path of a baseball while it is in the air
   d. None of the above

4. Which of the following is an example of linear motion?
   a. The path of a baseball while it is in the air
   b. A child performing a cartwheel
   c. A runner's leg motion during a 100-m race
   d. None of the above

5. Which of the following could be considered in a kinematic study?
   a. The force between a runner and the starting blocks
   b. The torque developed by the muscles crossing the knee joint in a runner
   c. The change in position of a runner over time
   d. None of the above

6. Which of the following could be considered in a kinetic study?
   a. The velocity of a runner during a race
   b. The acceleration of a runner during the start of a race
   c. The torque developed by the muscles crossing the knee joint in a runner
   d. None of the above

7. Which of the following are examples of a static analysis?
   a. A weight lifter lifting a barbell over his head
   b. The movement of a spaceship coasting through space
   c. The takeoff of a spaceship from earth
   d. None of the above

8. The unit of mass is _____.
   a. gram
   b. centimeter
   c. newton
   d. all of the above

9. A dynamic human movement implies _____.
   a. the velocity is zero
   b. the acceleration is zero
   c. the net forces are zero
   d. none of the above

10. Functional anatomy is _____.
    a. the science of the structure of the body
    b. the study of the body components needed to achieve or perform a human movement or function
    c. the arrangement of muscles, nerves, blood vessels, and bones
    d. none of the above

11. External rotation of a segment occurs in _____.
    a. a sagittal plane
    b. a transverse plane
    c. a frontal plane
    d. a longitudinal plane

12. Which motion occurs primarily in the frontal plane?
    a. Push-up
    b. Jumping jack
    c. Frisbee throw
    d. Squat

13. The sagittal plane bisects the body into _____.
    a. right and left portions
    b. top and bottom portions
    c. front and back portions
    d. none of the above

14. Compared to the anatomical position, the shoulder is more _____ in the fundamental position.
    a. flexed
    b. adducted
    c. abducted
    d. extended

15. The ankle joint is _____ to the knee joint.
    a. proximal
    b. medial
    c. distal
    d. anterior

16. The hip joint is _____ to the knee joint.
    a. proximal
    b. medial
    c. distal
    d. anterior

17. The right foot is _____ to the left foot.
    a. proximal
    b. contralateral
    c. inferior
    d. ipsilateral

18. A joint moving in the sagittal plane in which the relative angle extends past its zero position undergoes _____.
    a. hyperextension
    b. hyperflexion
    c. hyperadduction
    d. hyperabduction

19. Which of the following is not a movement of the scapula?
    a. Depression
    b. Lateral flexion
    c. Upward rotation
    d. Retraction

20. Radial flexion takes place on _____.
    a. the thumb side of the hand
    b. the little finger side of the hand
    c. the big toe side of the foot
    d. the little toe side of the foot

21. The axes of a relative reference system are aligned _____.
    a. horizontally and vertically
    b. with a segment of the body
    c. with the ground
    d. none of the above

22. Motion in the frontal plane takes place about which axis?
    a. Longitudinal
    b. Mediolateral
    c. Transverse
    d. Anteroposterior

23. Most human movements take place in _____.
    a. the sagittal plane
    b. the frontal plane
    c. the transverse plane
    d. multiple planes

24. The kinematic chain for a particular motion is _____.
    a. the flexion/extension, abduction/adduction, and internal/external rotation
    b. the positions, velocities, and accelerations
    c. the ligaments, bones, and muscles that restrict the degrees of freedom
    d. the summation of degrees of freedom in adjacent joints

25. A triaxial joint has how many degrees of freedom?
    a. 1
    b. 2
    c. 3
    d. More than 3

References

1. Hatze, H. (1974). The meaning of the term "biomechanics." *Journal of Biomechanics*, 7:189-190.
2. European Society of Biomechanics. *The founding and goals of the society.* Available at http://www.esbiomech.org/current/about_esb/index.html/.
3. Richie, D. H., et al. (1985). Aerobic dance injuries: A retrospective study of instructors and participants. *The Physician and Sportsmedicine*, 13:130-140.
4. Ulibarri, V. D., et al. (1987). Ground reaction forces in selected aerobics movements. *Biomechanics in Sport.* New York: Bioengineering Division of the American Society of Mechanical Engineering, pp. 19-21.

SKELETAL CONSIDERATIONS FOR MOVEMENT

OBJECTIVES

After reading this chapter, the student will be able to:

1. Define how the mechanical properties of a structure can be expressed in terms of its stress–strain relationship.

2. Define stress, strain, elastic region, plastic region, yield point, failure point, and elastic modulus.

3. Identify the elastic region, yield point, plastic region, and failure point on a stress–strain curve.

4. Describe the difference between elastic and viscoelastic materials.

5. Differentiate between brittle, stiff, and compliant materials.

6. List the functions of the bone tissue that makes up the skeletal system.

7. Describe the composition of bone tissue and the characteristics of cortical and cancellous bone.

8. Identify the types of bones found in the skeletal system, and describe the role each type of bone plays in human movement or support.

9. Describe how bone tissue forms and the differences between modeling and remodeling.

10. Discuss the impact of activity and inactivity on bone formation.

11. Define osteoporosis, and discuss the development of osteoporosis.

12. Discuss the strength and stiffness of bone as well as bone's anisotropic and viscoelastic properties.

13. Define the following types of loads that bone must absorb, and provide an example to illustrate each load on the skeletal system: compression, tension, shear, bending, and torsion.

14. Describe stress fractures and other common injuries to the skeletal system, and explain the load causing the injury.

15. Describe the types of cartilage and their functions in the skeletal system.

16. Describe the function of ligaments in the skeletal system.

17. Describe all of the components of the diarthrodial joint, factors that contribute to joint stability, and examples of injury to the diarthrodial joint.

18. List the seven different types of diarthrodial joints, and provide examples of each one.

19. Describe the characteristics of the synarthrodial and amphiarthrodial joints, and provide an example of each.

20. Define osteoarthritis, and discuss the development of osteoarthritis.

OUTLINE

Measuring the Mechanical Properties
of Body Tissues
    Basic Structural Analysis
Biomechanical Characteristics of Bone
    Bone Tissue Function
    Composition of Bone Tissue
    Macroscopic Structure of Bone
    Bone Formation
Mechanical Properties of Bone
    Strength and Stiffness of Bone
    Loads Applied to Bone
    Stress Fractures

Cartilage
    Articular Cartilage
    Fibrocartilage
Ligaments
Bony Articulations
    The Diarthrodial or Synovial Joint
    Other Types of Joints
    Osteoarthritis
Summary
Review Questions

 Measuring the Mechanical Properties of Body Tissues

Bone, tendon, ligament, and muscle are some of the basic structures that make up the human body. Of great interest to biomechanists are the mechanical properties of these tissues. Generally, when analyzing the mechanical properties of such structures, we discern the external forces that are applied to the structure and relate them to the resulting deformation of the structure. The ability of a structure to resist deformation depends on its material organization and overall shape. Therefore, this type of analysis is important because it provides information on the mechanical properties of the structure that may ultimately influence its function.

BASIC STRUCTURAL ANALYSIS

Stress and Strain

The force applied to deform a structure and the resulting deformation is referred to as *stress* and *strain*, respectively. To enable comparison of structures of different sizes, stress and strain are scaled quantities of the force applied and the deformation of the structure, respectively. The values of stress and strain are measured using a machine that can place either tension (pulling stress) or compression (pushing stress) on the structure. In Figure 2-1, the load cell measures the tension, or pulling force, applied to the tendon, and the extensiometer measures the length to which the tendon is stretched. The actuator is a motor that initiates the pull on the tendon. Figure 2-2 shows a similar setup to determine the compression stress on an amputated foot. The graph relating stress to strain is the **stress–strain curve** of a structure. A stress–strain analysis can be used to discern how a material changes with age, how materials react to different force applications, and how a material reacts to lack of everyday stress. Figure 2-3 illustrates the stress–strain relationships of bone vertebrae of normal rhesus monkeys versus those that have been immobilized. A stress–strain analysis can be performed

with pulling force (tension), pushing force (compression), or shear force (a push or pull along the surface of the material). This book deals only with tension and compression stress–strain relationships.

In this type of test, **stress** is defined as the force per unit area and is designated with the Greek letter sigma (σ). Stress is calculated thus:

$$\sigma = F/A$$

where F is the applied force and A is the unit area over which the force is applied. The force is applied perpendicular

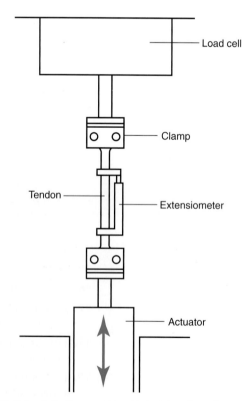

FIGURE 2-1 A testing machine that determines the stress–strain properties of a tendon. The actuator stretches the tendon. (Reprinted with permission from Alexander, R. M. [1992]. *The Human Machine.* New York: Columbia University Press.)

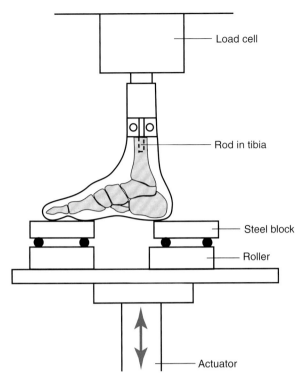

FIGURE 2-2 A testing machine that determines the stress–strain properties of an amputated foot. (Reprinted with permission from Alexander, R. M. [1992]. *The Human Machine*. New York: Columbia University Press.)

to the surface of the structure over a predetermined area. The unit in which a force is measured is the newton (N). The unit of area is the square meter (m²). Thus, the unit of stress is newton per square meter (N/m²), or the pascal (Pa).

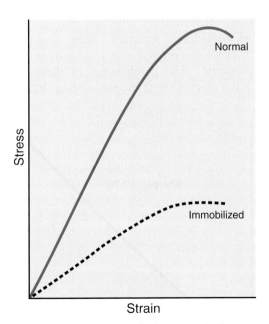

FIGURE 2-3 Stress–strain curves for bone vertebral segments from a normal and immobilized rhesus monkey. (Adapted from Kazarian, L. E., Von Gierke, H. E. [1969]. Bone loss as a result of immobilization and chelation. Preliminary results in *Macaca mulatta. Clinical Orthopaedics*, 65:67–75.)

Deformation or strain is also scaled to the initial length of the structure being tested. That is, the deformation caused by the applied stress is compared with the initial, or resting, length of the material, when no force is applied. Strain, designated by the Greek letter epsilon (ε), is therefore defined as the ratio of the change in length to the resting length. Thus,

$$\varepsilon = \Delta L / L$$

where ΔL is the change in length of the structure and L is the initial length. Because we are dividing a length by a length, there are no units, so strain is a dimensionless number.

A stress–strain curve is presented in Figure 2-4. A number of key points on this curve are important to the ultimate function of the structure. In this curve, the slope of the linear portion of the curve is the **elastic modulus**, or stiffness of the material. Stiffness is thus calculated as

$$k = \text{stress/strain} = \sigma/\varepsilon$$

As greater force is applied to the structure, the slope of the curve eventually decreases. At this point, the structure is said to **yield** or reach its yield point. Up to the yield point, the structure is said to be in the **elastic region**. If the stress is removed while the material is in this region, the material will return to its original length with no structural damage. After the yield point, the molecular components of the material are permanently displaced with respect to each other, and if the applied force is removed, the material will not return to its original length (Fig. 2-5). The difference between the original length of the material and the (resting) length resulting from stress into the plastic region is the **residual strain**.

The region after the yield point is the **plastic region**. For rigid materials, such as bone, the yield or plastic region is relatively small, but for other materials, it can be relatively large. If the applied force continues beyond the plastic region, the structure will eventually reach **failure**, at which point the stress quickly falls to zero. The maximal stress reached when failure occurs determines the failure strength and failure strain of the material.

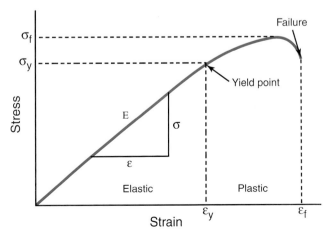

FIGURE 2-4 An idealized stress–strain curve showing the elastic and plastic regions and the elastic modulus.

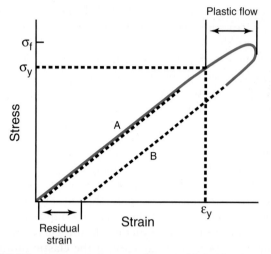

FIGURE 2-5 A stress–strain curve of a material that has been elongated into the plastic region. **(A)** The period of the applied load. **(B)** The period when the applied load is removed. The residual strain results because of the reorganization of the material at the molecular level.

In normal functional activities, the stress applied will not cause a strain that reaches the yield point. When structures are designed by an engineer, the engineer considers a **safety factor** when determining the stress–strain relationship of the structure. This safety factor is generally in the range of 5 to 10 times the stress that would normally be placed on the structure. That is, the applied force to reach the yield point is significantly greater than the force generally applied in everyday activities. It is obvious and has been suggested that biologic materials and biologic structures must have a significantly large safety factor. Needless to say, the stresses placed on a biologic structure in everyday activities are much less than the structure can handle. Figure 2-6 illustrates a stress–strain curve for a

human adult tibia and the actual stress–strain relationship during jogging.

When a structure is deformed by an applied force, the strain developed in the material relates to the mechanical energy absorbed by the material. The amount of mechanical energy stored is proportional to the area under the stress–strain curve (Fig. 2-7). That is, the stored mechanical energy is:

$$ME = 1/2\sigma\varepsilon$$

When the applied force is removed, the stored energy is released. For example, a rubber band can be stretched by pulling on both ends. When one end is released, the rubber band rebounds back to its original length but, in doing so, releases the energy stored during stretching. For practical purposes, this is the same concept as a trampoline. The weight of the person bouncing on it deforms the bed and stores energy. The trampoline rebounds and releases the stored energy to the person.

Types of Materials
Elastic
The idealized material described in Figure 2-4 is an **elastic material**. In this type of material, a linear relationship exists between the stress and strain. That is, when the material is deformed by the applied force, the amount of deformation is the same for a given amount of stress. When the applied load is removed, the material returns to its resting length as long as the material did not reach its yield point. In an elastic material, the mechanical energy that was stored is fully recovered.

Viscoelastic
As opposed to elastic structures, certain materials show stress–strain characteristics that are not strictly linear; these are **viscoelastic** materials. These structures have nonlinear or viscous properties in combination with linear elastic properties. The combination of these properties results in the magnitude of the stress being dependent on

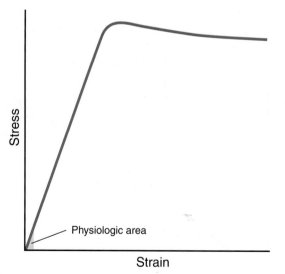

FIGURE 2-6 The *shaded area* represents the tension stress–strain values of an adult human tibia during jogging, and the *solid line* represents bone samples tested to failure. (Adapted with permission from Nordin, M., Frankel, V. H. [1989]. *Basic Biomechanics of the Musculoskeletal System*. Philadelphia, PA: Lea & Febiger.)

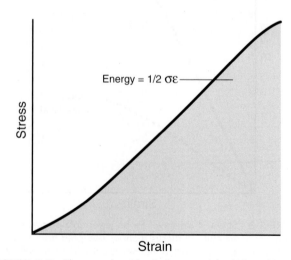

FIGURE 2-7 The stored mechanical energy (*shaded area*) is equal to the area under the stress–strain curve.

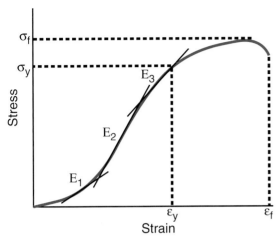

FIGURE 2-8 A stress–strain curve of a typical viscoelastic material. The elastic modulus (slope of the curve) varies according to the portion of the curve on which it is calculated.

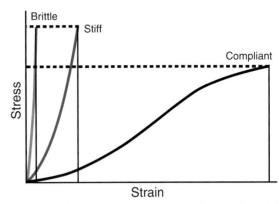

FIGURE 2-10 Stress–strain curves of compliant, stiff, and brittle materials. The elastic modulus is significantly different in the three materials.

the rate of loading, or how fast the load is applied. Nearly all biologic materials, such as tendon and ligament, show some level of viscoelasticity.

Figure 2-8 illustrates a viscoelastic material. On a stress–strain curve of a viscoelastic material, the terms *stiffness*, *yield point*, and *failure point* also apply. The elastic and plastic regions are defined similarly as in an elastic material. In contrast to an elastic structure, however, stiffness has several values that can be determined by where it was calculated on the curve. In Figure 2-8, the stiffness designated by E_1 is less than that of E_2. E_3, however, is certainly less than E_2. In addition, in a viscoelastic material, the stored mechanical energy is not completely returned when the applied load is removed. Thus, the energy returned is not equal to the energy stored. The energy that is lost is **hysteresis** (Fig. 2-9).

Materials, whether they are elastic or viscoelastic, are often referred to as stiff, compliant, or brittle, depending on the elastic modulus. Stress–strain curves of these materials are presented in Figure 2-10. A compliant

material has an elastic modulus less than that of a stiff material. The compliant material stores considerably more energy than a stiff material. On the other hand, a brittle material has a greater elastic modulus and stores less energy than a stiff material. Nonetheless, all of these terms are relative. Depending on the materials being tested, a brittle material may be considered stiff relative to one material and compliant relative to another. For example, bone is brittle relative to tendon but compliant relative to glass.

Biomechanical Characteristics of Bone

BONE TISSUE FUNCTION

The skeleton is built of bone tissue. Joints, or articulations, are the intersections between bones. Ligaments connect bones at the articulations, thus reinforcing the joints. The skeleton consists of approximately 20% of total body weight. The skeletal system is generally broken down into axial and appendicular skeletons. The major bones of the body are presented in Figure 2-11. Bone tissue performs many functions, including support, attachment sites, leverage, protection, storage, and blood cell formation.

Support
The skeleton provides significant structural support and can maintain a posture while accommodating large external forces, such as those involved in jumping. The bones increase in size from top to bottom in proportion to the amount of body weight they bear; thus, the bones of the lower extremities and the lower vertebrae and pelvic bones are larger than their upper extremity and upper torso counterparts. A visual comparison of the humerus and femur or the cervical vertebrae and lumbar vertebrae demonstrates these size relationships. Internally, bones also protect the internal organs.

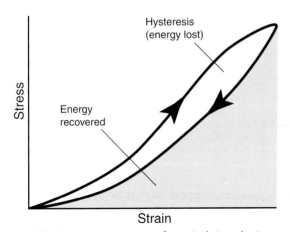

FIGURE 2-9 A stress–strain curve of a typical viscoelastic material showing the energy recovered when the material is allowed to return to its resting length. The hysteresis or energy lost is equal to the energy stored when the material is deformed minus the energy recovered.

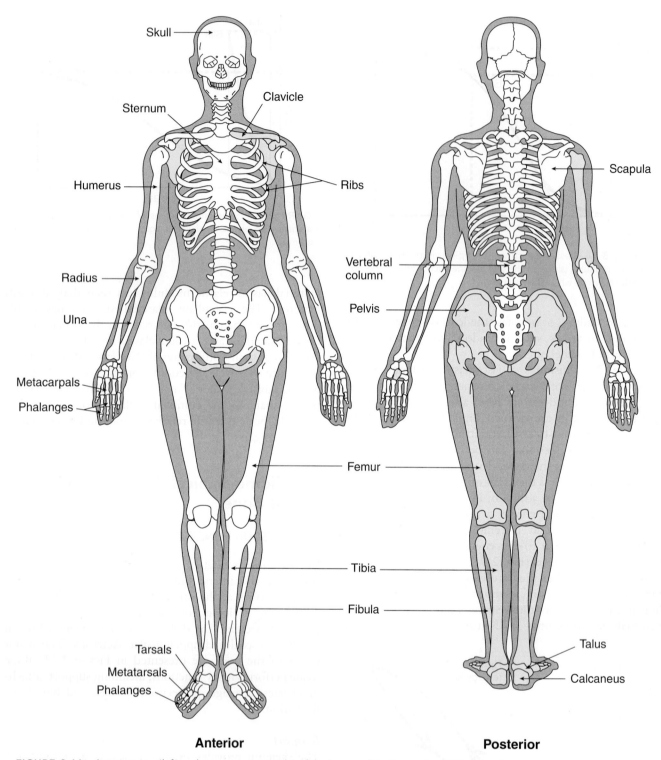

FIGURE 2-11 Anterior view (*left*) and posterior view (*right*) of the bones of the human body. (Reprinted with permission from Willis, M. C. [1996]. *Medical Terminology: The Language of Health Care.* Baltimore, MD: Lippincott Williams & Wilkins.)

Attachment Sites

Bones provide sites of attachment for tendons, muscles, and ligaments, allowing for the generation of movement through force applications to the bones through these sites. Knowledge of attachment sites on each bone provides good information about the movement potential of specific muscles, support offered by ligaments, and potential sites of injury.

Leverage

The skeletal system provides the levers and axes of rotation about which the muscular system generates the movements. A **lever** is a simple machine that magnifies the force, speed, or both of movements and consists of a rigid rod that is rotated about a fixed point or axis called the **fulcrum**. The rigid rod in a skeletal lever system is primarily one of the longer bones of the body, and the fixed point of rotation or axis is one of the joints where the bones meet. The skeletal lever system transmits movement generated by muscles or external forces. Chapter 10 provides an in-depth discussion of levers.

Other Functions

Three additional bone functions are not specifically related to movement: protection, storage, and blood cell formation. The bone protects the brain and internal organs. Bone also stores fat and minerals and is the main store of calcium and phosphate. Finally, blood cell formation, called hematopoiesis, takes place within the cavities of bone.

COMPOSITION OF BONE TISSUE

Bone, or **osseous** tissue, is a remarkable material with properties that make it ideal for its support and movement functions. It is light but it has high tensile and compressive strength and a significant amount of elasticity. Bone is also a very dynamic material with minerals moving in and out of it constantly. As much as a half a gram of calcium may enter or leave the adult skeleton every day, and humans recycle 5% to 7% of their bone mass every week. Bone can also be made to grow in different ways, and it is tissue that is continually being modified, reshaped, remodeled, and overhauled.

Osseous tissue is strong and is one of the body's hardest structures because of its combination of inorganic and organic elements. Bone is composed of a matrix of inorganic salts and **collagen**, an organic material found in all connective tissue. The inorganic minerals, calcium and phosphate, along with the organic collagen fibers, make up approximately 60% to 70% of bone tissue. Water constitutes approximately 25% to 30% of the weight of bone tissue (43). Collagen provides the bone with tensile strength and flexibility, and the bone minerals provide compressive strength and rigidity (38).

There are three types of bone cells (Fig. 2-12). **Osteoclasts** are large multinucleated cells that have properties similar to macrophages. They are created by the fusion of 15 to 20 individual cells, and they work to dissolve the bone in areas of microfracture. **Osteoblasts** have a single nucleus, and they produce a new bone called osteoid. These cells are also responsible for the calcification of the bone. Some of the osteoblasts are trapped within the formation of the new bone and develop into **osteocytes**. The osteocytes are similar to nerve cells because they have long extensions that reach out and connect to other osteocytes. They appear to have a mechanosensory and communicative function that allows them to sense strain and to direct osteoclast activity.

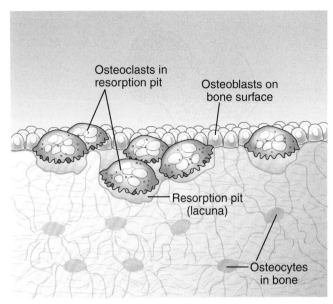

FIGURE 2-12   The three main types of bone cells are osteoclasts, osteoblasts, and osteocytes.

MACROSCOPIC STRUCTURE OF BONE

Bone is composed of two types of tissue: **cortical bone** and **cancellous bone**. The hard outer layer is cortical bone; internal to this is cancellous bone. A section of the femoral head presented in Figure 2-13 illustrates the architecture of the long bone. The architectural arrangement of bony tissue is remarkably well suited for the mechanical demands imposed on the skeletal system during physical activity.

Cortical Bone

Cortical bone is often referred to as compact bone and constitutes about 80% of the skeleton. Cortical bone looks solid, but closer examination reveals many passageways for blood vessels and nerves. The exterior layer of the bone is very dense and has a **porosity** less than 15% (48). Porosity is the ratio of pore space to the total volume; when porosity increases, bone mechanical strength deteriorates. Small changes in porosity can lead to significant changes in the stiffness and strength of the bone.

Cortical bone consists of a system of hollow tubes called **lamellae** that are placed inside one another. Lamellae are composed of collagen fibers, all running in a single direction. The collagen fibers of adjacent lamellae always run in different directions. A series of lamellae form an **osteon** or **Haversian system**. Osteons are pillar-like structures that are oriented parallel to the stresses that are placed on the bone. The arrangement of these weight-bearing pillars and the density of the cortical bone provide strength and stiffness to the skeletal system. Cortical bone can withstand high levels of weight bearing and muscle tension in the longitudinal direction before it fails and fractures (46).

Cortical bone is especially capable of absorbing tensile loads if the collagen fibers are parallel to the load.

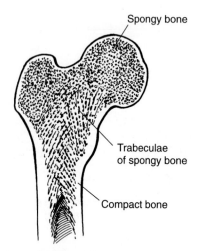

Spongy bone

Trabeculae
of spongy bone

Compact bone

FIGURE 2-13 Midsection of the proximal end of the femur showing both cortical bone and cancellous bone. The dense cortical bone lines the outside of the bone, continuing down to form the shaft of the bone. Cancellous bone is found in the ends and is distinguishable by its lattice-like appearance. Note the curvature in the trabeculae, which forms to withstand the stresses.

Typically, the collagen is arranged in layers running in longitudinal, circumferential, and oblique configurations. This offers resistance to tensile forces in different directions because the more layers there are, the greater the strength and stiffness the bone has. Also, where muscles, ligaments, and tendons attach to the skeleton, collagen fibers are arranged parallel to the insertion of the soft tissue, thereby offering greater tensile strength for these attachments.

A thick layer of cortical bone is found in the shafts of long bones, where strength is necessary to respond to the high loads imposed down the length of the bone during weight bearing or in response to muscular tension. The thickness is greater in the middle part of the long bones because of the increased bending and torsion forces (9). Thin layers of cortical bone are found on the ends of the long bones, the epiphyses, and covering the short and irregular bones.

Cancellous Bone

The bone tissue interior to the cortical bone is referred to as cancellous or spongy bone. Cancellous bone is found in the ends of the long bones, in the body of the vertebrae, and in the scapulae and pelvis. This type of bone has a lattice-like structure with a porosity greater than 70% (48). Cancellous bone structure, although quite rigid, is weaker and less stiff than the cortical bone. Cancellous bone is not as dense as cortical bone because it is filled with spaces. The small, flat pieces of bone that serve as small beams between the spaces are called **trabeculae** (Fig. 2-13). The trabeculae adapt to the direction of the imposed stress on the bone, providing strength without adding much weight (11). Collagen runs along the axis of the trabeculae and provides cancellous bone with both tensile and compressive resistance.

The high porosity gives cancellous bone high-energy storage capacity so that this type of bone becomes a crucial element in energy absorption and stress distribution when loads are applied to the skeletal structure (43). This type of bone is metabolically more active and responsive to stimuli than cortical bone (30). It has a much higher turnover rate than cortical bone, ending up with more remodeling along the lines of stress (38). Cancellous bone is not as strong as cortical bone, and there is a high incidence of fracture in the cancellous bone of elderly individuals. This is believed to be caused by the loss of compressive strength because of mineral loss (osteoporosis).

Anatomical Classification of Bones

The skeletal system has two main parts: the axial (skull, spine, ribs, and sternum) and the appendicular (shoulder and pelvic girdles and arms and legs) skeleton. Making up each section of the skeleton are four types of bones (Fig. 2-14). These include bones designated as long, short, flat, and irregular. Each type of bone performs specific functions.

Long Bones

The long bones are longer than they are wide. The long bones in the body are the clavicle, humerus, radius, ulna, femur, tibia, fibula, metatarsals, metacarpals, and phalanges. The long bone has a shaft, the **diaphysis**, a thick layer of cortical bone surrounding the bone marrow cavity (Fig. 2-15). The shaft widens toward the end into the section called the **metaphysis**. In the immature skeleton, the end of the long bone, the **epiphysis**, is separated from the diaphysis by a cartilaginous disk. The epiphyses consist of a thin outer layer of cortical bone covering cancellous inner bone. A thin white membrane, the **periosteum**, covers the outside of the bone with the exception of the parts covered by cartilage.

Long bones offer the body support and provide the interconnected set of levers and linkages that allow us to move. A long bone can act as a column by supporting loads along its long axis. Long bones typically are not straight; rather, they are beam shaped, which creates a stronger structure so bones can handle and minimize bending loads imposed on them. A long bone is strongest when it is stressed by forces acting along the long axis of the bone. Muscle attachment sites and protuberances are formed by tensile forces of muscles pulling on the bones.

Short Bones

The short bones, such as the carpals of the hand and tarsals of the foot, consist primarily of cancellous bone covered with a thin layer of cortical bone. These bones play an important role in shock absorption and the transmission of forces. A special type of short bone, the sesamoid bone, is embedded in a tendon or joint capsule. The patella is a sesamoid bone at the knee joint that is embedded in the tendon of the quadriceps. Other sesamoid bones can be

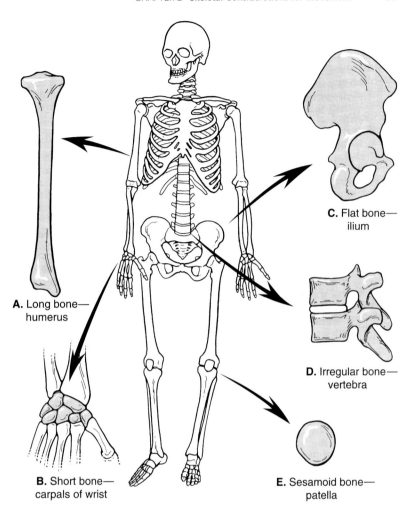

A. Long bone—
humerus

B. Short bone—
carpals of wrist

C. Flat bone—
ilium

D. Irregular bone—
vertebra

E. Sesamoid bone—
patella

FIGURE 2-14 Various types of bones serve specific functions. **(A)** Long bones serve as levers. **(B)** Short bones offer support and shock absorption. **(C)** Flat bones protect and offer large muscular attachment sites. **(D)** Irregular bones have specialized functions. **(E)** Sesamoid bones alter the angle of muscular insertion.

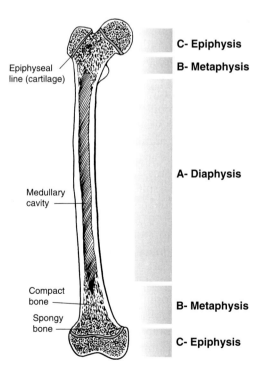

Epiphyseal
line (cartilage)

Medullary
cavity

Compact
bone

Spongy
bone

C- Epiphysis

B- Metaphysis

A- Diaphysis

B- Metaphysis

C- Epiphysis

FIGURE 2-15 The long bone has a shaft, or diaphysis **(A)**, which broadens out into the metaphysis **(B)** and the epiphysis **(C)**. Layers of cortical bone make up the diaphysis. The metaphysis and epiphysis are made up of cancellous bone inside a thin layer of cortical bone.

found at the base of the first metatarsal in the foot, where the bones are embedded in the distal tendon of the flexor hallucis brevis muscle, and at the thumb, where the bones are embedded in the tendon of the flexor pollicis brevis muscle. The role of the sesamoid bone is to alter the angle of insertion of the muscle and to diminish friction created by the muscle.

Flat Bones

A third type of bone, the flat bone, is represented by the ribs, ilium, sternum, and scapula. These bones consist of two layers of cortical bone with cancellous bone and marrow in between. Flat bones protect internal structures and offer broad surfaces for muscular attachment.

Irregular Bones

Irregular bones, such as those found in the skull, pelvis, and vertebrae, consist of cancellous bone with a thin exterior of cortical bone. These bones are termed irregular because of their specialized shapes and functions. The irregular bones perform a variety of functions, including supporting weight, dissipating loads, protecting the spinal cord, contributing to movement, and providing sites for muscular attachment.

BONE FORMATION

Bone is a highly adaptive material that is very sensitive to disuse, immobilization, or vigorous activity and high levels of loading. Across the life span, bone is continually optimized for its load-bearing role through functionally adaptive remodeling. Changes occur in whole bone architecture and bone mass as a functional adaptation occurs where the bone mass and architecture is matched to functional demand (40). In the appendicular skeleton, this is particularly important because of the load bearing. Adaptive changes are maximum in growing bone and decrease with aging but still occur to some level as the skeleton adapts to changes in mechanical use. Bone tissue is self-repairing and can alter its properties and configuration in response to mechanical demand. This was first determined by the German anatomist Julius Wolff, who provided the theory of bone development, termed Wolff's law. This law states: "Every change in the form and function of a bone or of their function alone is followed by certain definitive changes in their internal architecture and equally definite secondary alteration in their external conformation, in accordance with mathematical laws" (32).

Ossification, Modeling, and Remodeling

The formation of bone is a complex process that cannot be fully explored in detail here. Bone is always formed by the replacement of some preexisting tissue. In fetuses, much of the preexisting tissue is hyaline cartilage. **Ossification** is the formation of bone by the activity of the osteoblasts and osteoclasts. In fetuses, the cartilage is slowly replaced through this process so that at the time of birth, many of the bones have been at least partially ossified. In sites such as the flat bones of the skull, bone replaces a soft fibrous tissue instead of cartilage.

Long bones grow from birth through adolescence through activity at cartilage plates located between the shaft and the heads of the bones. These **epiphyseal plates** expand as new cells are formed and the bone is lengthened until the thicknesses of the plates diminish to reach what is called full ossification. This occurs in different bones at different ages but is usually complete by age 25 years.

Modeling occurs during growth to create new bone as bone resorption and bone formation (ossification) occur at different locations and rates to change the bone shape and size. In growing bone, bone properties are related to the growth-related demands on size and to changes in tensile and compressive forces acting on the body. Bone is deposited by osteoblasts while it is also being resorbed by osteoclasts. In the process of **resorption**, old bone tissue is broken down and digested by the body. This process is not the same in all bones or even in a single bone. For example, whereas the bone in the distal part of the femur is replaced every five to six months, the bone in the shaft is replaced much more slowly.

Living bone is always undergoing **remodeling** in which the bone matrix is constantly being removed and replaced.

The removal of bone by the osteoclasts is relatively quick—about three weeks—while formation of new bone by the osteoblasts takes about three months. The thickness and strength of bone must be continually maintained by the body, and this is done by an ongoing cycle of replacing old bone with new bone. A dynamic steady state is maintained by replacing a small amount of bone at the same site while leaving the size and shape of the remodeled bone basically the same. At least some new bone is being formed continually, and bone remodeling is the process through which bone mass adapts to the demands placed on it. After an individual is past the growing stage, the rate of bone deposit and resorption are equal to each other, keeping the total bone mass fairly constant. Through exercise, however, bone mass can be increased, even up through young adulthood. Bone deposits exceed bone resorption when greater strength is required or when an injury has occurred. Thus, weight lifters develop thickenings at the insertion of very active muscles, and bones are densest where stresses are greatest. The dominant arms of professional tennis players have cortical thicknesses that are 35% greater than the contralateral arm (32). The shape of bone also changes during fracture healing.

This ongoing rebuilding process continues up until age 40 years, when the osteoblastic activity slows and bones become more brittle. This remodeling process has two major benefits: The skeleton is reshaped to respond to gravitational, muscular, and external contact forces, and blood calcium levels are maintained for important physiologic functions.

Bone Tissue Changes across the Life Span

In immature bone, the fibers are randomly distributed, providing strength in multiple directions but lower overall strength. In mature bone, mineralization takes place, Haversian canals are created and lined with bone, and fibers are oriented in the primary load-bearing directions. Bone continues to reorganize throughout life to mend damage and to repair wear on the bone. In older bone, bone restoration still occurs, but the Haversian system is smaller, and the canals are larger because of slower bone deposits. There is some indication that this structural adjustment may be a result of decreased muscle strength, leading to partial disuse and subsequent bone remodeling that reduces strength (20).

Physical Activity and Inactivity and Bone Formation
Physical Activity

Bones require mechanical stress to grow and strengthen. Bones slowly add or lose mass and alter form in response to alterations in mechanical loading. Thus, physical activity is an important component of the development and maintenance of skeletal integrity and strength. Bone tissue must have a daily stimulus to maintain health. Muscle contraction in active movement coupled with external forces exerts the biggest pressure on bones. Not all exercises are equally effective. Overloading forces must be applied to the bone to stimulate and adapt force, and continued adaptation requires a progressive overload (33).

Generally, dynamic loading is better for bone formation than static loading, loading at higher frequencies is more effective, and prolonged exercise has diminishing returns (52). Repetitive, coordinated bone loading associated with habitual activity may have little role in preserving bone mass and may even reduce osteogenic potential because bone tissue becomes desensitized (40). Shorter periods of vigorous activity are more efficient in promoting an increase in bone mass (40). To stimulate an osteogenic effect in adult bone, four cycles a day of loading has been shown to be sufficient to stop bone loss (40). The daily applied loading history, comprising the number of loading cycles and the stress magnitude, influences the density of the bone. Again, it is recommended that one long session be broken into smaller sessions such as four sessions per day or three to five daily sessions per week (47,51).

The effect of physical activity on increasing bone mass varies across the life span. In the growing skeleton, loads applied to the skeleton provide a much greater stimulus than to the mature skeleton (52). In older adults with low bone mass, exercise is only moderately effective in bone building. The goal is to maximize the gain in **bone mineral density** in the first three decades of life and then minimize the decline after age 40 years (33). Bone mass reaches maximum levels between the ages of 18 and 35 years (9) and then decrease by about 0.5% per year after age 40 years (33) (Fig. 2-16). In adulthood, bone mass is the maximum bone mass minus the quantity lost, so exercise may be effective in just attenuating the rate of loss rather than increasing bone (33).

Which exercises are best?

There are several principles that help determine the best exercise for bone: 1) The forces and the rate of force development should be high (impact activities), 2) the number of impacts does not need to be great because bone becomes desensitized to mechanical stimulus rather quickly, 3) the sensitivity to mechanical stimulus recovers with rest, and 4) loading from different directions increases osteogenesis. Form groups to discuss and rank the following sport activities according to their potential for building bone: swimming, cycling, gymnastics, running, cross-country skiing, basketball, and soccer.

Turner C, Robling A. (2003) Designing exercise regimens to increase bone strength. *Exercise and Sport Sciences Reviews*, 31(1):45-50.

Inactivity

Bone loss after a decrease in the activity level may be significant (56). When under loaded in conditions such as fixation and bed rest, bone mass is resorbed, resulting in reduced bone mass and thinning. The skeletal system senses changes in load patterns and adapts to carry the load most efficiently using the least amount of bone mass. In microgravity conditions, astronauts, subjected to reduced activity and the loss of body weight influences, lose significant bone mass in relatively short periods. Some of the changes that occur to bone as a result of space travel include loss of rigidity, increased bending displacement, a decrease in bone length and cortical cross section, and slowing of bone formation (57).

What exercise prescription will facilitate bone growth in children and adolescents?

Exercise Mode	Exercise Intensity	Exercise Frequency	Exercise Duration
Impact activities (gymnastics, plyometrics, jumping); moderate-intensity resistance training; sports involving running and jumping	High; weight training, <60% of 1RM for safety purposes	At least 3 d per week	10–20 min

Source: Kohrt, W. M., et al. (2004) Physical activity and bone health. ACSM position stand. *Medicine & Science in Sports & Exercise*, 36:1985-1996.

What exercise prescription will help preserve bone health in adults?

Exercise Mode	Exercise Intensity	Exercise Frequency	Exercise Duration
Weight-bearing endurance activities (tennis, stair climbing, jogging); activities involving jumping (basketball, volleyball); resistance exercise	Moderate to high	Weight-bearing endurance activities 3–5 d per week; resistance training 2 d per week	30–60 min

Source: Kohrt, W. M., et al. (2004) Physical activity and bone health. ACSM position stand. *Medicine & Science in Sports & Exercise*, 36:1985-1996.

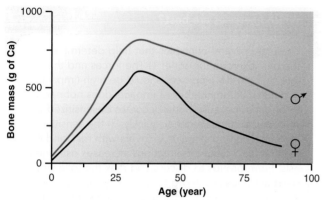

FIGURE 2-16 Peak bone mass occurs during the late third decade of life. Females have a lower peak bone mass and greater reductions in later life, especially after menopause.

Osteoporosis

In osteoporosis, bone resorption exceeds bone deposits. Osteoporosis is a disease of increasing bone fragility that is initially subtle, affecting only the trabeculae in cancellous bone, but leads to more severe examples in which one might experience an osteoporotic vertebral fracture just opening a window or rising from a chair (40,51). Bone fragility depends on the ultimate strength of the bone, the level of brittleness in the bone, and the amount of energy the bone can absorb (51). These factors are influenced by bone size, bone shape, bone architecture, and bone tissue quality. The symptoms of osteoporosis often begin to appear in elderly individuals, especially post-menopausal women. Osteoporosis may begin earlier in life, however, when bone mineral density decreases. When bone deposition cannot keep up with bone resorption, bone mineral mass decreases, resulting in reduced bone density accompanied by loss of trabecular integrity. The loss of bone mineral density means loss of the stiffness in the bone, and the loss of trabecular integrity weakens the structure. Both of these losses create the potential for a much greater incidence of fracture (12), ranging from 2.0% to 3.7% in nonosteoporotic individuals and almost doubling to 5% to 7% in osteoporotic individuals (28). A reason for higher fracture rates may be the higher strain in the osteoporotic bone under similar loading patterns. For example, the osteoporotic femoral head was shown to handle only 59% of the original external load in walking with strains 70% higher than normal and less uniformly distributed (53).

The exact causes of osteoporosis are not fully understood, but the condition has been shown to be related to genetics, hormonal factors, nutritional imbalances, and lack of exercise. Normal bone volume is 1.5 to 2 L, and the cortical diameter of bone is at its maximum between ages 30 and 40 years for both men and women (19,44). After age 30 years, a 0.2% to 0.5% yearly loss in the mineral weight of bone occurs (56), accelerating after menopause in women to bone loss

that is 50% greater than in men of similar age (44). It is speculated that a substantial proportion of this bone loss may be related to the accompanying reduction in activity level (56).

Lifestyle and activity habits seem to play an important role in the maintenance of bone health (13). In one study, the incidence of osteoporosis was 47% in a sedentary population compared with only 23% in a population whose occupations included hard physical labor (8). It is clear that elderly individuals may benefit from some form of weight-bearing exercise that is progressive and of at least moderate resistance.

Estrogen levels in anorexic women and amenorrheic female athletes have also been related to the presence of osteoporosis in this population. There is speculation that stress fractures in the femoral neck of female runners may be related to a noted loss of bone mineral density caused by osteoporosis (12). Elite female athletes in a variety of sports have had bone loss, usually associated with bouts of heavy training and associated menstrual irregularity. Some of these athletes have lost so much bone mass that their skeletal characteristics resemble those of elderly women.

Mechanical Properties of Bone

The mechanical properties of bone are as complex and varied as its composition. The measurement of bone strength, stiffness, and energy depends on both the material composition and the structural properties of bone. In addition, the mechanical properties also vary with age and gender and with the location of the bone, such as the humerus versus the tibia. Additional variation may result from other factors such as orientation of load applied to the bone, rate of load application, and type of load.

Bone must be capable of withstanding a variety of imposed forces simultaneously. In a static position, bone resists the force of gravity, supports the weight of the body, and absorbs muscular activity produced to maintain the static posture. In a dynamic mode such as running, the forces are magnified many times and become multi-directional.

STRENGTH AND STIFFNESS OF BONE

The behavior of any material under loading conditions is determined by its strength and stiffness. When an external force is applied to a bone or any other material, an internal reaction occurs. The strength can be evaluated by examining the relationship between the load imposed (external force) and the amount of deformation (internal reaction) occurring in the material.

As noted earlier, bone must be stiff yet flexible and strong yet light. Strength is necessary for load bearing, and lightness is necessary to allow movement. The strength in weight-bearing bones lies in their ability to resist bending by being stiff. Flexibility is needed to absorb high-impact forces, and the elastic properties of bone allow it to absorb energy by changing shape without failing and then return to its normal length. If the imparted energy exceeds the zone of elastic deformation, plastic deformation occurs at the price of microdamage to the bone. If both the elastic and plastic zones are exceeded, the imparted energy is released in the form of a fracture.

Strength

The strength of bone or any other material is defined by the failure point or the load sustained before failure. The overall ability of the bone to bear a load depends on having sufficient bone mass with adequate material properties and fiber arrangement that resist loading possibilities in different directions (40). The failure of bone depends on the type of load imposed (Fig. 2-17); there is actually no standardized strength value for bone because the measurement is so dependent on the type of bone and testing site. Failure of bone involves either a single traumatic event or the accumulation of micro-fractures. Thus, both fracture and fatigue behaviors of bone are important. Strength of bone is provided by the mineralization of its tissue: the greater the tissue mineral content, the stiffer and stronger the material. If bone becomes too mineralized, however, it becomes brittle and does not give during impact loading. Strength is assessed in terms of energy storage or the area under a stress–strain curve.

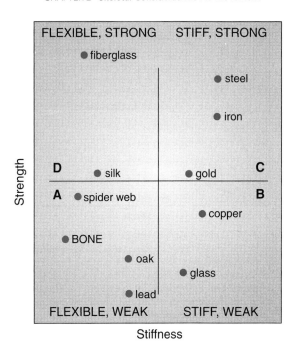

FIGURE 2-18 The strength and stiffness of a variety of materials are plotted in four quadrants representing material that is flexible and weak (**A**), stiff and weak (**B**), stiff and strong (**C**), and flexible and strong (**D**). Bone is categorized as being flexible and weak, along with other materials, such as spider web and oak wood. (Adapted with permission from Shipman, P., Walker, A., Bichell, D. [1985]. *The Human Skeleton.* Cambridge, MA: Harvard University Press.)

The compressive strength of cortical bone is greater than that of concrete, wood, or glass (Fig. 2-18). The strength of cortical bone in the middle of long bones is demonstrated in the ability to tolerate large impact loads and resist bending. Cancellous bone strength is less than that of cortical bone, but cancellous bone can undergo more deformation before failure.

> **What makes a skeleton stronger and less brittle?**
>
> 1. An increase in bone mass
> 2. Effective distribution of bone mass so more bone is present where mechanical demand is greater
> 3. Improved bone material properties

Stiffness

Stiffness, or the modulus of elasticity, is determined by the slope of the load deformation curve in the elastic response range and is representative of the material's resistance to load as the structure deforms. The stress–strain curve for ductile, brittle, and bone material is shown in Figure 2-19, and Figure 2-18 plots a variety of materials according to strength and stiffness (49). Metal is a type of ductile material that has high stiffness, and at stresses beyond its yield point, exhibits ductile behavior in which it undergoes

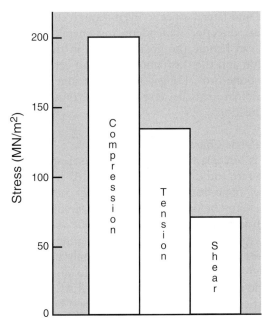

FIGURE 2-17 Ultimate stress for human adult cortical bone specimens. (Adapted with permission from Nordin, M., Frankel, V. H. (Eds.) [1989]. *Basic Biomechanics of the Musculoskeletal System.* Philadelphia, PA: Lea & Febiger.)

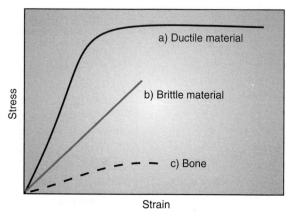

FIGURE 2-19 Stress–strain curves illustrating the differences in the behavior between ductile material **(A)**, brittle material **(B)**, and bone **(C)**, which has both brittle and ductile properties. When a load is applied, a brittle material responds linearly and fails or fractures before undergoing any permanent deformation. The ductile material enters the plastic region and deforms considerably before failure or fracture. Bone deforms slightly before failure.

large plastic deformation before failure. Glass is a brittle material that is stiff but fails early, having no plastic region. Bone is not as stiff as glass or metal, and unlike these materials, it does not respond in a linear fashion because it yields and deforms nonuniformly during the loading phase (43). Bone has a much lower level of stiffness than metal or glass and fractures after very little plastic deformation.

At the onset of loading, bone exhibits a linearly elastic response. When a load is first applied, a bone deforms through a change in length or angular shape. Bone deforms no more than approximately 3% (50). This is considered in the elastic region of the stress–strain curve because when the load is removed, the bone recovers and returns to its original shape or length. With continued loading, the bone tissue reaches its yield point, after which its outer fibers begin to yield, with microtears and debonding of the material in the bone. This is termed the *plastic region* of the stress–strain curve. The bone tissue begins to deform permanently and eventually fractures if loading continues in the plastic region. Thus, when the load is removed, the bone tissue does not return to its original length but stays permanently elongated. Although bone can exhibit a plastic response, normal loading remains well within the elastic region. Bone behaves largely like a brittle material, exhibiting very little permanent plastic deformation to failure.

Anisotropic Characteristics

Bone tissue is an anisotropic material, which means that the behavior of bone varies with the direction of the load application (Fig. 2-20). The differences between the properties of the cancellous and cortical bone contribute to the anisotropy of bone. The contribution of cancellous and cortical components of whole bone to overall strength varies with anatomical location because variable amounts of cortical and cancellous bone are present at every site. Cancellous bone provides bending

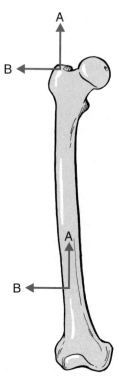

FIGURE 2-20 Bone is considered anisotropic because it responds differently if forces are applied in different directions. **(A)** Bone can handle large forces applied in the longitudinal direction. **(B)** Bone is not as strong in handling forces applied transversely across its surface.

strength, and cortical bone provides significant compressive strength. Within each bone, considerable variation exists, as seen in the femur, where the bone is weaker and less stiff in the anterior and posterior aspects than in the medial and lateral aspects. Even though the properties of bone are direction dependent, in general, tissue of long bones can handle the greatest loads in the longitudinal direction and the least amount of load across the surface of the bone (46). Long bones are stronger withstanding longitudinal loads because they are habitually loaded in that direction.

Viscoelastic Characteristics

Bone is viscoelastic, meaning that its response depends on the rate and duration of the load. At higher speeds of loading, bone becomes stiffer, tougher and it can absorb more energy before breaking. These strain rates are seen in high-impact situations involving falls or vehicular accidents. As shown in Figure 2-21, a bone loaded slowly fractures at a load that is approximately half of the load handled by the bone at a fast rate of loading.

Bone tissue is a viscoelastic material whose mechanical properties are affected by its deformation rate. The ductile properties of the bone are provided by its collagenous material. The collagen content gives the bone the ability to withstand tensile loads. Bone is also brittle, and its strength depends on the loading mechanism. The brittleness of bone is provided by the mineral constituents that provide bone with the ability to withstand compressive loads.

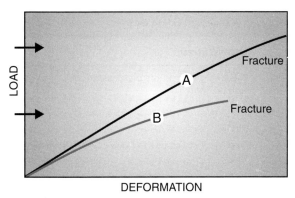

FIGURE 2-21 Bone is considered viscoelastic because it responds differently when loaded at different rates. **(A)** When loaded quickly, bone responds with more stiffness and can handle a greater load before fracturing. **(B)** When loaded slowly, bone is not as stiff or strong, fracturing under lower loads.

LOADS APPLIED TO BONE

The skeletal system is subject to a variety of applied forces as bone is loaded in various directions. Loads are produced by weight bearing, gravity, muscular forces, and external forces. Internally, loads can be applied to bones through the joints by means of ligaments or at tendinous insertions, and these loads are usually below any fracture level. Externally, bone accommodates multiple forces from the environment that have no limit on magnitude or direction.

Muscular activity can also influence the loads that bone can manage. Muscles alter the forces applied to the bone by creating compressive and tensile forces. These muscular forces may reduce tensile forces or redistribute the forces on the bone. Because most bones can handle greater compressive forces, the total amount of load can increase with the muscular contribution. If muscles fatigue during an exercise bout, their ability to alleviate the load on the bone diminishes. The altered stress distribution or increase in tensile forces leaves the athlete or performer susceptible to injury.

The stress and strain produced by forces applied to bones are responsible for facilitating the deposit of osseous material. Stress can be perpendicular to the plane of a cross section of the loaded object. This is termed **normal stress**. If stress is parallel to the plane of the cross section, it is termed **shear stress**. Each type of stress produces a strain. For example, whereas normal strain involves a change in the length of an object, shear strain is characterized by a change in the original angle of the object. An example of both normal strain and shear strain is the response of the femur to weight bearing. The femur shortens in response to normal strain and bends anteriorly in response to shear strain imposed by the body weight (46). Normal stress and shear stress, developed in response to tension applied to the tibia, are presented in Figure 2-22. Normal and shear strain, developed in response to compression of the femur, are also illustrated.

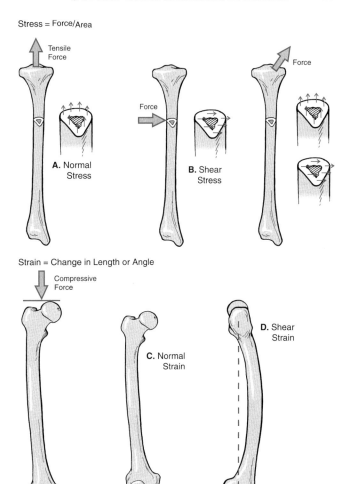

FIGURE 2-22 Stress, or force per unit area, can be perpendicular to the plane (normal stress) **(A)** or parallel to the plane (shear stress) **(B)**. Strain, or deformation of the material, is normal **(C)**, in which the length varies, or shear **(D)**, in which the angle changes.

Whether or not a bone incurs an injury as a result of an applied force is determined by the critical strength limits of the material and the loading history of the bone. External factors related to fracture include the magnitude, direction, and duration of the force coupled with the rate at which the bone is loaded. The ability of a bone to resist fracture is related to its energy-absorbing capacity. The ability of a bone to resist deformation varies through its length because of the different makeup of cortical and cancellous bone (38). Cancellous bone, depending on its architecture, can deform more and can absorb considerably more energy than cortical bone (38). These limits are primarily influenced by the loading on the bone. The loading of the bone can be increased or decreased by physical activity and conditioning, immobilization, and skeletal maturity of the individual. The rate of loading is also important because the response and tolerance of bone is rate sensitive. At high rates of loading, when bone tissue cannot deform fast enough, an injury can occur.

TABLE 2-1	Different Types of Loads Acting on Bone		
Load	Type of Force	Source	Stress/Strain
Compression	Presses ends of bones together to cause widening and shortening	Muscles, weight bearing, gravity or external forces	Maximal stress on the plane perpendicular to the applied load
Tension	Pulls or stretches the bone to cause lengthening and narrowing	Usually pull of contracting muscle tendon	Maximal stress on the plane perpendicular to the applied load
Shear	Force applied parallel to surface, causing internal deformation in an angular direction	Compressive or tension force application or external force	Maximum stress on the plane parallel to the applied load
Bending	Force applied to the bone having no direct support from the structure	Weight bearing or multiple forces applied at different points on the bone	Maximum tensile forces on the convex surface of the bent member and maximum compression forces on the concave side
Torsion	Twisting force	Force applied with one end of the bone fixed	Maximum shear stress on both the perpendicular and parallel to axes of bone with tension and compression forces also present at an angle across the surface

The five types of forces applying loads to bone are compression, tension, shear, bending, and torsion. These forces are summarized in Table 2-1 and illustrated in Figure 2-23.

Compression Forces

Compressive forces are necessary for development and growth in the bone. Specific bones need to be more suited to handle compressive forces. For example, the femur carries a large portion of the body's weight and needs to be stiff to avoid compression when loaded. The loads acting on the femur have been measured in the range of 1.8 to 2.7 body weight during one-leg standing and as high as 1.5 body weight in a leg lift in bed (2).

If a large compressive force is applied and if the load surpasses the stress limits of the structure, a compression fracture will occur. Numerous sites in the body are susceptible to compressive fractures. Compressive forces are responsible for patellar pain and softening and destruction of the cartilage underneath the patella. As the knee joint moves through a range of motion, the patella moves up and down in the femoral groove. The load between the patella and the femur increases and decreases to a point at which the compressive patellofemoral force is greatest at approximately 50° of flexion and least at full extension or hyperextension of the knee joint. The high-compressive force in flexion, primarily on the lateral patellofemoral surface, is the source of the destructive process that breaks down the cartilage and the underlying surface of the patella (17).

Compression is also the source of fractures to the vertebrae (18). Fractures to the cervical area have been reported in activities such as water sports, gymnastics, wrestling, rugby, ice hockey, and football. Normally the cervical spine is slightly extended with a curve anteriorly convex. If the head is lowered, the cervical spine will flatten out to approximately 30° of flexion. If a force is applied against the top of the head when it is in this position, the cervical vertebrae are loaded along the length of the cervical vertebrae by a compressive force, creating a dislocation or fracture–dislocation of the facets of the vertebrae. When spearing or butting during tackling with the head in flexion was outlawed in football, the number of cervical spine injuries was dramatically reduced (18).

Compression fractures in the lumbar vertebrae of weight lifters, football linemen, and gymnasts who load the vertebrae while the spine is held in hyperlordotic or sway-back position have also been reported (23). Figure 2-24 is a radiograph of a fracture to the lumbar vertebrae, demonstrating the shortening and widening effect of the compressive force. Finally, compression fractures are common in individuals with osteoporosis.

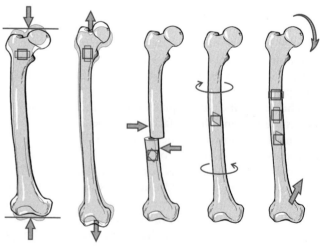

A. Compression   B. Tension   C. Shear   D. Torsion   E. Bending

FIGURE 2-23 The skeletal system is subjected to a variety of loads that alter the stresses in the bone. The *square* in the femur indicates the original state of the bone tissue. The *colored area* illustrates the effect of the force applied to the bone. **(A)** Compressive force causes shortening and widening. **(B)** Tensile force causes narrowing and lengthening. **(C)** and **(D)** Shear force and torsion create angular distortion. **(E)** Bending force includes all of the changes seen in compression, tension, and shear.

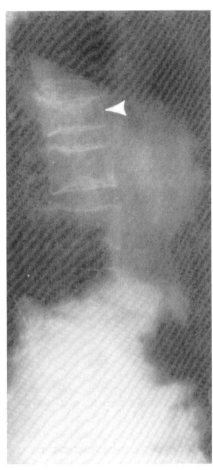

FIGURE 2-24 The lumbar vertebrae can incur compressive fracture (*arrow*), in which the body of the vertebra is shortened and widened. This type of fracture has been associated with loading of the vertebrae while maintaining a hyperlordotic position. (Reprinted with permission from Nordin, M., Frankel, V. H. [1989]. *Basic Biomechanics of the Musculoskeletal System.* (2nd Ed.). Philadelphia, PA: Lea & Febiger.)

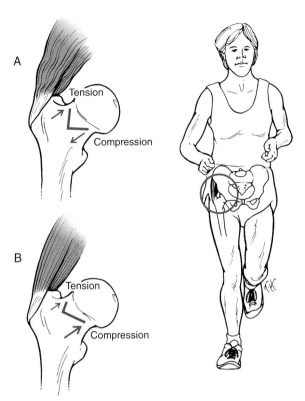

FIGURE 2-25 **(A)** During standing or in the stance phase of walking and running, bending force applied to the femoral neck creates a large compressive force on the inferior neck and tensile force on the superior neck. **(B)** If the gluteus medius contracts, the compressive force increases, and the tensile force decreases. This reduces the injury potential because injury is more likely to occur in tension.

Specific lifts in weight training result in spondylolysis, a stress fracture of the pars interarticularis section of the vertebra. Lifts that have a high incidence of this fracture are the clean and jerk and the snatch from the Olympic lifts and the squat and dead lift from power lifting (22,23). In gymnasts, it is associated with extreme extension positions in the lumbar vertebrae. This injury will be discussed in greater detail in Chapter 7, when the trunk is reviewed.

A compressive force at the hip joint can increase or decrease the injury potential of the femoral neck. The hip joint must absorb compressive forces of approximately three to seven times body weight during walking (43,46). Compressive forces are up to 15 to 20 times body weight in jumping (46). In a normal standing posture, the hip joint assumes approximately one-third of the body weight if both limbs are on the ground (43). This creates large compressive forces on the inferior portion of the femoral neck and a large pulling, or tensile, force on the superior portion of the neck. Figure 2-25 shows how this happens as the body pushes down on the femoral head, pushing the bottom of the femoral neck

together and pulling the top of the femoral neck apart as it creates bending.

The hip abductors, specifically the gluteus medius, contract to counteract the body weight during stance. As shown in Figure 2-25, they also produce a compressive load on the superior aspect of the femoral neck that reduces the tensile forces and injury potential in the femoral neck because bone usually fractures sooner with a tensile force (43). It is proposed that runners develop femoral neck fractures because the gluteus medius fatigues and cannot maintain its reduction of the high tensile force, producing the fracture (29,46). A femoral neck fracture can also be produced by a strong co-contraction of the hip muscles, specifically the abductors and adductors, creating excessive compressive forces on the superior neck.

Tension Forces

When muscle applies a tensile force to the system through the tendon, the collagen in the bone tissue arranges itself in line with the tensile force of the tendon. Figure 2-26 shows an example of collagen alignment at the tibial tuberosity. This figure also illustrates the influence of tensile forces on the development of apophyses. An **apophysis** is a bony outgrowth, such as a process, tubercle, and tuberosity. Figure 2-26 illustrates how an apophysis, the tibial tuberosity, is formed by tensile forces.

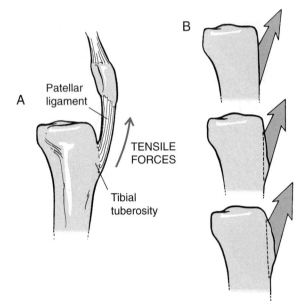

FIGURE 2-26 **A.** When tensile forces are applied to the skeletal system, the bone strengthens in the direction of the pull as collagen fibers align with the pull of the tendon or ligament. **B.** Tensile forces are also responsible for the development of apophyses, bony outgrowths such as processes, tubercles, and tuberosities.

Failure of the bone usually occurs at the site of muscle insertion. Tensile forces can also create ligament avulsions. A ligament avulsion, or an **avulsion fracture**, occurs when a portion of the bone at the insertion of the ligament is torn away. This occurs more frequently in children than in adults. Avulsion fractures occur when the tensile strength of the bone is not sufficient to prevent the fracture. This is typical of some of the injuries occurring in the high-velocity throwing motion of a little leaguer's pitching arm. The avulsion fracture in this case is commonly on the medial epicondyle as a result of tension generated in the wrist flexors.

Two other common tension-produced fractures are at the fifth metatarsal, caused by the tensile forces generated by the peroneal muscle group, and at the calcaneus, where the forces are generated by the triceps surae muscle group. The tensile force on the calcaneus can also be produced in the stance phase of gait as the arch is depressed and the plantar fascia covering the plantar surface of the foot tightens, exerting tensile force on the calcaneus. Some sites of avulsion fractures for the pelvic region, presented in Figure 2-27, include the anterior superior and inferior spines, lesser trochanter, ischial tuberosity, and pubic bone.

Tension forces are generally responsible for sprains and strains. For example, the typical ankle inversion sprain occurs when the foot is oversupinated. That is, the foot rolls over its lateral border, stretching the ligaments on the lateral side of the ankle. Tensile forces are also identified with shin splints. This injury occurs when the tibialis anterior pulls on its attachment site on the tibia and on the interosseous membrane between the tibia and the fibula.

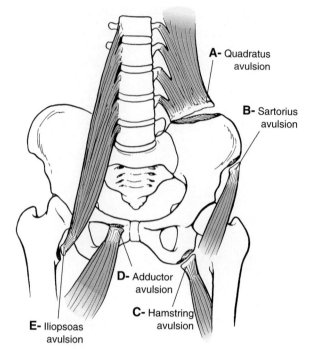

FIGURE 2-27 Avulsion fractures can result from tension applied by a tendon or a ligament. Sites of avulsion fractures in the pelvic region include the anterior superior spine **(A)**, anterior inferior spine **(B)**, ischial tuberosity **(C)**, pubic bone **(D)**, and lesser trochanter **(E)**.

Another site exposed to high tensile forces is the tibial tuberosity, which transmits very high tensile forces when the quadriceps femoris muscle group is active. This tensile force, if sufficient in magnitude and duration, may cause tendinitis or inflammation of the tendon in older participants. In younger participants, however, the damage usually occurs at the site of tendon–bone attachment and can result in inflammation, bony deposits, or an avulsion fracture of the tibial tuberosity. Osgood–Schlatter disease is characterized by inflammation and formation of bony deposits at the tendon–bone junction.

Bone responds to the demands placed upon it, as described by Wolff's law (32). Therefore, different bones and different sections in a bone respond to tension and compressive forces differently. For example, the tibia and femur participate in weight bearing in the lower extremity and are strongest when loaded with a compressive force. The fibula, which does not participate significantly in weight bearing but is a site for muscle attachment, is strongest when tensile forces are applied (43). An evaluation of the differences found in the femur has uncovered greater tensile strength capabilities in the middle third of the shaft, which is loaded through a bending force in weight bearing. In the femoral neck, the bone can withstand large compressive forces, and the attachment sites of the muscles have great tensile strength (43).

Shear Forces

Shear forces are responsible for some vertebral disk problems, such as spondylolisthesis, in which the vertebrae slip anteriorly over one another. In the lumbar vertebrae, shear force increases with increased swayback, or hyperlordosis (22). The pull of the psoas muscle on the lumbar vertebrae also increases shear force on the vertebrae. This injury is discussed in greater detail in Chapter 7.

Examples of fractures caused by shear forces are commonly found in the femoral condyles and the tibial plateau. The mechanism of injury for both is usually hyperextension in the knee through some fixation of the foot and valgus or medial force to the thigh or shank. In adults, this shear force can fracture a bone as well as injure the collateral or cruciate ligaments (37). In developing children, this shear force can create epiphyseal fractures, such as in the distal femoral epiphysis. The mechanism of injury and the resulting epiphyseal damage are presented in Figure 2-28. The effects of such a fracture in developing children can be significant because this epiphysis accounts for approximately 37% of the bone growth in length (15).

Compressive, tensile, and shear forces applied simultaneously to the bone are important in the development of bone strength. Figure 2-29 illustrates both compressive and tensile stress lines in the tibia and femur during running. Bone strength develops along these lines of stress.

Bending Forces

Bone is regularly subjected to large bending forces. For example, during gait, the lower extremity bones are subjected to bending forces caused by alternating tension and compression forces. During normal stance, both the femur and the tibia bend. The femur bends both anteriorly and laterally because of its shape and the manner of the force transmission caused by weight bearing. Also,

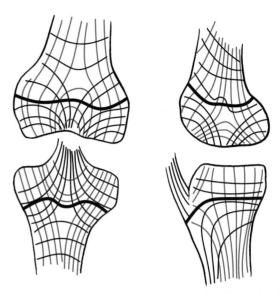

FIGURE 2-29 The lines of compressive stress (*bold black lines*) and tension stress (*lighter black and blue lines*) for the distal femur and proximal tibia during the stance phase of running.

weight bearing produces an anterior bend in the tibia. Although these bending forces are not injury producing, the bone is strongest in the regions where the bending force is greatest (46).

Typically, a bone fails and fractures on the convex side in response to high tensile forces because bone can withstand greater compressive forces than tensile forces (43). The magnitude of the compressive and tensile forces produced by bending increases with distance from the axis of the bone. Thus, the force magnitudes are greater on the outer portions of the bone.

Injury-producing bending loads are caused by multiple forces applied at different points on the bone. Generally, these situations are called three- or four-point force applications. A force is usually applied perpendicular to the bone at both ends of the bone and a force applied in the opposite direction at some point between the other two forces. The bone will break at the point of the middle force application, as is the case in a ski boot fracture shown in Figure 2-30. This fracture is produced as the skier falls over the top of the boot, with the ski and boot pushing in the other direction. The bone usually fractures on the posterior side because that is where the convexity and the tensile forces are applied. Ski boot fractures have been significantly reduced because of improvements in bindings, skis that turn more easily, well-groomed slopes, and a change in skiing technique that puts the weight forward on the skis. The reduction of tibial fractures through the improvement of equipment and technique, however, has led to an increase in the number of knee injuries, for the same reasons (14).

The three-point bending force is also responsible for injuries to a finger that is jammed and forced into hyperextension and to the knee or lower extremity when the foot is fixed in the ground and the lower body bends.

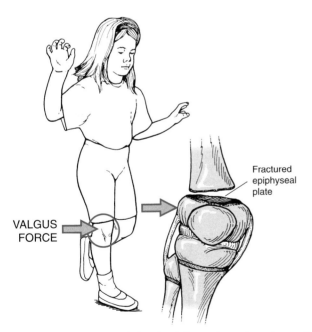

FIGURE 2-28 Fracture of the distal femoral epiphysis is usually created by shear force. This is commonly produced by a valgus force applied to the thigh or shank with the foot fixed and the knee hyperextended.

VALGUS FORCE

Fractured epiphyseal plate

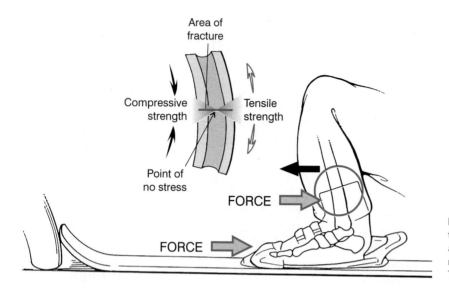

FIGURE 2-30   The ski boot fracture, created by a three-point bending load, occurs when the ski stops abruptly. Compressive force is created on the anterior tibia and tensile force on the posterior tibia. The tibia usually fractures on the posterior side.

Simply eliminating the long cleats in the shoes of football players and playing on good resurfaced fields reduced this type of injury by half (22). Three-point bending force applications are also used in bracing. Figure 2-31 presents two brace applications using the three-point force application to correct a postural deviation or stabilize a region.

A four-point bending load is two equal and opposite pairs of forces at each end of the bone. In the case of four-point bending, the bone breaks at its weakest point. This is illustrated in Figure 2-32 with the application of a four-point bending force to the femur.

Torsional Forces

Fractures resulting from torsional force can occur in the humerus when poor throwing technique creates a twist on the arm (46) and in the lower extremity when the foot is planted and the body changes direction. A spiral fracture is a result of torsional force. An example of the mechanism of spiral fracture to the humerus in a pitcher is shown in Figure 2-33. Spiral fractures usually begin on the outside of the bone parallel to the middle of the bone. Torsional loading of the lower extremity is also responsible for knee cartilage and ligament injuries (22) and can occur when the foot is caught while the body is spinning.

Combined Loading

Tension, compression, shear, bending, and torsion represent simple and pure modes of loading. It is more common to incur various combinations of loads acting simultaneously on the body. For example, the lower extremity bones are loaded in multiple directions during exercise. The mechanical loading provides the stimulus

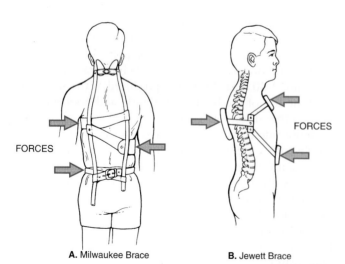

FIGURE 2-31   Three-point bending loads are used in many braces. **(A)** The Milwaukee brace, used for correction of lateral curvature of the spine, applies three-point bending force to the spine. **(B)** The Jewett brace applies three-point bending force to the thoracic spine to create spinal extension in that region.

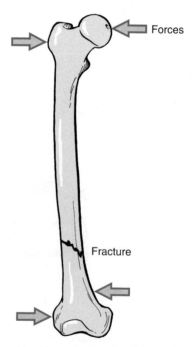

FIGURE 2-32   Hypothetical example of four-point bending load applied to the femur, creating a fracture or failure at the weakest point.

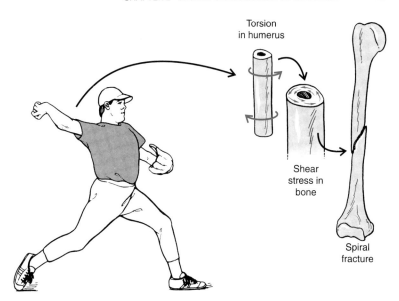

FIGURE 2-33 Example of torsion applied to the humerus, creating shear stress across the surface.

for bone adaptation and selection of exercises for this purpose becomes an important consideration. Because bone responds more stiffly at higher rates of loading, the strain rate also becomes important. In Figure 2-34, the bone strain in the tibia during the performance of the leg press, bicycle, stepmaster, and running is compared with baseline walking values (39). Whereas compression and shear values produced during the leg press, stepmaster, and running are higher than walking, tension strain values vary between exercise modes. Bicycling results in lower tension, compression, and shear values than walking. When the rate of loading is evaluated (Fig. 2-35), however, only running produces higher strain rates than walking.

STRESS FRACTURES

Injury to the skeletal system can be produced by a single high-magnitude application of one of these types of loads or by repeated application of a low-magnitude load over time. The former injury is referred to as a **traumatic fracture**. The latter type is a **stress fracture**, **fatigue fracture**, or **bone strain**. Figure 2-36 shows a radiograph of a stress fracture to the metatarsal. These fractures occur as a consequence of cumulative **microtrauma** imposed upon the skeletal system when loading of the system is so frequent that bone repair cannot keep up with the breakdown of bone tissue.

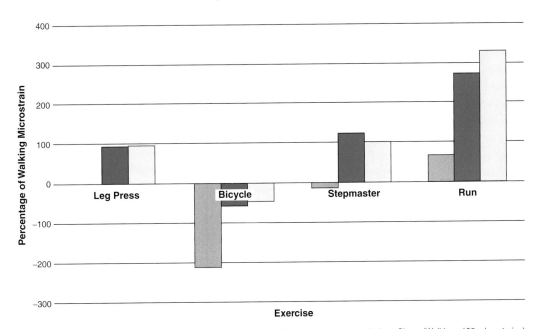

Principal Strain on Tibia during Exercise

☐ Tension (Walking = 840 microstrains)   ■ Compression (Walking = 454 microstrains)   ☐ Shear (Walking = 183 microstrains)

FIGURE 2-34 Comparison of in vivo tibial strain during four exercises compared with walking. (Adapted from Milgrom, C., et al. [2000]. *Journal of Bone & Joint Surgery*, 82-B:591-594.)

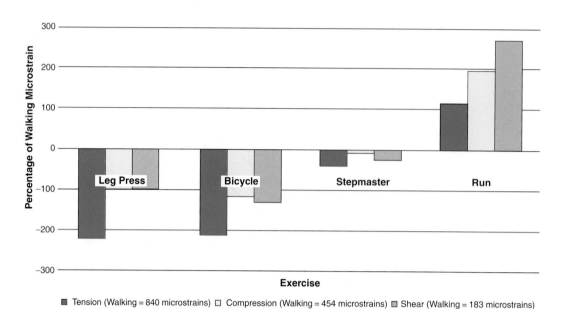

FIGURE 2-35   Comparison of in vivo tibial strain rates during four exercises compared with walking. (Adapted from Milgrom, C., et al. [2000]. *Journal of Bone & Joint Surgery*, 82-B:591-594.)

A stress fracture occurs when bone resorption weakens the bone too much and the bone deposit does not occur rapidly enough to strengthen the area. Stress fractures in the lower extremity can be attributed to muscle fatigue that reduces shock absorption and causes redistribution of forces to specific focal points in the bone. In the upper extremity, stress fractures result from repetitive muscular forces pulling on the bone. Stress fractures account for 10% of injuries to athletes (36).

The typical stress fracture injury occurs during a load application that produces shear or tensile strain and results in laceration, fracture, rupture, or avulsion. Bone tissue can also develop a stress fracture in response to compressive or tensile loading that overloads the system, either through excessive force applied one or a few times or through too frequent application of a low or moderate level of force (29,34,36). Fatigue microdamage occurs under cyclic loading that needs to be repaired before the bone progresses to failure, resulting in a stress fracture. The relationship between the magnitude and frequency of applications of load on bone is presented in Figure 2-37.

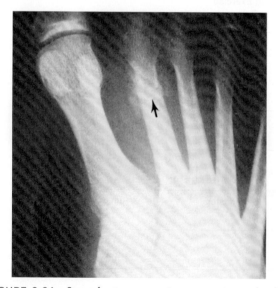

FIGURE 2-36   Stress fractures occur in response to overloading of the skeletal system so that cumulative microtraumas occur in the bone. A stress fracture to the second metatarsal, as shown in this radiograph (*arrow*), is caused by running on hard surfaces or in stiff shoes. It is also associated with persons with high arches and can be created by fatigue of the surrounding muscles. (Reprinted with permission from Fu, H. F., Stone, D. A. [1994]. *Sports Injuries*. Baltimore, MD: Williams & Wilkins.)

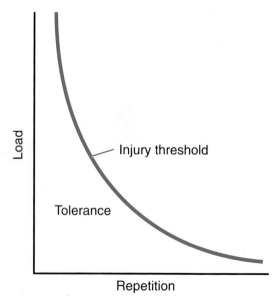

FIGURE 2-37   Injury can occur when a high load is applied a small number of times or when low loads are applied numerous times. It is important to remain within the injury tolerance range.

The tolerance of bone to injury is a function of the load and the cycles of loading.

Examples of injuries to the skeletal system are presented in Table 2-2. The activity associated with the injury, the type of load causing the injury, and the mechanism of injury are summarized. It is still not clear why some athletes participating in the same activity acquire a stress fracture injury and others do not. It has been suggested that other factors such as limb alignment and soft tissue dampening of imposed loads may play a role in influencing the risk of fracture (5).

Cartilage

Cartilage is a firm, flexible tissue made up of cells called **chondrocytes** surrounded by an extracellular matrix. The two main types of cartilage that will be discussed in this chapter are articular or hyaline cartilage and fibrocartilage.

ARTICULAR CARTILAGE

Articulating joints connect the different bones of the skeleton. In freely moving joints, the articulating ends of the

TABLE 2-2	Injuries to the Skeletal System		
Type of Injury	**Activity Examples**	**Load Causing Injury**	**Mechanism of Injury**
Tibial stress	Dancing, running, basketball	Compression	Poor conditioning, stiff footwear, unyielding surfaces, hypermobile foot (overpronation)
Medial epicondyle fracture	Gymnastics	Tension, compression	Too much work on floor exercise and tumbling
Stress fracture of the big toe	Sprinting, fencing, rugby	Tension	Toe extensors create bowstring effect on the big toe when up on the toes; primarily in individuals with hallux valgus
Stress fracture of the femoral neck	Running, gymnastics	Compression	Muscle fatigue, high-arched foot
Stress fracture in the calcaneus	Running, basketball, volleyball	Compression	Hard surface, stiff footwear
Stress fracture in the lumbar vertebrae	Weight lifting, gymnastics, football	Compression, tension	High loads with hyperlordotic low-back posture
Tibial plateau fractures	Skiing	Compression	Hyperextension and valgus of the knee, as in turning, with the force on the inside edge of the downhill ski, abruptly halted with heavy snow
Stress fracture to the medial malleolus	Running	Compression	Ankle sprain to the outside, causing compression between the talus and medial malleolus or excessive pronation because the medial malleolus rotates in with tibial rotation and pronation
Hamate fracture of the hand	Baseball, golf, tennis	Compression	Relaxed grip in the swing that stops suddenly at the end of the swing as the club hits the ground, the bat is forcefully checked, or the racket is out of control
Fracture of the tibia	Skiing	Bending, compression, tension	Three-point bending fall in which the body weight, boot, and ground bend the tibia posteriorly
Fracture of the femoral condyles	Skiing, football	Shear	Hyperextension of the knee with valgus force
Stress fracture in the fibula	Running, aerobics, jumping	Tension	Jumping or deep-knee bends with a walk; pull by soleus, tibialis posterior, peroneals, and toe flexors, pulling the tibia and fibula together
Meniscus tear of the knee	Basketball, football, jumping, volleyball, soccer	Compression, torsion	Turning on a weight-bearing limb or valgus force to the knee
Stress fracture in the metatarsal	Running	Compression	Hard surfaces, stiff footwear, high-arched foot, fatigue
Stress fracture in the femoral shaft	Running, triathlon	Tension	Excessive training and mileage; created by pull of the vastus medialis or adductor brevis

bones are covered with a connective tissue referred to as articular cartilage.

Articular or **hyaline cartilage** is an avascular substance consisting of 60% to 80% water and a solid matrix composed of collagen and proteoglycan. Collagen is a protein with the important mechanical properties of stiffness and strength. Proteoglycan is a highly hydrated gel. It is unclear how collagen and the proteoglycan gel interact during stress to the cartilage. However, the interaction between the two materials determines cartilage's mechanical properties. Cartilage has no blood supply and no nerves and is nourished by the fluid within the joint (41).

Articular cartilage is anisotropic, meaning it has different material properties for different orientations relative to the joint surface. The properties of cartilage make it well suited to resisting shear forces because it responds to load in a viscoelastic manner. It deforms instantaneously to a low or moderate load, and if rapidly loaded, it becomes stiffer and deforms over a longer period. The force distribution across the area in the joint determines the stress in the cartilage, and the distribution of the force depends on the cartilage's thickness.

> **What is the role of articular cartilage?**
>
> 1. Transmit compressive forces across the joint
> 2. Allows motion in the joint with minimal friction and wear
> 3. Redistributes contact stresses over a larger area
> 4. Protects the underlying bone

Cartilage is important to the stability and function of a joint because it distributes loads over the surface and reduces the contact stresses by half (50). Collagen fibers are arranged to withstand load bearing. For example, in the knee, the medial meniscus transmits 50% of the compression load. Removal of just a small part of the cartilage has been shown to increase the contact stress by as much as 350% (25). Several years ago, a cartilage tear would have meant removal of the whole cartilage, but today orthopedists trim the cartilage and remove only minimal amounts to maintain as much shock absorption and stability in the joint as possible.

Cartilage is 1 to 7 mm thick, depending on the stress and the incongruity of the joint surfaces (26). For example, in the ankle and the elbow joints, the cartilage is very thin, but at the hip and knee joints, it is thick. The cartilage is thin in the ankle because of the ankle's architecture. A substantial area of force distribution imposes less stress on the cartilage. Conversely, the knee joint is exposed to lower forces, but the area of force distribution is smaller, imposing more stress on the cartilage. Some of the thickest cartilage in the body, approximately 5 mm, lies on the underside of the patella (54).

Articular cartilage allows movement between two bones with minimal friction and wear. The joint surfaces have remarkably low coefficients of friction. Articular cartilage contributes significantly to this. The coefficient of friction in some joints has been reported to range from 0.01 to 0.04; the coefficient of friction of ice at 0°C is about 0.1. These almost frictionless surfaces allow the surfaces to glide over each other smoothly.

Cartilage growth across the life span is dynamic. At maturity, stabilization of articular cartilage thickness occurs, but ossification does not entirely cease (4). The interface between the cartilage and the underlying subchondral bone remains active and is responsible for the gradual change in joint shape during aging. The amount of cartilage growth is regulated by compressive stress, and the higher the joint contact pressures, the thicker the cartilage. In activities of daily living across the life span, the changes in joint use cause a change in cartilage, resulting in thinning or thickening.

FIBROCARTILAGE

Another type of cartilage is **fibrocartilage**, which is often found where articular cartilage meets a tendon or a ligament. Fibrocartilage acts as an intermediary between hyaline cartilage and the other connective tissues. Fibrocartilage is found where both tensile strength and the ability to withstand high pressures are necessary, such as in the intervertebral disks, the jaw, and the knee joint. A fibrocartilage structure is referred to as an articular disk, or **meniscus**. The menisci also improve the fit between articulating bones that have slightly different shapes. Meniscus tears usually occur during a sudden change of direction with the weight all on one limb. The resultant compression and tension on the meniscus tear the fibrocartilage. No pain is associated with the actual tear; rather, the peripheral attachment sites are the site of the irritation and resulting sensitivity.

Ligaments

A **ligament** is a short band of tough fibrous connective tissue that binds bone to bone and consists of collagen, elastin, and reticulin fibers (55). The ligament usually provides support in one direction and often blends with the capsule of the joint. Ligaments can be capsular, extracapsular, or intra-articular. **Capsular ligaments** are simply thickenings in the wall of the capsule, much like the glenohumeral ligaments in the front of the shoulder capsule. **Extracapsular ligaments** lie outside the joint itself. The collateral ligaments found in numerous joints are extracapsular (i.e., fibular collateral ligament of the knee). Finally, **intra-articular ligaments**, such as the cruciate ligaments of the knee and the capitate ligaments in the hip, are located inside a joint.

The maximum stress that a ligament can endure is related to its cross-sectional area. Ligaments exhibit visco-elastic behavior, which helps to control the dissipation of energy and controls the potential for injury (7). Ligaments respond to loads by becoming stronger and stiffer over time, demonstrating both a time-dependent and a non-linear stress–strain response. The collagen fibers in a ligament are arranged so the ligament can handle both tensile loads and shear loads; however, ligaments are best suited for tensile loading. An example of viscoelastic behavior is presented in Figure 2-38. The collagen fibers in a ligament have a nearly parallel configuration. When unloaded, they have a wavy or crimped configuration. At low stresses, the crimp in the collagen fibers of the ligament disappears. At this point, the ligament behaves almost linearly, with strains that are relatively small and within the physiologic limit. At greater stresses, the ligament tears, either partially or completely. Generally, when a tensile load is applied to a joint very quickly, the ligament can dissipate energy quickly and the chance of failure is more likely to be at the bone rather than in the ligament.

The strength of a ligament also diminishes rapidly with immobilization. A tensile injury to a ligament is termed a sprain. Sprains are rated 1, 2, or 3 in severity, depending on whether there is a partial tear of the fibers (rated 1), a tear with some loss of stability (rated 2), or a complete tear with loss of joint stability (rated 3) (26).

At the end of the range of motion for every joint, a ligament usually tightens up to terminate the motion. Ligaments provide passive restraint and transfer loads to the bone. A ligament can be subjected to extreme stress and damaged while overloaded when performing the role of restricting abnormal motion. Because the ligaments stabilize, control, and limit joint motion, any injury to a ligament influences joint motion. Ligament damage can result in joint instability, which in turn, can cause altered joint kinematics, resulting in altered load distribution and vulnerability to injury.

What is the function of a ligament?

1. To guide normal joint function
2. To restrict abnormal joint movement

Bony Articulations

THE DIARTHRODIAL OR SYNOVIAL JOINT

Movement potential of a segment is determined by the structure and function of the **diarthrodial** or **synovial joint**. The diarthrodial joint provides low-friction articulation capable of withstanding significant wear and tear. The characteristics of all diarthrodial joints are similar. For example, the knee has similar structures to the finger joints. Because of this similarity, it is worthwhile to look at the various components of the diarthrodial joint to gain general knowledge about joint function, support, and nourishment. Figure 2-39 shows the characteristics of the diarthrodial joint.

Characteristics of the Diarthrodial Joint
Covering the ends of the bones is the **articular end plate**, a thin layer of cortical bone over cancellous bone.

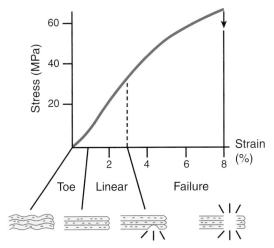

FIGURE 2-38 A stress–strain curve for a ligament. In the toe region, the collagen fibers of the ligament are wavy. The fibers straighten out in the linear region. In the plastic region, some of the collagen fibers tear. (Adapted with permission from Butler, D. L., et al. [1978]. Biomechanics of ligaments and tendons. *Exercise and Sports Sciences Reviews*, 6:125-181.)

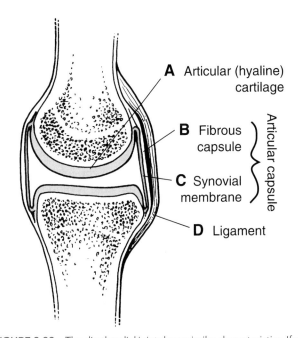

FIGURE 2-39 The diarthrodial joints have similar characteristics. If you study the knee, interphalanges, elbow, or any other diarthrodial joint, you will find the same structures. These include **(A)** articular or hyaline cartilage, **(B)** capsule, **(C)** synovial membrane, and **(D)** ligaments.

On top of the end plate is articular cartilage. This cartilage in the joint offers additional load transmission, stability, improved fit of the surfaces, protection of the joint edges, and lubrication.

Another important characteristic of the diarthrodial joint is the **capsule**, a fibrous white connective tissue made primarily of collagen. It protects the joint. Thickenings in the capsule, known as ligaments, are common where additional support is needed. The capsule basically defines the joint, creating the interarticular portion, or inside, of the joint, which has a joint cavity and a reduced atmospheric pressure (50). Although soft-tissue loads are difficult to compute, the capsule sustains some of the load imposed on the joint (27).

Any immobilization of the capsule alters the mechanical properties of the capsular tissue and may result in joint stiffness. Likewise, injury to the capsule usually results in the development of a thick or fibrous section that can be externally palpable (16).

On the inner surface of the joint capsule is the **synovial membrane**, a loose, vascularized connective tissue that secretes **synovial fluid** into the joint to lubricate and provide nutrition to the joint. The fluid, which has the consistency of an egg white, decreases in viscosity as shear rates increase. The consistency is similar to ketchup, hard to start but easy to move after it is going. When the joint moves slowly, the fluid is highly viscous, and the support is high. Conversely, when the joint moves rapidly, the fluid is elastic in its response, decreasing the friction in the joint (50).

A healthy joint provides effortless motion along preferred anatomical directions with an accompanying restriction of abnormal joint motion. Freedom of mobility is also provided by the lubricating action of articular cartilage. The healthy joint is also stable as a result of the interaction between bony connections, ligaments, and other soft tissue. Finally, the ligaments operate to guide and restrict motion, which defines the normal envelope of passive motion of the joint.

Any injury to the joint is noticeable in both a thickening in the membrane and a change in the consistency of the fluid. The fluid fills the capsular compartment and creates pain in the joint. Physicians drain the joint to relieve the pressure, often finding that the fluid is bloodstained.

Stability of the Diarthrodial Joint

Stability in a diarthrodial joint is provided by the structure—the ligaments surrounding the joints, the capsule, and the tendons spanning the joint—gravity, and the vacuum in the joint produced by negative atmospheric pressure. The hip is one of the most stable joints in the body because it has good muscular, capsular, and ligamentous support. The hip joint has congruency between the surfaces, with a high degree of bone-to-bone contact. Most of the stability in the hip, however, is derived from the effects of gravity and the vacuum in the joint (26). The negative pressure in the joint is sufficient to hold the femur in the joint if all other structures, such as supporting ligaments and muscles, are removed.

In contrast, the stability of the shoulder is supplied only by the capsule and the muscles surrounding the joint. Also, the congruency of the shoulder joint is limited, with only a small proportion of the head of the humerus making contact with the glenoid cavity.

Stability versus Mobility

Our joints are held together by muscle, ligaments, the capsule and the negative pressure within the capsule, and the bones themselves. In an ideal world we could restrict the linear movement of the bones (sliding and dislocating) while allowing unrestricted angular movement (rotation). In reality, there is often a compromise that is made between the necessary stability of a joint and the amount of mobility. We require greater mobility in the upper extremity joints so that we can manipulate objects in our environment. We require greater stability in the lower extremity joints so that we can withstand the large ground reaction forces caused by walking and running. Each of the stabilizing structures listed above have positive and negative attributes such as the amount of mobility allowed, the energy cost of providing the stability, the fatigability of the structure, the ability to recover from injury and the strength of the stabilizing influence. Discuss these attributes in reference to the stabilizing structures listed above.

Close-Packed versus Loose-Packed Positions

As movement through a range of motion occurs, the actual contact area varies between the articulating surfaces. When the joint position is such that the two adjacent bones fit together best and maximum contact exists between the two surfaces, the joint is considered to be in a **close-packed position**. This is the position of maximum compression of the joint, in which the ligaments and the capsule are tense and the forces travel through the joint as if it did not exist. Examples of close-packed positions are full extension for the knee, extension of the wrist, extension of the interphalangeal joints, and maximum dorsiflexion of the foot (50). All other joint positions are termed **loose-packed positions** because there is less contact area between the two surfaces and the contact areas are frequently changing. There is more sliding and rolling of the bones over one another in a loose-packed position. This position allows for continuous movement, reducing the friction in the joint. Although the loose-packed joint position is less stable than the close-packed position, it is not as susceptible to injury because of its mobility. The close- and loose-packed positions of the knee joint are presented in Figure 2-40. Note the greater contact area in the close-packed position.

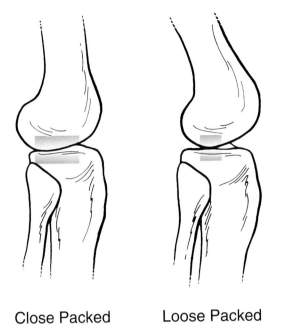

Close Packed   Loose Packed

FIGURE 2-40 In the close-packed position, contact between the two joint surfaces is maximal and mobility is minimal. In the loose-packed joint position, there is less contact between the surfaces in the joint and more mobility and movement between the two surfaces.

While in the close-packed position, the joint is very stable but vulnerable to injury because the structures are taut and the joint surfaces are pressed together. The joint is especially susceptible to injury if hit by an external force, such as hitting the knee when it is fully extended.

Types of Diarthrodial Joints

A classification system categorizes seven types of diarthrodial joints according to the differences in articulating surfaces, the directions of motion allowed by the joint, and the type of movement occurring between the segments. Figure 2-41 offers a graphic representation of these seven joints.

Plane or Gliding Joint

The first type of joint is the **plane** or **gliding joint**, found in the foot among the tarsals and in the hand among the carpals. Movement at this type of joint is termed nonaxial because it consists of two flat surfaces that slide over each other rather than around an axis.

In the hand, for example, the carpals slide over each other as the hand moves to positions of flexion, extension, radial deviation, and ulnar deviation. Likewise, in the foot, the tarsals shift during pronation and supination, sliding over each other in the process.

Hinge Joint

The **hinge joint** allows movement in one plane (flexion, extension); it is uniaxial. Examples of the hinge joint in the body are the interphalangeal joints in the foot and hand and the ulnohumeral articulation at the elbow.

Pivot Joint

The **pivot joint** also allows movement in one plane (rotation; pronation and supination) and is uniaxial. Pivot joints are found at the superior and inferior radioulnar joint and the atlantoaxial articulation at the base of the skull.

Condylar Joint

The **condylar joint** allows a primary movement in one plane (flexion and extension) with small amounts of movement in another plane (rotation). Examples are the metacarpals, interphalangeal, metacarpals, and the temporomandibular joint. The knee joint is also referred to as a condylar joint because of the articulation between the two condyles of the femur and the tibial plateau. However, because of the mechanical linkages created by the ligaments, the knee joint functions as a hinge and is referred to as a modified hinge joint in the literature for this reason.

Ellipsoid Joint

The **ellipsoid joint** allows movement in two planes (flexion and extension; abduction and adduction) and is biaxial. Examples of this joint are the radiocarpal articulation at the wrist and the metacarpophalangeal articulation in the phalanges.

Saddle Joint

The **saddle joint**, found only at the carpometacarpal articulation of the thumb, allows two planes of motion (flexion and extension; abduction and adduction) plus a small amount of rotation. It is similar to the ellipsoid joint in function.

Ball-and-Socket Joint

The last type of diarthrodial joint, the **ball-and-socket joint**, allows movement in three planes (flexion and extension; abduction and adduction; rotation) and is the most mobile of the diarthrodial joints. The hip and shoulder are examples of ball-and-socket joints. A summary of the major joints in the body is presented in Table 2-3.

OTHER TYPES OF JOINTS

Synarthrodial or Fibrous Joints

Other articulations are limited in movement characteristics but nonetheless play important roles in stabilization of the skeletal system. Some bones are held together by fibrous articulations such as those found in the sutures of the skull. These articulations, referred to as **synarthrodial joints**, allow little or no movement between the bones and hold the bones firmly together (Fig. 2-42).

Amphiarthrodial or Cartilaginous Joints

Cartilaginous or **amphiarthrodial joints** hold bones together with either hyaline cartilage, such as is found at the epiphyseal plates, or fibrocartilage, as in the pubic symphysis and the intervertebral articulations. The movement at these articulations is also limited, although not to the degree of the synarthrodial joints.

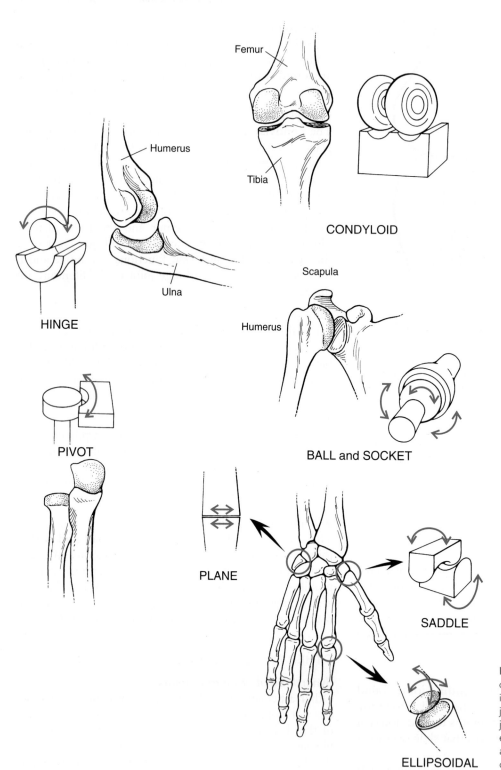

CONDYLOID

HINGE

PIVOT

BALL and SOCKET

PLANE

SADDLE

ELLIPSOIDAL

FIGURE 2-41 The seven types of diarthrodial joints. The nonaxial joint is the plane or gliding joint. Uniaxial joints include the hinge and pivot joints; biaxial joints are the condylar, ellipsoid, and saddle joint. The ball-and-socket joint is the only triaxial diarthrodial joint.

OSTEOARTHRITIS

Injury to the structures of the diarthrodial joint can occur during high load or through repetitive loading over an extended period. The articular cartilage in the joint is especially subject to wear during one's lifetime.

Osteoarthritis is a disease characterized by degeneration of the articular cartilage, which leads to fissures, fibrillation, and finally disappearance of the full thickness of the articular cartilage. Osteoarthritis is the leading chronic medical condition and is the leading cause of disability for persons aged 65 years and older (4).

TABLE 2-3	Major Joints of the Body	
Joint	Type	Degrees of Freedom
Vertebra	Amphiarthrodial	3
Hip	Ball-and-socket	3
Shoulder	Ball-and-socket	3
Knee	Condyloid	2
Wrist	Ellipsoid	2
Metacarpophalangeal	Ellipsoid	2 (fingers)
Carpometacarpal	Saddle	2 (thumb)
Elbow	Hinge	1
Radioulnar	Pivot	1
Atlantoaxial	Pivot	1
Ankle	Hinge	1
Interphalangeal	Hinge	1

A. Synarthrodial

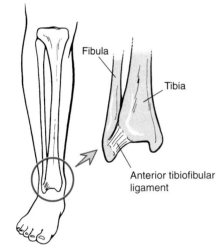

Distal tibiofibular joint

B. Amphiarthrodial

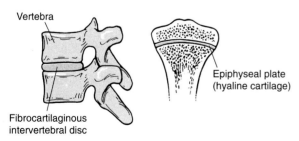

Intervertebral Disc   Epiphysis

FIGURE 2-42 **(A)** An example of the synarthrodial joint is the fibrous articulation at the distal tibiofibular joint. **(B)** The amphiarthrodial, or cartilaginous, joint can be found between the vertebrae or in the epiphyseal plate of a growing bone.

Osteoarthritis starts as a result of trauma to or repeated wear on the joint that causes a change in the articular substance to the point of removal of actual material by mechanical action. This results in diminished contact areas and erosion of the cartilage through development of rough spots in the cartilage. The rough spots develop into fissures and eventually go deep enough that only subchondral bone is exposed. Osteophytes or cysts form in and around the joint, and this is the beginning of degenerative joint disease, or osteoarthritis. The radiographs in Figure 2-43 show the areas of joint degeneration associated with osteoarthritis in the hip and vertebrae.

It is theorized that osteoarthritis develops first in the subchondral, or cancellous, bone underlying the joint (45). The cartilage overlying the bone in the joint is thin; consequently, the underlying subchondral bone absorbs the shock of loading. Repetitive loading or unequal loading in the joint causes microfractures in the subchondral bone. When the microfractures heal, the subchondral bone is stiffer and less able to absorb shock, passing this role on to the cartilage. The cartilage deteriorates as a consequence of this overloading, and the body lays down bone in the form of osteophytes to increase the contact area.

Osteoarthritis has been shown to have no relationship to hyperlaxity in the joint (6), levels of osteoporosis (24), or general physical activity (35). An injured joint deteriorates at a faster rate, however, making it more susceptible to the development of osteoarthritis. Additionally, the risk of osteoarthritis is increased by factors such as occupation, level of sports participation, and exercise intensity levels (21). Heavy loading and twisting are seen as contributing factors, but elevated physical activities do not appear to be a risk factor.

Osteoarthritis can also be created by joint immobilization because the joint and the cartilage require loading and compression to exchange nutrients and wastes (42). After only 30 days of immobilization, the fluid in the cartilage is increased, and an early form of osteoarthritis develops. Fortunately, this process can be reversed with a return to activity.

Injury to other structures in the diarthrodial joint can also be serious. An injury to the joint capsule results in formation of more fibrous tissue and possibly stretching of the capsule (16). Injury to the meniscus can create instability, loss of range of motion, and an increase in synovial effusion into the joint (swelling). Injury to the synovial membrane causes an increase in vascularity and produces gradual fibrosis of the tissue, eventually leading to chronic synovitis or inflammation of the membrane. Amazingly, many of these injury responses can also be reproduced through immobilization of the joint, which can produce adhesions, loss of range of motion, fibrosis, and synovitis.

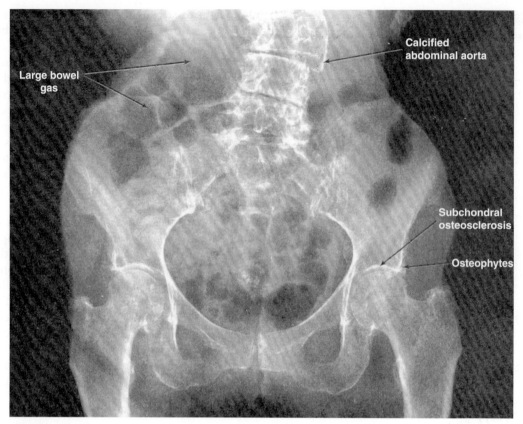

FIGURE 2-43   Osteoarthritis is characterized by physical changes in the joint consisting of cartilage erosion and formation of cysts and osteophytes. This radiograph shows osteoarthritis in the hip and vertebrae.

 **Summary**

Individual structures of the human body may be analyzed mechanically using a stress–strain curve to help determine its basic properties. Stress–strain curves illustrate the elastic and plastic regions and the elastic modulus of the structure. Structures and materials can be differentiated as elastic or viscoelastic based on their stress–strain curves. These basic mechanical properties may give insight into how a movement may take place.

The skeleton is composed of bones, joints, cartilage, and ligaments. It provides a system of levers that allows a variety of movements at the joints, provides a support structure, serves as a site for muscular attachment, protects the internal structures, stores fats and minerals, and participates in blood cell formation. Bone is an organ with blood vessels and nerves running through it.

The types of bones that compose the skeletal system (long, short, flat, and irregular) are shaped differently, perform different functions, and are made up of different proportions of cancellous and cortical bone tissue.

Bone tissue is one of the body's hardest structures because of its organic and inorganic components. Bone tissue continuously remodels through deposition and resorption of tissue. Modeling of bone is responsible for shaping both the shape and size of bone, and remodeling maintains bone mass by resorption and deposit at the

same site. Bone is sensitive to disuse and loading. Bone tissue is deposited in response to stress on the bone and removed through resorption when not stressed. One of the ways of increasing the strength and density of bone is through a program of physical activity. Osteoporosis occurs when bone resorption exceeds bone deposit and the bone becomes weak.

The study of the architecture of bone tissue has identified two types of bone, cortical and cancellous. Cortical bone, found on the exterior of bone and in the shaft of the long bones, is suited to handling high levels of compression and high tensile loads produced by the muscles. Cancellous bone is suited for high-energy storage and facilitates stress distribution within the bone.

Bone is both anisotropic and viscoelastic in its response to loads and responds differently to variety in the direction of the load and to the rate at which the load is applied. When first loaded, bone responds by deforming through a change in length or shape, known as the elastic response. With continued loading, microtears occur in the bone as it yields during the plastic phase. Bone is considered to be a flexible and weak material compared with other materials such as glass and steel.

The skeletal system is subject to a variety of loads and can handle larger compressive loads than tensile or shear loads. Commonly, bone is loaded in more than one direction, as with bending, in which both compression and tension are applied, and in torsion loads, in which shear, compression, and tensile loads are all produced. Injury to bone occurs when the applied load exceeds the strength of the material.

Two types of cartilage are found in the skeletal system. Articular or hyaline cartilage covers the ends of the bones at synovial joints. This cartilage is composed of water and a solid matrix of collagen and proteoglycan. Articular cartilage functions to attenuate shock in the joint, improve the fit of the joint, and provide minimal friction in the joint. Cartilage has viscoelastic properties in its response to loads. A second type of cartilage, fibrocartilage, offers additional load transmission and stability in a joint. Fibrocartilage is often referred to as an articular disk or meniscus.

Ligaments connect the bone to bone and are categorized as capsular, intracapsular, or extracapsular, depending on their location relative to the joint capsule. Ligaments exhibit viscoelastic behavior. They respond to loads by becoming stiffer as the load increases.

The movements of the long bones occur at a synovial joint, a joint with common characteristics such as articular cartilage, a capsule, a synovial membrane, and ligaments. The synovial joint can be injured through a sprain, in which the ligaments are injured. Joints are also susceptible to degeneration characterized by breakdown in the cartilage and bone. This degeneration is known as osteoarthritis.

The amount of motion between two segments is largely influenced by the type of synovial joint. For example, the planar joint allows simple translation between the joint surfaces; the hinge joint allows flexion and extension; the pivot joint allows rotation; the condylar joint allows flexion and extension with some rotation; the ellipsoid and the saddle joints allow flexion, extension, abduction, and adduction; and the ball-and-socket joint allows flexion, extension, abduction, adduction, and rotation. Other types of joints—synarthrodial and amphiarthrodial—allow little or no movement.

REVIEW QUESTIONS

True or False

1. ____ The portion of a stress–strain curve up to the yield point is called the plastic region.

2. ____ In a pure elastic material the mechanical energy is fully recovered after a deformation.

3. ____ The stiffness of a material can be determined by calculating the slope of the plastic portion of the stress–strain curve.

4. ____ A lever amplifies the force of movement.

5. ____ Hematopoiesis takes place within the cortical bone.

6. ____ Cancellous bone is also known as compact bone.

7. ____ Cancellous bone constitutes about 80% of the skeleton.

8. ____ The tarsals are short bones.

9. ____ The shaft of a long bone is called the diaphysis.

10. ____ Cortical bone is more porous than cancellous bone.

11. ____ Cortical bone is stronger than cancellous bone.

12. ____ The metaphysis is located between the epiphysis and the diaphysis of long bones.

13. ____ Flat bones make the best levers for muscles.

14. ____ Bone remodeling only takes place after age 40.

15. ____ Once bone is formed the shape cannot change.

16. ____ Bones use inactivity as a stimulus for bone growth.

17. ____ Joints have less friction than ice.

18. ____ Osteoporosis is an inflammation of the bone tissue.

19. ____ Bone is strongest when it is in tension.

20. ____ Bone becomes stiffer when the rate of loading is high.

21. ____ Shear stress is parallel to the plane of the cross section.

22. ____ A fatigue fracture is also known as a traumatic fracture.

23. ____ Articular cartilage has no blood supply.

24. ____ The meniscus of the knee is a type of fibrocartilage.

25. ____ Capsular ligaments are part of the joint capsule.

Multiple Choice

1. Which is not a part of the stress–strain curve?
   a. Elastic region
   b. Plastic region
   c. Yield point
   d. Nylon region

2. During failure the stress in a material will_____.
   a. rise quickly
   b. fall to zero
   c. remain equal to the yield stress
   d. None of the above

3. Stress is _____.
   a. the ratio of the change in length to the resting length
   b. the amount of force at a particular strain
   c. the force per unit area
   d. the stored mechanical energy

4. A viscoelastic material _____.
   a. has elastic and viscous properties
   b. exhibits nonlinear stress–strain behavior
   c. has multiple stiffnesses
   d. All of the above

5. These cells are responsible for forming new bone.
   a. Osteoclasts
   b. Osteopaths
   c. Osteocytes
   d. Osteoblasts

6. Which group contains examples of flat bones?
   a. Femur, humerus, skull
   b. Ribs, carpals, tarsals
   c. Ribs, skull, scapula
   d. Carpals, clavicle, vertebrae

7. The process of bone resorption by osteoclasts takes approximately 3 _____.
   a. hours
   b. days
   c. weeks
   d. months

8. These bone cells are responsible for sensing mechanical stress.
   a. Osteoclasts
   b. Osteopaths
   c. Osteoblasts
   d. Osteocytes

9. The bone in the distal part of the femur is replaced every _____.
   a. 5 to 6 months
   b. 10 to 12 months
   c. 2 years
   d. 4 years

10. Which is not a class of bones?
    a. Long
    b. Wide
    c. Short
    d. Flat

11. Cancellous bone is greater than _____% porous.
    a. 10
    b. 30
    c. 50
    d. 70

12. Building new bone at the same site that old bone is being removed is called _____.
    a. modeling
    b. remodeling
    c. resorption
    d. micromodeling

13. This structure improves the fit between articulating bones.
    a. Capsular ligament
    b. Articular fibrocartilage
    c. Fibrous capsule
    d. Ligament

14. This characteristic of bone suggests that the stiffness depends on the rate of loading.
    a. Isotropic
    b. Anisotropic
    c. Anisotonic
    d. Viscoelastic

15. Injury threshold _____ with repetition.
    a. increases
    b. decreases
    c. remains the same
    d. is eliminated

16. A standing person has _____ forces on the inferior portion and _____ forces on the superior portion of the femoral neck.
    a. torsion, tensile
    b. tensile, compressive
    c. compressive, tensile
    d. tensile, torsion

17. Peak bone mass occurs during the latter portion of the _____ decade of life.
    a. first
    b. second
    c. third
    d. fourth

18. Cartilage reduces contact forces by _____.
    a. 50%
    b. 60%
    c. 70%
    d. None of the above

19. Hyaline cartilage at the ends of long bones is called _____.
    a. articular cartilage
    b. fibrocartilage
    c. fibrous cartilage
    d. All of the above

20. Cartilage exhibits _____ characteristics.
    a. isotropic
    b. anisotropic
    c. Both a and b
    d. Neither a nor b

21. A diarthrodial joint is also known as a _____ joint.
    a. hinge
    b. condyloid
    c. synarthrodial
    d. synovial

22. Ellipsoidal joints have _____ degrees of freedom.
    a. 1
    b. 2
    c. 3
    d. 4

23. The elbow is an example of a(n) _____ joint.
    a. ellipsoid
    b. hinge
    c. condylar
    d. simple

24. The most mobile type of joint is the _____ joint.
    a. ball-and socket
    b. saddle
    c. pivot
    d. hinge

25. Osteoarthritis affects the _____.
    a. joint capsule
    b. ligaments
    c. articular cartilage
    d. articular fibrocartilage

References

1. An, K.N., et al. (1991). Pressure distribution on articular surfaces: Application to joint stability evaluation. *Journal of Biomechanics*, 23:1013.

2. Beaupre, G. S., et al. (2000). Mechanobiology in the development, maintenance and degeneration of articular cartilage. *Journal of Rehabilitation Research & Development*, 37: 145-151.

3. Bennell, K., et al. (2004). Ground reaction forces and bone parameters in females with tibial stress fractures. *Medicine & Science in Sports & Exercise*, 36:397-404.

4. Bird, H. A. (1986). A clinical review of the hyperlaxity of joints with particular reference to osteoarthrosis. *Engineering in Medicine*, 15:81.

5. Bonifasi-Lista, C., et al. (2005). Viscoelastic properties of the human medial collateral ligament under longitudinal, transverse and shear loading. *Journal of Orthopedic Research*, 23: 67-76.

6. Brewer, V., et al. (1983). Role of exercise in prevention of involutional bone loss. *Medicine & Science in Sports & Exercise*, 15:445.

7. Bubanj, S., Obradovic, B. (2002) Mechanical force and bones density. *Physical Education and Sport*, 9:37-50.

8. Choi, K., Goldstein, S. A. (1922). A comparison of the fatigue behavior of human trabecular and cortical bone tissue. *Journal of Biomechanics*, 25:1371.

9. Cook, S. D., et al. Trabecular bone density and menstrual function in women runners. *American Journal of Sports Medicine*, 15:503, 1987.

10. Cussler, E. C., et al. (2003). Weight lifted in strength training predicts bone changes in postmenopausal women. *Medicine & Science in Sports & Exercise*, 35:10-17.

11. Dolinar, J. (1990). Keeping ski injuries on the down slope. *The Physician and Sportsmedicine*. 19(2):120-123.

12. Downing, J. F., et al. (Eds.) (1991). Four complex joint injuries. *The Physician and Sportsmedicine* 19(10):80-97.

13. Egan, J. M. (1987). A constitutive model for the mechanical behavior of soft connective tissues. *Journal of Biomechanics*, 20:681-692,

14. Eisele, S. A. (1991). A precise approach to anterior knee pain. *The Physician and Sportsmedicine*, 19(6):126-130, 137-139.

15. Fine, K. M., et al. (1991). Prevention of cervical spine injuries in football. *The Physician and Sportsmedicine*, 19(10): 54-64.

16. Frost, H. M. (1985). The pathomechanics of osteoporoses. *Clinical Orthopaedics and Related Research*, 200:198.

17. Frost, H. M. (1997). On our age-related bone loss: Insights from a new paradigm. *Journal of Bone and Mineral Research*, 12:1539-1546.

18. Griffin, T. M., Guilak, F. (2005). The role of mechanical loading in the onset and progression of osteoarthritis. *Exercise and Sport Sciences Review*, 33:195-200.

19. Halpbern, B., et al. (1987). High school football injuries: Identifying the risk factors. *American Journal of Sports Medicine*, 15:316.

20. Halpbern, B. C., Smith, A. D. (1991). Catching the cause of low back pain. *The Physician and Sportsmedicine*, 19(6):71-79.

21. Healey, J. H., et al. (1985). The coexistence and characteristics of osteoarthritis and osteoporosis. *Journal of Bone & Joint Surgery*, 67A:586-592.

22. Henning, C. E. (1988). Semilunar cartilage of the knee: Function and pathology. In K. N. Pandolf (Ed.). *Exercise and Sport Sciences Reviews*. New York: Macmillan, 205-214.

23. Hettinga, D. L. (1985). In flammatory response of synovial joint structures. In J. Gould, G. J. Davies (Eds.). *Journal of Orthopaedics & Sports Physical Therapy*. St. Louis, MO: Mosby, 87-117.

24. Hoffman, A. H., Grigg, P. (1989). Measurement of joint capsule tissue loading in the cat knee using calibrated mechanoreceptors. *Journal of Biomechanics*, 22:787-791.

25. Iskrant, A. P., Smith, R. W. (1969). Osteoporosis in women 45 years and related to subsequent fractures. *Public Health Reports*, 84:33-38.

26. Jackson, D. L. (1990). Stress fracture of the femur. *The Physician and Sportsmedicine*, 19(7):39-44.

27. Jacobs, C. R. (2000). The mechanobiology of cancellous bone structural adaptation. *Journal of Rehabilitation Research & Development*, 37:209-216.

28. Kazarian, L. E., Von Gierke, H. E. (1969). Bone loss as a result of immobilization and chelation. Preliminary results in *Macaca mulatta*. *Clinical Orthopaedics*, 65:67-75.

29. Kohrt, W. M., et al. (2004) Physical activity and bone health. ACSM position stand. *Medicine & Science in Sports & Exercise*, 36:1985-1996.

30. Lakes, R. S., et al. (1990). Fracture mechanics of bone with short cracks. *Journal of Biomechanics*, 23:967-975.

31. Lane, N. E., et al. (1990). Running, osteoarthritis, and bone density: Initial 2-year longitudinal study. *American Journal of Medicine*, 88:452-459.

32. Matheson, G. O., et al. (1987). Stress fractures in athletes. *American Journal of Sports Medicine*, 15:46-58.

33. McConkey, J. P., Meeuwisse, W. (1988). Tibial plateau fractures in alpine skiing. *American Journal of Sports Medicine*, 16:159-164.

34. McGee, A. M., et al. (2004). Review of the biomechanics and patterns of limb fractures. *Trauma*, 6:29-40.

35. Milgrom, C., et al. (2000). In vivo strain measurements to evaluate strengthening potential of exercises on the tibial bone. *Journal of Bone & Joint Surgery*, 82-B:591-594.

36. Mosley, J. R. (2000). Osteoporosis and bone functional adaptation: Mechanobiological regulation of bone architecture in growing and adult bone, a review. *Journal of Rehabilitation Research & Development*, 37:189-199.

37. Mow, V. C., et al. (1989). Biomechanics of articular cartilage. In M. Nordin, V. H. Frankel (Eds.). *Basic Biomechanics of the Musculoskeletal System*. Philadelphia, PA: Lea & Febiger, 31-58.

38. Navarro, A. H., Sutton, J. D. (1985). Osteoarthritis IX: Biomechanical factors, prevention, and nonpharmacologic management. *Maryland Medical Journal*, 34:591-594.

39. Nordin, M., Frankel, V. H. (1989). Biomechanics of bone. In M. Nordin, V. H. Frankel (Eds.). *Basic Biomechanics of the Musculoskeletal System*. Philadelphia, PA: Lea & Febiger, 3-30.

40. Oyster, N., et al. (1984). Physical activity and osteoporosis in post-menopausal women. *Medicine & Science in Sports & Exercise*, 16:44-50.

41. Radin, E. L., et al. (1972). Role of mechanical factors in the pathogenesis of primary osteoarthritis. *Lancet*, 1:519-522.

42. Riegger, C. L. (1985). Mechanical properties of bone. In J. A. Gould, G. J. Davies (Eds.). *Journal of Orthopaedic and Sports Physical Therapy*. St. Louis, MO: Mosby, 3-49.

43. Robling, A. G., et al. (2002). Shorter, more frequent mechanical loading sessions enhance bone mass. *Medicine & Science in Sports & Exercise*, 34:196-202.

44. Schaffler, M. B., Burr, D. B. (1988). Stiffness of cortical bone: Effects of porosity and density. *Journal of Biomechanics*, 21:13-16.

45. Shipman, P., et al. (1985). *The Human Skeleton*. Cambridge, MA: Harvard University.

46. Soderberg, G. L. (1986). *Kinesiology: Application to Pathological Motion*. Baltimore, MD: Williams & Wilkins, 1986.

47. Turner, C. H. (2002). Biomechanics of bone: Determinants of skeletal fragility and bone quality. *Osteoporosis International*, 13:97-104.

48. Turner, C. H., et al. (2003) Designing exercise regimens to increase bone strength. *Exercise Sport Science Reviews*, 31:45-50.

49. van Rietbergen, B., et al. (2003). Trabecular bone tissue strains in the healthy and osteoporotic human femur. *Journal of Bone and Mineral Research* 18:1781-1788.

50. Wallace, L. A., et al. (1985). The knee. In J. A. Gould, G. J. Davies (Eds.). *Journal of Orthopaedic & Sports Physical Therapy*. St. Louis, MO: Mosby, 342-364.

51. Weiss, J. A., et al. (2001). Computational modeling of ligament mechanics. *Critical Reviews in Biomedical Engineering*, 29:1-70.

52. Whalen, R. T., et al. (1988). Influence of physical activity on the regulation of bone density. *Journal of Biomechanics*, 21:825-837.

53. Zernicke, R. F., et al. (1990). Biomechanical response of bone to weightlessness. In K. B. Pandolf, J. O. Holloszy (Eds.). *Exercise and Sport Sciences Reviews*. Baltimore, MD: Williams & Wilkins, 167-192.

MUSCULAR CONSIDERATIONS FOR MOVEMENT

OBJECTIVES

After reading this chapter, the student will be able to:

1. Define the properties, functions, and roles of skeletal muscle.

2. Describe the gross and microscopic anatomical structure of muscles.

3. Explain the differences in muscle fiber arrangement, muscle volume, and cross section as it relates to the output of the muscle.

4. Describe the difference in the force output between the three muscle fiber types (types I, IIa, and IIb).

5. Describe the characteristics of the muscle attachment to the bone, and explain the viscoelastic response of the tendon.

6. Discuss how force is generated in the muscle.

7. Describe how force is transmitted to the bone.

8. Discuss the role of muscle in terms of movement production or stability.

9. Compare isometric, concentric, and eccentric muscle actions.

10. Describe specific considerations for the two-joint muscles.

11. Discuss the interaction between force and velocity in the muscle.

12. Describe factors that influence force and velocity development in the muscle, including muscle cross section and length, the length–tension relationship, neural activation, fiber type, the presence of a prestretch, and aging.

13. Explain the physical changes that occur in muscles as a result of strength training and elaborate on how training specificity, intensity, and training volume influence strength training outcomes.

14. Describe types of resistance training, and explain how training should be adjusted for athletes and nonathletes.

15. Identify some of the major contributors to muscle injury, the location of common injuries, and means for prevention of injury to muscles.

OUTLINE

Muscle Tissue Properties
   Irritability
   Contractility
   Extensibility
   Elasticity
Functions of Muscle
   Produce Movement
   Maintain Postures and Positions

   Stabilize Joints
   Other Functions
Skeletal Muscle Structure
   Physical Organization of Muscle
Force Generation in the Muscle
   Motor Unit
   Muscle Contraction
   Transmission of Muscle Force to Bone

Mechanical Model of Muscle:
    The Musculotendinous Unit
Role of Muscle
    Origin versus Insertion
    Developing Torque
    Muscle Role versus Angle of
        Attachment
    Muscle Actions Creating, Opposing,
        and Stabilizing Movements
    Net Muscle Actions
    One- and Two-Joint Muscles
Force–Velocity Relationships in
Skeletal Muscle
    Force–Velocity and Muscle Action
        or Load

Factors Influencing Force and Velocity
    Generated by Skeletal Muscle
Strengthening Muscle
    Principles of Resistance Training
    Training Modalities
Injury to Skeletal Muscle
    Cause and Site of Muscle Injury
    Preventing Muscle Injury
    Inactivity, Injury, and Immobilization
        Effects on Muscle
Summary
Review Questions

Muscles exert forces and thus are the major contributors to human movement. Muscles are used to hold a position, to raise or lower a body part, to slow down a fast-moving segment, and to generate great speed in the body or in an object that is propelled into the air. A muscle only has the ability to pull and creates a motion because it crosses a joint. The tension developed by muscles applies compression to the joints, enhancing their stability. In some joint positions, however, the tension generated by the muscles can act to pull the segments apart and create instability.

Exercise programming for a young, healthy population incorporates exercises that push the muscular system to high levels of performance. Muscles can exert force and develop power to produce the desired movement outcomes. The same exercise principles used with young, active individuals can be scaled down for use by persons of limited ability. Using the elderly as an example, it is apparent that strength decrement is one of the major factors influencing efficiency in daily living activities. The loss of strength in the muscular system can create a variety of problems, ranging from inability to reach overhead or open a jar lid to difficulty using stairs and getting up out of a chair. Another example is an overweight individual who has difficulty walking any distance because the muscular system cannot generate sufficient power and the person fatigues easily. These two examples are really not different from the power lifter trying to perform a maximum lift in the squat. In all three cases, the muscular system is overloaded, with only the magnitude of the load and the output varying.

Muscle tissue is an excitable tissue and is either striated or smooth. Striated muscles include skeletal and cardiac muscles. Both cardiac and smooth muscles are under the control of the autonomic nervous system. That is, they are not under voluntary control. Skeletal muscle, on the other hand, is under direct voluntary control. Of primary interest in this chapter is the skeletal muscle. All aspects of muscle structure and function related to human movement and efficiency of muscular contribution are explored in this chapter. Because muscles are responsible for locomotion, limb movements, and posture and joint stability, a good understanding of the features and limitations of muscle action is necessary. Although it is not the function of this chapter to describe all of the muscles and their actions, it is necessary for the reader to have a good understanding of the location and action of the primary skeletal muscles. Figure 3-1 illustrates the surface skeletal muscles of the human body.

 ## Muscle Tissue Properties

Muscle tissue is very resilient and can be stretched or shortened at fairly high speeds without major damage to the tissue. The performance of the muscle tissue under varying loads and velocities is determined by the four properties of the muscle tissue: **irritability, contractility, extensibility**, and **elasticity**. A closer examination of these properties as they relate specifically to skeletal muscle tissue will enhance the understanding of skeletal muscle actions described later in the chapter.

IRRITABILITY

Irritability, or excitability, is the ability to respond to stimulation. In a muscle, the stimulation is provided by a motor neuron releasing a chemical neurotransmitter. Skeletal muscle tissue is one of the most sensitive and responsive tissues in the body. Only nerve tissue is more sensitive than skeletal muscle. As an excitable tissue, skeletal muscle can be recruited quickly, with significant control over how many muscle fibers and which ones will be stimulated for a movement.

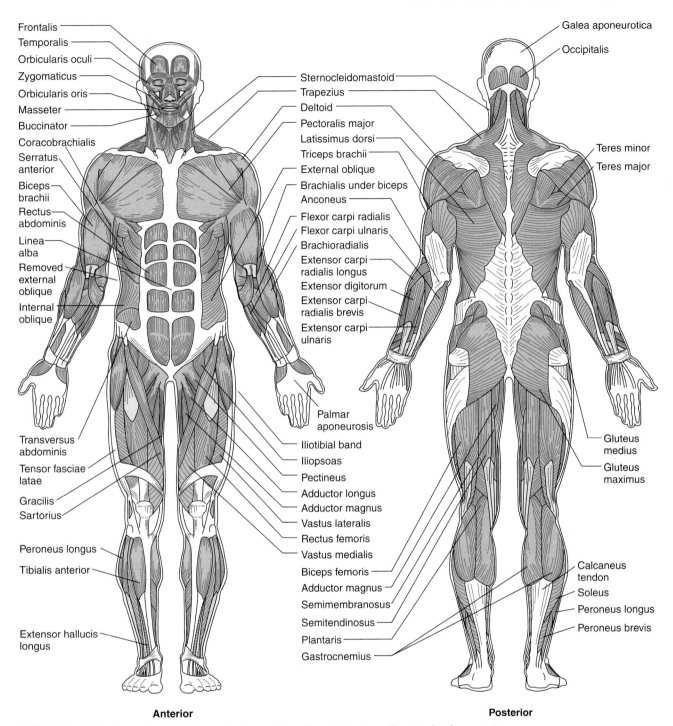

Frontalis
Temporalis
Orbicularis oculi
Zygomaticus
Orbicularis oris
Masseter
Buccinator
Coracobrachialis
Serratus anterior
Biceps brachii
Rectus abdominis
Linea alba
Removed external oblique
Internal oblique
Transversus abdominis
Tensor fasciae latae
Gracilis
Sartorius
Peroneus longus
Tibialis anterior
Extensor hallucis longus

Sternocleidomastoid
Trapezius
Deltoid
Pectoralis major
Latissimus dorsi
Triceps brachii
External oblique
Brachialis under biceps
Anconeus
Flexor carpi radialis
Flexor carpi ulnaris
Brachioradialis
Extensor carpi radialis longus
Extensor digitorum
Extensor carpi radialis brevis
Extensor carpi ulnaris
Palmar aponeurosis
Iliotibial band
Iliopsoas
Pectineus
Adductor longus
Adductor magnus
Vastus lateralis
Rectus femoris
Vastus medialis
Biceps femoris
Adductor magnus
Semimembranosus
Semitendinosus
Plantaris
Gastrocnemius

Galea aponeurotica
Occipitalis
Teres minor
Teres major
Gluteus medius
Gluteus maximus
Calcaneus tendon
Soleus
Peroneus longus
Peroneus brevis

Anterior

Posterior

FIGURE 3-1   Skeletal muscles of the human body: anterior and posterior views. (Reprinted with permission from Willis, M. C. [1986]. *Medical Terminology: The Language of Health Care.* Baltimore, MA: Lippincott Williams & Wilkins.)

CONTRACTILITY

Contractility is the ability of a muscle to generate tension and shorten when it receives sufficient stimulation. Some skeletal muscles can shorten as much as 50% to 70% of their resting length. The average range is about 57% of resting length for all skeletal muscles. The distance through which a muscle shortens is usually limited by the physical confinement of the body. For example, the sartorius muscle can shorten more

than half of its length if it is removed and stimulated in a laboratory, but in the body, the shortening distance is restrained by the hip joint and positioning of the trunk and thigh.

EXTENSIBILITY

Extensibility is the muscle's ability to lengthen, or stretch beyond the resting length. The skeletal muscle itself cannot produce the elongation; another muscle or an external force is required. Taking a joint through a passive range of motion, that is, pushing another's limb past its resting length, is a good example of elongation in muscle tissue. The amount of extensibility in the muscle is determined by the connective tissue surrounding and within the muscle.

ELASTICITY

Elasticity is the ability of the muscle fiber to return to its resting length after the stretch is removed. Elasticity in the muscle is determined by the connective tissue in the muscle rather than the fibrils themselves. The properties of elasticity and extensibility are protective mechanisms that maintain the integrity and basic length of the muscle. Elasticity is also a critical component in facilitating output in a shortening muscle action that is preceded by a stretch.

Using a ligament as a comparison makes it easy to see how elasticity benefits muscle tissue. Ligaments, which are largely collagenous, have little elasticity, and if they are stretched beyond their resting length, they will not return to the original length but rather will remain extended. This can create laxity around the joint when the ligament is too long to exert much control over the joint motion. On the other hand, muscle tissue always returns to its original length. If the muscle is stretched too far, it eventually tears.

FUNCTIONS OF MUSCLE

Skeletal muscle performs a variety of different functions, all of which are important to efficient performance of the human body. The three functions relating specifically to human movement are contributing to the production of skeletal movement, assisting in joint stability, and maintaining posture and body positioning.

PRODUCE MOVEMENT

Skeletal movement is created as muscle actions generate tensions that are transferred to the bone. The resulting movements are necessary for locomotion and other segmental manipulations.

MAINTAIN POSTURES AND POSITIONS

Muscle actions of a lesser magnitude are used to maintain postures. This muscle activity is continuous and results in small adjustments as the head is maintained in position and the body weight is balanced over the feet.

STABILIZE JOINTS

Muscle actions also contribute significantly to the stability of the joints. Muscle tensions are generated and applied across the joints via the tendons, providing stability where they cross the joint. In most joints, especially the shoulder and the knee, the muscles spanning the joint via the tendons are among the primary stabilizers.

OTHER FUNCTIONS

The skeletal muscles also provide four other functions that are not directly related to human movement. First, muscles support and protect the visceral organs and protect the internal tissues from injury. Second, tension in the muscle tissue can alter and control pressures within the cavities. Third, skeletal muscle contributes to the maintenance of body temperature by producing heat. Fourth, the muscles control the entrances and exits to the body through voluntary control over swallowing, defecation, and urination.

Skeletal Muscle Structure

PHYSICAL ORGANIZATION OF MUSCLE

Muscles and muscle groups are arranged so that they may contribute individually or collectively to produce a very small, fine movement or a very large, powerful movement. Muscles rarely act individually but rather interact with other muscles in a multitude of roles. To understand muscle function, the structural organization of muscle from the macroscopic external anatomy all the way down to the microscopic level of muscular action must be examined. A good starting point is the gross anatomy and external arrangement of muscles and the microscopic view of the muscle fiber.

Groups of Muscles

Groups of muscles are contained within compartments that are defined by **fascia**, a sheet of fibrous tissue. The compartments divide the muscles into functional groups, and it is common for muscles in a compartment to be innervated by the same nerve. The thigh has three compartments: the anterior compartment, containing the quadriceps femoris; the posterior compartment, containing the hamstrings; and the medial compartment, containing the adductors. Compartments for the thigh and the leg are illustrated in Figure 3-2.

The compartments keep the muscles organized and contained in one region, but sometimes the compartment is not large enough to accommodate the muscle or muscle groups. In the anterior tibial region, the compartment is small, and problems arise if the muscles are overdeveloped for the amount of space defined by the compartment. This is known as anterior compartment syndrome, and it can be serious if the cramped compartment impinges on nerves or blood supply to the leg and foot.

Muscle Architecture

Two major fiber arrangements are found in the muscle: **parallel** and **pennate**.

> Many of the more than 600 muscles in the body are organized in right and left pairs. About 70 to 80 pairs of muscles are responsible for the majority of movements.

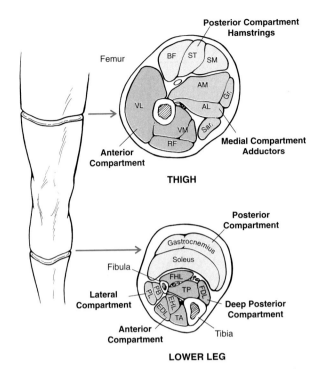

FIGURE 3-2 Muscles are grouped into compartments in each segment. Each compartment is maintained by fascial sheaths. The muscles in each compartment are functionally similar and define groups of muscles that are classified according to function, such as extensors and flexors. EDL, extensor digitorum longus; EHL, extensor hallucis longus; FDL, flexor digitorum longus; FHL, flexor hallucis longus; PB, peroneus brevis; PL, peroneus longus; TA, tibialis anterior; TP, tibialis posterior.

Parallel Fiber Arrangements

In the parallel fiber arrangement, the fascicles are parallel to the long axis of the muscle. The five different shapes of parallel fiber arrangements are **flat**, **fusiform**, **strap**, **radiate** or **convergent**, and **circular** (Fig. 3-3). The flat, parallel fiber arrangement is usually thin and broad and originates from sheet-like aponeuroses. Forces generated in the flat-shaped muscle can be spread over a larger area. Examples of flat muscles are the rectus abdominis and the external oblique. The fusiform muscle is spindle-shaped with a central belly that tapers to tendons on each ends. This muscle shape allows force transmission to small bony sites. Examples of fusiform muscles are the brachialis, biceps brachii, and brachioradialis. Strap muscles do not have muscle belly regions with a uniform diameter along the length of the muscle. This muscle shape allows for force transmission to targeted sites. The sartorius is an example of a strap-shaped muscle. The radiate or convergent muscle shape has a combined arrangement of flat and fusiform fiber shapes that originate on broad aponeuroses and converge onto a tendon. The pectoralis major and the trapezius muscles are examples of convergent muscle shapes. Circular muscles are concentric arrangements of strap muscles, and this muscle surrounds openings to close the openings upon contraction. The orbicularis oris surrounding the mouth is an example of a circular muscle.

The fiber force in a parallel muscle fiber arrangement is in the same direction as the musculature (23). This results in a greater range of shortening and yields greater movement velocity. This is basically because the parallel muscles are often longer than other types of muscles and the muscle fiber is longer than the tendon. The fiber length of the biceps brachii muscle (fusiform) is shown in Figure 3-4 and can be equal to the muscle length.

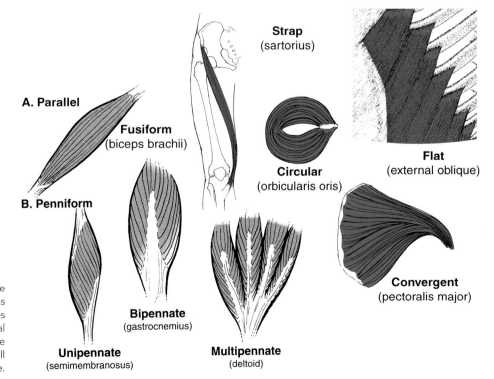

FIGURE 3-3 **(A)** Parallel muscles have fibers running in the same direction as the whole muscle. **(B)** Penniform muscles have fibers that run diagonally to a central tendon through the muscle. The muscle fibers of a penniform muscle do not pull in the same direction as the whole muscle.

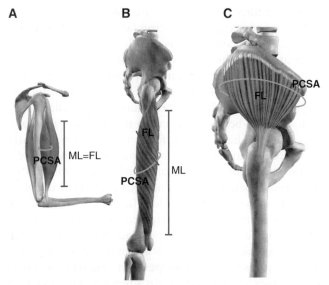

A   B   C

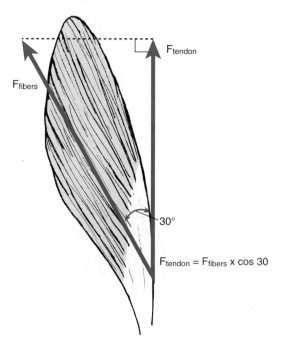

FIGURE 3-4 **(A)** The muscle length (ML) is equal to the fiber length (FL) in the biceps brachii, and it has a small physiologic cross-sectional area (PCSA), making it more suitable for a larger range of motion. **(B)** The vastus lateralis is capable of greater force production because it has a larger physiologic cross section. Additionally, the fiber length is shorter than the muscle length, making it less suitable for moving through a large distance. **(C)** The largest physiologic cross section is seen in the gluteus medius.

FIGURE 3-5 The pennation angle is the angle made between the fibers and the line of action of pull of the muscle.

Penniform Fiber Arrangements

In the second type of fiber arrangement, penniform, the fibers run diagonally with respect to a central tendon running the length of the muscle. The general shape of the penniform muscle is featherlike because the fascicles are short and run at an angle to the length of the muscle. Because the fibers of the penniform muscle run at an angle relative to the line of pull of the muscle, the force generated by each fiber is in a different direction than the muscle force (23). The fibers are shorter than the muscle, and the change in the individual fiber length is not equal to the change in the muscle length (23). The fibers can run diagonally off one side of the tendon, termed **unipennate** (e.g., biceps femoris, extensor digitorum longus, flexor pollicis longus, semimembranosus, and tibialis posterior); off both sides of the tendon, termed **bipennate** (e.g., rectus femoris, flexor hallucis longus, gastrocnemius, vastus medialis, vastus lateralis, and infraspinatus); or both, termed **multipennate** (e.g., deltoid and gluteus maximus).

Because the muscle fibers are shorter and run diagonally into the tendon, the penniform fibers create slower movements through a smaller range of motion than a fusiform muscle. The trade-off is that a penniform muscle has a much greater physiologic cross section that can generally produce more strength than can a fusiform muscle.

Pennation Angle

The **pennation angle** is the angle made by the fascicles and the line of action (pull) of the muscle (Fig. 3-5). The greater the angle of pennation, the smaller the amount of force transmitted to the tendon, and because the pennation angle increases with contraction, the force-producing capabilities will reduce. For example, the medial gastrocnemius working at the ankle joint is at a disadvantageous position when the knee is positioned at 90° because of pennation angles of approximately 60°, allowing only half of the force to be applied to the tendon (29). When the pennation angle is low, as with the quadriceps muscles, the pennation angle is not a significant factor.

Muscle Volume and Cross Section

A number of parameters can be calculated to describe the muscle potential in relationship to muscle architecture. Muscle mass, muscle length, and surface pennation angle can be measured directly after dissection of whole-muscle cadaver models. Ultrasonography and magnetic resonance imaging can also be used to collect some of these parameters. **Muscle volume** (cm³) can be calculated after these initial factors are known using the following equation:

$$MV = m/\rho$$

where m is the mass of the muscle (g) and ρ (g/cm³) is the density of the muscle (1.056 g/cm²). **Cross-sectional area** (cm²) can be calculated with the following equation:

$$CSA = MV/L$$

where MV is the muscle volume (cm³) and L is fiber length (cm). This is an estimate for the whole muscle. In one study (58), the largest muscle volume recorded in the thigh and the lower leg was in the vastus lateralis (1,505 cm³) and the soleus (552 cm³). Large cross-sectional areas were recorded in the gluteus maximus (145.7 cm²) and vastus medialis (63 cm²). A measurement of the cross-sectional area perpendicular to the longitudinal axis of the muscle is called the **anatomical cross section** and is only relevant to the site where the slice is taken.

The **physiologic cross section** is the sum total of all of the cross sections of fibers in the muscle in the plane perpendicular to the direction of the fibers. The formula for physiologic cross-sectional area (PCSA) is:

$$PCSA = m \cos \theta / \rho L$$

where m is the mass of the muscle, ρ is the density of the muscle (1.056 g/cm^2), L is the muscle length, and θ is the surface pennation angle. The soleus muscle has a PCSA of 230 cm^2, which is three to eight times larger than those of the medial gastrocnemius (68 cm^2) and the lateral gastrocnemius (28 cm^2), making its potential for force production greater (14). The PCSA is directly proportional to the amount of force generated by a muscle. Muscles such as the quadriceps femoris that have a large PCSA and short fibers (low fiber length/muscle length) can generate large forces. Conversely, muscles such as the hamstrings that have smaller PCSA and long fibers (high fiber length/muscle length) are more suited to developing high velocities. Figure 3-4 illustrates the difference between fiber length, muscle length, and physiologic cross section in fusiform (biceps brachii) and pennate muscles (vastus lateralis and gluteus medius).

Fiber Type

Each muscle contains a combination of fiber types that are categorized as **slow-twitch fibers** (type I) or **fast-twitch fibers** (type II). Fast-twitch fibers are further broken down into types IIa and IIb. Fiber type is an important consideration in muscle metabolism and energy consumption, and muscle fiber type is thoroughly studied in exercise physiology. Mechanical differences in the response of slow- and fast-twitch muscle fibers warrant an examination of the fiber type.

Slow-Twitch Fiber Types

Slow-twitch, or type I, fibers are oxidative. The fibers are red because of the high content of myoglobin in the muscle. These fibers have slow contraction times and are well suited for prolonged, low-intensity work. Endurance athletes usually have a high quantity of slow-twitch fibers.

Intermediate- and Fast-Twitch Fiber Types

Fast-twitch, or type II, fibers are further broken down into type IIa, oxidative–glycolytic, and type IIb, glycolytic. Type IIa fiber is a red muscle fiber known as the intermediate fast-twitch fiber because it can sustain activity for long periods or contract with a burst of force and then fatigue. The white type IIb fiber provides us with rapid force production and then fatigues quickly.

Most, if not all, muscles contain both fiber types. An example is the vastus lateralis, which is typically half fast-twitch and half slow-twitch fibers (31). The fiber type influences how the muscle is trained and developed and what techniques will best suit individuals with specific fiber types. For example, sprinters and jumpers usually have great concentrations of fast-twitch fibers. These fiber types are also found in high concentrations in muscles on which these athletes rely, such as the gastrocnemius. On the other hand, distance runners usually have greater concentrations of slow-twitch fibers.

Individual Muscle Structure

The anatomy of a skeletal muscle is presented in Figure 3-6. Each individual muscle usually has a thick central portion, the **belly** of the muscle. Some muscles, such as the biceps brachii, have very pronounced bellies, but other muscles, such as the wrist flexors and extensors, have bellies that are not as apparent.

Covering the outside of the muscle is another fibrous tissue, the **epimysium**. This structure plays a vital role in the transfer of muscular tension to the bone. Tension in the muscle is generated at various sites, and the epimysium transfers the various tensions to the tendon, providing a smooth application of the muscular force to the bone.

Each muscle contains hundreds to tens of thousands of muscle fibers, which are carefully organized into compartments within the muscle itself. Bundles of muscle fibers are called **fascicles**. Each fascicle may contain as many as 200 muscle fibers. A fascicle is covered with a dense connective sheath called the **perimysium** that protects the muscle fibers and provides pathways for the nerves and blood vessels. The connective tissue in the perimysium and the epimysium gives the muscle much of its ability to stretch and return to a normal resting length. The perimysium is also the focus of flexibility training because the connective tissue in the muscle can be stretched, allowing the muscle to elongate.

The fascicles run parallel to each other. Each fascicle contains the long, cylindrical, threadlike muscle **fibers**, the cells of skeletal muscles, where the force is generated. Muscle fibers are 10 to 100 m in width and 15 to 30 cm in length. Fibers also run parallel to each other and are covered with a membrane, the **endomysium**. The endomysium is a very fine sheath carrying the capillaries and nerves that nourish and innervate each muscle fiber. The vessels and the nerves usually enter in the middle of the muscle and are distributed throughout the muscle by a path through the endomysium. The endomysium also serves as an insulator for the neurologic activity within the muscle.

Directly underneath the endomysium is the **sarcolemma**. This is a thin plasma membrane surface that branches into the muscle. The neurologic innervation of the muscle travels through the sarcolemma and eventually reaches each individual contractile unit by means of a chemical neurotransmission.

At the microscopic level, a fiber can be further broken down into numerous **myofibrils**. These delicate rodlike strands run the total length of the muscle and contain the contractile proteins of the muscle. Hundreds or even thousands of myofibrils are present in each muscle fiber, and each fiber is filled with 80% myofibrils (5). The remainder of the fiber consists of the usual organelles, such as the mitochondria, the **sarcoplasm, sarcoplasmic**

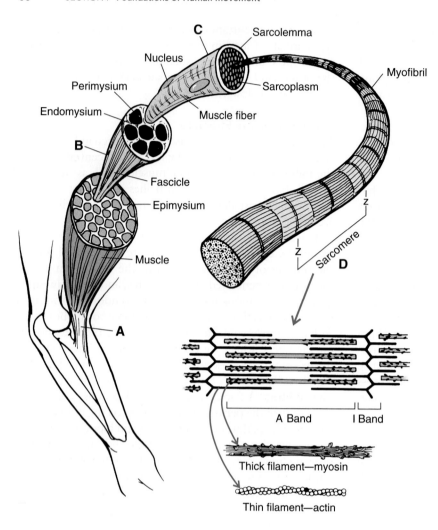

FIGURE 3-6   (A) Each muscle connects to the bone via a tendon or aponeurosis. (B) Within the muscle, the fibers are bundled into fascicles. (C) Each fiber contains myofibril strands that run the length of the fiber. (D) The actual contractile unit is the sarcomere. Many sarcomeres are connected in series down the length of each myofibril. Muscle shortening occurs in the sarcomere as the myofilaments in the sarcomere, actin, and myosin slide toward each other.

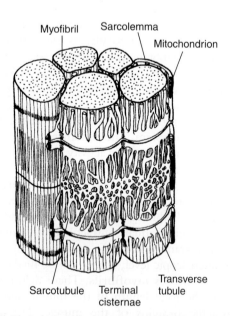

FIGURE 3-7   A portion of a skeletal muscle fiber illustrating the sarcoplasmic reticulum that surrounds the myofibril. (Adapted with permission from Pittman, M. I., Peterson, L. [1989]. Biomechanics of skeletal muscle. In M. Nordin, V. H. Frankel (Eds.). *Basic Biomechanics of the Musculoskeletal System* (2nd ed.). Philadelphia, PA: Lea & Febiger, 89-111.

reticulum, and the **t-tubules** (or **transverse tubules**). Myofibrils are 1 to 2 μm in diameter (about a 4 millionth of an inch wide) and run the length of the muscle fiber (5). Figure 3-7 illustrates muscle myofibrils and some of these organelles.

The myofibrils are cross-striated by light and dark filaments placed in an order that forms repeating patterns of bands. The dark banding is the thick protein **myosin,** and the light band is a thin polypeptide, **actin.** One unit of these bands is called a **sarcomere.** This structure is the actual contractile unit of the muscle that develops tension. Sarcomeres are in series along a myofibril. That is, sarcomeres form units along the length of the myofibril much like the links in a chain.

 Force Generation in the Muscle

MOTOR UNIT

Skeletal muscle is organized into functional groups called motor units. A **motor unit** consists of a group of muscle fibers that are innervated by the same motor neuron.

Motor units are discussed in more detail in Chapter 4, but it is important to discuss some aspects in this chapter. Motor units can consist of only a few muscle fibers (e.g., the optic muscles) or may have up to 2,000 muscle fibers (e.g., the gastrocnemius). The signal to contract that is transmitted from the motor neuron to the muscle is called an **action potential**. When a motor neuron is stimulated enough to cause a contraction, all muscle fibers innervated by that motor neuron contract. The size of the action potential and resulting muscle action are proportional to the number of fibers in the motor unit. An increase in output of force from the muscle requires an increase in the number of motor units activated.

MUSCLE CONTRACTION

The action potential from a motor neuron reaches a muscle fiber at a **neuromuscular junction** or **motor end plate** that lies near the center of the fiber. At this point, a synapse, or space, exists between the motor neuron and the fiber membrane. When the action potential reaches the synapse, a series of chemical reactions take place, and acetylcholine (ACh) is released. ACh diffuses across the synapse and causes an increase in permeability of the membrane of the fiber. The ACh rapidly breaks down to prevent continuous stimulation of the muscle fiber. The velocity at which the action potential is propagated along the membrane is the **conduction velocity**.

The muscle resting membrane potential inside is 270 to 295 mV with respect to the outside. At the threshold level of the membrane potential (approximately 250 mV), a change in potential of the fiber membrane or sarcolemma occurs. The action potential is characterized by a **depolarization** from the **resting potential** of the membrane so that the potential becomes positive (approximately +40 mV) and is said to overshoot. There is a hyperpolarized state (**hyperpolarization**) before returning to the resting potential. This is followed by a **repolarization**, or a return to the polarized state.

The wave of depolarization of the action potential moves along the nerve until it reaches the muscle fibers, where it spreads to the muscle membrane as calcium ions (Ca^{2+}) are released into the area surrounding the myofibrils. These Ca^{2+} ions promote cross-bridge formation, which results in an interaction between the actin and myosin filaments (see the discussion of sliding filament theory in the next section). When the stimulation stops, ions are actively removed from the area surrounding the myofibrils, releasing the cross-bridges. This process is **excitation–contraction coupling** (Fig. 3-8). The calcium ions link action potentials in a muscle fiber to contraction by binding to the filaments and turning on the interaction of the actin and myosin to start contraction of the sarcomere.

Muscle force production is achieved in two ways. First, muscle force can be increased by recruiting increasingly larger motor units. Initially, during a muscle contraction, smaller motor units are activated. As muscle force increases, more and larger motor units are engaged. This is the size principle (20). Second, a motor unit may be activated at any of the several frequencies. A single action potential that activates a fiber will cause the force to increase and decrease. This is referred to as a **twitch**. If a second stimulus occurs before the initial twitch has subsided, another twitch builds upon the first. With subsequent high frequency of stimulations, the force continues to build and forms a state called unfused **tetanus**. Finally, the force builds to a level in which there is no increase in the muscle force. At this point, the force level has reached tetanus. This scenario is illustrated in Figure 3-9. In a muscle contraction, both size recruitment and frequency of stimulation are simultaneously used to increase muscle force.

Sliding Filament Theory

How a muscle generates tension has been an area of much research. An explanation of the shortening of the sarcomere has been presented via the **sliding filament theory** presented by Huxley (26). This theory is the most widely accepted explanation of muscular contraction but certainly is not the only one. In the past, muscle contraction was thought, for example, to be similar to the principle of blood clotting, the behavior of India rubber, a chain of circular elastic rings, and a sliding movement caused by opposite electric charges in the different filaments (42).

In Huxley's sliding filament theory, when calcium is released into the muscle through neurochemical stimulation, the contracting process begins. The sarcomere contracts as the myosin filament walks along the actin filament, forming cross-bridges between the head of the myosin and a prepared site on the actin filament. In the contracted state, the actin and myosin filaments overlap along most of their lengths (Fig. 3-10).

The simultaneous sliding of many thousands of sarcomeres in series changes the length and force of the muscle (5). The amount of force that can be developed in the muscle is proportional to the number of cross-bridges formed. The shortening of many sarcomeres, myofibrils, and fibers develops tension running through the muscle and to the bone at both ends to create a movement.

TRANSMISSION OF MUSCLE FORCE TO BONE

Tendon versus Aponeurosis

A muscle attaches to bone in one of three ways: directly into the bone, via a **tendon**, or via an **aponeurosis**, a flat tendon. These three types of attachments are presented in Figure 3-11. Muscle can attach directly to the periosteum of the bone through fusion between the epimysium and the surface of the bone, such as the attachment of the trapezius (56). Muscle can attach via a tendon that is fused with the muscle fascia, such as in the hamstrings, biceps brachii, and flexor carpi radialis. Last, muscle can attach to a bone via a sheath of fibrous tissue known as an aponeurosis seen in the abdominals and the trunk attachment of the latissimus dorsi.

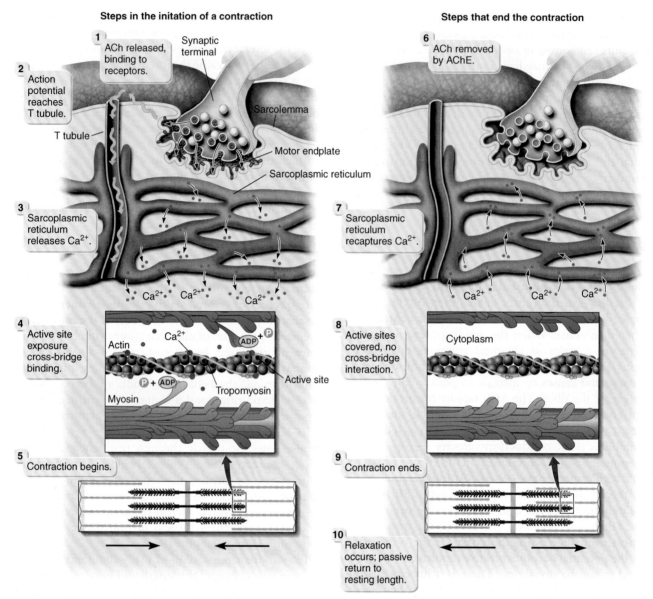

Steps in the initiation of a contraction

1 ACh released, binding to receptors.

Synaptic terminal

2 Action potential reaches T tubule.

T tubule

Sarcolemma

Motor endplate

Sarcoplasmic reticulum

3 Sarcoplasmic reticulum releases Ca²⁺.

Ca^{2+} Ca^{2+} Ca^{2+}

4 Active site exposure cross-bridge binding.

Ca^{2+}

Actin

ADP + P

P + ADP

Myosin

Tropomyosin

Active site

5 Contraction begins.

Steps that end the contraction

6 ACh removed by AChE.

7 Sarcoplasmic reticulum recaptures Ca²⁺.

Ca^{2+} Ca^{2+} Ca^{2+}

8 Active sites covered, no cross-bridge interaction.

Cytoplasm

9 Contraction ends.

10 Relaxation occurs; passive return to resting length.

FIGURE 3-8 Excitation–contraction coupling occurs when the action potential traveling down the motor neuron reaches the muscle fiber where acetylcholine (ACh) is released. This causes depolarization and the release of Ca^{2+} ions that promote cross-bridge formation between actin and myosin, resulting in shortening of the sarcomere. AChE, acetylcholinesterase; ADP, adenosine diphosphate.

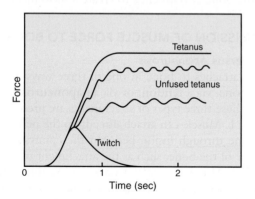

FIGURE 3-9 When a single stimulus is given, a twitch occurs. When a series of stimuli is given, muscle force increases to an uneven plateau or unfused tetanus. As the frequency of stimuli increases, the muscle force ultimately reaches a limit, or tetanus. (Adapted from McMahon, T. A. [1984]. *Muscles, Reflexes, and Locomotion.* Princeton, NJ: Princeton University Press.)

Characteristics of the Tendon

The most common form of attachment, the tendon, transmits the force of the associated muscle to the bone. The tendon connects to the muscle at the myotendinous junction, where the muscle fibers are woven in with the collagen fibers of the tendon. Tendons are powerful and carry large loads via connections where fibers perforate the surfaces of bones. Tendons can resist stretch, are flexible, and can turn corners running over cartilage, sesamoid bones, or bursae. Tendons can be arranged in a cord or in strips and can be circular, oval, or flat. Tendons consist of an inelastic bundle of collagen fibers arranged parallel to the direction of the force application of the muscle. Even though the fibers are inelastic, tendons can respond in an elastic fashion through recoiling and the elasticity of connective tissue. Tendons can withstand high tensile forces produced by the muscles, and they exhibit viscoelastic

MYOFIBRIL

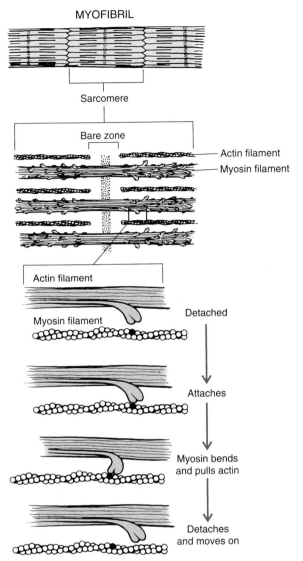

FIGURE 3-10 The sliding filament theory. Shortening of the muscle has been explained by this theory. Shortening takes place in the sarcomere as the myosin heads bind to sites on the actin filament to form a cross-bridge. The myosin head attaches and turns, moving the actin filament toward the center. It then detaches and moves on to the next actin site.

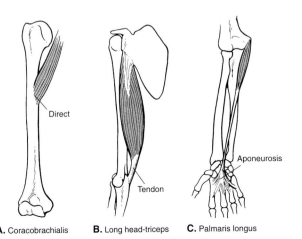

FIGURE 3-11 A muscle can attach directly into the bone (**A**) or indirectly via a tendon (**B**) or aponeurosis (**C**).

behavior in response to loading. The Achilles tendon has been reported to resist tensile loads to a degree equal to or greater than that of steel of similar dimensions.

The stress–strain response of a tendon is viscoelastic. That is, tendons show a nonlinear response and exhibit hysteresis. Tendons are relatively stiff and much stronger than other structures. Tendons respond very stiffly when exposed to a high rate of loading. This stiff behavior of tendons is thought to be related to their relatively high collagen content. Tendons are also very resilient and show relatively little hysteresis or energy loss. These characteristics are necessary for the function of tendons. Tendons must be stiff and strong enough to transmit force to bone without deforming much. Also, because of the low hysteresis of tendons, they are capable of storing and releasing elastic strain energy. The differences in the strength and performance characteristics of tendons versus muscles and bones are presented in Figure 3-12.

Tendons and muscles join at **myotendinous junctions**, where the actual myofibrils of the muscle fiber join the collagen fibers of the tendon to produce a multilayered interface (62). The tendon connection to the bone consists of fibrocartilage that joins to mineralized fibrocartilage and then to the lamellar bone. This interface blends with the periosteum and the subchondral bone.

Tendons and muscles work together to absorb or generate tension in the system. Tendons are arranged in series, or in line with the muscles. Consequently, tendons bear the same tension as muscles (46). The mechanical interaction between muscles and tendons depends on the amount of force that is being applied or generated, the speed of the muscle action, and the slack in the tendon.

Tendons are composed of parallel fibers that are not perfectly aligned, forming a wavy, crimped appearance. If tension is generated in the muscle fibers while the tendon is slack, there is initial compliance in the tendon as it straightens out. It will begin to recoil or spring back to its initial length (Fig. 3-13). As the slack in the tendon is taken up by the recoiling action, the time taken to stretch the tendon causes a delay in the achievement of the required level of tension in the muscle fibers (46).

Recoiling of the tendon also reduces the speed at which a muscle may shorten, which in turn increases the load a muscle can support (46). If the tendon is stiff and has no recoil, the tension will be transmitted directly to the muscle fibers, creating higher velocities and decreasing the load the muscle can support. The stiff response in a tendon allows for the development of rapid tensions in the muscle and results in brisk, accurate movements.

The tendon and the muscle are very susceptible to injury if the muscle is contracting as it is being stretched. An example is the follow-through phase of throwing. Here, the posterior rotator cuff stretches as it contracts to slow the movement. Another example is the lengthening and contraction of the quadriceps femoris muscle group during the support phase of running as the center of mass is lowered via knee flexion. The tendon picks up the initial

MUSCLE

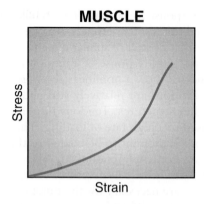

TENDON

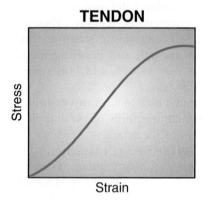

BONE

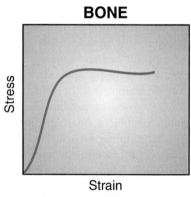

FIGURE 3-12   The stress–strain curves for muscle, tendon, and bone tissue. **Top.** Muscle is viscoelastic and thus deforms under low load and then responds stiffly. **Middle.** Tendon is capable of handling high loads. The end of the elastic limits of the tendon is also the ultimate strength level (no plastic phase). **Bottom.** Bone is a brittle material that responds stiffly and then undergoes minimal deformation before failure.

the angle of pull of the muscle and reducing the tension generated in the muscle. Examples of this can be found with the quadriceps femoris muscles and the patella and with the tendons of the hamstrings and the gastrocnemius as they travel over condyles on the femur. Some tendons are covered with synovial sheaths to keep the tendon in place and protect the tendon.

The tension in the tendons also produces the actual ridges and protuberances on the bone. The apophyses found on a bone are developed by tension forces applied to the bone through the tendon (see Chapter 2). This is of interest to physical anthropologists because they can study skeletal remains and make sound predictions about lifestyle and occupations of a civilization by evaluating prominent ridges, size of the trochanters and tuberosities, and basic size of the specimen.

Tendon Influences on Force Development (Force–Time Characteristics)

When a muscle begins to develop tension through the contractile component (CC) of the muscle, the force increases nonlinearly over time because the passive elastic components in the tendon and the connective tissue stretch and absorb some of the force. After the elastic components are stretched, the tension that the muscle exerts on the bone increases linearly over time until maximum force is achieved.

The time to achieve maximum force and the magnitude of the force vary with a change in joint position. In one joint position, maximum force may be produced very quickly, but in other joint positions, it may occur later in the contraction. This reflects the changes in tendon laxity, not changes in the tension-generating capabilities of the CCs. If the tendon is slack, the maximum force occurs later and vice versa.

MECHANICAL MODEL OF MUSCLE: THE MUSCULOTENDINOUS UNIT

A series of experiments by A. V. Hill gave rise to a behavioral model that predicted the mechanical nature of muscle. The Hill model has three components that act together in a manner that describes the behavior of a whole muscle (21,22). A schematic of configurations of the Hill model is presented in Figure 3-14. Hill used the techniques of a systems engineer to perform experiments that helped him identify key phenomena of muscle function. The model contained components referred to as the CC, **parallel elastic component** (PEC), and **series elastic component** (SEC). Because this is a behavioral model, it is inappropriate to ascribe these mechanical components to specific structures in the muscle itself.

However, the model has given great insight into how muscle functions to develop tension and is often used as a basis for many computer models of muscle.

The CC is the element of the muscle model that converts the stimulation of the nervous system into a force and reflects the shortening of the muscle through the actin and

stretch of the relaxed muscle, and if the muscle contracts as it is stretched, the tension increases steeply in both the muscle and the tendon (46).

When tension is generated in a tendon at a slow rate, injury is more likely to occur at the tendon–bone junction than other regions. At a faster rate of tension development, the actual tendon is the more common site of failure (54). For the total muscle–tendon unit, the likely site of injury is the belly of the muscle or the myotendinous junction.

Many tendons travel over bony protuberances that reduce some of the tension on the tendon by changing

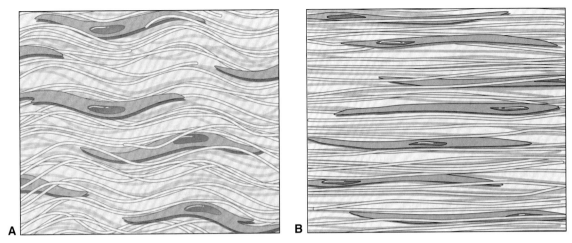

FIGURE 3-13 **(A)** In a relaxed state, the fibers in many tendons are slack and wavy. **(B)** When tension is applied, the tendon springs back to its initial length, causing a delay in the achievement of the required muscle tension.

myosin structures. The CC has mechanical characteristics that determine the efficiency of a contraction, that is, how well the signal from the nervous system translates into a force. We have already discussed the first of these mechanical characteristics, the relationship between stimulation and activation. Two others, the force–velocity and force–length relationships, are discussed later in this chapter.

The elasticity inherent in the muscle is represented by the SEC and the PEC. Because the SEC is in series with the CC, any force produced by the CC is also applied to the SEC. It first appears that the SEC is the tendon of the muscle, but the SEC represents the elasticity of all elastic elements in series with the force-generating structures of the muscle. The SEC is a highly nonlinearly elastic structure.

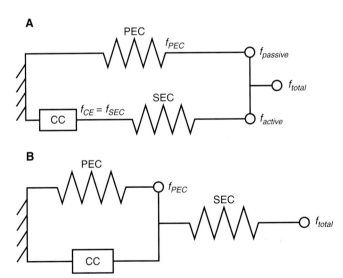

FIGURE 3-14 **(A)** The most common form of the Hill muscle model. **(B)** An alternative form. Because series elastic component (SEC) is usually stiffer than parallel elastic component (PEC) for most muscles, it generally does not matter which form of the model is used. (Adapted with permission from Winters, J. M. [2000]. Terminology and foundations of movement science. In J. M. Winters, P. E. Crago (Eds.). *Biomechanics and Neural Control of Posture and Movement.* New York: Springer-Verlag, 3-35.)

Muscle displays elastic behavior even when the CC is not producing force. An external force applied to a muscle causes the muscle to resist, but the muscle also stretches. This inactive elastic response is produced by structures that must be in parallel to the CC rather than in series to the CC. Thus, we have the PEC. The PEC is often associated with the connective tissue that surrounds the muscle and its compartments, but again, this is a behavioral model rather than a structural model, so this association cannot be made. The PEC, similar to the SEC, is highly nonlinear, and increases in stiffness as the muscle lengthens. Both the SEC and the PEC also behave like springs when acting quickly.

Role of Muscle

In the performance of a motor skill, only a small portion of the potential movement capability of the musculoskeletal system is used. Twenty to thirty degrees of freedom may be available to raise your arm above your head and comb your hair. Many of the available movements, however, may be inefficient in terms of the desired movement (e.g., combing the hair). To eliminate the undesirable movements and create the skill or desired movement, muscles or groups of muscles play a variety of roles. To perform a motor skill at a given time, only a small percentage of the potential movement capability of the motor system is used.

ORIGIN VERSUS INSERTION

A muscle typically attaches to a bone at both ends. The attachment closest to the middle of the body, or more proximal, is termed the **origin**, and this attachment is usually broader. The attachment farther from the midline, or more distal, is called the **insertion**; this attachment usually converges to a tendon. There can be more than one attachment site at both ends of the muscle. Traditional anatomy classes usually incorporate a study of the origins and insertions of the muscles. It is a common mistake to view the origin as the bony attachment that does not move when the muscle contracts. Muscle force is generated and applied to both skeletal connections, resulting in movement of one bone or both. The reason that both bones do not move when a muscle contracts is the stabilizing force of adjacent muscles or the difference in the mass of the two segments or bones to which the muscle is attached. Additionally, many muscles cross more than one joint and have the potential to generate multiple movements on more than one segment.

Numerous examples are available of a muscle shifting between moving one end of its attachment and the other, depending on the activity. One example is the psoas muscle, which crosses the hip joint. This muscle flexes the thigh, as in leg raises, or raises the trunk, as in a curl-up or sit-up (Fig. 3-15). Another example is the gluteus medius, which moves the pelvis when the foot is on the ground and the leg when the foot is off the ground. The effect of tension in a muscle should be evaluated at all attachment sites even if no movement is resulting from the force. Evaluating all attachment sites allows assessment of the magnitude of the required stabilizing forces and the actual forces applied at the bony insertion.

DEVELOPING TORQUE

A muscle controls or creates a movement through the development of **torque**. Torque is defined as the tendency of a force to produce rotation about a specific axis. In the case of a muscle, a force is generated in the muscle along the **line of action** of the force and applied to a bone, which causes a rotation about the joint (axis). The muscle's line of action or line of pull is the direction of the resultant muscle force running between the attachment sites on both ends of the muscle. The two components of torque are the magnitude of the force and the shortest or perpendicular distance from the pivot point to the line of action of the force, often termed the **moment arm**. Mathematically, torque is:

$$T = F \times r$$

where T is torque, F is the applied force in newtons, and r is the perpendicular distance in meters from the line of action of the force to the pivot point (moment arm). The amount of torque generated by the muscle is influenced by the capacity to generate force in the muscle itself and the muscle's moment arm. During any movement, both of these factors are changing. In particular, the moment arm increases or decreases depending on the line of pull of the muscle relative to the joint (Fig. 3-16). If the muscle's moment arm increases anywhere in the movement, the muscle can produce less force and still produce the same torque around the joint. Conversely, if the moment arm decreases, more muscle force is required to produce the same torque around the joint (Fig. 3-17). A more thorough discussion of torque is presented in a later chapter.

MUSCLE ROLE VERSUS ANGLE OF ATTACHMENT

The muscle supplies a certain amount of tension that is transferred via the tendon or aponeurosis to the bone. Not all of the tension or force produced by the muscle is put to use in generating rotation of the segment. Depending on the angle of insertion of the muscle, some

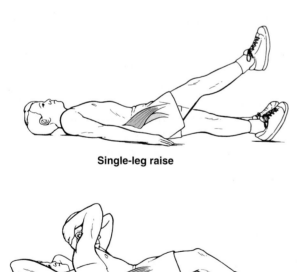

Single-leg raise

Sit-up

FIGURE 3-15   The origin of the psoas muscle is on the bodies of the last thoracic and all of the lumbar vertebrae, and the insertion is on the lesser trochanter of the femur. It is incorrect to assume that the origin remains stable in a movement. Here the psoas pulls on both the vertebrae and the femur. With the trunk stabilized, the femur moves (leg raise), and with the legs stabilized, the trunk moves (sit-up).

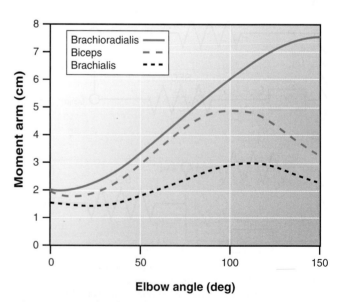

FIGURE 3-16   The elbow flexor muscle moment arms can change dramatically as the elbow is flexed. The brachioradialis muscle can produce nearly three times as much torque at 150° of flexion than it can at 0° of flexion with the same amount of force. The changes in moment arm are often nonlinear.

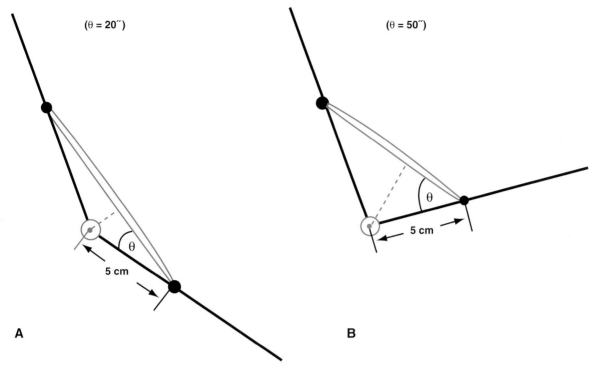

FIGURE 3-17 A muscle with a small moment arm (A) needs to produce more force to generate the same torque as a muscle with a larger moment arm (B).

force is directed to stabilizing or destabilizing the segment by pulling the bone into or away from the joint.

Muscular force is primarily directed along the length of the bone and into the joint when the tendon angle is acute or lying flat on the bone. When the forearm is extended, the tendon of the biceps brachii inserts into the radius at a low angle. Initiating an arm curl from this position requires greater muscle force than from other positions because most of the force generated by the biceps brachii is directed into the elbow rather than into moving the segments

around the joint. Fortunately, the resistance offered by the forearm weight is at a minimum in the extended position. Thus, the small muscular force available to move the segment is usually sufficient. Both the force directed along the length of the bone and that which is applied perpendicular to the bone to create joint movement can be determined by resolving the angle of the muscular force application into its respective parallel and rotary components. Figure 3-18 shows the parallel and rotary components of the biceps brachii force for various attachment angles.

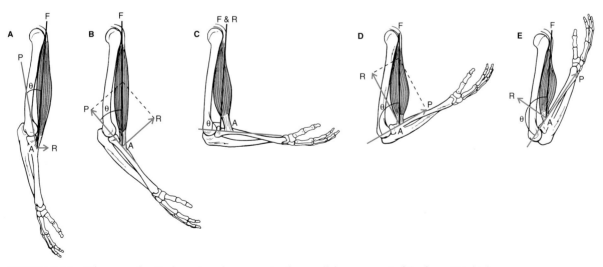

FIGURE 3-18 When muscle attachment angles are acute, the parallel component of the force (P) is highest and is stabilizing the joint. The rotatory component (R) is low (A). As the angle increases, the rotatory component also increases (B). The rotatory component increases to its maximum level at a 90° angle of attachment (C). Beyond a 90° angle of attachment, the rotatory component diminishes, and the parallel component increases to produce a dislocating force (D and E).

Even though muscular tension may be maintained during a joint movement, the rotary component and the torque varies with the angle of insertion. Many neutral starting positions are weak because most of the muscular force is directed along the length of the bone. As segments move through the midrange of the joint motion, the angle of insertion usually increases and directs more of the muscular force into moving the segment. Consequently, when starting a weight-lifting movement from the fully extended position, less weight can be lifted than if the person started the lift with some flexion in the joint. Figure 3-19 shows the isometric force output of the shoulder flexors and extensors for a range of joint positions.

In addition, at the end of some joint movements, the angle of insertion may move past 90°, the point at which the moving force again begins to decrease and the force along the length of the bone acts to pull the bone away from the joint. This dislocating force is present in the elbow and shoulder joints when a high degree of flexion is present in the joints.

The mechanical actions of broad muscles that have fibers attaching directly into bone over a large attachment site, such as the pectoralis major and trapezius, are difficult to describe using one movement for the whole muscle (56). For example, the lower trapezius attaches to the scapula at an angle opposite that of the upper trapezius; thus, these sections of the same muscle are functionally independent. When the shoulder girdle is elevated and abducted as the arm is moved up in front of the body, the lower portion of the trapezius may be inactive. This presents a complicated problem when studying the function of the muscle as a whole and requires multiple lines of action and effect (56).

MUSCLE ACTIONS CREATING, OPPOSING, AND STABILIZING MOVEMENTS

Agonists and Antagonists
The various roles of selected muscles in a simple arm abduction exercise are presented in Figure 3-20. Muscles

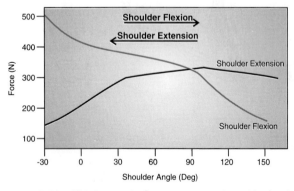

FIGURE 3-19 The isometric force output varies with the joint angle. As the shoulder angle increases, the shoulder extension force increases. The reverse happens with shoulder flexion force values, which decrease with an increase of the shoulder angle. (Adapted with permission from Kulig, K., et al. [1984]. Human strength curves. In R. L. Terjund (Ed.). *Exercise and Sport Sciences Reviews*, 12:417-466.)

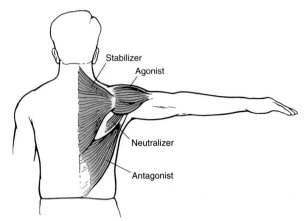

FIGURE 3-20 Muscles perform a variety of roles in movement. In arm abduction, the deltoid is the agonist because it is responsible for the abduction movement. The latissimus dorsi is the antagonistic muscle because it resists abduction. There are also muscles stabilizing in the region so the movement can occur. Here, the trapezius is shown stabilizing and holding the scapula in place. Last, there may be some neutralizing action: The teres minor may neutralize via external rotation any internal rotation produced by the latissimus dorsi.

creating the same joint movement are termed **agonists**. Conversely, muscles opposing or producing the opposite joint movement are called **antagonists**. The antagonists must relax to allow a movement to occur or contract concurrently with the agonists to control or slow a joint movement. Because of this, the most sizable changes in relative position of muscles occur in the antagonists (25). Thus, when the thigh swings forward and upward, the agonists producing the movement are the hip flexors, that is, the iliopsoas, rectus femoris, pectineus, sartorius, and gracilis muscles. The antagonists, or the muscles opposing the motion of hip flexion, are the hip extensors, hamstrings, and gluteus maximus. The antagonists combined with the effect of gravity slow down the movement of hip flexion and terminate the joint action. Both the agonists and antagonists are jointly involved in controlling or moderating movement.

When a muscle is playing the role of an antagonist, it is more susceptible to injury at the site of muscle attachment or in the muscle fiber itself. This is because the muscle is contracting to slow the limb while being stretched.

Stabilizers and Neutralizers
Muscles are also used as **stabilizers**, acting in one segment so that a specific movement in an adjacent joint can occur. Stabilization is important, for example, in the shoulder girdle, which must be supported so that arm movements can occur smoothly and efficiently. It is also important in the pelvic girdle and hip region during gait. When one foot is on the ground in walking or running, the gluteus medius contracts to maintain the stability of the pelvis so it does not drop to one side.

The last role muscles are required to play is that of **synergist**, or **neutralizer**, in which a muscle contracts to eliminate an undesired joint action of another muscle.

Forces can be transferred between two adjacent muscles and supplement the force in the target muscle (25). For example, the gluteus maximus is contracted at the hip joint to produce thigh extension, but the gluteus maximus also attempts to rotate the thigh externally. If external rotation is an undesired action, the gluteus minimus and the tensor fascia latae contract to produce a neutralizing internal rotation action that cancels out the external rotation action of the gluteus maximus, leaving the desired extension movement.

NET MUSCLE ACTIONS

Isometric Muscle Action

Muscle tension is generated against resistance to maintain position, raise a segment or object, or lower or control a segment. If the muscle is active and develops tension with no visible or external change in joint position, the muscle action is termed **isometric** (31). Examples of isometric muscle actions are illustrated in Figure 3-21. To bend over into 30° of trunk flexion and hold that position, the muscle action used to hold the position is termed isometric because no movement is taking place. The muscles contracting isometrically to hold the trunk in a position of flexion are the back muscles because they are resisting the force of gravity that tends to farther flex the trunk.

To take the opposite perspective, consider the movement in which the trunk is curled up to 30° and that position is held. To hold this position of trunk flexion, an isometric muscle action using the trunk flexors is produced. This muscle action resists the action of gravity that is forcing the trunk to extend.

Concentric Muscle Action

If a muscle visibly shortens while generating tension actively, the muscle action is termed **concentric** (31). In concentric joint action, the net muscle forces produc-ing movement are in the same direction as the change in joint angle, meaning that the agonists are the controlling muscles (Fig. 3-21). Also, the limb movement produced in a concentric muscle action is termed positive because the joint actions are usually against gravity or are the initiating source of movement of a mass.

Many joint movements are created by a concentric muscle action. For example, flexion of the arm or forearm from the standing position is produced by a concentric muscle action from the respective agonists or flexor muscles. Additionally, to initiate a movement of the arm across the body in a horizontal adduction movement, the horizontal adductors initiate the movement via a concentric muscle action. Concentric muscle actions are used to generate forces against external resistances, such as raising a weight, pushing off the ground, and throwing an implement.

Eccentric Muscle Action

When a muscle is subjected to an external torque that is greater than the torque generated by the muscle, the muscle lengthens, and the action is known as **eccentric** (31). The source of the external force developing the external torque that produces an eccentric muscle action is usually gravity or the muscle action of an antagonistic muscle group (5).

In eccentric joint action, the net muscular forces producing the rotation are in the opposite direction of the change in joint angle, meaning that the antagonists are the controlling muscles (Fig. 3-21). Also, the limb movement produced in eccentric muscle action is termed negative because the joint actions are usually moving down with gravity or are controlling rather than initiating the movement of a mass. In an activity such as walking downhill, the muscles act as shock absorbers as they resist the downward movement while lengthening.

Most movements downward, unless they are very fast, are controlled by an eccentric action of the antagonistic muscle

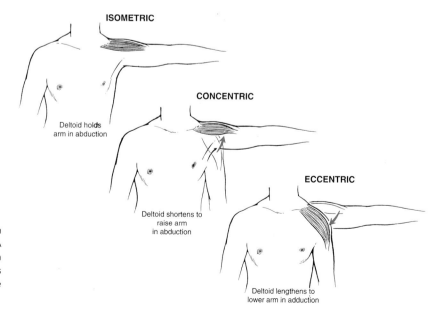

FIGURE 3-21 A muscle action is isometric when the tension creates no change in joint position. A concentric muscle action occurs when the tension shortens the muscle. An eccentric muscle action is generated by an external force when the muscle lengthens.

ISOMETRIC

Deltoid holds arm in abduction

CONCENTRIC

Deltoid shortens to raise arm in abduction

ECCENTRIC

Deltoid lengthens to lower arm in adduction

groups. To reverse the example shown in Figure 3-21, during adduction of the arm from the abducted position, the muscle action is eccentrically produced by the abductors or antagonistic muscle group. Likewise, lowering into a squat position, which involves hip and knee flexion, requires an eccentric movement controlled by the hip and knee extensors. Conversely, the reverse thigh and shank extension movements up against gravity are produced concentrically by the extensors.

From these examples, the potential sites of muscular imbalances in the body can be identified because the extensors in the trunk and the lower extremity are used to both lower and raise the segments. In the upper extremity, the flexors both raise the segments concentrically and lower the segments eccentrically, thereby obtaining more use.

Eccentric actions are also used to slow a movement. When the thigh flexes rapidly, as in a kicking action, the antagonists (extensors) eccentrically control and slow the joint action near the end of the range of motion. Injury can be a risk in a movement requiring rapid deceleration for athletes with impaired eccentric strength.

Eccentric muscle actions preceding concentric muscle actions increase the force output because of the contribution of elastic strain energy in the muscle. For example, in throwing, the trunk, lower extremity, and shoulder internal rotation are active eccentrically in the windup, cocking, and late cocking phases. Elastic strain energy is stored in these muscles, which enhances the concentric phase of the throwing motion (39).

Comparison of Isometric, Concentric, and Eccentric Muscle Actions

Isometric, concentric, and eccentric muscle actions are not used in isolation but rather in combination. Typically, isometric actions are used to stabilize a body part, and eccentric and concentric muscle actions are used sequentially to maximize energy storage and muscle performance. This natural sequence of muscle function, during which an eccentric action precedes a concentric action, is known as the **stretch–shortening cycle**, which is described later in this chapter.

These three muscle actions are very different in terms of their energy cost and force output. The eccentric muscle action can develop the same force output as the other two

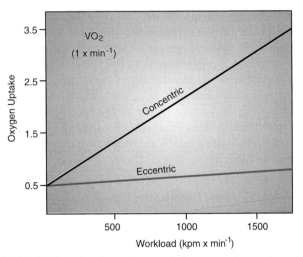

FIGURE 3-22 It has been illustrated that eccentric muscle action can produce high workloads at lower oxygen uptake levels than the same loads produced with concentric muscle action. (Adapted with permission from Asmussen, E. [1952]. Positive and negative muscular work. *Acta Physiologica Scandinavica*, 28:364–382.)

types of muscle actions with fewer muscle fibers activated. Consequently, eccentric action is more efficient and can produce the same force output with less oxygen consumption than the others (3) (Fig. 3-22).

In addition, the eccentric muscle action is capable of greater force output using fewer motor units than isometric or concentric actions (Fig. 3-23). This occurs at the level of the sarcomere, where the force increases beyond the maximum isometric force if the myofibril is stretched and stimulated (10,13).

Concentric muscle actions generate the lowest force output of the three types. Force is related to the number of cross-bridges formed in the myofibril. In isometric muscle action, the number of bridges attached remains constant. As the muscle shortens, the number of attached bridges is reduced with increased velocity (13). This reduces the level of force output generated by tension in the muscle fibers. A hypothetical torque output curve for the three muscle actions is presented in Figure 3-24.

An additional factor contributing to noticeable force output differences between eccentric and concentric muscle actions is present when the actions are producing vertical movements. In this case, the force output in both concentric and eccentric actions is influenced by torques

Examples of muscles and actions

Muscle	Movement	Muscle Action
Biceps brachii—elbow flexor	Elbow flexion in lifting	Concentric—shortening
Hamstrings—knee flexor	Knee extension in kicking	Eccentric—lengthening
Anterior deltoid—shoulder flexor	Shoulder flexion in handstand	Isometric—stabilization

What is the muscle action of the quadriceps femoris in the lowering action of a squat?
What is the muscle action of the posterior deltoid in the follow-through phase of a throw?

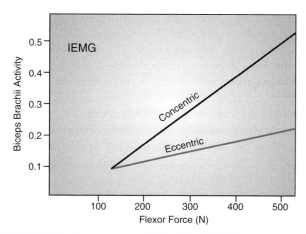

FIGURE 3-23 The integrated EMG activity (IEMG) in the biceps brachii muscle is higher as the same forces are generated using concentric muscle action compared with eccentric muscle action. (Adapted with permission from Komi, P. V. [1986]. The stretch–shortening cycle and human power output. In N. L. Jones, et al. (Eds.). *Human Muscle Power*. Champaign, IL: Human Kinetics, 27-40.)

created by gravity. The gravitational force creates torque that contributes to the force output in an eccentric action as the muscles generate torque that controls the lowering of the limb or body. The total force output in a lowering action is the result of both muscular torques and gravitational torques.

The force of gravity inhibits the movement of a limb upward, and before any movement can occur, the concentric muscle action must develop a force output that is greater than the force of gravity acting on the limb or body (weight). The total force output in a raising action is predominantly muscle force. This is another reason concentric muscle action is more demanding than the eccentric or isometric action.

This information is useful when considering exercise programs for unconditioned individuals or rehabilitation

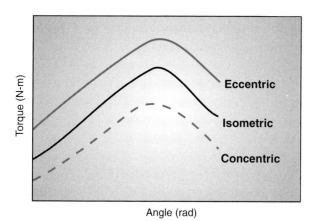

FIGURE 3-24 Eccentric muscle action can generate the greatest amount of torque through a given range of motion. Isometric muscle action can generate the next highest level of torque, and concentric muscle action generates the least torque (Adapted with permission from Enoka, R. M. [1988]. *Neuromuscular Basis of Kinesiology*. Champaign, IL: Human Kinetics.)

programs. Even the individual with the least amount of strength may be able to perform a controlled lowering of a body part or a small weight but may not be able to hold or raise the weight. A program that starts with eccentric exercises and then leads into isometric followed by concentric exercises may prove to be beneficial in the progression of strength or in rehabilitation of a body part. Thus, a person unable to do a push-up should start at the extended position and lower into the push-up, then receiving assistance on the up phase until enough strength is developed for the concentric portion of the skill. Factors to consider in the use of eccentric exercises are the control of the speed at which the limb or weight is lowered and control over the magnitude of the load imposed eccentrically because muscle injury and soreness can occur more readily with eccentric muscle action in high-load and high-speed conditions.

Concentric or eccentric contractions?

What muscles do you use to stand up and sit down? First note the action at the knee joint. While standing, the knee extends, and while sitting, the knee flexes. Does this mean you use the knee extensors to stand and the knee flexors to sit? No. You use knee extensors to stand and to sit. As you stand and then sit you can feel the tension in your knee extensors (quadriceps). The knee flexors (hamstrings) are relatively flaccid. Try this until you can convince yourself that it is predominantly the knee extensors that are active during standing up and sitting down. Use the knee extensors to slowly raise out of your seat. The knee extensors muscles are getting shorter but if you contract them with just a little less force, you start to lower yourself back into your seat. During this downward phase you find the quadriceps is lengthening and you are using your knee extensor muscles eccentrically to control the speed of the sitting motion. The actual force pulling you down is gravity.

ONE- AND TWO-JOINT MUSCLES

As stated earlier, one cannot determine the function or contribution of a muscle to a joint movement by simply locating the attachment sites. A muscle action can move a segment at one end of its attachment or two segments at both ends of its attachment. In fact, a muscle can accelerate and create movement at all joints, whether the muscle spans the joint or not. For example, the soleus is a plantarflexor of the ankle, but it can also force the knee into extension even though it does not cross the knee joint (61). This can occur in the standing posture. The soleus contracts and creates plantarflexion at the ankle. Because the foot is on the ground, the plantarflexion movement necessitates extension of the knee joint. In this manner, the soleus accelerates the knee joint twice as much as it accelerates the ankle, even though the soleus does not even span the knee.

Most muscles cross only one joint, so the dominating action of the one-joint muscle is at the joint it crosses. The two-joint muscle is a special case in which the muscle crosses two joints, creating a multitude of movements that often occur in opposite sequences to each other. For example, the rectus femoris is a two-joint muscle that creates both hip flexion and knee extension. Take the example of jumping. Hip extension and knee extension propel the body upward. Does the rectus femoris, a hip flexor and knee extensor, contribute to the extension of the knee, does it resist the movement of hip extension, or does it do both?

The action of a two-joint, or biarticulate, muscle depends on the position of the body and the muscle's interaction with external objects such as the ground (61). In the case of the rectus femoris, the muscle contributes primarily to the extension of the knee because of the hip joint position. This position results in the force of the rectus femoris acting close to the hip, thereby limiting the action of the muscle and its effectiveness in producing hip flexion (Fig. 3-25).

The perpendicular distance from the action line of the force of the muscle over to the hip joint is the moment arm, and the product of the force and the moment arm is the muscle torque. If the moment arm increases, torque at the joint increases, even if the applied muscle force is the same. Thus, in the case of a two-joint muscle, the muscle primarily acts on the joint where it has the largest moment arm or where it is farther from the joint. The hamstring group primarily creates hip extension rather than knee flexion because of the greater moment arm at the hip (Fig. 3-25). The gastrocnemius produces plantarflexion at the ankle rather than flexion at the knee joint because the moment arm is greater at the ankle.

For example, in vertical jumping, maximum height is achieved by extending the proximal joints first and then moving distally to where extension (plantarflexion) occurs in the ankle joint. By the time the ankle joint is involved in the sequence, very high joint moments and extension velocities are required (57). The role of the two-joint muscle becomes very important. The biarticular gastrocnemius muscle crosses both the knee and ankle joints. Its contribution to jumping is influenced by the knee joint. In jumping, the knee joint extends and effectively optimizes the length of the gastrocnemius (6). This keeps the contraction velocity in the gastrocnemius muscle low even when the ankle is plantarflexing very quickly. With the velocity lowered, the gastrocnemius is able to produce greater force in the jumping action.

The most important contribution of the two-joint muscle in the lower extremity is the reduction of the work required from the single-joint muscles. Two-joint muscles initiate a mechanical coupling of the joints that allows for a rapid release of stored elastic energy in the system (61).

Two-joint muscles save energy by allowing positive work at one joint and negative work at the adjacent joint. Thus, while the muscles acting at the ankle are producing a concentric action and positive work, the knee muscles can be eccentrically storing elastic energy through negative work (61).

The two-joint muscle actions for walking are presented in Figure 3-26. Two-joint muscles that work together in walking are the sartorius and rectus femoris at heel strike; the hamstrings and gastrocnemius at midsupport; the gastrocnemius and rectus femoris at toe-off; the rectus femoris, sartorius, and hamstrings at forward swing; and the hamstrings and gastrocnemius at foot descent (60). At heel strike, the sartorius, a hip flexor and a knee flexor, works with the rectus femoris, a hip flexor and knee extensor. As the heel strikes the surface, the rectus femoris performs negative work, absorbing energy at the knee as it moves into flexion. The sartorius, on the other hand, performs positive work as the knee and the hip both flex with gravity (60).

The two-joint muscle is limited in function at specific joint positions. When the two-joint muscle is constrained in elongation, it is termed **passive insufficiency**. This occurs when the antagonistic muscle cannot be elongated any

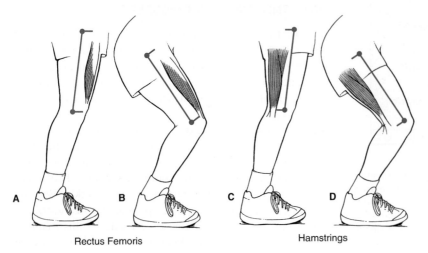

A   B   C   D

Rectus Femoris   Hamstrings

FIGURE 3-25 The rectus femoris moment arms at the hip and knee while standing **(A)** and in a squat **(B)** demonstrate why this muscle is more effective as an extender of the knee than as a flexor at the hip. Likewise, the hamstring moment arm while standing **(C)** and in a squat **(D)** demonstrates why the hamstrings are more effective as hip extensors than as knee flexors.

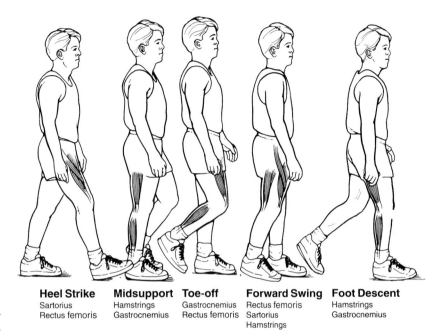

Heel Strike	**Midsupport**	**Toe-off**	**Forward Swing**	**Foot Descent**
Sartorius	Hamstrings	Gastrocnemius	Rectus femoris	Hamstrings
Rectus femoris	Gastrocnemius	Rectus femoris	Sartorius	Gastrocnemius
			Hamstrings	

FIGURE 3-26 Two-joint muscles work synergistically to optimize performance, shown here for walking.

farther and the full range of motion cannot be achieved. An example of passive insufficiency is the prevention of the full range of motion in knee extension by a tight hamstring. A two-joint muscle can also be restrained in contraction through **active insufficiency** where the muscle is slackened to the point where it has lost its ability to generate maximum tension. An example of active insufficiency is seen at the wrist where the finger flexors cannot generate maximum force in a grip when they are shortened by an accompanying wrist flexion movement.

Forceful wrist flexion extends fingers

Have you ever needed to remove something from a child's hand without hurting them? The concepts of active and passive insufficiency can help. By gently flexing the child's wrist, the extensor muscles will reach the limit that they can stretch and start to extend the fingers (passive insufficiency). At the same time the flexor muscles will shorten so much that they will not be able to produce much force (active insufficiency). The combination of passive and active insufficiencies will force the hand to open and release the contents. Demonstrate this with a partner.

Force–Velocity Relationships in Skeletal Muscle

Muscle fibers will shorten at a specific speed or velocity while concurrently developing a force used to move a segment or external load. Muscles create an active force to match the load in shortening, and the active force continuously adjusts to the speed at which the contractile system moves (10). When load is low, the active force is adjusted by increasing the speed of contraction. With greater loads, the muscle adjusts the active force by reducing the speed of shortening.

FORCE–VELOCITY AND MUSCLE ACTION OR LOAD

Force–Velocity Relationship in Concentric Muscle Actions

In concentric muscle action, velocity increases at the expense of a decrease in force and vice versa. The maximum force can be generated at zero velocity, and the maximum velocity can be achieved with the lightest load. An optimal force can be created at zero velocity because a large number of cross-bridges are formed. As the velocity of the muscle shortening increases, the cycling rate of the cross-bridges increases, leaving fewer cross-bridges attached at one time (24). This equates to less force, and at high velocities, when all of the cross-bridges are cycling, the force production is negligible (Fig. 3-27). This is opposite to what happens in a stretch in which an increase in the velocity of deformation of the passive components of the muscle results in higher force values. Maximum velocity in a concentric muscle action is determined by the cross-bridge cycling rates and the whole-muscle fiber length over which the shortening can occur.

Force–Velocity in the Muscle Fiber versus External Load

The force–velocity relationship relates to the behavior of muscle fiber, and it is sometimes confusing to relate this concept to an activity such as weight lifting. As an athlete increases the load in a lift, the speed of movement is likely to decrease. Although the force–velocity relationship is

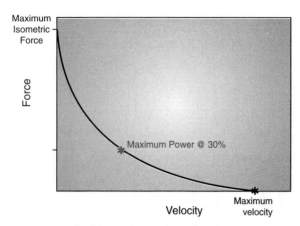

FIGURE 3-27 The force–velocity relationship in a concentric muscle action is inverse. The amount of tension or force-developing capability in the muscle decreases with an increase in velocity because fewer cross-bridges can be maintained. Maximum tension can be generated in the isometric or zero velocity condition, in which many cross-bridges can be formed. Maximum power can be generated in concentric muscle action with the velocity and force levels at 30% of maximum.

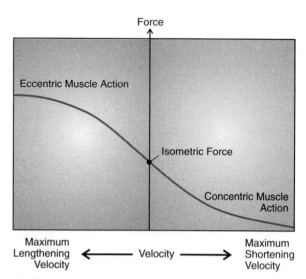

FIGURE 3-28 The relationship between force and velocity in eccentric muscle action is opposite to that of concentric muscle action. In eccentric muscle action, the force increases as the velocity of the lengthening increases. The force continues to increase until the eccentric action can no longer control lengthening of the muscle.

still present in the muscle fiber itself, the total system is responding to the increase in the external load or weight. The muscle may be generating the same amount of force in the fiber, but the addition of the weight slows the movement of the total system. In this case, the action velocity of the muscle is high, but the movement velocity of the high load is low (48). Muscles generate forces greater than the weight of the load in the early stages on the activity to move the weight and at the later stages of the lift, less muscle force may be required after the weight is moving.

Power

The product of force and velocity, **power**, is one of the major distinguishing features between successful and average athletes. Many sports require large power outputs, with the athlete expected to move his or her body or some external object very quickly. Because velocity diminishes with the increase of load, the most power can be achieved if the athlete produces one-third of maximum force at one-third of maximum velocity (43,44). In this way, the power output is maximized even though the velocities or the forces may not be at their maximum levels.

To train athletes for power, coaches must schedule high-velocity activities at 30% of maximum force (43). The development of power is also enhanced by fast-twitch muscle fibers, which are capable of generating four times more peak power than slow-twitch fibers.

Force–Velocity Relationships in Eccentric Muscle Actions

The force–velocity relationship in an eccentric muscle action is opposite to that in the shortening or concentric action. An eccentric muscle action is generated by antagonistic muscles, gravity, or some other external force. When a load greater than the maximum isometric strength value is applied to a muscle fiber, the fiber begins to lengthen eccentrically. At the initial stages of

lengthening, when the load is slightly greater than the isometric maximum, the speed of lengthening and the length changes in the sarcomeres are small (10).

If a load is as high as 50% greater than the isometric maximum, the muscle elongates at a high velocity. In eccentric muscle action, the tension increases with the speed of lengthening because the muscle is stretching as it contracts (Fig. 3-28). The eccentric force–velocity curve ends abruptly at some lengthening velocity when the muscle can no longer control the movement of the load.

FACTORS INFLUENCING FORCE AND VELOCITY GENERATED BY SKELETAL MUSCLE

Many factors influence how much force, or how fast of a contraction, a muscle can produce. Muscle contraction force is opposed by many factors, including the passive internal resistance of the muscle and tissue, the opposing muscles and soft tissue, and gravity or the effect of the load being moved or controlled. Let's examine some of the major factors that influence force and velocity development. These include muscle cross section, muscle length, muscle fiber length, preloading of the muscle before contraction, neural activation of the muscle, fiber type, and the age of the muscle.

What are some of the factors that determine force production in the muscle?

1. The number of cross-bridges formed at the sarcomere level
2. The cross section of the individual muscle fiber
3. The cross section of the total muscle
4. Muscle fiber arrangement

What are some of the factors that determine velocity production in the muscle?

1. Muscle length
2. Shortening rate per sarcomere per fiber
3. Muscle fiber arrangement

n	⎍⎍⎍ 1 sarcomere	⎍⎍⎍•⎍⎍⎍•⎍⎍⎍ 3 sarcomeres in series	3 sarcomeres in parallel
Force	f = 1	f = 1	nf = 3
Range of motion	x = 1	nx = 3	x = 3
Contraction time	t = 1	t = 1	t = 1
Velocity	x/t = 1	x/t = 3	x/t = 1

FIGURE 3-29 This table shows the effect of arranging springs (or sarcomeres) in series or in parallel. They have an advantage in producing large ranges of motion and velocity when arranged in series. They have an advantage in producing force when arranged in parallel.

Muscle Cross Section and Whole Muscle Length

Muscle architecture determines whether the muscle can generate large amounts of force and whether it can change its length significantly to develop higher velocity of movement. In the case of the latter, the shortening ability of a muscle is reflected by changes in both length and speed, depending on the situation.

Generally speaking, the strength of a muscle and the potential for force development are determined mainly by its size. Muscle can produce a maximum contractile force between 25 and 35 N per square centimeter of cross section, so a bigger muscle produces more force.

In the penniform muscle, the fibers are typically shorter and not aligned with the line of pull. An increased number of sarcomeres are aligned in parallel, which enhances the force-producing capacity. With an increased PCSA, the penniform muscle is able to exert more force than a similar mass of parallel fibers.

Parallel fibers with longer fiber lengths typically have a longer working range, producing a larger range of motion and a higher contraction velocity. With the fibers aligned parallel to the line of pull, an increased number of sarcomeres are attached end-to-end in series. This results in the increased fiber lengths and the capacity to generate greater shortening velocity.

A muscle with a greater ratio of muscle length to tendon length has the potential to shorten over a greater distance. Consequently, muscles attaching to the bone with a short tendon (e.g., the rectus abdominis) can move through a greater shortening distance than muscles with longer tendons (e.g., the gastrocnemius) (19). Great amounts of shortening also occur because skeletal muscle can shorten up to approximately 30% to 50% of its resting length. Similarly, a muscle having less pennation can also shorten over a longer distance and generate higher velocities (i.e., hamstrings, dorsiflexor). In contrast, a muscle having greater pennation (e.g., the gastrocnemius) can generate larger forces. The advantages of arranging sarcomeres in parallel (relatively shorter fibers with increased PCSA as in pennate muscle) or in series (relatively longer fibers as in fusiform muscle) are illustrated in Figure 3-29.

Muscle Fiber Length

The magnitude of force produced by a muscle during a contraction is also related to the length at which the muscle is held (10). Muscle length may increase, decrease, or remain constant during a contraction depending on the external opposing forces. Muscle length is restricted by the anatomy of the region and the attachment to the bone. The maximum tension that can be generated in the muscle fiber occurs when a muscle is activated at a length slightly greater than resting length, somewhere between 80% and 120% of the resting length. Fortunately, the length of most muscles in the body is within this maximum force production range. Figure 3-30 shows the **length–tension relationship** and demonstrates the contribution of active and passive components in the muscle during an isometric contraction.

Tension at Shortened Lengths

The tension-developing capacity drops off when the muscle is activated at both short and elongated lengths. The optimal length at the sarcomere level is when there

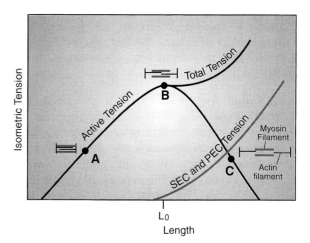

FIGURE 3-30 Muscle fibers cannot generate high tensions in the shortened state **(A)** because the actin and myosin filaments are maximally overlapped. The greatest tension in the muscle fiber can be generated at a length slightly greater than resting length **(B)**. In the elongated muscle **(C)**, the fibers are incapable of generating tension because the cross-bridges are pulled apart. The total muscle tension increases, however, because the elastic components increase their tension development. PEC, parallel elastic component; SEC, series elastic component.

is maximum overlap of myofilaments, allowing for the maximal number of cross-bridges. When a muscle has shortened to half its length, it is not capable of generating much more contractile tension. At short lengths, less tension is present because the filaments have exceeded their overlapping capability, creating an incomplete activation of the cross-bridges because fewer of these can be formed (10) (Fig. 3-30). Thus, at the end of a joint movement or range of motion of a segment, the muscle is weak and incapable of generating large amounts of force.

Tension at Elongated Lengths

When a muscle is lengthened and then activated, muscle fiber tension is initially greater because the cross-bridges are pulled apart after initially joining (49). This continues until the muscle length is increased slightly past the resting length. When the muscle is further lengthened and contracted, the tension generated in the muscle drop off because of slippage of the cross-bridges, resulting in fewer cross-bridges being formed (Fig. 3-30).

Contribution of the Elastic Components

The CC is not the only contributor to tension at different muscle lengths. The tension generated in a shortened muscle is shared by the SEC, that is, most tension develops in the tendon. The tension in the muscle is equal to the tension in the SEC when the muscle contracts in a shortened length.

As the tension-developing characteristics of the active components of the muscle fibers diminish with elongation, tension in the total muscle increases because of the contribution of the passive elements in the muscle. The SEC is stretched, and tension is developed in the tendon and the cross-bridges as they are rotated back (24). Significant tension is also developed in the PEC as the connective tissue in the muscle offers resistance to the stretch. As the muscle is lengthened, passive tension is generated in these structures, so that the total tension is a combination of contractile and passive components (Fig. 3-30). At extreme muscle lengths, the tension in the muscle is almost exclusively elastic, or passive, tension.

Optimal Length for Tension

The optimal muscle length for generating muscle tension is slightly greater than the resting length because the CCs are optimally producing tension and the passive components are storing elastic energy and adding to the total tension in the unit (18). This relationship lends support for placing the muscle on a stretch before using the muscle for a joint action. One of the major purposes of a windup or preparatory phase is to put the muscle on stretch to facilitate output from the muscle in the movement.

Neural Activation of the Muscle

The amount of force generated in the muscle is determined by the number of cross-bridges formed at the sarcomere level. The nature of stimulation of the motor units and the types of motor units recruited both affect

> **How does the tension in the active and passive components contribute to force generation in the muscle?**
>
> 1. At less than 50% of resting length, the muscle cannot develop contractile force.
> 2. At normal resting length, the active tension generated in the muscle contributes the most to the muscle force. Some slight passive elastic tension also contributes.
> 3. Beyond resting length, the passive tension offsets some of the decrement in the active muscle force.
> 4. With additional stretching of the muscle, the passive tension accounts for most of the force generation.

force output. Force output increases from light to higher levels as motor unit recruitment expands from type I slow-twitch to type IIa and then type IIb fast-twitch fibers. Recruitment of additional motor units or recruitment of fast-twitch fibers increases force output.

Fiber Type

At any given velocity of movement, the force generated by the muscle depends on the fiber type. A fast-twitch fiber generates more force than a slow-twitch fiber when the muscle is lengthening or shortening. Type IIb fast-twitch fibers produce the highest maximal force of all the fiber types. At any given absolute force level, the velocity is also greater in muscles with a greater percentage of fast-twitch fibers. Fast-twitch fibers generate faster velocities because of a quicker release of Ca^{2+} and higher ATPase activity. Slow-twitch fibers, which are recruited first, are the predominately active fiber type in low-load situations, and their maximal shortening velocity is slower than that of fast-twitch fibers.

Preloading of the Muscle before Contraction

If concentric, or shortening, muscle action is preceded by a prestretch through eccentric muscle action, the resulting concentric action is capable of generating greater force. Termed **stretch-contract** or stretch-shortening cycle, the stretch on the muscle increases its tension through storage of potential elastic energy in the SEC of the muscle (32) (Fig. 3-31). When a muscle is stretched, there is a small change in the muscle and tendon length (33) and maximum accumulation of stored energy. Thus, when a concentric muscle action follows, an enhanced recoil effect adds to the force output through the muscle–tendon complex (24).

A concentric muscle action beginning at the end of a prestretch is also enhanced by the stored elastic energy in the connective tissue around the muscle fibers. This contributes to a high-force output at the initial portion of the concentric muscle action as these tissues return to their normal length.

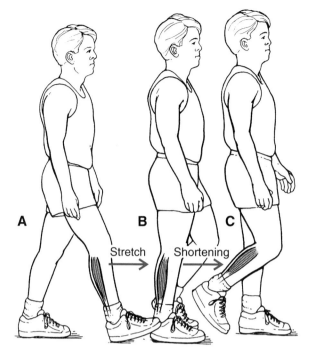

FIGURE 3-31 If a stretch of the muscle **(A–B)** precedes a concentric muscle action **(B–C)**, the resulting force output is greater. The increased force output is attributable to contributions from stored elastic energy in the muscle, tendon, and connective tissue and through some neural facilitation.

If the shortening contraction of the muscle occurs within a reasonable time after the stretch (up to 0.9 seconds), the stored energy is recovered and used. If the stretch is held too long before the shortening occurs, the stored elastic energy is lost through conversion to heat (31).

Neural Contributions

The stretch preceding the concentric muscle action also initiates a stimulation of the muscle group through reflex potentiation. This activation accounts for only approximately 30% of the increase in the concentric muscle action (31). The remaining increase is attributed to stored energy. The actual process of proprioceptive activation through the reflex loop is presented in the next chapter.

Use of the Prestretch

A short-range or low-amplitude prestretch occurring over a short time is the best technique to significantly improve the output of concentric muscle action through return of elastic energy and increased activation of the muscle (4,31). To get the greatest return of energy absorbed in the negative or eccentric action, the athlete should go into the stretch quickly but not too far. Also, the athlete should not pause at the end of the stretch but move immediately into the concentric muscle action. In jumping, for example, a quick counterjump from the anatomical position, featuring a drop–stop–pop action, lowering only through 8 to 12 inches, is much more effective than a jump from a squat position or a jump from a height that forces the limbs into more flexion (4). The influence

of this type of jumping technique on the gastrocnemius muscle is presented in Figure 3-32.

Slow- and fast-twitch fibers handle a prestretch differently. Muscles with predominantly fast-twitch fibers benefit from a very high-velocity prestretch over a small distance because they can store more elastic energy (31). The fast-twitch fibers can handle a fast stretch because myosin cross-bridging occurs quickly. In slow-twitch fibers, the cross-bridging is slower (17).

In a slow-twitch fiber, the small-amplitude prestretch is not advantageous because the energy cannot be stored fast enough and the cross-bridging is slower (17,31). Therefore, slow-twitch fibers benefit from a prestretch that is slower and advances through a greater range of motion. Some athletes with predominantly slow-twitch fibers should be encouraged to use longer prestretches of the muscle to gain the benefits of the stretch. For most athletes, however, the quick prestretch through a small range of motion is the preferred method.

Plyometrics

The use of a quick prestretch is part of a conditioning protocol known as **plyometrics**. In this protocol, the muscle is put on a rapid stretch, and a concentric muscle action is initiated at the end of the stretch. Single-leg bounding, depth jumps, and stair hopping are all plyometric activities for the lower extremity. Surgical tubing or elastic bands are also used to produce a rapid stretch on muscles in the upper extremity. Plyometrics is covered in greater detail in Chapter 4.

Age of Muscle

Sarcopenia is the term for loss of muscle mass and decline in muscle quality seen in aging. Sarcopenia results in a loss of muscle force that impacts bone density, function, glucose intolerance, and a number of other factors leading to disability in the elderly. Both anatomical and biochemical changes occur in the aging muscle to lead to sarcopenia. Anatomically, a number of changes take place in the aging muscle, including decreased muscle mass and cross section, more fat and connective tissue, decreases in type II fiber size, decrease in the number of both type I and II fibers, changes at the sarcomere level, and a decreased number of motor units (28). Biochemically, a reduction in protein synthesis, some impact on enzyme activity, and changes in muscle protein expression take place.

Muscle force decreases with aging at the rate of about 12% to 15% per decade after the age of 50 years (28). The rate of strength loss increases with age and is related to many factors, some of which are anatomical, biochemical, nutritional, and environmental. Progressive resistance training is the best intervention to slow or reverse the effects of aging on the muscle.

Other Factors Influencing Force and Velocity Development

A number of other factors can influence the development of force and velocity in the skeletal muscle. Muscle fatigue

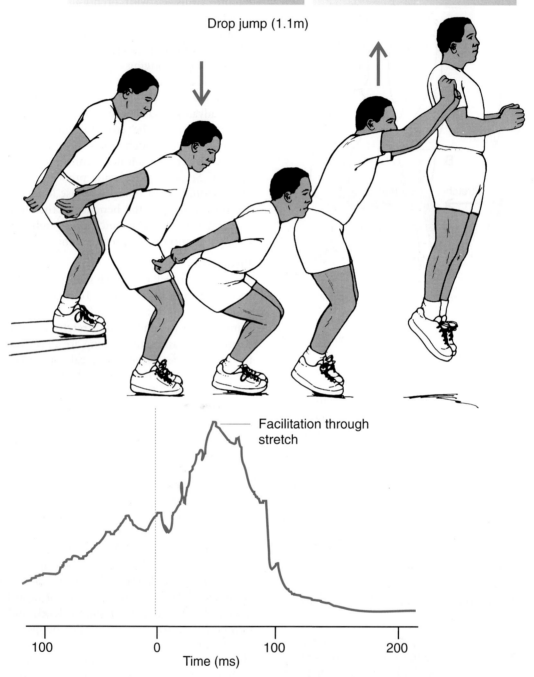

Stretch through eccentric muscle action

Concentric muscle action

Drop jump (1.1m)

Facilitation through stretch

100   0   100   200

Time (ms)

FIGURE 3-32 Neural facilitation in the gastrocnemius. In trained jumpers, the prestretch is used to facilitate the neural activity of the lower extremity muscles. Neural facilitation coupled with the recoil effect of the elastic components adds to the jump if it is performed with the correct timing and amplitude. (Adapted with permission from Sale, D. G. [1986]. Neural adaptation in strength and power training. In N. L. Jones, et al. (Eds.). *Human Muscle Power.* Champaign, IL: Human Kinetics, 289–308.)

can influence force development as the muscle becomes progressively weaker, the shortening velocity is reduced, and the rate of relaxation slows. Gender differences and psychological factors can also influence force and velocity development.

 Strengthening Muscle

Strength is defined as the maximum amount of force produced by a muscle or muscle group at a site of attachment on the skeleton (38). Mechanically, strength is equal to maximum isometric torque at a specific angle. Strength, however, is usually measured by moving the heaviest possible external load through one repetition of a specific range of motion. The movement of the load is not performed at a constant speed because joint movements are usually done at speeds that vary considerably through the range of motion. Many variables influence strength measurement. Some of these include the muscle action (eccentric, concentric, and isometric) and the speed of the limb movement (30). Also, length–tension, force–angle, and force–time characteristics influence strength measurements as strength varies throughout the range of motion. Strength measurements are limited by the weakest joint position.

Training of the muscle for strength focuses on developing a greater cross-sectional area in the muscle and on developing more tension per unit of the cross-sectional area (59). This holds true for all people, both young and old. Greater cross section, or **hypertrophy**, associated with weight training is caused by an increase in the size of the actual muscle fibers and more capillaries to the muscle, which creates greater mean fiber area in the muscle (32,40). The size increase is attributed to increase in size of the actual myofibrils or separation of the myofibrils, as shown in Figure 3-33. Some researchers speculate that the actual muscle fibers may split (Fig. 3-33), but this has not been experimentally substantiated in humans (40). The increase in tension per unit of cross section reflects the neural influence on the development of strength (47). In the early stages of strength development, the nervous system adaptation accounts for a significant portion of the strength gains

through improvement in motor unit recruitment, firing rates, and synchronization (37). Hypertrophy follows as the quality of the fibers improves. Figure 3-34 illustrates the strength progression.

What are the components of a resistance-training program?

1. The type of muscle actions that will be used (concentric, eccentric, and isometric)
2. Exercise selection (single- or multiple-joint exercise)
3. Exercise order and workout schedule (total body versus upper/lower body versus split workouts)
4. Loading (amount of weight to be lifted – % of one repetition maximum)
5. Training volume (number of sets and repetitions in a session)
6. Rest intervals (30–40 s to 2–3 min)
7. Repetition velocity (slow versus fast lifting)
8. Frequency (1–6 d/wk)

Source: Kraemer, W. J., Ratamess, N. A. (2004). Fundamentals of resistance training: progression and exercise prescription. *Medicine & Science in Sports & Exercise*, 36:674–688.

PRINCIPLES OF RESISTANCE TRAINING

Training Specificity

Training specificity, relating to the specific muscles, is important in strength training. Only the muscles used in a specific movement pattern gain strength. This principle, specific adaptation to imposed demands, should direct the choice of lifts toward movement patterns related to the sport or activity in which the pattern might be used (59). This training specificity has a neurologic basis, somewhat like learning a new motor skill—one is usually clumsy until the neurologic patterning is established. Figure 3-35 shows two sport skills, football lineman drives and basketball rebounding, along with lifts specific to the movement. Decisions concerning muscle actions, speed of movement, range of motion, muscle groups, and intensity and volume are all important in terms of training specificity (Table 3-1) (37).

A learning process takes place in the early stages of strength training. This process continues into the later stages of training, but it has its greatest influence at the beginning of the program. In the beginning stages of a program, the novice lifter demonstrates strength gains as a consequence of learning the lift rather than any noticeable increase in the physical determinants of strength, such as increase in fiber size (15,59). This is the basis for using submaximal resistance and high-repetition lifting at the beginning of a strength-training program, so that the lift can first be learned safely.

A

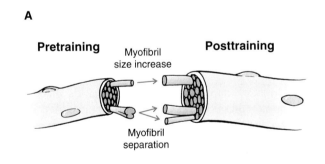

Pretraining   Myofibril size increase   **Posttraining**

Myofibril separation

B   Training

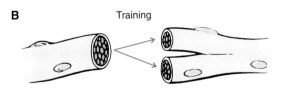

FIGURE 3-33 **(A)** During strength training, the muscle fibers increase in cross section as the myofibrils become larger and separate. **(B)** It has been hypothesized that the fibers may also actually split, but this has yet to be demonstrated in humans. (Adapted with permission from MacDougall, J. D. [1992]. Hypertrophy or hyperplasia. In P. Komi (Ed.). *Strength and Power in Sport*. Boston, MA: Blackwell Scientific, 230–238.)

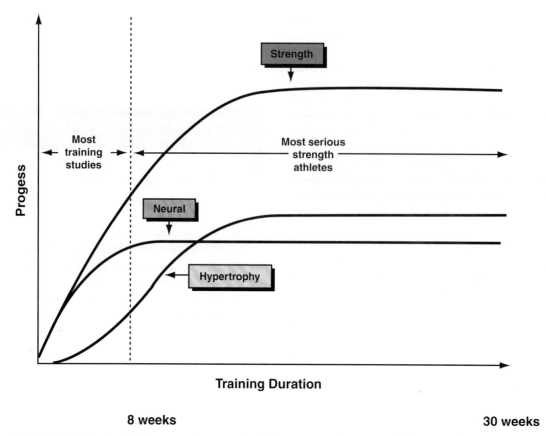

FIGURE 3-34 In the initial stages of a strength training program, the majority of the strength gain is because of neural adaptation, which is followed by hypertrophy of the muscle fibers. Both of these changes contribute to the overall increase in strength.

In addition to the specificity of the pattern of joint movement, specificity of training of the muscle also relates to the speed of training. If a muscle is trained at slow speeds, it will improve strength at slow speeds but may not be strengthened at higher speeds, although training at a faster speed of lifting can promote greater strength gains (53). It is important that if power is the ultimate goal for an athlete, the strength-training routine should contain movements focusing on force and velocity components to maximize and emulate power. After a strength base is established, power is obtained with high-intensity loads and a low number of repetitions (48).

Intensity

The intensity of the training routine is another important factor to monitor in the development of strength. Strength gains are directly related to the tension produced in the muscle. A muscle must be overloaded to a particular threshold before it will respond and adapt to the training (60). The amount of tension in the muscle rather than the number of repetitions is the stimulus for strength. The amount of overload is usually determined as a percentage of the maximum amount of tension a muscle or muscle group can develop.

Athletes attempt to work at the highest percentage of their maximal lifting capability to increase the magnitude of their strength gains. If the athlete trains regularly using a high number of repetitions with low amounts of tension per repetition, the strength gains will be minimal because the muscle has not been overloaded beyond its threshold. The greatest strength gains are achieved when the muscle is worked near its maximum tension before it reaches a fatigue state (two to six repetitions).

The muscle adapts to increased demands placed on it, and a systematic increase through **progressive overload** can lead to positive improvements in strength, power, and local muscular endurance (36). Overload of the muscle can be accomplished by increasing the load, increasing the repetitions, altering the repetition speed, reducing the rest period between exercises, and increasing the volume (37).

Rest

The quality and success of a strength development routine are also directly related to the rest provided to the muscles between sets, between days of training, and before competition. Rest of skeletal muscle that has been stressed through resistive training is important for the recovery and rebuilding of the muscle fiber. As the skeletal muscle fatigues, the tension-development capability deteriorates, and the muscle is not operating at optimal overload.

FIGURE 3-35   Weight-lifting exercises should be selected so that they reproduce some of the movements used in the sport. For football lineman (**A–C**), the dead lift and the power clean include similar joint actions. Likewise, for basketball players who use a jumping action, the squat and heel-raising exercises are helpful (**D–F**).

Volume

The volume of work that a muscle performs may be the important factor in terms of rest of the muscle. Volume of work on a muscle is the sum of the number of repetitions multiplied by the load or weight lifted (59). Volume can be computed per week, month, or year and should include all of the major lifts and the number of lifts. In a week, the volume of lifting for two lifters may be the same even though their regimens are not the same. For example, one lifter lifts three sets of 10 repetitions at 100 lb for a volume of 3,000 lb, and another lifts three sets of two repetitions at 500 lb, also for a volume of 3,000 lb.

Considerable discussion has focused on the number of sets that are optimal for strength development. Some evidence suggests that similar strength gains can be obtained with single rather than multiple sets (7). On the other side, an increasing amount of evidence supports considerably higher strength gains with three sets as compared to single sets (53).

TABLE 3-1	Sample Weight-Training Cycle			
Phase	Preparation Hypertrophy	Transition Basic Strength	Competition Strength/Power	Transition (Active Rest) Peak/Maintain
Sets	3–10	3–5	3–5	1–3
Repetitions	8–12	4–6	2–3	1–3
Days/week	1–3	1–3	1–2	1
Times/day	1–3	1–3	1–2	1
Intensity/cycle[a]	2:1–3:1	2:1–4:1	2:1–3:1	—
Intensity	Low	High	High	Very high to low
Volume	High	Moderate to high	Low	Very low

[a]Ratio of heavy training weeks to light training weeks.
Source: NSCA 1986, 8(6), 17–24.

At the beginning of a weight-training program, the volume is usually high, with more sessions per week, more lifts per session, more sets per exercise, and more repetitions per set taking place than later in the program (15). As one progresses through the training program, the volume decreases. This is done by lifting fewer times per week, performing fewer sets per exercise, increasing the intensity of the lifts, and performing fewer repetitions.

The yearly repetition recommendation is 20,000 lifts, which can be divided into monthly and weekly volumes as the weights are increased or decreased (15). In a month or in a week, the volume of lifting varies to offer higher and lower volume days and weeks.

A lifter performing heavy-resistance exercises with a low number of repetitions must allow five to 10 minutes between sets for the energy systems to be replenished (59). If the rest is less than three minutes, a different energy system is used, resulting in lactic acid accumulation in the muscle.

Bodybuilders use the short-rest and high-intensity training to build up the size of the muscle at the expense of losing some strength gains achieved with a longer rest period. If a longer rest period is not possible, it is believed that a high-repetition, low-resistance form of circuit training between the high-resistance lifts may reduce the buildup of lactic acid in the muscle. Bodybuilders also exercise at loads less than those of power lifters and weight lifters (six to 12 RM). This is the major reason for the strength differences between the weight lifter (greater strength) and the bodybuilder (less strength).

The development of strength for performance enhancement usually follows a detailed plan that has been outlined in the literature for numerous sports and activities. The long-term picture usually involves some form of **periodization** during which the loads are increased and the volume of lifting is decreased over a period of months. Variation through periodization is important for long-term progression to overcome plateaus or strength decrements caused by slowed physical adaptations to the loads. As the athlete heads into a performance season, the lifting volume may be reduced by as much as 60%, which will actually increase the strength of the muscles. If an athlete stops lifting in preparation for a performance, strength can be maintained for at least five days and may be even higher after a few days of rest (59).

Strength Training for the Nonathlete

The principles of strength or resistance training have been discussed using the athlete as an example. It is important to recognize that these principles are applicable to rehabilitation situations, the elderly, children, and unconditioned individuals. Strength training is now recommended as part of one's total fitness development. The American College of Sports Medicine recommends at least one set of resistance training two days a week and including eight to 12 exercises for adults (1). The untrained responds favorably to most protocols demonstrating high rates of improvement compared with the trained (37).

Strength training is recognized as an effective form of exercise for elderly individuals. A marked strength decrement occurs with aging and is believed to be related to reduced activity levels (27). Strength training that is maintained into the later years may counteract atrophy of bone tissue and moderate the progression of degenerative joint alterations. Eccentric training has also been shown to be effective in developing strength in the elderly (39). The muscle groups identified for special attention in a weight-training program for the elderly include the neck flexors, shoulder girdle muscles, abdominals, gluteals, and knee extensors.

Only the magnitude of the resistance should vary in weight training for athletes, elderly individuals, young individuals, and others. Whereas a conditioned athlete may perform a dumbbell lateral raise with a 50-lb weight in the hand, an elderly person may simply raise the arm to the side using the arm weight as the resistance. High-resistance weight lifting must be implemented with caution, especially with young and elderly individuals. Excessive loading of the skeletal system through high-intensity lifting can fracture bone in elderly individuals, especially in the individual with osteoporosis.

The epiphyseal plates in young people are also susceptible to injury under high loads or improper lifting technique; thus, high-intensity programs for children are not recommended. If safety is observed, however, children and adolescents can achieve training-induced strength gains (11). Regular participation in a progressive resistance-training program by children and adolescents has many potential benefits, including increased bone strength, weight control, injury reduction, sports performance enhancement, and increased muscle endurance (11).

TRAINING MODALITIES

Isometric Exercise

There are various ways of loading the muscle, all of which have advantages and disadvantages in terms of strength development. Isometric training loads the muscle in one-joint position so the muscle torque equals the resistance torque and no movement results (2). Individuals have demonstrated moderate strength gains using isometric exercises, and power lifters may use heavy-resistance isometric training to enhance muscle size.

Isometric exercise is also used in rehabilitation and with unconditioned individuals because it is easier to perform than concentric exercise. The major problem associated with isometric exercise is that there is minimal transfer to the real world because most real-world activities involve eccentric and concentric muscle actions. Furthermore, isometric exercise only enhances the strength of the muscle group at the joint angle in which the muscle is stressed, which limits development of strength throughout the range of motion.

Isotonic Exercise

The most popular strength-training modality is isotonic exercise. An exercise is considered **isotonic** when the segment moves a specified weight through a range of motion. Although the weight of the barbell or body segment is constant, the actual load imposed on the muscle varies throughout the range of motion. In an isotonic lift, the initial load or resistance is overcome and then moved through the motion (2). The resistance cannot be heavier than the amount of muscle torque developed by the weakest joint position because the maximum load lifted is only as great as this position. Examples of isotonic modalities are the use of free weights and multijoint machines, such as universal gyms, in which the external resistance can be adjusted (Fig. 3-36).

The use of free weights versus machines has generated considerable discussion. Free weights include dumbbells, barbells, weighted vests, medicine balls, and other added loads that allow the lifter to generate normal movements with the added weight. The advocates for free weights promote stabilization and control as major benefits to the use of free weights. Machines apply a resistance in a guided or restricted manner and are seen as requiring less overall control. Both training techniques can generate strength,

FIGURE 3-36 Two forms of isotonic exercises for the upper extremity. **(A)** The use of free weights (bench press). **(B)** The use of a machine.

power, hypertrophy, or endurance, so the choice should be left to the individual. Free weights may be preferable to enhance specificity of training, but correct technique is mandatory.

An isotonic movement can be produced with an eccentric or concentric muscle action. For example, the squat exercise involves eccentrically lowering a weight and concentrically raising the same weight. Even though the weight in an isotonic lift is constant, the torque developed by the muscle is not. This is because of the changes in length–tension or force–angle or to the speed of the lift. To initiate flexion of the elbow while holding a 2.5-kg weight, a person generates maximum tension in the flexors at the beginning of the lift to get the weight moving.

Remember that this is also one of the weakest joint positions because of the angle of attachment of the muscle. Moving through the midrange of the motion requires reduced muscular tension because the weight is moving and the musculoskeletal lever is more efficient. The resistive torque also peaks in this stage of the movement.

The isotonic lift may not adequately overload the muscle in the midrange, where it is typically the strongest. This is especially magnified if the lift is performed very quickly. If the person performs isotonic lifts with a constant speed (no acceleration) so that the midrange is exercised, the motive torque created by the muscle will match the load offered by the resistance. Strength assessment using isotonic lifting is sometimes difficult because specific joint actions are hard to isolate. Most isotonic exercises involve action or stabilization of adjacent segments.

Isokinetic Exercise

A third training modality is the **isokinetic** exercise, an exercise performed at a controlled velocity with varying resistance. This exercise must be performed on an isokinetic dynamometer, allowing for isolation of a limb; stabilization of adjacent segments; and adjustment of the speed of movement, which typically ranges from 0° to 600°/second (Fig. 3-37).

When an individual applies a muscular force against the speed-controlled bar of the isokinetic device, an attempt is made to push the bar at the predetermined speed. As the individual attempts to generate maximum tension at the specific speed of contraction, the tension varies because of changes in leverage and muscular attachment throughout the range of motion. Isokinetic testing has been used for quantifying strength in the laboratory and in the rehabilitation setting. An extensive body of literature presents a wide array of norms for isokinetic testing of different joints, joint positions, speeds, and populations.

The velocity of the devices significantly influences the results. Therefore, testing must be conducted at a variety of speeds or at a speed close to that which will be used in activity. This is often the major limitation of isokinetic dynamometers. For example, the isokinetic strength of the shoulder internal rotators of a baseball pitcher may be assessed at 300°/second on the isokinetic dynamometer, but the actual speed of the movement in the pitch has been shown to average 6000°/second (8). Isokinetic testing allows for a quantitative measurement of power that has previously been difficult to measure in the field.

Using isokinetic testing and training has some drawbacks. The movement at a constant velocity is not the type of movement typically found in the activities of daily living or in sport, and the cost of most isokinetic systems and lack of mass usage make isokinetic training or testing prohibitive for many.

Closed and Open Kinetic Chain Exercise

Although most therapists still use isokinetic testing for assessment, many have discontinued its use for training and have gone to closed-chain training, in which individuals use body weight and eccentric and concentric muscle actions. A **closed-chain exercise** is an isotonic exercise in which the end of the chain is fixed, as in the case of a foot or hand on the floor. An example of a closed-chain exercise for the quadriceps is a simple squat movement with the feet on the floor (Fig. 3-38). It is believed that this form of exercise is more effective than an **open-chain exercise**, such as a knee extension on the isokinetic dynamometer or knee extension machine, because it uses body weight, maintains muscle relationships, and is more transferable to normal human function. The use of closed-chain kinetic exercise for the knee joint has been shown to promote a more balanced quadriceps activation than open-chain exercise (51). With the promotion of accelerated rehabilitation after anterior cruciate ligament surgery, the trend in physical therapy is toward the use of closed-chain kinetic exercises. Research has shown no difference in the strains produced at the anterior cruciate ligament in open- versus closed-chain exercise, however, even with more anterior tibial translation seen in the open-chain exercise (12).

Functional Training

The final training modality presented is functional training, a specialized training protocol for specific purposes. With the goal of enhancement of specificity of training, functional training uses different equipment to individualize training for each functional purpose. This training typically

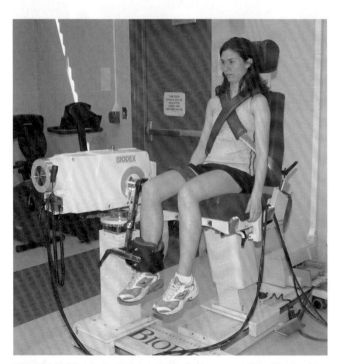

FIGURE 3-37   An isokinetic exercise for knee extension. The machine is the Biodex isokinetic dynamometer.

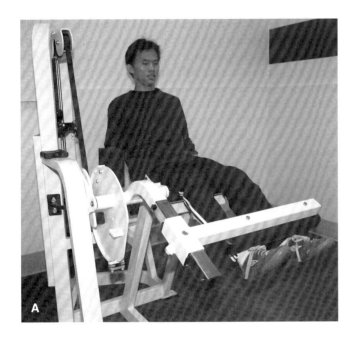

FIGURE 3-38 **(A)** An open-chain exercise for the same muscles (leg extension). **(B)** A closed-chain exercise for the quadriceps femoris muscles (squat).

incorporates balance and coordination into each exercise so that stability is inherent in the movement. The use of medicine balls, stability balls, Bosu®, rubber tubing, and pulley systems are examples of various tools used in functional training. Examples of functional training exercises include throwing a medicine ball, performing an overhead press while sitting on a stability ball, applying variable resistance to an exercise by using rubberized tubing, standing on a balance board or Bosu® while performing an exercise, and applying resistance via a cable system during a diagonal movement pattern. A specific type of functional training, multivector strength training, is resistance training in which the individual must coordinate muscle action occurring in three directions or planes of motion at the same time.

Whatever form of exercise selected, the training should simulate the contraction characteristics of the activity. Improved strength alone does not necessarily transfer to better functional performance (55). After improvement in strength, muscle stiffness and physical changes take place as well as changes in neural input, which require more coordination. Thus, improvement in function does not always occur because of strength improvements.

Injury to Skeletal Muscle

CAUSE AND SITE OF MUSCLE INJURY

Injury to the skeletal muscle can occur through a bout of intense exercise, exercising a muscle over a long duration, or in eccentric exercise. The actual injury is usually a microinjury with small lesions in the muscle fiber. The result of a muscle strain or microtear in the muscle is manifested by pain or muscle soreness, swelling, possible anatomical deformity, and athletic dysfunction.

Muscles at greatest risk for strain are two-joint muscles, muscles limiting the range of motion, and muscles used eccentrically (16). The two-joint muscles are at risk because they can be put on stretch at two joints (Fig. 3-39). Extension at the hip joint with flexion at the knee joint puts the rectus femoris on extreme stretch and renders it very vulnerable to injury.

Eccentric exercise has been identified as a primary contributor to muscle strain (50). After a prolonged concentric or isometric exercise session, the muscles are fatigued, but it is usually a temporary state. After an unaccustomed session of eccentric exercise, the muscles remain weak longer and are also stiff and sore (45). The muscle damage process brought on by eccentric exercise starts with initial damage at the sarcomere level followed by a secondary adaptation to protect the muscle from further damage (45). It has also been documented that rest may play a big role in determining the force decrement after eccentric muscle actions. Shorter work–rest cycles (10 seconds vs. five minutes) have been shown to result in more force decrement two days after exercise (9). Although it is known that high forces in muscles working eccentrically can cause tissue damage, some believe that this type of eccentric contraction may actually promote positive adaptations in the muscle and tendon, resulting in increased size and strength (39).

A. Lowering

B. Support phase of running

C. Hurdling

FIGURE 3-39 Muscles undergoing an eccentric muscle action are at increased risk for injury. **(A)** The quadriceps femoris performing an eccentric muscle action in lowering as they control the knee flexion on the way down. **(B)** The quadriceps and gastrocnemius eccentrically acting during the support phase of running. Two-joint muscles are also placed in injury-prone positions, making them more susceptible to strain. **(C)** The hamstrings on extreme stretch when the hip is flexed and the knee is extended during hurdling.

How is muscle postulated to be damaged during eccentric exercise?

1. During eccentric exercise, the muscle can be overstretched, disrupting the sarcomeres.

2. The membrane is damaged, resulting in an uncontrolled release of Ca^2+.

3. The result is a shift in optimal length, a decrease in active tension, an increase in passive tension, and delayed soreness and swelling.

Source: Proske, U., Allen, T. J. (2005). Damage to skeletal muscle from eccentric exercise. *Exercise and Sport Sciences Reviews*, 33:98–104.

Muscles used to terminate a range of motion are at risk because they are used to eccentrically slow a limb moving very quickly. Common sites where muscles are strained as they slow a movement are the hamstrings as they slow hip flexion and the posterior rotator cuff muscles as they slow the arm in the follow-through phase of throwing (16).

Although the muscle fiber itself may be the site of damage, it is believed that the source of muscle soreness immediately after exercise and strain to the system is the connective tissue. This can be in the muscle sheaths, epimysium, perimysium, or endomysium, or it can be injury to the tendon or ligament (49). In fact, a common site of muscle strain is at the muscle–tendon junction because of the high tensions transmitted through this region. Injuries at this site are common in the gastrocnemius, pectoralis major, rectus femoris, adductor longus, triceps brachii, semimembranosus, semitendinosus, and biceps femoris muscles (16).

It is important to identify those who are at risk for muscle strain. First, the chance of injury increases with muscular fatigue as the neuromuscular system loses its ability to control the forces imposed on the system. This commonly results in an alteration in the mechanics of movement and a shifting of shock-absorbing load responsibilities. Repetitive muscle strain can occur after the threshold of mechanical activity has been exceeded. Practice times should be controlled, and events late in the practice should not emphasize maximum load or stress conditions.

Second, an individual can incur a muscle strain at the onset of practice if it begins with muscles that are weak from recent usage (49). Muscles should be given ample time to recover from heavy usage. After extreme bouts of exercise, rest periods may have to be one week or more, but normally, a muscle can recover from moderate usage within one or two days.

Third, if trained or untrained individuals perform a unique task for the first time, they will probably have pain, swelling, and loss of range of motion after performing the exercise. This swelling and injury are most likely to occur in the passive elements of the muscle and generally lessen or be reduced as the number of practices increase (49).

Last, an individual with an injury is susceptible to a recurrence of the injury or development of an injury elsewhere in the system resulting from compensatory actions. For example, if the gastrocnemius is sore from a minor muscle strain, an individual may eccentrically load the lower extremity with a weak and inflexible gastrocnemius. This forces the person to pronate more during the support phase and run more on the balls of the feet, indirectly producing knee injuries or metatarsal fractures. With every injury, a functional substitution happens elsewhere in the system; this is where the new injury will occur.

PREVENTING MUSCLE INJURY

Conditioning of the connective tissue in the muscle can greatly reduce the incidence of injury. Connective tissue responds to loading by becoming stronger, although the rate of strengthening of connective tissue lags behind the rate of strengthening of the muscle. Therefore, base work with low loads and high repetitions should be instituted for three to four weeks at the beginning of a strength and conditioning program to begin the strengthening process of the connective tissue before muscle strength is increased (53).

Different types of training influence the connective tissue in different ways. Endurance training has been shown to increase the size and tensile strength of both ligaments and tendons. Sprint training improves ligament weight and thickness, and heavy loading strengthens the muscle sheaths by stimulating the production of more collagen. When a muscle produces a maximum voluntary contraction, only 30% of the maximum tensile strength of the tendon is used (53). The remaining tensile strength serves as an excess to be used for very high dynamic loading. If this margin is exceeded, muscle injury occurs.

Other important considerations in preventing muscle injury are a warm-up before beginning exercise routines, a progressive strength program, and attention to strength and flexibility balance in the musculoskeletal system. Finally, early recognition of signs of fatigue also helps prevent injury if corrective actions are taken.

INACTIVITY, INJURY, AND IMMOBILIZATION EFFECTS ON MUSCLE

Changes in the muscle with disuse or immobilization can be dramatic. Atrophy is one of the first signs of immobilization of a limb, showing as much as a 20% to 30% decrease in cross-sectional area after eight weeks of cast immobilization (52). Disuse or inactivity leads to atrophy because of muscle remodeling, resulting in loss of proteins and changes in the muscle metabolism. The level of atrophy appears to be muscle specific where lower extremity muscles lose more cross section than back or upper extremity muscles (52). The greatest change occurs in the initial weeks of disuse, and this should be a focus of attention in rehabilitation and exercise.

Muscle regrowth after inactivity or immobilization varies between young, adult, and elderly individuals (41). Regrowth in young muscle is more successful than in the aging muscle, and the regrowth process varies between fast and slow muscles. Also, when successfully rebuilding cross section of the atrophied muscle, the force output of the muscle lags behind (52).

When a muscle is injured, the force-producing capabilities usually decrease. Compensation occurs where other muscles change in function to make up for the injured muscle or the motion can be changed to minimize the use of the injured muscle (34). For example, injury to a hip flexor can cause a large reduction of force in the soleus, an ankle muscle, because of its role in propelling the trunk forward via pushoff in plantarflexion. Injury to the gluteus maximus (hip extensor) can shift duties of hip extension over to the gluteus medius and hamstrings. Loss of function in one muscle can impact all of the joints in the linked segments such as the lower extremity, so the whole musculoskeletal system should be the focus of retraining efforts.

 ## Summary

Skeletal muscle has four properties: irritability, contractility, extensibility, and elasticity. These properties allow the muscle to respond to stimulation, shorten, lengthen beyond resting length, and return to resting length after a stretch, respectively.

Muscles can perform a variety of functions, including producing movement, maintaining postures and positions, stabilizing joints, supporting internal organs, controlling pressures in the cavities, maintaining body temperature, and controlling entrances and exits to the body.

Groups of muscles are contained in compartments that can be categorized by common function. The individual muscles in the group are covered by an epimysium and usually have a central portion called the belly. The muscle can be further divided internally into fascicles covered by the perimysium; the fascicles contain the actual muscle fibers covered by the endomysium. Muscle fibers can be organized in a parallel arrangement, in which the fibers run parallel and connect to a tendon at both ends, or in a penniform arrangement, in which the fibers run diagonally to a tendon running through the muscle. In penniform muscle, the anatomical cross section, situated at right angles to the direction of the fibers, is less than the physiologic cross section, the sum of all of the cross sections in the fiber. In parallel muscle, the anatomical and physiologic cross sections are equal. Muscle volume and physiologic cross section are larger in the penniform muscle. The force applied in the penniform muscle is influenced by the pennation angle, where a smaller force is applied to the tendon at greater pennation angles.

Each muscle contains different fiber types that influence the muscle's ability to produce tension. Slow-twitch fiber types have slow contraction times and are well suited for prolonged, low-intensity workouts. Intermediate- and fast-twitch fiber types are better suited for higher force outputs over shorter periods.

A motor unit is a group of muscle fibers innervated by a single motor neuron. Muscle contraction occurs as the action potential traveling along the axon reaches the muscle fiber and stimulates a chemical transmission across the synapse. Once at the muscle, excitation–contraction coupling occurs as the release of Ca^{2+} ions promotes cross-bridge formation.

Each muscle fiber contains myofibrils that house the contractile unit of the muscle fiber, the sarcomere. It is at the sarcomere level that cross-bridging occurs between the actin and myosin filaments, resulting in shortening or lengthening of the muscle fiber.

A muscle attaches to the bone via an aponeurosis, or tendon. Tendons can withstand high tensile forces and respond stiffly to high rates of loading and less stiffly at lower loading rates. Tendons recoil during muscle contraction and delay the development of tension in the muscle. This recoiling action increases the load that a muscle can support. Tendons and muscles are more prone to injury during eccentric muscle actions.

A mechanical model of muscular contraction breaks the muscle down into active and passive components. The active component includes the CCs found in the myofibrils and cross-bridging of the actin and myosin filaments. The passive or elastic components are in the tendon and the cross-bridges and in the sarcolemma and the connective tissue.

Muscles perform various roles, such as agonist or antagonist and stabilizer or neutralizer. Torque is generated in a muscle, developing tension at both ends of the muscle. The amount of tension is influenced by the angle of attachment of the muscle.

Muscle tension is generated to produce three types of muscle actions: isometric, concentric, and eccentric. The isometric muscle action is used to stabilize a segment, the concentric action creates a movement, and the eccentric muscle action controls a movement. The concentric muscle actions generate the lowest force output of the three, and the eccentric muscle action generates the highest.

Two-joint muscles are unique in that they act at two adjacent joints. Their effectiveness at one joint depends on the positioning of the other joint, the moment arms at each joint, and the muscle synergies in the movement.

Numerous factors influence the amount of force that can be generated by a muscle, including the angle of attachment of the tendon, muscle cross section, laxity or stiffness in the tendon that influences the force–time relationship, fiber type, neural activation, length of the muscle, contributions of the elastic component, age of the muscle, and velocity of the muscle action.

Greater force can be developed in a concentric muscle action if it is preceded by an eccentric muscle action, or prestretch (stretch–shortening cycle). The muscle force is increased by facilitation via stored elastic energy and neurologic facilitation. A quick, short-range prestretch is optimal for developing maximum tension in fast-twitch fibers, and a slow, larger range prestretch is beneficial for tension development in slow-twitch fibers.

The development of strength in a muscle is influenced by genetic predisposition, training specificity, training intensity, muscle rest during training, and total training volume. Training principles apply to all groups, including conditioned and unconditioned individuals, and only the magnitude of the resistance needs to be altered. Muscles can be exercised isometrically, isotonically, isokinetically,

or through specific functional training. Another important exercise consideration should be the decision on the use of open- or closed-chain exercises.

Muscle injury is common and occurs most frequently in two-joint muscles and during eccentric muscle action. To prevent muscle injury, proper training and conditioning principles should be followed.

REVIEW QUESTIONS

True or False

1. ____ Muscles shorten by 50% to 70% of their resting length.

2. ____ When muscle is stretched it returns to its resting length because of its contractility prosperities.

3. ____ Muscle compartments are separated by fascia.

4. ____ Eccentric muscle actions are used to raise a load from the floor.

5. ____ Fusiform fiber arrangement is a type of pennate architecture.

6. ____ Pennate fiber arrangements have a pennation angle.

7. ____ Muscle force is related to the physiologic cross-sectional area.

8. ____ Type II fibers are oxidative.

9. ____ The largest tensions can be developed in a muscle through eccentric contractions.

10. ____ Each muscle fiber is covered by perimysium.

11. ____ The dark banding seen in muscle tissue is the result of actin.

12. ____ The sarcoplasm is the actual contractile unit of the muscle.

13. ____ Lengthening of a muscle before contraction increases the force in a concentric contraction.

14. ____ The fiber force in a pennate muscle is in the same direction as the tendon.

15. ____ A muscle can have more muscle fibers than motor units.

16. ____ The physiologic cross section of a fusiform muscle is the same as the anatomical cross section.

17. ____ Depolarization occurs in both the nerve and the muscle.

18. ____ Calcium ions must be present for cross-bridges to form.

19. ____ Calcium ions are stored within the synaptic terminal.

20. ____ All muscles attach to the bone through a tendon.

21. ____ The muscle moment arm does not affect the torque a muscle can produce.

22. ____ Faster concentric actions can produce more force than slow ones.

23. ____ Maximum muscle power occurs during an isometric contraction.

24. ____ Maximum muscle tension can be generated at a muscle length slightly less than the resting length.

25. ____ Sarcopenia is the loss of muscle mass seen in aging.

Multiple Choice

1. The ability to shorten when stimulated is called _____.
   a. contractility
   b. extensibility
   c. irritability
   d. flexibility

2. What is the muscle structure from smallest to largest?
   a. Myofilaments, myofibrils, fascicles, fibers, muscle
   b. Myofibrils, myofilaments, fascicles, fibers, muscle
   c. Myofibrils, myofilaments, fibers, fascicles, muscle
   d. Myofilaments, myofibrils, fibers, fascicles, muscle

3. Maximal force is generated in the muscle fibers during _____.
   a. concentric contractions
   b. isometric contractions
   c. eccentric contractions
   d. None of the above

4. The stretch-shortening cycle consists of _____ in the same muscle.
   a. an isometric contraction followed by an eccentric contraction
   b. a concentric contraction followed by an eccentric contraction
   c. two short eccentric contractions separated by less than 60 ms
   d. an eccentric contraction followed by a concentric contraction

5. The perimysium covers the _____.
   a. muscle fibers
   b. muscle fascicles
   c. muscle fibrils
   d. muscle filaments

6. A motor unit is _____.
   a. one neuron and one muscle fiber
   b. all of the neurons and muscle fibers in a muscle
   c. one neuron and all of the muscle fibers it connects to
   d. one fiber and all of the neurons that connect to it

7. The series elastic component in a muscle is hypothesized to be located in the _____.
   a. fascia
   b. tendon
   c. cross-bridges
   d. tendon and cross-bridges

8. Strength gains during the first eight weeks of an exercise program are primarily due to _____.
   a. neural factors
   b. hypertrophy of muscle fibers
   c. an increase in the number of muscle fibers
   d. conversion of type I fibers to type II fibers

9. Functional training _____.
   a. incorporates balance and coordination into each exercise
   b. always includes closed-chain exercise
   c. only uses diagonal patterns of movement
   d. uses high speeds

10. Muscle fiber depolarization occurs when _____.
    a. ATP attaches to the myosin head
    b. ACh binds to the muscle fiber
    c. calcium ions are released from the sarcoplasmic reticulum
    d. calcium ions are pumped into the sarcoplasmic reticulum

11. The amount of force that can be produced in a muscle is most closely related to _____.
    a. anatomical cross section
    b. physiologic cross section
    c. average muscle fiber length
    d. muscle volume

12. During which of the following maximal contractions is the muscle tension greatest?
    a. Isometric contraction with the length greater than Lo
    b. Eccentric contraction with the length greater than Lo
    c. Concentric contraction with the length less than Lo
    d. Eccentric contraction with the length less than Lo

13. Muscles with lots of sarcomeres in parallel can achieve high _____.
    a. velocity
    b. force
    c. range of motion
    d. None of the above

14. Compared with a pennate muscle of the same volume, a fusiform muscle will have _____.
    a. more sarcomeres in parallel
    b. a greater PCSA
    c. greater maximal force production
    d. more sarcomeres in series

15. As a person sits from a standing position, what is the predominant contraction type in the lower extremity?
    a. Concentric
    b. Eccentric
    c. Isotonic
    d. Isometric

16. Muscle can produce the greatest tension when the length is _____.
   a. slightly less than resting length
   b. at resting length
   c. slightly greater than resting length
   d. Length does not matter

17. In this role, muscle is active to secure a bone so that movement in and adjacent segment can occur.
   a. stabilizer
   b. assistant mover
   c. agonist
   d. neutralizer

18. Actin–myosin cross-bridges form when _____.
   a. ATP attaches to the myosin head
   b. ACh binds to the muscle fiber
   c. calcium ions cause active sites to be exposed
   d. None of the above

19. Which scenario describes an eccentric contraction?
   a. The knee is flexing as the knee flexor muscle group is active
   b. The knee is flexing as the knee extensor muscle group is active
   c. The knee is extending as the knee extensor muscle group is active
   d. None of these describe an eccentric contraction

20. The order of maximum potential force output by contraction type is _____.
   a. concentric, eccentric, isometric
   b. eccentric, isometric, concentric
   c. eccentric, concentric, isometric
   d. there is no difference in maximum potential force among contraction types

21. Muscle torque differs from muscle force because _____.
   a. torque does not consider the velocity of contraction
   b. force does not consider the distance to the axis of rotation
   c. force does not consider the duration of the contraction
   d. torque does not consider the fiber type

22. Both active and passive tension can be generated at muscle lengths _____.
   a. greater than Lo
   b. equal to Lo
   c. less than Lo
   d. that are changing

23. A muscle contracts predominantly _____ when you throw an object.
   a. isokinetically
   b. eccentrically
   c. concentrically
   d. isometrically
   e. Both c and d

24. The connective tissue component of a skeletal muscle that surrounds fibers is called the _____.
   a. perimysium
   b. epimysium
   c. endomysium
   d. tendomysium

25. Power is _____.
   a. the rate of change of velocity
   b. the product of length and tension
   c. the product of torque and acceleration
   d. the product of force and velocity

References

1. ACSM Position Stand. (1990). The recommended quantity and quality of exercise for developing and maintaining cardiorespiratory and muscular fitness in healthy adults. *Medicine & Science in Sports & Exercise*, 22:265-274.
2. Ariel, G. (1984). Resistive exercise machines. In J. Terauds, et al. (Eds.). *Biomechanics*. Eugene, OR: Microform Publications, 21-26.
3. Asmussen, E. (1952). Positive and negative muscular work. *Acta Physiologica Scandinavica*, 28:364-382.
4. Asmussen, E., Bonde-Petersen, F. (1974). Apparent efficiency and storage of elastic energy in human muscles during exercise. *Acta Physiologica Scandinavia*, 92:537-545.
5. Billeter, R., Hoppeler, H. (1992). Muscular basis of strength. In P. Komi (Ed.). *Strength and Power in Sport*. Boston, MA: Blackwell Scientific, 39-63.
6. Bobbert, M. F., van Ingen Schenau, G. J. (1988). Coordination in vertical jumping. *Journal of Biomechanics*, 21:249-262.
7. Carpinelli, R. N. (2002). Berger in retrospect: effect of varied weight training programmes on strength. *British Journal of Sports Medicine*, 36:319-324.
8. Cook, E. E., et al. (1987). Shoulder antagonistic strength ratios: A comparison between college-level baseball pitchers and nonpitchers. *Journal of Orthopaedic & Sports Physical Therapy*, 8:451-461.
9. Cutlip, R. G., et al. (2005). Impact of stretch-shortening cycle rest interval on in vivo muscle performance. *Medicine & Science in Sports & Exercise*, 37:1345-1355.
10. Edman, K. A. P. (1992). Contractile performance of skeletal muscle fibers. In P. Komi (Ed.). *Strength and Power in Sport*. Boston: Blackwell Scientific, 96-114.
11. Faigenbaum, A. D. (2003). Youth resistance training. *President's Council on Physical Fitness and Sports Research Digest*, 4:1-8.
12. Fleming, B. C., et al. (2005). Open or closed-kinetic chain exercises after anterior cruciate reconstruction? *Exercise and Sport Sciences Reviews*, 33:134-140.
13. Fuglevand, A. J., et al. (1993). Impairment of neuromuscular propagation during human fatiguing contractions at submaximal forces. *Journal of Physiology*, 460:549-572.

14. Fukunaga, T., et al. (1992). Physiological cross-sectional area of human leg muscles based on magnetic resonance imaging. *Journal of Orthopedic Research*, 10:928-934.

15. Garhammer, J., Takano, B. (1992). Training for weight-lifting. In P. Komi (Ed.). *Strength and Power in Sport*. Boston, MA: Blackwell Scientific, 357-369.

16. Garrett, W. E. (1991). Muscle strain injuries: Clinical and basic aspects. *Medicine & Science in Sports & Exercise*, 22: 436-443.

17. Goldspink, G. (1992). Cellular and molecular aspects of adaptation in skeletal muscle. In P. Komi (Ed.). *Strength and Power in Sport*. Boston, MA: Blackwell Scientific, 211-229.

18. Gowitzke, B. A. (1984). Muscles alive in sport. In M. Adrian, H. Deutsch (Eds.). *Biomechanics*. Eugene, OR: Microform Publications, 3-19.

19. Hay, J. G. (1992). Mechanical basis of strength expression. In P. Komi (Ed.). *Strength and Power in Sport*. Boston, MA: Blackwell Scientific, 197-207.

20. Henneman, E., et al. (1965). Excitability and inhibitability of motor neurons of different sizes. *Journal of Neurobiology*, 28:599-620.

21. Hill, A. V. (1938). Heat and shortening and the dynamic constants of muscle. *Proceedings of the Royal Society of London (Biology)*, 126:136-195.

22. Hill, A. V. (1970). *First and Last Experiments in Muscle Mechanics*. Cambridge: Cambridge University Press.

23. Huijing, P. A. (1992). Mechanical muscle models. In P. Komi (Ed.). *Strength and Power in Sport*. Boston, MA: Blackwell Scientific, 130-150.

24. Huijing, P. A. (1992). Elastic potential of muscle. In P. Komi (Ed.). *Strength and Power in Sport*. Boston, MA: Blackwell Scientific, 151-168.

25. Huijing, P.A. (2003). Muscular force transmission necessitates a multilevel integrative approach to the analysis of function of skeletal muscle. *Exercise and Sport Sciences Reviews*, 31: 167-175.

26. Huxley, A. F. (1957). Muscle structure and theories of contraction. *Progress in Biophysics and Biophysical Chemistry*, 7:255-318.

27. Israel, S. (1992). Age-related changes in strength and special groups. In P. Komi (Ed.). *Strength and Power in Sport*. Boston, MA: Blackwell Scientific, 319-328.

28. Kamel, H. K. (2003). Sarcopenia and aging. *Nutrition Reviews*, 61:157-167.

29. Kawakami, Y, et al. (1998). Architectural and functional features of human triceps surae muscles during contraction. *Applied Physiology*, 85:398-404.

30. Knuttgen, H. G., Komi, P. (1992). Basic definitions for exercise. In P. Komi (Ed.). *Strength and Power in Sport*. Boston, MA: Blackwell Scientific, 3-6.

31. Komi, P. V. (1984). Physiological and biomechanical correlates of muscle function: Effects of muscle structure and stretch–shortening cycle on force and speed. In R. L. Terjund (Ed.). *Exercise and Sport Sciences Reviews*, 12:81-121.

32. Komi, P. V. (1986). The stretch–shortening cycle and human power output. In N. L. Jones et al. (Eds.). *Human Muscle Power*. Champaign, IL: Human Kinetics, 27-40.

33. Komi, P. V. (1992). Stretch–shortening cycle. In P. Komi (Ed.). *Strength and Power in Sport*. Boston, MA: Blackwell Scientific, 169-179.

34. Komura, T. Nagano, A. (2004). Evaluation of the influence of muscle deactivation on other muscles and joints during gait motion. *Journal of Biomechanics*, 37:425-436.

35. Kornecki, S. (1992). Mechanism of muscular stabilization process in joints. *Journal of Biomechanics*, 25:235-245.

36. Kraemer, W. J., et al. (2002). Progression models in resistance training for healthy adults. *Medicine & Science in Sports & Exercise*, 34:364-380.

37. Kraemer, W. J., Ratamess, N. A. (2004). Fundamentals of resistance training: progression and exercise prescription. *Medicine & Science in Sports & Exercise*, 36:674-688.

38. Kulig, K., et al. (1984). Human strength curves. In R. L. Terjund (Ed.). *Exercise and Sport Sciences Reviews*, 12:417-466.

39. LaStayo, P. C., et al. (2003). Eccentric muscle contractions: Their contribution to injury, prevention, rehabilitation and sport. *Journal of Orthopaedic & Sports Physical Therapy*, 33:557-571.

40. MacDougall, J. D. (1992). Hypertrophy or hyperplasia. In P. Komi (Ed.). *Strength and Power in Sport*. Boston, MA: Blackwell Scientific, 230-238.

41. Machida, S, Booth, F. W. (2004). Regrowth of skeletal muscle atrophied from inactivity. *Medicine & Science in Sports & Exercise*, 36:52-59.

42. McMahon, T. A. (1984). *Muscles, Reflexes, and Locomotion*. Princeton, NJ: Princeton University Press, 3-25.

43. Moritani, T. (1992). Time course of adaptations during strength and power training. In P. Komi (Ed.). *Strength and Power in Sport*. Boston, MA: Blackwell Scientific, 226-278.

44. Munn, J., et al. (2005). Resistance training for strength: Effect of number of sets and contraction speed. *Medicine & Science in Sports & Exercise*, 37:1622-1626.

45. Perrine, J. J. (1986). The biophysics of maximal muscle power outputs: Methods and problems of measurement. In N. L. Jones, et al. (Eds.). *Human Muscle Power*. Champaign, IL: Human Kinetics, 15-46.

46. Proske, U., Allen, T. J. (2005). Damage to skeletal muscle from eccentric exercise. *Exercise and Sport Sciences Reviews*, 33:98-104.

47. Proske, U., Morgan, D. L. (1987). Tendon stiffness: Methods of measurement and significance for the control of movement. *Journal of Biomechanics*, 20:75-82.

48. Sale, D. G. (1986). Neural adaptation in strength and power training. In N. L. Jones et al. (Eds.). *Human Muscle Power*. Champaign, IL: Human Kinetics, 289-308.

49. Schmidtbleicher, D. (1992). Training for power events. In P. Komi (Ed.). *Strength and Power in Sport*. Boston, MA: Blackwell Scientific, 381-395.

50. Stauber, W. T. (1989). Eccentric action of muscles: Physiology, injury, and adaptation. In K. Pandolf (Ed.). *Exercise and Sports Sciences Reviews*, 17: 157-185.

51. Stensdotter, A., et al. (2003) Quadriceps activation in closed and in open kinetic chain exercise. *Medicine & Science in Sports & Exercise*, 35:2043-2047.

52. Stevens, J. E., et al. (2004). Muscle adaptations with immobilization and rehabilitation after ankle fracture. *Medicine & Science in Sports & Exercise*, 36:1695-1701.

53. Stone, M. H. (1990). Muscle conditioning and muscle injuries. *Medicine & Science in Sports & Exercise*, 22:457-462.

54. Stone, M. H. (1992). Connective tissue and bone response to strength training. In P. Komi (Ed.). *Strength and Power in Sport*. Boston, MA: Blackwell Scientific, 279-290.

55. Toumi, H. T., et al. (2004). Muscle plasticity after weight and combined (weight + jump) training. *Medicine & Science in Sports & Exercise*, 36:1580-1588.

56. Vanderhelm, F. C. T., Veenbaas, R. (1991). Modeling the mechanical effect of muscles with large attachment sites: Application to the shoulder mechanism. *Journal of Biomechanics*, 24:1151-1163.

57. Van Soest, A. J., et al. (1993). The influence of the biarticularity of the gastrocnemius muscle on vertical jumping achievement. *Journal of Biomechanics*, 26:1-8.

58. Voronov, A. V. (2003). Anatomical cross-section areas and volumes of the muscles of the lower extremities. *Human Physiology*, 29:210-211.

59. Weiss, L. W. (1991). The obtuse nature of muscular strength: The contribution of rest to its development and expression. *Journal of Applied Sport Science Research*, 5:219-227.

60. Wells, R. P. (1988). Mechanical energy costs of human movement: An approach to evaluating the transfer possibilities of two-joint muscles. *Journal of Biomechanics*, 21:955-964.

61. Zajac, F. E., Gordon, M. E. (1989). Determining muscle's force and action in multi-articular movement. In K. B. Pandolf (Ed.). *Exercise and Sports Sciences Reviews*, 17:187-230.

62. Zernicke, R. F., Loitz, B. J. (1992). Exercise-related adaptations in connective tissue. In P. Komi (Ed.). *Strength and Power in Sport*. Boston, MA: Blackwell Scientific, 77-95.

NEUROLOGIC CONSIDERATIONS FOR MOVEMENT

OBJECTIVES

After reading this chapter, the student will be able to:

1. Describe the anatomy of a motor unit, including central nervous system pathways, the neuron structure, the neuromuscular junction, and the ratio of fibers to neurons that are innervated.

2. Explain the differences between the three motor unit types (I, IIa, and IIb).

3. Discuss the characteristics of the action potential, emphasizing how a twitch or tetanus develops, and the influence of local graded potentials.

4. Describe the pattern of motor unit contribution to a muscle contraction through discussion of the size principle, synchronization, recruitment, and rate coding of the motor unit activity.

5. Discuss the components of a reflex action, and provide examples.

6. Describe the anatomy of the muscle spindle and the functional characteristics of the spindle during a stretch of the muscle or during gamma motoneuron influence.

7. Describe the anatomy of the Golgi tendon organ (GTO), and explain how the GTO responds to tension in the muscle.

8. Discuss the effect of exercise and training on neural input and activation levels in the muscle.

9. Identify the factors that influence flexibility, and provide examples of specific stretching techniques that are successful in enhancing flexibility.

10. Discuss the components of proprioceptive neuromuscular facilitation.

11. Describe a plyometric exercise, detailing the neurologic and structural contributions to the exercise.

12. Explain what electromyography is, how increasing muscle force affects it, how to record it, and its limitations.

OUTLINE

General Organization of the
  Nervous System
Motoneurons
  Structure of the Motoneuron
  The Motor Unit
  Neural Control of Force Output
Sensory Receptors and Reflexes
  Muscle Spindle
  Golgi Tendon Organ
  Tactile and Joint Sensory
    Receptors

Effect of Training and Exercise
  Flexibility Exercise
  Plyometric Exercise
Electromyography
  The Electromyogram
  Recording an Electromyographic Signal
  Factors Affecting the Electromyogram
  Analyzing the Signal
  Application of Electromyography
  Limitations of Electromyography
Summary
Review Questions

Human movement is controlled and monitored by the nervous system. The nature of this control is such that many muscles may have to be activated to perform a vigorous movement such as sprinting, or only a few muscles may have to be activated to push a doorbell or make a phone call. The nervous system is responsible for identifying the muscles that will be activated for a particular movement and then generating the stimulus to develop the level of force that will be required from that muscle.

Many human movements require stabilization of adjacent segments while a fine motor skill is performed. This requires a great deal of coordination on the part of the nervous system to stabilize such segments as the arm and forearm while very small, coordinated movements are created with the fingers, as in the act of writing.

Accuracy of movement is another task with which the nervous system is faced. The nervous system coordinates the muscles to throw a baseball with just the right amount of muscular force so that the throw is successful. Recognizing the difficulty of being accurate with a physical movement contributes to an appreciation of the complexity of neural control.

The neural network is extensive because each muscle fiber is individually innervated by a branch of the nervous system. Information exits the muscle and provides input to the nervous system, and information enters the muscle to initiate a muscle activity of a specific nature and magnitude. Through this loop system, which is interconnected with many other loops from other muscles and with central nervous control, the nervous system is able to coordinate the activity of many muscles at once. Specific levels of force may be generated in several muscles simultaneously so that a skill such as kicking may be performed accurately and forcefully. Knowledge of the nervous system is helpful in improving muscular output, refining a skill or task, rehabilitating an injury, and stretching a muscle group.

 # General Organization of the Nervous System

The nervous system consists of two parts, the central nervous system and the peripheral nervous system, both illustrated in Figure 4-1. The **central nervous system** consists of the brain and the spinal cord and should be

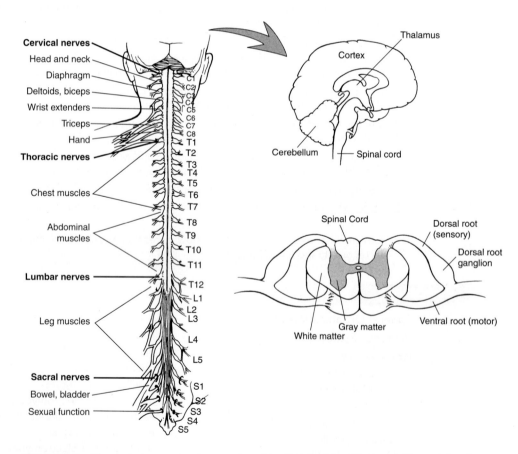

FIGURE 4-1   The central nervous system consists of the brain and the spinal cord. The peripheral nervous system consists of all of the nerves that lie outside the spinal cord. The 31 pairs of spinal nerves exit and enter the spinal cord at the various vertebral levels servicing specific regions of the body. Motor information leaves the spinal cord through the ventral root (anterior), and sensory information enters the spinal cord through the dorsal root (posterior).

Anterior View

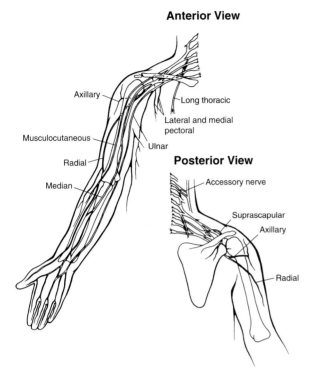

Posterior View

FIGURE 4-2 The upper extremity nerves. Nine nerves innervate the muscles of the upper extremity.

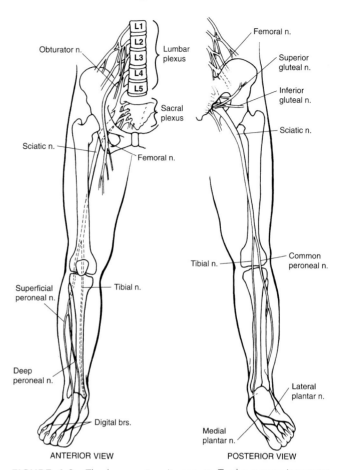

FIGURE 4-3 The lower extremity nerves. Twelve nerves innervate the muscles of the lower extremity.

viewed as the means by which human movement is initiated, controlled, and monitored.

The **peripheral nervous system** consists of all of the branches of nerves that lie outside the spinal cord. The peripheral nerves primarily responsible for muscular action are the **spinal nerves**, which enter on the posterior, or dorsal, side of the vertebral column and exit on the anterior, or ventral, side at each vertebral level of the spinal cord. Eight pairs of nerves enter and exit the cervical region, 12 pairs at the thoracic region, five at the lumbar region, five in the sacral region, and one in the coccygeal region. The pathways of the nerves are presented for the upper and lower extremities in Figures 4-2 and 4-3, respectively.

The nerves entering the spinal cord on the dorsal, or back, side of the cord are called **sensory neurons** because they transmit information into the system from the muscle. This pathway is termed the afferent pathway and carries all incoming information. The nerves exiting on the ventral, or front, side of the body are called **motoneurons** because they carry impulses away from the system to the muscle. This pathway is termed the efferent pathway and carries all outgoing information. Nerves from the dorsal and ventral roots join together as they exit so that sensory and motor neurons are mixed together to form a spinal nerve that can carry information in and out of the spinal cord.

Areas of the body supplied by the spinal nerves

Spinal Nerve	Area Supplied
Cervical—eight pairs	Back of the head, neck and shoulders, arms and hands, and diaphragm
Thoracic—12 pairs	Chest, some muscles of the back, and parts of the abdomen
Lumbar—five pairs	Lower parts of the abdomen and back, the buttocks, some parts of the external genital organs, and parts of the legs
Sacral—five pairs	Thighs and lower parts of the legs, the feet, most of the external genital organs, and the area around the anus

⚙️ Motoneurons

STRUCTURE OF THE MOTONEURON

The **neuron** is the functional unit of the nervous system that carries information to and from the nervous system. The structure of a neuron, specifically the motoneuron, warrants examination to clarify the process of muscular contraction. Figure 4-4 shows a close-up view of the neuron and the neuromuscular junction.

The motoneuron consists of a cell body containing the nucleus of the nerve cell. The **cell body**, or **soma**, of a motoneuron is usually contained within the gray matter of the spinal cord or in bundles of cell bodies just outside the cord, referred to as **ganglia**. The cell bodies are arranged in pools spanning one to three levels of the spinal cord and innervate portions of a single muscle or selected synergists.

Projections on the cell body, called **dendrites**, serve as receivers and bring information into the neuron from

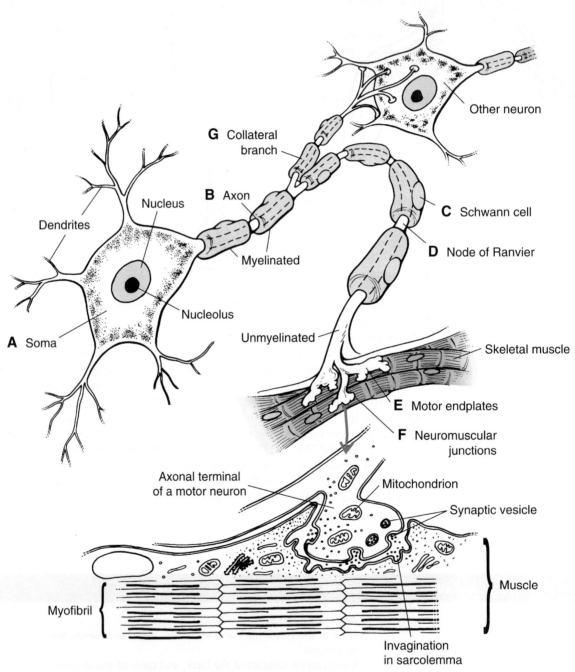

FIGURE 4-4   The cell body, or soma (**A**), of the neuron is in or just outside the spinal cord. Traveling from the soma is the axon (**B**), which is myelinated by Schwann cells (**C**), separated by gaps, the nodes of Ranvier (**D**). On the ends of each axon, the branches become unmyelinated to form the motor endplates (**E**) that terminate at the neuromuscular junction (**F**) on the muscle. Neurons receive information from other neurons through collateral branches (**G**).

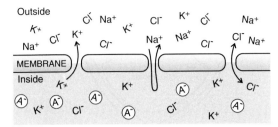

Exchange of Ions across Membrane

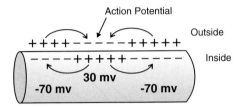

Action Potential Generated through Change in Electrical Potential

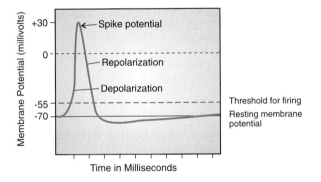

Recording of Action Potential

FIGURE 4-5 The action potential travels down the nerve as the permeability of the nerve membrane changes, allowing an exchange of sodium (Na^+) and potassium (K^+) ions across the membrane. This creates a voltage differential that is negative on the outside of the membrane. This negative voltage, or action potential, travels down the nerve until it reaches the muscle and stimulates a muscle action potential that can be recorded.

other neurons. The dendrites are bunched to form small bundles. A bundle contains dendrites from other neurons and may consist of dendrites from different spinal cord levels or different neuron pools. The composition of the bundle changes as dendrites are added and subtracted. This arrangement facilitates cross-talk between neurons.

A large nerve fiber, the **axon**, branches out from the cell body and exits the spinal cord via the ventral root, where it is bundled together with other peripheral nerves. The axon of the motoneuron is fairly large, making it capable of transmitting nerve impulses at high velocities, up to 100 m/sec. This large and rapidly transmitting motoneuron is also called an **alpha motoneuron**. The axon of the motoneuron is **myelinated**, or covered with an insulated shell. The myelination is sectioned, with **Schwann cells**

insulating and enveloping a specific length along the axon, followed by a gap, termed the **node of Ranvier**, and then a repeat of the insulated Schwann cell covering.

When the myelinated motoneuron approaches a muscle fiber, it breaks off into unmyelinated terminals, or branches, called **motor endplates**, which embed into fissures, or clefts, near the center of the muscle fiber. This site is called the **neuromuscular junction**. The neuron does not make contact with the actual muscle fiber; instead, a small gap, termed the synaptic gap or **synapse**, exists between the terminal branch of the neuron and the muscle. This is the reason muscular contraction involves a chemical transmission—the only way for a nerve impulse to reach the actual muscle fiber is some type of chemical transmission across the gap.

The nerve impulse travels down the axon in the form of an action potential (Fig. 4-5). As reviewed in Chapter 3, each action potential generates a twitch response in the muscle. If action potentials are in close enough sequence, the tensions generated by one muscle twitch are summed with other twitches to form a tetanus, or constant tension in the muscle fiber (see Fig. 3-9). This level of tension declines as the motor unit becomes incapable of regenerating the individual twitch responses fast enough.

The action potential is a propagated impulse, meaning that the amplitude of the impulse remains the same as it travels down the axon to the motor endplate. At the motor endplate, the action potential traveling down through the nerve becomes a muscle action potential traveling through the muscle. Externally, these two action potentials are indistinguishable. Eventually, the muscle action potential initiates the development of the cross-bridging and shortening within the muscle sarcomere. The total process is referred to as excitation–contraction coupling (see Chapter 3).

THE MOTOR UNIT

The structure of the **motor unit** was introduced in Chapter 3, in which we concentrated on the action of the muscles in the motor unit. In this section, we concentrate on the nervous system portion of the motor unit. The neuron, cell body, dendrites, axon, branches, and muscle fibers constitute the motor unit (Fig. 4-6). A neuron may terminate on as many as 2,000 fibers in muscles, as in the gluteus maximus, or as few as five or six fibers, as in the orbicularis oculi of the eye. The typical ratio of neurons to muscle fibers is 1:10 for the eye muscles, 1:1,600 for the gastrocnemius, 1:500 for the tibialis anterior, 1:1,000 for the biceps brachii, 1:300 for the dorsal interossei in the hand, and 1:96 for the lumbricals in the hand (4). The average number of fibers per neuron is between 100 and 200 (4,53). The number of fibers controlled by one neuron is termed the **innervation ratio**. Whereas fibers with a small innervation ratio are capable of exerting fine motor control, those with a large innervation ratio are only capable of gross motor control. The fibers innervated by each motor unit are not bunched together and are not all in the same fascicle; rather, they are spread throughout the muscle.

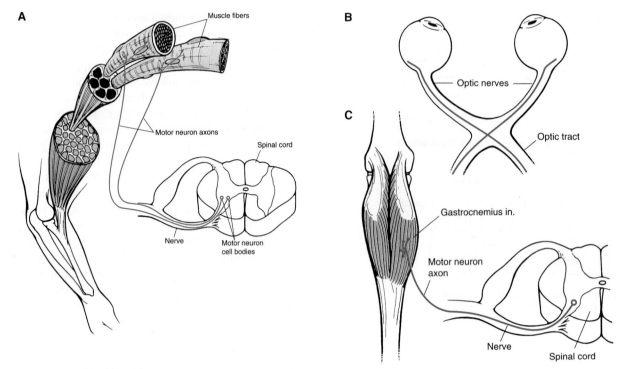

FIGURE 4-6   **(A)** The motor unit consists of a neuron and all of the fibers innervated by that neuron. The motoneurons exit the anterior side of the spinal cord and branch out, terminating on a muscle fiber. **(B)** Fine motor movements can occur when the motor unit services only a small number of muscle fibers, such as in the eye. **(C)** When the motor unit terminates on large numbers of muscle fibers, such as in the gastrocnemius, finer movement capabilities are lost at the gain of more overall muscle activity.

Precision of muscular contraction

When a motor unit is recruited, all of the fibers that the neuron innervates are activated at once. If a muscle were to consist of a single motor unit you would have very poor control over that muscle. On the other hand if a muscle were divided into many different motor units you could recruit any submaximal number of motor units depending on the needed force level. Discuss which of the following muscles would be capable of contracting more precisely: 93,000 fibers with 378 motoneurons or 68,000 fibers with 300 motoneurons? Consider the innervation ratio.

When a motor unit is activated sufficiently, all of the muscle fibers belonging to it contract within a few milliseconds. This is referred to as the **all-or-none principle**. A muscle that has motor units with very low ratios of nerve to fiber, such as is seen in eye and hand movements, allows finer control of the movement characteristics. Many lower extremity muscles have large neuron-to-fiber ratios suitable to functions in which large amounts of muscular output are required, such as in weight bearing and walking.

Muscle fibers of different motor units are intermingled so that the force applied to the tendon remains constant even when different muscle fibers are contracting or relaxing. Muscle tone is maintained in the resting muscle as random motor units contract.

The activity in the motor unit is determined from all of the inputs it receives. These include motor commands causing excitation via the alpha motoneuron and excitatory and inhibitory inputs the motor unit receives from other neurons. This is discussed later in the section on receptors.

Motor Unit Types

Three different types of motor units exist, corresponding to the three fiber types discussed in the previous chapter: slow-twitch oxidative (type I or S), fast-twitch oxidative (type IIa or FR), and fast-twitch glycolytic (type IIb or FF). Performance and size differences are illustrated in Figure 4-7. All three types of muscle fibers are found in all muscles, but the proportion of fiber types within a muscle varies. Whereas certain muscles, such as the soleus, consist primarily of type I muscle fibers and motor units, muscles such as the vastus lateralis are approximately 50% type I and the remainder type II.

All of the muscle fibers in a motor unit are of the same type. The fast-twitch glycolytic motor units (type IIb) are innervated by very large alpha motoneurons that conduct the impulses at very fast velocities (100 m/s), creating rapid contraction times in the muscle (approximately 30 to 40 ms) (13). As a result, these large motor units generate muscular activity that contracts fast, develops high

Events in the action potential

REST Voltage = −70 to −90 mV	Unequal distribution of charged particles on the inside and outside of the membrane. Inside = negative because of the presence of negatively charged proteins. Outside = positive relative to inside because of large number of positively charged ions attracted to the outer surface by the negative charge on the inside.	More Na^+ outside of the cell than inside and more K^+ inside the cell than outside
DEPOLARIZATION Voltage = +30 mV	If the threshold stimulus to the neuron is strong enough (>10 mV), the neuron fires an action potential. The action potential is propagated as the interior becomes more positively charged.	1. Large numbers of voltage-activated Na^+ channels open 2. Rapid movement of Na^+ into the cell 3. The interior becomes more positively charged 4. Stops when the Na^+ channels inactivate as a result of the voltage change
REPOLARIZATION	Sodium (Na^+) channel inactivation and opening up of potassium (K^+) channels	Rapid outward movement of K^+
HYPERPOLARIZATION	K^+ channels open and close slowly; requires a greater than normal stimulus to activate another action potential	More K^+ leaves the cell than is necessary to repolarize the membrane

tensions, and fatigues quickly. These motor units usually have large neuron-to-fiber ratios and are found in some of the largest muscles in the body, such as the quadriceps femoris group. These motor units are useful in activities such as sprinting, jumping, and weight lifting.

The fast-twitch oxidative motor units (type IIa) also have fast conduction speeds (80 to 90 m/s) and short contraction times (30 to 50 ms), but they have the advantage over fast-twitch glycolytic motor units because they are more fatigue-resistant (13). These moderately sized

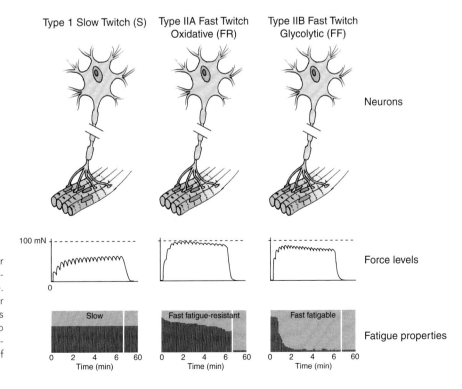

FIGURE 4-7 **(A)** Type I slow-twitch (S) motor unit is smaller and is capable of generating sustained contractions and lower levels of force. **(B)** Type IIa fast-twitch oxidative (FR) motor unit can also generate sustained contractions at higher force levels than type Ia. **(C)** Type IIb fast-twitch glycolytic (FF) cannot sustain a contraction for any length of time but is capable of generating the highest force levels.

Motor unit properties

Properties	Type 1 Slow-Twitch (S)	Type IIa Fast-Twitch Oxidative (FR)	Type IIb Fast-Twitch Glycolytic (FF)
Contraction speed	Slow	Fast	Fast
Number of fibers	Few	Many	Many
Motoneuron size	Small	Large	Large
Fiber diameter	Moderate	Large	Large
Force of unit	Low	High	High
Fatigability	Low	Medium	High
Excitability	High	Low	Low
Metabolic type	Oxidative	Intermediate	Glycolytic
Mitochondrial density	High	Medium	Low
Myosin ATPase activity	Low	High	High

motor units are capable of generating moderate tensions over longer periods. The activity from these motor units is useful in activities such as swimming and bicycling and in job tasks in factories and among longshoremen.

The slow-twitch oxidative motor units (type I) transmit the impulses slowly (70 to 80 m/s), generating slow contraction times in the muscle (70 ms) (13). These motor units are capable of generating very little tension but can sustain this tension over a long time. Type I fibers are more efficient than the other two fiber types. Consequently, the slow-twitch motor units, the smallest of the three types, are useful in maintaining postures, stabilizing joints, and doing repetitive activities such as typing and gross muscular activities such as jogging.

NEURAL CONTROL OF FORCE OUTPUT

Chapter 3 explored a number of factors such as muscle cross section that determines maximal force produced by a muscle. We also stated that the force exerted by a motor unit is determined by the number of fibers innervated by the motor unit and the rate at which the motor unit discharges the impulse or action potential (19). When a muscle is producing its maximal force, all motor units are activated and all muscle fibers are active.

Motor Pool

Groups of neurons in the spinal cord that innervate a single muscle are termed a **motor pool**. Pool sizes range from a few hundred to a thousand depending on the size of the muscle. Motor neurons in the pool vary in electrical properties, amplitude of the input they receive, and in contractile properties (e.g., speed, force generation, and fatigue resistance) (19).

Recruitment

The tension or force generated by a muscle is determined by the number of motor units actively stimulated at the same time and by the frequency at which the motor units are firing. **Recruitment**, the term used to describe the

order of activation of the motor units, is the prime mechanism for force production in the muscle. Force produced by a muscle can be increased by increasing the number of active motor units to increase the active cross-sectional area of the muscle. Recruitment usually follows an orderly pattern in which pools of motor units are sequentially recruited (14). There is a functional pool of motor units for each task, whereby separate recruitment sequences can be initiated to stimulate the three different types of motor units (types I, IIa, and IIb) for the performance of different actions within the same muscle.

The sequence of motor unit recruitment usually follows the **size principle**, whereby the small, slow-twitch motoneurons are recruited first, followed by recruitment of the fast-twitch oxidative and finally the large, fast-twitch glycolytic motor units (14). This is because the small motoneurons have lower thresholds than the large ones. Thus, the small motoneurons are used over a broad tension range before the moderate or large fibers are recruited.

In walking, for example, the low-threshold motor units are used for most of the gait cycle, except for some brief recruitment of the intermediate motor units during peak activation times. The high-threshold, fast-twitch motor units are not usually recruited unless a rapid change of direction or a stumble takes place.

In running, more motor units are recruited, with some high-threshold units recruited for the peak output times in the cycle. Furthermore, the low-threshold units are recruited for activities such as walking and jogging, and the fast-twitch fibers are recruited in activities such as weight lifting (14,25). Recruitment sequences for walking and for different exercise intensities are presented in Figure 4-8.

The motor units are recruited **asynchronously**, whereby the activation of a motor unit is temporally spaced but is summed with the preceding motor unit activity. If the tension is held isometrically over a long time, some of the larger motoneurons are activated. Likewise, in vigorous, rapid movements, both small and large motoneurons are activated.

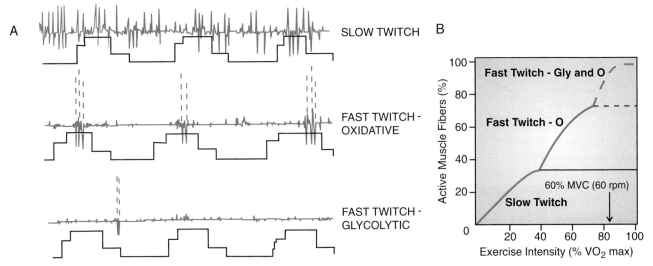

FIGURE 4-8   The order of activation of the motor units, termed recruitment, usually follows the size principle: The small slow-twitch fibers are recruited first, followed by the fast-twitch oxidative and last by the
fast-twitch glycolytic fibers. **(A)** The muscle activity for the three muscle types for three support phases in
walking. Slow-twitch fibers are used for most of the gait cycle, with some recruitment of fast-twitch fibers
at peak activation times. **(B)** Similar recruitment pattern, with slow-twitch fibers recruited for up to 40% of
the exercise intensity, at which point the fast-twitch oxidative fibers are recruited. It is not until 80% of exercise intensity is reached that the fast-twitch glycolytic fibers are recruited. (Reprinted with permission from
(A) Grimby, L. (1986). Single motor unit discharge during voluntary contraction and locomotion. In N. L.
Jones, et al. (Eds.). *Human Muscle Power*. Champaign, IL: Human Kinetics, 111–129; **(B)** Sale, D. G. (1987).
Influence of exercise and training on motor unit activation. *Exercise and Sport Sciences Reviews*, 16:95-151.)

The motor unit recruitment pattern proceeds from small to large motoneurons, slow to fast, small force to large force, and fatigue-resistant to fatigable muscles. After a motor unit is recruited, it will remain active until the force declines, and when the force declines, the motor units are deactivated in reverse order of activation, with the large motoneurons going first. Also, the motor unit recruitment pattern is established in the muscle for a specific movement pattern (58). If the joint position changes and a new pattern of movement is required, the recruitment pattern changes because different motor units are recruited, although the order of recruitment from small to large remains the same. The force developed during recruitment does not increase in a jerky manner because the larger motoneurons are not brought into action until the muscle is already developing a large amount of force. In fact, the fractional increase in force is constant such that the larger the tension already in the muscle, the larger the size of motor units recruited.

Rate Coding

The frequency of motor unit firing can also influence the amount of force or tension developed by the muscle. This is known as **frequency coding** or **rate coding** and involves intermittent high-frequency bursts of action potentials or impulses ranging from three to 120 impulses per second (53). With constant tension or slow increases in tension, the firing frequency is in the range of 15 to 50 impulses per second. This frequency rate can increase to a range of 80 to 120 impulses per second during fast

contraction velocities. With increased rate coding, the rate of impulses increases in a linear fashion and only after all of the motor units are recruited (7).

In the small muscles, all of the motor units are usually recruited and activated when the external force of the muscle is at levels of only 30% to 50% of the maximum voluntary contraction level. Beyond this level, the force output in the muscle is increased through increases in rate coding, allowing for the production of a smooth, precise contraction.

In the large muscles, recruitment of motor units takes place all through the total force range, so that some muscles are still recruiting more motor units at 100% of maximum voluntary contraction. The deltoid and the biceps brachii are examples of muscles still recruiting motor units at 80% to 100% of maximum output of the muscle.

The rate coding also varies with fiber type and changes with the type of movement. Examples of the rate coding of both high- and low-threshold fibers in two muscle contractions is illustrated in Figure 4-9. In ballistic movements, the higher-threshold fast-twitch motor units fire at higher rates than the slow-twitch units. To produce rapid accelerations of the segments, the fast-twitch motor units increase the firing rates more than the slow-twitch motor units (25). The high-threshold fast-twitch fibers cannot be driven for any considerable length of time, but it is believed that trained athletes can drive the high-threshold units longer by maintaining the firing rates, resulting in the ability to produce a vigorous contraction for a limited time. Eventually, the frequency of motor unit firing

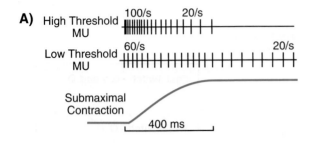

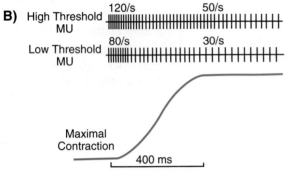

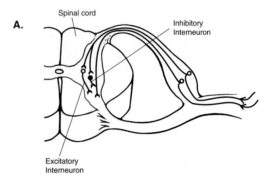

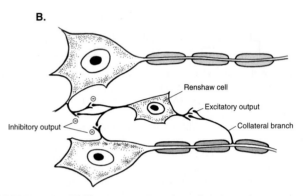

FIGURE 4-9 **(A)** Tension development in the muscle is influenced by the frequency at which a motor unit is activated, termed *rate coding*. In a submaximal muscle contract and hold, the high-threshold fast-twitch fibers increase firing rates in the ramp phase more than the low-threshold units. The frequency of motor unit firing drops off during the hold phase, and the high-threshold units cease firing. **(B)**. In a more vigorous contract and hold, the rate coding increases and is maintained further into the contraction by both the high- and low-threshold motor units. (Reprinted with permission from Sale, D. G. (1987). Influence of exercise and training on motor unit activation. *Exercise and Sport Sciences Reviews*, 16:95-151.)

FIGURE 4-10 **(A)** The action potential traveling through a motor unit can be altered by input from interneurons, which are small connecting nerve branches that generate a local graded potential; the potential may or may not institute a change in the connecting neuron. The interneurons may produce an excitatory local graded potential, which facilitates the action potential, or an inhibitory local graded potential sufficient to inhibit the action potential. **(B)** A special interneuron, the Renshaw cell, receives excitatory information from a collateral branch of another neuron, stimulating an inhibitory local graded potential.

decreases during any continuous muscular contraction, whether vigorous or mild.

The action potential in a motor unit can be facilitated or inhibited by the input it receives from the many neurons that are connecting to it within the spinal cord. As shown in Figure 4-10 a motor unit receives synaptic input from other neurons and from **interneurons**, which are connecting branches that can be both excitatory and inhibitory. The input is in the form of a **local graded potential** that, unlike the action potential, does not maintain its amplitude as it travels along. Thus, the stimulus has to be sufficient to reach its destination on another neuron and be large enough to generate a response in the neuron with which it has interfaced.

The alpha motoneuron has many collateral branches interacting with other neurons, and the number of collateral branches is highest in the distal muscles. An inhibitory interneuron receiving input from these collaterals is the **Renshaw cell**, also in the spinal cord. The Renshaw cell is considered one of the key elements in organizing muscular response in the agonists, antagonists, and synergists when it is stimulated sufficiently by a collateral branch (29,30).

Some evidence suggests that alternative recruitment patterns may be initiated by input from the excitatory and inhibitory pathways. This is done through interneurons that alter the threshold response of the slow- and fast-twitch units. The threshold level of the fast-twitch motor unit can be lowered via excitatory interneurons.

In ballistic movements involving rapid alternating movements, there appears to be **synchronous** or concurrent activation of the motor unit pool whereby large motor units are recruited along with the small motoneurons. This synchronous firing has also been shown to occur as a result of weight training. It is believed that in athletic performance requiring a wide range of muscular output, the neuromuscular sequence may actually be reversed, with the fast-twitch fibers recruited first in vigorous muscle actions (8,14).

Sensory Receptors and Reflexes

The body requires an input system to provide feedback on the condition and changing characteristics of the musculoskeletal system and other body tissues, such as the skin. Sensors collect information on such events as stretch in the muscle, heat or pressure on the muscle, tension in

CHAPTER 4 Neurologic Considerations for Movement **109**

the muscle, and pain in the extremity. These sensors send information to the spinal cord, where the information is processed and used by the central nervous system in the adjustment or initiation of motor output to the muscles. These sensors are connected to the spinal cord via sensory neurons.

When the sensory information from one of these receptors brings information into the cord, triggering a predictable motor response, it is termed a **reflex**. A reflex is an involuntary neural response to a specific sensory stimulus and is a stereotypical behavior in both time and space. A simple reflex arc is shown in Figure 4-11. In the case of tendon jerk reflex at the knee joint (**stretch reflex**), the magnitude of the reflex contraction of the quadriceps muscle resulting in a sudden involuntary extension of the leg is proportional to the intensity of the tap stimulus applied to the patellar tendon. Most reflexes are not that simple and can be modified with input from different areas. For example, a **flexor reflex** initiates a quick withdrawal response after receiving sensory information indicating pain such as from touching something hot. Compare this reflex with an anticipatory situation in which you are told that you are going to be jerked by the hand and need to maintain your balance by resisting with arm flexion. When the jerk is applied, there is a reflex response, but it is different from the flexor reflex because the context is different, and even though the response occurs at the spinal cord level, the circuitry has been reset to respond differently. Each reflex is influenced by the state of many interneurons, which receive input both from segmental and descending systems (47).

Reflexes that bring information into the spinal cord and are processed through both sides and different levels of the spinal cord are termed **propriospinal**. An example of this type of reflex is the **crossed extensor reflex**, which is initiated by receiving or expecting to receive a painful stimulus, such as stepping on a nail. This sensory information is processed in the spinal cord by creating a flexor and

withdrawal response in the pained limb and an increase or excitation in the extension muscles of the other limb.

Another propriospinal reflex is the **tonic neck reflex**, which is stimulated by movements of the head that create a motor response in the arms. When the head is rotated to the left, this reflex stimulates an asymmetric response of extension of the same-side arm (left) and a flexion of the opposite arm (right). Also, when the head flexes or extends, this reflex initiates flexion or extension of the arms, respectively.

Another type of reflex is the **supraspinal reflex**, which brings information into the spinal cord and processes it in the brain. The result is a motor response. The **labyrinthine righting reflex** is an example of this type of reflex. This reflex is stimulated by leaning, being upside down, or falling out of an upright posture. The response from the upper centers is to stimulate a motor response from the neck and limbs to maintain or move to an upright position. This complex reflex involves many levels of the spinal cord and the upper centers of the nervous system. Examples of these various reflex actions are presented in Figure 4-12.

MUSCLE SPINDLE

Proprioceptors are sensory receptors in the musculoskeletal system that transform mechanical distortion in the muscle or joint, such as any change in joint position, muscle length, and muscle tension, into nerve impulses that enter the spinal cord and stimulate a motor response (63). The **muscle spindle** is a proprioceptor found in higher abundance in the belly of the muscle lying parallel to the muscle fibers and actually connecting into the fascicles via connective tissue (Fig. 4-13). The fibers of the muscle spindle are termed intrafusal compared with muscle fibers that are termed extrafusal. The **intrafusal fibers** of the spindle are contained within a capsule, forming a spindle shape, hence the name muscle spindle. Some muscles, such as those of the eye, hand, and upper back, have hundreds of spindles; other muscles, such as the latissimus dorsi and other shoulder muscles, may have only a handful (63). Every muscle has some spindles. However, the muscle spindle is absent from some of type IIb fast-twitch glycolytic muscle fibers within some muscles.

Each spindle capsule may contain as many as 12 intrafusal fibers, which can be either of two types: nuclear bag or nuclear chain (63). Both types of fibers have noncontractile centers that contain the nuclei of the fiber in addition to sensory nerve fibers that take information into the system through the dorsal root of the spinal cord. The spindle also has contractile ends that can be innervated by a **gamma motoneuron**, creating shortening upon receipt of motor input. The gamma or fusimotor motoneuron is intermingled with the alpha motoneuron in the ventral horn of the spinal column. Smaller than the alpha motoneuron, each gamma motoneuron innervates multiple muscle spindles.

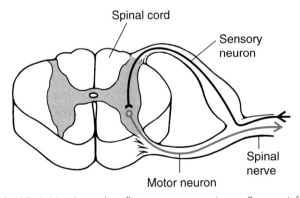

FIGURE 4-11 A simple reflex or monosynaptic arc. Sensory information from receptors is brought into the cord, where it initiates a motor response sent back out to the extremities. The stretch reflex is a reflex arc that sends sensory information into the cord in response to stretch of the muscle; the cord sends back motor stimulation to the same muscle, causing a contraction.

Spinal cord
Sensory neuron
Spinal nerve
Motor neuron

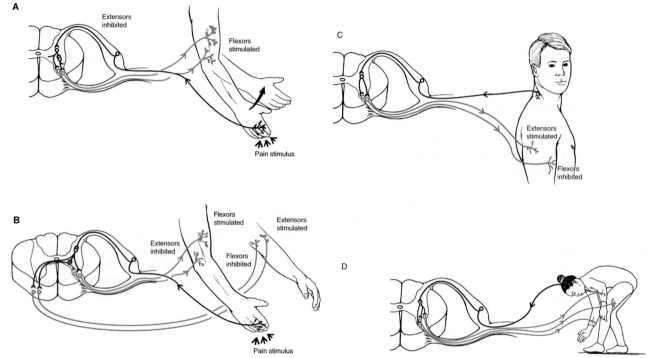

FIGURE 4-12 Examples of reflex actions (a reflex is a motor response developed in the central nervous system after sensory input is received). **(A)** The flexor reflex is triggered by sensory information registering pain, which facilitates a quick flexor withdrawal from the pain source. **(B)** The crossed extensor reflex is also initiated by pain; it works with the flexor reflex to create flexion on the stimulated limb and extension on the contralateral limb. **(C)** The tonic neck reflex is stimulated by head movements; it creates flexion or extension of the arms, depending on the direction of the neck movement. **(D)** The labyrinthine righting reflex is stimulated by body positioning; it causes movements of the limbs and neck to maintain a balanced, upright posture.

The **nuclear bag fiber** has a large cluster of nuclei in its center. It is also thicker, and its fibers connect to the capsule and to the actual connective tissue of the muscle

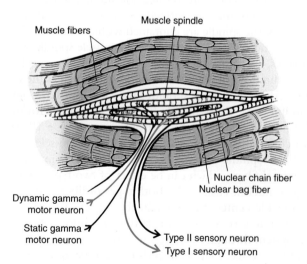

FIGURE 4-13 The muscle spindle lies parallel with the muscle fibers. Within each spindle capsule are the spindle fibers, which can be either of two types: nuclear chain or nuclear bag fibers. Both types have contractile ends that are innervated by gamma motoneurons. Sensory information responding to stretch leaves the middle portion of both the chain and bag fibers through type Ia sensory neuron and from the ends of the nuclear chain fibers via type II sensory neuron.

fiber itself. The **nuclear chain fiber** is smaller, with the nuclei arranged in rows in the equatorial region. The nuclear chain fiber does not connect to the actual muscle fiber but only makes connection with the spindle capsule.

Exiting from the equatorial region of both types of spindle fibers, **type Ia primary afferent** neuron is stimulated by a change in length of the middle of the spindle. Information from the sensory endings sends information into the dorsal horn and makes a **monosynaptic**, or direct, connection with a motoneuron, resulting in a contraction of the same muscle. Because the spindle lies parallel to the muscle fibers, it is subject to the same stretch as the muscle. The other mechanism of "stretching" the middle portion of the spindle is through contraction of the ends of the spindle via gamma motoneuron innervation. Both the nuclear bag and nuclear chain fibers are innervated by their own gamma motoneuron, the dynamic and static gamma efferents, respectively. The shortening of the ends of the spindle fibers through gamma innervation allows tuning of the muscle spindle to meet the needs of the movement parameters (Fig. 4-14).

From the polar ends of the nuclear chain fiber, additional sensory information is transmitted via **type II secondary afferent** sensory neuron. This sensory neuron is medium sized and is stimulated by stretch in the

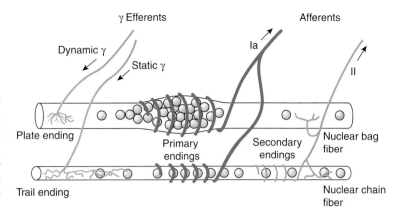

FIGURE 4-14 Two afferent pathways bring information from the spindle into the spinal cord. Type Ia primary afferent pathway exiting from the equatorial regions of both the nuclear chain and bag fibers provides sensory information about muscle length and velocity of stretch. Type II secondary afferent pathway exiting from the ends of the nuclear chain fibers provides information about muscle length. Both fiber types receive motor innervation of the contractile ends via gamma efferent neurons.

muscle, responding at a higher threshold of stretch than type I sensory neuron. There are generally one or two type II sensory neurons per muscle spindle, although some muscle spindles and even some muscles (10% to 20%) have none (63).

The primary afferents are sensitive to the rate of change of the stretch of the muscle and act as velocity sensors. The sensitivity of the primary afferents is nonlinear and is very sensitive to small changes in length and rate of change (short-range stretch), but falls off with slower

or larger changes in length. The secondary afferents are muscle length sensors with some sensitivity to rate of change in length. Figure 4-15 illustrates the response of the primary and secondary afferents in the absence of any gamma innervation with stretch of the muscle, a quick tap of the muscle, a cyclic stretch and release, and with the release of the stretch.

When a stretch is imposed on the muscle, the equatorial region of the intrafusal fibers deforms the nerve endings and type I sensory neuron sends impulses into the spinal

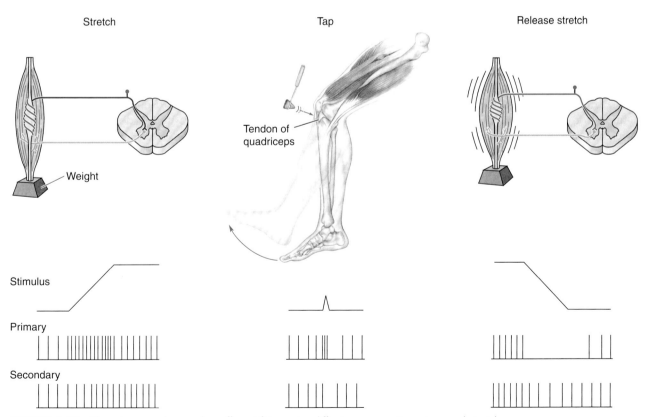

FIGURE 4-15 The primary and secondary afferent firing rates differ in response to an imposed stretch or relaxation of the muscle. The responses of both the primary and secondary afferents are shown for three different stretch conditions with the influence of any gamma innervation removed. The primary afferent responds to a stretch imposed on the muscle and fires at higher rates when a rapid stretch is imposed on the muscle in the case of a tap. When the stretch is removed, the primary afferent ceases firing. The secondary afferent fires at a more consistent rate to reflect the length of the muscle.

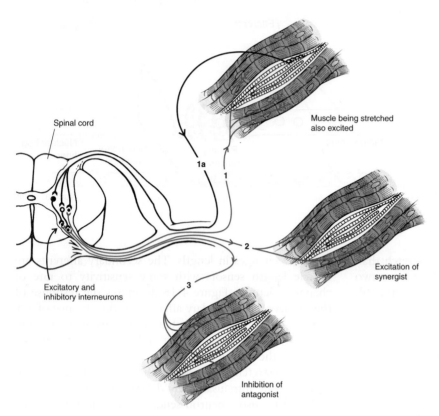

Spinal cord

Muscle being stretched
also excited

1a

1

2

Excitation of
synergist

Excitatory and
inhibitory interneurons

3

Inhibition of
antagonist

FIGURE 4-16   Type Ia loop initiated by a stretch of the muscle. Responding proportionally to the rate of stretch, the muscle spindle sends impulses to the spinal cord via type Ia sensory neuron. Within the cord, connections with interneurons produce a local, graded potential that inhibits the antagonistic muscles and excites the synergists and the muscle in which the stretch occurred. This is the typical stretch reflex response, also termed *autogenic facilitation*.

cord. Sensory action potentials connect with interneurons, generating an excitatory local graded potential that is sent back to the muscle being stretched. If the stretch is vigorous enough, a local graded impulse is sent back to the same muscle with sufficient magnitude to initiate a contraction via the alpha motoneurons. Sensory information enters and motor information leaves the spinal cord at the same level, creating a **monosynaptic reflex arc** in which the sensory input connects directly on the motoneuron. An example of this reflex is the stretch or **myotatic reflex**, which is stimulated by sensory neurons responding to stretch in the muscle, which in turn, initiates an increase in the motor input to the same muscle (63). Type Ia loop is illustrated in Figure 4-16. It is also termed **autogenic facilitation** because of the facilitation of the alpha motoneurons of the same muscle. The stretch reflex primarily recruits slow-twitch muscle fibers.

The information coming into the spinal cord via type I sensory neuron is also sent to the cerebellum and cerebral sensory areas to be used as feedback on muscle length and velocity. Additional connections are made in the spinal cord with inhibitory interneurons, creating a **reciprocal inhibition**, or relaxation of the antagonistic muscles (29). Other excitatory interneuron connections are made with the alpha motoneurons of synergistic muscles to facilitate their muscle activity along with the agonist.

When type II, or secondary, afferent neuron is stimulated, it has a different response from that of type I sensory neuron. It produces a sensory input in response to stretch or change in length in the muscle, and it is a

good feedback indicator of the actual length in the muscle because its sensory impulses do not diminish when the muscle is held in a stationary position.

The innervation of the ends of the spindle fibers by the gamma motoneuron alters the response of the muscle spindle considerably. The first important effect of gamma innervation of the spindle is that it does not allow spindle discharge to cease when a muscle is shortened. If the muscle shortened with no alpha–gamma coactivation, the spindle activity would be silenced by the removal of the external stretch on the muscle. The alpha–gamma coactivation keeps the spindle taut and allows it to continue to provide position and length information despite shortening of the muscle (63). There is some indication that this is only true for slow movements and for movements under load but is not true for fast movements. In fast movements, the stretching activity in the spindles of the antagonistic muscle may provide the length and position information.

The second major input from gamma motoneuron innervation of the muscle spindle is an indirect enhancement of the motor impulses being sent to the muscle via the alpha neuron pathways. This adds to the impulses coming down through the system, alters the gain, and increases the potential for full activation via the alpha pathways. It is a main contributor to coordinating the output and patterning of the alpha motoneurons.

In anticipation of lifting something heavy, the alpha and gamma motoneurons establish a certain level of excitability in the system for accommodating the heavy resistance. If the object lifted is much lighter than anticipated,

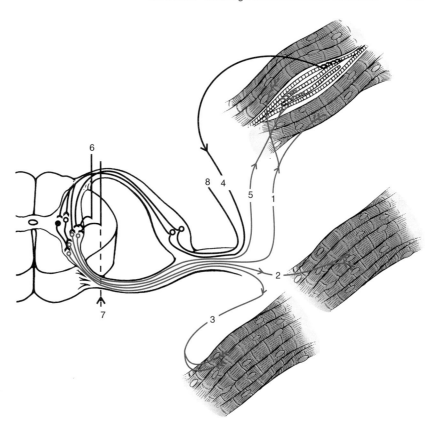

FIGURE 4-17 Type Ia loop in which information is sent from the spindle (4), causing inhibition (3) and excitation of synergists and agonists (2, 1). It is facilitated by input from the gamma motoneuron (5), which initiates a contraction of the ends of the spindle fibers, creating an internal stretch of the spindle fibers. The gamma motoneuron receives input via the upper centers or other interneurons in the spinal cord (6, 7, 8).

the gamma system acts to reduce the output of the type I afferent, making a quick adjustment in the alpha motoneuron output to the muscle and reduce the number of motor units activated.

Finally, the gamma motoneuron is activated at a lower threshold than the alpha motoneuron and can therefore initiate responses to postural changes by resetting the spindle and activating the alpha output (27). The afferent pathways, gamma pathways, and alpha pathways are all part of the **gamma loop**, which is shown in Figure 4-17.

GOLGI TENDON ORGAN

Another important proprioceptor significantly influencing muscular action is the **Golgi tendon organ** (GTO). This structure monitors force or tension in the muscle. As illustrated in Figure 4-18, the GTO lies at the musculoskeletal junction. It is a spindle-shaped collection of collagen fascicles surrounded by a capsule that continues inside the fascicles to create compartments. The collagen fibers of the GTO are connected directly to **extrafusal fibers** from the muscles (63).

Two sensory neurons exit from a site between the collagen fascicles. When the collagen is compressed through a stretch or contraction of the muscle fibers, type Ib nerve endings of the GTO generate a sensory impulse proportional to the amount of deformation created in them. The response to the load and the rate of change in the load are linear. Several muscle fibers insert in one GTO, and any tension generated in any of the muscles will generate a response in the GTO.

In a stretch of the muscle, the tension in the individual GTO is generated along with all other GTOs in the tendon. Consequently, the GTO response is more sensitive in tension than in stretch. This is because the GTO measures load bearing in series with the muscle fibers but is parallel to the tension developed in the passive elements during stretch (33). Thus, contraction has a lower threshold than stretch.

The GTO generates an inhibitory local graded potential in the spinal cord known as the **inverse stretch reflex**. If the graded potential is sufficient, relaxation or autogenic inhibition is produced in the muscle fibers connected in series with the GTO stimulated. The alpha motoneuron output to muscles undergoing a high-velocity stretch or producing a high-resistance output is reduced.

The GTO is very sensitive to small changes in tension, so it is used to modulate changes in force. It assists with providing information on force so that the individual applies just the right amount of force to overcome a load. The GTO is reliable in signaling whole-muscle tension whether it is active or passive tension, even after a fatiguing routine (24). The GTO can generate an inhibitory response via type Ib pathway to reduce contraction strength in a muscle experiencing a rapid increase in force. Alternately, the GTO can actually provide excitatory input in an activity such as walking, during which the GTO detects tension in the support muscles and stimulates an extensor reflex. Again, with input from upper neural centers, the context changes and circuits are adjusted accordingly.

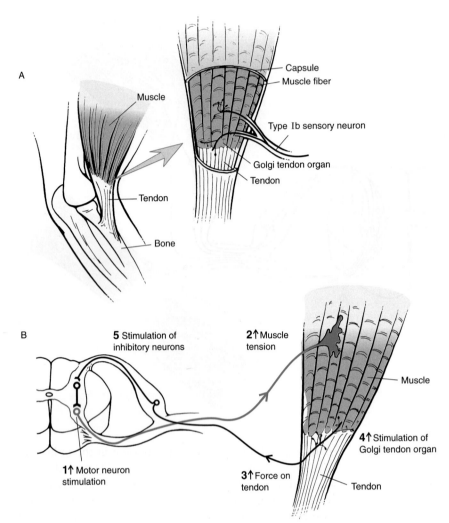

A

Muscle

Capsule
Muscle fiber

Type Ib sensory neuron

Golgi tendon organ

Tendon

Tendon

Bone

B

5 Stimulation of inhibitory neurons

2↑ Muscle tension

Muscle

4↑ Stimulation of Golgi tendon organ

1↑ Motor neuron stimulation

3↑ Force on tendon

Tendon

FIGURE 4-18 **(A)** The Golgi tendon organ (GTO) is at the muscle–tendon junction. **(B)** When tension is at this site, the GTO sends information into the spinal cord via type Ib sensory neurons. The sensory input from the GTO facilitates relaxation of the muscle via stimulation of inhibitory interneurons. This response is as the inverse stretch reflex, or autogenic inhibition.

TACTILE AND JOINT SENSORY RECEPTORS

There is limited information on the sensory neuron input from the tactile and joint receptors placed in and around the synovial joints (Fig. 4-19). One such tactile receptor, the **Ruffini ending,** lies in the joint capsule and responds to change in joint position and velocity of movement of the joint (54). The **pacinian corpuscle** is another tactile receptor in the capsule and connective tissue that responds to pressure created by the muscles and to pain within the joint (54). These joint receptors, as well as other receptors in the ligaments and tendons, provide continuous input to the nervous system about the conditions in and around the joint.

Effect of Training and Exercise

During training of the muscular system, a neural adaptation modifies the activation levels and patterns of the neural input to the muscle. In strength training, for example, significant strength gains can be demonstrated after approximately four weeks of training. This strength

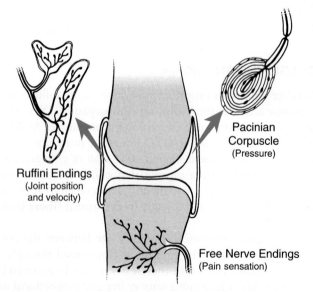

Pacinian Corpuscle
(Pressure)

Ruffini Endings
(Joint position and velocity)

Free Nerve Endings
(Pain sensation)

FIGURE 4-19 A number of other sensory receptors send information into the central nervous system. In the joint capsules and connective tissue are found the pacinian corpuscle, which responds to pressure, and the Ruffini endings, which respond to changes in joint position. Also, free nerve endings around the joints create pain sensations.

gain is not attributable to an increase in muscle fiber size but is rather a learning effect in which neural adaptation has occurred (59), resulting in increases in factors such as firing rates, motoneuron output, motor unit synchronization, and motoneuron excitability (1).

The effect of the neural adaptation is an improved muscular contraction of higher quality through coordination of motor unit activation. The neural input to the muscle, as a consequence of maximal voluntary contractions, is increased to the agonists and synergists, and inhibition of the antagonists is greater. This neural adaptation, or learning effect, levels off after about four to five weeks of training and is typically the result of an increase in the frequency of motor unit activation. Increases in strength beyond this point are usually attributable to structural changes and physical increases in the cross section of the muscle. The influence of training on both the electromechanical delay (EMD) and the amount of electromyographic (EMG) activity is presented from the work of Hakkinen and Komi (26) in Figure 4-20.

Specificity of training is important for enhancement of neural input to the muscles. If one limb is trained at a time, greater force production can be attained with more neural input to the muscles of that limb than if two limbs are trained at once. The loss of both force and neural input to the muscles through bilateral training is termed **bilateral deficit** (5,14). In fact, training of one limb neurologically enhances the activity and increases the voluntary strength in the other limb.

When working with athletes who use the limbs asymmetrically, as in running or throwing, the trainer should incorporate some unilateral limb movements into the conditioning program. Participants in sports or activities that use both limbs together, such as weight lifting, should train bilaterally.

Specificity of training also determines the fiber type that is enhanced and developed. Through resistive training, type II fibers can be enhanced through reduction in central inhibition and increased neural facilitation. This may serve to resist fatigue in short-term, high-intensity exercise in which the fatigue is brought on by the inability to maintain optimal nerve activation.

Even a short warm-up (five to 10 minutes) preceding an event or performance influences neural input by increasing the motor unit activity (38). Another factor that enhances the neural input to the muscle is the use of an antagonistic muscle contraction that precedes the contraction of the agonist, such as seen in preparatory movements in a skill (e.g., backswing and lowering). This diminishes the inhibitory input to the agonist and allows for more neural input and activation in the agonist contraction.

A stretch of a muscle before it contracts produces some neural stimulation of the muscle via the stretch reflex arc. Athletes who must produce power, such as jumpers and sprinters, have been shown to have excitable systems in which the reflex potentiation is high (38).

When fatigue occurs during exercise, a reduction occurs in the maximal force capacity of the muscle, caused

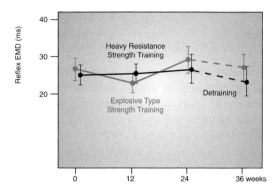

A. Influence of Training on Electromechanical Delay

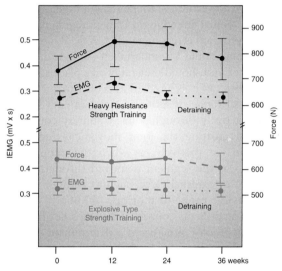

B. Influence of Training on IEMG

FIGURE 4-20 **(A)** Explosive strength training has been shown to decrease the electromechanical delay (EMD) in the muscle contraction after 12 weeks of training. The EMD increases again, however, if training continues to 24 weeks and drops off slightly with detraining. The influence of heavy resistance training on EMD is negligible. **(B)** The IEMG increases in the early weeks during heavy resistance training but not with explosive training. It is believed that some neural adaptation occurs in the early stages of specific types of resistance training, which facilitates an early increase in force production. (Reprinted with permission from Hakkinen, K., Komi, P. V. [1986]. Training-induced changes in neuromuscular performance under voluntary and reflex conditions. *European Journal of Applied Physiology,* 55:147-155.)

by impairment of both muscular and neural mechanisms (31). A decline in force output involves multiple neural mechanisms, including the influence of afferent feedback, descending inputs, and spinal circuitry influence on the output of the motor pool. The mechanism contributing to fatigue depends on the task, and variations in contraction intensity create differences in the balance of excitatory and inhibitory inputs to the pool.

In summary, the neural input to the muscle can be enhanced through training to increase the number of active motor units contributing, alter the pattern of firing, and increase the reflex potentiation of the system (52). Likewise, immobilization of the muscle can create the opposite response by lowering neural input to the muscle and decreasing reflex potentiation.

FLEXIBILITY EXERCISE

Flexibility is viewed by many to be an essential component of physical fitness and is seen as an important component of performance in sports such as gymnastics and dance. Flexibility can be increased with a stretching program.

No solid evidence demonstrates that increased flexibility is important for injury reduction or that it is a protection against injury (67). In fact, stretching before a sport performance may even have a negative influence by reducing force production and power output (46) and is only recommended for activities that require high levels of flexibility (37). Nevertheless, to maintain a functional range of motion, a regular stretching routine integrated into a conditioning program is recommended. A flexibility conditioning program should be undertaken daily or at least three times a week and should preferably take place after exercise.

Flexibility, as it is used in this section, is defined as the terminal range of motion of a segment. This can be obtained actively through some voluntary contraction of an agonist creating the joint movement (**active range of motion**) or passively, as when the agonist muscles are relaxed as the segment is moved through a range of motion by an external force, such as another person and object (**passive range of motion**) (60,68).

Many components contribute to one's flexibility or lack thereof. First, joint structure is a determinant of flexibility because it limits the range of motion in some joints and produces the termination or end point of the movement. This is true in a joint such as the elbow, in which the movement of extension is terminated by bony contact between the olecranon process and fossa on the back of the joint. A person who can hyperextend the forearm at the elbow is not one who is exceptionally flexible but is someone who has either a deep olecranon fossa or a small olecranon process. Bony restrictions to range of motion are present in a variety of joints in the body, but this type of restriction is not the main mechanism that limits or enhances joint flexibility.

Soft tissue around the joint is another factor contributing to flexibility. As a joint nears the ends of the range of

motion, the soft tissue of one segment is compressed by the soft tissue of the adjacent segment. This compression between adjacent tissue components eventually contributes to the termination of the range of motion. This means that obese individuals and individuals with large amounts of muscle mass or hypertrophy usually demonstrate lower levels of flexibility. An individual with hypertrophied muscles, however, can obtain good flexibility in a joint by applying a greater force at the end of the range of motion, which compresses the restrictive soft tissue to a greater degree. An obese individual who lacks strength is definitely limited in flexibility because of an inability to produce the force necessary to achieve the greater range of motion.

Ligaments restrict range of motion and flexibility by offering maximal support at the end of the range of motion. For example, the ligaments of the knee terminate the extension of the leg. An individual who can hyperextend the knees is commonly called double jointed but actually has slightly long ligaments that allow more than the usual joint motion.

The main factors that influence flexibility are the actual physical length of the antagonistic muscle or muscles; the viscoelastic characteristics of the muscles, ligaments, and other connective tissues; and the level of neurologic innervation in a muscle being stretched. Soft-tissue extensibility is related to the resistance of the tissue when it lengthens, and stretching overcomes the passive resistance in the tissue (57). All of these factors can be influenced by specific types of flexibility training.

When a muscle is stretched, neurologic mechanisms can influence the range of motion. In a rapid stretch, type Ia primary afferent sensory neuron in the muscle spindle initiates the stretch reflex, creating increased muscular activity through alpha motoneuron innervation. This response is proportional to the rate of stretch. Thus, the faster the stretch, the more the same muscle contracts. After the stretch is completed, type Ia sensory neurons decrease to a lower firing level, reducing the level of motoneuron activation or resistance in the muscle. A flexibility technique that enhances this response is **ballistic stretching**, in which the segments are bounced to achieve the terminal range of motion. This type of stretching is not recommended for the improvement of flexibility because of the stimulation of type Ia neurons and the increase in the resistance in the muscle. Slow rates of elongation permit greater stress relaxation and generate lower tissue forces. However, ballistic stretching is a component of many common movements, such as a preparatory windup in baseball or the end of the follow-through in a kick.

A better stretching technique for the improvement of range of motion is **static stretching**, in which the limb is moved slightly beyond the terminal position slowly and then maintained in that position for 10 to 30 seconds (9). Moving the limb slowly decreases the response of type Ia sensory neuron, and holding the position at the end reduces type Ia input, allowing minimal interference to the joint movement.

Mechanisms Restricting Joint Range of Motion	Example
(1) Joint structures—bony contact	Elbow extension
(2) Soft tissue—fat, muscle mass	Elbow flexion
(3) Ligaments	Knee extension
(4) Length of antagonists	Hip flexion
(5) Neurologic mechanisms	Ballistic throwing motion at the shoulder

Stretching methods

Method	Technique
Passive	Slow, sustained stretching with partner; 10–30 s
Dynamic	Slow, cyclical, elongating, static stretch, and shortening of a muscle; 30 s/ 6 repetitions of 5-s stretches
Static	Slow lengthening with a hold at end of range of motion; 10–30 s
Isometric	Static stretch against an immovable force or object such as a wall or floor; 10–30 s
Ballistic	Rapid lengthening of muscle using jerky or bouncing movements
Proprioceptive neuromuscular facilitation	Alternating passive muscle lengthening with partner after the antagonistic muscle contracts against resistance (contract–relax; hold–relax; slow-reversal hold–relax); 6-s contraction followed by 10- to 30-s assisted stretch

The primary restriction to the stretch of a muscle is found in the connective tissue and tendons in and around the muscle (23,50). This includes the fascia, epimysium, perimysium, endomysium, and tendons. The actual muscle fibers do not play a significant role in the elongation of a muscle through flexibility training. To understand how the connective tissue responds to a stretch, it is necessary to examine the stress–strain characteristics of the muscle unit.

When a stretch is first imposed, the muscle creates a linear response to the load through elongation in all parts of the muscle. This is the elastic phase of external stretch. If the external load is removed from the muscle during this phase of stretching, it will return to its original length within a few hours, and no residual or long-term increase in muscle length will remain. The stretching techniques working the elastic response of the muscle are common; they include short-duration, repetitive joint movements. These stretches, usually preceding an activity, produce some increase in muscle length for use in the practice or game but do not produce any long-term improvement in flexibility.

If a muscle is put in a terminal position and maintained in the position for an extended period, the tissue enters the plastic region of response to the load, elongating and undergoing plastic deformation (66). This plastic deformation is a long-term increase in the length of the muscle and carries over from day to day (41). A model describing the behavior of the elastic and plastic elements acting in a stretch is presented in Figure 4-21.

To create increases in length through plastic or long-term elongation, the muscle should be stretched while it is warm, and the stretch should be maintained for a long time under a low load (54,56). Thus, to gain long-term benefits from stretching, the stretch should occur after a practice or workout, and individual stretches should be held in the terminal joint positions for an extended period.

Cooling of a warm muscle enhances the permanent elongation of the tissues in that muscle. The joint positions should be held for at least 30 seconds and ideally up to one minute. In muscles that are inflexible and require extra attention, however, stretching should occur for longer times, six to 10 minutes (61). To avoid any significant tissue damage, stretching with pain should not take place.

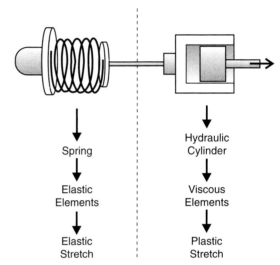

FIGURE 4-21 When a repetitive stretch of short duration is applied to the muscle, the connective tissue and muscle respond like a spring, with a short-term elongation of the tissue but a return to the original length after a short time. In a long-term sustained stretch, especially while the muscle is warm, the tissues behave more hydraulically, as a long-term deformation of the tissues takes place. (Reprinted with permission from Sapega, A. A., et al. [1981]. Biophysical factors in range of motion exercises. *The Physician and Sportsmedicine*, 9:57-64.)

Proprioceptive Neuromuscular Facilitation

Proprioceptive neuromuscular facilitation (PNF) is a technique used to stimulate relaxation of the stretched muscle so that the joint can be moved through a greater range of motion (36). This technique, used in rehabilitation settings, can also be put to good use with athletes and individuals who have limited flexibility in certain muscle groups, such as the hamstrings (56).

PNF incorporates various combination sequences using relaxation and contraction of the muscles being stretched. A simple PNF exercise is passive movement of an individual's limb into the terminal range of motion, have him or her contract back isometrically against the manual resistance applied by a partner, and then relax and move further into the stretch (contract–relax). Repeating this cycle can achieve a significant increase in the terminal range of motion (20). This procedure increases the range of motion because the input from type Ia afferent from the muscle spindle is reduced by the resetting of the spindle (27). PNF exercises are usually diagonal and in line with the fiber direction of the muscle. Stretching in an oblique pattern is closer to the actions found in common movements (39).

The process can be enhanced even more if a contraction of the agonist occurs at the end of the range of motion. This sets up an increase in the relaxation of the antagonist or the muscle being stretched. For example, passively move the foot into plantarflexion to stretch the dorsiflexors. Contract the dorsiflexors isometrically against resistance applied by a partner on the top of the foot. Move the foot farther into plantarflexion and then contract the plantarflexors. Both of these techniques produce the greatest increase in the range of motion. In a hold–relax PNF exercise, the GTO is stimulated so that reflex inhibition is produced, making the subsequent passive stretch easier. In a slow-reversal hold–relax PNF exercise, the GTO is also stimulated in the isometric "hold" phase. The antagonists generate a slow-reversal movement to elongate the target muscle, activating the muscle spindles and desensitizing the spindle during the follow-up passive elongation. Examples of PNF exercises for the muscles of the hip and shoulder joint are presented in Figure 4-22.

PLYOMETRIC EXERCISE

The purpose of **plyometric training** is to improve the velocity and power output in a performance. Plyometric training has been effective in increasing power output in athletes in sports such as volleyball, basketball, high jumping, long jumping, throwing, and sprinting. Plyometrics builds on the idea of specificity of training, whereby a muscle trained at higher velocities will function better at those velocities.

A plyometric exercise consists of rapidly stretching a muscle and immediately following with a contraction of the same muscle (5). The stretch–contract principle behind plyometric exercise was discussed in the previous chapter and shown to be an effective stimulator of force output. For example, a countermovement jump can make a 2- to 4-cm difference in the height of a vertical jump compared with a squat jump that does not include the stretch–contract sequence (11). Plyometric exercises improve power output in the muscle through facilitation of the neurologic input to the muscle and through increased muscle tension generated in the elastic components of the muscle.

The neurologic basis for plyometrics is the input from the stretch reflex via the type Ia sensory neuron. Rapid stretching of the muscle produces excitation of the alpha motoneurons contracting that muscle. This excitation is increased with the velocity of the stretch and is at its maximum level at the conclusion of a rapid stretch, after which the excitation levels decrease. Thus, if a muscle can be rapidly stretched and immediately contracted with no pause at the end of the stretch, this reflex loop produces maximum facilitation. If an individual pauses at the end of the stretch, this myoneural input is greatly diminished. The myoelectric enhancement of the muscle being

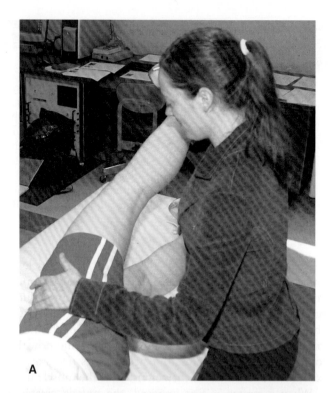

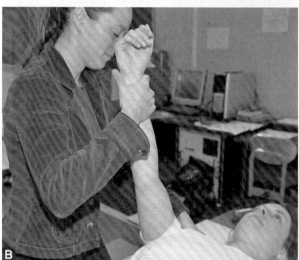

FIGURE 4-22 Examples of PNF exercises. **(A)** At the hip, the thigh moves through a diagonal pattern, with manual resistance applied at the foot and thigh. **(B)** In the shoulder, the arm moves into flexion, with manual resistance offered at the hand.

stretched accounts for approximately 25% to 30% of the increase of the force output in the plyometric stretch–contract sequence (40).

The factor accounting for most of the increases in output (70% to 75%) as a consequence of plyometric exercise is the restitution of elastic energy in the muscle (40). At the end of the stretch phase in a plyometric exercise, the muscle initiates an eccentric muscle action that increases the force and stiffness in the musculotendinous unit, resulting in storage of elastic energy. When a muscle is stretched, elastic potential energy is stored in the connective tissue and tendon and in the cross-bridges as they are rotated back with the stretch (2). With a vigorous short-term stretch, maximal recovery of the elastic potential energy is returned to the succeeding contraction of that same muscle. The net result of this short-range prestretch with a small time period between the stretch and the contraction is that larger forces can be produced for any given velocity, enhancing the power output of the system (12). Implementation of this technique suggests that a quick stretch through a limited range of motion should be followed immediately by a vigorous contraction of the same muscle.

Plyometric Examples

A plyometric exercise program includes a series of exercises imposing a rapid stretch followed by a vigorous contraction. Because the muscle is undergoing a vigorous eccentric contraction, attention should be given to the number of exercises and the load imposed through the eccentric contraction (16,45). It is suggested that plyometric exercises be done on yielding surfaces and not more than two days a week. Injury rates are higher in the use of plyometric training if these factors are not taken into account. Furthermore, plyometric training should be used very conservatively when the participants lack strength in the muscles being trained. A strength base should be developed first. It is suggested that an individual be able to squat 60% of body weight five times in five seconds before beginning plyometrics (15). This is done to see if eccentric and concentric muscle actions can be reversed quickly.

Lower extremity plyometric exercises include activities such as single-leg bounds, depth jumps from various heights, stair hopping, double-leg speed hops, split jumps, bench jumps, and quick countermovement jumping. The height from which the plyometric jump is performed is an important consideration. Heights can range from 0.25 to 1.5 m and should be based on the fitness level of the participant. A height is too high if a quick, vigorous rebound cannot be achieved shortly after landing.

Plyometric exercises can be done one to two times per week by a conditioned athlete. A sample plyometric workout may include three to five low-intensity exercises (10 to 20 repetitions), such as jumping in place or double-leg hops; three to four moderate-intensity exercises (five to 10 repetitions) including single-leg hops, double-leg hops over a hurdle, or bounding; and two to three high-intensity exercises, including depth jumping (five to 10 repetitions). In the beginning, the height of the box for depth jumping should be limited to avoid injury because the amount of force to be absorbed and controlled will increase with each height increase.

Upper extremity activities can best be implemented with surgical tubing or material that can be stretched. The muscle can be pulled into a stretch by the surgical tubing, after which the muscle can contract against the resistance offered by the tubing. For example, hold surgical tubing in a diagonal position across the back and simulate a

throwing motion with the right hand while holding the left hand in place. The arm will generate a movement against the surgical tube resistance and then be drawn back into a quick stretch by the tension generated in the tubing. These resistive tubes or straps can be purchased in varying resistances, offering compatibility with a variety of different strength levels.

Other forms of upper extremity plyometrics include catching a medicine ball and immediately throwing it. This puts a rapid stretch on the muscle in the catch, which is followed by a concentric contraction of the same muscles in the throw. Figure 4-23 shows specific plyometric exercises.

FIGURE 4-23 Plyometric exercises can be developed for any sport or region of the body by use of a stretch-contract cycle in exercise. Examples for the lower extremity include bounding **(A)** and depth **(B)** jumps. For the upper extremity, the use of surgical tubing **(C)** and medicine ball throws **(D)** are good exercises.

A form of combined training is called **complex training** in which strength is combined with speed work to enhance multiple components of the muscle. For example, a squat exercise could be paired with depth jumps. The squat will facilitate concentric performance via strength training, and the depth jump will facilitate eccentric performance and rate of force development through plyometrics (18).

Electromyography

The electrical activity in the muscle can be measured with EMG. This allows for the measurement of the change in the membrane potential as the action potentials are transmitted along the fiber. The study of muscle from this perspective can be valuable in providing information concerning the control of voluntary and reflexive movements. The study of muscle activity during a particular task can yield insight into which muscles are active and when the muscles initiate and cease their activity. In addition, the magnitude of the electrical response of the muscles during the task can be quantified. EMG has limitations, however, and these must be clearly understood if it is to be used correctly.

THE ELECTROMYOGRAM

The **electromyogram** is the profile of the electrical signal detected by an electrode on a muscle, that is, it is the measure of the action potential of the sarcolemma. The EMG signal is complex and is the composite of multiple action potentials of all active motor units superimposed on each other. Figure 4-24 illustrates the complexity of the signal. Note that the raw signal has both positive and negative components.

The amplitude of the EMG signal varies with a number of factors (discussed in the section characteristics of the electromyogram). Although the amplitude increases as the intensity of the muscular contraction increases, this does not mean that a linear relationship exists between EMG amplitude and muscle force. In fact, increases in EMG activity do not necessarily indicate an increase in muscle force (65). Only in isometric contractions are muscle electrical activity and muscle force closely associated (22).

RECORDING AN ELECTROMYOGRAPHIC SIGNAL

Electrodes

The EMG signal is recorded using an electrode. An electrode, which acts like an antenna, may be either indwelling or on the surface. The **indwelling electrode**, which may be either a needle or fine wire, is placed directly in the muscle. These electrodes are used for deep or small muscles. **Surface electrodes** are placed on the skin over a muscle and thus are mainly used for superficial muscles; they should not be used for deep muscles. The surface electrode is most often used in biomechanics, so most of the following discussion addresses surface electrodes.

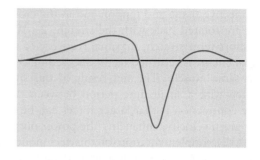
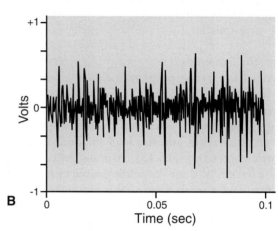

FIGURE 4-24   Electromyography (EMG) records. **(A)** A single action potential. **(B)** A single EMG record containing many action potentials. The duration of the single action potential is much shorter than that of the EMG signal in **(B)**.

Surface electrodes can be placed in either a monopolar or bipolar arrangement (Fig. 4-25). In a monopolar mode, one electrode is placed directly over the muscle in question, and a second electrode goes over an electrically neutral site, such as a bony prominence. Monopolar recordings are nonselective relative to bipolar recordings, and although they are used in certain situations, such as static contractions, they are poor in nonisometric movements. Bipolar electrodes are much more commonly used in biomechanics. In this case, two electrodes with a diameter of about 8 mm are placed over the muscle about 1.5 to 2.0 cm apart, and a third electrode is placed at an electrically neutral site. This arrangement uses a differential amplifier, which records the difference between the two recording electrodes. This differential technique removes any signal that is common to the inputs from the two recording electrodes.

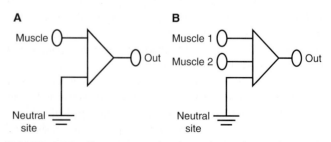

FIGURE 4-25   Electromyography electrodes can have either monopolar **(A)** or bipolar **(B)** configuration.

The correct placement of electrodes is critical to a good recording. It is obvious that the electrodes must be placed so that the action potentials from the underlying muscle can be recorded. Therefore, electrodes should not be placed over tendinous areas of the muscle or over the motor point, that is, the point at which the nerve enters the muscle. Because action potentials propagate in both directions along the muscle from the motor point, signals recorded above the motor point have the potential to be attenuated because of cancellation of signals from both electrodes. Various sources describe the standard locations for electrode placement (42).

Electrodes must also be oriented correctly, that is, parallel to the muscle fiber. The EMG signal is greatly affected when the electrodes are perpendicular rather than parallel to the fiber.

When using surface electrodes, the resistance of the skin must be taken into consideration. For an electrical signal to be detected, this resistance should be very low. To obtain a low skin resistance, the skin must be thoroughly prepared by shaving the site, abrading the skin, and cleaning the skin with alcohol. When this is done, the electrodes can be placed properly.

Amplification of the Signal

The EMG signal is relatively small, varying from 10 to 5 mV. It is therefore imperative that the signal be amplified, generally up to a level of 1 V. The usual type is the differential amplifier, which can amplify the EMG signal linearly without amplifying the noise or error in the signal. The noise in the EMG signal can be from sources other than the muscle, such as power line hum, machinery, or the amplifier itself. In addition, the amplifier must have high input impedance (resistance) and good frequency response and must be able to eliminate common noise from the signal.

FACTORS AFFECTING THE ELECTROMYOGRAM

Any of a number of factors, both physiologic and technical, can influence the interpretation of an EMG signal (35) (Fig. 4-26). It is essential to fully understand these factors before a knowledgeable interpretation of the EMG signal

can be made. Some, such as muscle fiber diameter, number of fibers, number of active motor units, muscle fiber conduction velocity, muscle fiber type and location, motor unit firing rate, muscle blood flow, distance from the skin surface to the muscle fiber, and tissue surrounding the muscle, may appear obvious because they all relate to the muscle itself. Others, including electrode–skin interface, signal conditioning, and electrode spacing, essentially relate to how the data are collected. These factors are amplified when measuring a dynamic contraction, and additional factors, such as a nonstationary EMG signal, shifting of the electrodes relative to the action potential origins, and changes in tissue conductivity characteristics, also become concerns (21).

ANALYZING THE SIGNAL

Except under special circumstances, it is difficult to record a single action potential. Thus, we are left with a signal made up of numerous action potentials from many motor units. Researchers are often interested in quantifying the EMG signal, and employ several procedures to do so (44). Most often, biomechanists first rectify the signal. **Rectification** involves taking the absolute value of the raw signal, that is, making all values in the signal positive. At this point, a **linear envelope** may be determined. This involves filtering out the high-frequency content of the signal to produce a smooth pattern that represents the volume of the activity. An alternative technique to the linear envelope is to integrate the rectified signal. When the signal is integrated, the EMG activity is summed over time so that the total accumulated activity can be determined over the chosen time period. Rectification, linear enveloping, and integration can be accomplished using electronic hardware, although they can also be done by computer. Figure 4-27 illustrates the results of these procedures.

In the procedures just described, the EMG signal was presented as a function of time or in the **time domain**. The EMG signal has also been analyzed in the **frequency domain** so that the frequency content of the signal can be determined. In this case, the power of the signal is plotted as a function of the frequency of the signal (Fig. 4-28). This profile is referred to as a frequency spectrum.

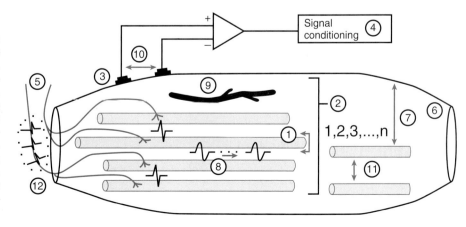

FIGURE 4-26 Some of the influences on the electromyographic signal. (1) Muscle fiber diameter. (2) Number of muscle fibers. (3) Electrode–skin interface. (4) Signal conditioning. (5) Number of active motor units. (6) Tissue. (7) Distance from skin surface to muscle fiber. (8) Muscle fiber conduction velocity. (9) Muscle blood flow. (10) Interelectrode spacing. (11) Fiber type and location. (12) Motor unit firing rate. (Adapted with permission from Kamen, G., Caldwell, G. E. [1996]. Physiology and interpretation of the electromyogram. *Journal of Clinical Neurophysiology*, 13:366-384.)

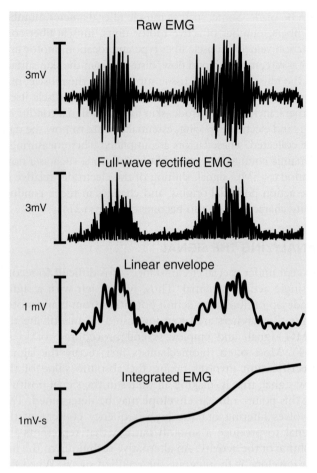

FIGURE 4-27 A raw electromyography (EMG) signal, full-wave rectified EMG signal, linear envelope, and integrated EMG signal.

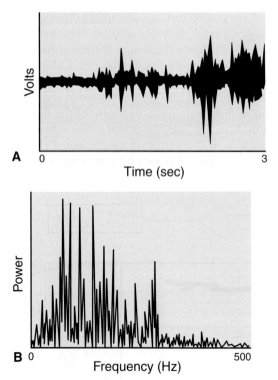

FIGURE 4-28 A raw EMG signal in the time domain (A) and frequency domain (B).

Electromechanical Delay

When a muscle is activated by a signal from the nervous system, the action potential must travel the length of the muscle before tension can be developed in the muscle. Thus, a temporal disassociation or delay is seen between the onset of the EMG signal and the onset of the development of force in the muscle. This is referred to as the EMD. Tension develops at some time after the signal is detected because chemical events need to occur before the contraction takes place. The EMD portion of the EMG signal represents the activation of the motor units and the shortening of the series elastic component of the muscle and can be affected by mechanical factors that change the rate of series elastic shortening. These factors include the initial muscle length and muscle loading. It has been reported that athletes who have a high percentage of fast-twitch muscle fibers exhibit a short EMD (34). The actual duration of this delay is not known, and values in the literature range from 50 to 200 ms. Figure 4-29 illustrates this concept.

APPLICATION OF ELECTROMYOGRAPHY

Muscle Force–Electromyography Relationship

In isometric conditions, the relationship between muscle force and EMG activity is relatively linear (32,43). That is, for a given increment in muscle force, there is a concomitant increase in EMG amplitude. These increases in EMG amplitude are probably produced by a combination of motor unit recruitment and an increase in motor unit firing rate. Many relationships, however, including both linear and curvilinear, between EMG and force have been suggested for different muscles (6) (Fig. 4-30).

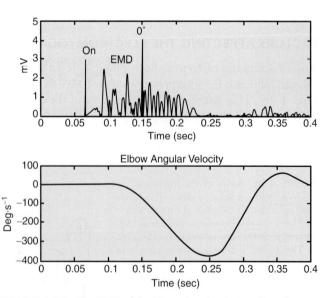

FIGURE 4-29 The EMD of the biceps brachii during elbow flexion. **Top.** The EMG activity of the biceps brachii. **Bottom.** The elbow angular velocity profile. (Reprinted with permission from Gabriel, D. A., Boucher, J. P. [1998]. Effects of repetitive dynamic contractions upon electromechanical delay. *European Journal of Physiology*, 79:37-40.)

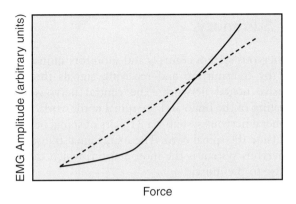

FIGURE 4-30 A linear relationship between EMG amplitude and external muscle force is frequently observed (*dotted line*). Numerous exceptions often result in a curvilinear relationship. (Adapted from a drawing by G. Kamen, University of Massachusetts at Amherst.)

In terms of concentric or eccentric contractions, descriptions of EMG–force relationships are controversial. The methodologies of the studies that report such relationships are often questioned because of the predominant use of isokinetic dynamometers that constrain the joint velocity. In the literature, only a few studies have attempted to relate EMG and force during unconstrained movements (28,55).

Muscle Fatigue

EMG has greatly enhanced the study of muscle fatigue. Fatigue can result from either peripheral (muscular) or central (neural) mechanisms, although EMG cannot directly determine the exact site of the fatigue. This section briefly discusses local muscle fatigue. When a motor unit fatigues, the frequency content and the amplitude of the EMG signal change (3). The signal in the frequency domain shifts toward the low end of the frequency scale, and the amplitude increases (Fig. 4-31). A number of physiologic explanations have been proposed for

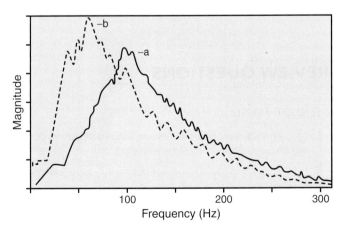

FIGURE 4-31 The frequency and amplitude changes during a sustained isometric contraction of the first dorsal interosseous muscle. (Adapted from Basmajian, J. V., DeLuca, C. J. [1985]. *Muscles Alive: Their Functions Revealed by Electromyography*, 5th ed. Baltimore, MD: Lippincott Williams & Wilkins, 205.)

these changes, including motor unit recruitment, motor unit synchronization, firing rate, and motor unit action potential rate. Basically, the force capacity in the muscles diminishes because of the impairment in both the neural and muscular mechanisms (31). The shifts in the frequency domain are recoverable after sufficient rest, with the amount of rest dependent on the type and duration of loading. The recovery in the frequency spectrum of the signal, however, does not appear to correspond to mechanical or physiologic recovery of the muscle (51).

Clinical Gait Analysis

In the clinical setting, a gait analysis often involves EMG to determine which muscle group is used at a particular phase of the gait cycle (55). Generally, the raw or rectified EMG signal is used to determine when the muscles are active and when they are inactive, that is, to determine the activation order. Onsets and offsets of muscles should not be evaluated from any type of signal other than the raw or rectified signal because further processing, such as filtering the data, distorts the onset or offset. More often, however, linear envelopes of EMG signals are used after appropriate scaling to determine amplitudes. Figure 4-32 illustrates typical EMG activity of lower extremity muscle groups during walking.

Ergonomics

EMG has been used in ergonomics for many applications. For example, studies have used EMG to investigate the effects of sitting posture, hand and arm movement, and armrests on the activity of neck and shoulder muscles of electronics assembly-line workers (62); to investigate the shoulder, back, and leg muscles during load carrying by varying load magnitude and the duration of load carrying (10); to study the erector spinae muscles of persons sitting in chairs with inclined seat pans (64); and to study postal letter sorting (17).

A particularly interesting use of EMG in ergonomics has been in the study of the low back in industry (48,49). This research has focused on proper lifting techniques and rehabilitation of workers who have low back problems. In addition, EMG has been used to study low back mechanics in exercise and during weight lifting.

LIMITATIONS OF ELECTROMYOGRAPHY

At best, EMG is a semiquantitative technique because it gives only indirect information regarding the strength of the contraction of muscles. Although many attempts have been made to quantify EMG, they have been largely unsuccessful. A second limitation is that it is difficult to obtain satisfactory recordings of dynamic EMG during movements such as walking and running. The EMG recording, therefore, is an indication of muscle activity only. One positive aspect of EMG recordings, however, is that they do reveal when a muscle is active and when it is not.

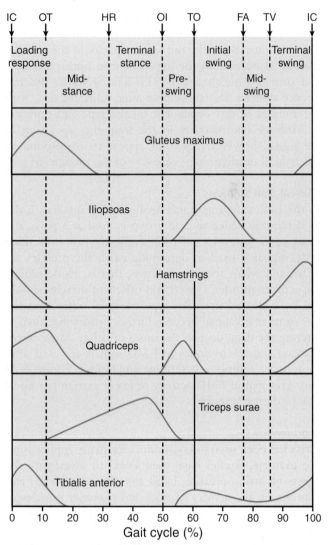

IC   OT   HR   OI   TO   FA   TV   IC

Loading response

Mid-stance

Terminal stance

Pre-swing

Initial swing

Mid-swing

Terminal swing

Gluteus maximus

Iliopsoas

Hamstrings

Quadriceps

Triceps surae

Tibialis anterior

Gait cycle (%)

FIGURE 4-32 Typical EMG activity of the major lower extremity muscle groups during a stride cycle of walking. FA, feet adjacent; HR, heel raise; IC, initial contact; OI, opposite initial contact; OT, opposite toe-off; TO, toe-off; TV, tibial vertical. (Reprinted with permission from Whittle, M. W. [1996]. *Gait Analysis,* 2nd ed. Oxford: Butterworth-Heinemann, 68.)

Electromyography and force

When a muscle contraction is required your nervous system sends a signal to the muscle that results in tension. EMG is the neural input to the muscle and force is the output. During dynamic muscular contractions the relationship between the input and output is altered by a number of factors. For instance, a given amount of neural input will produce more tension in a muscle that is at its optimal length than a shorter length. Discuss some other factors that might change the amount of force produced by muscle for a given amount of EMG. What implications do these factors have on the linearity (or one-to-one correspondence) of the relationship between muscle input and output?

 **Summary**

The nervous system controls and monitors human movement by transmitting and receiving signals through an extensive neural network. The central nervous system, consisting of the brain and the spinal cord, works with the peripheral nervous system via 31 pairs of spinal nerves that lie outside the spinal cord. The main signal transmitter of the nervous system is the motoneuron, which carries the impulse to the muscle.

The nerve impulse travels to the muscle as an action potential, and when it reaches the muscle, a similar action potential develops in the muscle, eventually initiating a muscle contraction. The actual tension in the muscle is determined by the number of motor units actively stimulated at one time.

Sensory neurons play an important role in the nervous system by providing feedback on the characteristics of the muscle or other tissues. When a sensory neuron brings information into the spinal cord and initiates a motor response, it is termed a *reflex*. The main sensory neurons for the musculoskeletal system are the proprioceptors. One proprioceptor, the muscle spindle, brings information into the spinal cord about any change in the muscle length or velocity of a muscle stretch. Another important proprioceptor is the GTO, which responds to tension in the muscle.

Flexibility, an important component of fitness, is influenced by a neurologic restriction to stretching that is produced by the proprioceptive input from the muscle spindle. Another area of training that uses the neurologic input from the sensory neurons is plyometrics. A plyometric exercise is one that involves a rapid stretch of a muscle that is immediately followed by a contraction of the same muscle.

EMG is a technique whereby the electrical activity of muscle can be recorded. From EMG, we can gain insight into which muscles are active and when the muscles initiate and cease their activity. A number of concepts must be understood, however, if one is to clearly interpret EMG signals.

REVIEW QUESTIONS

True or False

1. ____ When a motoneuron fires and sends a signal, all of the fibers in the motor unit will contract.

2. ____ The soleus muscle consists of primarily type II fibers.

3. ____ The sensory nerves enter the spinal cord on the anterior side.

4. ____ The amplitude of an action potential decreases the closer it gets to the muscle.

5. ____ Schwann cells cover the dendrites of an alpha motoneuron.

6. ____ Nerve impulse can reach velocities up to 100 m/s.

7. ____ Dendrites act as receivers for information transmitted by other neurons.

8. ____ A single action potential causes a muscle twitch.

9. ____ Slow-twitch fibers are recruited first.

10. ____ A muscle with a small innervation ratio has fine control over tension in the muscle.

11. ____ A nerve action potential becomes a muscle action potential at the node of Ranvier.

12. ____ All of the fibers within a muscle have the same fiber type.

13. ____ A motor pool is a group of neurons that innervate a single muscle.

14. ____ All of the motor units in a muscle are activated at the same time.

15. ____ The firing frequency of a motor unit does not change during a contraction.

16. ____ During a contraction of increasing strength, recruitment will take place before rate coding.

17. ____ Motor units start producing force at a stimulation frequency of 60/s.

18. ____ The muscle spindle detects both velocity of stretch and tension of the muscle fiber.

19. ____ The stretch reflex utilizes autogenic facilitation.

20. ____ The Golgi tendon organ senses muscle tension.

21. ____ Strength gain during the first four weeks of a strength training program is typically the result of an increased physiologic cross section.

22. ____ The most effective stretching technique uses ballistic actions.

23. ____ EMG measures the electrical activity in the heart.

24. ____ The EMG signal varies from 5 to 10 V.

25. ____ EMG activity is loosely related to muscle tension.

Multiple Choice

1. Sensory neurons ____.
   a. send signals to the spinal cord
   b. send signals to the muscle
   c. are within the spinal cord
   d. None of the above

2. A single neuron and all the muscle cells it innervates is ____.
   a. a peripheral nerve
   b. a spinal nerve
   c. a motor pool
   d. a motor unit

3. Muscle spindles ____.
   a. make monosynaptic connections with motoneurons
   b. are stimulated by extrafusal muscle contractions
   c. send information to the cerebral cortex
   d. Both b and c
   e. Both a and c
   f. All of the above

4. Projections on the cell body receive information called ____.
   a. dendrites
   b. ganglia
   c. Schwann cells
   d. None of the above

5. Two ways of controlling the amount of tension in a muscle are ____.
   a. recruitment and rate coding
   b. number of motor units and frequency of firing
   c. size principle and rate coding
   d. the type of muscle fiber and the number of motor units

6. A quick stretch to the muscle results in ____.
   a. the stretch reflex
   b. relaxation of the antagonists
   c. contraction of the agonists
   d. Both a and c
   e. All of the above

7. The all-or-none principle refers to ____.
   a. a muscle
   b. a muscle fiber
   c. a muscle fascicle
   d. a motor unit

8. The synaptic gap occurs ____.
   a. at a node of Ranvier
   b. in the soma
   c. at the neuromuscular junction
   d. at a collateral branch of the nerve

9. A single motor unit may innervate ____ fibers.
   a. type I and type II
   b. type IIa and type I
   c. Both a and b
   d. None of the above

10. The size principle refers to ____.
    a. the threshold for when motor units are recruited
    b. the size of the muscle cross section for which muscles are recruited
    c. the fact that larger muscles contain more sensory neurons
    d. None of the above

11. At rest, the electrical potential on the inside of the nerve has a value of ____.
    a. 60 mV
    b. −260 mV
    c. 290 mV
    d. −70 mV

12. The sequence of motor unit recruitment is usually ____.
    a. IIb, IIa, I
    b. IIb, I, IIa
    c. IIa, IIb, I
    d. I, IIa, IIb

13. Interneurons are ____.
    a. inhibitory
    b. excitatory
    c. both inhibitory and excitatory
    d. neither inhibitory nor excitatory

14. A flexor reflex ____.
    a. results from a quick stretch to the muscle
    b. results from high tension in a muscle
    c. results from pain such as heat
    d. None of the above

15. Fibers of the muscle spindle are called ____.
    a. intrafusal
    b. extrafusal
    c. supraspinal
    d. propriospinal

16. Type I primary afferents of the muscle spindle transmit a signal related to ____.
    a. force
    b. length
    c. velocity
    d. None of the above

17. EMG can give ____.
    a. information about muscle activation
    b. information about the level of force output of the muscle
    c. information about the velocity of stretch of the muscle
    d. Both a and b

18. When a motor unit fatigues, ____.
    a. the EMG amplitude increases and force decreases
    b. the EMG amplitude decreases and force increases
    c. the EMG amplitude remains the same and force decreases
    d. None of the above

19. The GTO is sensitive to ____.
    a. muscle length
    b. movement
    c. joint position
    d. muscle tension

20. The Ruffini endings respond to ____ and ____.
    a. joint position, velocity
    b. velocity, pressure
    c. pressure, pain
    d. pain, velocity

21. The main factors that influence flexibility are ____.
    a. length of the antagonistic muscles
    b. viscoelasticity
    c. neurologic innervation of the stretched muscle
    d. All of the above
    e. Both a and b

22. Electromechanical delay is ____.
    a. onset of EMG to onset of force
    b. onset of force to onset of movement
    c. onset of nerve action potential to onset of muscle action potential
    d. onset of nerve action potential to onset of movement

23. Which form of EMG can have both positive and negative values?
    a. raw
    b. full-wave rectified
    c. linear envelope
    d. integrated

24. The axon of motoneurons is fairly large, making it capable of transmitting impulses up to ____ m/s.
    a. 30
    b. 100
    c. 300
    d. 500

25. Which is not a method of stretching?
    a. Passive
    b. Static
    c. Ballistic
    d. Antagonistic
    e. Proprioceptive neuromuscular facilitation

References

1. Aagaard, P. (2003). Training induced changes in neural function. *Exercise and Sport Sciences Reviews*, 31:61-67.
2. Asmussen, E., Bonde-Peterson, F. (1974). Storage of elastic energy in skeletal muscles in man. *Acta Physiologica Scandinavia*, 91:385-392.
3. Basmajian, J. V., DeLuca, C. J. (1985). *Muscles Alive: Their Functions Revealed by Electromyography*, 5th ed. Baltimore, MD: Lippincott Williams & Wilkins.
4. Basmajian, J. V. (1978). *Muscles Alive: Their Functions Revealed by Electromyography*, 4th ed. Baltimore, MD: Lippincott Williams & Wilkins.
5. Bedi, J. F., et al. (1987). Increase in jumping height associated with maximal vertical depth jumps. *Research Quarterly for Exercise and Sport*, 58(1):11-15.
6. Bigland-Ritchie, B. (1980). EMG/force relations and fatigue of human voluntary contractions. *Exercise and Sport Sciences Reviews*, 8:75-117.
7. Bigland-Ritchie, B., et al. (1983). Changes in motor neuron firing rates during sustained maximal voluntary contractions. *Journal of Physiology*, 340:335-346.
8. Billeter, R., Hoppeler, H. (1992). Muscular basis of strength. In P. Komi (Ed.). *Strength and Power in Sport*. Boston, MA: Blackwell Scientific, 39-63.
9. Blanke, D. (1982). Flexibility training: Ballistic, static, or proprioceptive neuromuscular facilitation. *Archives of Physical Medicine Rehabilitation*, 63:261-263.
10. Bobet, J., Norman, R. W. (1982). Use of the average electromyogram in design evaluation investigation of a whole-body task. *Ergonomics*, 25:1155-1163.
11. Bobbert, M. F., Casius, L. J. R. (2005). Is the effect of a countermovement on jump height due to active state development? *Medicine & Science in Sports & Exercise*, 37:440-446.

12. Bosco, C., et al. (1982). Neuromuscular function and mechanical efficiency of human leg extensor muscles during jumping exercises. *Acta Physiologica Scandinavia*, 114:543-550.
13. Burke, R. E. (1981). Motor units: Anatomy, physiology, and functional organization. In J. M. Brookhart, V. B. Mountcastle (Eds.). *Handbook of Physiology: The Nervous System*. Bethesda, MD: American Physiological Society, 345-422.
14. Burke, R. E. (1986). The control of muscle force: Motor unit recruitment and firing patterns. In N. L. Jones, et al. (Eds.). *Human Muscle Power*. Champaign, IL: Human Kinetics, 97-109.
15. Chu, D. (1983). Plyometrics: The link between strength and speed. *National Strength and Conditioning Association Journal*, 5:20-21.
16. Chu, D., Plummer, L. (1985). The language of plyometrics. *National Strength and Conditioning Association Journal*, 6:30-31.
17. DeGroot, J. P. (1987). Electromyographic analysis of a postal sorting task. *Ergonomics*, 30:1079-1088.
18. Ebben, W., et al. (2000). Electromyographic and kinetic analysis of complex training variables. *Journal of Strength & Conditioning Research*, 14:451-456.
19. Enoka, R. (2005). Central modulation of motor unit activity. *Medicine & Science in Sports & Exercise*, 37:2111-2112.
20. Entyre, B. R., Abraham, L. D. (1986). Reflex changes during static stretching and two variations of proprioceptive neuromuscular facilitation techniques. *Electroencephalography & Clinical Neurophysiology*, 63:174-179.
21. Farina, D. (2006). Interpretation of the surface electromyogram in dynamic contractions. *Exercise and Sport Sciences Reviews*, 34:121-127.
22. Fuglevand, A. J., et al. (1993). Impairment of neuromuscular propagation during human fatiguing contractions at submaximal forces. *Journal of Physiology*, 460:549-572.
23. Garrett, W. E., et al. (1987). Biomechanical comparison of stimulated and nonstimulated skeletal muscle pulled to failure. *American Journal of Sports Medicine*, 15:448-454.
24. Gregory, J. E., et al. (2002). Effect of eccentric muscle contractions on Golgi tendon organ responses to passive and active tension in the cat. *Journal of Physiology*, 538:209-218.
25. Grimby, L. (1986). Single motor unit discharge during voluntary contraction and locomotion. In N. L. Jones, et al. (Eds.). *Human Muscle Power*. Champaign, IL: Human Kinetics, 111-129.
26. Hakkinen, K., Komi, P. V. (1986). Training-induced changes in neuromuscular performance under voluntary and reflex conditions. *European Journal of Applied Physiology*, 55:147-155.
27. Hardy, L., Jones, D. (1986). Dynamic flexibility and proprioceptive neuromuscular facilitation. *Research Quarterly for Exercise and Sport*, 51:625-635.
28. Hof, A. L., van den Berg, J. (1981). EMG to force processing: 1. An electrical analog of the Hill muscle model. *Journal of Biomechanics*, 14:747-758.
29. Hultborn, H., et al. (1971). Recurrent inhibition of interneurons monosynaptically activated from group Ia afferents. *Journal of Physiology*, 215:613-636.
30. Hultborn, H. (1972). Convergence on interneurones in the reciprocal Ia inhibitory pathway to motoneurones. *Acta Physiologica Scandinavica*, 85(s375), 1-42.
31. Hunter, S. K., et al. (2004). Muscle fatigue and the mechanisms of task failure. *Exercise and Sport Sciences Reviews*, 32:44-49.
32. Jacobs, R., van Ingen Schenau, G. J. (1992). Control of an external force in leg extensions in humans. *Journal of Physiology*, 457:611-626.
33. Jansen, J. K., Rudford, T. (1964). On the silent period and Golgi tendon organs of the soleus muscle of the cat. *Acta Physiologica Scandinavica*, 62:364-379.
34. Kamen, G., et al. (1981). Fractionated reaction time in power trained athletes under conditions of fatiguing isometric exercise. *Journal of Motor Behavior*, 13:117-129.
35. Kamen, G., Caldwell, G. E. (1996). Physiology and interpretation of the electromyogram. *Journal of Clinical Neurophysiology*, 13:366-384.
36. Knot, M., Voss, D. E. (1968). *Proprioceptive Neuromuscular Facilitation: Patterns and Techniques*, 2nd ed. New York: Harper & Row.
37. Knudson, D. V., Magnusson, P., McHugh, M. (2000). Current issues in flexibility fitness. *President's Council of Physical Fitness and Sport Research Digest*, 3:1-8.
38. Koceja, D. M., Kamen, G. (1992). Segmental reflex organization in endurance-trained athletes and untrained subjects. *Medicine & Science in Sports & Exercise*, 24(2):235-241.
39. Kofotolis, N., Kellis, E. (2006). Facilitation programs on muscle endurance, flexibility, and functional performance in women with chronic low back pain. *Physical Therapy*, 86:1001-1012.
40. Komi, P. V. (1986). The stretch-shortening cycle and human power output. In N. L. Jones, et al. (Eds.). *Human Muscle Power*. Champaign, IL: Human Kinetics, 27-42.
41. Kottke, F. J., et al. (1966). The rationale for prolonged stretching for correction of shortening of connective tissue. *Archives of Physical Medicine and Rehabilitation*, 47:345-352.
42. LeVeau, B., Andersson, G. (1992). Output forms: Data analysis and applications. In G. L. Soderberg (Ed.). *Selected Topics in Surface Electromyography for the Use in the Occupational Setting*. Washington, DC: National Institute for Occupational Safety and Health, U.S. Public Health Service, 70-102.
43. Lippold, O. C. J. (1952). The relation between integrated action potential in human muscle and its isometric tension. *Journal of Physiology*, 117:492-499.
44. Loeb, G. E., Gans, C. (1986). *Electromyography for Experimentalists*. Chicago, IL: University of Chicago Press.
45. Lundin, P. (1985). A review of plyometric training. *National Strength and Conditioning Association Journal*, 7(3):69-74.
46. Marek, S. M., et al. (2005). Acute strength effects of static and proprioceptive neuromuscular facilitation stretching on muscle strength and power output. *Journal of Athletic Training*, 40:94-103.
47. McCrea, D. A. (1992). Can sense be made of spinal interneuron circuits? *Behavioral and Brain Sciences*, 15:633-643.
48. McGill, S. M. (1991). Electromyographic activity of the abdominal and low back musculature during the generation of isometric and dynamic axial trunk torque: Implications for lumbar mechanics. *Journal of Orthopaedic Research*, 9:91-103.
49. McGill, S. M., Sharrat, M. T. (1990). The relationship between intra-abdominal pressure and trunk EMG. *Clinical Biomechanics*, 5:59-67.
50. McHugh, M. P., et al. (1992). Viscoelastic stress relaxation in human skeletal muscle. *Medicine & Science in Sports & Exercise*, 24(12):1375-1382.

51. Mills, K. R. (1982). Power spectral analysis of electromyogram and compound muscle action potential during muscle fatigue and recovery. *Journal of Physiology*, 326:401-409.

52. Moritani, T. (1993). Neuromuscular adaptations during the acquisition of muscle strength, power, and motor tasks. *Journal of Biomechanics*, 26:95-107.

53. Moritani, T., DeVries, H. A. (1979). Neural factors versus hypertrophy in the time course of muscle strength gain. *American Journal of Physical Medicine*, 58(3):115-130.

54. Newton, R. A. (1982). Joint receptor contributions to reflexive and kinesthetic responses. *Physical Therapy*, 62(1):23-29.

55. Olney, S. J., Winter, D. A. (1985). Predictions of knee and ankle moments of force in walking from EMG and kinematic data. *Journal of Biomechanics*, 18:9-20.

56. Osternig, L. R., et al. (1990). Differential responses to proprioceptive neuromuscular facilitation (PNF) stretch techniques. *Medicine & Science in Sports & Exercise*, 22:106-111.

57. Reid, D., McNair, P. J. (2004). Passive force, angle and stiffness changes after stretching of hamstring muscles. *Medicine & Science in Sports & Exercise*, 36:1944-1948.

58. Sale, D. G. (1987). Influence of exercise and training on motor unit activation. In K. B. Pandolf (Ed.). *Exercise and Sport Sciences Reviews*, 16:95-151.

59. Sale, D. G. (1988). Neural adaptation to resistance training. *Medicine & Science in Sport & Exercise*, 20:S135-S145.

60. Sandy, S. P., et al. (1982). Flexibility training: Ballistic, static, or proprioceptive neuromuscular facilitation? *Archives of Physical Medicine and Rehabilitation*, 6:132-138.

61. Sapega, A. A., et al. (1981). Biophysical factors in range of motion exercises. *The Physician and Sportsmedicine*, 9:57.

62. Schuldt, K., et al. (1986). Effects of sitting work posture on static neck and shoulder muscle activity. *Ergonomics*, 29:1525-1537.

63. Smith, J. L. (1976). Fusimotor loop properties and involvement during voluntary movement. In J. Keogh, R. S. Hutton (Eds.). *Exercise and Sport Sciences Reviews*, 4:297-333.

64. Soderberg, G. L., et al. (1986). An EMG analysis of posterior trunk musculature during flat and anteriorly inclined sitting. *Human Factors*, 28:483-491.

65. Solomonow, M., et al. (1990). Electromyogram power spectra frequencies associated with motor unit recruitment strategies. *Journal of Applied Physiology*, 68:1177-1185.

66. Taylor, D. C., et al. (1990). Viscoelastic properties of muscle-tendon units: The biomechanical effects of stretching. *American Journal of Sports Medicine*, 18:300-309.

67. Thacker, S. B et al (2004). The impact of stretching on sports injury risk: A systematic review of the literature. *Medicine & Science in Sports & Exercise*, 36:371-378.

68. Wallin, D. V., et al. (1985). Improvement of muscle flexibility: A comparison between two techniques. *American Journal of Sports Medicine*, 13:263-268.

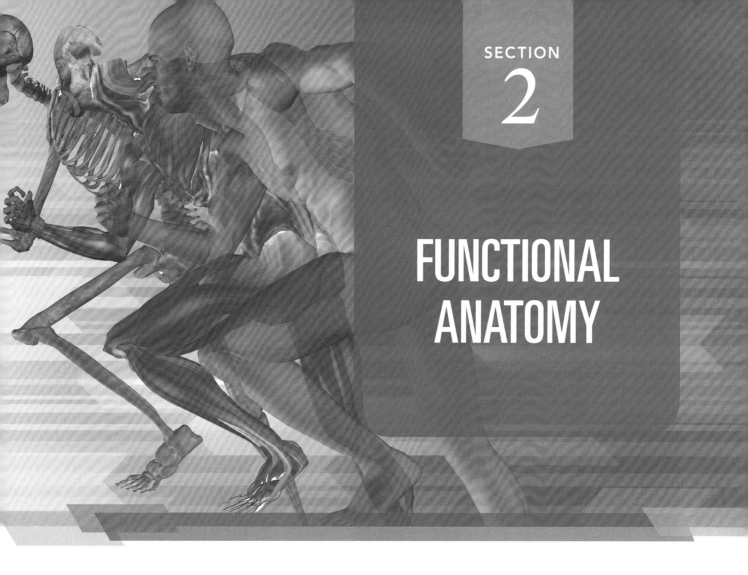

FUNCTIONAL ANATOMY

CHAPTER 5
Functional Anatomy of the Upper Extremity

CHAPTER 6
Functional Anatomy of the Lower Extremity

CHAPTER 7
Functional Anatomy of the Trunk

FUNCTIONAL ANATOMY OF THE UPPER EXTREMITY

OBJECTIVES

After reading this chapter, the student will be able to:

1. Describe the structure, support, and movements of the joints of the shoulder girdle, shoulder joint, elbow, wrist, and hand.

2. Describe the scapulohumeral rhythm in an arm movement.

3. Identify the muscular actions contributing to shoulder girdle, elbow, wrist, and hand movements.

4. Explain the differences in muscle strength across the different arm movements.

5. Identify common injuries to the shoulder, elbow, wrist, and hand.

6. Develop a set of strength and flexibility exercises for the upper extremity.

7. Describe some common wrist and hand positions used in precision or power.

8. Identify the upper extremity muscular contributions to activities of daily living (e.g., rising from a chair, throwing, swimming, and swinging a golf club).

OUTLINE

The Shoulder Complex
    Anatomical and Functional Characteristics of the Joints of the Shoulder
    Combined Movement Characteristics of the Shoulder Complex
    Muscular Actions
    Strength of the Shoulder Muscles
    Conditioning
    Injury Potential of the Shoulder Complex

The Elbow and Radioulnar Joints
    Anatomical and Functional Characteristics of the Joints of the Elbow
    Muscular Actions
    Strength of the Forearm Muscles
    Conditioning
    Injury Potential of the Forearm

The Wrist and Fingers
    Anatomical and Functional Characteristics of the Joints of the Wrist and Hand
    Combined Movements of the Wrist and Hand
    Muscular Actions
    Strength of the Hand and Fingers
    Conditioning
    Injury Potential of the Hand and Fingers

Contribution of Upper Extremity Musculature to Sports Skills or Movements
    Overhand Throwing
    The Golf Swing

External Forces and Moments Acting at Joints in the Upper Extremity

Summary

Review Questions

The upper extremity is interesting from a functional anatomy perspective because of the interplay among the various joints and segments necessary for smooth, efficient movement. Movements of the hand are made more effective through proper hand positioning by the elbow, shoulder joint, and shoulder girdle. Also, forearm movements occur in concert with both hand and shoulder movements (47). These movements would not be half as effective if the movements occurred in isolation. Because of the heavy use of our arms and hands, the shoulder needs a high degree of structural protection and a high degree of functional control (4).

 ## The Shoulder Complex

The shoulder complex has many articulations, each contributing to the movement of the arm through coordinated joint actions. Movement at the shoulder joint involves a complex integration of static and dynamic stabilizers. There must be free motion and coordinated actions between all four joints: the **scapulothoracic**, **sternoclavicular**, **acromioclavicular (AC)**, and **glenohumeral joints** (63,75). Although it is possible to create a small amount of movement at any one of these articulations in isolation, movement usually is generated at all of these joints concomitantly as the arm is raised or lowered or if any other significant arm action is produced (88).

ANATOMICAL AND FUNCTIONAL CHARACTERISTICS OF THE JOINTS OF THE SHOULDER

Sternoclavicular Joint

The only point of skeletal attachment of the upper extremity to the trunk occurs at the sternoclavicular joint. At this joint, the **clavicle** is joined to the manubrium of the sternum. The clavicle serves four roles by serving as a site of muscular attachment, providing a barrier to protect underlying structures, acting as a strut to stabilize the shoulder and prevent medial displacement when the muscles contract, and preventing an inferior migration of the shoulder girdle (75). The large end of clavicle articulating with a small surface on the sternum at the sternoclavicular joint requires significant stability from the ligaments (75). A close view of the clavicle and the sternoclavicular joint is shown in Figure 5-1. This gliding synovial joint has a fibrocartilaginous disk (89). The joint is reinforced by three ligaments: the interclavicular, costoclavicular, and sternoclavicular ligaments, of which the costoclavicular ligament is the main support for the joint (73) (Fig. 5-2). The joint is also reinforced and supported by muscles, such as the short, powerful subclavius. Additionally, a strong joint capsule contributes to making the joint resilient to **dislocation** or disruption.

Movements of the clavicle at the sternoclavicular joint occur in three directions, giving it 3 degrees of freedom.

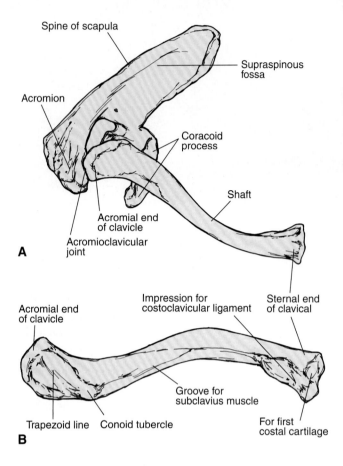

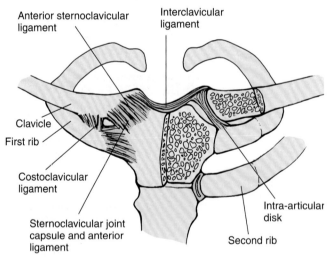

FIGURE 5-1 The clavicle articulates with the acromion process on the scapula to form the acromioclavicular joint **(A)**. An S-shaped bone **(B)**, the clavicle also articulates with the sternum to form the sternoclavicular joint **(C)**.

The clavicle can move superiorly and inferiorly in movements referred to as **elevation** and **depression**, respectively. These movements take place between the clavicle and the meniscus in the sternoclavicular joint and have a range of motion of approximately 30° to 40° (75,89).

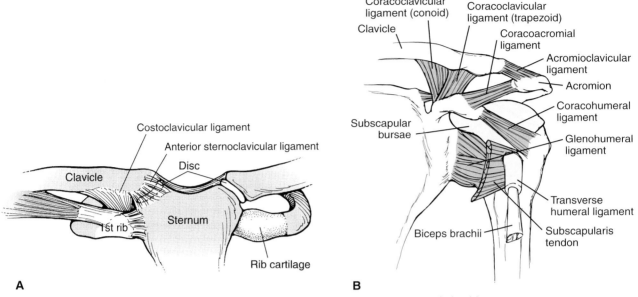

FIGURE 5-2 Ligaments of the shoulder region. Anterior aspects of the sternum **(A)** and shoulder **(B)** are shown.

The clavicle can also move anteriorly and posteriorly via movements in the transverse plane termed **protraction** and **retraction**, respectively. These movements occur between the sternum and the meniscus in the joint through a range of motion of approximately 30° to 35° in each direction (75). Finally, the clavicle can rotate anteriorly and posteriorly along its long axis through approximately 40° to 50° (75,89).

Acromioclavicular Joint

The clavicle is connected to the scapula at its distal end via the AC joint (Fig. 5-1). This is a small, gliding synovial joint that is the size of 9 mm × 19 mm in adults (75) and it frequently has a fibrocartilaginous disk similar to the sternoclavicular joint (73). At this joint, most of the movements of the scapula on the clavicle occur, and the joint handles large contact stresses as a result of high axial loads that are transmitted through the joint (75).

The AC joint lies over the top of the humeral head and can serve as a bony restriction to arm movements above the head. The joint is reinforced with a dense capsule and a set of **AC** ligaments lying above and below the joint (Fig. 5-2). The AC ligaments primarily support the joint in low load and small movement situations. Close to the AC joint is the important **coracoclavicular** ligament, which assists scapular movements by serving as an axis of rotation and by providing substantial support in movements requiring more range of motion and displacement. The shoulder girdle is suspended from the clavicle by this ligament and serves as the primary restraint to vertical displacement (75).

Another ligament in the region that does not cross a joint is the **coracoacromial** ligament. This ligament

protects underlying structures in the shoulder and can limit excessive superior movement of the humeral head.

Scapulothoracic Joint

The scapula interfaces with the thorax via the scapulothoracic joint. This is not a typical articulation, connecting bone to bone. Rather, it is a physiologic joint (89) containing neurovascular, muscular, and bursal structures that allows for a smooth motion of the scapula on the thorax (75). The scapula actually rests on two muscles, the serratus anterior and the subscapularis, both connected to the scapula and moving across each other as the scapula moves. Underneath these two muscles lies the thorax.

Seventeen muscles attach to or originate on the scapula (75). As shown in Figure 5-3, the scapula is a large, flat, triangular bone with five thick ridges (glenoid, spine, medial and lateral border, and coracoid process) and two thin, hard, and laminated surfaces (infraspinous and supraspinous fossas) (27). It serves two major functions relative to shoulder motion. First, the scapulothoracic articulation offers another joint so that the total rotation of the humerus with respect to the thorax increases (27). This increases the range of motion beyond the 120° generated solely in the glenohumeral joint. As the arm elevates at the glenohumeral joint, there is 1° of scapulothoracic elevation for every 2° of glenohumeral elevation (75).

The second function of the scapula is facilitating a large lever for the muscles attaching to the scapula. Because of its size and shape, the scapula provides large movements around the AC and the sternoclavicular joints. Small muscles in the region can provide a sufficient amount of torque to be effective at the shoulder joint (27).

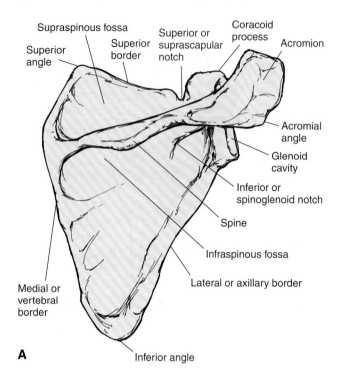

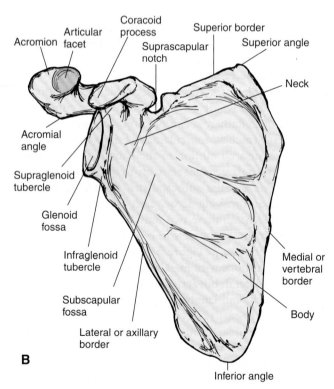

FIGURE 5-3 The scapula is a flat bone that serves as a site of muscular attachment for many muscles. The dorsal **(A)** and ventral **(B)** surfaces of the scapula on the right side are shown.

The scapula moves across the thorax as a consequence of actions at the AC and the sternoclavicular joints, giving a total range of motion for the scapulothoracic articulation of approximately 60° of motion for 180° of arm abduction or flexion. Approximately 65% of this range of motion occurs at the sternoclavicular joint, and 35%

occurs as a result of AC joint motion (89). The clavicle acts as a crank for the scapula, elevating and rotating to elevate the scapula.

The movement of the scapula can occur in three directions, as shown in Figure 5-4. The scapula can move anteriorly and posteriorly about a vertical axis; these motions are known as **protraction** or **abduction** and **retraction** or **adduction**, respectively. Protraction and retraction occur as the acromion process moves on the meniscus in the joint and as the scapula rotates about the medial coracoclavicular ligament. They can be anywhere from 30° to 50° of protraction and retraction of the scapula (73).

The second scapular movement occurs when the base of the scapula swings laterally and medially in the frontal plane. These actions are termed *upward* and *downward rotation*. This movement occurs as the clavicle moves on the meniscus in the joint and as the scapula rotates about the trapezoid portion of the lateral coracoclavicular ligament. This movement can occur through a range of motion of approximately 60° (89).

The third and final movement potential, or degree of freedom, is the scapular movement up and down, termed *elevation* and *depression*. This movement occurs at the AC joint and is not assisted by rotations about the coracoclavicular

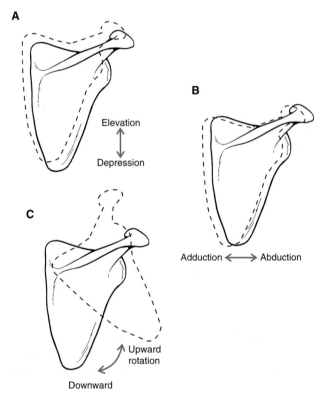

FIGURE 5-4 Scapular movements take place in three directions. **(A)** Elevation and depression of the scapula occur with a shoulder shrug or when the arm raises. **(B)** Abduction (protraction) and adduction (retraction) occur when the scapulae are drawn away from or toward the vertebrae, respectively, or when the arm is brought in front or behind the body, respectively. **(C)** The scapula also rotates upward and downward as the arm raises and lowers, respectively.

ligament. The range of motion at the AC joint for elevation and depression is approximately 30° (73,89).

The scapular movements also depend on the movement and position of the clavicle. The movements at the sternoclavicular joint are opposite to the movements at the AC joint for elevation, depression, protraction, and retraction. For example, as elevation occurs at the AC joint, depression occurs at the sternoclavicular joint and vice versa. This is not true for rotation because the clavicle rotates in the same direction along its length. The clavicle does rotate in different directions to accommodate the movements of the scapula: anteriorly with protraction and elevation and posteriorly with retraction and depression.

Glenohumeral Joint

The final articulation in the shoulder complex is the shoulder joint, or the glenohumeral joint, illustrated in Figure 5-5. Motions at the shoulder joint are represented by the movements of the arm. This is a synovial ball-and-socket joint that offers the greatest range of motion and movement potential of any joint in the body.

The joint contains a small, shallow socket called the **glenoid fossa**. This socket is only one-quarter the size of the humeral head that must fit into it. One of the reasons the shoulder joint is suited for extreme mobility is because of the size difference between the humeral head and the small glenoid fossa on the scapula (4). At any given time, only 25% to 30% of the humeral head is in contact with the glenoid fossa, but this does not necessarily lead to excessive movement because in the normal shoulder, the head of the humerus is constrained to within 1 to 2 mm of the center of the glenoid cavity by muscles (75).

Shoulder Joint Stability

Because there is minimal contact between the glenoid fossa and the head of the humerus, the shoulder joint largely depends on the ligamentous and muscular structures for stability. Stability is provided by both static and dynamic components, which provide restraint and guide and maintain the head of the humerus in the glenoid fossa (4,75).

The passive, static stabilizers include the articular surface, glenoid labrum, joint capsule, and ligaments (15,75). The articular surface of the glenoid fossa is slightly flattened and has thicker articular cartilage at the periphery, creating a surface for interface with the humeral head. The joint is also fully sealed, which provides suction and resists a dislocating force at low loads (75).

The joint cavity is deepened by a rim of fibrocartilage referred to as the **glenoid labrum**. This structure receives supplementary reinforcement from the surrounding ligaments and tendons. The labrum varies from individual to individual and is even absent in some cases (68). The glenoid labrum increases the contact area to 75% and deepens the concavity of the joint by 5 to 9 mm (75).

The joint capsule has approximately twice the volume of the humeral head, allowing the arm to be raised through a considerable range of motion (29). The capsule tightens in various extreme positions and is loose in the mid-range of motion (75). For example, the inferior capsule tightens in extreme abduction and external rotation seen in throwing (32). Likewise, the anterosuperior capsule works with the muscles to limit inferior and posterior translation of the humeral head and the posterior capsule limits posterior humeral translation when the arm is flexed and internally rotated (15).

The final set of passive stabilizers consists of the ligaments (Fig. 5-2). The **coracohumeral** ligament is taut when the arm is adducted, and it constrains the humeral head on the glenoid in this position (75) by restraining inferior translation. It also prevents posterior translation of the humerus during arm movements and supports the weight of the arm. The three **glenohumeral** ligaments reinforce the capsule, prevent anterior displacement of the humeral head, and tighten up when the shoulder externally rotates.

Dynamic support of the shoulder joint occurs primarily in the mid-range of motion and is provided by the muscles as they contract in a coordinated pattern to compress the humeral head in the glenoid cavity (15). The posterior rotator cuff muscles provide significant posterior stability, the subscapularis muscle provides anterior stability, the long head of the biceps brachii prevents anterior and superior humeral head translation, and the deltoid and the other scapulothoracic muscles position the scapula to provide maximum glenohumeral stability (15). When all of the rotator cuff muscles contract, the humeral head is compressed into the joint, and with an asymmetric contraction of the rotator cuff, the humeral head is steered to the correct position (75). This muscle group also rotates and depresses the humeral head during arm elevation to keep the humeral head in position. These muscles are examined more closely in a later section.

On the anterior side of the joint, support is provided by the capsule, the glenoid labrum, the glenohumeral ligaments, three reinforcements in the capsule, the coracohumeral ligament, fibers of the subscapularis, and the pectoralis major (78). These muscles blend into the joint capsule (29). Both the coracohumeral and the middle glenohumeral ligament support and hold up the relaxed arm. They also offer functional support through abduction, external rotation, and extension (43,73). Posteriorly, the joint is reinforced by the capsule, the glenoid labrum, and fibers from the teres minor and infraspinatus, which also blend into the capsule.

The superior aspect of the shoulder joint is often termed the *impingement area*. The glenoid labrum, the coracohumeral ligament, and the muscles support the superior portion of the shoulder joint, and the supraspinatus and the long head of the biceps brachii reinforce the capsule. Above the supraspinatus muscle lie the **subacromial bursae** and the coracoacromial ligament. These form an arch underneath the AC joint. This area and a typical impingement position are presented in Figure 5-6.

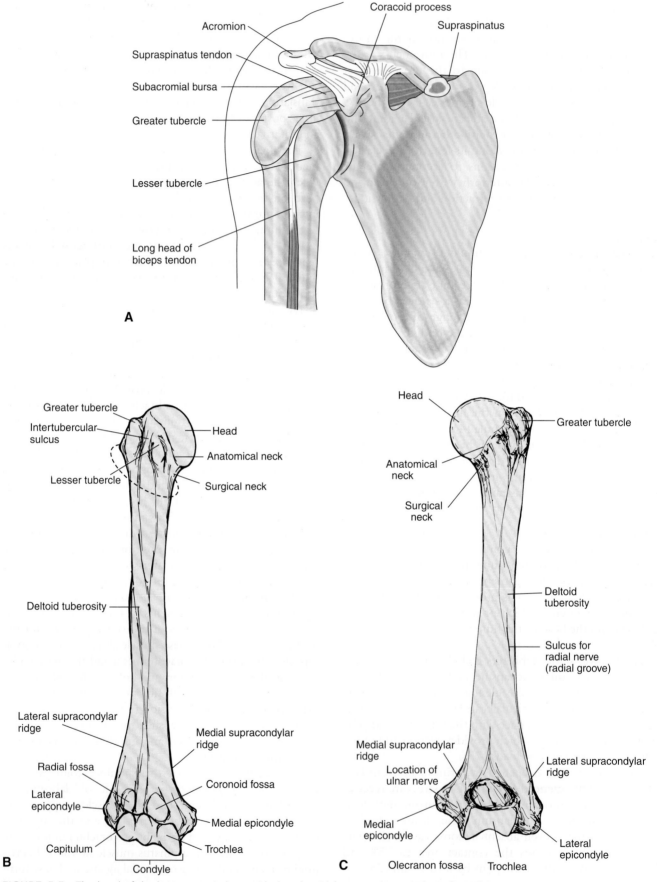

FIGURE 5-5 The head of the humerus articulates with the glenoid fossa on the scapula to form the glenohumeral joint. The landmarks of the shoulder complex **(A)** and the anterior **(B)** and posterior **(C)** surfaces of the humerus are shown.

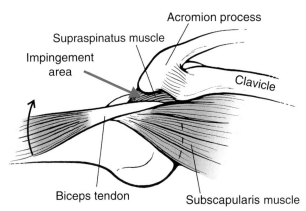

FIGURE 5-6 The impingement area of the shoulder contains structures that can be damaged with repeated overuse. The actual impingement occurs in the abducted position with the arm rotated.

A **bursa** is a fluid-filled sac found at strategic sites around the synovial joints that reduces the friction in the joint. The supraspinatus muscle and the bursae in this area are compressed as the arm rises above the head and can be irritated if the compression is of sufficient magnitude or duration. The inferior portion of the shoulder joint is minimally reinforced by the capsule and the long head of the triceps brachii.

Movement Characteristics

The range of motion at the shoulder joint is considerable for the aforementioned structural reasons (Fig. 5-7). The arm can move through approximately 165° to 180° of flexion to approximately 30° to 60° of hyperextension in the sagittal plane (11,89). The amount of flexion can be limited if the shoulder joint is also externally rotated. With the joint in maximal external rotation, the arm can be flexed through only 30° (11). Also, during passive flexion and extension, there is accompanying anterior and posterior translation, respectively, of the head of the humerus on the glenoid (30).

The arm can also abduct through 150° to 180°. The abduction movement can be limited by the amount of internal rotation occurring simultaneously with abduction. If the joint is maximally rotated internally, the arm can produce only about 60° of abduction (11), but a

certain amount of rotation is needed to reach 180°. As the arm adducts down to the anatomical or neutral position, it can continue past the neutral position for approximately 75° of hyperadduction across the body.

The arm can rotate both internally and externally 60° to 90° for a total of 120° to 180° of rotation (29). Rotation is limited by abduction of the arm. In an anatomical position, the arm can rotate through the full 180°, but in 90° of abduction, the arm can rotate only through 90° (11). Finally, the arm can move across the body in an elevated position for 135° of **horizontal flexion** or adduction and 45° of **horizontal extension** or abduction (89).

COMBINED MOVEMENT CHARACTERISTICS OF THE SHOULDER COMPLEX

The movement potential of each joint was examined in the previous section. This section examines the movement of the shoulder complex as a whole, sometimes referred to as **scapulohumeral rhythm**.

As stated earlier, the four joints of the shoulder complex must work together in a coordinated action to create arm movements. Any time the arm is raised in flexion or abduction, accompanying scapular and clavicular movements take place. The scapula must rotate upward to allow full flexion and abduction at the shoulder joint, and the clavicle must elevate and rotate upward to allow the scapular motion. A posterior view of the relationship between the arm and scapular movements is shown in Figure 5-8.

In the first 30° of abduction or the first 45° to 60° of flexion, the scapula moves either toward the vertebral column or away from the vertebral column to seek a position of stability on the thorax (73). After stabilization has been achieved, the scapula moves laterally, anteriorly, and superiorly in the movements described as upward rotation, protraction or abduction, and elevation. The clavicle also rotates posteriorly, elevates, and protracts as the arm moves through flexion or abduction (20).

In the early stages of abduction or flexion, the movements are primarily at the glenohumeral joint except for the stabilizing motions of the scapula. Past 30° of abduction or 45° to 60° of flexion, the ratio of glenohumeral to scapular movements becomes 5:4. That is, there is 5° of humeral movement for every 4° of scapular movement

Necessary Range of Motion at the Shoulder and Elbow

Activity	Shoulder Range of Motion	Elbow Range of Motion
Combing hair	20° to 100° of elevation with 37.7° of rotation	115° of flexion
Eating with a spoon	36° of elevation	116° of flexion with 33° of pronation
Reading	57.5° of elevation with 5° of rotation	20° of flexion with 102° of pronation

Source: Magermans, D. J., et al. (2005). Requirements for upper extremity motions during activities of daily living. *Clinical Biomechanics*, 20:591-599.

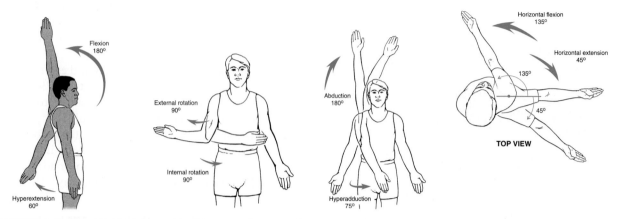

FIGURE 5-7 The shoulder has considerable range of motion. The arm can move through 180° of flexion or abduction, 60° of hyperextension, 75° of hyperadduction, 90° of internal and external rotation, 135° of horizontal flexion, and 45° of horizontal extension.

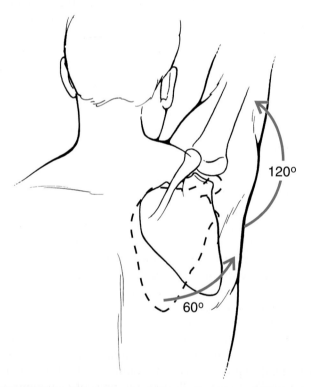

FIGURE 5-8 The movement of the arm is accompanied by movements of the shoulder girdle. The working relationship between the two is known as the scapulohumeral rhythm. The arm can move through only 30° of abduction and 45° to 60° of flexion with minimal scapular movements. Past these points, the scapula movements occur concomitantly with the arm movements. For 180° of flexion or abduction, approximately 120° of motion occurs in the glenohumeral joint and 60° of motion occurs as a result of scapular movement on the thorax.

on the thorax (67,73). For the total range of motion through 180° of abduction or flexion, the glenohumeral to scapular movement ratio is 2:1; thus, the 180° range of motion is produced by 120° of glenohumeral motion and 60° of scapular motion (29). The contributing joint actions to the scapular motion are 20° produced at the AC

joint, 40° produced at the sternoclavicular joint, and 40° of posterior clavicular rotation (20).

As the arm abducts to 90°, the greater tuberosity on the humeral head approaches the coracoacromial arch, compression of the soft tissue begins to limit further abduction, and the tuberosity makes contact with the acromion process (20). If the arm is externally rotated, 30° more abduction can occur as the greater tuberosity is moved out from under the arch. Abduction is limited even more and can occur through only 60° with arm internal rotation because the greater tuberosity is held under the arch (20). External rotation accompanies abduction up through about 160° of motion. Also, full abduction cannot be achieved without some extension of the upper trunk to assist the movement.

MUSCULAR ACTIONS

The insertion, action, and nerve supply for each individual muscle of the shoulder joint and shoulder girdle are outlined in Figure 5-9. Most muscles in the shoulder region stabilize and execute movements. Special interactions between the muscles are presented in this section.

The muscles contributing to shoulder abduction and flexion are similar. The deltoid generates about 50% of the muscular force for elevation of the arm in abduction or flexion. The contribution of the deltoid increases with increased abduction. The muscle is most active from 90° to 180° (66). However, the deltoid has been shown to be most resistant to fatigue in the range of motion from 45° to 90° of abduction, making this range of motion most popular for arm-raising exercises.

When the arm elevates, the **rotator cuff** (teres minor, subscapularis, infraspinatus, and supraspinatus) also plays an important role because the deltoid cannot abduct or flex the arm without stabilization of the humeral head (89). The rotator cuff as a whole is also capable of generating flexion or abduction with about 50% of the force normally generated in these movements (29).

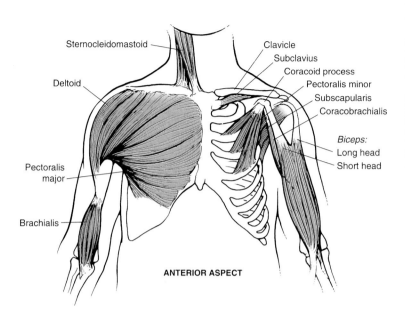

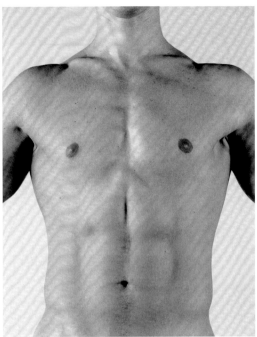

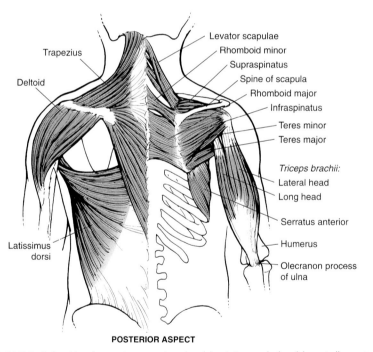

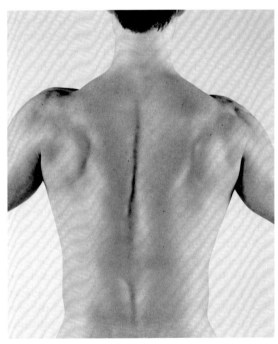

FIGURE 5-9 Muscles acting on the shoulder joint and shoulder girdle, anterior (*top*) and posterior (*bottom*) aspects. Along with insertion and nerve supply, the muscles responsible for the noted movements (PM) and the assisting muscles (Asst) are included in the table on the next page.

In the early stages of arm flexion or abduction, the deltoid's line of pull is vertical, so it is assisted by the supraspinatus, which produces abduction while at the same time compressing the humeral head and resisting the superior motion of the humeral head by the deltoid. The rotator cuff muscles contract as a group to compress the humeral head and maintain its position in the glenoid fossa (65). The teres minor, infraspinatus, and subscapularis muscles stabilize the humerus in elevation by applying a downward force. The latissimus dorsi also contracts eccentrically to assist with the stabilization of the humeral head and increases in activity as the angle increases (42). The interaction between the deltoid and the rotator cuff in abduction and flexion is shown in Figure 5-10. The inferior and medial force of the rotator cuff allows the deltoid to elevate the arm.

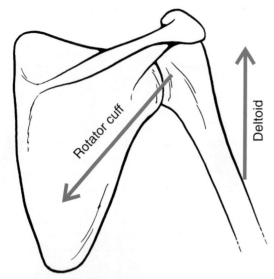

FIGURE 5-10   For efficient flexion or abduction of the arm, the deltoid muscle and the rotator cuff work together. In the early stages of abduction and flexion through 90°, the rotator cuff applies a force to the humeral head that keeps the head depressed and stabilized in the joint while the deltoid muscle applies a force to elevate the arm.

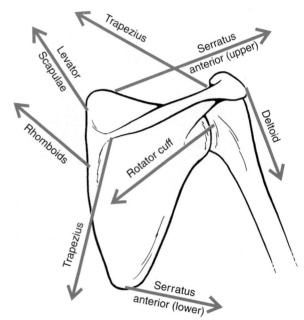

FIGURE 5-11   The direction of pull of various shoulder girdle muscles, the deltoid, and the rotator cuff for the resting arm. Note the line of pull of the trapezius and the serratus anterior, which work together to produce abduction, elevation, and upward rotation of the scapula necessary in arm flexion or abduction. Likewise, note the pull of the levator scapulae and the rhomboid, which also assist in elevation of the scapula.

Above 90° of flexion or abduction, the rotator cuff force decreases, leaving the shoulder joint more vulnerable to injury (29). However, one of the rotator cuff muscles, the supraspinatus, remains a major contributor above 90° of flexion or abduction. In the upper range of motion, the deltoid begins to pull the humeral head down and out of the joint cavity, thus creating a subluxating force (73). Motion through 90° to 180° of flexion or abduction requires external rotation in the joint. If the humerus externally rotates 20° or more, the biceps brachii can also abduct the arm (29).

When the arm is abducted or flexed, the shoulder girdle must protract or abduct, elevate, and upwardly rotate with posterior clavicular rotation to maintain the glenoid fossa in the optimal position. As shown in Figure 5-11, the serratus anterior and the trapezius work as a force couple to create the lateral, superior, and rotational motions of the scapula (29). These muscle actions take place after the deltoid and the teres minor have initiated the elevation of the arm and continue up through 180°, with the greatest muscular activity through 90° to 180° (66). The serratus anterior is also responsible for holding the scapula to the thorax wall and preventing any movement of the medial border of the scapula off the thorax.

If the arm is slowly lowered, producing adduction or extension of the arm with accompanying retraction, depression, and downward rotation of the shoulder girdle with forward clavicular rotation, the muscle actions are eccentric. Therefore, the movement is controlled by the muscles previously described in the arm abduction and flexion section. If the arm is forcefully lowered or if it is lowered against external resistance, such as a weight machine, the muscle action is concentric.

In a concentric adduction or extension against external resistance, such as in a swimming stroke, the muscles responsible for creating these joint actions are the latissimus dorsi, teres major, and sternal portion of the pectoralis major. The teres major is active only against a resistance, but the latissimus dorsi has been shown to be active in these movements even when no resistance is offered (13).

As the arm is adducted or extended, the shoulder girdle retracts, depresses, and downwardly rotates with forward clavicular rotation. The rhomboid muscle downwardly rotates the scapula and works with the teres major and the latissimus dorsi in a force couple to control the arm and scapular motions during lowering. Other muscles actively contributing to the movement of the scapula back to the resting position while working against resistance are the pectoralis minor (depresses and downwardly rotates the scapula) and the middle and lower portions of the trapezius (retract the scapula with the rhomboid). These muscular interactions are illustrated in Figure 5-12.

Two other movements of the arm, internal and external rotation, are very important in many sport skills and in the efficient movement of the arm above 90° (measured from arm at the side). An example of both external and internal rotation in a throwing action is shown in Figure 5-13. External rotation is an important component of the preparatory, or cocking, phase of an overhand throw, and internal rotation is important in the force application and follow-through phase of the throw.

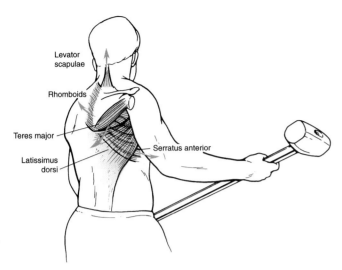

FIGURE 5-12 Lowering the arm against a resistance uses the latissimus dorsi and teres major working as a force coupled with the rhomboid. Other muscles that contribute to the lowering action are the pectoralis major, pectoralis minor, levator scapulae, and serratus anterior.

FIGURE 5-13 Shoulder joint rotation is an important contributor to the overhand throw. In the preparatory, or cocking, phase, the arm externally rotates to increase the range of motion and the distance over which the ball will travel. Internal rotation is an active contributor to the force application phase. The movement continues in the follow-through phase as the arm slows down.

External rotation, which is necessary when the arm is above 90°, is produced by the infraspinatus and the teres minor muscles (73). The activity of both of these muscles increases with external rotation in the joint (36). Because the infraspinatus is also an important muscle in humeral head stabilization, it fatigues early in elevated arm activities.

Internal rotation is produced primarily by the subscapularis, latissimus dorsi, teres major, and portions of the pectoralis major. The teres major is an active contributor to internal rotation only when the movement is produced against resistance. The muscles contributing to the internal rotation joint movement are capable of generating a large force, yet the internal rotation in most upper extremity actions never requires or uses much internal rotation force (73).

The shoulder girdle movements accompanying internal and external rotation depend on the position of the arm. In an elevated arm position, the shoulder girdle movements described in conjunction with abduction and flexion are necessary. Rotation produced with the arm in the neutral or anatomical position requires minimal shoulder girdle assistance. It is also in this position that the full range of rotation through 180° can be obtained. This is because as the arm is raised, muscles used to rotate the humerus are also used to stabilize the humeral head, which is restrained in rotation in the upper range of motion. Specifically, internal rotation is difficult in elevated arm positions because the tissue under the acromion process is very compressed by the greater tuberosity (66).

Two final joint actions that are actually combinations of elevated arm positions are horizontal flexion or adduction and horizontal extension or abduction. Because the arm is elevated, the same muscles described earlier for abduction and flexion also contribute to these movements of the arm across the body.

Muscles contributing more significantly to horizontal flexion are the pectoralis major and the anterior head of the deltoid. This movement brings the arms across the body in the elevated position and is important in power movements of upper extremity skills. Horizontal extension in which the arm is brought back in the elevated position is produced primarily by the infraspinatus, teres minor, and posterior head of the deltoid. This joint action is common in the backswing and preparatory actions in upper extremity skills (89).

Internal/External Shoulder Rotation

Internally and externally rotate your shoulder joint. Is this a motion you use a lot in everyday life? Think of some activities that use this motion. Abduct your shoulder 90° and flex your elbow 90°. Now perform internal and external rotation at the shoulder. Describe this motion and note some activities that use a motion similar to this.

STRENGTH OF THE SHOULDER MUSCLES

In a flexed position, the shoulder muscles can generate the greatest strength output in adduction when muscle fibers of the latissimus dorsi, teres major, and pectoralis major contribute to the movement. The adduction strength of the shoulder muscles is twice that for abduction, even though the abduction movement and muscle group are used more frequently in activities of daily living and sports (89).

The movement capable of generating the next greatest level of strength after the adductors is an extension movement that uses the same muscles that contribute to arm adduction. The extension action is slightly stronger than its opposite movement, flexion. After flexion, the next strongest joint action is abduction, illustrating the fact that shoulder joint actions are capable of generating greater force output in the lowering phase using the adductors and extensors than in the raising phase, when the flexors and abductors are used. These strength relationships change, however, when the shoulder is held in a neutral or slightly hyperextended position because the isometric force development is greater in the flexors than in the extensors. This reversal in strength differences is related to the length–tension relationship created by the starting point.

The weakest joint actions in the shoulder are rotational, with external rotation being weaker than internal rotation. The strength output of the rotators is influenced by arm position, and the greatest internal rotation strength can be obtained with the arm in the neutral position. The greatest external rotation strength can be obtained with the shoulder in 90° of flexion. With the arm elevated to 45°, however, both internal and external rotation strength outputs are greater in 45° of abduction than 45° of flexion (28). External rotation is important in the upper 90° of arm elevation, providing stability to the joint. Internal rotation creates instability in the joint, especially in the upper elevation levels, as it compresses the soft tissue in the joint.

Muscle strength imbalance is accentuated in athletic populations because of use patterns. For example, swimmers, water polo players, and baseball pitchers have been found to have relatively stronger adductors and internal rotators (14). In paraplegic wheelchair athletes, the adductors are relatively weaker than the abductors, and this is more pronounced in athletes with shoulder injuries (14).

CONDITIONING

The shoulder muscles are easy to stretch and strengthen because of the mobility of the joint. The muscles usually work in combination, making it difficult to isolate a specific muscle in an exercise. Examples of stretching, manual resistance, and weight training for the shoulder abductors and flexors are presented in Figure 5-14.

Some resistance exercises may irritate the shoulder joint and should be avoided by individuals with specific injuries. Any lateral dumbbell raise using the deltoid may cause impingement in the coracoacromial area. This impingement is magnified if the shoulder is internally rotated. A solution for those wishing to avoid impingement or who have injuries in this area is to rotate the arm externally and then perform the lateral raise (20). It is important to recognize that when an adjustment like this is made, the muscle activity and the forces generated internally also change. External rotation during a lateral raise alters the activity of the deltoid and facilitates activity in the internal rotators.

Exercises such as the bench press and push-ups should be avoided by individuals with instability in the anterior or posterior portion of the shoulder joint caused by adduction and internal rotation. Likewise, stress on the anterior portion of the capsule is produced by the pullover exercise that moves from an extreme flexed, abducted, and externally rotated position. Other exercises to be avoided by individuals with anterior capsule problems are behind-the-neck pull-downs, incline bench press, and rowing exercises. The risks in these three exercises can be minimized if no external rotation is maintained or even if some internal rotation is maintained in the joint. The external rotation position produces strain on the anterior portion of the shoulder (20). In an exercise such as the squat, which uses the lower extremity musculature, the position of the shoulder in external rotation may even prove to be harmful because of the strain on the anterior capsule created by weights held in external rotation. Attempts should be made to minimize this joint action by balancing a portion of the weight on the trapezius or using alternative exercises, such as the dead lift.

Finally, if an individual is having problems with rotator cuff musculature, heavy lifting in an abduction movement should be minimized or avoided. This is because the rotator cuff muscles must generate a large force during the abduction action to support the shoulder joint and complement the activity of the deltoid. Heavy weight lifting above the head should be avoided to reduce strain on the rotator cuff muscles (49).

INJURY POTENTIAL OF THE SHOULDER COMPLEX

The shoulder complex is subject to a wide variety of injuries that can be incurred in two ways. The first type of injury is through trauma. This type of injury usually occurs when contact is made with an external object, such as the ground or another individual. The second type of injury is through repetitive joint actions that create inflammatory sites in and around the joints or muscular attachments.

Many injuries to the shoulder girdle are traumatic, a result of impacts during falls or contact with an external object. The sternoclavicular joint can **sprain** or dislocate anteriorly if an individual falls on the top of the shoulder in the area of the middle deltoid. An individual with a sprain to this joint has pain in horizontal extension movements of the shoulder, such as in the golf swing or the backstroke in swimming (85). Anterior **subluxations** of this joint in adolescents have also occurred spontaneously during throwing because they have greater mobility in this joint than adults. A posterior dislocation or subluxation of the sternoclavicular joint can be quite serious because the trachea, esophagus, and numerous veins and arteries lie below this structure. This injury occurs as a consequence of force to the sternal end of the clavicle. The individual may have symptoms such as choking, shortness of breath, and difficulty in swallowing (85). Overall, the sternoclavicular joint is well reinforced with ligaments, and fortunately, injury in the form of sprains, subluxations, and dislocations is not common.

The clavicle is frequently a site of injury by direct trauma received through contact in football and some other sports. The most common injury is a **fracture** to

Muscle Group	Sample Stretching Exercise	Sample Strengthening Exercise	Other Exercises
Shoulder flexors/ abductors		Front dumbell raise  Side dumbell raise 	Military press Shoulder press Upright row Front lateral raise Seated dumbell press
Shoulder extensors/ adductors		Lat pull down 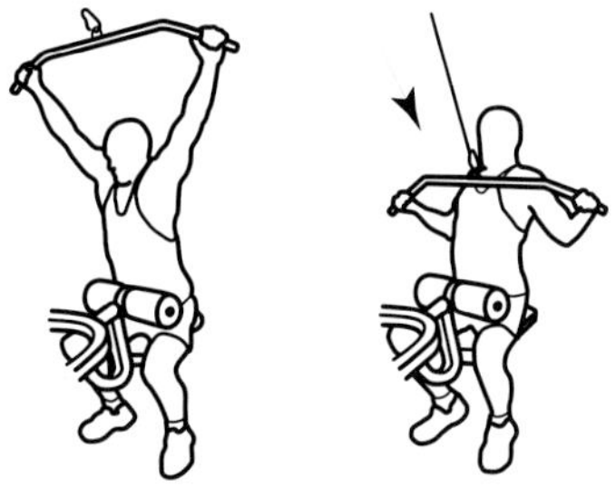 Straight arm pull down 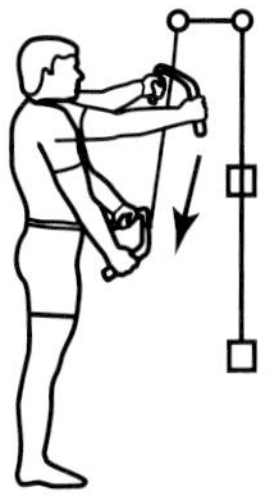	Pull up Chin up Cable pull down
Shoulder rotators	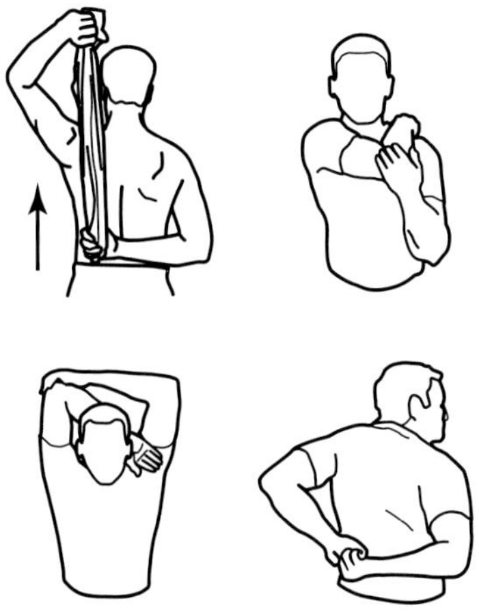	External rotation  Internal rotation 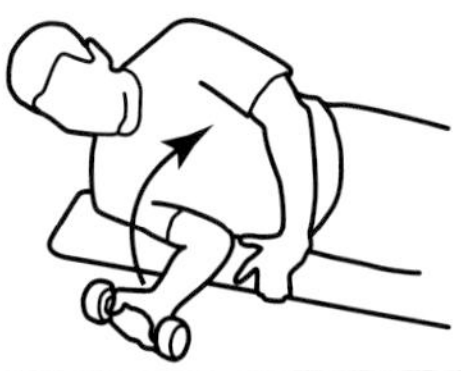	Backscratch Cable external and internal rotation

FIGURE 5-14   Sample stretching and strengthening exercises for selected muscle groups.

Muscle Group	Sample Stretching Exercise	Sample Strengthening Exercise	Other Exercises
Shoulder girdle elevators		Dumbell shrugs  Barbell upright row 	Barbell shrug Seated shrug Cable upright rows
Shoulder girdle abductors		Dumbell pullover 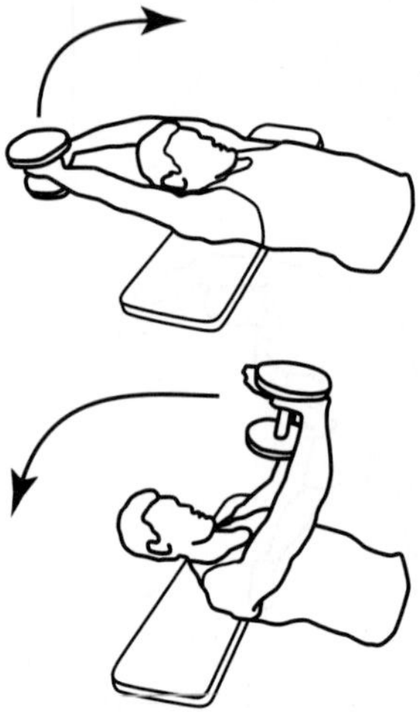	Push-up Standing fly Punch
Shoulder girdle adductors		V pull 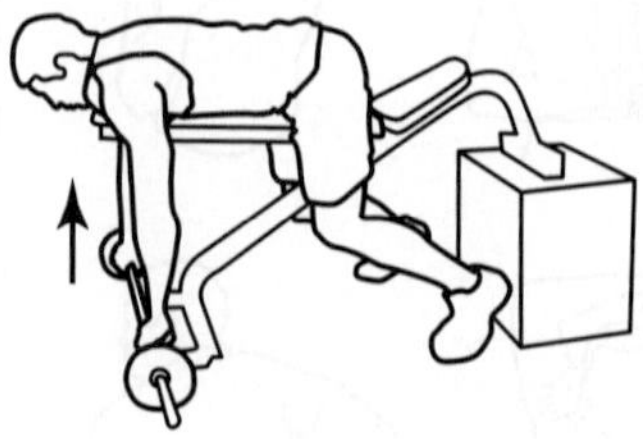	Seated rows Bent over dumbbell rows

FIGURE 5-14   (continued)

the middle third of the clavicle. This injury is incurred by falling on the shoulder or outstretched arm or receiving a blow on the shoulder so that a force is applied along the shaft of the clavicle. Other less common fractures occur to the medial clavicle as a result of direct trauma to the lateral end of the clavicle or as a result of direct trauma to the tip of the shoulder (85). Clavicular fractures in adolescents heal quickly and effectively; but in adults, the healing and repair process is not as efficient or effective. This is related to the differences in the level of skeletal maturation. In adolescents, new bone is being formed at a much faster rate than in mature individuals.

Injuries to the AC joint can cause a considerable amount of disruption to shoulder movements. Again, if an individual falls on the point of the shoulder, the AC joint can subluxate or dislocate. This can also occur because of a fall on the elbow or on an outstretched arm. This joint is also frequently subjected to overuse injuries in sports using the overhand pattern, such as throwing, tennis, and swimming. Other sports that repeatedly load the joint in the overhead position, such as weight lifting and wrestling, may also cause the overuse syndrome. The consequences of overuse of the joint are capsule injury, an **ectopic calcification** in the joint, and possible **degeneration** of the cartilage (85).

The scapula rarely receives sufficient force to cause an injury. If an athlete or an individual falls on the upper back, however, it is possible to fracture the scapula and bruise the musculature so that arm abduction is quite painful. Another site of fracture on the scapula is the **coracoid process**, which can be fractured with separation of the AC joint. Throwers can also acquire **bursitis** at the inferomedial border of the scapula, causing pain as the scapula moves through the cocking and acceleration phases in the throw. The pain is diminished in the follow-through phase. Bursitis is the inflammation of the bursa, a fluid-filled sac found at strategic sites around the synovial joints that reduces the friction in the joint.

Activities such as weight lifting (bench press and push-ups), lifting above the head, playing tennis, and carrying a backpack can produce trauma to the brachial nerve plexus by means of a traction force (i.e., a pulling force). If the long thoracic nerve is impinged, isolated paralysis of the serratus anterior can cause movement of the medial border of the scapula away from the thorax and a decreased ability to abduct and flex at the shoulder joint (85).

The shoulder joint is commonly injured either through direct trauma or repeated overuse. Dislocation or subluxation in the glenohumeral joint is frequent because of the lack of bony restraint and the dependence on soft tissue for restraint and support of the joint. Dislocation occurs most frequently in collision sports such as ice hockey (15). The glenoid fossa faces anterolaterally, creating more stability in the posterior joint than the anterior. Thus, the most common direction of dislocation is anterior. Anterior and inferior dislocations account for 95% of dislocations (59).

The usual cause of the dislocation is contact or some force applied to the arm when it is abducted and externally rotated overhead. This drives the humeral head anteriorly, possibly tearing the capsule or the glenoid labrum. The rate of recurrence of dislocation depends on the age of the individual and the magnitude of the force producing the dislocation (33). The recurrence rate for the general population is 33% to 50%, increasing to 66% to 90% in individuals younger than 20 years of age (66). In fact, the younger the age at the first dislocation, the more likely is a recurrent dislocation. Also, if a relatively small amount of force created the dislocation, a recurrent dislocation is more likely.

Recurrent dislocations also depend on the amount of initial damage and whether the glenoid labrum was also damaged (64). A tear to the glenoid labrum, similar to tearing the meniscus in the knee, results in clicking and pain with the arm overhead (88). An anterior dislocation also makes it difficult to rotate the arm internally, so that the contralateral shoulder cannot be touched with the hand on the injured side.

Posterior dislocations of the shoulder are rare (2%) and are usually associated with a force applied with an adducted and internally rotated arm with the hand below shoulder level (88). The clinical signs of a posterior dislocation are inability to abduct and externally rotate the arm.

Soft-tissue injuries at the shoulder joint are numerous and are most often associated with overhead motions of the arm, such as in throwing, swimming, and racket sports. Because of the extreme range of motions and high velocities in throwing, the dynamic stabilizing structures of the shoulder joint are at great risk for injury (52). Injuries in this category include examples such as posterior and anterior instability, impingement, and glenoid labrum damage. The rotator cuff muscles, which are active in controlling the humeral head and motion during the overhand pattern, are highly susceptible to injury.

In an upper extremity throwing pattern, when the arm is in the preparatory phase with the shoulder abducted and externally rotated, the anterior capsule—specifically, the subscapularis muscle—is susceptible to strain or tendinitis at the insertion on the lesser tuberosity (72). In late cocking and early acceleration phase, the posterior portion of the capsule and posterior labrum are susceptible to injury as the anterior shoulder is tightened, driving the head of the humerus backward (10). In the follow-through phase, when the arm is brought horizontally across the body at a very high speed, the posterior rotator cuff, infraspinatus, and teres minor are very susceptible to muscle strain or tendinitis on the greater tuberosity insertion site as they work to decelerate the arm (19).

The most common mechanism of injury to the rotator cuff occurs when the greater tuberosity pushes against the underside of the acromion process. This subacromial **impingement syndrome** occurs during the acceleration phase of the overhand throwing pattern when the arm is internally rotating while still maintained in the abducted position. Impingement can also occur in the lead arm of golfers and in a variety of other activities that use the overhead pattern (40). The rotator cuff, subacromial bursa, and biceps tendon are compressed against the anterior undersurface of the acromion and coracoacromial ligament (51) (Fig. 5-15). The impingement has been seen as the main source of soft-tissue injury, although others point to tension overload, overuse, and traumatic injury as other competing sources of injury to the rotator cuff (51). Impingement occurs in the range of 70° to 120° of flexion or abduction and is most common in such activities as the tennis serve, throwing, and the butterfly and crawl strokes in swimming (29). If an athlete maintains the shoulder joint in an internally rotated position, impingement is more likely to occur. It is also commonly injured in wheelchair athletes and in individuals transferring from a wheelchair to a bed or chair (9,14). The supraspinatus muscle, lying in the subacromial space, is compressed and can be torn with impingement, and with time, calcific deposits can be laid down in the muscle or tendon. This irritation can occur with any overhead activity, creating a painful arc of arm motion through 60° to 120° of abduction or flexion (73).

The Subacromial Arch

The functional abduction/adduction range of motion at the shoulder joint depends on the amount of shoulder rotation. As you internally and externally rotate at the shoulder you move the greater tubercle. Try internally rotating the shoulder while in anatomical position and then abducting the shoulder as much as possible. In this position, the greater tubercle comes into contact with the roof of the subacromial arch. If you externally rotate the shoulder and then abduct you will get about 30° more abduction. Overhead motions that require internal and external rotation at the shoulder can impinge the soft tissue and irritate the subacromial bursa.

Another injury that is a consequence of impingement is **subacromial bursitis**. This injury results from an irritation of the bursae above the supraspinatus muscle and underneath the acromion process (29). It also develops in wheelchair propulsion because of greater-than-normal pressures in the joint and abnormal distribution of stress in the subacromial area (9).

Finally, the tendon of the long head of the biceps brachii can become irritated when the arm is forcefully abducted and rotated. **Bicipital tendinitis** develops as the biceps tendon is subluxated or irritated within the bicipital groove. In throwing, the arm externally rotates to 160° in the cocking phase, and the elbow moves through 50° of motion. Because the biceps brachii acts on the shoulder

and is responsible for decelerating the elbow in the final 30° of extension, it is often maximally stressed (6). In a rapid throw, the long head of the biceps brachii may also be responsible for tearing the anterosuperior portion of the glenoid labrum. Irritation to the biceps tendon is manifested in a painful arc syndrome similar to that of the rotator cuff injury.

In summary, the shoulder complex has the greatest mobility of any region in the body, but as a consequence of this great mobility, it is an unstable area in which numerous injuries may occur. Despite the high probability of injury, successful rehabilitation after surgery is quite common. It is important to maintain the strength and flexibility of the musculature surrounding the shoulder complex because there is considerable dependence on the musculature and soft tissue for support and stabilization.

The Elbow and Radioulnar Joints

The role of forearm movement, generated at the elbow or **radioulnar joint**, is to assist the shoulder in applying force and in controlling the placement of the hand in space. The combination of shoulder and elbow–radioulnar joint movements affords the capacity to place the hand in many positions, allowing tremendous versatility. Whether you are working above your head, shaking someone's hand, writing a note, or tying your shoes, hand position is important and is generated by the working relationship between the shoulder complex and the forearm. The

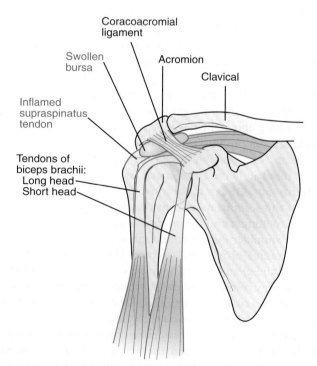

FIGURE 5-15 Impingement of soft tissue between the greater tubercle and the acromion/coracoacromial ligament can cause damage to the rotator cuff, subacromial bursa, or the biceps tendon.

elbow joint also works as a fulcrum for the forearm, allowing both powerful grasping and fine hand motion (24).

ANATOMICAL AND FUNCTIONAL CHARACTERISTICS OF THE JOINTS OF THE ELBOW

The elbow is considered a stable joint, with structural integrity, good ligamentous support, and good muscular support (41). The elbow has three joints allowing motion between the three bones of the arm and forearm (humerus, radius, and ulna). Movement between the forearm and the arm takes place at the ulnohumeral and radiohumeral articulations, and movements between the radius and the ulna take place at the radioulnar articulations (73). Landmarks on the radius and ulna and the ulnohumeral, radiohumeral, and proximal radioulnar articulations are shown in Figure 5-16.

Ulnohumeral Joint
The **ulnohumeral joint** is the articulation between the ulna and the humerus and is the major contributing joint to flexion and extension of the forearm. The joint is the union between the spool-like **trochlea** on the distal end of the humerus and the **trochlear notch** on the ulna. On the front of the ulna is the **coronoid process**, which makes contact in the **coronoid fossa** of the humerus, limiting flexion in the terminal range of motion. Likewise, on the posterior side of the ulna is the **olecranon process**, which makes contact with the **olecranon fossa** on the humerus, terminating extension. An individual who can hyperextend at the elbow joint may have a small olecranon process or a large olecranon fossa, which allows more extension before contact occurs.

The trochlear notch of the ulna fits snugly around the trochlea, offering good structural stability. The trochlea is covered with articular cartilage over the anterior, inferior, and posterior surfaces and is asymmetrical, with an oblique posterior projection (87). In the extended position, the asymmetrical trochlea creates an angulation of the ulna laterally referred to as a valgus position. This is termed the **carrying angle** and ranges from 10° to 15° in males and 15° to 25° in females (58,87). Measurement of the carrying angle is shown in Figure 5-17. As the forearm flexes, this valgus position is reduced and may even result in a varus position with full flexion (24).

Radiohumeral Joint
The second joint participating in flexion and extension of the forearm is the **radiohumeral joint**. At the distal end of the humerus is the articulating surface for this joint, the **capitulum**, which is spheroidal and covered with cartilage on the anterior and inferior surfaces. The top of the round radial head butts up against the capitulum, allowing radial movement around the humerus during flexion and extension. The capitulum acts as a buttress for lateral

compression and other rotational forces absorbed during throwing and other rapid forearm movements.

Radioulnar Joint
The third articulation, the radioulnar joint, establishes movement between the radius and the ulna in **pronation** and **supination**. There are actually two radioulnar articulations, the superior in the elbow joint region and the inferior near the wrist. Also, midway between the elbow and the wrist is another fibrous connection between the radius and the ulna, recognized by some as a third radioulnar articulation.

The superior or proximal radioulnar joint consists of the articulation between the radial head and the radial fossa on the side of the ulna. The radial head rotates in a fibrous osseous ring and can turn both clockwise and counterclockwise, creating movement of the radius relative to the ulna (12). In the neutral position, the radius and ulna lie next to each other, but in full pronation, the radius has crossed over the ulna diagonally. As the radius crosses over in pronation, the distal end of the ulna moves laterally. The opposite occurs during supination.

An **interosseous membrane** connecting the radius and ulna runs the length of the two bones. This fascia increases the area for muscular attachment and ensures that the radius and ulna maintain a specific relationship with each other. Eighty percent of compressive forces are typically applied to the radius, and the interosseous membrane transmits forces received distally from the radius to the ulna. The membrane is taut in a semiprone position (12).

Two final structural components in the elbow region are the **medial** and **lateral epicondyles**. These are prominent landmarks on the medial and lateral sides of the humerus. The lateral epicondyle serves as a site of attachment for the lateral ligaments and the forearm supinator and extensor muscles, and the medial epicondyle accommodates the medial ligaments and the forearm flexors and pronators (1). These extensions of the humerus are also common sites of overuse injury.

Ligaments and Joint Stability
The elbow joint is supported on the medial and lateral sides by collateral ligaments. The medial, or ulnar, collateral ligament (MCL) connects the ulna to the humerus and offers support and resistance to valgus stresses imposed on the elbow joint. Support in the valgus direction is very important in the elbow joint because most forces are directed medially, creating a valgus force. The anterior band of the MCL is taut in extension, and the posterior band is relaxed in extension but increases in tension in flexion (1,69). Consequently, the MCL is taut in all joint positions. If the MCL is injured, the radial head becomes important in providing stability when a valgus force is applied (4). The flexor–pronator muscles originating on the medial epicondyle also provide dynamic stabilization to the medial elbow (70).

A set of collateral ligaments on the lateral side of the joint is termed the lateral or radial collateral ligaments.

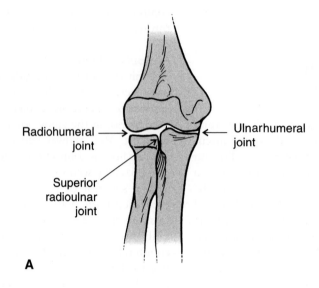

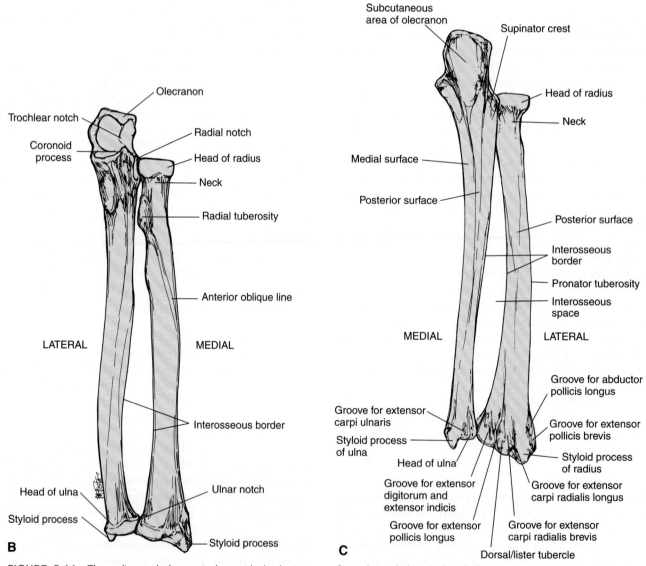

FIGURE 5-16   The radius and ulnar articulate with the humerus to form the radiohumeral and ulnar humeral joints. Shown are the elbow joint complex **(A)** and the anterior **(B)** and posterior **(C)** surfaces of the radius and ulna.

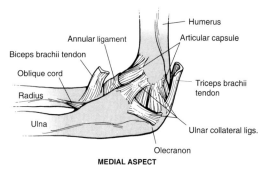

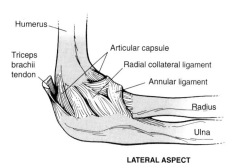

FIGURE 5-18 The elbow ligaments.

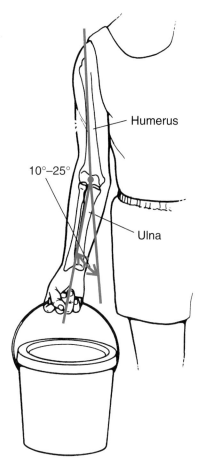

FIGURE 5-17 In the extended position, the ulna and humerus form the carrying angle because of asymmetry in the trochlea. The carrying angle is measured as the angle between a line describing the long axis of the ulna and a line describing the long axis of the humerus. The angle ranges from 10° to 25°.

The radial collateral is taut throughout the entire range of flexion (1,69), but because varus stresses are rare, these ligaments are not as significant in supporting the joint (89). The small anconeus muscle provides dynamic stabilization to the lateral elbow (70).

A ligament that is important for the function and support of the radius is the **annular ligament**. This ligament wraps around the head of the radius and attaches to the side of the ulna. The annular ligament holds the radius in the elbow joint while still allowing it to turn in pronation and supination. The elbow ligaments and their actions can be reviewed in Figure 5-18.

Movement Characteristics
The three joints of the elbow complex do not all reach a close-packed position (i.e., position of maximum joint surface contact and ligamentous support) at the same point in the range of motion. A close-packed position for the radiohumeral is achieved when the forearm is flexed to 80° and in the semipronated position (12). The fully extended position is the close-packed position for the ulnohumeral joint. Thus, when the ulnohumeral articulation is most stable in the extended position, the

radiohumeral articulation is loose packed and least stable. The proximal radioulnar joint is in its close-packed position in the semipronated position, complementing the close-packed position of the radiohumeral (12).

The range of motion at the elbow in flexion and extension is approximately 145° of active flexion, 160° of passive flexion, and 5° to 10° of hyperextension (12). An extension movement is limited by the joint capsule and the flexor muscles. It is also terminally restrained by bone-on-bone impact with the olecranon process.

Flexion at the joint is limited by soft tissue, the posterior capsule, the extensor muscles, and the bone-on-bone contact of the coronoid process with its respective fossa. A significant amount of hypertrophy or fatty tissue will limit the range of motion in flexion considerably. Approximately 100° to 140° of flexion and extension is required for most daily activities, but the total range of motion is 30° to 130° of flexion (53).

The range of motion for pronation is approximately 70°, limited by the ligaments, the joint capsule, and soft tissue compressing as the radius and ulna cross. Range of motion for supination is 85° and is limited by ligaments, the capsule, and the pronator muscles. Approximately 50° of pronation and 50° of supination are required to perform most daily activities (89).

MUSCULAR ACTIONS

Twenty-four muscles cross the elbow joint. Some of them act on the elbow joint exclusively; others act at the wrist and finger joints (3). Most of these muscles are capable of producing as many as three movements at the elbow, wrist,

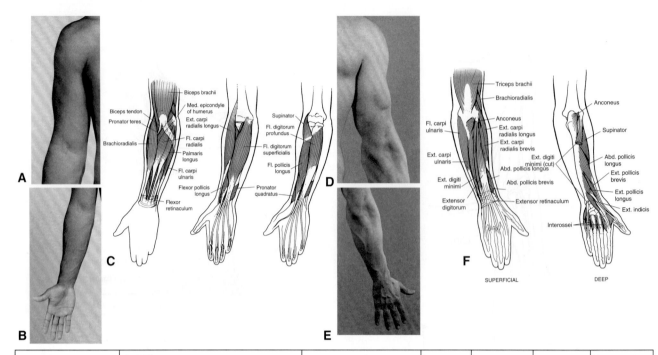

Muscle	Insertion	Nerve Supply	Flexion	Extension	Pronation	Supination
Anconeus	Lateral epicondyle of humerus TO olecranon process on ulna	Radial nerve; C7, C8		Asst		
Biceps brachii	Supraglenoid tubercle; corocoid process TO radial tuberosity	Musculocutaneous nerve: C5, C6	PM			PM
Brachialis	Anterior surface of lower humerus TO coronoid process on ulna	Musculocutaneous nerve: C5, C6	PM			
Brachioradialis	Lateral supracondylar ridge of humerus TO styloid process of radius	Radial nerve; C6, C7	PM			
Extensor carpi radialis brevis	Lateral epicondyle of humerus TO base of third metacarpal	Radial nerve; C6, C7	Asst			
Extensor carpi radialis longus	Lateral supracondylar ridge of humerus TO base of second metacarpal	Radial nerve; C6, C7	Asst			
Extensor carpi ulnaris	Lateral epicondyle of humerus TO base of fifth metacarpal	Posterior interosseous nerve; C6–C8		Asst		
Flexor carpi radialis	Medial epicondyle of humerus TO base of second, third metacarpal	Median nerve; C6, C7	Asst			
Flexor carpi ulnaris	Medial epicondyle TO pisiform; hamate base; base of fifth metecarpal	Ulnar nerve; C8, T1	Asst			
Palmaris longus	Medial epicondyle TO palmar aponeurosis	Median nerve; C6, C7	Asst			
Pronator quadratus	Distal anterior surface of ulnar TO distal anterior surface of radius	Anterior interosseous nerve; C8, T1			PM	
Pronator teres	Medial epicondyle of humerus, coronoid process on ulna TO midlateral surface of radius	Median nerve; C6, C7	Asst		PM	
Supinator	Lateral epicondyle of humerus TO upper lateral side of radius	Posterior interosseous nerve; C5, C6				PM

FIGURE 5-19   Elbow and forearm muscles. The anterior surface of the arm **(A)** and forearm **(B)** are shown with the anterior muscles **(C)**. The posterior surface of the arm **(D)** and forearm **(E)** are shown with the corresponding posterior muscles **(F)**.

or phalangeal joints. One movement is usually dominant, however, and it is the movement with which the muscle or muscle group is associated. There are four main muscle groups, the anterior flexors, posterior extensors, lateral extensor–supinators, and medial flexor–pronators (1). The locations, actions, and nerve supplies of the muscles acting at the elbow joint can be found in Figure 5-19.

The elbow flexors become more effective as elbow flexion increases because their mechanical advantage increases with an increase in the magnitude of the moment arm (3,58). The brachialis has the largest cross-sectional area of the flexors but has the poorest mechanical advantage. The biceps brachii also has a large cross section with better mechanical advantage, and the brachioradialis has a smaller cross section but the best mechanical advantage (Fig. 5-20). At 100° to 120° of flexion, the mechanical advantage of the flexors is maximal because the moment arms are longer (brachioradialis = 6 cm; brachialis = 2.5 to 3.0 cm; biceps brachii = 3.5 to 4.0 cm) (58).

Each of the three main elbow flexors is limited in its contribution to the elbow flexion movement depending on the joint position or mechanical advantage. The brachialis is active in all forearm positions but is limited by its

poor mechanical advantage. The brachialis plays a bigger role when the forearm is in the pronated position. The biceps brachii can be limited by actions at both the shoulder and the radioulnar joints. Because the long head of the biceps crosses the shoulder joint, flexion of the shoulder joint generates slack in the long head of the biceps brachii, and extension of the shoulder generates more tension. Because the biceps tendon attaches to the radius, the insertion can be moved in pronation and supination. The influence of pronation on the tendon of the biceps brachii is illustrated in Figure 5-21. Because the tendon wraps around the radius in pronation, the biceps brachii is most effective as a flexor in supination. Finally, the brachioradialis is a muscle with a small volume and very long fibers; it is a very efficient muscle, however, because of its excellent mechanical advantage. The brachioradialis flexes the elbow most effectively when the forearm is in midpronation, and it is heavily recruited during rapid movements. It is well positioned to contribute to elbow flexion in the semiprone position.

In the extensor muscle group is the powerful triceps brachii, the strongest elbow muscle. The triceps brachii has great strength potential and work capacity because of its muscle volume (3). The triceps brachii has three portions: the long head, medial head, and lateral head.

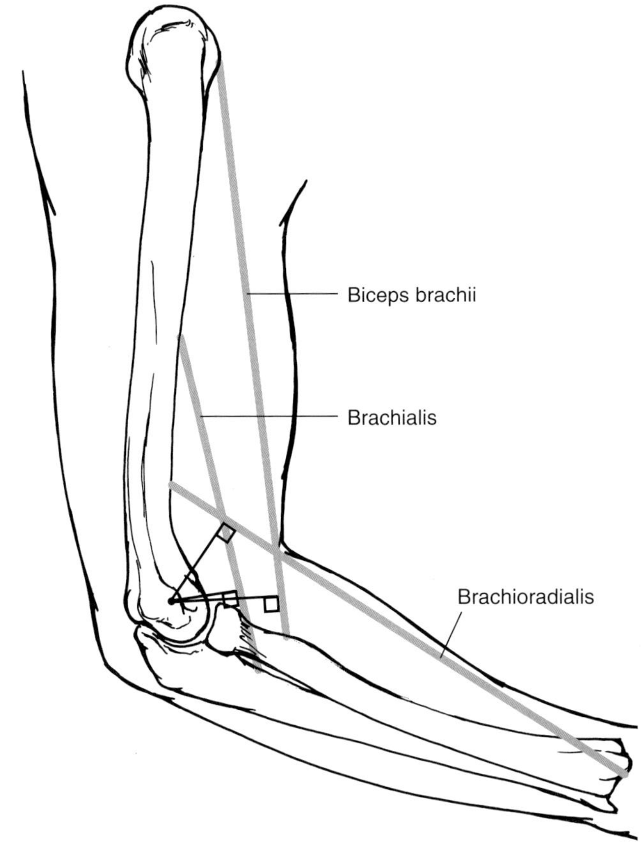

Biceps brachii

Brachialis

Brachioradialis

FIGURE 5-20 The line of action of the three forearm muscles. The brachialis (BRA) is a large muscle, but it has the smallest moment arm, giving it the poorest mechanical advantage. The biceps brachii (BIC) also has a large cross section and has a longer moment arm, but the brachioradialis (BRD), with its smaller cross section, has the longest moment arm, giving it the best mechanical advantage in this position.

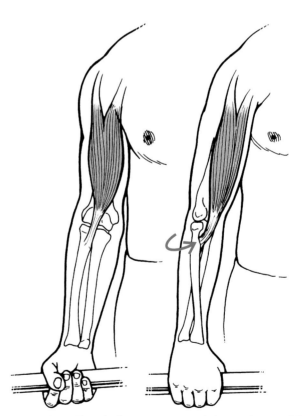

FIGURE 5-21 When the forearm is pronated, the attachment of the biceps brachii to the radius is twisted underneath. This position interferes with the flexion-producing action of the biceps brachii, which is more efficient in producing flexion when the forearm is supinated and the tendon is not twisted under the radius.

Of these three, only the long head crosses the shoulder joint, making it dependent partially on shoulder position for its effectiveness. The long head is the least active of the triceps. However, it can be increasingly more involved with shoulder flexion as its insertion on the shoulder is stretched.

The medial head of the triceps brachii is considered the workhorse of the extension movement because it is active in all positions, at all speeds, and against maximal or minimal resistance. The lateral head of the triceps brachii, although the strongest of the three heads, is relatively inactive unless movement occurs against resistance (73). The output of the triceps brachii is not influenced by forearm positions of pronation and supination.

The medial flexor–pronator muscle group originating on the medial epicondyle includes the pronator teres and three wrist muscles (flexor carpi radialis, flexor carpi ulnaris, and palmaris longus). The pronator teres and the three wrist muscles assist in elbow flexion, and the pronator teres and the more distal pronator quadratus are primarily responsible for forearm pronation. The pronator quadratus is more active regardless of forearm position, whether the activity is slow or fast or working against a resistance or not. The pronator teres is called on to become more active when the pronation action becomes rapid or against a high load. The pronator teres is most active at 60° of forearm flexion (74).

The final muscle group at the elbow is the extensor–supinator muscles originating on the lateral epicondyle, which includes the supinator and three wrist muscles (extensor carpi ulnaris, extensor carpi radialis longus, and extensor carpi radialis brevis). The wrist muscles can assist with elbow flexion. Supination is produced by the supinator muscle and by the biceps brachii under special circumstances. The supinator is the only muscle that contributes to a slow, unresisted supination action in all forearm positions. The biceps brachii can supinate during rapid or rested movements when the elbow is flexed. The flexion action of the biceps brachii is neutralized by actions from the triceps brachii, allowing contribution to the supination action. At 90° of flexion, the biceps brachii becomes a very effective supinator.

Many of the muscles acting at the elbow joint create multiple movements, and a large number of two-joint muscles also generate movements at two joints. Where an isolated movement is desired, synergistic actions are required to neutralize the unwanted action. For example, the biceps brachii flexes the elbow and supinates the radioulnar joint. To provide a supination movement without flexion, synergistic action from an elbow extensor must occur. Likewise, if flexion is the desired movement, a supination synergist must be recruited. Another example is the biceps brachii action at the shoulder joint where it generates shoulder flexion. To eliminate a shoulder movement during elbow flexion, there must be action from the shoulder extensors. A final example is the triceps brachii action at the shoulder where it creates shoulder extension.

If a strong extension is required at the elbow in pushing and throwing actions, shoulder flexors must be engaged to eliminate the shoulder extension movement. If an adjacent joint is to remain stationary, appropriate changes in muscle activity must occur and are usually proportional to the velocity of the movement (26).

STRENGTH OF THE FOREARM MUSCLES

The flexor muscle group is almost twice as strong as the extensors at all joint positions, making us better pullers than pushers. The joint forces created by a maximum isometric flexion in an extended position that is equal to approximately two times body weight.

The semiprone elbow position is the position at which maximum strength in flexion can be developed, followed by the supine position, and finally, the pronated position (62). The supine position generates about 20% to 25% more strength than the pronation position. The semiprone position is most commonly used in daily activities. Semiprone flexion exercises should be included in a conditioning routine to take advantage of the strong position of the forearm.

Extension strength is greatest from a position of 90° of flexion (89). This is a common forearm position for daily living activities and for power positions in upper extremity sport skills. Finally, pronation and supination strength is greatest in the semiprone position, with the torque dropping off considerably at the fully pronated or fully supinated position.

CONDITIONING

The effectiveness of exercises used to strengthen or stretch depends on the various positions of the arm and the forearm. In stretching the muscles, the only positions putting any form of stretch on the flexors and extensors must incorporate some hyperextension and flexion at the shoulder joints. Stretching these muscles while the arm is in the neutral position is almost impossible because of the bony restrictions to the range of motion.

The position of the forearm is important in forearm strengthening activities. The forearm position in which the flexors and extensors are the strongest is semiprone. For the flexors specifically, the biceps brachii can be brought more or less into the exercise by supinating or pronating, respectively. Numerous exercises are available for both the flexors and extensors; examples of which are provided in Figure 5-22.

The pronators and supinators offer a greater challenge in the prescription of strength or resistive exercises (Fig. 5-22). Stretching these muscle groups presents no problem because a maximal supination position can adequately stretch the pronation musculature and vice versa. Also, low-resistance exercises can be implemented by applying a force in a turning action (e.g., to a doorknob or some other immovable object). High-resistance exercises

Muscle Group	Sample Stretching Exercise	Sample Strengthening Exercise	Other Exercises
Flexors	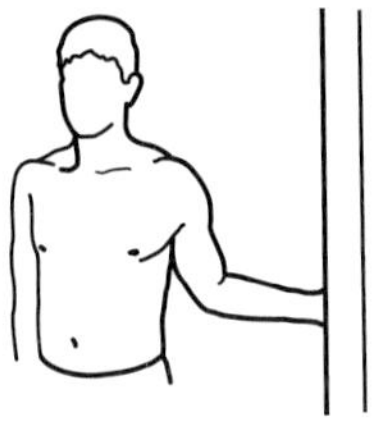	Dumbell biceps curls  Machine biceps curl 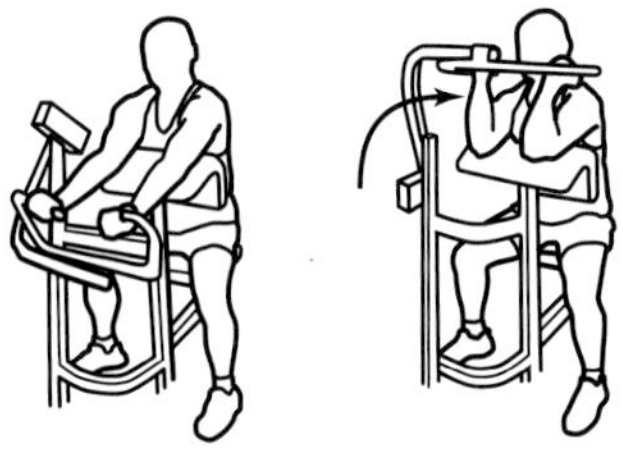	Pull-up Upright row
Extensors	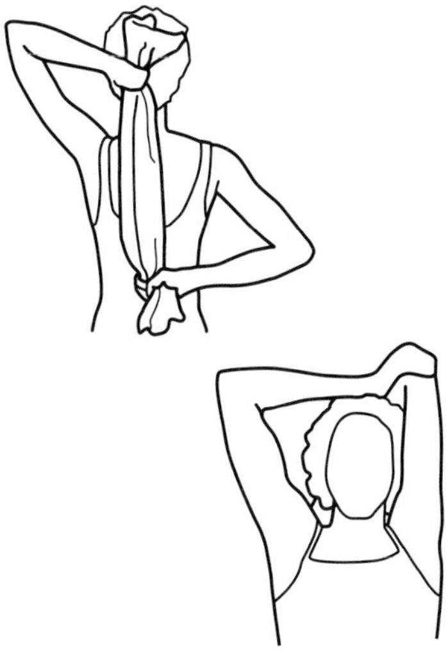	Triceps extensions 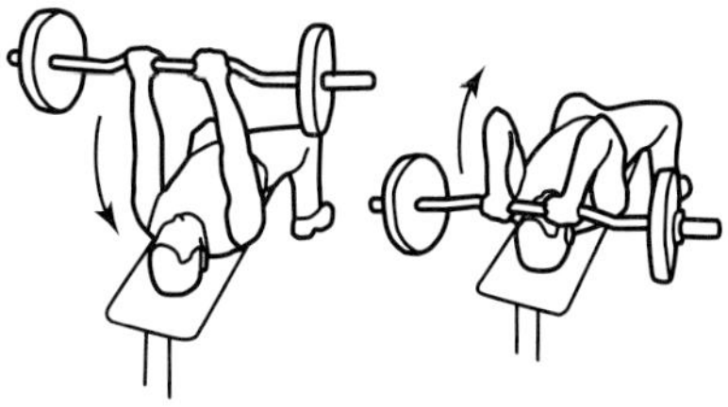 Triceps push-down 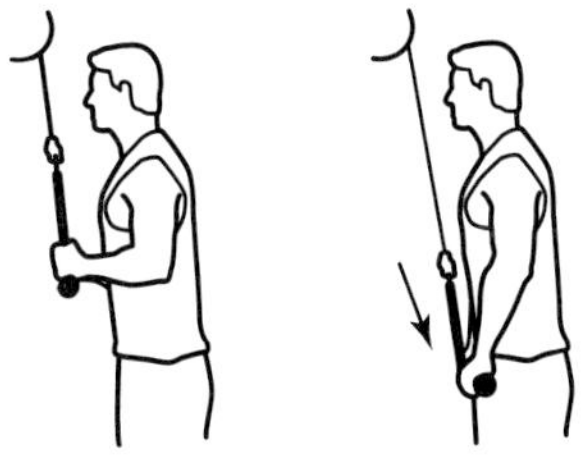	Push-up Tricep press
Pronators/ supinators	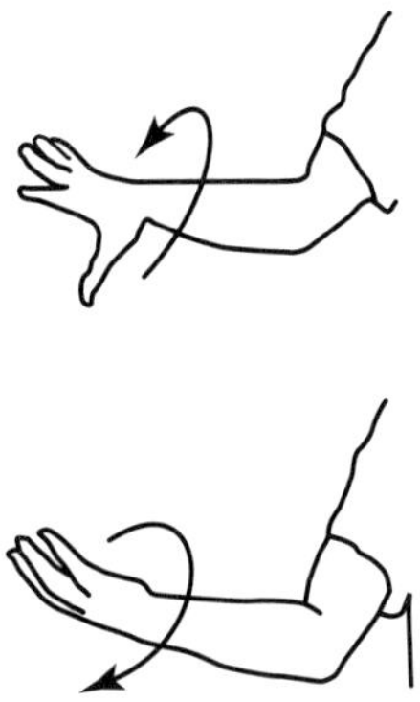	Forearm pronation   Forearm supination	Side-to-side Dumbbells Rice bucket Grabs

FIGURE 5-22 Sample stretching and strengthening exercises for selected muscle flexors, extensors, pronators, and supinators.

necessitate the use of creativity, however, because there are no standardized sets of exercises for these muscles.

INJURY POTENTIAL OF THE FOREARM

There are two categories of injuries at the elbow joint: traumatic or high-force injuries and repetitive or over-use injuries. The elbow joint is subjected to traumatic injuries caused by the absorption of a high force, such as in falling, but most of the injuries at the elbow joint result from repetitive activities, such as throwing and throwing-type actions. The high-impact or traumatic injuries are presented first, followed by the more common overuse injuries.

One of the injuries occurring as a consequence of absorbing a high force is a dislocation. These injuries usually occur in sports such as gymnastics, football, and wrestling. The athlete falls on an outstretched arm, causing a posterior dislocation (35). With the dislocation, a fracture in the medial epicondyle or the coronoid process may occur. The elbow is the second most common dislocated joint in the body (46). Other areas that may fracture with a fall include the olecranon process; the head of the radius; and the shaft of the radius, the ulna, or both. Additionally, spiral fractures of the humerus can be incurred through a fall.

Direct blows to any muscle can culminate in a condition known as myositis ossificans. In this injury, the body deposits **ectopic bone** in the muscle in response to the severe bruising and repeated stress to the muscle tissue. Although it is most common in the quadriceps femoris in the thigh, the brachioradialis muscle in the forearm is the second most common area of the body to develop this condition (35).

A high muscular force can create a **rupture** of the long head of the biceps brachii, commonly seen in adults. The joint movements facilitating this injury are arm hyperextension, forearm extension, and forearm pronation. If these three movements occur concomitantly, the strain on the biceps brachii may be significant. Finally, falling on the elbow can irritate the olecranon bursa, causing **olecranon bursitis**. This injury looks very disabling because of the swelling but is actually minimally painful (12).

The repetitive or overuse injuries occurring at the elbow can be associated with throwing or some overhead movement, such as the tennis serve. Throwing places stringent demands on the medial side of the elbow joint. Through the high-velocity actions of the throw, large tensile forces develop on the medial side of the elbow joint, compressive forces develop on the lateral side of the joint, and shear forces occur on the posterior side of the joint. A maximal valgus force is applied to the medial side of the elbow during the latter part of the cocking phase and through the initial portion of the acceleration phase. The elbow joint is injured because of the change in a varus to a valgus angle, greater forces, smaller contact areas, and contact areas that move more to the periphery as the joint moves through the throwing action (17).

The valgus force is responsible for creating the **medial tension syndrome**, or pitcher's elbow (35,89). This excessive valgus force is responsible for sprain or rupture of the ulnar collateral ligaments, medial **epicondylitis**, **tendinitis** of the forearm or wrist flexors, avulsion fractures to the medial epicondyle, and **osteochondritis dissecans** to the capitulum or olecranon (35,89). The biceps and the pronators are also susceptible to injury because they control the valgus forces and slow down the elbow in extension (45).

Medial epicondylitis is an irritation of the insertion site of the wrist flexor muscles attached to the medial epicondyle. They are stressed with the valgus force accompanied by wrist actions. This injury is seen in the trailing arm during the downswing in golf, in the throwing arm, and as a result of spiking in volleyball. Osteochondritis dissecans, a lesion in the bone and articular cartilage, commonly occurs on the capitulum as a result of compression during the valgus position that forces the radial head up against the capitulum. During the valgus overload, coupled with forearm extension, the olecranon process can be wedged against the fossa, creating an additional site for osteochondritis dissecans and breakdown in the bone. Additionally, the olecranon is subject to high tensile forces and can develop a **traction apophysitis**, or bony outgrowth, similar to that seen with the patellar ligament of the quadriceps femoris group (35).

The lateral overuse injuries to the elbow usually occur as a consequence of overuse of the wrist extensors at their attachment site on the lateral epicondyle. The overuse of the wrist extensors occurs as they eccentrically slow down or resist any flexion movement at the wrist. Lateral epicondylitis, or tennis elbow, is associated with force overload resulting from improper technique or use of a heavy racket. If the backhand stroke in tennis is executed with the elbow leading or if the performer hits the ball consistently off center, the wrist extensors and the lateral epicondyle will become irritated (44). Also, a large racket grip or tight strings may increase the load on the epicondyle by the extensors. Lateral epicondylitis is common in individuals working in occupations such as construction, food processing, and forestry in which repetitive pronation and supination of the forearm accompanies forceful gripping actions. Lateral epicondylitis is seven to 10 times more common than medial epicondylitis (86).

The Wrist and Fingers

The hand is primarily used for manipulation activities requiring very fine movements incorporating a wide variety of hand and finger postures. Consequently, there is much interplay between the wrist joint positions and efficiency of finger actions. The hand region has many stable yet very mobile segments, with complex muscle and joint actions.

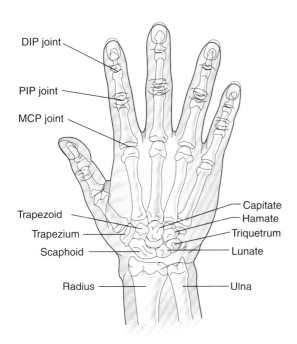

FIGURE 5-23 The wrist and hand can perform both precision and power movements because of numerous joints controlled by a large number of muscles. Most of the muscles originate in the forearm and enter the hand as tendons. DIP, distal interphalangeal; MCP, metacarpophalangeal; PIP, proximal interphalangeal.

ANATOMICAL AND FUNCTIONAL CHARACTERISTICS OF THE JOINTS OF THE WRIST AND HAND

Beginning with the most proximal joints of the hand and working distally to the tips of the fingers offers the best perspective on the interaction between segments and joints in the hand. All of the joints of the hand are illustrated in Figure 5-23. Ligaments and muscle actions for the wrist and hand are illustrated in Figures 5-24 and 5-25, respectively (also see Fig. 5-19).

Radiocarpal Joint

The wrist consists of 10 small carpal bones but can be functionally divided into the **radiocarpal** and the **midcarpal joints**. The radiocarpal joint is the articulation where movement of the whole hand occurs. The radiocarpal joint involves the broad distal end of the radius and two carpals, the scaphoid and the lunate. There is also minimal contact and involvement with the triquetrum. This ellipsoid joint allows movement in two planes: flexion–extension and radial–ulnar flexion. It should be noted that wrist extension and radial and ulnar flexion primarily occur at the radiocarpal joint but a good portion of the wrist flexion is developed at the midcarpal joints.

Distal Radioulnar Joint

Adjacent to the radiocarpal joint but not participating in any wrist movements is the distal radioulnar articulation. The ulna makes no actual contact with the carpals and is separated by a fibrocartilage disk. This arrangement is important so that the ulna can glide on the disk in pronation and supination while not influencing wrist or carpal movements.

Midcarpal and Intercarpal Joints

To understand wrist joint function, it is necessary to examine the structure and function at the joints between the carpals. There are two rows of carpals, the proximal row, containing the three carpals that participate in wrist joint function (lunate, scaphoid, and triquetrum), and the pisiform bone, which sits on the medial side of the hand, serving as a site of muscular attachment. In the distal row, there are also four carpals: the trapezium interfacing with the thumb at the saddle joint, the trapezoid, the capitate, and the hamate.

The articulation between the two rows of carpals is called the midcarpal joint, and the articulation between a pair of carpal bones is referred to as an **intercarpal**

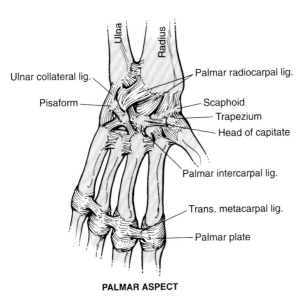

PALMAR ASPECT

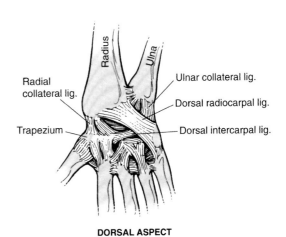

DORSAL ASPECT

FIGURE 5-24 Ligaments of the wrist and hand.

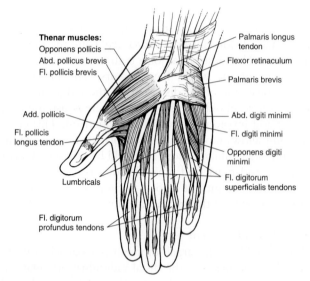

Thenar muscles:
Opponens pollicis
Abd. pollicus brevis
Fl. pollicis brevis

Add. pollicis

Fl. pollicis
longus tendon

Lumbricals

Fl. digitorum
profundus tendons

Palmaris longus
tendon

Flexor retinaculum

Palmaris brevis

Abd. digiti minimi

Fl. digiti minimi

Opponens digiti
minimi

Fl. digitorum
superficialis tendons

FIGURE 5-25 Muscles of the wrist and hand. Along with insertion and nerve supply, the muscles responsible for the noted movements (PM) and the assisting muscles (Asst) are included in the table on page 161.

joint. All of these are gliding joints in which translation movements are produced concomitantly with wrist movements. However, the proximal row of carpals is more mobile than the distal row (82). A concave transverse arch runs across the carpals, forming the carpal arch that determines the floor and walls of the carpal tunnel, through which the tendons of the flexors and the median nerve travel.

The scaphoid may be one of the most important carpals because it supports the weight of the arm, transmits forces received from the hand to the bones of the forearm, and is a key participant in wrist joint actions. The scaphoid supports the weight of the arm and transmits forces when the hand is fixed and the forearm weight is applied to the hand. Because the scaphoid interjects into the distal row of carpals, it sometimes moves with the proximal row and at other times with the distal row.

When the hand flexes at the wrist joint, the movement begins at the midcarpal joint. This joint accounts for 60% of the total range of flexion motion (86), and 40% of wrist flexion is attributable to movement of the scaphoid and lunate on the radius. The total range of motion for wrist flexion is 70° to 90°, although it is reported that only 10° to 15° of wrist flexion is needed for most daily activities involving the hand (89). Wrist flexion range of motion is reduced if flexion is performed with the fingers flexed because of the resistance offered by the finger extensor muscles.

Wrist extension is also initiated at the midcarpal joint, where the capitate moves quickly and becomes close packed with the scaphoid. This action draws the scaphoid into movements of the second row of carpals. This reverses the role of the midcarpal and radiocarpal joints to the extension movement, with more than 60%

of the movement produced at the radiocarpal joint and more than 30% at the midcarpal joint (73). This switch is attributed to the fact that the scaphoid moves with the proximal row of carpals in the flexion movement and with the distal row of carpals in extension. The range of motion for extension is approximately 70° to 80°, with approximately 35° of extension needed for daily activities (82). The range of motion of wrist extension is reduced if the extension is performed with the fingers extended.

The hand can also move laterally in radial and ulnar flexion or deviation. These movements are created as the proximal row of carpals glides over the distal row. In the radial flexion movement, the proximal carpal row moves toward the ulna and the distal row moves toward the radius. The opposite occurs for ulnar flexion. The range of motion for radial flexion is approximately 15° to 20° and for ulnar flexion is about 30° to 40° (89).

The close-packed position for the wrist, in which maximal support is offered, is in a hyperextended position. The close-packed position for the midcarpal joint is radial flexion. Both of these positions should be considered when selecting positions that maximize stability in the hand. For example, in racket sports, the wrist is most stable in a slightly hyperextended position. Also, when one falls on the hand with the arm outstretched and the wrist hyperextended, the wrist—specifically, the scaphoid carpal bone—is especially susceptible to injury because it is in the close-packed position.

Carpometacarpal Joints

Moving distally, the next articulation is the carpometacarpal (CMC) joint, which connects the carpals with each of the five fingers via the metacarpals. Each metacarpal and phalanx is also called a ray. They are numbered from the thumb to the little finger, with the thumb being the first ray and the little finger the fifth. The CMC articulation is the joint that provides the most movement for the thumb and the least movement for the fingers.

For the four fingers, the CMC joint offers very little movement, being a gliding joint that moves directionally with the carpals. The movement is very restricted at the second and third CMC but increases to allow as much as 10° to 30° of flexion and extension at the CMC joint of the ring and little fingers (89). There is also a concave transverse arch across the metacarpals of the fingers similar to that of the carpals. This arch facilitates the gripping potential of the hand.

The CMC joint of the first ray, or thumb, is a saddle joint consisting of the articulation between the trapezium and the first metacarpal. It provides the thumb with most of its range of motion, allowing for 50° to 80° of flexion and extension, 40° to 80° of abduction and adduction, and 10° to 15° of rotation (74). The thumb sits at an angle of 60° to 80° to the arch of the hand and has a wide range of functional movements (34).

The thumb can touch each of the fingers in the movement of opposition and is very important in all gripping and prehension tasks. Opposition can take place through a range of motion of approximately 90°. Without the thumb, specifically the movements allowed at the CMC joint, the function of the hand would be very limited.

Metacarpophalangeal Joints

The metacarpals connect with the phalanges to form the **metacarpophalangeal (MCP) joints**. Again, the function of the MCP joints of the four fingers differs from that of the thumb. The MCP joints of the four fingers are condyloid joints allowing movements in two planes: flexion–extension and abduction–adduction. The joint is well reinforced on the dorsal side by the dorsal hood of the fingers, on the palmar side by the palmar plates that span the joint, and on the sides by the collateral ligaments or deep transverse ligaments.

The fingers can flex through 70° to 90°, with most flexion in the little finger and least in the index finger (73). Flexion, which determines grip strength, can be more effective and produces more force when the wrist joint is held in 20° to 30° of hyperextension, a position that increases the length of the finger flexors.

Extension of the fingers at the MCP joints can take place through about 25° of motion. The extension can be limited by the position of the wrist. That is, finger extension is limited with the wrist hyperextended and enhanced with the wrist flexed.

The fingers spread in abduction and are brought back together in adduction at the MCP joint. Approximately 20° of abduction and adduction is allowed (82). Abduction is extremely limited if the fingers are flexed because the collateral ligaments become very tight and restrict movement. Thus, the fingers can be abducted when extended and then cannot be abducted or adducted when flexed around an object.

The MCP for the thumb is a hinge joint allowing motion in only one plane. The joint is reinforced with collateral ligaments and the palmar plates but is not connected with the other fingers via the deep transverse ligaments. Approximately 30° to 90° of flexion and 15° of extension can take place at this joint (82).

Interphalangeal Joints

The most distal joints in the upper extremity link are the interphalangeal (IP) articulations. Each finger has two IP joints, the proximal interphalangeal (PIP) and the **distal interphalangeal (DIP) joints**. The thumb has one IP joint and consequently has only two sections or phalanges, the proximal and distal phalanges. The fingers, however, have three phalanges, the proximal, middle, and distal. The IP joints are hinge joints allowing for movement in one plane only (flexion and extension), and they are reinforced on the lateral sides of the joints by collateral ligaments that restrict movements other than flexion and extension. The range of motion in flexion of the fingers is 110° at the PIP joint and 90° at the DIP joint and the IP joint of the thumb (82,89).

As with the MCP joint, the flexion strength at these joints determines grip strength. It can be enhanced with the wrist hyperextended by 20° and is impaired if the wrist is flexed. Various finger positions can be obtained through antagonistic and synergistic actions from other muscles so that all fingers can flex or extend at the same time. There can also be extension of the MCP with flexion of the IP and vice versa. There is usually no hyperextension allowed at the IP joints unless an individual has long ligaments that allow extension because of joint laxity.

COMBINED MOVEMENTS OF THE WRIST AND HAND

The wrist position influences the position of the metacarpal joints, and the metacarpal joints influence the position of the IP joints. This requires a balance between muscle groups. The wrist movements are usually reverse to those of the fingers because the extrinsic muscle tendons are not long enough to allow the full range of motion at the wrist and fingers (76,77). Thus, complete flexion of the fingers is generally only possible if the wrist is in slight extension, and the extension of the fingers is facilitated with synergistic action from the wrist extensors.

MUSCULAR ACTIONS

Most of the muscles that act at the wrist and finger joints originate outside the hand in the region of the elbow joint and are termed extrinsic muscles (see Fig. 5-25). These muscles enter the hand as tendons that can be quite long, as in the case of some finger tendons that eventually terminate on the distal tip of a finger. The tendons are held in place on the dorsal and palmar wrist area by extensor and flexor retinacula. These are bands of fibrous tissue running transversely across the distal forearm and wrist that keep the tendons close to the joint. During wrist and finger movements, the tendons move through considerable distances but are still maintained by the retinacula. Thirty-nine muscles work the wrist and hand, and no muscle works alone; antagonists and agonists work in pairs. Even the smallest and simplest movement requires antagonistic and agonistic action (76). The extrinsic muscles provide considerable strength and dexterity to the fingers without adding muscle bulk to the hand.

In addition to the muscles originating in the forearm, intrinsic muscles originating within the hand create movement at the MCP and IP joints. The four intrinsic muscles of the thumb form the fleshy region in the palm known as the **thenar eminence**. Three intrinsic muscles of the little finger form the smaller **hypothenar eminence**, the fleshy ridge on the little finger side of the palm.

The wrist flexors (flexor carpi ulnaris, flexor carpi radialis, and palmaris longus) are all fusiform muscles originating in the vicinity of the medial epicondyle on the humerus. These muscles run about halfway along the forearm before becoming a tendon. The flexor carpi radialis and flexor carpi ulnaris contribute the most to wrist flexion. The palmaris longus is variable and may be as small as a tendon or even absent in about 13% of the population (73). The strongest flexor of the group, the flexor carpi ulnaris, gains some of its power by encasing the pisiform bone and using it as a sesamoid bone to increase mechanical advantage and reduce the overall tension on the tendon. Because most activities require the use of a small amount of wrist flexion, attention should always be given to the conditioning of this muscle group.

The wrist extensors (extensor carpi ulnaris, extensor carpi radialis longus, and extensor carpi radialis brevis) originate in the vicinity of the lateral epicondyle. These muscles become tendons about one-third of the way along the forearm. The wrist extensors also act and create movements at the elbow joint. Thus, elbow joint position is important for wrist extensor function. The extensor carpi radialis longus and extensor carpi radialis brevis create flexion at the elbow joint and thus can be enhanced as a wrist extensor with extension at the elbow. The extensor carpi ulnaris creates extension at the elbow and is enhanced as a wrist extensor in elbow flexion. Also, wrist extension is an important action accompanying and supporting a gripping action using finger flexion. Thus, the wrist extensor muscles are active with this activity.

The wrist flexors and extensors pair up to produce ulnar and radial flexion. Ulnar flexion is produced by the ulnaris wrist muscles, consisting of the flexor carpi ulnaris and the extensor carpi ulnaris. Likewise, radial flexion is produced by the flexor carpi radialis, extensor carpi radialis longus, and extensor carpi radialis brevis. Radial flexion joint movement, although it has just half the range of motion of ulnar flexion, is important in many racket sports because it creates the close-packed position of the wrist, thus stabilizing the hand (82).

Finger flexion is performed primarily by the flexor digitorum profundus and flexor digitorum superficialis. These extrinsic muscles originate in the vicinity of the medial epicondyle. The flexor digitorum profundus cannot independently flex each finger. Thus, flexion at the middle, ring, and little fingers usually occurs together because the flexor tendons all arise from a common tendon and muscle. Because of the separation of the flexor digitorum profundus muscle and tendon for this digit, the index finger can independently flex.

The flexor digitorum superficialis is capable of flexing each finger independently. The fingers can be independently flexed at the PIP but not at the DIP joint. Flexion of the little finger is also assisted by one of the intrinsic muscles, the flexor digiti minimi brevis. Flexion of the fingers at the MCP articulation is produced by the lumbricals and the interossei, two sets of intrinsic muscles that lie in the palm and between the metacarpals. These muscles also produce extension at the IP joints because they attach to the fibrous extensor hood running the length of the dorsal surface of the fingers. Consequently, to achieve full flexion of the MCP, PIP, and DIP joints, the long finger flexors

must override the extension component of the lumbricals and interossei. This is easier if tension is taken off the extensors by some wrist extension.

Extension of the fingers is created primarily by the extensor digitorum muscle. This muscle originates at the lateral epicondyle and enters the hand as four tendon slips that branch off at the MCP articulation. The tendons create a main slip that inserts into the extensor hood and two collateral slips that connect into adjacent fingers. The extensor hood, formed by the tendon of the extensor digitorum and fibrous connective tissue, wraps around the dorsal surface of the phalanges and runs the total length of the finger to the distal phalanx. The structures in the finger are shown in Figure 5-26.

Because the lumbricals and interossei connect into this hood, they also assist with extension of the PIP and DIP joints. Their actions are facilitated as the extensor digitorum contracts, applying tension to the extensor hood and stretching these muscles (82).

Abduction of fingers two, three, and four is performed by the dorsal interossei. The dorsal interossei consist of four intrinsic muscles lying between the metacarpals. They connect to the lateral sides of digits two and four and to both sides of digit three. The little finger, digit five, is abducted by one of its intrinsic muscles, the abductor digiti minimi brevis.

The three palmar interossei, lying on the medial side of digits two, four, and five, pull the fingers back into adduction. The middle finger is adducted by the dorsal interossei, which is connected to both sides of the middle finger. Abduction and adduction movements are necessary for grasping, catching, and gripping objects. When the fingers

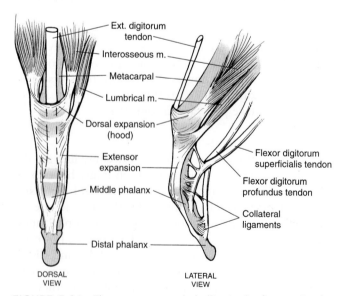

FIGURE 5-26 There are no muscle bellies in the fingers. On the dorsal surface of the fingers are the extensor expansion and the extensor hood, to which the finger extensors attach. Tendons of the finger flexors travel the ventral surface of the fingers. The fingers flex and extend as tension is generated in the tendons via muscular activity in the upper forearm.

are flexed, abduction is severely limited by the tightening of the collateral ligament and the limited length–tension relationship in the interossei, which are also flexors of the MCP joint.

The thumb has eight muscles that control and generate an expansive array of movements. The muscles of the thumb are presented in Figure 5-25. Opposition is the most important movement of the thumb because it provides the opportunity to pinch, grasp, or grip an object by bringing the thumb across to meet any of the fingers. Although all of the hypothenar muscles contribute to opposition, the main muscle responsible for initiating the movement is the opponens pollicis. The little finger is also assisted in opposition by the opponens digiti minimi.

STRENGTH OF THE HAND AND FINGERS

Strength in the hand is usually associated with grip strength, and there are many ways to grasp or grip an object. Whereas a firm grip requiring maximum output uses the extrinsic muscles, fine movements, such as a pinch, use more of the intrinsic muscles to fine-tune the movements.

In a grip, the fingers flex to wrap around an object. If a **power grip** is needed, the fingers flex more, with the most powerful grip being the fist position with flexion at all three finger joints, the MCP, PIP, and DIP. If a fine **precision grip** is required, there may be only limited flexion at the PIP and DIP joints, and only one or two fingers may be involved, such as in pinching and writing (89). Examples of both power and precision grips are shown in Figure 5-27. The thumb determines whether a fine preci-

sion position or power position is generated. If the thumb remains in the plane of the hand in an adducted position and the fingers flex around an object, a power position is created. An example of this is the grip used in the javelin throw and in the golf swing. This power position still allows for some precision, which is important in directing the golf club or the javelin.

Power in the grip can be enhanced by producing a fist with the thumb wrapped over the fully flexed fingers. With this grip, there is minimal, if any, precision. In activities that require precise actions, the thumb is held more perpendicular to the hand and moved into opposition, with limited flexion at the fingers. An example of this type of position is in pitching, writing, and pinching. In a pinch or prehensile grip, greater force can be generated if the pulp of the thumb is placed against the pulps of the index and long fingers. This pinch is 40% stronger than the pinch grip with the tips of the thumb and fingers (39).

The strength of a grip can be enhanced by the position of the wrist. Placing the wrist in slight extension and ulnar flexion increases the finger flexion strength. The least finger strength can be generated in a flexed and radial flexed wrist position. Grip strength at approximately 40° of wrist hyperextension is more than three times that of grip strength measured in 40° of wrist flexion (89). The strength of the grip may increase with specific wrist positioning, but the incidence of strain or impingement on structures around the wrist also increases. The neutral position of the wrist is the safest position because it reduces strain on the wrist structures.

The strongest muscles in the hand region, capable of the greatest work capacity, in order from high to low are the flexor digitorum profundus, flexor carpi ulnaris, extensor digitorum, flexor pollicis longus, extensor carpi ulnaris, and extensor carpi radialis longus. Two muscles that are weak and capable of little work capacity are the palmaris longus and the extensor pollicis longus.

CONDITIONING

There are three main reasons why people condition the hand region. First, the fingers can be strengthened to enhance the grip strength in athletes who participate in racket sports, individuals who work with implements, and individuals who lack the ability to grasp or grip objects.

Second, the muscles acting at the wrist joint are usually strengthened and stretched to facilitate a wrist position for racket sports or to enhance wrist action in a throwing or striking event, such as volleyball. Wrist extension draws the hand back, and wrist flexion snaps the hand forward in activities such as serving and spiking in volleyball, dribbling in basketball, and throwing a baseball. Even though the speed of the flexion and extension movement may be determined by contributions from adjacent joints, strengthening the wrist flexor and extensor muscles enhances the force production. Commonly, the wrist is maintained in a position so that an efficient force application

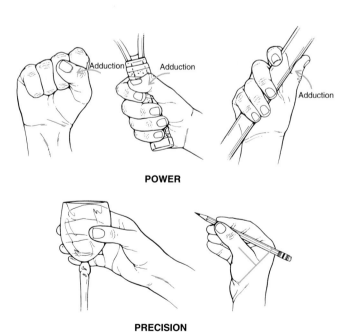

POWER

PRECISION

FIGURE 5-27 If power is needed in a grip, the fingers flex at all three joints to form a fist. Also, if the thumb adducts, the grip is more powerful. A precision grip usually involves slight flexion at a small number of finger joints with the thumb perpendicular to the hand.

can occur. In tennis and racket sports, for example, the wrist is held either in the neutral position or in a slightly radially flexed position. If the wrist is held stationary, the force applied to the ball by the racket will not be lost through movements occurring at the wrist. This position is maintained by both the wrist flexor and wrist extensor muscles. Another example of maintaining wrist position is in the volleyball underhand pass, in which the wrist is maintained in an ulnar flexed position. This opens up a broader area for contact and locks the elbows so they maintain an extended position upon contact. The wrist must be maintained in a stable, static position to achieve maximal performance from the fingers. Thus, while playing a piano or typing, the wrist must be maintained in the optimal position for finger usage. This is usually a slight hyperextended position via the wrist extensors.

The final reason for conditioning the hand region is to reduce or prevent injury. The tension developed in the hand and finger flexor and extensor muscles places considerable strain on the medial and lateral aspect of the elbow joint. Some of this strain can be reduced through stretching and strengthening exercises.

Overall, the conditioning of the hand region is relatively simple and can be done in a very limited environment with minimal equipment. Examples of some flexibility and resistance exercises for the wrist flexors and extensors and the fingers are presented in Figure 5-28. Wrist curls and tennis ball gripping exercises are the most popular for this region.

INJURY POTENTIAL OF THE HAND AND FINGERS

Many injuries can occur to the hand as a result of absorbing a blunt force, as in impact with a ball, the ground, or another object. Injuries of this type in the wrist region are usually associated with a fall, forcing the wrist into extreme flexion or extension. In this case, extreme hyperextension is the most common injury. This can result in a sprain of the wrist ligaments, a strain of the wrist muscles, a fracture of the scaphoid (70%) or other carpals (30%), a fracture of the distal radius, or a dislocation between the carpals and the wrist or other carpals (48).

The distal end of the radius is one of the most frequently fractured areas of the body because the bone is not dense and the force of the fall is absorbed by the radius. A common fracture of the radius, Colles fracture, is a diagonal fracture that forces the radius into more radial flexion and shortens it. These injuries are associated mainly with activities such as hockey, fencing, football, rugby, skiing, soccer, bicycling, parachuting, mountain climbing, and hang gliding in which the chance of a blunt macrotrauma is greater than in other activities.

Examples of injuries to the fingers and the thumb as a result of blunt impact are fractures, dislocations, and tendon avulsions. The thumb can be injured by jamming it or forcing it into extension, causing severe strain of the thenar muscles and the ligaments surrounding the MCP joint. **Bennett fracture** is a common fracture to the

thumb at the base of the first metacarpal. Thumb injuries caused by jamming by the pole are common in skiing (83). Thumb injuries are also common in biking (71).

The fingers are also frequently fractured or dislocated by an impact on the tip of the finger, forcing it into extreme flexion or extension. Fractures are relatively common in the proximal phalanx and rare in the middle phalanx. High-impact collisions with the hand, such as in boxing and the martial arts, result in more fractures or dislocations of the ring and little fingers because they are least supported in a fist position.

Finger flexor or extensor mechanisms can be disrupted with a blow, forcing the finger into extreme positions. **Mallet finger** is an avulsion injury to the extensor tendon at the distal phalanx caused by forced flexion, resulting in the loss of the ability to extend the finger. **Boutonniere deformity**, caused by avulsion or stretching of the middle branch of the extensor mechanism, creates a stiff and immobile PIP articulation (73). Avulsion of the finger flexors is called **jersey finger** and is caused by forced hyperextension of the distal phalanx. The finger flexors can also develop nodules, a **trigger finger**. This results in snapping during flexion and extension of the fingers. These finger and thumb injuries are also commonly associated with the sports and activities listed above because of the incidence of impact occurring to the hand region.

There are also overuse injuries associated with repetitive use of the hand in sports, work, or other activities. **Tenosynovitis** of the radial flexors and thumb muscles is common in activities such as canoeing, rowing, rodeo, tennis, and fencing. Tennis and other racket sports, golf, throwing, javelin, and hockey, in which the wrist flexors and extensors are used to stabilize the wrist or create a repetitive wrist action, are susceptible to tendinitis of the wrist muscles inserting into the medial and lateral epicondyles. Medial or lateral epicondylitis may also result from this overuse. Medial epicondylitis is associated with overuse of the wrist flexors, and lateral epicondylitis is associated with overuse of the wrist extensors.

A disabling overuse injury to the hand is **carpal tunnel syndrome**. Next to low-back injuries, carpal tunnel syndrome is one of the most frequent work injuries reported by the medical profession. The floor and sides of the carpal tunnel are formed by the carpals, and the top is formed by the transverse carpal ligament. Traveling through this tunnel are all of the wrist flexor tendons and the median nerve (Fig. 5-29). Through repetitive actions at the wrist, usually repeated wrist flexion, the wrist flexor tendons may be inflamed to the point where there is pressure and constriction of the median nerve. The median nerve innervates the radial side of the hand, specifically the thenar muscles of the thumb. Impingement of this nerve can cause pain, atrophy of the thenar muscles, and tingling sensations in the radial side of the hand.

To eliminate this condition, the source of the irritation must be removed by examining the workplace environment; a wrist stabilizing device can be applied to reduce

Muscle Group	Sample Stretching Exercise	Sample Strengthening Exercise	Other Exercises
Flexors		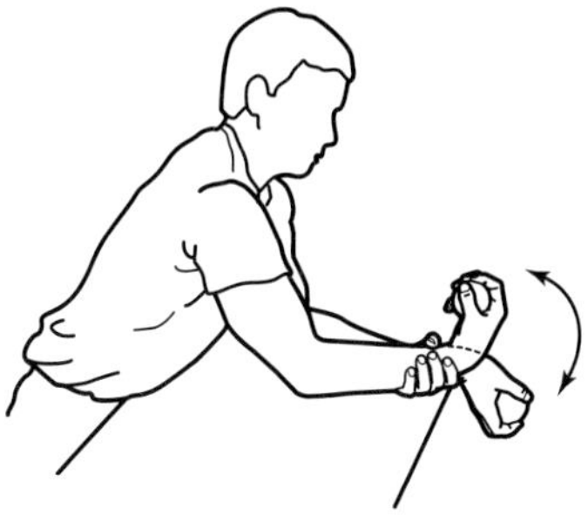	Rice bucket grabs Manual resistance
Extensors	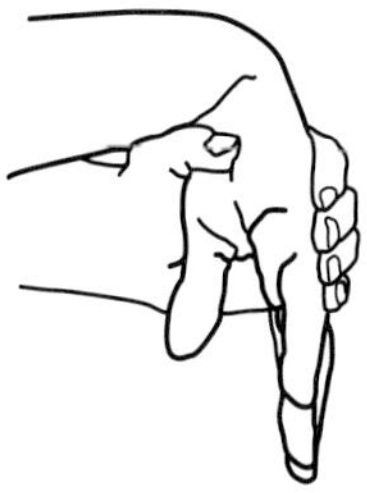	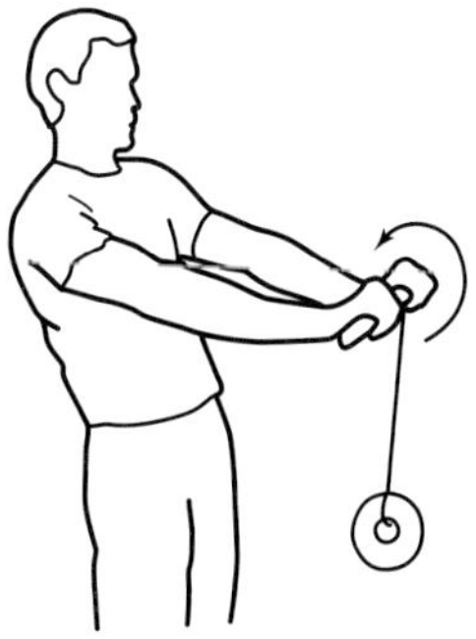	Fold hands together Reverse curl
Finger flexors	Straighten all fingers Spread all fingers 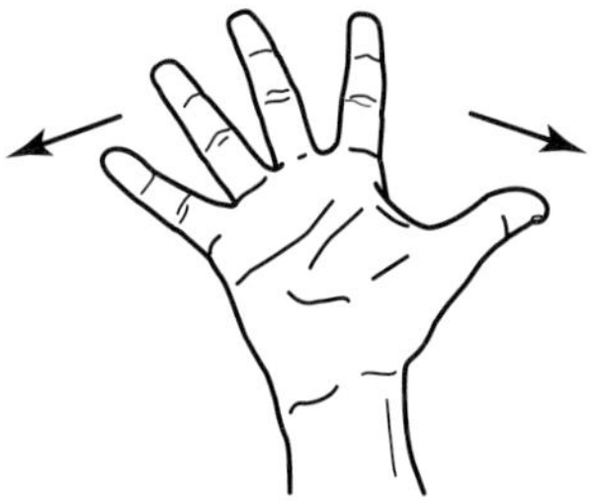  Large circle with thumb	Ball squeeze 	Make a fist

FIGURE 5-28 Sample stretching and strengthening exercises for wrist flexors and extensors and the finger flexors.

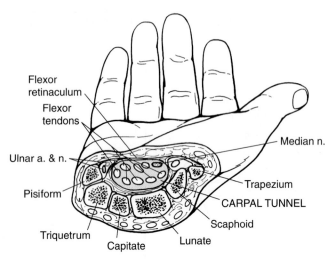

Flexor
retinaculum

Flexor
tendons

Median n.

Ulnar a. & n.

Trapezium

Pisiform

CARPAL TUNNEL

Scaphoid

Triquetrum

Capitate   Lunate

FIGURE 5-29 The floor and sides of the carpal tunnel are formed by the carpals, and the top of the tunnel is covered by ligament and the flexor retinaculum. Within the tunnel are wrist flexor tendons and the median nerve. Overuse of the wrist flexors can impinge the median nerve, causing carpal tunnel syndrome.

the magnitude of the flexor forces; or a surgical release can be administered. It is recommended that the wrist be maintained in a neutral position while performing tasks in the workplace to avoid carpal tunnel syndrome.

Ulnar nerve injuries can also result in loss of function to the ulnar side of the hand, specifically the ring and little fingers. Damage to this nerve can occur as a result of trauma to the elbow or shoulder region. Ulnar neuropathy is associated with activities such as cycling (56).

Contribution of Upper Extremity Musculature to Sport Skills or Movements

To fully appreciate the contribution of a muscle or muscle group to an activity, the activity or movement of interest must be evaluated and studied. This provides an understanding of the functional aspect of the movement, ideas for training and conditioning of the appropriate musculature, and a better comprehension of injury sites and mechanisms. The upper extremity muscles are important for the completion of many daily activities. For example, pushing up to get out of a chair or a wheelchair places a tremendous load on the upper extremity muscles because the full body weight is supported in the transfer from a sitting to a standing position (5,25). If you simply push up out of a chair or wheelchair, the muscle primarily used is the triceps brachii, followed by the pectoralis major, with some minimal contribution from the latissimus dorsi.

Upper extremity muscles are important contributors to a variety of physical activities. For example, in freestyle swimming, propelling forces are generated by the motion of the arms through the water. Internal rotation and

adduction are the primary movements in the propulsion phase of swimming and use the latissimus dorsi, teres major, and pectoralis major muscles (55,61). Also, as the arm is taken out of the water to prepare for the next stroke, the supraspinatus and infraspinatus (abduction and external rotation of the humerus), middle deltoid (abduction), and serratus anterior (very active in the hand lift as it rotates the scapula) are active. Swimming incorporates a high amount of upper extremity muscle actions.

A more thorough review of muscular activity is provided for the overhand throw and the golf swing. These are examples of a functional anatomy description of a movement and are gathered primarily from electromyographic research. Each activity is first broken down into phases. Next, the level of activity in the muscle is described as being low, moderate, or high. Finally, the action of the muscle is identified along with the movement it is concentrically generating or eccentrically controlling. It is important to note that these examples may not include all of the muscles that might be active in these activities but only the major contributing muscles.

OVERHAND THROWING

Throwing places a great deal of strain on the shoulder joint and requires significant upper extremity muscular action to control and contribute to the throwing movement even though the lower extremity is a major contributor to the power generation in a throw.

The throwing action described in this section is a pitch in baseball from the perspective of a right-hand thrower (Fig. 5-30). From the windup through the early cocking phases, the front leg strides forward and the hand and ball are moved as far back behind the body as possible. In late cocking, the trunk and legs rotate forward as the arm is maximally abducted and externally rotated (21,22,55). In these phases, the deltoid and the supraspinatus muscles are active in producing the abduction of the arm. The infraspinatus and the teres minor are also active, assisting with abduction and initiating the external rotation action. The subscapularis is also minimally active to assist during the shoulder abduction. During the late cocking phase, the latissimus dorsi and the pectoralis major muscles demonstrate a rapid increase in activity as they eccentrically act to slow the backward arm movement and concentrically act to initiate forward movement.

In the cocking phase of the throw, the biceps brachii and the brachialis are active as the forearm flexes and the arm is abducted. The activity of the triceps brachii begins at the end of the cocking phase, when the arm is in maximum external rotation and the elbow is maximally flexed. There is a co-contraction of the biceps brachii and the triceps brachii at this time. Additionally, the forearm is pronated to 90° at the end of the cocking phase via the pronator teres and pronator quadratus (37,54).

Muscles previously active in the early portion of the cocking phase also change their level of activity as the

FIGURE 5-30 Upper extremity muscles involved in the overhead throw showing the level of muscle activity (low, moderate, and high) and the type of muscle action (concentric [CON] and eccentric [ECC]) with the associated purpose.

arm nears the completion of this phase. Teres minor and infraspinatus activities increase at the end of the cocking phase to generate maximal external rotation. The activity of the supraspinatus increases as it maintains the abduction late into the cocking phase. Subscapularis activity also increases to maximum levels in preparation for the acceleration of the arm forward. The deltoid is the only muscle whose activity diminishes late in the cocking phase (55).

At the end of the cocking phase, the external rotation motion is terminated by the anterior capsule and ligaments and the actions of the subscapularis, pectoralis major, triceps brachii, teres major, and latissimus dorsi muscles. Consequently, in this phase of throwing, the anterior capsule and ligaments and the tissue of the specified muscles are at greatest risk for injury (21,55). Examples of injuries developing in this phase are tendinitis of the insertion of the subscapularis and strain of the pectoralis major, teres major, and latissimus dorsi muscle.

The acceleration phase is an explosive action characterized by the initiation of elbow extension, arm internal rotation with maintenance of 90° of abduction, scapula protraction or abduction, and some horizontal flexion as the arm moves forward. The muscles most active in the acceleration phase are those that act late in the cocking phase, including the subscapularis, latissimus dorsi, teres major, and pectoralis major, which generate the horizontal flexion and the internal rotation movements; the serratus anterior, which pulls the scapula forward into protraction or abduction; and the triceps brachii, which initiates and controls the extension of the forearm. Sites of irritation and strain in this phase of the throw are found at the sites of the muscular attachment and in the subacromial area.

This area is subjected to compression during adduction and internal rotation in this phase.

The last phase of throwing is the follow-through or deceleration phase. In this phase, the arm travels across the body in a diagonal movement and eventually stops over the opposite knee. This phase begins after the ball is released. In the early portion of this phase, after maximum internal rotation in the joint is achieved, a very quick muscular action takes place, resulting in external rotation and horizontal flexion of the arm. After this into the later stages of the follow-through are trunk rotation and replication of the shoulder and scapular movements of the cocking phase. This includes an increase in the activity of the deltoid as it attempts to slow the horizontally flexed arm; the latissimus dorsi as it creates further internal rotation; the trapezius, which creates slowing of the scapula; and the supraspinatus, to maintain the arm abduction and continue to produce internal rotation (37,55). There is also a very rapid increase in the activity of the biceps brachii and the brachialis in the follow-through phase as these muscles attempt to reduce the tensile loads on the rapidly extending forearm. In this phase of throwing, the posterior capsule and the corresponding muscles and the biceps brachii (6) are at risk for injury because they are rapidly stretched.

THE GOLF SWING

The golf swing presents a more complicated picture of the shoulder muscle function than throwing because the left and right arms must work in concert (Fig. 5-31). That is, the arms produce opposite movements and use opposing muscles. In the backswing for a right-handed golfer, the club is brought up and back behind the body as the left arm comes across the body and the right arm abducts minimally (38). The shoulder's muscular activity in this phase is minimal except for moderate subscapularis activity on the target arm to produce internal rotation and marked activity from the supraspinatus on the trailing arm to abduct the arm (50,55). In the shoulder girdle, all parts of the trapezius of the trailing side work together with the levator scapula and rhomboid to elevate and adduct the scapula. On the target side, the serratus anterior protract the scapula.

In the forward swing, movement of the club is initiated by moderate activity from the latissimus dorsi and subscapularis muscles on the target side. On the trailing side, there is accompanying high activity from the pectoralis major, moderate activity from the latissimus dorsi and subscapularis, and minimal activity from the supraspinatus and deltoid. In the shoulder girdle, the trapezius, rhomboid, and levator scapula of the target arm are active as the scapula is adducted. The serratus anterior is also active in the trailing limb as the scapula is abducted. This phase brings the club around to shoulder level through continued internal rotation of the left arm and the initiation of internal rotation with some adduction of the right arm.

The acceleration phase begins when the arms are at approximately shoulder level and continues until the club makes contact with the ball. On the target side, there is substantial muscular activity in the pectoralis major, latissimus dorsi, and subscapularis as the arm is extended and maintained in internal rotation. On the trailing side, there is even greater activity from these same three muscles as the arm is brought vigorously downward (50,55).

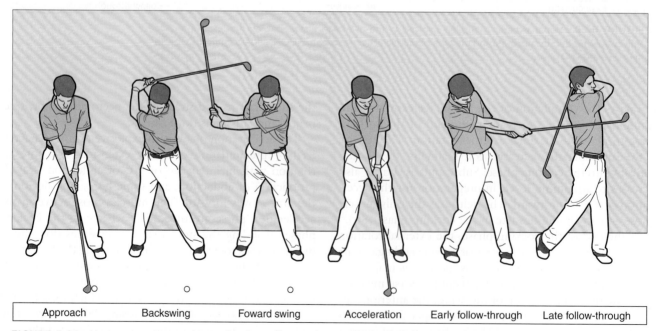

| Approach | Backswing | Foward swing | Acceleration | Early follow-through | Late follow-through |

FIGURE 5-31 Upper extremity muscles used in the golf swing showing the level of muscle activity (low, moderate, and high) and the type of muscle action (concentric [CON] and eccentric [ECC]).

As soon as contact with the ball is made, the follow-through phase begins with continued movement of the arm and club across the body to the target side. This action must be decelerated. In the follow-through phase, the target side has high activity in the subscapularis and moderate activity in the pectoralis major, latissimus dorsi, and infraspinatus as the upward movement of the arm is curtailed and slowed (55). It is here, in the follow-through phase, that considerable strain can be placed on the posterior portion of the trailing shoulder and the anterior portion of the target shoulder during the rapid deceleration.

External Forces and Moments Acting at Joints in the Upper Extremity

Muscle activity in the shoulder complex generates high forces in the shoulder joint itself. The rotator cuff muscle group as a whole, capable of generating a force 9.6 times the weight of the limb, generates maximum forces at 60° of abduction (89). Because each arm constitutes approximately 7% of body weight, the rotator cuff generates a force in the shoulder joint equal to approximately 70% of body weight. At 90° of abduction, the deltoid generates a force averaging eight to nine times the weight of the limb, creating a force in the shoulder joint ranging from 40% to 50% of body weight (89). In fact, the forces in the shoulder joint at 90° of abduction have been shown to be close to 90% of body weight. These forces can be significantly reduced if the forearm is flexed to 90° at the elbow.

In throwing, compressive forces have been measured in the range of 500 to 1,000 N (1,23,52,84) with anterior forces ranging from 300 to 400 N (52). In a tennis serve, forces at the shoulder have been recorded to be 423 and 320 N in the compressive and mediolateral directions, respectively (60). As a comparison, lifting a block to head height has been shown to generate 52 N of force (57), and crutch and cane walking have generated forces at the shoulder of 49 and 225 N, respectively (7,31).

The load-carrying capacity of the elbow joint is also considerable. In a push-up, the peak axial forces on the elbow joint average 45% of body weight (2,18). These forces depend on hand position, with the force reduced to 42.7% of body weight with the hands farther apart than normal and increased to 65% of body weight in the one-handed push-up (16). Radial head forces are greatest from 0° to 30° of flexion and are always higher in pronation. Joint forces at the ulnohumeral joint can range from one to three body weights (~750 to 2500 N) with strenuous lifting (24).

Sample Upper Extremity Joint Torques

Activity	Joint	Moments
Cane walking (7)	Shoulder	24.0 Nm
Lifting a 5-kg box from floor to shoulder height (7)	Shoulder	21.8 Nm
Lifting and walking with a 10-kg suitcase (7)	Shoulder	27.9 Nm
Lifting a block to head height (57)	Shoulder	14 Nm
	Elbow	5.8 Nm
Push-up (18)	Elbow	24.0 Nm
Rock climbing crimp grip (81)	Fingers (DIP)	26.4 Nm
Sit to stand (7)	Shoulder	16.2 Nm
Stand to sit (7)	Shoulder	12.3 Nm
Tennis serve (60)	Shoulder	94 Nm internal rotation torque
	Elbow	106 Nm varus torque
Follow-through phase of throwing (84)	Elbow	55 Nm flexion torque
Late cock phase of throwing (1,23,84)	Elbow	54–120 Nm of varus torque
Weight lifting (8)	Shoulder	32–50 Nm
Wheelchair propulsion (79)	Shoulder	50 Nm
Wheelchair propulsion (80)	Shoulder	−7.2 Nm level propulsion (paraplegia)
		−14.6 Nm propulsion up slope (paraplegia)
	Elbow	−3.0 Nm level propulsion (paraplegia)
		5.7 Nm propulsion up slope (paraplegia)

⚙ Summary

The upper extremity is much more mobile than the lower extremity, even though the extremities have structural similarities. There are similarities in the connection into girdles, the number of segments, and the decreasing size of the bones toward the distal end of the extremities.

The shoulder complex consists of the sternoclavicular joint, the AC joint, and the glenohumeral joint. The sternoclavicular joint is very stable and allows the clavicle to move in elevation and depression, protraction and retraction, and rotation. The AC joint is a small joint that allows the scapula to protract and retract, elevate and depress, and rotate up and down. The glenohumeral joint provides movement of the humerus through flexion and extension, abduction and adduction, medial and lateral rotation, and combination movements of horizontal abduction and adduction and circumduction. A final articulation, the scapulothoracic joint, is called a *physiologic joint* because of the lack of connection between two bones. It is here that the scapula moves on the thorax.

There is considerable movement of the arm at the shoulder joint. The arm can move through 180° of abduction, flexion, and rotation because of the interplay between movements occurring at all of the articulations. The timing of the movements between the arm, scapula, and clavicle is termed the *scapulohumeral rhythm*. Through 180° of elevation (flexion or abduction), there is approximately 2:1 degrees of humeral movement to scapular movement.

The muscles that create movement of the shoulder and shoulder girdle are also important for maintaining stability in the region. In abduction and flexion, for example, the deltoid produces about 50% of the muscular force for the movement, but it requires assistance from the rotator cuff (teres minor, subscapularis, infraspinatus, and supraspinatus) to stabilize the head of the humerus so that elevation can occur. Also, the shoulder girdle muscles contribute as the serratus anterior and the trapezius assist to stabilize the scapula and produce accompanying movements of elevation, upward rotation, and protraction.

To extend the arm against resistance, the latissimus dorsi, teres major, and pectoralis major act on the humerus and are joined by the rhomboid and the pectoralis minor, which retract, depress, and downwardly rotate the scapula. Similar muscular contributions are made by the infraspinatus and teres minor in external rotation of the humerus and the subscapularis, latissimus dorsi, teres major, and pectoralis major in internal rotation.

The shoulder muscles can generate considerable force in adduction and extension. The next strongest movement is flexion, and the weakest movements are abduction and rotation. The muscles surrounding the shoulder joint are capable of generating high forces in the range of eight to nine times the weight of the limb.

Conditioning of the shoulder muscles is relatively easy because of the mobility of the joint. Numerous strength and flexibility exercises are used to isolate specific muscle groups and to replicate an upper extremity pattern used in a skill. Special exercise considerations for individuals with shoulder injuries should exclude exercises that create impingement in the joint.

Injury to the shoulder complex can be acute in the case of dislocations of the sternoclavicular or glenohumeral joints and fractures of the clavicle or humerus. Injuries can also be chronic, as with bursitis and tendinitis. Common injuries associated with impingement of the shoulder joint are subacromial bursitis, bicipital tendinitis, and tears in the supraspinatus muscle.

The elbow and the radioulnar joints assist the shoulder in applying force and placing the hand in a proper position for a desired action. The joints that make up the elbow joint are the ulnohumeral and radiohumeral joints, where flexion and extension occur, and the superior radioulnar joint, where pronation and supination of the forearm occur. The region is well supported by ligaments and the interosseous membrane running between the radius and the ulna. The joint structures allow approximately 145° to 160° of flexion and 70° to 85° of pronation and supination.

Twenty-four muscles span the elbow joint, and these can be further classified into flexors (biceps brachii, brachioradialis, brachialis, pronator teres, and extensor carpi radialis), extensors (triceps brachii and anconeus), pronators (pronator quadratus and pronator teres), and supinators (biceps brachii and supinator). The flexor muscle group is considerably stronger than the extensor group. Maximum flexion strength can be developed from the semiprone forearm position. Extension strength is maximum in a flexion position of 90°. Pronation and supination strength is also maximum from the semiprone position.

The elbow and forearm are vulnerable to injury as a result of falling or repetitive overuse. In absorbing high forces, the elbow can dislocate or fracture or muscles can rupture. Through overuse, injuries such as medial or lateral tension syndrome can produce epicondylitis, tendinitis, or avulsion fractures.

The wrist and hand consist of complex structures that work together to provide fine movements used in a variety of daily activities. The main joints of the hand are the radiocarpal joint, inferior radioulnar joint, midcarpal and intercarpal joints, CMC joints, MCP joints, and IP joints. The hand is capable of moving through 70° to 90° of wrist flexion, 70° to 80° of extension, 15° to 20° of radial flexion, and 30° to 40° of ulnar flexion. The fingers can flex through 70° to 110°, depending on the actual joint of interest (MCP or IP), 20° to 30° of hyperextension, and 20° of abduction. The thumb has special structural and functional characteristics that are related to the role of the CMC joint.

The extrinsic muscles that act on the hand enter the region as tendons. The muscles work in groups to produce wrist flexion (flexor carpi ulnaris, flexor carpi radialis, and palmaris longus), extension (extensor carpi ulnaris, extensor carpi radialis longus, and extensor carpi radialis brevis), ulnar flexion (flexor carpi ulnaris and extensor carpi ulnaris), and radial flexion (flexor carpi radialis, extensor carpi radialis longus, and extensor carpi radialis brevis). Finger flexion is produced by the flexor digitorum profundus and flexor digitorum superficialis, and extension is produced primarily by the extensor digitorum. The fingers are abducted by the dorsal interossei and adducted by the palmar interossei.

Strength in the fingers is important in activities and sports in which a firm grip is essential. Grip strength can be enhanced by placing the thumb in a position parallel with the fingers (fist position). When precision is required, the thumb should be placed perpendicular to the fingers. The muscles of the hand can be exercised via a series of exercises that incorporates various wrist and finger positions.

The fingers and hand are frequently injured because of their vulnerability, especially when performing activities such as catching balls. Sprains, strains, fractures, and dislocations are common results of injuries sustained by the fingers or hands in the absorption of an external force. Other common injuries in the hand are associated with overuse, including medial or lateral tendinitis or epicondylitis and carpal tunnel syndrome.

The upper extremity muscles are very important contributors to specific sport skills and movements. In the push-up, for example, the pectoralis major, latissimus dorsi, and triceps brachii are important contributors. In swimming, the latissimus dorsi, teres major, pectoralis major, supraspinatus, infraspinatus, middle deltoid, and serratus anterior make important contributions. In throwing, the deltoid, supraspinatus, infraspinatus, teres minor, subscapularis, trapezius, rhomboid, latissimus dorsi, pectoralis major, teres major, and deltoid all contribute. In the forearm, the triceps brachii is an important contributor to raising from a chair, wheelchair activities, and throwing. Likewise, the biceps brachii and the pronator muscles are important in various phases of throwing.

The upper extremity is subject to a variety of loads, and loads as high as 90% of body weight can be applied to the shoulder joint as a result of muscle activity and other external forces. At the elbow, forces as high as 45% of body weight have been recorded. These forces are increased and decreased with changing joint positions and muscular activity.

REVIEW QUESTIONS

True or False

1. ____ The brachialis muscle is the elbow flexor with the smallest moment arm at 100°.

2. ____ The shoulder complex is composed of three joints.

3. ____ The acromioclavicular joint connects the upper extremity to the trunk.

4. ____ In elevation and depression the clavicle moves anteriorly and posteriorly.

5. ____ The scapula is connected to the humerus at the scapulothoracic joint.

6. ____ Adduction strength of the shoulder is greater than abduction strength.

7. ____ The scapulothoracic joint is a physiologic joint.

8. ____ The glenoid labrum is a ligament that crosses the glenohumeral joint.

9. ____ The glenoid fossa is the socket of a ball-and-socket joint.

10. ____ The coracohumeral ligament is taut when the arm is adducted.

11. ____ Scapulohumeral rhythm refers to the movement of the shoulder complex as a whole.

12. ____ External rotation is the strongest joint action in the shoulder.

13. ____ The carrying angle at the elbow is lesser in females.

14. ____ The lateral epicondyle is a site of injury because of tension in the wrist extensors.

15. ____ The clavicle is frequently injured by direct trauma.

16. ____ The biceps brachii is most effective as a flexor when the forearm is supinated.

17. ____ Typical carrying angles are 120° to 180°.

18. ____ There are two radioulnar articulations in each forearm.

19. ____ The annular ligament wraps around the radius at the distal end.

20. ____ The distal end of the radius is the most frequently fractured area of the body.

21. ____ The weakness in carpal tunnel syndrome is caused by impingement of the ulnar nerve.

22. ____ The last phase of throwing is the deceleration phase.

23. ____ Torques in the elbow can reach 24 Nm during a push-up.

24. ____ Placing the wrist in slight extension can increase finger flexion strength.

25. ____ Abduction at the shoulder joint is limited when the arm is externally rotated.

Multiple Choice

1. Which joint does the coracoacromial ligament cross?
   a. Glenohumeral
   b. Sternoclavicular
   c. Acromioclavicular
   d. None

2. Which muscle is not a rotator cuff muscle?
   a. Teres minor
   b. Teres major
   c. Supraspinatus
   d. Subscapularis

3. Which muscle is biarticular?
   a. Biceps brachii
   b. Brachialis
   c. Brachioradialis
   d. Coracobrachialis

4. Which structure does not play a role in the impingement area of the shoulder?
   a. Coracoacromial ligament
   b. Supraspinatus muscle
   c. Subacromial bursa
   d. Suprascapular notch

5. Which is not a scapular motion?
   a. Downward rotation
   b. Depression
   c. Retraction
   d. Pronation

6. The scapulothoracic joint is a ____.
   a. ball-and-socket joint
   b. physiologic joint
   c. saddle joint
   d. pivot joint

7. The biceps brachii can develop the most force ____.
   a. when the forearm is pronated
   b. when the forearm is supinated
   c. when the forearm is in the neutral position
   d. when the shoulder is flexed

8. Which structure is not on the scapula?
   a. Inferior angle
   b. Coracoid process
   c. Glenoid fossa
   d. Acromion
   e. Radial notch

9. The motion that takes place at the radioulnar joint is ____.
   a. pronation
   b. flexion
   c. abduction
   d. adduction

10. Stability in the glenohumeral joint is derived primarily from the ____.
    a. joint contact area
    b. ligaments and muscles
    c. vacuum in the joint
    d. none of the above

11. Which muscle does not cross the glenohumeral joint?
    a. Latissimus dorsi
    b. Pectoralis major
    c. Teres minor
    d. Rhomboid

12. Slowly lowering the arm in the sagittal plane would use the ____ muscle group.
    a. shoulder flexor
    b. shoulder extensor
    c. shoulder abductor
    d. shoulder adductor

13. The arm can rotate through
    a. 60° to 90°
    b. 90° to 110°
    c. 120° to 180°
    d. 250° to 280°

14. The greatest strength output in the shoulder is generated in ____.
    a. extension
    b. flexion
    c. abduction
    d. adduction

15. To lift the arm against gravity you would use a(n) ____ contraction.
    a. eccentric
    b. isometric
    c. concentric

16. Impingement at the shoulder can be minimized by ____ motion.
    a. shoulder abduction
    b. shoulder flexion
    c. shoulder internal rotation
    d. shoulder external rotation

17. Rotator cuff problems can be exacerbated by shoulder ____.
    a. flexion
    b. extension
    c. abduction
    d. adduction

18. A clicking sound in the shoulder may be due to an injury to the ____.
    a. glenoid labrum
    b. subacromial bursa
    c. supraspinatus muscle
    d. lesser tubercle

19. Ulnar flexion takes place at the ____ joint.
    a. radioulnar
    b. ulnohumeral
    c. midcarpal
    d. radiocarpal

20. There are ____ carpal bones.
    a. 4
    b. 6
    c. 8
    d. 10

21. Most of the muscles acting at the wrist and fingers are considered ____.
    a. concentric
    b. eccentric
    c. intrinsic
    d. extrinsic

22. The joints of the fingers are called ____ joints.
    a. interphalangeal
    b. digital
    c. carpal
    d. tarsal

23. Which is not an injury to the hand?
    a. Bennett fracture
    b. Boutonniere deformity
    c. Mallet finger
    d. Olecranon bursitis

24. The structure that connects the radius to the ulna is the _____.
    a. interosseous membrane
    b. annular ligament
    c. ulnar collateral ligament
    d. radial collateral ligament

25. The ratio of glenohumeral movement to scapular movement through 180° of abduction or flexion is _____.
    a. 2:1
    b. 5:4
    c. 1:3
    d. 2:5

References

1. Alcid, J. G. (2004). Elbow anatomy and structural biomechanics. *Clinical Sports Medicine*, 23:503-517.
2. An, K. N., et al. (1992). Intersegmental elbow joint load during pushup. *Biomedical Scientific Instrumentation*, 28:69-74.
3. An, K. N., et al. (1981). Muscles across the elbow joint: A biomechanical analysis. *Journal of Biomechanics*, 14:659-669.
4. Anders, C., et al. (2004). Activation of shoulder muscles in healthy men and women under isometric conditions. *Journal of Electromyography & Kinesiology*, 14:699-707.
5. Anderson, D. S., et al. (1984). Electromyographic analysis of selected muscles during sitting pushups. *Physical Therapy*, 64:24-28.
6. Andrews, J. R., et al. (1985). Glenoid labrum tears related to the long head of the biceps. *The American Journal of Sports Medicine*, 13:337-341.
7. Anglin, C., Wyss, U. P. (2000). Arm motion and load analysis of sit-to-stand, stand-to-sit, cane walking and lifting. *Clinical Biomechanics*, 15:441-448.
8. Arborelius, U. P., et al. (1986). Shoulder joint load and muscular activity during lifting. *Scandinavian Journal of Rehabilitative Medicine*, 18:71-82.
9. Bayley, J. C., et al. (1987). The weight-bearing shoulder. *The Journal of Bone & Joint Surgery*, 69(suppl A):676-678.
10. Blackburn, T. A., et al. (1990). EMG analysis of posterior rotator cuff exercises. *Athletic Training*, 25(1):40-45.
11. Blakely, R. L., Palmer, M. L. (1984). Analysis of rotation accompanying shoulder flexion. *Physical Therapy*, 64:1214-1216.
12. Bowling, R. W., Rockar, P. (1985). The elbow complex. In J. Gould, G. J. Davies (Eds.). *Orthopaedics and Sports Physical Therapy*. St. Louis, MO: C.V. Mosby, 476-496.
13. Broome, H. L., Basmajian, J. V. (1970). The function of the teres major muscle: An electromyographic study. *Anatomical Record*, 170:309-310.
14. Burnham, R. S., et al. (1993). Shoulder pain in wheelchair athletes. *The American Journal of Sports Medicine*, 21:238-242.
15. Chen, F. S., et al. (2005). Shoulder and elbow injuries in the skeletally immature athlete. *The Journal of the American Academy of Orthopaedic Surgeons*, 13:172-185.

16. Chou, P. H., et al. (2002). Elbow load with various forearm positions during one-handed pushup exercise. *International Journal of Sports Medicine*, 23:457-462.
17. David, T. S. (2003). Medial elbow pain in the throwing athlete. *Orthopedics*, 26:94-106.
18. Donkers, M. J., et al. (1993). Hand position affects elbow joint load during push-up exercise. *Journal of Biomechanics*, 26:625-632.
19. Duda, M. (1985). Prevention and treatment of throwing arm injuries. *The Physician and Sportsmedicine*, 13:181-186.
20. Einhorn, A. R. (1985). Shoulder rehabilitation: Equipment modifications. *Journal of Orthopaedic & Sports Physical Therapy*, 6:247-253.
21. Fleisig, G. S., et al. (1991). A biomechanical description of the shoulder joint during pitching. *Sports Medicine Update*, 6:10-24.
22. Fleisig, G. S., et al. (1996). Kinematic and kinetic comparison between baseball pitching and football passing. *Journal of Applied Biomechanics*, 12:207-224.
23. Fleisig, G. S., et al. (1996). Biomechanics of overhand throwing with implications for injury. *Sports Medicine*, 21:421-437.
24. Fornalski, S., et al. (2003) Anatomy and biomechanics of the elbow joint. *Techniques in Hand & Upper Extremity Surgery*, 7:168-178.
25. Gellman, H., et al. (1988). Late complications of the weight-bearing upper extremity in the paraplegic patient. *Clinical Orthopaedics and Related Research*, 233:132-135.
26. Gribble, P. L., Ostry, D. J. (1999). Compensation for interaction torques during single and multijoint limb movement. *Journal of Neurophysiology*, 82:2310-2326.
27. Gupta, S., van der Helm, F. C. T. (2004). Load transfer across the scapula during humeral abduction. *Journal of Biomechanics*, 37:1001-1009.
28. Hageman, P. A., et al. (1989). Effects of position and speed on eccentric and concentric isokinetic testing of the shoulder rotators. *Journal of Orthopaedic & Sports Physical Therapy*, 11:64-69.
29. Halbach, J. W., Tank, R. T. (1985). The shoulder. In J. A. Gould, G. J. Davies (Eds.). *Orthopaedic and Sports Physical Therapy*. St. Louis, MO: C.V. Mosby, 497-517.
30. Harryman, D. T., et al. (1990). Translation of the humeral head on the glenoid with passive glenohumeral motion. *The Journal of Bone & Joint Surgery*, 72(suppl A):1334-1343.
31. Haubert, L. L., et al. (2006). A comparison of shoulder joint forces during ambulation with crutches versus a walker in persons with incomplete spinal cord injury. *Archives of Physical Medicine and Rehabilitation*, 87:63-70.
32. Heinrichs, K. I. (1991). Shoulder anatomy, biomechanics and rehabilitation considerations for the whitewater slalom athlete. *National Strength and Conditioning Association Journal*, 13:26-35.
33. Henry, J. H., Genung, J. A. (1982). Natural history of glenohumeral dislocation—revisited. *The American Journal of Sports Medicine*, 10:135-137.
34. Imaeda, T., et al. (1992). Functional anatomy and biomechanics of the thumb. *Hand Clinics*, 8:9-15.
35. Ireland, M. L., Andrews, J. R. (1988). Shoulder and elbow injuries in the young athlete. *Clinics in Sports Medicine*, 7: 473-494.
36. Jiang, C. C., et al. (1987). Muscle excursion measurements and moment arm determinations of rotator cuff muscles. *Biomechanics in Sport*, 13:41-44.

37. Jobe, F. W., et al. (1984). An EMG analysis of the shoulder in pitching. *The American Journal of Sports Medicine*, 12:218-220.

38. Jobe, F. W., et al. (1996). Rotator cuff function during a golf swing. *The American Journal of Sports Medicine*, 14:388-392.

39. Jones, L. A. (1989). The assessment of hand function: A critical review of techniques. *Journal of Hand Surgery*, 14A:221-228.

40. Kim, D. H., et al. (2004). Shoulder injuries in golf. *The American Journal of Sports Medicine*, 32:1324-1330.

41. King, G. J., et al. (1993). Stabilizers of the elbow. *Journal of Shoulder and Elbow Surgery*, 2:165-174.

42. Kronberg, M., et al. (1990). Muscle activity and coordination in the normal shoulder. *Clinical Orthopaedics and Related Research*, 257:76-85.

43. Kuhn, J. E., et al. (2005). External rotation of the gleno-humeral joint: ligament restraints and muscle effects in the neutral and abducted positions. *Journal of Shoulder and Elbow Surgery/American Shoulder and Elbow Surgeons*, 14:39S-48S.

44. Kulund, D. N., et al. (1979). The long-term effects of playing tennis. *The Physician and Sportsmedicine*, 7:87-91.

45. Lachowetz, T., et al. (1998). The effect of an intercollegiate baseball strength program on reduction of shoulder and elbow pain. *Journal of Strength and Conditioning Research*, 12:46-51.

46. Lippe, C. N., Williams, D. P. (2005) Combined posterior and convergent elbow dislocations in an adult. A case report and review of the literature. *The Journal of Bone & Joint Surgery*, 87:1597-1600.

47. Magermans, D. J., et al. (2005). Requirements for upper extremity motions during activities of daily living. *Clinical Biomechanics*, 20:591-599.

48. Mayfield, J. K. (1980). Mechanism of carpal injuries. *Clinical Orthopaedics and Related Research*, 149:45-54.

49. McCann, P. D., et al. (1993). A kinematic and electromyo-graphic study of shoulder rehabilitation exercises. *Clinical Orthopaedics and Related Research*, 288:179-188.

50. McHardy, A., Pollard, H. (2005). Muscle activity during the golf swing. *British Journal of Sports Medicine*, 39, 799-804.

51. McClure, P. W., et al. (2006). Shoulder function and 3-dimensional scapular kinematics in people with and without shoulder impingement syndrome. *Physical Therapy*, 86:1075-1090.

52. Meister, K. (2000). Injuries to the shoulder in the throwing athlete: Part one: Biomechanics/pathophysiology/classification of injury. *The American Journal of Sports Medicine*, 28:265-275.

53. Morrey, B. F., et al. (1981). A biomechanical study of normal functional elbow motion. *The Journal of Bone & Joint Surgery*, 63(suppl A):872-877.

54. Morris, M., et al. (1989). Electromyographic analysis of elbow function in tennis players. *The American Journal of Sports Medicine*, 17:241-247.

55. Moynes, D. R., et al. (1986). Electromyography and motion analysis of the upper extremity in sports. *Physical Therapy*, 66:1905-1910.

56. Munnings, F. (1991). Cyclist's palsy. *The Physician and Sportsmedicine*, 19:113-119.

57. Murray, I. A., Johnson, G. R. (2004). A study of external forces and moments at the shoulder and elbow while performing everyday tasks. *Clinical Biomechanics*, 19:586-594.

58. Murray, W. M., et al. (1995) Variation of muscle moment arms with elbow and forearm position. *Journal of Biomechanics*, 28:513-525.

59. Nitz, A. J. (1986). Physical therapy management of the shoulder. *Physical Therapy*, 66:1912-1919.

60. Noffal, G. J., Elliott, B. (1998). Three-dimensional kinetics of the shoulder and elbow joints in the high performance tennis serve: Implications for injury. In Kibler W. B., Roetert E. P. (Eds). *4th International Conference on Sports Medicine and Science in Tennis*. Key Biscayne, FL: USTA.

61. Nuber, G. W., et al. (1986). Fine wire electromyography analysis of muscles of the shoulder during swimming. *The American Journal of Sports Medicine*, 14:7-11.

62. Ober, A. G. (1988). An electromyographic analysis of elbow flexors during sub-maximal concentric contractions. *Research Quarterly for Exercise and Sport*, 59:139-143.

63. Oizumi, N., et al. (2006). Numerical analysis of cooperative abduction muscle forces in a human shoulder joint. *Journal of Shoulder and Elbow Surgery/American Shoulder and Elbow Surgeons*, 15:331-338.

64. Pappas, A. M., et al. (1983). Symptomatic shoulder instability due to lesions of the glenoid labrum. *The American Journal of Sports Medicine*, 11:279-288.

65. Parsons, I. M., et al. (2002). The effect of rotator cuff tears on reaction forces at the glenohumeral joint. *Journal of Orthopaedic Research*, 20:439-446.

66. Peat, M., Graham, R. E. (1977). Electromyographic analysis of soft tissue lesions affecting shoulder function. *American Journal of Physical Medicine*, 56:223-240.

67. Poppen, N. K., Walker, P. S. (1976). Normal and abnormal motion of the shoulder. *The Journal of Bone & Joint Surgery*, 58-A:195-200.

68. Prodromos, C. C., et al. (1990). Histological studies of the glenoid labrum from fetal life to old age. *The Journal of Bone & Joint Surgery*, 72-A:1344-1348.

69. Regan, W. D., et al. (1991). Biomechanical study of ligaments around the elbow joint. *Clinical Orthopaedics and Related Research*, 271:170-179.

70. Safran, M. R. (1995). Elbow injuries in athletes. *Clinical Orthopaedics and Related Research*, 310:257-277.

71. Shea, K. G., et al. (1991). Shifting into wrist pain. *The Physician and Sportsmedicine*, 19:59-63.

72. Simon, E. R., Hill, J. A. (1989). Rotator cuff injuries: An update. *Journal of Orthopaedic & Sports Physical Therapy*, 10:394-398.

73. Soderberg, G. L. (1986). *Kinesiology: Application to Pathological Motion*. Baltimore, MD: Lippincott Williams & Wilkins, pp. 109-128.

74. Stewart, O. J., et al. (1981). Influence of resistance, speed of movement, and forearm position on recruitment of the elbow flexors. *American Journal of Physical Medicine*, 60(4):165-179.

75. Terry, G. C., Chopp, T. M. (2000). Functional anatomy of the shoulder. *Journal of Athletic Training*, 35:248-255.

76. Tubiana R., Chamagne, P. (1988). Functional anatomy of the hand. *Medical Problems of Performing Artists*, 3:83-87.

77. Tubiana R. (1988). Movements of the fingers. *Medical Problems of Performing Artists*, 3:123-128.

78. Turkel, S. J., et al. (1981). Stabilizing mechanisms preventing anterior dislocation of the glenohumeral joint. *The Journal of Bone & Joint Surgery*, 63(8):1208-1217.

79. Van der Helm, F. T., Veeger, H. J. (1996). Quasi-static analysis of muscle forces in the shoulder mechanism during wheelchair propulsion. *Journal of Biomechanics*, 29:32-50.

80. Van Drongelen, S. A., et al. (2005). Mechanical load on the upper extremity during wheelchair activities. *Archives of Physical Medicine and Rehabilitation*, 86:1214-1220.

81. Vigouroux, L., et al. (2006). Estimation of finger muscle tendon tensions and pulley forces during specific sport-climbing techniques. *Journal of Biomechanics*, 39:2583-25892.

82. Wadsworth, C. T. (1985). The wrist and hand. In J. A. Gould, G. J. Davies (Eds.). *Orthopaedic and Sports Physical Therapy*. St. Louis, MO: C.V. Mosby, 437-475.

83. Wadsworth, L. T. (1992). How to manage skier's thumb. *The Physician and Sportsmedicine*, 20:69-78.

84. Werner, S. L., et al. (1993) Biomechanics of the elbow during baseball pitching. *Journal of Sports Physical Therapy*, 17:274-278.

85. Whiteside, J. A., Andrews, J. R. (1992). On-the-field evaluation of common athletic injuries: 6. Evaluation of the shoulder girdle. *Sports Medicine Update*, 7:24-28.

86. Wilson, J. J. Best, T. M. (2005). Common overuse tendon problems: A review and recommendations for treatment. *American Family Physician*, 72:811-819.

87. Yocum, L. A. (1989). The diagnosis and nonoperative treatment of elbow problems in the athlete. *Office Practice of Sports Medicine*, 8:437-439.

88. Zarins, B., Rowe, R. (1984). Current concepts in the diagnosis and treatment of shoulder instability in athletes. *Medicine & Science in Sports & Exercise*, 16:444-448.

89. Zuckerman, J. D., Matsea III, F. A. (1989). Biomechanics of the shoulder. In M. Nordin, V. H. Frankel (Eds.). *Biomechanics of the Musculoskeletal System*. Philadelphia, PA: Lea & Febiger, 225-248.

FUNCTIONAL ANATOMY OF THE LOWER EXTREMITY

OBJECTIVES

After reading this chapter, the student will be able to:

1. Describe the structure, support, and movements of the hip, knee, ankle, and subtalar joints.

2. Identify the muscular actions contributing to movements at the hip, knee, and ankle joints.

3. List and describe some of the common injuries to the hip, knee, ankle, and foot.

4. Discuss strength differences between muscle groups acting at the hip, knee, and ankle.

5. Develop a set of strength and flexibility exercises for the hip, knee, and ankle joints.

6. Describe how alterations in the alignment in the lower extremity influence function at the knee, hip, ankle, and foot.

7. Discuss the structure and function of the arches of the foot.

8. Identify the lower extremity muscular contributions to walking, running, stair climbing, and cycling.

9. Discuss various loads on the hip, knee, ankle, and foot in daily activities.

OUTLINE

The Pelvis and Hip Complex
Pelvic Girdle
Hip Joint
Combined Movements of the Pelvis and Thigh
Muscular Actions
Strength of the Hip Joint Muscles
Conditioning of the Hip Joint Muscles
Injury Potential of the Pelvic and Hip Complex

The Knee Joint
Tibiofemoral Joint
Patellofemoral Joint
Tibiofibular Joint
Movement Characteristics
Muscular Actions

Combined Movements of the Hip and Knee
Strength of the Knee Joint Muscles
Conditioning of the Knee Joint Muscles
Injury Potential of the Knee Joint

The Ankle and Foot
Talocrural Joint
Subtalar Joint
Midtarsal Joint
Other Articulations of the Foot
Arches of the Foot
Movement Characteristics
Combined Movements of the Knee and Ankle/Subtalar
Alignment and Foot Function

Muscle Actions
Strength of the Ankle and Foot
   Muscles
Conditioning of the Foot and Ankle
   Muscles
Injury Potential of the Ankle and Foot
Contribution of Lower Extremity
   Musculature to Sports Skills or
   Movements
   Stair Ascent and Descent

Locomotion
Cycling
Forces Acting on Joints in the
   Lower Extremity
   Hip Joint
   Knee Joint
   Ankle and Foot
Summary
Review Questions

The lower extremities are subject to forces that are generated via repetitive contacts between the foot and the ground. At the same time, the lower extremities are responsible for supporting the mass of the trunk and the upper extremities. The lower limbs are connected to each other and to the trunk by the pelvic girdle. This establishes a link between the extremities and the trunk that must always be considered when examining movements and the muscular contributions to movements in the lower extremity.

Movement in any part of the lower extremity, pelvis, or trunk influences actions elsewhere in the lower limbs. Thus, a foot position or movement can influence the position or movement at the knee or hip of either limb, and a pelvic position can influence actions throughout the lower extremity (23). It is important to evaluate movement and actions in both limbs, the pelvis, and the trunk rather than focus on a single joint to understand lower extremity function for the purpose of rehabilitation, sport performance, and exercise prescription.

For example, in a simple kicking action, it is not just the kicking limb that is critical to the success of the skill. The contralateral limb plays a very important role in stabilization and support of body weight (BW). The pelvis establishes the correct positioning for the lower extremity, and trunk positioning determines the efficiency of the lower extremity musculature. Likewise, in evaluating a limp in walking, attention should not be focused exclusively on the limb in which the limp occurs because something happening in the other extremity may cause the limp.

The Pelvis and Hip Complex

PELVIC GIRDLE

The pelvic girdle, including the hip joint, plays an integral role in supporting the weight of the body while offering mobility by increasing the range of motion in the lower extremity. The pelvic girdle is a site of muscular attachment for 28 trunk and thigh muscles, none of which are positioned to act solely on the pelvic girdle (129). Similar to the shoulder girdle, the pelvis must be oriented to place

the hip joint in a favorable position for lower extremity movement. Therefore, concomitant movement of the pelvic girdle and the thigh at the hip joint is necessary for efficient joint actions.

The pelvic girdle and hip joints are part of a closed kinetic chain system whereby forces travel up from the lower extremity through the hip and the pelvis into the trunk or down from the trunk through the pelvis and the hip to the lower extremity. Finally, pelvic girdle and hip joint positioning contribute significantly to the maintenance of balance and standing posture by using continuous muscular action to fine-tune and ensure equilibrium.

The pelvic region is one area of the body where there are noticeable differences between the sexes in the general population. As illustrated in Figure 6-1, women generally have pelvic girdles that are lighter, thinner, and wider than their counterparts in men (65). The female pelvis flares out more laterally in the front. The female **sacrum** is also

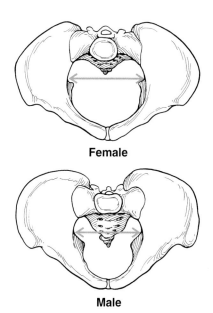

Female

Male

FIGURE 6-1   The pelvis of a female is lighter, thinner, and wider than that of a male. The female pelvis also flares out in the front and has a wider sacrum in the back.

wider in the back, creating a broader pelvic cavity than in men. This skeletal difference is discussed later in this chapter because it has a direct influence on muscular function in and around the hip joint.

The bony attachment of the lower extremity to the trunk occurs via the pelvic girdle (Fig. 6-2). The pelvic girdle consists of a fibrous union of three bones: the superior **ilium**, the posteroinferior **ischium**, and the anteroinferior **pubis**. These are separate bones connected by hyaline cartilage at birth but are fully fused, or ossified, by age 20 to 25 years.

The right and left sides of the pelvis connect anteriorly at the **pubic symphysis**, a cartilaginous joint that has a fibrocartilage disk connecting the two pubic bones. The ends of each pubic bone are also covered with hyaline cartilage. This joint is firmly supported by a **pubic ligament** that runs along the anterior, posterior, and superior sides of the joint. Movement at this joint is limited, maintaining a firm connection between right and left sides of the pelvic girdle.

The pelvis is connected to the trunk at the **sacroiliac joint**, a strong synovial joint containing fibrocartilage and powerful ligamentous support (Fig. 6-2). The articulating surface on the sacrum faces posteriorly and laterally and articulates with the ilium, which faces anteriorly and medially (164).

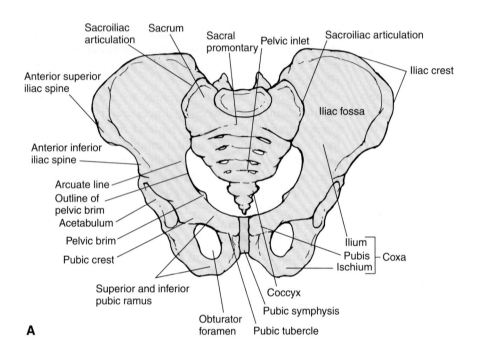

A

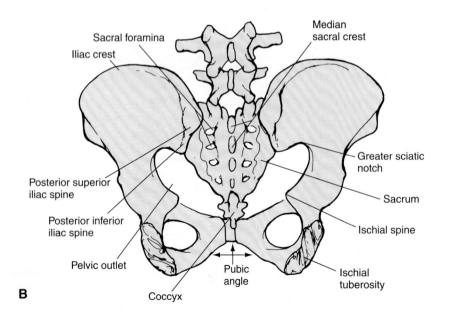

B

FIGURE 6-2 The pelvic girdle supports the weight of the body, serves as an attachment site for numerous muscles, contributes to the efficient movements of the lower extremity, and helps maintain balance and equilibrium. The girdle consists of two coxal bones, each created through the fibrous union of the ilium, ischium, and pubic bones. The right and left coxal bones are joined anteriorly at the pubic symphysis **(A)**, and connect posteriorly **(B)** via the sacrum and the two sacroiliac joints.

The sacroiliac joint transmits the weight of the body to the hip and is subject to loads from the lumbar region and from the ground. It is also an energy absorber of shear forces during gait (129). Three sets of ligaments support the left and right sacroiliac joints, and these ligaments are the strongest in the body (Fig. 6-3).

Even though the sacroiliac joint is well reinforced by very strong ligaments, movement occurs at the joint. The

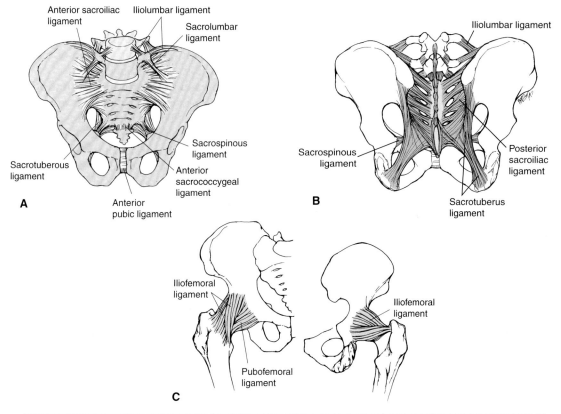

Ligament	Insertion	Action
Anterior pubic	Transverse fiber from body of pubis TO body of pubis	Maintain relationship between right and left pubic bones
Anterior sacrococcygeal	Anterior surface of sacrum TO front of coccyx	Maintain relationship between sacrum and coccyx
Anterior sacroiliac	Thin; pelvic surface of sacrum TO pelvic surface of ilium	Maintains relationship between sacrum and ilium
Iliofemoral	Anterior, inferior iliac spine TO intertrochanteric line of femur	Supports anterior hip; resists in movements of extension, internal rotation, external rotation
Iliolumbar	Transverse process of L5 TO iliac crest	Limits lumbar motion in flexion, rotation
Interosseous (SI)	Tuberosity of ilium TO tuberosity of sacrum	Prevents downward displacement of sacrum caused by body weight
Ischiofemoral	Posterior acetabulum TO iliofemoral ligament	Resists adduction and internal rotation
Ligament of head	Acetabular notch and transverse ligament TO pit of head of femur	Transmits vessel to head of femur; no mechanical function
Posterior sacroiliac	Posterior, inferior spine of ilium TO pelvic surface of sacrum	Maintains relationship between sacrum and ilium
Pubofemoral	Pubic part of acetabulum; superior rami TO intertrochanteric line	Resists abduction and external rotation
Sacrospinous	Spine of ischium TO lateral margins of the sacrum and coccyx	Prevent posterior rotation of ilia respect to the sacrum
Sacrotuberous	Posterior ischium TO sacral tubercles, inferior margin of sacrum, & upper coccyx	Prevents the lower part of the sacrum from titling upward and backward under the weight of the rest of the vertebral column

FIGURE 6-3 Ligaments of the pelvis and hip region shown for the anterior **(A)** and posterior **(B)** perspectives and for the hip joint **(C)**.

amount of movement allowed at the joint varies considerably between individuals and sexes. Males have thicker and stronger sacroiliac ligaments and consequently do not have mobile sacroiliac joints. In fact, three in 10 men have fused sacroiliac joints (164).

In females, the sacroiliac joint is more mobile because there is greater laxity in the ligaments supporting the joint. This laxity may increase during the menstrual cycle, and the joint is extremely lax and mobile during pregnancy (59).

Another reason the sacroiliac joint is more stable in males is related to positioning differences in the center of gravity. In the standing position, BW forces the sacrum down, tightening the posterior ligaments and forcing the sacrum and ilium together. This provides stability to the joint and is the close-packed position for the sacroiliac joint (129). In females, the center of gravity is in the same plane as the sacrum, but in males, the center of gravity is more anterior. Thus, in males, a greater load is placed on the sacroiliac joint, which in turn creates a tighter and more stable joint (164).

Motion at the sacroiliac joint can best be described by sacral movements. The movements of the sacrum that accompany each specific trunk movement are presented in Figure 6-4. The triangular sacrum is actually five fused vertebrae that move with the pelvis and trunk. The top of the sacrum, the widest part, is the base of the sacrum, and when this base moves anteriorly, it is termed **sacral flexion** (129). Clinically, this is also referred to as **nutation**. This movement occurs with flexion of the trunk and with bilateral flexion of the thigh.

Sacrum extension, or **counternutation**, occurs as the base moves posteriorly with trunk extension or bilateral thigh extension. The sacrum also rotates along an axis running diagonally across the bone. Right rotation is designated if the anterior surface of the sacrum faces to the right and left rotation if the anterior surface faces to the left. This sacral torsion is produced by the piriformis muscle in a side-bending exercise of the trunk (129). Additionally, in the case of asymmetrical movement such as standing on one leg, there can be asymmetrical movement at the sacroiliac joint, which results in torsion of the pelvis.

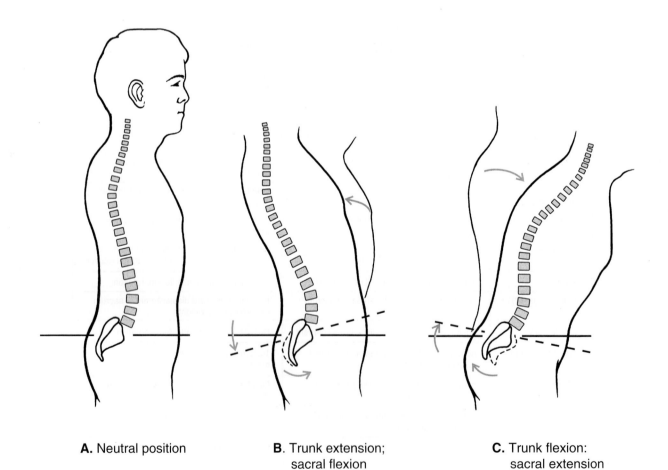

A. Neutral position

B. Trunk extension;
sacral flexion

C. Trunk flexion:
sacral extension

FIGURE 6-4 **(A)** In the neutral position, the sacrum is placed in the close-packed position by the force of gravity. The sacrum responds to movements of both the thigh and the trunk. **(B)** When the trunk extends or the thigh flexes, the sacrum flexes. Flexion of the sacrum occurs when the wide base of the sacrum moves anteriorly. **(C)** During trunk flexion or thigh extension, the sacrum extends as the base moves posteriorly. The sacrum also rotates to the right or left with lateral flexion of the trunk (not shown).

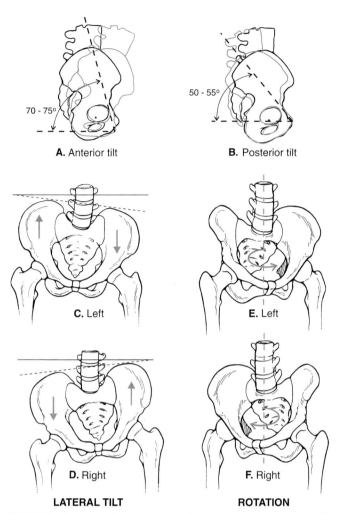

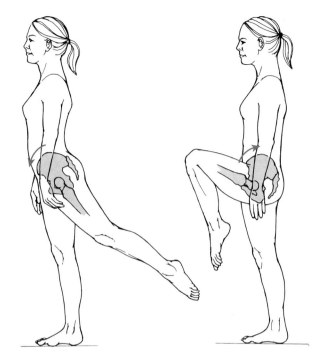

FIGURE 6-6 The pelvis can assist with movements of the thigh by tilting anteriorly to add to hip extension (*left*) or tilt posteriorly to add to hip flexion (*right*).

LATERAL TILT   **ROTATION**

FIGURE 6-5 The pelvis moves in six directions in response to a trunk or thigh movement. Anterior tilt of the pelvis accompanies trunk flexion or thigh extension (**A**). Posterior tilt accompanies trunk extension or thigh flexion (**B**). Left (**C**) and right (**D**) lateral tilt accompany weight bearing on the right and left limbs, respectively, or lateral movements of the thigh or trunk. Left (**E**) and right (**F**) rotation accompany left and right rotation of the trunk, respectively, or unilateral leg movement.

In addition to the movement between the sacrum and the ilium, there is movement of the pelvic girdle as a whole. These movements, shown in Figure 6-5, accompany trunk and thigh movements to facilitate positioning of the hip joint and the lumbar vertebrae. Although muscles facilitate the movements of the pelvis, no one set of muscles acts on the pelvis specifically; thus, pelvic movements occur as a consequence of movements of the thigh or the lumbar vertebrae.

Movements of the pelvis are described by monitoring the ilium, specifically, the anterior superior, and anterior inferior iliac spines on the front of the ilium. In a closed-chain weight-bearing movement, the pelvis moves about a fixed femur, and anterior tilt of the pelvis occurs when the trunk flexes and the hip flexes. In an open-chain position such as hanging, the femur moves on the pelvis, and anterior tilt occurs with extension of the thighs. This anterior tilt can be created by protruding the abdomen

and creating a swayback position in the low back. Both anterior tilt and posterior tilt in an open-chain movement can substitute for hip extension and hip flexion, respectively (Fig. 6-6). In a closed-chain movement, posterior tilt is created through trunk extension or flattening of the low back and hip extension. In the open chain, posterior tilt occurs with flexion of the thigh.

The pelvis can also tilt laterally and naturally tries to move through a right lateral tilt when weight is supported by the left limb. In the closed-chain weight-bearing position, if the right pelvis elevates, adduction of the hip is produced on the weight-bearing limb and abduction of the hip is produced on the opposite side to which the pelvis drops. This movement is controlled by muscles, particularly the gluteus medius, so that it is not pronounced unless the controlling muscles are weak. Thus, right and left lateral tilts occur with weight bearing and any lateral movement of the thigh or trunk (Fig. 6-7).

One Giant Step for Humankind

Have a partner start with their left foot forward and take one slow giant step with the right foot. Note the motion of the pelvis during this step. Does the pelvis have any lateral tilt? What direction? Why does it tilt in this direction? Does the pelvis rotate? What direction? Why does it rotate this direction? Is there any rotation of the hip joint? Internal or external? Why?

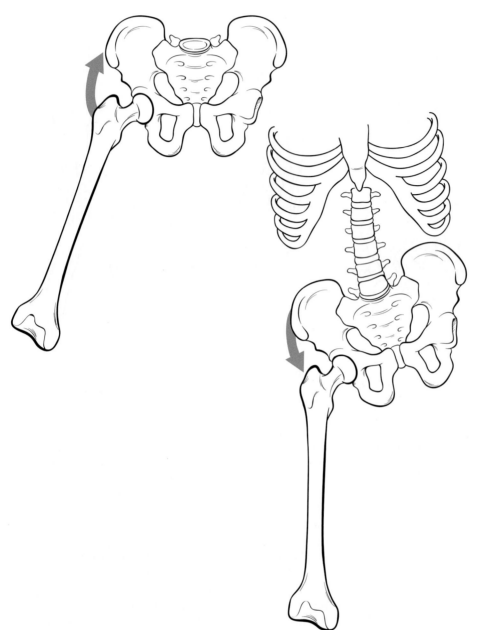

FIGURE 6-7 In the lower extremity, segments interact differently depending on whether an open- or closed-chain movement is occurring. As shown on the *left*, the hip abduction movement in the open chain occurs as the thigh moves up toward the pelvis. In the closed-chain movement shown on the *right*, abduction occurs as the pelvis lowers on the weight-bearing side.

Finally, the pelvic girdle rotates to the left and right as unilateral leg movements take place. As the right limb swings forward in a walk, run, and kick, the pelvis rotates to the left. Hip external rotation accompanies the forward pelvis, and hip internal rotation accompanies the backward pelvic side.

HIP JOINT

The final joint in the pelvic girdle complex is the hip joint, which can be generally characterized as stable yet mobile. The hip, which has 3 degrees of freedom (df), is a ball-and-socket joint consisting of the articulation between the **acetabulum** on the pelvis and the head of the femur. The structure of the hip joint and femur is illustrated in Figure 6-8.

The acetabulum is the concave surface of the ball and socket, facing anteriorly, laterally, and inferiorly (118,132). Interestingly, the three bones forming the pelvis—the ilium, ischium, and pubis—make their fibrous connections with each other in the acetabular cavity. The cavity is lined with articular cartilage that is thicker at the edge and thickest on the top part of the cavity (76,118). There is no cartilage on the underside of the acetabulum. As with the shoulder, a rim of fibrocartilage called the **acetabular labrum** encircles the acetabulum. This structure serves to deepen the socket and increase stability (151).

The spherical head of the femur fits snugly into the acetabular cavity, giving the joint both congruency and a large surface contact area. Both the femoral head and the acetabulum have large amounts of spongy trabecular bone that facilitates the distribution of the forces absorbed by

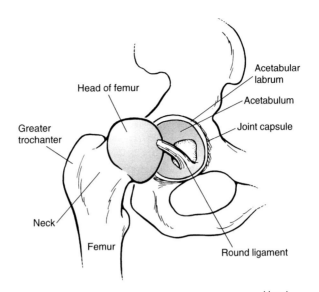

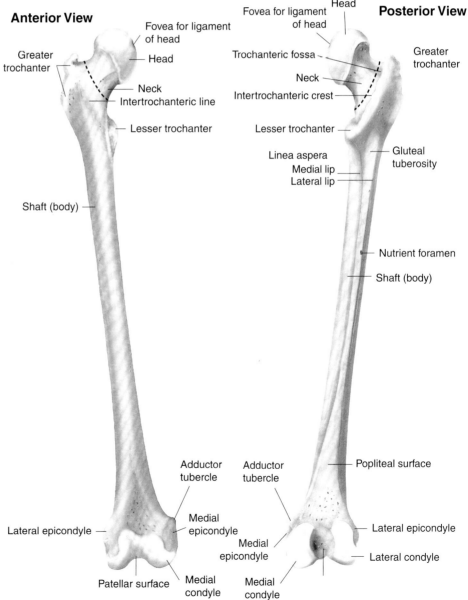

FIGURE 6-8 The hip is a stable joint with considerable mobility in three directions. It is formed by the concave surface of the acetabulum on the pelvis and the large head of the femur. The femur is one of the strongest bones in the body.

the hip joint (118). The head is also lined with articular cartilage that is thicker in the middle central portions of the head, where most of the load is supported. The cartilage on the head thins out at the edges, where the acetabular cartilage is thick (118). Approximately 70% of the head of the femur articulates with the acetabulum compared with 20% to 25% for the head of the humerus with the glenoid cavity.

Surrounding the whole hip joint is a loose but strong capsule that is reinforced by ligaments and the tendon of the psoas muscle and encapsulates the entire femoral head and a good portion of the femoral neck. The capsule is densest in the front and top of the joint, where the stresses are the greatest, and it is quite thin on the back side and bottom of the joint (142).

Three ligaments blend with the capsule and receive nourishment from the joint (Fig. 6-3). The **iliofemoral ligament**, or Y-ligament, is strong and supports the anterior hip joint in the standing posture, resisting extension, external rotation, and some adduction (151). This ligament is capable of supporting most of the BW and plays an important role in standing posture (122). Also, hyperextension may be so limited by this ligament that it may not actually occur in the hip joint itself but rather as a consequence of anterior pelvic tilt.

The second ligament on the front of the hip joint, the **pubofemoral ligament**, primarily resists abduction, with some resistance to external rotation and extension. The final ligament on the outside of the joint is the **ischiofemoral ligament**, on the posterior capsule, where it resists extension, adduction, and internal rotation (151). None of the ligaments surrounding the hip joint resist during flexion movements, and all are loose during flexion. This makes flexion the movement with the greatest range of motion.

The femur is held away from the hip joint and the pelvis by the femoral neck. The neck is formed by cancellous trabecular bone with a thin cortical layer for strength.

The cortical layer is reinforced on the lower surface of the neck, where greater strength is required in response to greater tension forces. Also, the medial femoral neck is the portion responsible for withstanding ground reaction forces. The lateral portion of the neck resists compression forces created by the muscles (118).

The femoral neck joins up with the shaft of the femur, which slants medially down to the knee. The shaft is very narrow in the middle, where it is reinforced with the thickest layer of cortical bone. Also, the shaft bows anteriorly to offer the optimal structure for sustaining and supporting high forces (142).

The femoral neck is positioned at a specific angle in both the frontal and transverse planes to facilitate congruent articulation within the hip joint and to hold the femur away from the body. The **angle of inclination** is the angle of the femoral neck with respect to the shaft of the femur in the frontal plane. This angle is approximately 125° (142) (Fig. 6-9). This angle is larger at birth by almost 20° to 25°, and it gets smaller as the person matures and assumes weight-bearing positions. It is also believed that the angle continues to reduce by approximately 5° in later adult years.

The range of the angle of inclination is usually within 90° to 135° (118). The angle of inclination is important because it determines the effectiveness of the hip abductors, the length of the limb, and the forces imposed on the hip joint (Fig. 6-10). An angle of inclination greater than 125° is termed **coxa valga**. This increase in the angle of inclination lengthens the limb, reduces the effectiveness of the hip abductors, increases the load on the femoral head, and decreases the stress on the femoral neck (151). **Coxa vara**, in which the angle of inclination is less than 125°, shortens the limb, increases the effectiveness of the hip abductors, decreases the load on the femoral head, and increases stress on the femoral neck. This **varus** position gives the hip abductors a mechanical advantage needed to counteract the forces produced by BW. The result is

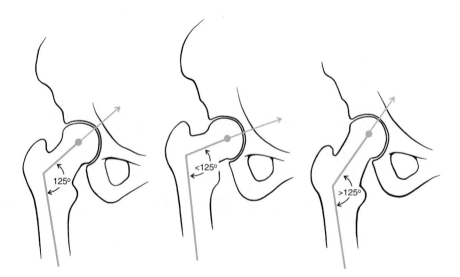

FIGURE 6-9 The angle of inclination of the neck of the femur is approximately 125°. If the angle is less than 125°, it is termed coxa vara. When the neck angle is greater than 125°, it is termed coxa valga.

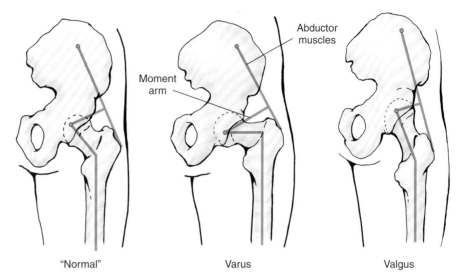

FIGURE 6-10 The femoral neck inclination angle influences both load on the femoral neck and the effectiveness of the hip abductors. When the angle is reduced in coxa vara, the limb is shortened and the abductors are more effective because of a longer moment arm resulting in less load on the femoral head but more load on the femoral neck. The coxa valgus position lengthens the limb, reduces the effectiveness of the abductors because of a shorter moment arm, increases the load on the femoral head, and decreases the load on the neck.

reductions in the load imposed on the hip joint and in the amount of muscular force needed to counteract the force of BW (142). There is a higher prevalence of coxa vara in athletic females than males (121).

The angle of the femoral neck in the transverse plane is termed the angle of **anteversion** (Fig. 6-11). Normally, the femoral neck is rotated anteriorly 12° to 14° with respect to the femur (151). Anteversion in the hip increases the mechanical advantage of the gluteus maximus, making it more effective as an external rotator (132). Conversely,

there is reduced efficiency of the gluteus medius and vastus medialis, resulting in a loss of control of motion in the frontal and transverse planes (121). If there is excessive anteversion in the hip joint, in which it rotates beyond 14° to the anterior side, the head of the femur is uncovered, and a person must assume an internally rotated posture or gait to keep the femoral head in the joint socket. The toeing-in accompanying excessive femoral anteversion is illustrated in Figure 6-12. Other accompanying lower extremity adjustments to excessive anteversion include

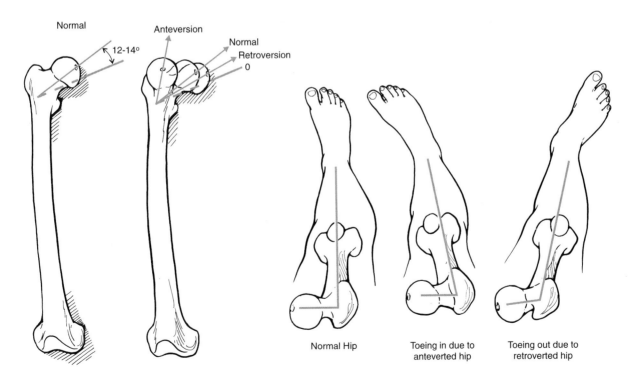

FIGURE 6-11 The angle of the femoral neck in the frontal plane is called the angle of anteversion. The normal angle is approximately 12° to 14° to the anterior side. If this angle increases, a toe-in position is created in the extremity. If the angle of anteversion is reversed so the femoral neck moves posteriorly, it is termed retroversion. Retroversion causes toeing-out.

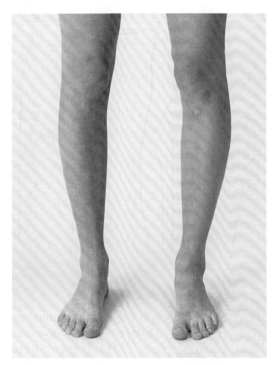

FIGURE 6-12 Individuals who have excessive femoral anteversion compensate by rotating the hip medially so that the knees face medially in stance. There is also usually an adaptation in the tibia that develops external tibial torsion to reorient the foot straight ahead.

an increase in the **Q-angle**, patellar problems, long legs, more **pronation** at the **subtalar joint**, and an increase in lumbar curvature (118,142). Excessive anteversion has also been associated with increased hip joint contact forces and higher bending moments (62) as well as higher patellofemoral joint contact pressures (121).

If the angle of anteversion is reversed so that it moves posteriorly, it is termed **retroversion** (Fig. 6-11). Retroversion creates an externally rotated gait, a supinated foot, and a decrease in the Q-angle (142).

The hip is one of the most stable joints in the body because of powerful muscles, the shape of the bones, the labrum, and the strong capsule and ligaments (122). The hip is a stable joint even though the acetabulum is not deep enough to cover all of the femoral head. The acetabular labrum deepens the socket to increase stability, and the joint is in a close-packed position in full extension when the lower body is stabilized on the pelvis. The joint is stabilized by gravity during stance, when BW presses the femoral head against the acetabulum (142). There is also a difference in atmospheric pressure in the hip joint, creating a vacuum and suction of the femur up into the joint. Even if all of the ligaments and muscles were removed from around the hip joint, the femur would still remain in the socket (74).

Strong ligaments and muscular support in all directions support and maintain stability in the hip joint. At 90° of flexion with a small amount of rotation and abduction, there is maximum congruence between the femoral head and the socket. This is a stable and comfortable position and is common in sitting. A position of instability for the hip joint is in flexion and adduction, as when the legs are crossed (74).

Movement Characteristics

The hip joint allows the thigh to move through a wide range of motion in three directions (Fig. 6-13). The thigh can move through 120° to 125° of flexion and 10° to 15° of hyperextension in the sagittal plane (56,118). These measurements are made with respect to a fixed axis and vary considerably if measured with respect to the pelvis (7). Also, if thigh extension is limited or impaired, compensatory joint actions at the knee or in the lumbar vertebrae accommodate the lack of hip extension.

Hip flexion range of motion is limited primarily by the soft tissue and can be increased at the end of the range of

FIGURE 6-13 The thigh can move through a wide range of motion in three directions. The thigh moves through approximately 120° to 125° of flexion, 10° to 15° of hyperextension, 30° to 45° of abduction, 15° to 30° of adduction, 30° to 50° of external rotation, and 30° to 50° of internal rotation.

motion if the pelvis tilts posteriorly. Hip flexion occurs freely with the knees flexed but is severely limited by the **hamstrings** if the flexion occurs with knee extension (74).

Extension is limited by the anterior capsule, the strong hip flexors, and the iliofemoral ligament. Anterior tilt of the pelvis contributes to the range of motion in hip extension.

The thigh can abduct through approximately 30° to 45° and can adduct 15° to 30° beyond the anatomical position (74). Most activities require 20° of abduction and adduction (74). Abduction is limited by the adductor muscles, and adduction is limited by the tensor fascia latae muscle.

Finally, the thigh can internally rotate through 30° to 50° and externally rotate through 30° to 50° from the anatomical position (74,130). The range of motion for rotation at the hip can be enhanced by the position of the thigh. Both internal and external rotation ranges of motion can be increased by flexing the thigh (74). Both internal and external rotations are limited by their antagonistic muscle group and the ligaments of the hip joint. Range of motion in the hip joint is usually lower in older age groups, but the difference is not that substantial and is usually in the range of 3° to 5° (136).

COMBINED MOVEMENTS OF THE PELVIS AND THIGH

The pelvis and the thigh commonly move together unless the trunk restrains pelvic activity. The coordinated movement between the pelvis and the hip joint is termed the **pelvifemoral rhythm**. In hip flexion movements in an open chain (leg raise), the pelvis rotates posteriorly in the first degrees of motion. In a leg raise with the knees flexed or extended, 26% to 39% of the hip flexion motion is attributed to pelvic rotation, respectively (36). At the end of the range of motion in hip flexion, additional posterior pelvic rotation can contribute to more hip flexion. Anterior pelvic tilt accompanies hip extension when the limb is off the ground. In running, the average anterior tilt of the swing limb has been shown to be approximately 22°, which increases if there is limited hip extension flexibility (144). There is more pelvic motion in non-weight-bearing motions.

In a closed-chain, weight-bearing, standing position, the pelvis moves anteriorly on the femur, and pelvic motion during hip flexion has been shown to contribute only 18% to the change in hip motion (109). Posterior pelvic motion in weight bearing contributes to hip extension.

In the frontal plane, pelvic orientation is maintained or adjusted in response to single-limb weight bearing seen in walking or running. When weight is taken onto one limb, there is a mediolateral shift toward the nonsupport limb that requires abduction and adduction muscle torque to shift the pelvis toward the stance foot (72). This elevation of the nonsupport side pelvis creates hip adduction on the support side and abduction on the nonsupport side.

In the transverse plane during weight bearing, a rotation forward of the pelvis on one side creates lateral rotation on the front hip and medial rotation on the back hip.

Activity	Hip Range of Motion	Knee Range of Motion	Ankle/Foot Range of Motion
Walking	• 35°–40° of flexion during late swing (118) • Full extension at heel lift • 12° of abduction and adduction (max abduction after toe-off; max adduction in stance) (75,143) • 8°–10° of external rotation in swing phase of gait (74) • 4°–6° of internal rotation before heel strike and through the support phase	• 5°–8° of knee flexion at heel strike (156) • 60°–88° of knee flexion during swing phase (77,156) • 17°–20° of flexion during support (77,156) • 12°–17° of rotation during swing phase (77,156) • 8°–11° of valgus during swing phase (77,156) • 5°–8° of knee flexion at heel strike (156) • 17°–20° of flexion during support (77,82,156) • 5°–7° of internal rotation during support (77,82,156) • 7°–14° of external rotation during support (77,82,156) • 3°–7° of varus during support (77,82,156)	• 20°–40° of total ankle movement • 10° of plantarflexion at heel strike (127) • 5°–10° of dorsiflexion in midstance (127) • 20° of plantarflexion at toe-off (127) • Dorsiflexion back to the neutral position in the swing phase (167) • 4° of calcaneal inversion at toe-off (88) • 6°–7° of calcaneal eversion in midstance (88) • 2°–3° of supination at heel strike (38) • 3°–10° of pronation at midstance (8,31,156) • 3°–10° of supination up until heel-off (61)

Activity	Hip Range of Motion	Knee Range of Motion	Ankle/Foot Range of Motion
Running		• 80° of knee flexion during swing phase (54) • 36° of flexion during support (54) • 8° of valgus during swing phase (54) • 19° of varus during support (54) • 8° of internal rotation during support (54) • 11° of external rotation during support (54)	• 10° of dorsiflexion prior to contact (156) • As much as 50° of dorsiflexion in midstance (156) • 25° of plantarflexion at toe-off (156) • 8°–15° of pronation in midstance (8,31,156)
Lowering into or raising out of a chair	• 80°–100° of flexion (64)	• 93° of flexion, 15° of abduction/adduction and 14° of rotation (84)	
Climbing stairs	• 63° of flexion for ascent; 24°–30° for descent (64,143)	• 83° of flexion, 17° of abduction, and 16° of rotation for ascent (84) • 83° of flexion, 14° of abduction/adduction and 15° of rotation for descent (84)	
Bending down and picking up an object	• 18°–20° of abduction (142) • 10°–15° of external rotation (142)		
Tying a shoe while seated		• 106° of flexion, 20° of abduction/adduction, 18° of rotation	

MUSCULAR ACTIONS

The insertion, action, and nerve supply for each individual muscle in the lower extremity are outlined in Figure 6-14. Thigh flexion is used in walking and running to bring the leg forward. It is also an important movement in climbing stairs and walking uphill and is forcefully used in kicking. Little emphasis is placed on training the hip joint for flexion movements because most consider flexion at the hip to play a minor role in activities. However, hip flexion is very important for sprinters, hurdlers, high jumpers, and others who must develop quick leg action. Elite athletes in these activities usually have proportionally stronger hip flexors and abdominal muscles than do less skilled athletes. Recently, more attention has been given to training of the hip flexors in long distance runners as well because it has been shown that fatigue in the hip flexors during running may alter gait mechanics and lead to injuries that may be avoidable with better conditioning of this muscle group.

The strongest hip flexor is the iliopsoas muscle, which consists of the psoas major, psoas minor, and iliacus (142). The iliopsoas is a two-joint muscle that acts on both the lumbar spine of the trunk and the thigh. If the trunk is stabilized, the iliopsoas produces flexion at the hip joint that is slightly facilitated with the thigh abducted and externally rotated. If the thigh is fixed, the iliopsoas produces hyperextension of the lumbar vertebrae and flexion of the trunk.

The iliopsoas is highly activated in hip flexion exercises where the whole upper body is lifted or the legs are lifted (6). In sit-ups with the hips flexed and the feet held in place, the hip flexors are more active. Also, double-leg lifts result in much higher activity in the iliopsoas than single-leg lifts (6).

The rectus femoris is another hip flexor whose contribution depends on knee joint positioning. This is also a two-joint muscle because it acts as an extensor of the knee joint as well. It is called the kicking muscle because it is in maximal position for output at the hip during the preparatory phase of the kick, when the thigh is drawn back into hyperextension and the leg is flexed at the knee. This position puts the rectus femoris on stretch and into an optimal length–tension relationship for the succeeding joint action, in which the rectus femoris makes a powerful contribution to both hip flexion and knee extension. During the kicking action, the rectus femoris is very susceptible to injury and avulsion at its insertion site, the anterior inferior spine on the ilium. Loss of function of the rectus femoris diminishes thigh flexion strength as much as 17% (95).

The three other secondary flexors of the thigh are the sartorius, pectineus, and tensor fascia latae (see Fig. 6-14). The sartorius is a two-joint muscle originating at the anterior superior iliac spine and crossing the knee joint to the medial side of the proximal tibia. It is a weak fusiform

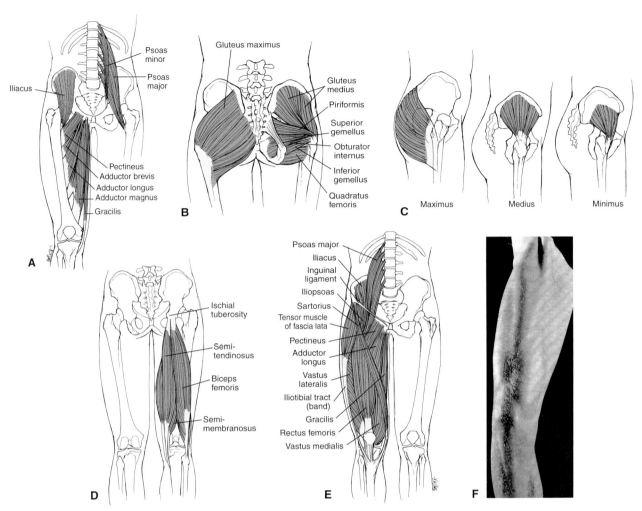

Muscle Group	Insertion	Nerve Supply	Flexion	Extension	Abduction	Adduction	Medial Rotation	Lateral Rotation
Adductor brevis	Inferior rami of pubis TO upper half of posterior femur	Anterior obturator nerve; L3, L4				PM		
Adductor longus	Inferior rami of pubis TO middle third of posterior femur	Anterior obturator nerve; L3, L4				PM	Asst	
Adductor magnus	Anterior pubis, ischial tuberosity TO linea aspera on posterior femur, adductor tubercle	Posterior obturator, sciatic; L3, L4		Asst		PM	Asst	
Biceps femoris	Ischial tuberosity TO lateral condyle of tibia, head of fibula	Tibial, peroneal portion of sciatic nerve; L5, S1–S3		PM				Asst
Gemellus inferior	Ischial tuberosity TO greater trochanter on femur	Sacral plexus; L4, L5, S1 sacral nerve						Asst
Gemellus superior	Ischial spine TO greater trochanter	Sacral plexus; L5, S1, S2 sacral nerve						Asst
Gluteus maximus	Posterior ilium, sacrum, coccyx TO gluteal tuberosity; iliotibial band	Inferior gluteal nerve; L5, S1, S2		PM				PM
Gluteus medius	Anterior, lateral ilium TO lateral surface of greater trochanter	Superior gluteal nerve; L4, L5, S1			PM		PM	
Gluteus minimus	Outer, lower ilium TO front of greater trochanter	Superior gluteal nerve; L4, L5, S1			PM		PM	
Gracilis	Inferior rami of pubis TO medial tibis (pes anserinus)	Anterior obturator nerve; L3, L4				PM	Asst	

FIGURE 6-14 Muscles acting on the hip joint, including the adductors and flexors **(A)**, the external rotators **(B)**, abductors **(C)**, and extensors **(D)**. A combination of knee and hip joint muscles comprise the anterior thigh region **(E, F)**. *(continued)*

Muscle Group	Insertion	Nerve Supply	Flexion	Extension	Abduction	Adduction	Medial Rotation	Lateral Rotation
Iliacus	Inner surface of ilium, sacrum TO lesser trochanter	Femoral nerve; L2, L3	PM					Asst
Obturator externus	Sciatic notch, margin of obturator foramen TO greater trochanter	Sacral plexus; L5, S1, S2						PM
Obturator internus	Pubis, ischium, margin of obturator formamen TO upper posterior femur	Obturator nerve; L3, L4						PM
Pectineus	Pectineal line on pubis TO below lesser trochanter	Femoral nerve; L2–L4	PM			PM		
Piriformis	Anterior lateral sacrum TO superior greater trochanter	S1, S2, L5			Asst			PM
Psoas	Transverse processes, body of L1–L5, T12 TO lesser trochanter	Femoral nerve; L1–L3	PM					Asst
Quadratus femoris	Ischial tuberosity TO greater trochanter	Sacral plexus; L4, L5, S1						PM
Rectus femoris	Anterior inferior iliac spine TO patella, tibial tuberosity	Femoral nerve; L2, L3, L4	PM		Asst			
Sartorius	Anterior superior iliac spine TO medial tibia (pes anserinus)	Femoral nerve; L2, L3	PM		Asst			PM
Semimembranosus	Ischial tuberosity TO medial condyle of tibia	Tibial portion of sciatic nerve: L5, S1, S2		PM			Asst	
Semitendinosus	Ischial tuberosity TO medial tibia (pes anserinus)	Tibial portion of sciatic nerve: L5, S1, S2		PM			Asst	
Tensor fascia latae	Anterior superior iliac spine TO ilitibial tract	Superior gluteal nerve: L4, L5, S1	Asst		Asst		PM	

FIGURE 6-14 (continued)

muscle producing abduction and external rotation in addition to the flexion action of the hip.

The pectineus is one of the upper groin muscles. It is primarily an adductor of the thigh except in walking, actively contributing to thigh flexion. It is accompanied by the tensor fascia latae, which is generally an internal rotator. During walking, however, the tensor fascia latae aids thigh flexion. The tensor fascia latae is considered a two-joint muscle because it attaches to the fibrous band of fascia, the **iliotibial band**, running down the lateral thigh and attaching across the knee joint on the lateral aspect of the proximal tibia. Thus, this muscle is stretched in knee extension.

During thigh flexion, the pelvis is pulled anteriorly by these muscles unless stabilized and counteracted by the trunk. The iliopsoas muscle and tensor fascia latae pull the pelvis anteriorly. If either of these muscles is tight, pelvic torsion, pelvic instability, or a functional short leg may occur.

Extension of the thigh is important in the support of the BW in stance because it maintains and controls the hip joint actions in response to gravitational pull. Thigh extension also assists in propelling the body up and forward in walking, running, or jumping by producing hip joint actions that counteract gravity. The extensors attach to the pelvis and consequently play a major role in stabilizing the pelvis in the anterior and posterior directions.

The muscles contributing in all conditions of extension at the hip joint are the hamstrings. The two medial hamstrings—the semimembranosus and the semitendinosus—are not as active as the lateral hamstring, the biceps femoris, which is considered the workhorse of extension at the hip.

Because all of the hamstrings cross the knee joint, producing both flexion and rotation of the lower tibia, their effectiveness as hip extensors depends on positioning at the knee joint. With the knee joint extended, the hamstrings are put on stretch for optimal action at the hip. The hamstring output also increases with increasing amounts of thigh flexion; however, the hamstrings can be lengthened to a position of muscle **strain** if the leg is extended with the thigh in maximal flexion.

The hamstrings also control the pelvis by pulling down on the ischial tuberosity, creating a posterior tilt of the pelvis. In this manner, the hamstrings are responsible for maintaining an upright posture. Tightness in the hamstrings can create significant postural problems by flattening the low back and producing a continuous posterior tilt of the pelvis.

In level walking or in low-output hip extension activities, the hamstrings are the predominant muscles that contributed to the extension movement in the weight-bearing positions. Loss of function in the hamstrings produces significant impairment in hip extension.

If the resistance in extension is increased or if a more vigorous hip extension is needed, the gluteus maximus is recruited as a major contributor (151). This occurs in running up hills, climbing stairs, rising out of a deep squat, sprinting, and rising from a chair. It also occurs in an optimal length–tension position with thigh hyperextension and external rotation (151).

The gluteus maximus appears to dominate the pelvis during gait rather than contribute significantly to the generation of extension forces. Because the thigh is almost extended during the walking cycle, the function of the gluteus maximus is more trunk extension and posterior tilt of the pelvis. At foot strike when the trunk flexes, the gluteus maximus prevents the trunk from pitching forward. Because the gluteus maximus also externally rotates the thigh, internal rotation places the muscle on stretch. Loss of function of the gluteus maximus muscle does not significantly impair the extension strength of the thigh because the hamstrings dominate the production of extension strength (95).

Finally, because the flexors and extensors control the pelvis anteroposteriorly, it is important that they are balanced in both strength and flexibility so that the pelvis is not drawn forward or backward as a result of one group being stronger or less flexible.

Abduction of the thigh is an important movement in many dance and gymnastics skills. During gait, the abduction muscles are more important in their role as stabilizers of the pelvis and thigh. The abductors can raise the thigh laterally in the frontal plane, or if the foot is on the ground, they can move the pelvis on the femur in the frontal plane. When abduction occurs, such as in doing splits on the ground, both hip joints displace the same number of degrees in abduction, even though only one limb may have moved. The relative angle between the thigh and the trunk is the same in both hip joints in abduction because of the pelvic shift in response to abduction initiated in one hip joint.

The main abductor of the thigh at the hip joint is the gluteus medius. This multipennate muscle contracts during the stance in a walk, run, and jump to stabilize the pelvis so that it does not drop to the nonstance limb. This is important for all the joints and segments in the lower extremity because a weak gluteus medius can lead to changes such as contralateral pelvis drop and increased femoral adduction and internal rotation, which can lead to increased knee valgus, excessive lateral tracking of the patella, and increased tibial rotation and pronation in the foot (43). The effectiveness of the gluteus medius muscle is determined by its mechanical advantage. It is more effective if the angle of inclination of the femoral neck is less than 125°, taking the insertion farther away from the hip joint, and it is also more effective for the same reason in the wider pelvis (132). As the mechanical advantage of the gluteus medius increases, the stability of the pelvis in gait will also improve.

The gluteus minimus, tensor fascia latae, and piriformis also contribute to abduction of the thigh, with the gluteus minimus being the most active of the three. A 50% reduction in the function of the abductors results in a slight to moderate impairment in abduction function (95). If the abductors are weak, there will be an excessive tilt in the frontal plane, with a higher pelvis on the weaker side (87). The abductors on the support side work to keep the pelvis level and to avoid any tilting. Additionally, the shear forces across the sacroiliac joint will greatly increase, and the individual will walk with greater side-to-side sway.

The adductor muscle group works to bring the thigh across the body, as seen commonly in dance, soccer, gymnastics, and swimming. The adductors, similar to the abductors, also work to maintain the pelvic position during gait. The adductors as a group constitute a large muscle mass, with all of the muscles originating on the pubic bone and running down the inner thigh. Although the adductors are important in specific activities, it has been shown that a 70% reduction in the function of the thigh adductors results in only a slight or moderate impairment in hip function (95).

The adductor muscles include the gracilis, on the medial side of the thigh; the adductor longus, on the anterior side of the thigh; the adductor brevis, in the middle of the thigh; and the adductor magnus, on the posterior side of the inner thigh. High in the groin is the pectineus, previously discussed briefly in its role as hip flexor. The adductors are active during the swing phase of gait as they work to swing the limb through (151), and if they are tight, a scissors gait can result, leading to a crossover plant.

The adductors work with the abductors to balance the pelvis. The abductors on one side of the pelvis work with the adductors on the opposite side to maintain pelvic positioning and prevent tilting. As shown earlier, the abductors and adductors must be balanced in strength and flexibility so that the pelvis can be balanced side to side. Figure 6-15 illustrates how imbalances in abduction and adduction can tilt the pelvis. If the abductors overpower the adductors through contracture or a strength imbalance, the pelvis will tilt to the side of the strong, contracted abductor. Adductor contracture or strength imbalances produce a similar effect in the opposite direction. The adductors also work with the hip flexors and extensors to maintain limb position and to counteract the rotation of the pelvis when the front limb is flexed and the back limb is extended in the double-support phase of walking (122).

External rotation of the thigh is important in preparation for power production in the lower extremity because it follows the trunk during rotation. The muscles primarily responsible for external rotation are the gluteus maximus, obturator externus, and quadratus femoris. The obturator internus, inferior and superior gemellus, and piriformis contribute to external rotation when the thigh is extended. The piriformis also abducts the hip when the hip is flexed and creates the movement on lifting the leg into abduction with the toes pointing upward in external rotation. Because most of these muscles attach to the

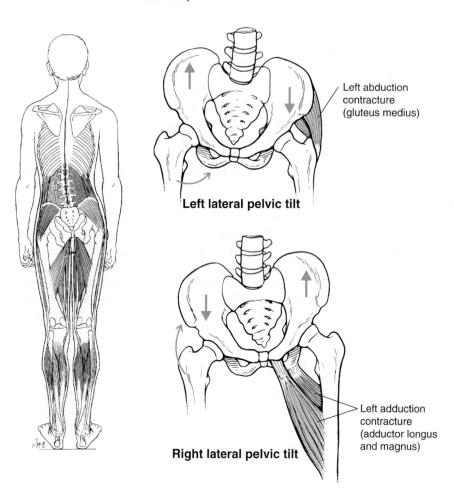

Left abduction
contracture
(gluteus medius)

Left lateral pelvic tilt

Left adduction
contracture
(adductor longus
and magnus)

Right lateral pelvic tilt

FIGURE 6-15 The abductors and adductors work in pairs to maintain pelvic height and levelness. For example, the left abductors work with the right adductors and lateral trunk flexors to create a left lateral tilt. If an abductor or adductor muscle group is stronger than the contralateral group, the pelvis will tilt to the strong side. This also happens with contracture of the muscle group.

anterior face of the pelvis, they also exert considerable control over the pelvis and sacrum.

Internal rotation of the thigh is basically a weak movement. It is a secondary movement for all of the muscles contracting to produce this joint action. The two muscles most involved in internal rotation are the gluteus medius and the gluteus minimus. Internal rotation is also aided by contractions of the gracilis, adductor longus, adductor magnus, tensor fascia latae, semimembranosus, and semitendinosus.

The muscles of the trunk, pelvis, and hip also work together to control the pelvic posture. The pelvis serves as a link between the lumbar vertebrae and the hip and must be stabilized by the trunk or thigh musculature to maintain its position (134). For example, at the beginning of lifting, the gluteus maximus contracts to stabilize the pelvis so that the spinal extensors can extend the trunk in the lift. The gluteus maximus also stabilizes the pelvis in trunk rotation (111). In upright standing, the pelvis is maintained in a vertical position but can also assume a variety of tilt postures. The rectus femoris and the erector spinae muscles can pull the pelvis anteriorly, and the gluteals and abdominals can pull the pelvis posteriorly if the pelvis is in a position out of the neutral vertical position (42,134).

STRENGTH OF THE HIP JOINT MUSCLES

The hip muscles generate the greatest strength output in extension. The most massive muscle in the body, the gluteus maximus, combines with the hamstrings to produce hip extension. Extension strength is maximum with the hip flexed to 90° and diminishes by about half as the hip flexion angle approaches the 0° or neutral position (151). Extension strength also depends on knee position because the hamstrings cross the knee joint. The hamstrings' contribution to hip extension strength is enhanced with the knees extended (74).

Many muscles contribute to hip flexion strength, but many of the muscles do so secondarily to other main roles. Hip flexion strength is primarily generated with the powerful iliopsoas muscle, although its strength diminishes with trunk flexion. Additionally, the flexion strength of the thigh can be enhanced if flexion at the knee joint increases the contribution of the rectus femoris to flexion strength. Abduction strength is maximal from the neutral position and diminishes more than half at 25° of abduction (151). This reduction is associated with decreases in muscle length even though the ability of the gluteus medius to abduct the leg improves as a consequence of improving the direction of the pull of the muscle. The strength output of the abduction movement can also be increased if it is performed with the thigh flexed (151). Abduction strength has also been shown to be greater in the dominant limb than in the nondominant limb (70,114).

The potential for the development of adduction strength is substantial because the muscles contributing to the movement are massive as a group and adductors can develop more force output than the abductors (96). Adduction, however, is not the primary contributor to many movements or sport activities, so it is minimally loaded or strengthened through activity. Adduction strength values are greater from a position of slight abduction as a stretch is placed on the muscle group.

The strength of the external rotators is 60% greater than that of the internal rotators except in hip flexion, when the internal rotators are slightly stronger (151). The strength output of both the internal and external rotators is greater in a seated position than in a supine one.

CONDITIONING OF THE HIP JOINT MUSCLES

The muscles surrounding the hip joint receive some form of conditioning during walking, rising from or lowering into a chair, and performing other common daily activities, such as climbing stairs. The hip musculature should be balanced so that the extensors do not overpower the flexors and the abductors are equivalent to the adductors. This ensures sufficient control over the pelvis. Sample stretching and strengthening exercises for the hip joint muscles are provided in Figure 6-16.

Because the hip muscles are used in all support activities, it is best to design exercises using a closed kinetic chain. In this type of activity, the foot or feet are in contact with a surface (i.e., the ground), and forces are applied to the system at the foot or feet. An example of a closed-chain exercise is a squat lift in weight training. An example of an open kinetic chain exercise is one using a machine, in which the muscle group moves the limb through a prescribed arc of motion. Finally, many two-joint muscles act at the hip joint, so careful attention should be paid to adjacent joint positioning to maximize a stretch or strengthening exercise.

The flexors are best exercised in the supine or hanging position so that the thigh can be raised against gravity or in lifting the whole upper body. The hip flexors are minimally used in a lowering activity, such as a squat, when there is flexion of the thigh, because the extensors control the movement eccentrically. Because the hip flexors attach on the trunk and across the knee joint, their contribution to flexion can be enhanced with the trunk extended. Flexion at the knee also enhances thigh flexion. It is easy to stretch the flexors with both the trunk and thigh placed in hyperextension. The rectus femoris can be placed in a very strenuous stretch with thigh hyperextension and maximal knee flexion.

The success of conditioning the extensors depends on the trunk and knee joint positioning. The greater the

knee flexion, the less the hamstrings contribute to extension, requiring a greater contribution from the gluteus maximus. For example, in a quarter-squat activity with the extensors used eccentrically to lower the body and concentrically to raise the body, the hamstrings are the most active contributors. In a deep squat, with the amount of knee flexion increased to 90° and beyond, the gluteus maximus is used more because the hamstrings are incapacitated by their reduced length.

Trunk positioning is also important, and the activity of the hamstrings is enhanced with trunk flexion because trunk flexion increases the length of both the hamstrings and the gluteus maximus. The extensors are best exercised in a standing, weight-supported position because they are used in this position in most cases and are one of the propulsive muscle groups in the lower extremity.

The extensors can be stretched to maximum levels with hip flexion accompanied by full extension at the knee. The stretch on the gluteus maximus can be increased with thigh internal rotation and adduction.

The abductors and adductors are difficult to condition because they influence balance and pelvic position so significantly. In a standing position, the thigh can be abducted against gravity, but it will shift the pelvis dramatically so that the person loses balance. The adductors present an even greater problem. It is very difficult to place the adductors so that they work against gravity because the abductors are responsible for lowering the limb to the side after abducting it. Consequently, the supine position is best for strengthening and stretching the abductors and adductors. Resistance can be offered manually or through an exercise machine with external resistance to the movement.

The abductors and adductors can be exercised from the side-lying position so that they can work against gravity. This position requires stabilization of the pelvis and low back. It is hard to exercise the abductors or adductors on one side without working the other side as well; both sides are affected equally because of the action of the pelvis. For example, 20° of abduction at the right hip joint results in 20° of abduction at the left hip joint because of the pelvic tilt accompanying the movement.

The rotators of the thigh are the most challenging in terms of conditioning because it is so difficult to apply resistance to the rotation. The seated position is recommended for strengthening the rotators because the rotators are strong in this position and resistance to the rotation can easily be applied to the leg either with surgical tubing or manually. Because the internal rotators lose effectiveness in the extended supine position, they should definitely be exercised with the person seated. Both muscle groups can be stretched in the same way they are strengthened, using the opposite joint action for the stretch. These exercises may be contraindicated, however, for individuals with knee pain, particularly patellofemoral pain.

Muscle Group	Sample Stretching Exercise	Sample Strengthening Exercise	Other Exercises
Hip flexors	 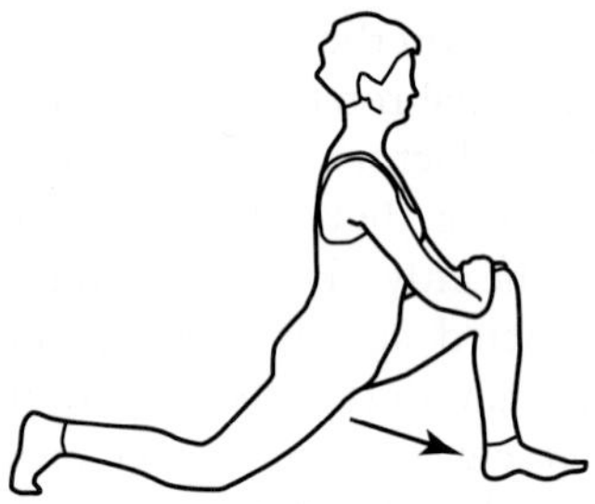	Hanging leg lifts  Hip flexor machine 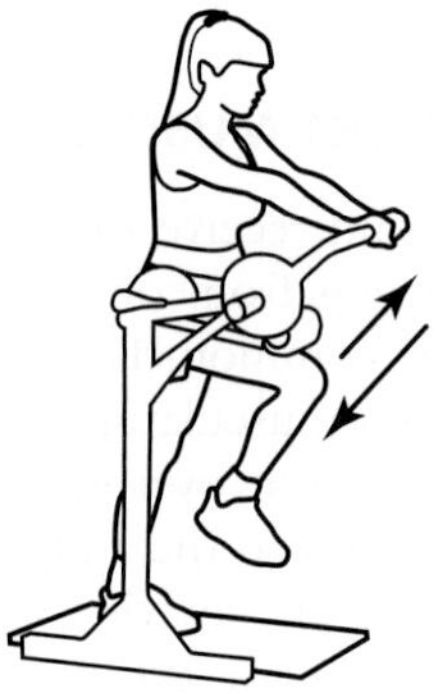	Manual resistance Band or tube hip flexion
Hip extensors	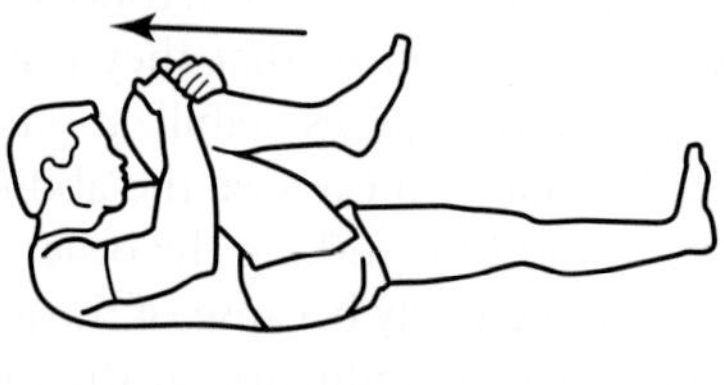 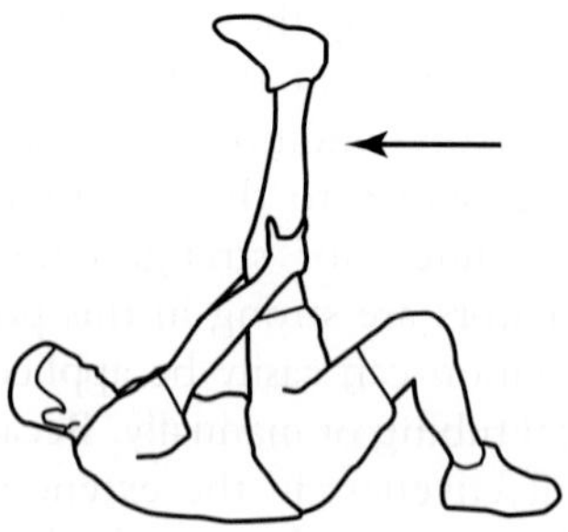	Glutal leg lift 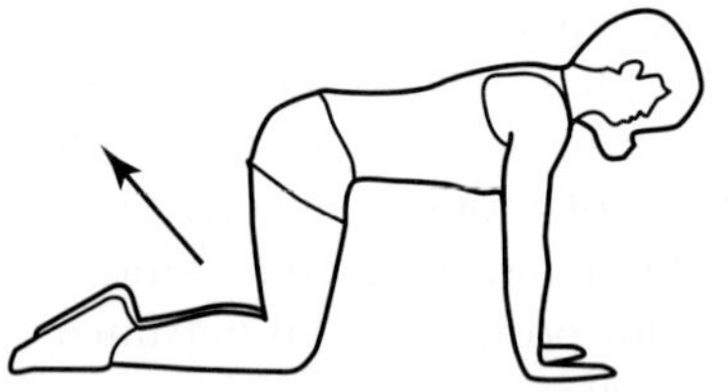 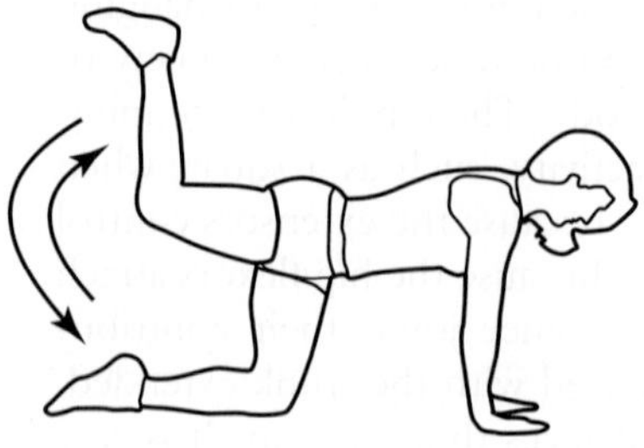	Leg curl squat

FIGURE 6-16   Sample stretching and strengthening exercises for selected muscle groups.

Muscle Group	Sample Stretching Exercise	Sample Strengthening Exercise	Other Exercises
Hip abductors	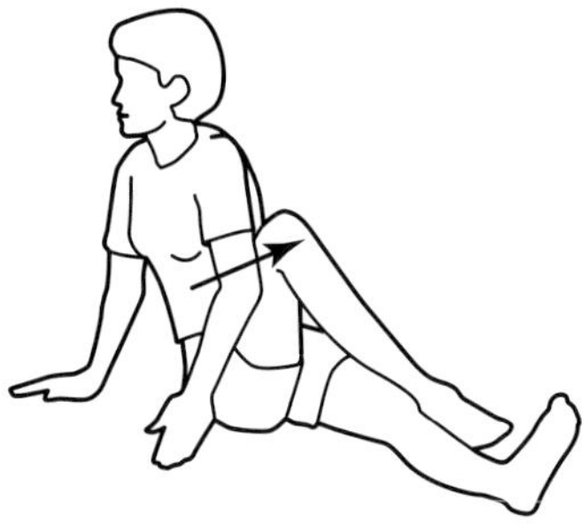	Abductor machine 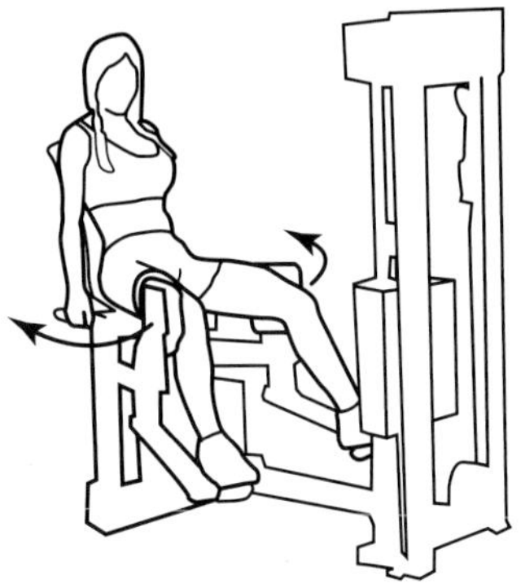	Leg swings (tubing)
Hip adductors		Adductor machine 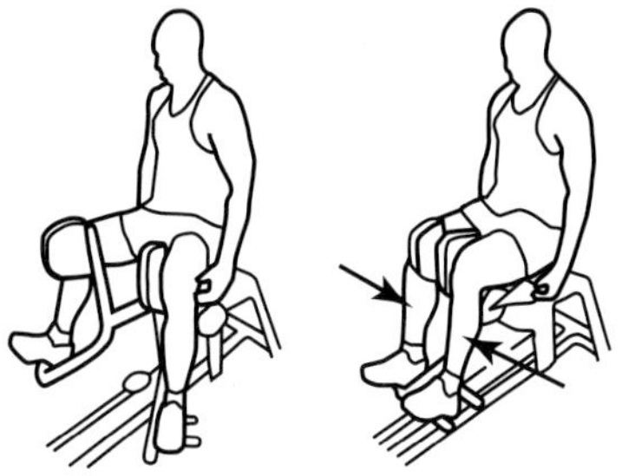	Leg swings (tubing) Ball sqeeze
Hip rotators	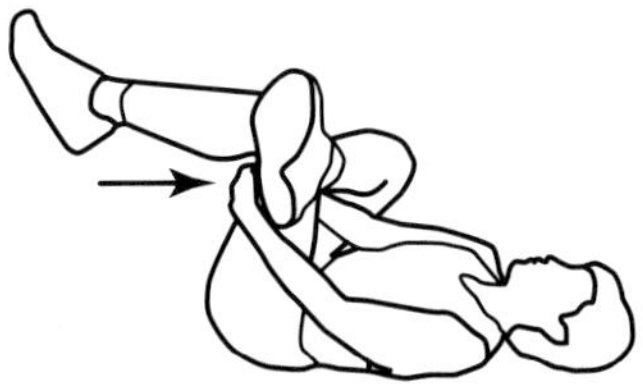 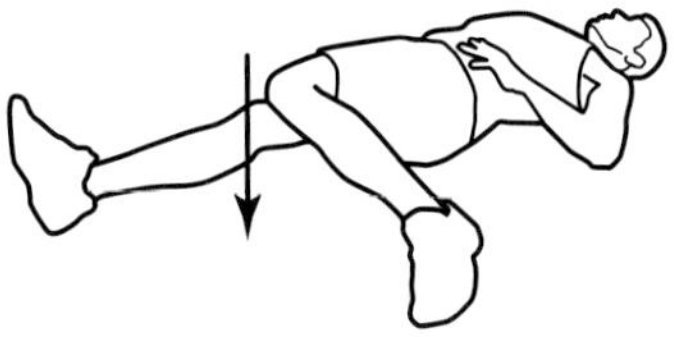	 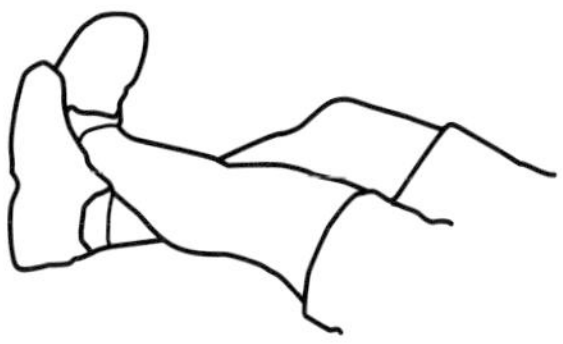	

FIGURE 6-16   (*continued*)

INJURY POTENTIAL OF THE PELVIC AND HIP COMPLEX

Injuries to the pelvis and hip joint are a small percentage of injuries in the lower extremity. In fact, overuse injuries to this area account for only 5% of the total for the whole body (128). This may be attributable to the strong ligamentous support, significant muscular support, and solid structural characteristics of the region.

Injuries to the pelvis primarily occur in response to abnormal function that excessively loads areas of the pelvis. This can result in an irritation at the site of muscular attachment, and in adolescents, a more common type of injury might be an avulsion fracture at the apophysis, or bony outgrowth. **Iliac apophysitis** is an example of such an injury, in which excessive arm swing in gait causes excessive rotation of the pelvis, creating stress on the attachment site of the gluteus medius and tensor fascia latae on the iliac crest (128). This can also occur at the iliac crest as a result of direct blow or as a result of a sudden, violent contraction of the abdominals (3,103). A hip pointer results when the anterior iliac crest is bruised as a result of a direct blow. **Apophysitis**, an inflammation of an apophysis, can also develop into a **stress fracture**.

Another site in the pelvis subjected to apophysitis or stress fracture is the anterior superior iliac spine, where the sartorius attaches (103) and high tensions develop in activities such as sprinting where there is vigorous hip extension and knee flexion. At the anterior inferior iliac spine, the rectus femoris can produce the same type of injury in an activity such as kicking.

A stress fracture in the pubic rami can be produced by strong contractions from the adductors, often associated with overstriding in a run (158). Finally, the hamstrings can exert enough force to create an avulsion fracture on the ischial tuberosity. Commonly called the hurdler's fracture, this ischial tuberosity injury is also common to waterskiing (5). All of these injuries are most common in activities such as sprinting, jumping, soccer, football, basketball, and figure skating, in which sudden bursts of motion are required (103).

The sacrum and sacroiliac joint can dysfunction as a result of injury or poor posture. If one assumes a round-shouldered, forward-head posture, the center of gravity of the body moves forward. This increase in the curvature of the lumbar spine produces a ligamentous laxity in the dorsal sacroiliac ligaments and stress on the anterior ligaments (39). Also, any skeletal asymmetry, such as a short leg, produces a ligament laxity in the sacroiliac joint (129).

With excessive mobility, large forces are transferred to the sacroiliac joint, producing an inflammation of the joint known as **sacroiliitis**. Inflammation of the joint may occur in an activity such as long jumping, in which the landing is absorbed with the leg extended at the knee. At the same time, the hip is flexed or there is extreme flexion of the trunk combined with lateral flexion (129). The sacroiliac joint also becomes very mobile in pregnant women, making them more susceptible to sacroiliac **sprain** (59).

The functional positions of the sacrum and the pelvis are also important for maintaining an injury-free lower extremity. A functional short leg can be created by posterior rotation of the ipsilateral ilium, anterior ilium rotation of the opposite side, superior ilium movement on the same side, forward or backward sacral torsion to the same side, or sacral flexion of the opposite side (129). A functional short leg requires adjustments in the whole limb, creating stress at the sacroiliac joint, knee, and foot.

The hip joint can withstand large loads, but when muscle imbalance develops with high forces, injury can result. For example, in a high-force situation involving flexion, adduction, and internal rotation, a dislocation posteriorly can occur. Falling on an adducted limb with the knee flexed or an abrupt stop over the weight-bearing limb can push the femoral head to the posterior rim of the acetabulum, resulting in a hip subluxation (5). Activities more prone to a posterior dislocation of the hip are stooping activities, leg-crossing activities, and rising from a low seat (112). Anterior dislocations or subluxations are uncommon.

Also, a number of age-related hip conditions must be considered when working with children or older adults. In children 3 to 12 years old, the condition known as **Legg–Calvé–Perthes disease** may appear (142). In this condition, also called **coxa plana**, the femoral head degenerates, and the proximal femoral epiphysis is damaged. This disorder strikes boys five times more frequently than girls and usually occurs to only one limb (2). It is caused by trauma to the joint, synovitis or inflammation to the capsule, or some vascular condition that limits blood supply to the area.

Slipped capital femoral epiphysitis is another disorder that can affect children aged 10 to 17 years. It is usually caused by some traumatic event that forces the femoral neck into external rotation, or it can be caused by failure of the cartilaginous growth plates (2). This tilts the femoral head back and medially and tilts the growth plate forward and vertically, producing a nagging pain on the front of the thigh. An individual with this disorder walks with an externally rotated gait and has limited internal rotation with the thigh flexed and abducted (142). Such slippage may occur in a baseball player who rounds a base with the left foot fixed in internal rotation while the trunk and pelvis rotate in the opposite direction.

The final major childhood disorder to the hip joint is **congenital hip dislocation**, a disorder that affects girls more often than boys (142). This condition is usually diagnosed early as the infant assumes weight on the lower extremity. The hip joint subluxates or dislocates for no apparent reason. The thigh cannot abduct, the limb shortens, and a limp is usually present. Fortunately, this condition is easily corrected with an abduction orthotic.

An age-related disorder of the hip joint seen commonly in elderly individuals is **osteoarthritis**. This condition results in degeneration of the joint cartilage and the underlying subchondral bone, narrowing of the joint space, and the growth of osteophytes in and around the joint. This affliction strikes millions of elderly people, creating a significant amount of pain and discomfort during weight support and gait activities. To reduce the pain in the joint, individuals often assume a position of flexion, adduction, and external rotation or whichever position results in the least tension for the hip.

More than 60% of injuries to the hip occur in the soft tissue (87). Of these injuries, 62% occur in running, 62% are associated with a varum alignment in the lower extremity, and 30% are associated with a leg length discrepancy (87). These types of injuries are usually muscle strains, **tendinitis** of the muscle insertions, or **bursitis** (25).

The most common soft-tissue injury to the hip region is gluteus medius tendinitis, which occurs more frequently in women as a result of excessive pull by the gluteus medius during running (21,87). A hamstring strain is also common and is seen in activities such as hurdling, in which the lower limb is placed in a position of maximum hip flexion and knee extension. It can also occur with speed or hill running and in individuals performing with poor flexibility or conditioning in this muscle group.

Iliopsoas strain can occur in activities such as sprinting, in which a rapid forceful flexion taxes the muscle or the muscle is used eccentrically to slow a rapid extension at the hip. The adductors are often strained in an activity such as soccer, in which the lower extremity is rapidly abducted and externally rotated in preparation for contact with the ball. Strain to the rectus femoris can occur in a rapid forceful flexion of the thigh, such as is seen in sprinting, and in a vigorous hyperextension of the thigh, such as in the preparatory phase of a kick.

A piriformis strain may be caused by excessive external rotation and abduction when the thigh is being flexed. This creates pain in adduction, flexion, and internal rotation of the thigh. A piriformis syndrome can develop. This is an impingement of the sciatic nerve aggravated by internal and external rotation movements of the thigh during walking (21,87). The syndrome can also be created by a functional short leg that lengthens the piriformis and then stretches it as the pelvis drops to the shorter leg. The irritation of the sciatic nerve causes pain in the buttock area that can travel down the posterior surface of the thigh and leg.

Other soft-tissue injuries to the hip region are seen in the bursae. The most common of these is greater trochanteric bursitis, which is caused by hyperadduction of the thigh. This can be produced by running with too much leg crossover in each stride, imbalance between the abductors and adductors, running on banked surfaces, having a leg length difference, or remaining on the outside of the foot during the support phase of a walk or run (22,128). It is especially prevalent in runners with a wide pelvis, a large Q-angle, and an imbalance between the abductors and adductors (22,128).

Because the right hip adductors work with the left hip abductors and vice versa, any imbalance causes asymmetrical posture. For example, a weak right abductor creates a lateral pelvic tilt, with the right side high and the left side low. This places stress on the lateral hip, setting up the conditions for bursitis. Pain on the outside of the hip is accentuated with trochanteric bursitis when the legs are crossed.

Ischial bursitis can develop with prolonged sitting and is aggravated by walking, stair climbing, and flexion of the thigh. Finally, iliopectineal bursitis may develop in reaction to a tight iliopsoas muscle or osteoarthritis of the hip (142).

Two remaining soft-tissue injuries seen in dancers and distance runners are lateral hip pain created by **iliotibial band syndrome** and **snapping hip syndrome**. The strain to the iliotibial band is created because dancers warm up with the hip abducted and externally rotated. They have very few flexion and extension routines in warm-up and dance routines. The stress to the iliotibial band occurs with thigh adduction and internal rotation, movements that are extremely limited in dancers by technique (135). Iliotibial band syndrome can also be caused by excess tension in the tensor fascia latae in abducting the hip in single-stance weight bearing. The snapping hip also commonly produces a click as the hip capsule moves or the iliopsoas tendon snaps over a bony surface.

The bony or osseous injuries to the hip are usually a result of a strong muscular contraction that creates an avulsion fracture. Stress fractures can develop in the hip region and are common in endurance athletes, particularly women (25). Stress fractures to the femoral neck account for 5% to 10% of all stress fractures (97). A stress fracture to the inferior medial aspect of the femoral neck is seen more often in younger patients and is caused by high compression forces. In older adults, stress fractures to the femoral neck are seen more often on the superior side and are caused by high tension forces (3). The abductors can create an avulsion fracture on the greater trochanter, and the iliopsoas can pull hard enough to produce an avulsion fracture at the **lesser trochanter** (3,142). Stress fractures can also appear in the femoral neck. It is believed that these stress fractures may be related to some type of vascular necrosis in which the blood supply is limited or to some hormonal deficiency that reduces the bone density in the neck (87). Stress fracture at this site produces pain in the groin area.

The Knee Joint

The knee joint supports the weight of the body and transmits forces from the ground while allowing a great deal of movement between the femur and the tibia. In the extended position, the knee joint is stable because of its vertical alignment, the congruency of the joint surfaces, and the effect of gravity. In any flexed position, the knee joint is mobile and requires special stabilization from the powerful capsule, ligaments, and muscles surrounding the joint (147). The joint is vulnerable to injury because of the mechanical demands on it and the reliance on soft tissue for support.

The ligaments surrounding the knee support the joint passively as they are loaded in tension only. The muscles support the joint actively and are also loaded in tension, and bone offers support and resistance to compressive loads (100). Functional stability of the joint is derived from the passive restraint of the ligaments, joint geometry, the active muscles, and the compressive forces pushing the bones together.

There are three articulations in the region known as the knee joint: the **tibiofemoral joint**, the **patellofemoral joint**, and the superior **tibiofibular joint** (165). The bony landmarks of the knee joint and tibia and fibula are illustrated in Figure 6-17.

TIBIOFEMORAL JOINT

The tibiofemoral joint, commonly referred to as the actual knee joint, is the articulation between the two longest and strongest bones in the body, the femur and the tibia (Fig. 6-17). It has been referred to as a double condyloid joint or a modified hinge joint that combines a hinge and a pivot joint. In this joint, flexion and extension occur similar to flexion and extension at the elbow joint. In the knee joint, however, flexion is accompanied by a small but significant amount of rotation (147).

At the distal end of the femur are two large convex surfaces, the medial and lateral **condyles**, separated by the **intercondylar notch** in the posterior and the patellar, or trochlear, groove in the anterior (147) (Fig. 6-17). It is important to review the anatomical characteristics of these two condyles because their differences and the corresponding differences on the tibia account for the rotation in the knee joint. The lateral condyle is flatter, has a larger surface area, projects more posteriorly, is more prominent anteriorly to hold the patella in place, and is basically aligned with the femur (165). The medial condyle projects more distally and medially, is longer in the anteroposterior direction, angles away from the femur in the rear, and is aligned with the tibia (165). Above the condyles on both sides are the **epicondyles**, which are the sites of capsule, ligament, and muscular attachment.

The condyles rest on the condyle **facet** or **tibial plateau**, a medial and a lateral surface separated by a ridge of bone termed the **intercondylar eminence**. This ridge of bone serves as an attachment site for ligaments, centers the joint, and stabilizes the bones in weight bearing (165). The medial surface of the plateau is oval, larger, longer in the anteroposterior direction, and slightly concave to accept the convex condyle of the femur. The lateral tibial plateau is circular and slightly convex (165). Consequently,

the medial tibia and femur fit fairly snugly together, but the lateral tibia and femur do not fit together well because both surfaces are convex (147). This structural difference is one of the determinants of rotation because the lateral condyle has a greater excursion with flexion and extension at the knee.

Two separate fibrocartilage menisci lie between the tibia and the femur. As shown in Figure 6-18, the lateral **meniscus** is oval, with attachments at the anterior and posterior horns (52,165). It also receives attachments from the quadriceps femoris anteriorly and the popliteus muscle and **posterior cruciate ligament** (PCL) posteriorly. The lateral meniscus occupies a larger percentage of the area in the lateral compartment than the medial meniscus in the medial compartment. Also, the lateral meniscus is more mobile, capable of moving more than twice the distance of the medial meniscus in the anteroposterior direction (165).

The medial meniscus is larger and crescent shaped, with a wide base of attachment on both the anterior and posterior horns via the coronary ligaments (Fig. 6-18). It is connected to the quadriceps femoris and the anterior cruciate ligament (ACL) anteriorly, the tibial collateral ligament laterally, and the semimembranosus muscle posteriorly (165).

Both menisci are wedge shaped because of greater thickness at the periphery. The menisci are connected to each other at the anterior horns by a **transverse ligament**. The menisci have blood supply to the horns at the anterior and posterior ends of the arcs of each meniscus but have no blood supply to the inner portion of the fibrocartilage. Thus, if a tear occurs in the periphery of the menisci, healing can occur, unlike with tears to the thinner inner portion of the menisci.

The menisci are important in the knee joint. The menisci enhance stability in the joint by deepening the contact surface on the tibia. They participate in shock absorption by transmitting half of the weight-bearing load in full extension and a significant portion of the load in flexion (169). In flexion, the lateral meniscus carries the greater portion of the load. By absorbing some of the load, the menisci protect the underlying articular cartilage and subchondral bone. The menisci transmit the load across the surface of the joint, reducing the load per unit of area on the tibiofemoral contact sites (52). The contact area in the joint is reduced by two-thirds when the menisci are absent. This increases the pressure on the contacting surfaces and increases the susceptibility to injury (115). During low-load situations, the contact is primarily on the menisci, but in high-load situations, the contact area increases, with 70% of the load still on the menisci (52). The lateral meniscus carries a significantly greater percentage of the load.

The menisci also enhance lubrication of the joint. By acting as a space-filling mechanism, they allow dispersal of more synovial fluid to the surface of the tibia and the femur. It has been demonstrated that a 20% increase in

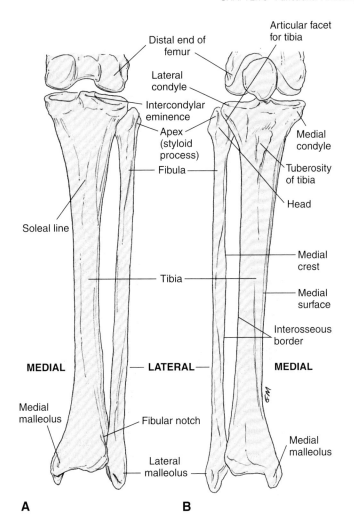

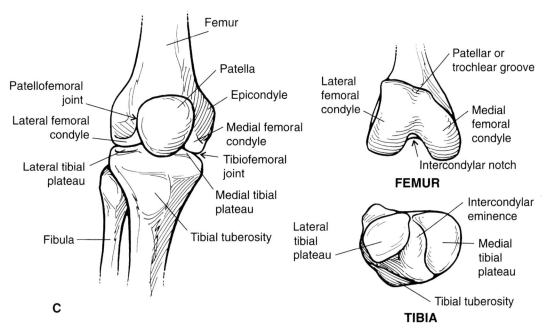

FIGURE 6-17 The structure of the knee joint is complex with asymmetrical condyles on the distal end of the femur articulating with asymmetrical facets on the tibial plateau. The patella moves in the trochlear groove on the femur. The anterior **(A)** and posterior **(B)** views of the lower leg and a close-up view of the knee joint **(C)** illustrate the complexity of the joints.

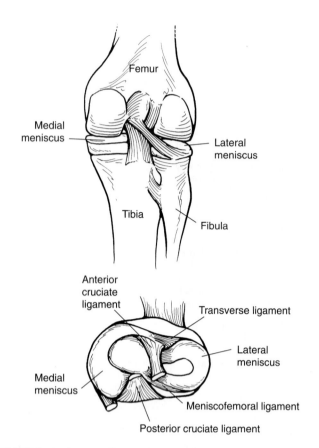

FIGURE 6-18 Two fibrocartilage menisci lie in the lateral and medial compartments of the knee. The medial meniscus is crescent shaped, and the lateral meniscus is oval to match the surfaces of the tibial plateau and the differences in the shape of the femoral condyles. Both menisci serve important roles in the knee joint by offering shock absorption, stability, and lubrication and by increasing the contact area between the tibia and the femur.

friction within the joint occurs with the removal of the meniscus (169).

Finally, the menisci limit motion between the tibia and femur. In flexion and extension, the menisci move with the femoral condyles. As the leg flexes, the menisci move posteriorly because of the rolling of the femur and muscular action of the popliteus and semimembranosus muscles (169). At the end of the flexion movement, the menisci fill up the posterior portion of the joint, acting as a space-filling buffer. The reverse occurs in extension. The quadriceps femoris and the patella assist in moving the menisci forward on the surface. Additionally, the menisci follow the tibia during rotation.

The tibiofemoral joint is supported by four main ligaments, two collateral and two cruciate. These ligaments assist in maintaining the relative position of the tibia and femur so that contact is appropriate and at the right time. See Figure 6-19 for insertions, actions, and illustration of these ligaments. They are the passive load-carrying structures of the joint and serve as a backup to the muscles (100).

On the sides of the joint are the collateral ligaments. The **medial collateral ligament** (MCL) is a flat, triangular ligament that covers a large portion of the medial side of the joint. The MCL supports the knee against any **valgus** force (a medially directed force acting on the lateral side of the knee) and offers some resistance to both internal and external rotations (117). It is taut in extension and reduces in length by approximately 17% in full flexion (166). The MCL offers 78% of the total valgus restraint at 25° of knee flexion (119).

The **lateral collateral ligament** (LCL) is thinner and rounder than the MCL. It offers the main resistance to varus force (a lateral force acting on the medial side) at the knee. This ligament is also taut in extension and reduces its length by approximately 25% in full flexion (166). The LCL offers 69% of the varus restraint at 25° of knee flexion (119) and offers some support in lateral rotation.

In full extension, the collateral ligaments are assisted by tightening of the posteromedial and posterolateral capsules, thus making the extended position the most stable (99). Both collaterals are taut in full extension even though the anterior portion of the MCL is also stretched in flexion.

The cruciate ligaments are intrinsic, lying inside the joint in the intercondylar space. These ligaments control both anteroposterior and rotational motions in the joint. The ACL provides the primary restraint for anterior movement of the tibia relative to the femur. It accounts for 85% of the total restraint in this direction (119). The ACL is 40% longer than its counterpart, the PCL. It elongates by about 7% as the knee moves from extension to 90° of flexion and maintains the same length up through maximum flexion (166). If the joint is internally rotated, the insertion of the ACL moves anteriorly, elongating the ligament slightly more. With the joint externally rotated, the ACL does not elongate up through 90° of knee flexion but elongates up to 10% from 90° to full flexion (166). Different parts of the ACL are taut in different knee positions. The anterior fibers are taut in extension, the middle fibers are taut in internal rotation, and the posterior fibers are taut in flexion. The ACL as a whole is considered to be taut in the extended position (Fig. 6-20).

The PCL offers the primary restraint to posterior movement of the tibia on the femur, accounting for 95% of the total resistance to this movement (119). This ligament decreases in length and slackens by 10% at 30° of knee flexion and then maintains that length throughout flexion (166). The PCL increases in length by about 5% with internal rotation of the joint up to 60° of flexion and then decreases in length by 5% to 10% as flexion continues. The PCL is not affected by external rotation in the joint, maintaining a fairly constant length. It is maximally strained through 45° to 60° of flexion (166) (Fig. 6-20). As with the ACL, the fibers of the PCL participate in different functions. The posterior fibers are taut in extension, the anterior fibers are taut in midflexion, and the posterior fibers are taut in full flexion; however, as a whole, the PCL is taut in maximum knee flexion.

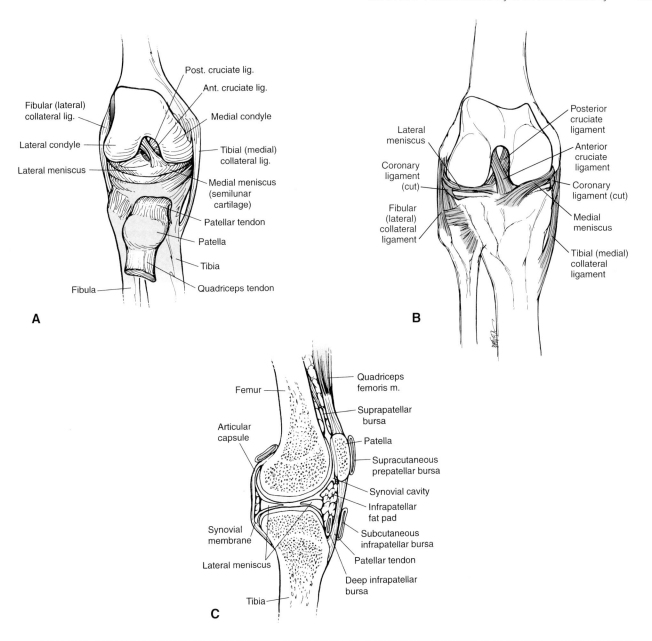

Ligament	Insertion	Action
Anterior cruciate	Anterior intercondylar area of tibia TO medial surface of lateral condyle	Prevents anterior tibial displacement; resists extension, internal rotation, flexion
Arcuate	Lateral condyle of femur TO head of fibula	Reinforces back of capsule
Coronary	Meniscus TO tibia	Holds menisci to tibia
Medial collateral	Medial epicondyle of femur TO medial condyle of tibia and medial meniscus	Resists valgus forces; taut in extension; resists internal, external rotation
Lateral collateral	Lateral epicondyle of femur TO head of fibula	Resists varus forces; taut in extension
Patellar	Inferior patella TO tibial tuberosity	Transfers force from quariceps to tibia
Posterior cruciate	Posterior spine of tibia TO inner condyle of femur	Resists posterior tibial movement; resists flexion and rotation
Posterior oblique	Expansion of semimembranosus muscle	Supports posterior, medial capsule
Transverse	Medial meniscus TO lateral meniscus in front	Connects menisci to each other

FIGURE 6-19 Ligaments of the knee joint shown from the anterior (A), posterior (B), and medial (C) perspectives.

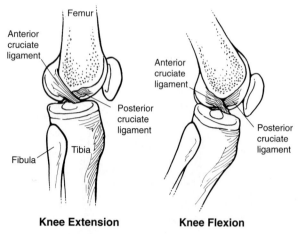

Knee Extension   **Knee Flexion**

FIGURE 6-20 The anterior cruciate ligament provides anterior restraint of the movement of the tibia relative to the femur. The posterior cruciate ligament offers restraint to posterior movement of the tibia relative to the femur.

Both the cruciate ligaments stabilize, limit rotation, and cause sliding of the condyles over the tibia in flexion. They both also offer some stabilization against varus and valgus forces. In a standing posture, with the tibial shaft vertical, the femur is aligned with the tibia and tends to slide posteriorly. A hyperextended position to 9° of flexion is unstable because the femur tilts posteriorly and is minimally restricted (100). At a 9° tilt of the tibia, the femur slides anteriorly to a position where it is more stable and supported by the patella and the quadriceps femoris.

Another important support structure surrounding the knee is the joint capsule. One of the largest capsules in the body, it is reinforced by numerous ligaments and muscles, including the MCL, the cruciate ligaments, and the arcuate complex (165). In the front, the capsule forms a substantial pocket that offers a large patellar area and is filled with the infrapatellar fat pad and the **infrapatellar bursa**. The fat pad offers a stopgap in the anterior compartment of the knee.

The capsule is lined with the largest synovial membrane in the body, which forms embryonically from three separate pouches (18). In 20% to 60% of the population, a permanent fold, called a **plica**, remains in the synovial membrane (19). The common location of the plica is medial and superior to the patella. It is soft and pliant and passes over the femoral condyle in flexion and extension. If injured, it can become fibrous and create both resistance and pain in motion (19). There are also more than 20 bursae in and around the knee, reducing friction between the muscle, tendon, and bone (165).

PATELLOFEMORAL JOINT

The second joint in the region of the knee is the patellofemoral joint, consisting of the articulation of the patella with the trochlear groove on the femur. The patella is a triangular sesamoid bone encased by the tendons of the quadriceps femoris. The primary role of the patella is to increase the mechanical advantage of the quadriceps femoris (18).

The posterior articulating surface of the patella is covered with the thickest cartilage found in any joint in the body (147). A vertical ridge of bone separates the underside of the patella into medial and lateral facets, each of which can be further divided into superior, middle, and inferior facets. A seventh facet, the odd facet, lies on the far medial side of the patella (165). The structure of the patella and the location of these facets are presented in Figure 6-21. During normal flexion and extension, five of these facets typically make contact with the femur.

The patella is connected to the tibial tuberosity via the strong **patellar tendon**. It is connected to the femur and tibia by small patellofemoral and patellotibial ligaments that are actually thickenings in the extensor retinaculum surrounding the joint (18).

Positioning of the patella and alignment of the lower extremity in the frontal plane is determined by measuring the Q-angle (quadriceps angle). Illustrated in

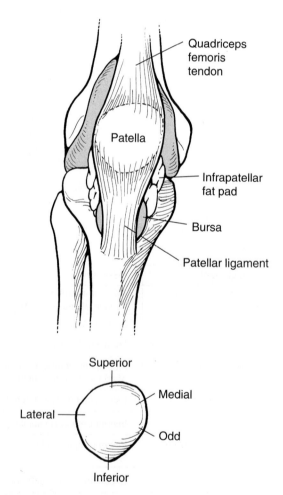

FIGURE 6-21 The patella increases the mechanical advantage of the quadriceps femoris muscle group. The patella has five facets, or articulating surfaces: the superior, inferior, medial, lateral, and odd facets.

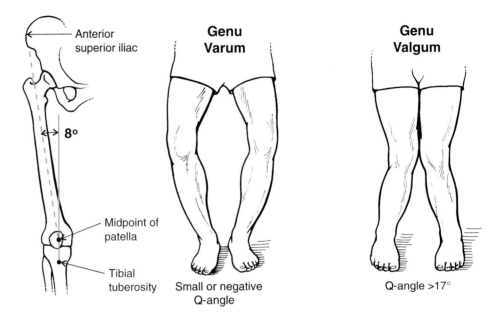

FIGURE 6-22 The Q-angle is measured between a line from the anterior superior iliac spine to the middle of the patella and the projection of a line from the middle of the patella to the tibial tuberosity. Q-angles range from 10° to 14° for males and 15° to 17° for females. Very small Q-angles create a condition known as genu varum, or bowleggedness. Large Q-angles create genu valgum, or knock-kneed position.

Figure 6-22, the Q-angle is formed by drawing one line from the anterior superior spine of the ilium to the middle of the patella and a second line from the middle of the patella to the tibial tuberosity. The Q-angle forms because the two condyles sit horizontal on the tibial plateau and because the medial condyle projects more distally, the femur angles laterally. In a normal alignment, the hip joint should still be vertically centered over the knee joint even though the anatomical alignment of the femur angles out. The most efficient Q-angle for quadriceps femoris function is one close to 10° of valgus (91). Whereas males typically have Q-angles averaging 10° to 14°, females average 15° to 17°, speculated to be primarily because of their wider pelvic basins (91). However, a recent evaluation of the Q-angle in males and females suggests that the positioning of the anterior superior iliac spine is not significantly positioned more laterally in females, and the differences in values between males and females are attributable to height differences (58).

The Q-angle represents the valgus stress acting on the knee, and if it is excessive, many patellofemoral problems can develop. Any Q-angle over 17° is considered to be excessive and is termed **genu valgum**, or knock-knees (91). A very small Q-angle constitutes bowleggedness, or **genu varum**.

Mediolaterally, the patella should be centered in the trochlear notch, and if the patella deviates medially or laterally, abnormal stresses can develop on the underside. The vertical position of the patella is determined primarily by the length of the patellar tendon measured from the distal end of the patella to the tibia. Patella alta is an alignment in which the patella is high and has been associated with higher levels of patellar subluxations. Patella baja is when the patella is lower than normal.

TIBIOFIBULAR JOINT

The third and final articulation is the small, superior tibiofibular joint, shown in Figure 6-23. This joint consists of the articulation between the head of the fibula and the posterolateral and inferior aspect of the tibial condyle. It is a gliding joint moving anteroposteriorly, superiorly, and inferiorly and rotating in response to rotation of the tibia and the foot (131). The fibula externally rotates and moves externally and superiorly with **dorsiflexion** of the

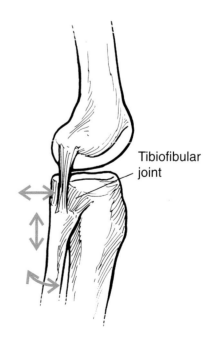

FIGURE 6-23 The tibiofibular joint is a small joint between the head of the fibula and the tibial condyle. It moves anteroposteriorly, superiorly, and inferiorly and rotates in response to movements of the tibia or the foot.

foot and accepts approximately 16% of the static load applied to the leg (131).

The primary functions of the superior tibiofibular joint are to dissipate the torsional stresses applied by the movements of the foot and to attenuate lateral tibial bending. Both the tibiofibular joint and the fibula absorb and control tensile rather than compressive loads applied to the lower extremity. The middle part of the fibula has more ability to withstand tensile forces than any other part of the skeleton (131).

MOVEMENT CHARACTERISTICS

The function of the knee is complex because of its asymmetrical medial and lateral articulations and the patellar mechanics on the front. When flexion is initiated in the closed-chain or weight-bearing position, the femur rolls backward on the tibia and laterally rotates and abducts with respect to the tibia. In an open-chain movement such as kicking, flexion is initiated with movement of the tibia on the femur, resulting in tibial forward motion, medial rotation, and adduction. The opposite occurs in extension with the femur rolling forward, medially rotating, and adducting in a closed-chain movement and the tibia rolling backward, laterally rotating, and abducting in an open-chain activity. The femoral contact with the tibia moves posteriorly during flexion and anteriorly during extension. Through 120° of extension, the anterior movement is 40% of the length of the tibial plateau (165). It has also been suggested that after the rolling is complete in the flexion movement that the femur finishes off in maximal flexion by just sliding anteriorly. These movements are illustrated in Figure 6-24.

Rotation at the knee is created partly by the greater movement of the lateral condyle on the tibia through almost twice the distance. Rotation can occur only with the joint in some amount of flexion. Thus, there is no rotation in the extended, locked position. Internal tibial rotation also occurs with dorsiflexion and pronation at the foot. Roughly 6° of subtalar motion results in roughly 10° of internal rotation (140). External rotation of the tibia also accompanies **plantarflexion** and **supination** of the foot. With 34° of supination, there is a corresponding 58° of external rotation (140).

The rotation occurring in the last 20° of extension has been termed the **screw-home mechanism**. The screw-home mechanism is the point at which the medial and lateral condyles are locked to form the close-packed position for the knee joint. The screw-home mechanism moves the tibial tuberosity laterally and produces a medial shift at the knee. Some of the speculative causes of the screw-home movement are that the lateral condyle surface is covered first and a rotation occurs to accommodate the larger surface of the medial condyle or that the ACL becomes taut just before rotation, forcing rotation of the femur on the tibia (148). Finally, it is speculated that the cruciate ligaments become taut in early extension and pull the condyles in opposite directions, causing the rotation. The screw-home mechanism is disrupted with injury to the ACL because the tibia moves more anteriorly on the femur. It is not significantly disrupted with loss of the PCL, indicating that the ACL is the main controller (148).

The normal range of motion at the knee joint is approximately 130° to 145° in flexion and 1° to 2° of hyperextension. It has been reported that there are 6° to

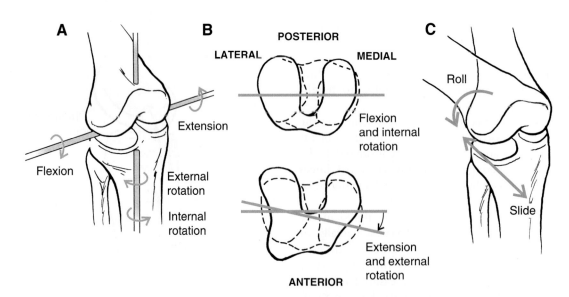

FIGURE 6-24   (A) The movements at the knee joint are flexion and extension and internal and external rotation. **(B)** When the knee flexes, there is an accompanying internal rotation of the tibia on the femur (non-weight bearing). In extension, the tibia externally rotates on the femur. **(C)** There are also translatory movements of the femur on the tibial plateau surface. In flexion, the femur rolls and slides posteriorly.

30° of internal rotation through 90° of flexion at the joint around an axis passing through the medial intercondylar tubercle of the tibial plateau (77,118). External rotation of the tibia is possible through approximately 45° (74). The range of motion in varus or abduction and valgus or adduction is small and in the range of 5°.

When the knee flexes, the patella moves distally through a distance more than twice its length, entering the intercondylar notch on the femur (74) (Fig. 6-25). In extension, the patella returns to its resting position high and lateral on the femur, where it is above the trochlear groove and resting on the suprapatellar fat pad. The patella is free to move in the extended position and can be shifted in multiple directions. Patellar movement is restricted in the flexed position because of the increased contact with the femur.

The movement of the patella is most affected by the joint surface and the length of the patellar tendon and minimally affected by the quadriceps femoris. In the first 20° of flexion, the tibia internally rotates and the patella is drawn from its lateral position down into the groove, where first contact is made with the inferior facets (165). The stability offered by the lateral condyle is most important because most subluxations and dislocations of the patella occur in this early range of motion.

The patella follows the groove to 90° of flexion, at which point contact is made with the superior facets of the patella (Fig. 6-25). At that time, the patella again moves laterally over the condyle. If flexion continues to 135°, contact is made with the odd facet (165). In flexion, the linear and translatory movements of the patella are posterior and inferior, but the patella also has some angular movements that affect its position. During knee flexion, the patella also flexes, abducts, and externally rotates, and these movements reverse in extension (extension,

adduction, and internal rotation). Flexion and extension of the patella occur about a mediolateral axis running through a fixed axis in the distal femur, with flexion representing the upward tilt and extension representing the downward tilt about this axis. Likewise, patellar abduction and adduction involve movement of the patella away from and toward midline in the frontal plane, respectively. External and internal rotations are rotations of the patella outward and inward about a longitudinal axis, respectively (81).

MUSCULAR ACTIONS

Knee extension is a very important contributor to the generation of power in the lower extremity for any form of human projection or translation. The musculature producing extension is also used frequently to contract eccentrically and decelerate a rapidly flexing knee joint. Fortunately, the quadriceps femoris muscle group, the producer of extension at the knee, is one of the strongest muscle groups in the body; it may be as much as three times stronger than its antagonistic muscle group, the hamstrings, because of its involvement in negatively accelerating the leg and continuously contracting against gravity (74).

The quadriceps femoris is a muscle group that consists of the rectus femoris and vastus intermedius forming the middle part of the muscle group, the vastus lateralis on the lateral side, and the vastus medialis on the medial side (19). The specific insertions, actions, and nerve supply are presented in Figure 6-26.

The quadriceps femoris connects to the tibial tuberosity via the patellar tendon and contributes somewhat to the stability of the patella. As a muscle group, it also pulls the menisci anteriorly in extension via the meniscopatellar ligament. When it contracts, it also reduces the strain in the MCL and works with the PCL to prevent posterior displacement of the tibia. It is antagonistic to the ACL.

The largest and strongest of the quadriceps femoris is the vastus lateralis, a muscle applying lateral force to the patella. Pulling medially is the vastus medialis. The vastus medialis has two portions referred to as the vastus medialis longus and the vastus medialis oblique, and the boundary of these two portions of the vastus medialis is located at the medial rim of the patella. The direction of the muscle fibers in the more proximal vastus medialis longus runs more vertical, and the fibers of the lower vastus medialis oblique run more horizontal (123). Although the vastus medialis as a whole is an extensor of the knee, the vastus medialis oblique is also a medial stabilizer of the patella (165).

It has been noted in the literature that the vastus medialis was selectively activated in the last few degrees of extension. This has been proved not to be true. No selective activation of the vastus medialis muscles occurs in the last degrees of extension, and the quadriceps muscles contract equally throughout the range of motion (85).

The only two-joint muscle of the quadriceps femoris group, the rectus femoris, does not significantly contribute to knee extension force unless the hip joint is in

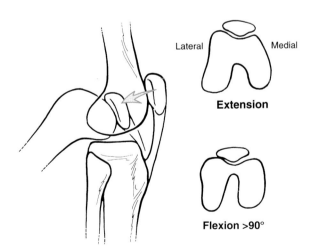

FIGURE 6-25 When the knee flexes, the patella moves inferiorly and posteriorly over two times its length. The patella sits in the groove and is held in place by the lateral condyle of the femur. If the knee continues into flexion past 90°, the patella moves laterally over the condyle until at approximately 135° of flexion, when contact is made with the odd facet.

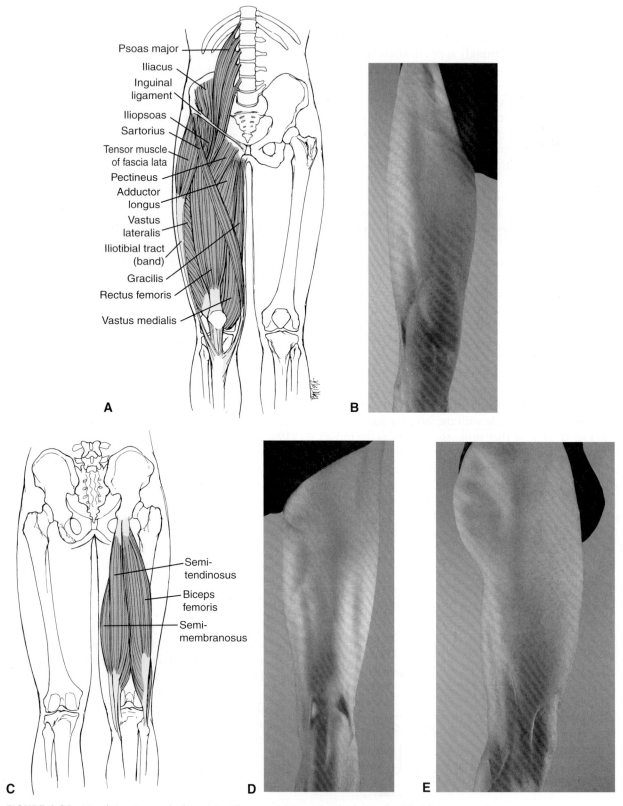

Psoas major
Iliacus
Inguinal ligament
Iliopsoas
Sartorius
Tensor muscle of fascia lata
Pectineus
Adductor longus
Vastus lateralis
Iliotibial tract (band)
Gracilis
Rectus femoris
Vastus medialis

A

B

Semi-tendinosus
Biceps femoris
Semi-membranosus

C

D

E

FIGURE 6-26   Muscles acting on the knee joint. Shown are the anterior thigh muscles **(A)** with corresponding surface anatomy **(B)**, the posterior thigh muscles **(C)** and posterior **(D)** and lateral **(E)** surface anatomy, and other supporting anterior and posterior muscles **(F)**.

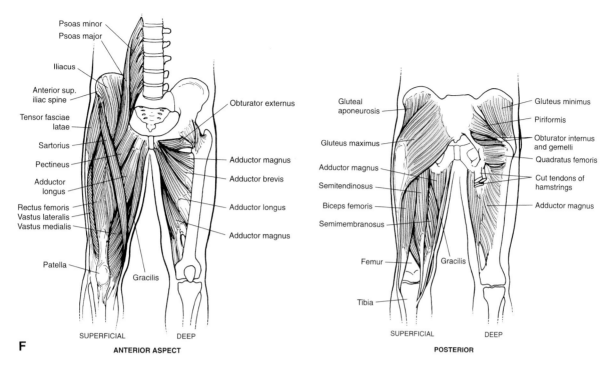

F   ANTERIOR ASPECT

POSTERIOR

Muscle	Insertion	Nerve Supply	Flexion	Extension	Pronation	Supination
Biceps femoris	Ischial tuberosity TO lateral condyle of tibia, head of fibula	Tibial, peroneal portion of sciatic nerve; L5, S1–S3	PM			PM
Gastrocnemius	Medial, lateral condyles of femur TO calcaneus	Tibial nerve; S1, S2	Asst			
Gracilis	Inferior rami of pubis TO medial tibial (pes anserinus)	Anterior obturator nerve; L3, L4	Asst		PM	
Popliteus	Lateral condyle of femur TO proximal tibia	Tibial nerve	Asst		PM	
Rectus femoris	Anterior inferior iliac spine TO patella, tibial tuberosity	Femoral nerve; L2–L4		PM		
Sartorius	Anterior superior iliac spine TO medial tibia (pes anserinus)	Femoral nerve; L2, L3	Asst			PM
Semimembranous	Ischial tuberosity TO medial condyle of tibia	Tibial portion of sciatic nerve: L5, S1, S2	PM		PM	
Semitendinosus	Ischial tuberosity TO medial tibia (pes anserinus)	Tibial portion of sciatic nerve: L4, S1, S2	PM		PM	
Vastus intermedius	Anterior lateral femur TO patella, tibial tuberosity	Femoral nerve; L2–L4		PM		
Vastus lateralis	Intertrochanteric line; linea aspera TO patella, tibial tuberosity	Femoral nerve; L2–L4		PM		
Vastus medialis	Linea apsera; trochanteric line TO patella, tibial tuberosity	Femoral nerve; L2–L4		PM		

FIGURE 6-26 (continued)

a favorable position. It is limited as an extensor of the knee if the hip is flexed and is facilitated as a knee extensor if the hip joint is extended, lengthening the rectus femoris. In walking and running, the rectus femoris contributes to the extension force in the toe-off phase when the thigh is extended. Likewise, in kicking, rectus femoris activity is maximized in the preparatory phase as the thigh is brought back into hyperextension with the leg in flexion.

Flexion of the leg at the knee joint occurs during support, when the body lowers toward the ground; however, this downward movement is controlled by the extensors so that buckling does not occur. The flexor muscles are very active with the limb off the ground, working frequently to slow a rapidly extending leg.

The major muscle group that contributes to knee flexion is the hamstrings, consisting of the lateral biceps femoris and the medial semimembranosus and semitendinosus (see Fig. 6-26). The action of the hamstrings can be quite complex because they are two-joint muscles that work to extend the hip. The hamstrings work with the ACL to resist anterior tibial displacement. They are also rotators of the knee joint because of their insertions on the sides of the knee. As flexors, the hamstrings can generate the greatest force from a flexion position of 90° (120).

Flexion strength diminishes with extension because of an acute tendon angle that reduces the mechanical advantage. At full extension, flexion strength is reduced by 50% compared with 90° of flexion (120).

The lateral hamstring, the biceps femoris, has two heads connecting on the lateral side of the knee and offering lateral support to the joint. The biceps femoris also produces external rotation of the lower leg.

The semimembranosus bolsters the posterior and medial capsule. In flexion, it pulls the meniscus posteriorly (165). This medial hamstring also contributes to the production of internal rotation in the joint. The other medial hamstring, the semitendinosus, is part of the **pes anserinus** muscular attachment on the medial surface of the tibia. It is the most effective flexor of the pes anserinus muscle group, contributing 47% to the flexion force (165). The semitendinosus works with both the ACL and the MCL in supporting the knee joint. It also contributes to the generation of internal rotation.

The hamstrings operate most effectively as knee flexors from a position of hip flexion by increasing the length and tension in the muscle group. If the hamstrings become tight, they offer greater resistance to extension of the knee joint by the quadriceps femoris. This imposes a greater workload on the quadriceps femoris muscle group.

The two remaining pes anserinus muscles, the sartorius and the gracilis, also contribute 19% and 34% to the flexion strength, respectively (120). The popliteus is a weak flexor that supports the PCL in deep flexion and draws the meniscus posteriorly. Finally, the two-joint gastrocnemius contributes to knee flexion, especially when the foot is in the neutral or dorsiflexed position.

Internal rotation of the tibia is produced by the medial muscles: sartorius, gracilis, semitendinosus, semimembranosus, and popliteus (see Fig. 6-26). Internal rotation force is greatest at 90° of knee flexion and decreases by 59% at full extension (124,125). The internal rotation force can be increased by 50% if it is preceded by 15° of external rotation. Of the three pes anserinus muscles, the sartorius and the gracilis are the most effective rotators, accounting for 34% and 40% of the pes anserinus force in rotation (120). The semitendinosus contributes 26% of the pes anserinus rotation force. The pes anserinus muscle group also contributes significantly to medial knee stabilization. Only one muscle, the biceps femoris, contributes significantly to the generation of external rotation of the tibia. Both internal and external rotations are necessary movements associated with the function of the knee joint.

COMBINED MOVEMENTS OF THE HIP AND KNEE

Many lower extremity movements require coordinated actions at the hip and knee joint, and this is complicated by the number of two-joint muscles that span both joints. Coactivation of both monoarticular and biarticular agonists and antagonists is required to produce motion with appropriate direction and force. This coordination is required for uninterrupted transitions between extension and flexion. For example, in walking, coactivation of the gluteus maximus (monoarticular) and the rectus femoris (knee extensor) is necessary to generate forces for the simultaneous extension of both the hip and the knee (143,162,171). Additionally, coactivation of the iliopsoas and the hamstrings facilitates knee flexion by cancelling out the motion at the hip joint.

Positioning of the hip changes the effectiveness of the muscles acting at the knee joint. For example, changing the hip joint angle has a large effect on increasing the moment arm of the biceps femoris. It is the opposite for the rectus femoris, which is more influenced by a change in the knee angle (163). The range of motion at the knee also changes with a change in hip positioning. For example, the knee flexes through approximately 145° with the thigh flexed and 120° with the thigh hyperextended (74). This difference in range is attributable to the length–tension relationship in the hamstring muscle group.

STRENGTH OF THE KNEE JOINT MUSCLES

The extensors at the knee joint are usually stronger than the flexors throughout the range of motion. Peak extension strength is achieved at 50° to 70° of knee flexion (115). The position of maximum strength varies with the speed of movement. For example, if the movement is slow, peak extension strength occurs in the first 20° of knee extension from the 90° flexed position. Flexion strength is greatest in the first 20° to 30° of flexion from the extended position (126). This position also fluctuates with the speed of movement. Greater knee flexion torques can be obtained if the hips are flexed because the hamstring length–tension relationship is improved.

It is common in sports medicine to evaluate the isokinetic strength of the quadriceps femoris and the hamstrings to construct a hamstring-to-quadriceps ratio. A generally acceptable ratio is 0.5, with the hamstrings at least half as strong as the quadriceps femoris. It has been suggested that anything below this ratio indicates a strength imbalance between the quadriceps femoris and the hamstrings that predisposes one to injury. Caution must be observed when using this ratio because it applies only to slow, isokinetic testing speeds.

At faster testing speeds, when the limbs move through 200° to 300°/s, the ratio approaches 1 because the efficiency of the quadriceps femoris decreases at higher speeds. Even at the isometric testing level, the hamstring-to-quadriceps ratio is 0.7. Thus, a ratio of 0.5 between the hamstrings and the quadriceps femoris is not acceptable at fast speeds and indicates a strength imbalance between the two groups, but at a slower speed, it would not indicate an imbalance (110).

Internal and external rotation torques are both greatest with the knee flexed to 90° because a greater range of rotation motion can be achieved in that position. Internal rotation strength increases by 50% from 45° of knee flexion to 90° (125). The position of the hip joint also influences internal rotation torque, with the greatest strength developed at 120° of hip flexion, at which point the gracilis and the hamstrings are most efficient (124). At low hip flexion angles and in the neutral position, the sartorius is the most effective lateral rotator. Peak rotation torques occur in the first 5° to 10° of rotation. The internal rotation torque is greater than the external rotation torque (125).

CONDITIONING OF THE KNEE JOINT MUSCLES

The extensors of the leg are easy to exercise because they are commonly used to both lower and raise the body. Examples of stretching and strengthening exercises for the extensors are presented in Figure 6-27. The squat is used to strengthen the quadriceps femoris. When one lowers into a squat, the force coming through the joint, directed vertically in the standing position, is partially directed across the joint, creating a shear force. This shear force increases as knee flexion increases. Thus, in a deep squat position, most of the original compressive force is directed posteriorly, creating a shear force. With the ligaments and muscles unable to offer much protection in the posterior direction at the full squat position, this is considered a vulnerable position. This position of maximum knee flexion is contraindicated for beginner and unconditioned lifters.

An experienced and conditioned lifter who has strong musculature and uses good technique at the bottom of the lift will most likely avoid any injury when in this position. Good technique involves control over the speed of descent and proper segmental positioning. For example, if the trunk is in too much flexion, the low back will be excessively loaded and the hamstrings will perform more of the work and the quadriceps femoris less, focusing control on the posterior side.

The quadriceps femoris group may also be exercised in an open-chain activity, as in a leg extension machine. Starting from 90° of flexion, one can exert considerable force because the quadriceps femoris muscles are very efficient throughout the early parts of the extension action. Near full extension, the quadriceps femoris muscles become inefficient and must exert greater force to move the same load. Thus, quadriceps activity in an open-chain leg extension is higher near full extension in the squat, but there is more activity in the quadriceps near full flexion at the bottom of the squat (48).

The terminal extension exercise is good for individuals who have patellar pain because the quadriceps femoris works hard with minimal patellofemoral compression force. This kind of exercise should be avoided, however, in early rehabilitation of an ACL injury because the anterior shear force is so large in this position. To minimize the stress on the ACL, no knee extension exercise should be used at any angle less than 64° (168). Coactivation from the hamstrings increases as the knee reaches full extension, and this also minimizes the stress on the ACL by preventing anterior displacement (40). However, any knee extension exercise for individuals with ACL injuries should be done from a position of considerable knee flexion. Also, the terminal extension exercise does not selectively exercise the medial quadriceps more than the lateral quadriceps (44).

The flexors of the knee are not actively recruited in the performance of a flexion action with gravity because the quadriceps femoris muscles control the flexion action via eccentric muscle activity. Fortunately, the hamstrings are extensors of the hip as well as flexors of the knee joint. Thus, they are active during a squat exercise by virtue of their influence at the hip because hip flexion in lowering is controlled eccentrically by the hip extensors. The squat generates twice as much activity in the hamstrings as a leg press on a machine (48). If it were not for the hamstrings' role as extensors at the hip, the hamstrings group would be considerably weaker than the quadriceps femoris.

The knee flexors are best isolated and exercised in a seated position using a leg curl apparatus. The seated position places the hip in flexion, thus optimizing their performance. The knee flexors, especially the hamstrings and the pes anserinus muscles, are important for knee stability because they control much of the rotation at the knee. As presented earlier in this chapter, the hamstrings should be half as strong as the quadriceps femoris groups for slow speeds and should be as strong as the quadriceps femoris group at fast speeds. It is also important to maintain flexibility in the hamstrings because if they are tight, the quadriceps femoris muscles must work harder and the pelvis will develop an irregular posture and function.

Muscle Group	Sample Stretching Exercise	Sample Strengthening Exercise	Other Exercises
Knee flexors	 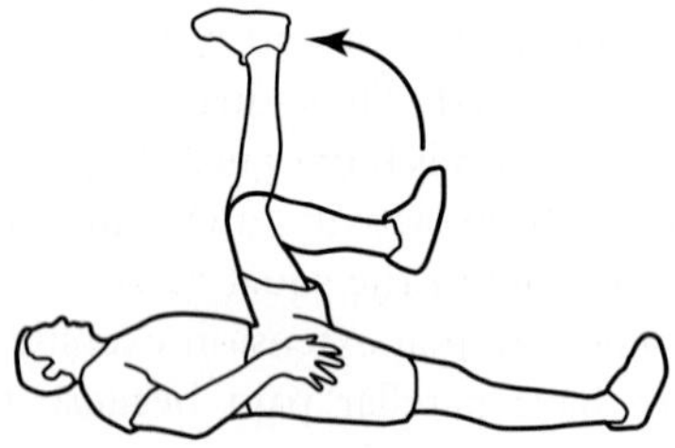 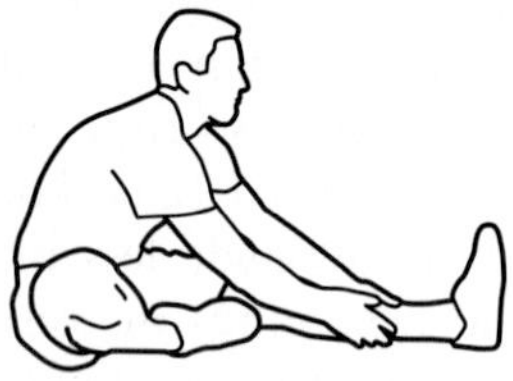	Leg curl 	Stability ball curls Squats
Knee extensors	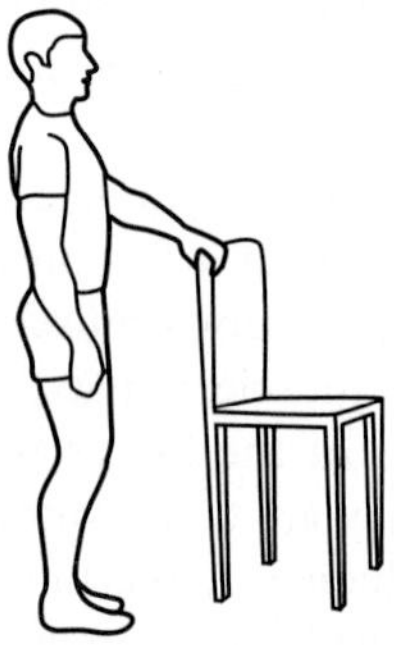 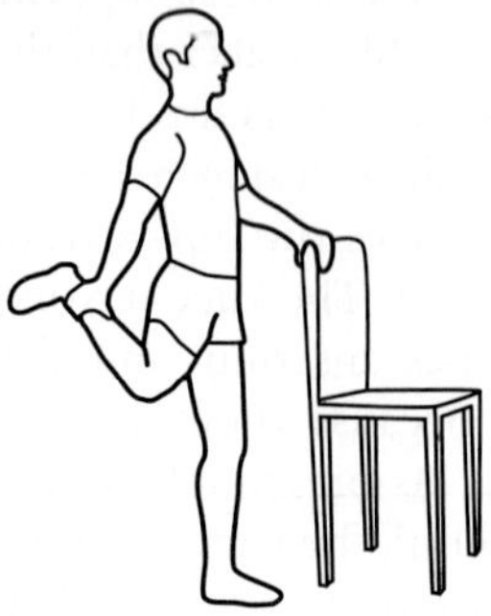	Leg extension 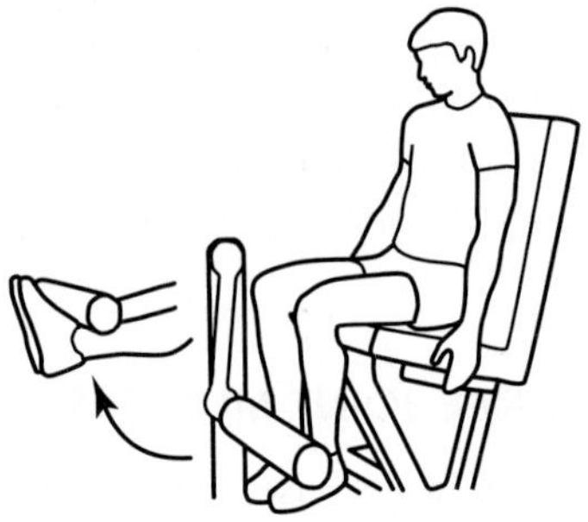 Squat 	Leg press Lunge

FIGURE 6-27   Sample stretching and strengthening exercises for selected muscle groups.

The rotators of the knee, because they are all flexor muscles, are exercised along with the flexion movements. If the rotators are to be selectively stretched or strengthened as they perform the rotation, it is best to do the exercise from a seated position with the knee flexed to 90° and the rotators in a position of maximum effectiveness. Toeing in the foot contracts the internal rotators and stretches the external rotators. Different levels of resistance can be added to this exercise through the use of elastic bands or cables.

There continues to be debate over the use of closed-chain versus open-chain exercises for the rehabilitation after ACL repair at the knee joint. Some surgeons and physical therapists advocate using only closed-chain exercises (28). The reason behind this is that closed-chain exercises have been shown to produce significantly less posterior shear force at all angles and less anterior shear force at most angles (90). This occurs because of higher compressive loads and muscular coactivation. Recently, there has been added support for the inclusion of open-chain exercises in an ACL rehabilitation protocol (15). Knee extension exercises at angles of 60° to 90° have been shown to be very effective for isolation of quadriceps and do not negatively influence healing of the ACL graft (50). Studies have shown that anterior tibial translation is less in a closed-chain exercise (80), giving support for their use. In other studies, however, maximum ACL strains have been shown to be similar in both open- and closed-chain exercises (16), supporting the inclusion of both types of exercise in the rehabilitation protocol.

Extension exercises for individuals with patellofemoral pain also vary between closed- and open-chain exercises. In the open-chain knee extension, the patellofemoral force increases with extension with the quadriceps force high from 90° to 25° of knee flexion (44). In a closed-chain squat, it is opposite, with the patellofemoral force zero at full extension and increasing with increases in knee flexion and with load (14).

INJURY POTENTIAL OF THE KNEE JOINT

The knee joint is a frequently injured area of the body, depending on the sport, accounting for 25% to 70% of reported injuries. In a 10-year study of athletic knee injuries in which 7,769 injuries related to the knee joint were documented, the majority of the knee injuries occurred in males and in the age group of 20 to 29 years (93). Activities associated with most of the injuries were soccer and skiing.

The cause of an injury to the knee can often be related to poor conditioning or training or to an alignment problem in the lower extremity. Injuries in the knee have been attributable to hindfoot and forefoot varus or valgus, tibial or femoral varus or valgus, limb length differences, deficits in flexibility, strength imbalances between agonists and antagonists, and improper technique or training.

A number of knee injuries are associated with running or jogging because the knee and the lower extremity are subjected to a force equivalent to approximately three times BW at every foot contact. It is clear that if 1,500 foot contacts are made per mile of running, the potential for injury is high.

Traumatic injuries to the knee usually involve the ligaments. Ligaments are injured as a result of application of a force causing a twisting action of the knee. High-friction or uneven surfaces are usually associated with increased ligamentous injury. Any movement fixing the foot while the body continues to move forward, such as often occurs in skiing, will likely produce a ligament sprain or tear. Simply, any turn on a weight-bearing limb leaves the knee vulnerable to ligamentous injury.

The ACL is the most common site of ligament injuries, which are usually caused by a twisting action while the knee is flexed, internally rotated, and in a valgus position while supporting weight. It can also be damaged with a forced hyperextension of the knee. If the trunk and thigh rotate over a lower extremity while supporting the body's weight, the ACL can be sprained or torn because the lateral femoral condyle moves posteriorly in external rotation (60). The quadriceps can also be responsible for ACL sprain by producing anterior displacement of the tibia when eccentrically controlling knee motion when there is limited hamstring coactivation (32). If the hamstrings are co-contracting, they resist the anterior translation of the tibia. Examples from sport in which this ligament is often injured are skiers catching the edge of the ski; a football player being blocked from the side; a basketball player landing off balance from a jump, cutting, or rapid deceleration; and a gymnast landing off balance from a dismount (124).

Loss of the ACL creates valgus laxity and single-plane or rotatory instability (30). Whereas planar instability is usually anterior, rotatory instabilities can occur in a variety of directions, depending on the other structures injured (22). Instability created by an inefficient or missing ACL places added stress on the secondary stabilizers of the knee, such as the capsule, collateral ligaments, and iliotibial band. There is an accompanying deficit in quadriceps femoris musculature. The "side effects" of an ACL injury are often more debilitating in the long run.

Injury to the PCL is less common than the ACL. The PCL is injured by receiving an anterior blow to a flexed or hyperextended knee or by forcing the knee into external rotation when it is flexed and supporting weight. Hitting the tibia up against the dashboard in a car crash or falling on a bent knee in soccer or football can also damage the PCL. Damage to the PCL results in anterior or posterior planar instability.

The collateral ligaments on the side are injured upon receipt of a force applied to the side of the joint. The MCL, torn in an application of force in the direction of the medial side of the joint, can also sprain or tear with a violent external rotation or tibial varus (35,150). The MCL is typically injured when the foot is fixed and slightly flexed. A change in direction with the person moving away from the support limb, as when running the bases in baseball, is a common event leading to an MCL injury. The MCL is usually injured at the proximal end, resulting in tenderness on the femoral side of the knee joint.

The LCL is injured upon receipt of a lateral force that is usually applied when the foot is fixed and the knee is in slight flexion (35). Injury to the MCL and LCL creates medial and lateral planar instabilities, respectively. A forceful varus or valgus force can also create a **distal femoral epiphysitis** as the collateral ligaments forcefully pull on their attachment site (83).

Damage to the menisci occurs much the same way as ligament damage. The menisci can be torn through compression associated with a twisting action in a weight-bearing position. They can also be torn in kicking and other violent extension actions. Tearing the meniscus by compression is a result of the femur grinding into the tibia and ripping the menisci. A meniscal tear in rapid extension is a result of the meniscus getting caught and torn as the femur moves rapidly forward on the tibia.

Tears to the medial meniscus are usually incurred during moves incorporating valgus, knee flexion, and external rotation in the supported limb or when the knee is hyperflexed (147). Lateral meniscus tears have been associated with a forced axial movement in the flexed position; a forced lateral movement with impact on the knee in extension; a forceful rotational movement; a movement incorporating varus, flexion, and internal rotation of the support limb; and the hyperflexed position (147).

Many injuries to the knee are a result of less traumatic noncontact forces. Muscle strains to the quadriceps femoris or the hamstrings muscle groups occur frequently. Strain to the quadriceps femoris usually involves the rectus femoris because it can be placed in a very lengthened position with hip hyperextension and knee flexion. It is commonly injured in a kicking action, especially if the kick is mistimed. A hamstring strain is usually associated with inflexibility in the hamstrings or a stronger quadriceps femoris that pulls the hamstrings into a lengthened position. Sprinting when the runner is not in condition to handle the stresses of sprinting can lead to a hamstring strain.

On the lateral side of the knee is the iliotibial band, which is frequently irritated as the band moves over the lateral epicondyle of the femur in flexion and extension. Iliotibial band syndrome is seen in individuals who run on cambered roads, specifically affecting the downhill limb. It has also been identified in individuals who run more than 5 miles per session, in stair climbing and downhill running, and in individuals who have a varum alignment in the lower extremity (57). Medial knee pain can be associated with many structures, such as tendinitis of the pes anserinus muscle attachment and irritation of the semimembranosus, parapatellar, or pes bursae (57).

Posterior knee pain is likely associated with popliteus tendinitis, which causes posterior lateral pain. This is often brought on by hill running. Posterior pain can also be associated with strain or tendinitis of the gastrocnemius

muscle insertion or by collection of fluid in the bursae, called a Baker's or popliteal cyst.

Anterior knee pain accounts for most overuse injuries to the knee, especially in women. **Patellofemoral pain syndrome** is pain around the patella and is often seen in individuals who exhibit valgum alignments or femoral anteversion in the extremity (34). Patellofemoral pain is aggravated by going down hills or stairs or squatting.

Stress on the patella is associated with a greater Q-angle because of increased stress on the patella. Patellar injury may be caused by abnormal tracking, which in addition to an increased Q-angle, can be created by a functional short leg, tight hamstrings, tight gastrocnemius, a long patellar tendon (termed **patella alta**), a short patellar tendon (termed **patella baja**), a tight lateral retinaculum or iliotibial band, or excessive pronation at the foot. Weak hip abductor muscles can allow excessive motion of the pelvis in the frontal plane and cause a characteristic gait called Trendelenburg gait (Fig. 6-28). This condition as well as compensations for this condition can affect the Q-angle and patellar tracking, and it may be associated with excessive pronation.

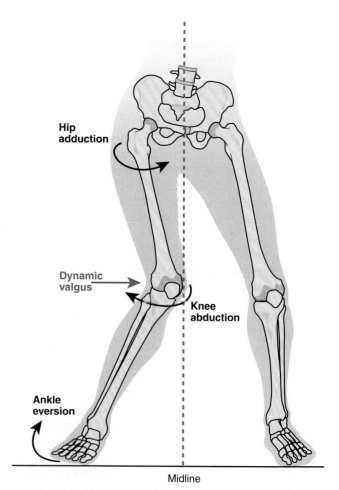

FIGURE 6-28 Weak hip abductors can cause the hip to drop when the contralateral leg is raised. The resulting hip adduction and internal rotation gives a characteristic gait pattern that can result in poor patellar tracking and excessive pronation.

Some patellofemoral pain syndromes are associated with cartilage destruction, in which the cartilage underneath the patella becomes soft and fibrillated. This condition is known as **chondromalacia patellae**. Patellar pain similar to that of patellar pain syndrome or chondromalacia patellae is also seen with medial retinaculitis, in which the medial retinaculum is irritated in running (165).

A subluxated or dislocated patella is common in individuals with predisposing factors. These are patella alta, ligamentous laxity, a small Q-angle with outfacing patella, external tibial torsion, and an enlarged fat pad with patella alta (165). Dislocation of the patella may be congenital. The dislocation occurs in flexion as a result of a faulty knee extension mechanism.

The attachment site of the quadriceps femoris to the tibia at the tibial tuberosity is another site for injury and the development of anterior pain. The tensile force of the quadriceps femoris can create tendinitis at this insertion site. This is commonly seen in athletes who do vigorous jumping, such as in volleyball, basketball, and track and field (105). In children aged 8 to 15 years, a tibial tubercle epiphysitis can develop. This is referred to as **Osgood–Schlatter disease**. This disease is an avulsion fracture of the growing tibial tuberosity that can also avulse the epiphysis. Bony growths can develop on the site. The cause of both of these conditions is overuse of the extensor mechanism (105).

Overuse of the extensor mechanism can also cause irritation of the plica. Plica injury can also result from a direct blow, a valgus rotary force applied to the knee, or weakness in the vastus medialis oblique. The plica becomes thick, inelastic, and fibrous with injury, making it difficult to sit for long periods and creating pain on the superior knee (19). The medial patella may snap and catch during flexion and extension with injury to the plica.

The Ankle and Foot

The foot and ankle make up a complex anatomical structure consisting of 26 irregularly shaped bones, 30 synovial joints, more than 100 ligaments, and 30 muscles acting on the segments. All of these joints must interact harmoniously and in combination to achieve a smooth motion. Most of the motion in the foot occurs at three of the synovial joints: the talocrural, the subtalar, and the midtarsal joints (102). The foot moves in three planes, with most of the motion occurring in the rear foot.

The foot contributes significantly to the function of the whole lower limb. The foot supports the weight of the body in both standing and locomotion. The foot must be a loose adapter to uneven surfaces at contact. Also, upon contact with the ground, it serves as a shock absorber, attenuating the large forces resulting from ground contact. Late in the support phase, it must be a rigid lever for effective propulsion. Finally, when the foot is fixed during stance, it must absorb the rotation of the lower extremity.

These functions of the foot all occur during a closed kinetic chain as it is receiving frictional and reaction forces from the ground or another surface (102).

The foot can be divided into three regions (Fig. 6-29). The rearfoot, consisting of the talus and the calcaneus; the midfoot, including the navicular, cuneiforms, and the cuboid; and the forefoot, containing the metatarsals and the phalanges. These structures are shown in Figure 6-30.

TALOCRURAL JOINT

The proximal joint of the foot is the **talocrural joint**, or ankle joint (Fig. 6-30). It is a uniaxial hinge joint formed by the tibia and fibula (tibiofibular joint) and the tibia and talus (**tibiotalar joint**). This joint is designed for stability rather than mobility. The ankle is stable when large forces are absorbed through the limb, when stopping and turning, and in many of the lower limb movements performed on a daily basis. If any of the anatomical support structures around the ankle joint are injured, however, the joint can become very unstable (60).

The tibia and fibula form a deep socket for the trochlea of the talus, creating a mortise. The medial side of the mortise is the inner side of the medial malleolus, a projection on the distal end of the tibia. On the lateral side is the inner surface of the lateral malleolus, a distal projection on the fibula. The lateral malleolus projects more distally than the medial malleolus and protects the lateral ligaments of the ankle. It also acts as a bulwark against any lateral displacement. Because the lateral malleolus projects more distally, it is also more susceptible to fracture with an **inversion** sprain to the lateral ankle.

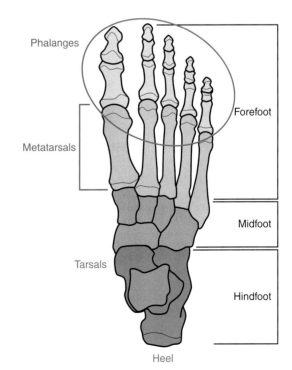

FIGURE 6-29   Division of the foot into functional regions.

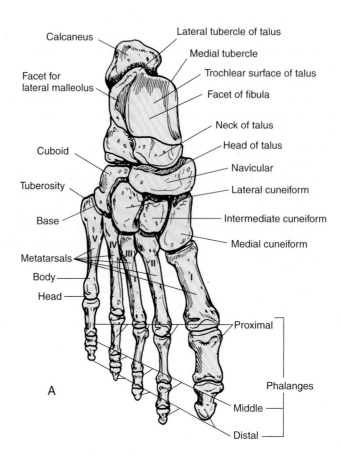

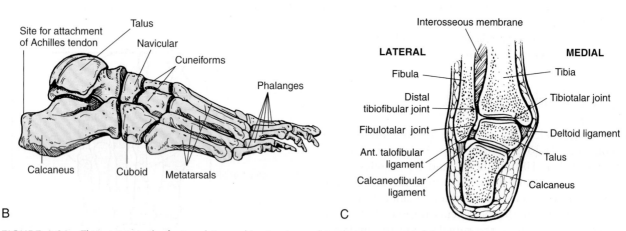

FIGURE 6-30 Thirty joints in the foot work in combination to produce the movements of the rear foot, midfoot, and forefoot. The subtalar and midtarsal joints contribute to pronation and supination. The intertarsal, tarsometatarsal, metatarsophalangeal, and interphalangeal joints contribute to movements of the forefoot and the toes. Joints are shown from the superior **(A)**, lateral **(B)**, and posterior **(C)** view.

The tibia and fibula fit snugly over the trochlea of the talus, a bone that is wider anteriorly than posteriorly (74). The difference in width of the talus allows for some abduction and adduction of the foot. The close-packed position for the ankle is the dorsiflexed position when the talus is wedged in at its widest spot.

The ankle has excellent ligamentous support on the medial and lateral sides. The location and actions of the ligaments are presented in Figure 6-31. The ligaments that surround the ankle limit plantarflexion and dorsiflexion, anterior and posterior movement of the foot, tilting of the talus, and inversion and eversion (155). Each of the

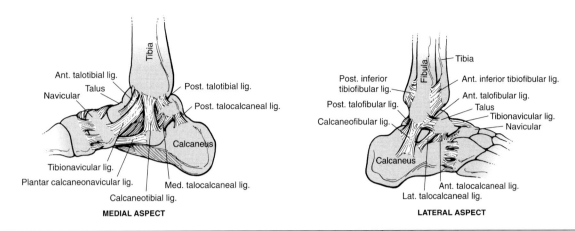

FIGURE 6-31 Ligaments of the foot and ankle.

Ligament	Insertion	Action
Anterior talofibular	Lateral malleolus TO neck of talus	Limits anterior displacement of foot or talar tilt; limits plantarflexion and inversion
Anterior talotibial	Anterior margin of tibia TO front margin on talus	Limits plantarflexion and abduction of foot
Calcaneocuboid	Calcaneus TO cuboid on dorsal surface	Limits inversion of foot
Calcaneofibular	Lateral malleolus TO tubercle on outer calcaneus	Resists backward displacement of foot; resists inversion
Deltoid	Medial malleolus TO talus, navicular, calcaneus	Resists valgus forces to ankle; limits plantarflexion, dorsiflexion, eversion, abduction of foot
Dorsal (tarsometatarsal)	Tarsals TO metatarsals	Supports arch; maintains relationship between tarsals and metatarsal
Dorsal calcaneocuboid	Calcaneus TO cuboid on dorsal side	Limits inversion
Dorsal talonavicular	Neck of talus TO superior surface of navicular	Supports talonavicular joint; limits inversion
Interosseous (intertarsal)	Connects adjacent tarsals	Supports arch of foot, intertarsal joints
Interosseous (talocalcaneal)	Undersurface of talus TO upper surface of calcaneus	Limits pronation, supination, abduction, adduction, dorsiflexion, plantarflexion
Plantar calcaneocuboid	Undersurface of calcaneus TO undersurface of cuboid	Supports arch
Plantar calcaneonavicular	Anterior margin of calcaneus TO undersurface on navicular	Supports arch; limits abduction
Posterior talofibular	Inner, back lateral malleolus TO posterior surface of talus	Limits plantarflexion, dorsiflexion, inversion; supports lateral ankle
Posterior talotibial	Tibial TO talus behind articulating facet	Limits planatarflexion; supports medial ankle
Talocalcaneal	Connecting anterior/posterior, medial, lateral talus TO calcaneus	Supports subtalar joint

lateral ligaments has a specific role in stabilizing the ankle depending on the position of the foot (63).

The stability of the ankle depends on the orientation of the ligaments, the type of loading, and the position of the ankle at the time of stress. The lateral side of the ankle

joint is more susceptible to injury, accounting for 85% of ankle sprains (155).

The axis of rotation for the ankle joint is a line between the two malleoli, running oblique to the tibia and not in line with the body (33). Dorsiflexion occurs at the ankle

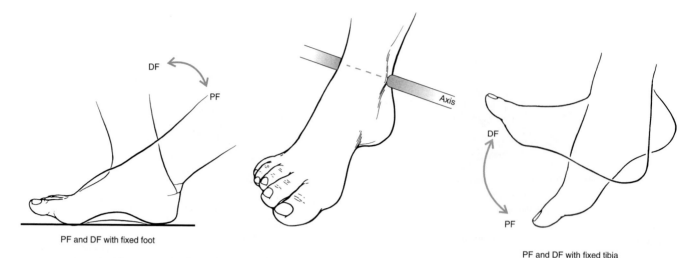

FIGURE 6-32 Plantarflexion (PF) and dorsiflexion (DF) occur about a mediolateral axis running through the ankle joint. The range of motion for plantarflexion and dorsiflexion is approximately 50° and 20°, respectively. Plantarflexion and dorsiflexion can be produced with the foot moving on a fixed tibia or with the tibia moving on a fixed foot.

joint as the foot moves toward the leg (e.g., when lifting the toes and forefoot off the floor) or as the leg moves toward the foot (e.g., in lowering down with the foot flat on the floor). These actions are illustrated in Figure 6-32.

SUBTALAR JOINT

Moving distally from the talocrural joint is the subtalar, or talocalcaneal, joint, which consists of the articulation between the talus and the calcaneus. All of the joints in the foot, including the subtalar joint, are shown in Figure 6-30. The talus and the calcaneus are the largest of the weight-bearing bones in the foot and form the hindfoot. The talus links the tibia and fibula to the foot and is called the keystone of the foot. No muscles attach to the talus. The calcaneus provides a moment arm for the Achilles tendon and must accommodate large impact loads at heel strike and high tensile forces from the gastrocnemius and soleus muscles.

The talus articulates with the calcaneus at three sites, anteriorly, posteriorly, and medially, where the convex surface of the talus fits into a concave surface on the calcaneus. The subtalar joint is supported by five short and powerful ligaments that resist severe stresses in lower extremity movements. The location and action of these ligaments are presented in Figure 6-31. The ligaments that support the talus limit the motions of the subtalar joint.

The axis of rotation for the subtalar joint runs obliquely from the posterior lateral plantar surface to the anterior dorsal medial surface of the talus (Fig. 6-33). It is tilted vertically 41° to 45° from the horizontal axis in the sagittal plane and is slanted 16° to 23° medially from the longitudinal axis of the tibia in the frontal plane (151). Because the axis of the subtalar joint is oblique through the sagittal, frontal, and transverse planes of the foot, triplanar motion can occur.

The triplanar movements at the subtalar joint are termed *pronation* and *supination*. Pronation, occurring in an open-chain system with the foot off the ground, consists of calcaneal eversion, abduction, and dorsiflexion (145). **Eversion** is the movement in the frontal plane in which the lateral border of the foot moves toward the leg in non-weight bearing or the leg moves toward the foot in weight

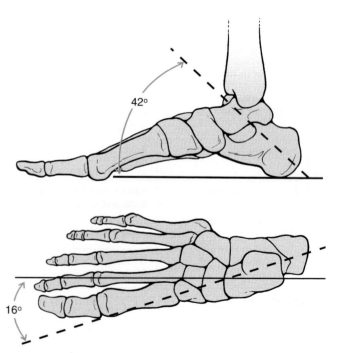

FIGURE 6-33 The axis of rotation for the subtalar joint runs diagonally from the posterolateral plantar surface to the anteromedial dorsal surface. The axis is approximately 42° in the sagittal plane (*top*) and 16° in the transverse plane. The *solid line* bisects the posterior surface of the calcaneus and the distal anteromedial corner of the calcaneus; the *dashed line* bisects the talus.

Right foot

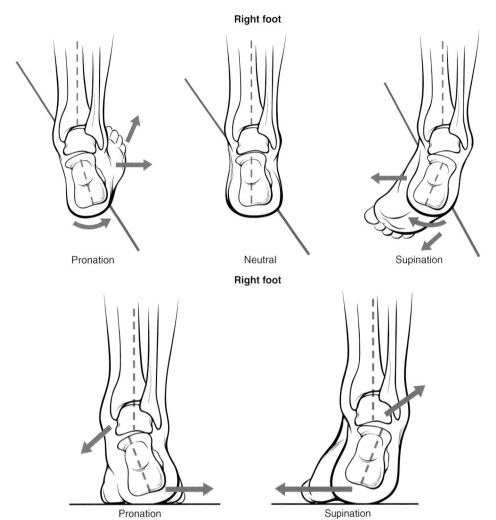

FIGURE 6-34 Top. With the foot off the ground, the foot moves on a fixed tibia, and the subtalar movement of pronation is produced by eversion, abduction, and dorsiflexion. Supination in the open chain is produced by inversion, adduction, and plantarflexion. Bottom. In a closed kinetic chain with the foot on the ground, much of the pronation and supination are produced by the weight of the body acting on the talus. In this weight-bearing position, the tibia moves on the talus to produce pronation and supination.

bearing (Fig. 6-34). The transverse plane movement is abduction, or pointing the toes out. It occurs with external rotation of the foot on the leg and lateral movement of the calcaneus in the non-weight-bearing position or internal rotation of the leg with respect to the calcaneus and medial movement of the talus in weight bearing. The sagittal plane movement of dorsiflexion occurs as the calcaneus moves up on the talus in non-weight bearing or as the talus moves down on the calcaneus in weight bearing. An illustration of differences in subtalar movements between open- and closed-chain positioning is shown in Figure 6-34.

Supination is just the opposite of pronation, with calcaneal inversion, adduction, and plantarflexion in the non-weight-bearing position and calcaneal inversion and talar abduction and dorsiflexion in the weight-bearing position (101). The frontal plane movement of inversion occurs as the medial border of the foot moves toward the medial leg in non-weight bearing or as the medial aspect of the leg moves toward the medial foot in weight bearing as the calcaneus lies on the lateral surface. In the transverse plane, adduction, or toeing-in, occurs as the foot internally rotates on the leg in non-weight bearing, and the calcaneus moves medially or the leg externally rotates on

the foot in weight bearing and the talus moves laterally. The plantarflexion movements in the sagittal plane occur as the calcaneus moves distally while non-weight bearing or as the talus moves proximally while weight bearing.

The prime function of the subtalar joint is to absorb the rotation of the lower extremity during the support phase of gait. With the foot fixed on the surface and the femur and tibia rotating internally at the beginning of stance and externally at the end of stance, the subtalar joint absorbs the rotation through the opposite actions of pronation and supination (71). Pronation is a combination of dorsiflexion, abduction, and eversion, and supination is a combination of plantarflexion, adduction, and inversion. The subtalar joint absorbs rotation by acting as a mitered hinge, allowing the tibia to rotate on a weight-bearing foot (159). Inversion and eversion are also used as corrective motions in postural adjustments to keep the foot stable under the center of gravity (159).

A second function of the subtalar joint is shock absorption. This may also be accomplished by pronation. The subtalar movements also allow the tibia to rotate internally faster than the femur, facilitating unlocking at the knee joint.

MIDTARSAL JOINT

Of the remaining articulations in the foot, the midtarsal, or transverse tarsal, joint has the greatest functional significance (Fig. 6-30). It actually consists of two joints, the **calcaneocuboid joint** on the lateral side and the **talonavicular joint** on the medial side of the foot. In combination, they form an S-shaped joint with two axes, oblique and longitudinal (151). Five ligaments support this region of the foot (see Fig. 6-32). Motion at these two joints contributes to the inversion and eversion, abduction and adduction, and dorsiflexion and plantarflexion at the subtalar and ankle joints.

Movement at the midtarsal joint depends on the subtalar joint position. When the subtalar joint is in pronation, the two axes of the midtarsal joint are parallel, which unlocks the joint, creating hypermobility in the foot (118). This allows the foot to be very mobile in absorbing the shock of contact with the ground and also in adapting to uneven surfaces. When the axes are parallel, the forefoot is also allowed to flex freely and extend with respect to the **rear foot**. The motion at the midtarsal joint is unrestricted from heel strike to foot flat as the foot bends toward the surface.

During supination of the subtalar joint, the two axes run through the midtarsal joint converge. This locks in the joint, creating rigidity in the foot necessary for efficient force application during the later stages of stance (118). The midtarsal joint becomes rigid and more stable from foot flat to toe-off in gait as the foot supinates. It is usually stabilized, creating a rigid lever, at 70% of the stance phase (101). At this time, there is also a greater load on the midtarsal joint, making the articulation between the talus and the navicular more stable. Figure 6-34 depicts these actions.

OTHER ARTICULATIONS OF THE FOOT

The other articulations in the **midfoot**, the intertarsal articulations, between the cuneiforms and the navicular and cuboid and intercuneiform, are gliding joints (Fig. 6-30). At the articulation between the cuneiforms and the navicular and cuboid, small amounts of gliding and rotation are allowed (74). At the intercuneiform articulations, a small vertical movement takes place, thus altering the shape of the **transverse arch** in the foot (37). These joints are supported by strong **interosseous ligaments**.

The forefoot consists of the metatarsals and the phalanges and the joints between them. The function of the forefoot is to maintain the transverse metatarsal arch, maintain the medial **longitudinal arch**, and maintain the flexibility in the first metatarsal. The plane of the forefoot at the metatarsal head is formed by the second, third, and fourth metatarsals. This plane is perpendicular to the vertical axis of the heel in normal forefoot alignment. This is the neutral position for the forefoot (Fig. 6-35). If the plane is tilted so that the medial side lifts, it is termed

forefoot supination or varus (71). If the medial side drops below the neutral plane, it is termed forefoot pronation or valgus. **Forefoot valgus** is not as common as **forefoot varus** (Fig. 6-36). Also, if the first metatarsal is below the plane of the adjacent metatarsal heads, it is considered to be a **plantarflexed first ray** and is commonly associated with high-arched feet (71).

The base of the metatarsals is wedge shaped, forming a mediolateral or transverse arch across the foot. The tarsometatarsal articulations are gliding or planar joints, allowing limited motion between the cuneiforms and the first, second, and third metatarsals and the cuboid and the fourth and fifth metatarsals (74). The **tarsometatarsal joint** movements change the shape of the arch. When the first metatarsal flexes and abducts as the fifth metatarsal flexes and adducts, the arch deepens, or increases in curvature. Likewise, if the first metatarsal extends and adducts and the fifth metatarsal extends and abducts, the arch flattens.

Flexion and extension at the tarsometatarsal articulations also contribute to inversion and eversion of the foot. Greater movement is allowed between the first metatarsal and the first cuneiform than between the second metatarsal and the cuneiforms (101). Mobility is an important factor in the first metatarsal because it is significantly involved in weight bearing and propulsion. The limited mobility at the second metatarsal is also significant because it is the peak of the plantar arch and a continuation of the long axis of the foot. The tarsometatarsal joints are supported by the medial and lateral dorsal ligaments.

The **metatarsophalangeal joints** are biaxial, allowing both flexion and extension and abduction and adduction (Fig. 6-30). These joints are loaded during the propulsive phase of gait after heel-off and the initiation of plantarflexion and phalangeal flexion (60). Two sesamoid bones lie under the first metatarsal and reduce the load on one of the hallucis muscles in the propulsive phase. The movements at the metatarsophalangeal joints are similar to those seen in the same joints in the hand except that greater extension occurs in the foot as a result of requirements for the propulsive phase of gait.

The **interphalangeal joints** are similar to those found in the hand (Fig. 6-30). These uniaxial hinge joints allow for flexion and extension of the toes. The toes are much smaller than the fingers. They are also less developed, probably because of continual wearing of shoes (74). The toes are less functional than the fingers because they lack an opposable structure like the thumb.

ARCHES OF THE FOOT

The tarsals and metatarsals of the foot form three arches, two running longitudinally and one running transversely across the foot. This creates an elastic shock-absorbing system. In standing, half of the weight is borne by the heel and half by the metatarsals. One-third of the weight

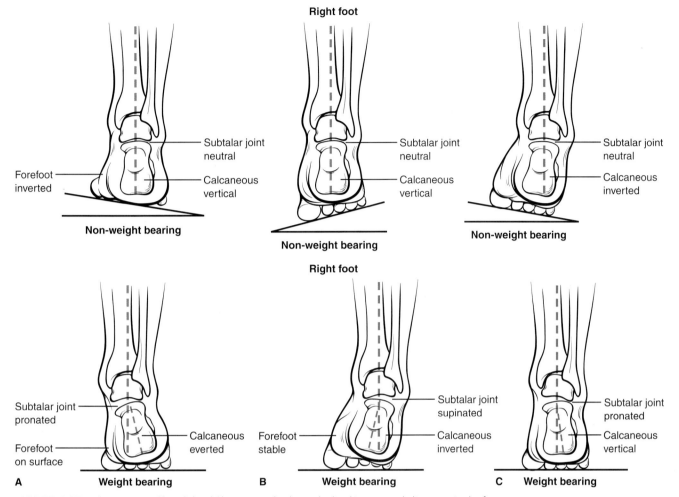

FIGURE 6-35 The metatarsal head should be perpendicular to the heel in a normal alignment in the foot. There are many variations in this alignment, including forefoot valgus **(B)**, in which the medial side of the forefoot drops below the neutral plane; forefoot varus **(A)**, in which the medial side lifts; and rear foot varus **(C)**, in which the calcaneus is inverted. In weight bearing, these alignments occur with different movements.

borne by the metatarsals is on the first metatarsal, and the remaining load is on the other metatarsal heads (60). The arches form a concave surface that is a quarter of a sphere (74). The arches are shown in Figure 6-36.

The lateral longitudinal arch is formed by the calcaneus, cuboid, and fourth and fifth metatarsals. It is relatively flat and limited in mobility (60). Because it is lower than the medial arch, it may make contact with the ground and bear some of the weight in locomotion, thus playing a support role in the foot.

The more dynamic medial longitudinal arch runs across the calcaneus to the talus, navicular, cuneiforms, and first three metatarsals. It is much more flexible and mobile than the lateral arch and plays a significant role in shock absorption upon contact with the ground. At heel strike, part of the initial force is attenuated by compression of a fat pad positioned on the inferior surface of the calcaneus. This is followed by a rapid elongation of the medial arch that continues to maximum elongation at toe contact with the ground. The medial arch shortens at midsupport and then slightly elongates and

again rapidly shortens at toe-off (60). Flexion at the transverse tarsal and tarsometatarsal joints increases the height of the longitudinal arch as the metatarsophalangeal joints extend at push-off (146). The movement of the medial arch is important because it dampens impact by transmitting the vertical load through deflection of the arch.

Even though the medial arch is very adjustable, it usually does not make contact with the ground unless a person has functional flat feet. The medial arch is supported by the keystone navicular bone, the **calcaneonavicular ligament**, the long plantar ligament, and the **plantar fascia** (37,61).

The plantar fascia, illustrated in Figure 6-37, is a strong, fibrous plantar aponeurosis running from the calcaneus to the metatarsophalangeal articulation. It supports both arches and protects the underlying neurovascular bundles. The plantar fascia can be irritated as a result of ankle motion through extreme ranges of motion because the arch is flattened in dorsiflexion and increased in plantarflexion. These actions place a wide range of tensions on

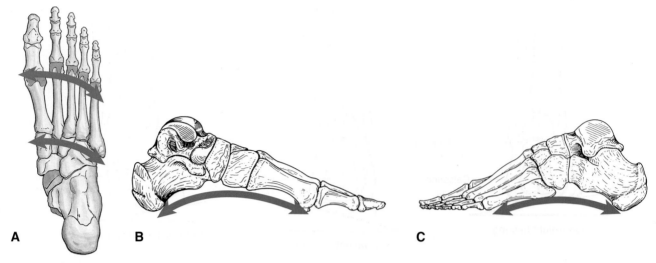

FIGURE 6-36   Three arches are formed by the tarsals and metatarsals: the transverse arches **(A)**, which support a significant portion of the body weight during weight bearing; the medial longitudinal arch **(B)**, which dynamically contributes to shock absorption; and the lateral longitudinal arch **(C)**, which participates in a support role function during weight bearing.

the fascial attachments (37). Also, if the plantar fascia is short, the arch is likely to be higher.

The digital slips of the plantar fascia extend beyond the metatarsal-phalangeal joints (Fig. 6-37). In a process termed the **windlass effect**, hyperextension of these joints tightens the plantar fascia and helps to stiffen the medial longitudinal arch. Sesamoid bones embedded within the fascia increase the mechanical advantage and the tension. This mechanism is an ingenious way of allowing the foot to be a mobile adapter when it is striking the ground and a stiff platform that will efficiently transmit forces during push-off.

The transverse arch is formed by the wedging of the tarsals and the base of the metatarsals. The bones act as beams for support of this arch, which flattens with weight bearing and can support three to four times BW (151). The flattening of this arch causes the forefoot to spread considerably in a shoe, indicating the importance of sufficient room in shoes to allow for this spread.

Volleyball and the Windlass Mechanism

Would you rather play volleyball on a sand court or on a solid surface court? It might depend on what you value most from playing volleyball. When you land after a high flying block you would probably value landing on a soft surface like sand. On the other hand, if you are jumping up to spike the ball you would get more height from a solid surface. There is a trade-off between protecting your skeleton (a soft surface may be best) and performance (a hard surface may be best). Discuss how the windlass mechanism in the foot gives you the best of both surfaces. How is it like jumping off of a hard surface and landing on a soft surface? How is the windlass mechanism similar to the dual axes of the midtarsal joint?

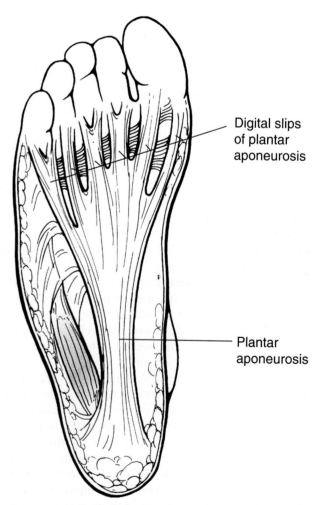

Digital slips of plantar aponeurosis

Plantar aponeurosis

FIGURE 6-37   The plantar fascia is a strong fibrous aponeurosis that runs from the calcaneus to the base of the phalanges. It supports the arches and protects structures in the foot.

Individuals can be classified according to the height of the medial arch into foot types that are normal, high-arched or **pes cavus**, and flat-footed or **pes planus**. They can be further classified as being rigid or flexible. The midfoot of the high-arched rigid foot does not make any contact with the ground and usually has little or no inversion or eversion in stance. It is a foot type that has poor shock absorption. The flat foot, on the other hand, is usually hypermobile, with most of the plantar surface making contact in stance. This weakens the medial side. It is a foot type usually associated with excessive pronation throughout the support phase of gait.

MOVEMENT CHARACTERISTICS

The range of motion at the ankle joint varies with the application of loads to the joint. The range of motion in dorsiflexion is limited by the bony contact between the neck of the talus and the tibia, the capsule and the ligaments, and the plantar flexor muscles. The average range of dorsiflexion is 20°, although approximately 10° of dorsiflexion is required for efficient gait (24). More dorsiflexion can be attained up through 40° plus when performing a full squat movement using BW. Healthy elderly individuals typically exhibit less passive dorsiflexion range of motion but more dorsiflexion in gait than their younger counterparts.

Any arthritic condition in the ankle also reduces passive dorsiflexion range of motion and increases active dorsiflexion range of motion. The increase in dorsiflexion in the arthritic joint is primarily because of a decrease in flexibility in the gastrocnemius or a weakness in the soleus. With the maintenance of the knee flexion angle during the support period of gait, a collapse into greater dorsiflexion is observed (88). With increased dorsiflexion and knee flexion, more weight is maintained on the heel.

Plantarflexion is movement of the foot away from the leg (e.g., rising up on the toes) or moving the leg away from the foot (such as in leaning back and away from the front of the foot) (Fig. 6-32). Plantarflexion is limited by the talus and the tibia, the ligaments and the capsule, and the dorsiflexor muscles. The average range of motion for plantarflexion is 50°, with 20° to 25° of plantarflexion used in gait (24,29,108).

In an arthritic or pathologic gait, plantarflexion range of motion is less for both passive and active measurements. The reduction of plantarflexion in gait is substantial because of weak calf muscles. Healthy elderly people do not demonstrate substantial loss in either passive or active plantarflexion range of motion (88).

In the rear foot, subtalar eversion and inversion can be measured by the angle formed between the leg and the calcaneus. In the closed-chain weight-bearing movement, the talus moves on the calcaneus, and in the open chain, the calcaneus moves on the talus. Calcaneal inversion and eversion are the same regardless of weight-bearing or open-chain motion. This makes calcaneal inversion and eversion measurements very useful in quantifying subtalar motion (Fig. 6-4). Subtalar inversion is possible through 20° to 32° of motion in young healthy individuals and 18° in healthy elderly individuals (88,104). Inversion is greatly reduced in individuals with osteoarthritis in the ankle joint. Eversion, measured passively, averages 5° and 4° for healthy young and elderly individuals, respectively (88). In 84% of arthritic patients, excessive calcaneal eversion creates what is known as a hindfoot valgus deformity.

COMBINED MOVEMENTS OF THE KNEE AND ANKLE/SUBTALAR

Movements at the knee and foot need to be coordinated to maximize absorption of forces and minimize strain in the lower extremity linkage. For example, during the support phase of gait, pronation and supination in the foot should correspond with rotation at the knee and hip. At heel strike, the foot typically makes contact with the ground in a slightly supinated position, and the foot is lowered to the ground in plantarflexion (38). The subtalar joint begins to immediately pronate, accompanying internal rotation and flexion at the knee and hip joints (61). The talus rotates medially on the calcaneus, initiating pronation as a result of lateral heel strike and putting stress on the medial side (139). Pronation continues until it reaches a maximum at approximately 35% to 50% of the stance phase (9,154), and this corresponds to the achievement of maximum flexion and internal rotation at the knee.

At the stage of foot flat in stance, the knee joint begins to externally rotate and extend, and because the forefoot is still fixed on the ground, these movements are transmitted to the talus (61). The subtalar joint should begin to supinate in response to the external rotation and extension that occurs up through heel-off. Many injuries of the lower extremity are thought to be associated with a lack of synchrony between these movements at the knee and subtalar joint.

Excessive pronation has been speculated to be a major cause of injury, but it is not necessarily the maximum degree of pronation but rather the percentage of support in which pronation is present and the synchronization with the knee joint movements. Pronation can be present for as much as 55% to 85% of stance, creating problems when the lower limb moves into external rotation and extension as the subtalar joint is still pronating (103). The lack of synchrony between the subtalar and knee joint motions has been shown to increase with increasing velocities (154) and increases in stride length (153).

ALIGNMENT AND FOOT FUNCTION

Foot function can be altered significantly with any variation in alignment in the lower extremity or as a result of abnormal motion in the lower extremity linkage. Typically, any varum alignment in the lower extremity increases the pronation at the subtalar joint in stance (66). A Q-angle at the knee greater than 20°, tibial varum greater than 5°, rear foot varum (calcaneus inversion) greater than 2°, and forefoot varum (forefoot adduction) greater than 3° are all deemed to be significant enough to produce an increase in subtalar pronation (88).

Rear foot varus is usually a combination of subtalar varus and tibial varum in which the calcaneus inverts and the lower third of the tibia deviates in the direction of inversion. Forefoot varus, the most common cause of excessive pronation, is the inversion of the forefoot on the rear foot with the subtalar joint in the neutral position (24). It is caused by the inability of the talus to derotate, leaving the foot pronated at heel lift and preventing any supination. This shifts the BW to the medial side of the foot, creating a hypermobile midtarsal joint and an unstable first metatarsal.

Both rear foot and forefoot varus double the amount of pronation in midstance compared with normal foot function and continue pronation into late stance (66). In some cases, the pronation continues until the very end of the support period. This is a major injury-producing mechanism because the continued pronation is contrary to the external rotation being produced in the leg. It is the primary cause of discomfort and dysfunction in the foot and leg. The transverse rotation being produced by the hypermobile foot, still in pronation late in stance, is absorbed at the knee joint and can create lateral hip pain through an anterior tilt of the pelvis or strain the invertor muscles (38).

A plantarflexed first ray can also produce excessive pronation (66). The first ray is usually plantarflexed by the pull of the peroneus longus muscle and is commonly seen in both rear foot and forefoot varus alignments. This alignment causes the medial side of the foot to load prematurely, with greater than normal loads limiting forefoot inversion and creating supination in midstance. However, sudden pronation is generated at heel-off, developing high shear forces across the forefoot, especially at the first and fifth metatarsals (66).

Hypermobility of the first ray is generated because the peroneus longus muscle cannot stabilize the first metatarsal. During pronation, the medial side is hypermobile, placing a large load and shear force on the second metatarsal. This is a common cause of stress fracture of the second metatarsal and subluxation of the first metatarsophalangeal joint (1,24).

Although it is not common, a person may have a forefoot valgus alignment. This may be caused by a bony deformity in which the plantar surfaces of the metatarsals evert relative to the calcaneus with the subtalar joint in the neutral position (24). Forefoot valgus causes the forefoot to be prematurely loaded in gait, creating supination at the subtalar joint. This alignment is typically seen in high-arched feet.

Foot type, as mentioned previously, can also affect the amount of pronation or supination. In the normal foot

with a subtalar axis of 42° to 45°, the internal rotation of the leg is equal to the internal rotation of the foot (69). In a high-arched foot, the axis of the subtalar joint is more vertical and is greater than 45°, so that for any given internal rotation of the leg, there is less internal rotation of the foot, creating less pronation for any given leg rotation.

In the flat foot, the subtalar joint axis is less than 45°, that is, closer to the horizontal. This has the opposite effect to an axis that is greater than 45°. Thus, for any given internal rotation of the leg, there is greater internal rotation of the foot, creating greater pronation (69).

A final alignment consideration is the equine foot, in which the Achilles tendon is short, creating a significant limitation of dorsiflexion in gait. The **equinus** deviation can be reproduced with a tight and inflexible gastrocnemius and soleus. Because the tibia is unable to move forward on the talus in midsupport, the talus moves anteriorly and pronates excessively to compensate (38). An early heel rise and toe walking are symptoms of this disorder.

MUSCLE ACTIONS

Twenty-three muscles act on the ankle and the foot, 12 originating outside the foot and 11 inside the foot. All of the 12 extrinsic muscles, except for the gastrocnemius, soleus, and plantaris, act across both the subtalar and midtarsal joints (49). The insertion, actions, and nerve supply of all of these muscles are presented in Figure 6-38.

The muscles of the foot play an important role in sustaining impacts of very high magnitude. They also generate and absorb energy during movement. The ligaments and tendons of the muscles store some of the energy for later return. For example, the Achilles tendon can store 37 joules (J) of elastic energy, and the ligaments of the arch can store 17 J as the foot absorbs the forces and BW (141).

Plantarflexion is used to propel the body forward and upward, contributing significantly to the other propelling forces generated in heel-off and toe-off. Plantar flexor muscles are also used eccentrically to slow down a rapidly dorsiflexing foot or to assist in the control of the forward movement of the body, specifically the forward rotation of the tibia over the foot.

Plantarflexion is a powerful action created by muscles that insert posterior to the transverse axis running through the ankle joint. The majority of the plantarflexion force is produced by the gastrocnemius and the soleus, which together are referred to as the triceps surae muscle group. Because the gastrocnemius also crosses the knee joint and can act as a knee flexor, it is more effective as a plantar flexor with the knee extended and the quadriceps femoris activated.

In a sprint racing start, the gastrocnemius is maximally activated with the knee extended and the foot placed in full dorsiflexion. The soleus, called the workhorse of plantarflexion, is flatter than the gastrocnemius (37). It is also the predominant plantarflexor during a standing posture.

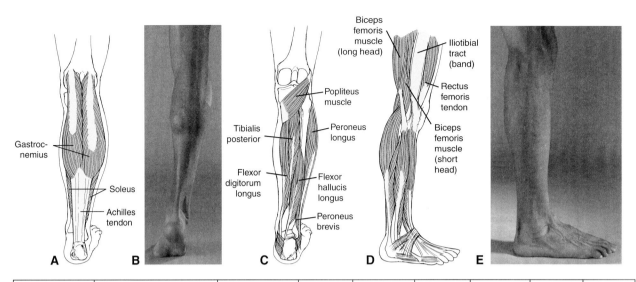

Muscle	Insertion	Nerve Supply	Flexion/ Dorsiflexion	Extension/ Plantarflexion	Abduction	Adduction	Inversion	Eversion
Abductor digiti minimi	Lateral calcaneus TO base of proximal phalanx of fifth toe	Lateral plantar nerve			PM: Little toe			
Abductor hallucis	Medial calcaneus TO medial base of proximal phalanx of first toe	Medial plantar nerve			PM: Big toe			
Adductor hallucis	Second, third, fourth metatarsal TO lateral side of proximal phalanx of big toe	Lateral plantar nerve				PM: Big toe		
Dorsal interossei	Sides of metatarsals TO lateral side of proximal phalanx	Lateral plantar nerve	PM: proximal phalanx		PM: Toes 2–4	PM: Second toe		
Extensor digitorum brevis	Lateral calcaneus TO proximal phalanx of first, second, third toes	Deep peroneal nerve		PM: Toes 1–4				
Extensor digitorum longus	Lateal condyle of tibia; fibula; interosseus membrane TO dorsal expansion of toes 2–5	Deep peroneal nerve	Asst: Ankle DF	PM Toe 2–5				PM
Extensor hallucis longus	Anterior fibula; interosseous membrane TO distal phalanx of big toe	Deep peroneal nerve	Asst: Ankle DF	PM: Big toe		PM: Forefoot		
Flexor digiti minimi brevis	Fifth metatarsal TO proximal phalanx of little toe	Lateral plantar nerve	PM: Little toe					
Flexor digitorum brevis	Medial calcaneus TO middle phalanx of toes 2–5	Medial plantar nerve	PM: Toe 2–5					
Flexor digitorum longus	Posterior tibia TO distal phalanx of toes 2–5	Tibial nerve	PM: Toe 2–5	Asst: Ankle PF			Asst	
Flexor hallucis brevis	Cuboid TO medial side of proximal phalanx of big toe	Medial plantar nerve	PM: Big toe					
Flexor hallucis longus	Lower 2–3 of posterior fibula, interosseous membrane	Tibial nerve	PM: Big toe	Asst: Ankle PF		PM: Forefoot	Asst	

FIGURE 6-38   Muscles acting on the ankle joint and foot, including superficial posterior muscles **(A)** and surface anatomy **(B)** of posterior lower leg; deep posterior muscles of the lower leg **(C)**, muscles **(D)** and surface anatomy of the lower leg **(E)**; anterior muscle **(F)** and surface anatomy **(G)**; surface anatomy of the foot and ankle **(H)**; and muscles in the dorsal **(I, J, K)** and ventral **(L)** surface of the foot. (*continued*)

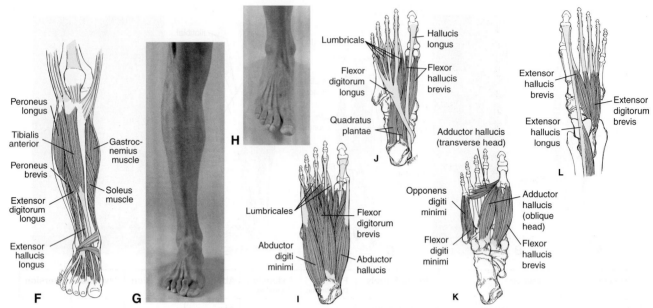

Muscle	Insertion	Nerve Supply	Flexion/ Dorsiflexion	Extension/ Plantarflexion	Abduction	Adduction	Inversion	Eversion
Gastrocnemius	Medial, lateral condyles of femur TO calcaneus	Tibial nerve; S1, S2		PM: Ankle PF				
Lumbricales	Tendon of flexor digitorum longus TO base of proximal phalanx of toes 2–5	Medial, lateral planter nerve	PM: proximal phalanx 2–5					
Peroneus brevis	Lower lateral fibula TO fifth metatarsal	Superficial peroneal nerve		Asst: Ankle PF				PM
Peroneus longus	Lateral condyle of tibia, upper lateral fibula TO first cuneiform; lateral first metatarsal	Superficial peroneal nerve		Asst: Ankle PF	PM: Forefoot			PM
Peroneus tertius	Lower anterior fibula; interosseous membrane TO base of fifth metatarsal	Deep peroneal nerve	PM					PM
Plantar interossei	Medial side of 3–5 meta-tarsal TO medial side of proximal phalanx of toes 3–5	Lateral plantar nerve				PM: Toes 3–5		
Plantaris	Linea aspera on femur TO calcaneus	Tibial nerve		Asst: Ankle PF				
Quadratus plantae	Medial lateral inferior calcaneus TO flexor digitorum tendon	Lateral plantar nerve	PM: Toes 2–5					
Soleus	Upper posterior tibia, fibula, interosseous membrane TO calcaneus	Tibial nerve		PM: Ankle PF				
Tibialis anterior	Upper lateral tibia, intero-sseous membrane TO medial plantar surface of first cuneiform	Deep peroneal nerve	PM: Ankle DF				PM	
Tibialis posterior	Upper posterior tibia, fibula, interosseous mem-brane TO inferior navicular	Tibial nerve		Asst: Ankle PF			PM	

FIGURE 6-38   (continued)

A tight soleus can create a functional short leg, often seen in the left leg of people who drive a car a great deal. As explained in an earlier section, an inflexible or tight soleus limits dorsiflexion and facilitates compensatory pronation that creates the functionally shorter limb.

The action of these plantarflexor muscles is mediated through a stiff subtalar joint, allowing for an efficient transfer of the muscular force. The gastrocnemius and possibly the soleus have also been shown to produce supination when the forefoot is on the floor during the later stages of the stance phase of gait. Plantarflexion is usually accompanied by both supination and adduction.

The other plantar flexor muscles produce only 7% of the remaining plantarflexor force (37). Of these, the peroneus longus and the peroneus brevis are the most significant, with minimal plantarflexor contribution from the plantaris, flexor hallucis longus, flexor digitorum longus, and tibialis posterior. The plantaris is an interesting muscle, similar to the palmaris longus in the hand, in that it is absent in some individuals, very small in others, and well developed in yet others. Overall, its contribution is usually insignificant.

Dorsiflexion at the ankle is actively used in the swing phase of gait to help the foot clear the ground and in the stance phase of gait to control lowering of the foot to the floor after heel strike. Dorsiflexion is also present in the middle part of the stance phase as the body lowers and the tibia travels over the foot, but this action is controlled eccentrically by the plantarflexor muscles (45). The dorsiflexor muscles are those that insert anterior to the transverse axis running through the ankle (49) (see Fig. 6-38).

The most medial dorsiflexor is the tibialis anterior, whose tendon is farthest from the joint, thus giving it a significant mechanical advantage and making it the most powerful dorsiflexor (37). The tibialis anterior has a long tendon that begins halfway down the leg. It is also the largest muscle and provides additional support to the medial longitudinal arch. Assisting the tibialis anterior in dorsiflexion is the extensor digitorum longus and the extensor hallucis longus. These muscles pull the toes up in extension. The peroneus tertius also contributes to the dorsiflexion force.

Eversion is created primarily by the peroneal muscle group. These muscles lie lateral to the long axis of the tibia. They are known as pronators in the non-weight-bearing position because they evert the calcaneus and abduct the **forefoot**. The peroneus longus is an everter that is also responsible for controlling the pressure on the first metatarsal and some of the finer movements of the first metatarsal and big toe, or hallux.

The lack of stabilization of the first metatarsal by the peroneus longus leads to hypermobility of the medial side of the foot. The peroneus brevis also contributes through the production of eversion and forefoot abduction, and the peroneus tertius contributes through dorsiflexion and eversion. Both the peroneus tertius and peroneus brevis stabilize the lateral aspect of the foot. Pronation

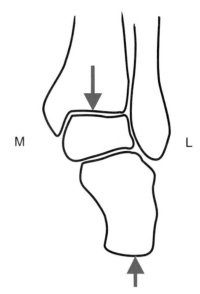

FIGURE 6-39 When the heel strikes the ground on the lateral **(L)** aspect, a vertical force is directed on the outside of the foot. The force of body weight is acting down through the ankle joint. Because these two forces do not line up, the talus is driven medially **(M)**, producing the pronation movement.

in the weight-bearing position is primarily generated by weight bearing on the lateral side of the foot in heel strike. This drives the talus medially, producing pronation. Figure 6-39 shows how pronation is produced through weight bearing.

The inverters of the foot are the muscles lying medial to the long axis of the tibia. These muscles generate inversion of the calcaneus and adduction of the forefoot (37). Inversion is created primarily by the tibialis anterior and the tibialis posterior, with assistance from the toe flexors, flexor digitorum longus, and flexor hallucis longus. The extensor hallucis longus works with the flexor hallucis longus to adduct the forefoot during inversion.

The intrinsic muscles of the foot work as a group and are very active in the support phase of stance. They basically follow supination and are more active in the later portions of stance to stabilize the foot in propulsion (69). In a foot that excessively pronates, they are also more active as they work to stabilize the midtarsal and subtalar joints. There are 11 intrinsic muscles, and 10 of these are on the plantar surface arranged in four layers. Figure 6-39 has a full listing of these muscles.

STRENGTH OF THE ANKLE AND FOOT MUSCLES

The strongest movement at the ankle or foot is plantarflexion. This is because of the larger muscle mass contributing to the movement. It is also related to the fact that the plantarflexors are used more to work against gravity and maintain an upright posture, control lowering to the ground, and add to propulsion. Even standing, the plantarflexors, specifically the soleus, contract to control dorsiflexion in the standing posture.

Plantarflexion strength is greatest from a position of slight dorsiflexion. A starting dorsiflexion angle of 105° increases plantarflexion strength by 16% from the neutral 90° position. Plantarflexion strength measured from 75° and 60° of plantarflexion is reduced by 27% and 42%, respectively, compared with strength measured in the neutral position (151). Additionally, plantarflexion strength can be increased if the knee is maintained in an extended position, placing the gastrocnemius at a more advantageous muscle length.

Dorsiflexion is incapable of generating a large force because of its reduced muscle mass and because it is minimally used in daily activities. The strength of the dorsiflexor muscles is only about 25% that of the plantarflexor muscles (151). Dorsiflexion strength can be enhanced by placing the foot in a few degrees of plantarflexion before initiating dorsiflexion.

CONDITIONING OF THE FOOT AND ANKLE MUSCLES

Both stretching and strengthening exercises for selected movements at the foot and ankle are presented in Figure 6-40. The plantarflexor muscles are exercised to a great extent in daily living activities: They are used to walk, get out of chairs, go up stairs, and drive a car. Strengthening the plantarflexors by using resistive exercises is also relatively easy. Any heel-raising exercise offers a significant amount of resistance because BW is lifted by this muscle group. With the weight centered over the foot, the leverage of the plantarflexors is very efficient for handling large loads; thus, a heel-raise activity with weight on the shoulders can usually be done with a considerable amount of weight. This exercise is perfect for the gastrocnemius because the strength of this muscle is enhanced with the knee extended and the quadriceps femoris contracting.

To specifically strengthen the soleus, a seated position is best. This position flexes the knee and reduces the contribution of the gastrocnemius significantly. Weight or resistance can be placed on the thigh as plantarflexion is produced.

It is important to maintain flexibility in the plantarflexors because any inflexibility in this muscle group can create an early heel raise and excessive pronation in gait. Inflexibility in the plantarflexors is common in women who wear high heels much of the time (74). In fact, both men and women are susceptible to strain in the plantarflexors when going from a higher heel to a lower heel in either exercise or activities of daily living. It is better to maintain the flexibility in the muscle group through stretching with the knee extended and the ankle in maximum dorsiflexion.

Flexibility in the gastrocnemius and the soleus can be somewhat isolated. Flexibility of the gastrocnemius can best be tested with the knee extended, and flexibility of the soleus is best tested with the knees flexed to 35°.

The strength of the dorsiflexors is limited, but it should be maintained so that fatigue does not set in during a long walk or run. Fatigue in the muscle group leads to foot drop in swing and slapping of the foot on the surface following heel strike. To strengthen the muscle group, a seated position works best so that resistance can be applied below the foot with sandbags, weights, or surgical tubing (Fig. 6-41). Also, ankle machines are available that allow a full range of dorsiflexion and high-resistance training of this movement. Flexibility of dorsiflexion can also be best achieved in the seated position through maximum plantarflexion activities.

Strength and flexibility of the inverters and everters of the ankle are important for athletes participating in activities in which ankle injuries are common. This includes basketball, volleyball, football, soccer, tennis, and a wide variety of other activities. Stretching and strengthening the inversion and eversion muscles can be done with the foot flat on the floor on a towel or attached to surgical tubing. Weight can be put on the towel, which can then be pulled toward the foot in either inversion or eversion depending on which side of the weights the foot is placed. Circumduction and figure-eight tracing are good flexibility exercises.

The intrinsic muscles of the foot are usually atrophied and weak because we regularly wear shoes. Because the intrinsic muscles support the arch of the foot and stabilize the foot during the propulsive phase of gait, it is worthwhile to give these muscles some conditioning. The best way to exercise the intrinsic muscle group as a whole is to forego shoes and go barefoot. The movement potential of the foot is best illustrated by individuals who have upper extremity disabilities and must use their feet to perform daily functions. These individuals can become very versatile and adept at using the feet to perform a wide range of functions.

During walking or running, impact is the same either with shoes or barefoot; it is the manner in which the forces are absorbed that is different between the two. With a shoe, the foot is more rigid during the shock absorption phase of support and depends on the shoe for support and protection. During shock absorption in barefoot gait, the foot is more mobile, with more arch deflection upon loading (137). This does not necessarily mean that shoes should not be worn—the injury rate in barefoot running would initially be high because of the significant change imposed by removing the shoes. There is also a danger associated with barefoot activity and the possibility of injury from sharp objects. Going barefoot in the summer, however, is one way of improving the condition of the intrinsic muscles.

Attesting to the benefits of barefoot activity is the low injury rate in populations that remain largely barefoot. The incidence of injury to barefoot runners is much lower than among the shoe population (137). Finally, the intrinsic musculature in a person with a flat mobile foot is much more developed than in a person with a high-arched, rigid foot because of the difference in movement characteristics in loading of the foot.

Muscle Group	Sample Stretching Exercise	Sample Strengthening Exercise	Other Exercises
Ankle plantarflexors	 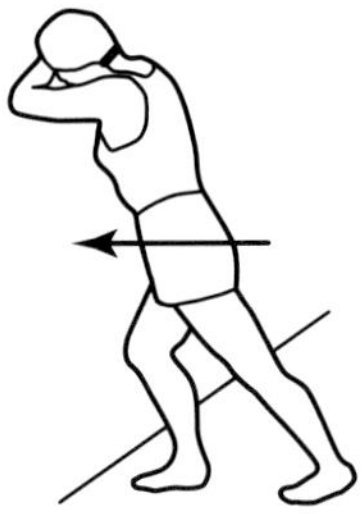	Standing calf raise  Seated calf raise 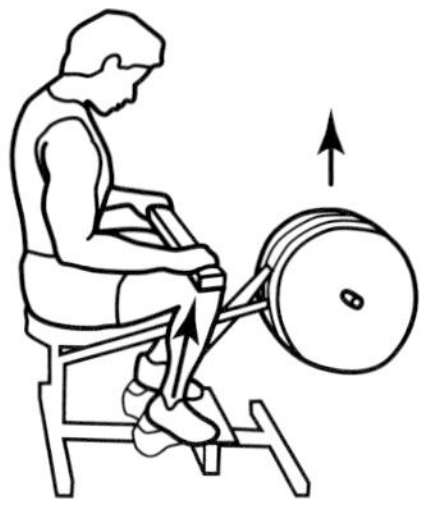	Dumbbell heel raise Barbell heel raise
Ankle dorsiflexors	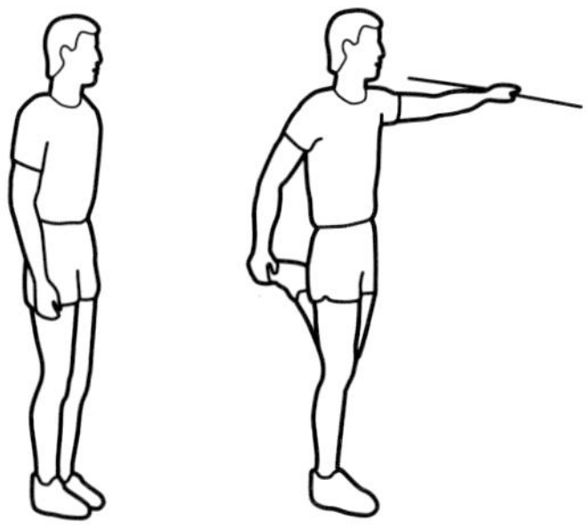	Dorsiflexion with tubing 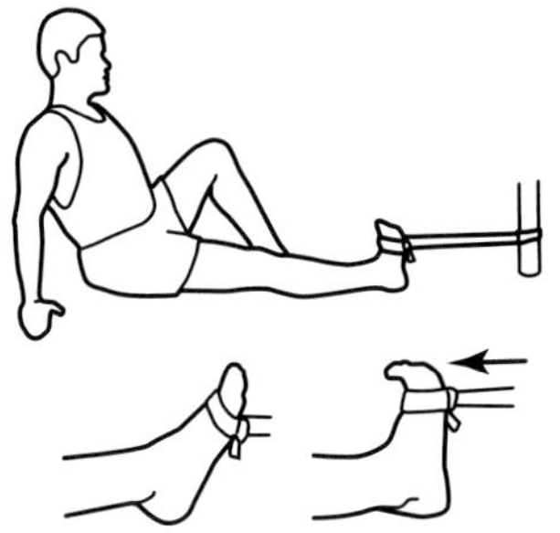	
Ankle eversion/inversion	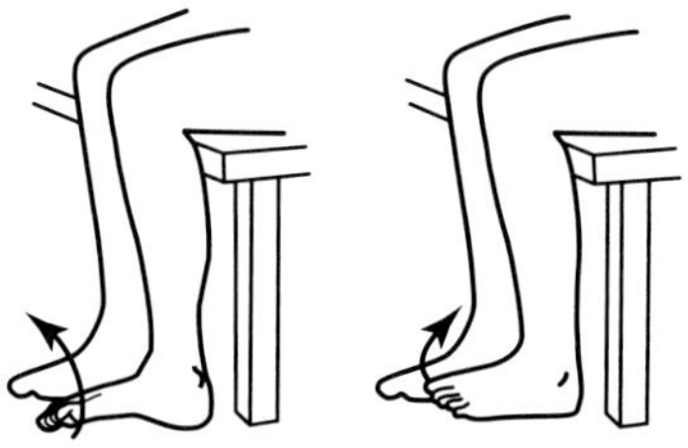 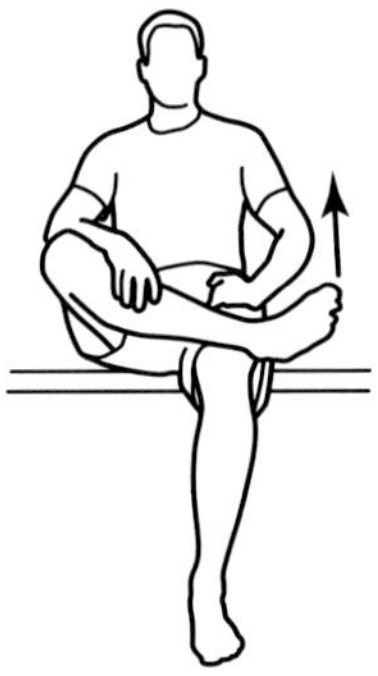	Eversion with elastic band 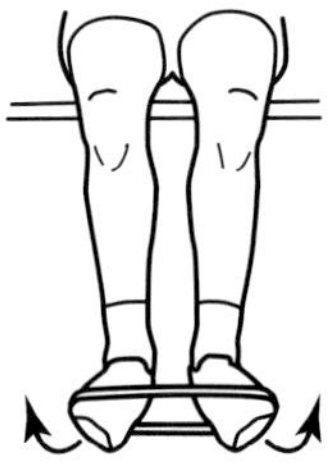 Ankle tubing inversion 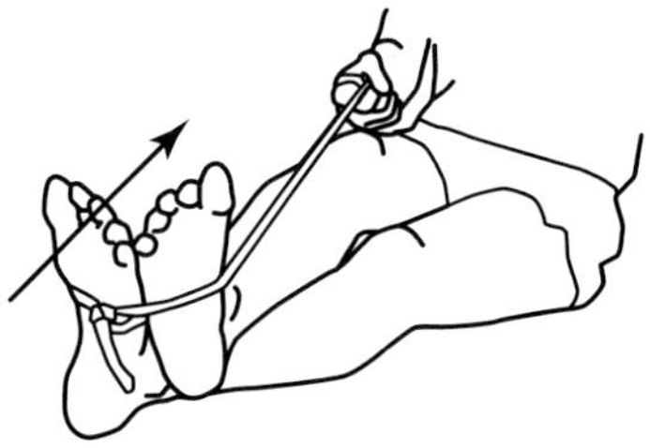	

FIGURE 6-40 Sample stretching and strengthening exercises for selected muscle groups are illustrated.

INJURY POTENTIAL OF THE ANKLE AND FOOT

Injuries to the foot and ankle account for a large portion of the injuries to the lower extremity. In some sports or activities, such as basketball, the ankle joint is the most frequently injured part of the lower extremity. Whereas injuries to the hindfoot usually occur as a result of vertical compression, injuries to the midfoot occur with excessive lateral movement or range of motion in the foot (37). Injuries to the forefoot occur similarly to injuries in long bones elsewhere in the body. In this area of the foot, both compressive and tensile forces create the injury.

Most injuries to the ankle joint and the foot occur as a result of overtraining or an excessive training bout. The ankle joint is injured frequently in activities such as running, during which the foot is loaded suddenly and repeatedly (137). Foot and ankle injuries are also associated with anatomical factors; a greater incidence of injury is seen in individuals who overpronate and in those with cavus alignment in the lower extremity. Functional ankle instability can also be related to a number of factors, including peroneal tendon weakness, rotational talar instability, subtalar instability, tibiofibular instability, or hindfoot misalignment (63).

One of the most common injuries to the foot is ankle sprain. Sprains most commonly occur in the lateral complex of the ankle during inversion. The mechanism of injury is a movement of the tibia laterally, posteriorly, anteriorly, or rotating while the foot is firmly fixed on the surface. Stepping into a hole in the ground, walking off a curb, and losing one's balance in high heels are other instances in which the ankle can be sprained. The factors associated with ankle sprain differ between men and women. Women with increased tibial varum and calcaneal eversion range of motion and men with increased talar tilt are more susceptible to ankle ligament injury (17).

Most ankle sprains in athletes are seen during the cutting maneuver, when the cut is made with the foot opposite to the direction of the run (49) or when landing on another player's foot. For example, the left foot is sprained as it drives in plantarflexion and inversion to the right. The plantarflexion and inversion action is the cause of sprain to the lateral ligamentous structure, with the anterior **talofibular ligament** most likely to be sprained (68). If the cut is made with greater foot inversion, the **calcaneofibular ligament** is the next ligament that may be damaged (68). The injury is created with a talar tilt as the talus moves forward out of the ankle mortise. Any talar tilt greater than 5° will likely cause ligament damage to the lateral ankle (41). With injury to the lateral ankle complex, an anterior subluxation of the talus and talar tilt may occur, creating great instability in the ankle and foot complex.

The medial ligaments of the ankle are not often sprained because of the support of the strong **deltoid ligament** and the bulwark created by the longer lateral malleolus. The powerful deltoid ligament can be sprained if the foot is planted and pronated and incurs a blow on the lateral side of the leg.

Although not at the ankle, the ligaments holding the tibiofibular joint together can be sprained with a forceful external rotation and dorsiflexion or a forceful inversion or eversion. The talus pushes the tibia and fibula apart, spraining the ligaments.

Many other soft-tissue injuries to the foot and ankle are typically associated with overuse or some other functional malalignment. Posterior or **medial tibial syndrome**, previously referred to as shin splints, generates pain above the medial malleolus (71). This condition usually involves the insertion site of the tibialis posterior and can be tendinitis of the tibialis posterior tendon. It may also be **periostitis**, in which the insertion of the tibialis posterior pulls on the interosseous membrane and periosteum on the bone, causing an inflammation. This muscle is usually irritated through excessive pronation, which places a great deal of tension and stretch on the muscle. **Lateral tibial syndrome** causes pain on the anterior lateral aspect of the leg and is an overuse condition similar to that of the tibialis anterior muscle.

The Achilles tendon is another frequently strained area of the foot that can be injured as a result of overtraining. A tight Achilles tendon can also lead to a number of conditions, including pain in the calf, heel, lateral or medial ankle, and plantar surface. Multiple vigorous contractions of the gastrocnemius that overstretch the muscle group, as in hill running or in moving to a low-heel shoe from a higher heel, may strain this tendon (12). The Achilles tendon can also be irritated if there is loss of absorption in the heel pad on the calcaneus. This creates a higher amplitude shock at heel strike during locomotion that is compensated for by an increase in soleus activity. The increased muscle activity produces a corresponding increase in the loading on the Achilles tendon. Achilles tendinitis can be very painful and difficult to heal because immobilization of the area is difficult. The Achilles tendon can also rupture as a result of a vigorous muscle contraction. For example, a vigorous forward push-off after a move backward can rupture a tendon. Another method of rupture is by stepping into a hole or off a curb.

A condition that mimics the pain associated with Achilles tendinitis is **retrocalcaneal bursitis**. This is an inflammation of the bursae lying superior to the Achilles attachment. It is generally caused by ill-fitting shoes (12).

Plantar fasciitis, an inflammation of the plantar fascia on the underside of the foot, is another common soft-tissue injury to the foot (137). Irritation usually develops on the medial plantar fascial attachment to the calcaneus and may be caused by training adjustments that increase hill running or mileage. It may also be caused by stepping into a hole or off a curb. Plantar fasciitis is most prevalent in high-arched foot types and in individuals with a tight Achilles tendon or leg length discrepancy (78). More tension placed on the plantar fascia in pronation predisposes this area to this type of injury. The plantar fascia can rupture with forceful plantarflexion, such as is seen in descending stairs or during rapid acceleration.

At the site of the irritation of the plantar fascia on the calcaneus, adolescents can develop a **calcaneal apophysitis**, an irritation of the epiphysis of the calcaneus (78). Adults may develop a similar irritation at the same site, where heel bone spurs develop in response to the pull of the plantar fascia.

Although osteoarthritis at the ankle joint occurs at a lower incidence than that seen at the hip or knee, it can be seen in younger patients (160). This is different than the degenerative arthritis commonly seen at the hip and knee joint. Recurring injury to the ankle ligaments or a single severe ankle sprain are predisposing factors for the development of ankle osteoarthritis.

Forefoot pain can be related to conditions such as metatarsal stress fractures, metatarsalgia, and Morton neuroma. Strain to the metatarsals, referred to as **metatarsalgia**, creates a dull burning sensation in the forefoot. Morton neuroma is an inflammation of a nerve, usually between the third and fourth metatarsals at the balls of the feet. Symptoms include sharp pain, burning, and numbness. Irritation to the ligaments or soft tissue in the forefoot is usually associated with running on a hard surface. Injuries to the metatarsal are more prevalent in overpronating feet.

Nerve compression can occur at various sites in the leg and foot. **Anterior compartment syndrome** is a case in which nerve and vascular compression occur as a result of hypertrophy in the anterior tibial muscles. The muscles hypertrophy to the point that they impinge on the nerves and blood vessels in the muscle compartment. Impingement can create tingling sensations or atrophy in the foot.

Injury to the osseous components of the foot typically occurs with overuse or pathologic function. Metatarsal fractures are typically found in the middle of the shaft of the second or third metatarsal. This fracture is associated with tight dorsiflexor muscles or forefoot varus. A stress fracture can also develop in the metatarsals on the lateral side of the foot as a result of a tight gastrocnemius. The tight gastrocnemius prevents dorsiflexion in gait, creating compensatory pronation, an unlocked subtalar joint, and more flexibility in the first metatarsal, with the lateral metatarsal absorbing the force. A person who lacks sufficient dorsiflexion in gait is almost five times more likely than usual to acquire a stress fracture.

Fractures to the metatarsals occur with a fall on the foot, avulsion by a muscle such as the site of the attachment of the peroneus brevis on the fifth metatarsal, or as a consequence of compression. Fractures have also been associated with loss of heel pad compressive ability, requiring greater force absorption by the foot. An example of a compression injury is a fracture of the tibia or talus on the medial side that accompanies a lateral ankle sprain. This jamming of the inner ankle can also loosen bony fragments, a condition known as **osteochondritis dissecans**.

An **osteochondral fracture** of the talus is a shearing type of fracture that occurs with a dorsiflexion–eversion action of the foot in which the talus impinges on the fibula during a crouch.

 # Contribution of Lower Extremity Musculature to Sport Skills or Movements

The lower extremity is involved primarily with weight bearing, walking, posture, and most gross motor activities. This section summarizes the lower extremity muscular contribution to a sample of movements. A more thorough review of muscular activity is provided for walking and cycling. These are examples of a functional anatomy description of a movement derived from electromyographic research.

Very few movements or sport skills do not require the use and contribution of the lower extremity muscles. For example, in landing from a jump or other airborne event, the weight of the body is decelerated over the lower extremity using the trunk, hip, and lower leg muscles (53). In a cut maneuver, the gluteus medius and the sartorius modify the foot position in the air through internal and external rotation of the hip, and in the stance phase of the cutting action, there is increased force from the gastrocnemius and the quadriceps muscles to generate more force for the change in direction (133).

STAIR ASCENT AND DESCENT

Going up a flight of stairs is first initiated with a limb lift via vigorous contraction of the iliopsoas, which pulls the limb up against gravity to the next stair (98) (Fig. 6-41). The rectus femoris becomes active in this phase as it assists in the thigh flexion and eccentrically slows the knee flexion. Next, the foot is placed on the next step. At this point, there is activity in the hamstrings, primarily working to slow down the extension at the knee joint (98). As the foot makes contact with the next step, weight acceptance involves some activity in the extensors of the thigh. The next phase is pull-up, in which the limb placed on the upper step is extended to bring the body up to that step. Most of the extension is generated at the knee joint by the quadriceps. The lower leg moves posteriorly via plantarflexion at the ankle to increase the vertical position and the primary ankle muscle producing this motion is the soleus, with some contribution from the gastrocnemius. There is minimal contribution from the hip other than contraction by the gluteus medius to pull up the trunk over the limb (98). Finally, in the forward propelling stage in which the limb on the lower step pushes up to the next step, there is minimal activity at the hip, with the ankle joint generating the most of the force. The greatest ankle power is generated in this phase as the individual

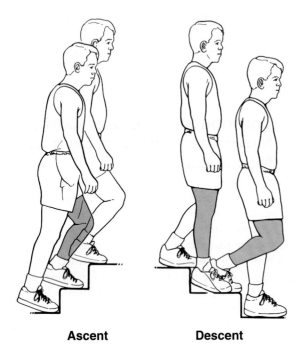

Ascent   **Descent**

FIGURE 6-41 In stair ascent with the left limb leading, there is significant contribution from the quadriceps, with assistance from the plantarflexors and the iliopsoas. In descent with the right limb leading, the same muscles control the movement eccentrically. For stair climbing as a whole, there is less contribution from the hip muscles than in walking or running.

continues on to the next step. At this point, the ankle pushes off, with the plantarflexors active as the body is pushed up to the next step (98).

Going downstairs, or descent, requires minimal hip muscular activity. In the limb pull phase, the hip flexors are active, followed by hamstring activity in the foot placement phase, when the limb is lowered to the step surface (98). As the limb makes contact with the next step in weight acceptance, the hip is minimally involved because most of the weight is eccentrically absorbed at the knee and ankle joints. The muscles acting at the knee joint are primarily responsible for generating the forces in the forward propelling phase. The plantarflexor muscles act eccentrically to absorb foot–surface contact (51,100). There is also co-contraction of the soleus and the tibialis anterior muscles early in the absorption phase to stabilize the ankle joint. As the person steps down, there is a small eccentric muscular activity in the soleus muscle as it contributes to the controlled drop and forward movement of the body. In the final phase of support, the controlled lowering phase, the body is lowered onto the step primarily through eccentric muscle activity at the knee joint. There is a minimum amount of hip extensor activity at the end of this phase.

LOCOMOTION

Several terms are used in gait studies to describe the timing of the key events. This terminology is necessary to understand the actions of the lower extremity in walking and running. In locomotion studies, a walking or running cycle is generally defined as the period from the contact of one foot on the ground to the next contact of the same foot. A gait cycle is usually broken down into two phases, referred to as the stance or support phase and the swing phase. In the stance or support phase, the foot is in contact with the ground. The support phase can also be broken down into subphases. The first half of the support phase is the braking phase, which starts with a loading or heelstrike phase and ends at midsupport. The second half of the support phase is the propulsion phase, which starts at midstance and continues to terminal stance and then to preswing as the foot prepares to leave the ground. The swing or noncontact phase is the period when the foot is not in contact with the ground, and it can be further subdivided into the initial swing phase, the midswing and the terminal swing subphases. Essentially, this phase represents the recovery of the limb in preparation for the next contact with the ground. These events are illustrated in Figure 6-42 for running and in Figure 6-43 for walking.

Running

There is considerable muscle activity in multiple muscles during running, and the joint motions typically occur over a greater range of motion than in walking. The exception is hyperextension, which is greater in walking because of the increased stance time. The muscular activity in running, however, is similar to that seen in walking. In running, there are 800 to 2,000 foot contacts with the ground per mile, and two to three times the BW is absorbed by the foot, leg, thigh, pelvis, and spine (22).

At the hip joint, the gluteus maximus controls flexion of the trunk on the stance side and decelerates the swing leg. The stance-side gluteus maximus also eccentrically controls flexion of the hip with the hamstrings (86,94). The gluteus medius and the tensor fascia latae are active in the initial braking portion of the support phase to control the pelvis in the frontal plane to keep it from tilting to the opposite side (94). During the propulsive portion of the support phase in running, the

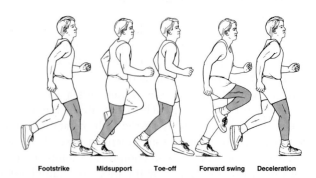

Footstrike   Midsupport   Toe-off   Forward swing   Deceleration

FIGURE 6-42 In running, there is high level of muscular activity in the hamstrings, gluteus minimus, gluteus maximus, quadriceps femoris group, and intrinsic muscles of the foot during the right support phase of the activity. During the swing phase, there is substantial activity in the iliopsoas and the tensor fascia latae.

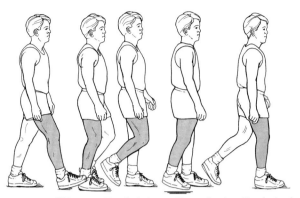

Loading response Mid stance Terminal stance Forward swing Terminal swing

Muscles	Loading Response Heel strike to foot flat			Mid-stance (10%–30%) Foot flat to midstance			Terminal Stance (30%–60%) Mid-stance to toe-off			Forward Swing (50%–80%) Toe-off to acceleration to midswing			Terminal Swing (85%–100%) Midswing to deceleration		
	Level	Action	Purpose	Level	Action	Purpose	Level	Action	Purpose	Level	Action	Purpose	Level	Action	Purpose
Gluteus medius and minimus	MOD	ISO	Control hip flexion	MOD	ISO	Opposing hip adduciton to stop contra-lateral pelvic drop	LOW	ISO	Opposing hip adduction to stop contra-lateral pelvic drop						
Iliospoas							MOD	ECC	Control of hip extension	HIGH	CON	Hip flexion	MOD	ECC	Control of hip extension
Tensor fascia latae	MOD	CON	Controls drop of contra-lateral pelvis	LOW	ISO	Stop contra-lateral pelvic drop	LOW	ISO	Stop contra-lateral pelvic drop	MOD	CON	Hip flexion			
Hip adductors							MOD	CON	Assist with hip flexion	HIGH	CON	Assist hip flexion and adduct thigh			
							LOW	ISO	Stabilize weight shift to other limb at toe-off						
Hamstrings	MOD	ECC	Control of hip flexion				MOD	CON	Hip extension at toe-off	LOW	CON	Knee flexion	HIGH	ECC	Decelerate knee extension
			Hip extension							MOD	ECC	Control of knee extension in midswing			
Quadriceps	MOD	ECC	Control of knee flexion	MOD	ECC	Control of knee flexion until COG over base of support	MOD	CON	Knee extension at toe-off	MOD	ISO	Limit knee flexion and augment hip flexioin	MOD	CON	Initiate knee extension
Plantar-flexors				HIGH	ECC	Control of ankle dorsi-flexion	HIGH	CON	Plantarflexion						
Dorsiflexors	HIGH	ECC	Control of lowering of foot into plantarflexion							HIGH	CON	Dorsiflexion so forefoot clears ground	MOD	ISO	Ankle dorsi-flexion for landing
Intrinsic foot muscles				HIGH	CON	To make foot rigid	LOW	ISO	Cause foot to be rigid as heel raises from floor						

Source: Gage, J. R. (1990). An overview of normal walking. *Instructional Course Lectures*, 39:291–303.
   Krebs, D. E., et al. (1998). Hip biomechanics during gait. *Journal of Sports Physical Therapy*, 28:51–59.
   Zajac, F. E. (2002). Understanding muscle coordination of the human leg with dynamical simulations. *Journal of Biomechanics*, 35:1011–1018.

FIGURE 6-43 Lower extremity muscles involved in walking showing the level of muscle activity (low, moderate, and high) and the type of muscle action (concentric [CON] and eccentric [ECC]) with the associated purpose.

hamstrings are very active as the thigh extends. The gluteus maximus also contributes to extension during late stance while also generating external rotation until toe-off.

At the knee joint, both the quadriceps femoris muscles and the hamstrings are active during various portions of the stance phase (106). At the instant of heel strike in running, a brief concentric contraction of the hamstrings flexes the knee to decrease the horizontal or braking force

being absorbed at impact. This is followed by activation of the quadriceps femoris. Initially, the quadriceps femoris acts eccentrically to slow the negative vertical of the body velocity. This action lasts until midsupport. The quadriceps femoris then acts concentrically to produce positive vertical velocity of the body. The hamstrings are also active with the quadriceps femoris to generate extension at the hip (92). The period from heel strike to midsupport represents more than half of the energy costs in running.

In the propelling portion of support, the quadriceps femoris is eccentrically active as the heel lifts off and then becomes concentrically active up through toe-off. The hamstrings are also concentrically active at toe-off.

Plantarflexor activity also increases sharply after heel strike and dominates through the total stance period (94). In the braking portion of stance, the plantarflexor muscles work not at the ankle but to eccentrically halt the vertical descent of the body over the foot. This continues into the propelling portion of support, when the plantarflexors shift to a concentric contraction, adding to the driving force of the run (94).

As soon as the foot leaves the ground to begin the swing phase, the limb is brought forward by the iliopsoas and rectus femoris, slowing the thigh in hyperextension and moving the thigh forward into flexion. The rectus femoris is the most important muscle for forward propulsion of the body because it accounts for the large range of motion in the lower extremity. It initiates the flexion movement so vigorously that the iliopsoas action also contributes to knee extension. The iliopsoas is active for more than 50% of the swing phase in running (94). In the early part of the swing phase, there is activity in the adductors that, as in walking, are working with the abductors to control the pelvis.

At the end of the swing phase, a great amount of eccentric muscular activity takes place in the gluteus maximus and the hamstrings as they begin to decelerate the rapidly flexing thigh. As the speed of the run increases, the activity of the gluteus maximus increases as it assumes more of the responsibility for slowing the thigh in preparation for foot contact in descent. Also, in the later portion of the swing phase, the abductors become active again as they lower the thigh eccentrically to produce adduction.

During the initial portion of the swing phase, the quadriceps femoris is active eccentrically to slow rapid knee flexion. In the later part of the swing phase, the hamstrings become active to limit both knee extension and hip flexion (94).

When running at faster speeds, the lower extremity muscles must generate considerable power. If muscle groups are weak, the running stride can be affected. For example, a weak gluteus maximus can slow down the leg transition between recovery and swing. Weak hamstrings can result in a failure to control hip flexion and knee extension in late recovery and weaken the hip extension force in the support phase (20).

Walking

During walking, the muscles around the pelvis and the hip joint contribute minimally to the actual propulsion in walking and are more involved with control of the pelvis (89). The muscular contribution of the lower extremity muscles active in walking is summarized in Figure 6-43.

At heel strike, moderate activity in the gluteus medius and minimus of the weight-bearing limb keeps the pelvis balanced against the weight of the trunk. The abduction

muscle force balances the trunk and the swing leg about the supporting hip (92). This activity continues until midsupport and then drops off in late stance (142). The adductors also work concurrently with both of these muscles to control the limb during support. The gluteus maximus is active at heel strike to assist with the movement of the body over the leg. Finally, the tensor fascia latae are active from heel strike to midsupport to assist with frontal plane control of the pelvis (142).

The hamstrings reach their peak of muscular activity as they attempt to arrest movement at the hip joint at heel strike. The quadriceps femoris then begins to contract to control the load (i.e., weight) being imposed on the knee joint by the body and the reaction force coming up from the ground. The knee also moves into flexion eccentrically controlled by the quadriceps femoris. A co-contraction of the hamstrings and the quadriceps femoris continues until the foot is flat on the ground, at which time the activity of the hamstrings drops off. The activity of the quadriceps femoris diminishes at approximately 30% of stance and is silent through midsupport and into the initial phases of propulsion.

In the propelling portion of the support phase of walking, the quadriceps femoris becomes active again around 85% to 90% of stance, when it is used to propel the body upward and forward. The hamstrings become active at approximately the same time to add to the forward propulsion.

At the ankle, there is maximum activity in the dorsiflexor muscles during heelstrike to eccentrically control the lowering of the foot to the ground in plantarflexion. The most muscle activity is seen in the tibialis anterior, extensor digitorum longus, and extensor hallucis longus (147). The activity in this muscle group decreases but maintains activity throughout the total stance phase.

There is little activity in the gastrocnemius and soleus at heel strike. They begin to activate after foot flat and continue into the propulsive phase as they control the movement of the tibia over the foot and generate propelling forces. The intrinsic muscles of the foot are inactive in this portion of stance.

In the propelling portion of support, the dorsiflexor muscles are still active, generating a second peak in the stance phase right before toe-off. The gastrocnemius and soleus reach a peak of muscular activity just before toe-off. The intrinsic muscles of the foot are active in the propelling phase of stance as they work to make the foot rigid and stable and control depression of the arch. Activity in the gastrocnemius, soleus, and intrinsic muscles ceases at toe-off.

At the beginning of the swing phase, the limb must be swung forward rapidly. This movement is initiated by a vigorous contraction of the iliopsoas, sartorius, and tensor fascia latae. The thigh adducts in the middle of the swing phase and internally rotates just after toe-off. The adductors are active at the beginning of the swing phase and continue into the stance phase. At the end of the swing phase, activity from the hamstrings and the gluteus maximus decelerates the limb (142).

In the swing phase, the hamstrings are active after toe-off and again at the end of the swing just before foot contact. Similar activity is seen in the quadriceps femoris, which slows knee flexion after toe-off and initiates knee extension prior to heel strike.

During the swing phase, the dorsiflexor muscles generate the only significant muscular activity in the ankle and foot. They hold the foot in a dorsiflexed position so that the foot clears the ground while the limb is swinging through.

CYCLING

In cycling, the key events are determined from the rotation of the crank of the bicycle. The motion of the crank forms a circle. A cycle is one revolution of this circle with 0° at the 12 o'clock position, 90° at 3 o'clock, 180° at 6 o'clock, and 270° at 9 o'clock. The end of the cycle occurs at 360° (or 0°), back at the 12 o'clock position. The 12 o'clock position is also referred to as top dead center, and the 6 o'clock position is referred to as bottom dead center. These events are presented in Figure 6-44.

The direction of the forces applied to the pedal changes during knee leg extension and coactivation of

agonists and antagonists occurs throughout the cycle. The quadriceps femoris muscles are the primary force producers with assistance from the other lower extremity muscles. At the hip joint, the gluteus maximus becomes active in the downstroke after about 30° into the cycle and continues through approximately 150° to extend the hip. As the activity from the gluteus maximus begins to decline, the activity of the hamstrings increases in the second quadrant and continues from approximately 130° to 250° as they extend the hip and begin to flex the knee (47). At the top of the crank cycle, from 0 to 90°, the quadriceps femoris is very active. The rectus femoris is active through the arc of 200° to 130° of the next cycle. The vastus medialis is active from 300° to 135°, and the vastus lateralis is active from 315° through 130° of the next cycle (11).

In the middle of the cycle, from 90° to 270°, the hamstrings contribute more to power production, with the biceps femoris active from 5° to 265° and the semimembranosus active from 10° to 265° (11). There is co-contraction of the quadriceps femoris and the hamstrings throughout the entire cycle but in different and changing proportions. In the last portion of the cycle, from 270° to 360°, the rectus femoris is actively involved as the leg is brought back up into the top position.

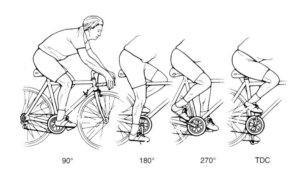

90°   180°   270°   TDC

Muscles	TDC-90 From top center to 90°			90–80 From 90° to 180°			180–270 From 180° to 270°			270-TDC From 270° to top center		
	Level	Action	Purpose	Level	Action	Purpose	Level	Action	Purpose	Level	Action	Purpose
Gluteus maximus	HIGH	CON	Hip extension	LOW	CON	Hip extension	LOW	ECC	Control hip flexion	LOW	ECC	Control hip flexion
Hamstrings	MOD	CON	Hip extension	HIGH	CON	Hip extension to knee extension	LOW	CON	Knee flexion	MOD	CON	Knee flexion
Vastus lateralis/ medialis	HIGH	CON	Knee extension	LOW	CON	Knee extension	LOW	ECC	Control knee flexion	MOD	CON	Knee extension
Rectus femoris	HIGH	CON	Knee extension	LOW	CON	Knee extension	MOD	CON	Hip flexion	HIGH	CON	Hip flexion
Soleus	HIGH	CON	Plantarflexion	MOD	CON	Plantarflexion						
Gastrocnemius	HIGH	CON	Plantarflexion	HIGH	CON	Plantarflexion	LOW	CON	Plantarflexion			
Tibialis anterior	LOW	CON	Dorsiflexion	LOW	CON	Dorsiflexion	LOW	CON	Dorsiflexion	HIGH	CON	Dorsiflexion

Source: Baum, B. S., Li, L. (2003). Lower extremity muscle activities during cycling are influence by load and frequency. *Journal of Electromyography & Kinesiology*, 13:181–190.
Jorge, M., Hull, M. L. (1986). Analysis of EMG measurements during bicycle pedaling. *Journal of Biomechanics*, 19:683–694.
Neptune, R. R., Kautz, S. A. (2000). Knee joint loading in forward versus backward pedaling: Implications for rehabilitation strategies. *Clinical Biomechanics*, 15:528–535.
van Ingen Schenau, G. J., et al. (1995). The control of mono-articular muscles in multijoint leg extensions in man. *Journal of Physiology*, 484:247–254.

FIGURE 6-44 Lower extremity muscles involved in cycling showing the level of muscle activity (low, moderate, and high) and the type of muscle action (concentric [CON] and eccentric [ECC]) with the associated purpose.

At the ankle, the gastrocnemius contributes through most of the power portion of the cycle and is active from 30° to 270° in the revolution. When the activity of the gastrocnemius ceases, the tibialis anterior becomes active from 280° until slightly past top dead center, thus contributing to the lift of the pedal. Again, when the tibialis anterior activity ceases, the gastrocnemius becomes active. Unlike the knee and hip, the ankle does not co-contract (73).

 ## Forces Acting on Joints in the Lower Extremity

The lower extremity joints can be subjected to high forces generated by muscles, BW, and ground reaction forces. The ground reaction forces generated in basic activities such as walking and stair climbing are 1.1 to 1.3 BW and 1.2 to 2.0 BW, respectively (152). Landing vertical forces are even higher (2.16 to 2.67 BW), and drop landings have been shown to generate maximum ground reaction forces in the range of 8.5 BW in children (10,75,149,170).

HIP JOINT

Even just standing on two limbs loads the hip joint with a force equivalent to 30% of BW (151). This force is generated primarily by the BW above the hip joint and is shared by right and left joints. When a person stands on one limb, the force imposed on the hip joint increases significantly, to approximately 2.5 to 3 times BW (142,151). This is mainly the result of the increase in the amount of BW previously shared with the other limb and a vigorous muscular contraction of the abductors. The increased muscular force of the abductors generates high hip joint forces as they work to counter the effects of gravity and control the pelvis.

In stair climbing, hip joint forces can reach levels of 3 times BW and are on an average 23% higher than walking (13); in walking, the forces range from 2.5 to 7 times BW; and in running, the forces can be as high as 10 times BW (67,79,118,142,151). In one study, hip joint forces (5.3 BW) were higher in running compared with skiing on long turns and a flat slope (4.1 BW), but short turns and steep slope skiing generated the highest hip joint forces (7.8 BW) (161). Cross-country skiing loaded the hip joint with 4.6 BW, which was less than walking (161). Fortunately, the hip joint can withstand 12 to 15 times BW before fracture or breakdown in the osseous component occurs (142).

KNEE JOINT

The knee is also subject to very high forces during most activities, whether generated in response to gravity, as a result of the absorption of the force coming up from the ground, or as a consequence of muscular contraction. The muscles generate considerable force, with the quadriceps femoris tension force being as high as 1 to 3 times BW in walking, 4 times BW in stair climbing, 3.4 times BW in climbing, and 5 times BW in a squat (26).

The tibiofemoral compression force can also be quite high in specific activities. For example, in a knee extension exercise, muscle forces applied against a low resistance (40 Nm [newton-meters]) can create tibiofemoral compression forces of 1100 N during knee extension acting through knee angles of 30° to 120°. This force increases to 1,230 N when extension occurs from the fully extended position (115). Tibiofemoral compression force in the extended position is greater, partly because the quadriceps femoris group loses mechanical advantage at the terminal range of motion and thus has to exert a greater muscular force to compensate for the loss in leverage.

The tibiofemoral shear force is maximum in the last few degrees of knee extension. The direction of the shear force changes with the amount of flexion in the joint, changing direction between 50° and 90° of flexion. Operating against the same 40-Nm resistance in extension, there is posterior shear of 200 N at 120° of flexion and 600 N of anterior shear in extension (115). This is partially because when nearing extension, the patellar tendon pulls the tibia anteriorly relative to the femur, but in flexion, it pulls the tibia posteriorly. The anterior force in the last 30° of extension places a great deal of stress on the ACL, which takes up 86% of the anterior shear force. By moving the contact pad closer to the knee in an extension exercise, the shear force can be directed posteriorly, taking the strain off of the ACL (115).

Even though tibiofemoral compression forces are greater in the extended position, the contact area is large, which reduces the pressure. There is 50% more contact area at the extended position than in 90° of flexion. Thus, in the extended position, the compression forces are high, but the pressure is less by 25% (115). The forces for women are 20% higher because of a decreased mechanical advantage associated with a shorter moment arm. Because women also have less contact area in the joint, greater pressure is created, accounting for the higher rate of osteoarthritis in the knees of women, an occurrence not seen in the hip.

Tibiofemoral compression forces during isokinetic knee extension (180°/s) has been found to be very high with a maximum of 6,300 N or 9 times BW (116). During cycling, the tibiofemoral compressive force has been recorded at 1.0 to 1.2 BW (46,113). Whereas tibiofemoral compression forces for walking are in the range of 2.8 to 3.1 BW (4,107,157), the tibiofemoral shear forces are in the range of 0.6 BW (157). In stair ascent, tibiofemoral compression and shear forces have been recorded as high as 5.4 and 1.3 BW, respectively (157).

The patellofemoral compressive force approximates 0.5 to 1.5 times BW in walking, 3 to 4 times BW in climbing, and 7 to 8 times BW in a squat exercise (115).

The patellofemoral joint absorbs compressive forces from the femur and transforms them into tensile forces in the quadriceps and patellar tendon. In vigorous activities, in which there are large negative acceleration forces, the patellofemoral force is also large. This force increases with flexion because the angle between the quadriceps femoris and the patella decreases, requiring greater quadriceps femoris force to resist the flexion or produce an extension.

The patellofemoral compressive force is maximum at 50° of flexion and declines at extension, approaching zero as the patella almost comes off the femur. The largest area of contact with the patella is at 60° to 90° of knee flexion. Of the patellar surface, 13% to 38% bears the force in joint loading (115). Fortunately, there is a large contact area when the patellofemoral compressive forces are large, which reduces the pressure. In fact, there is considerable pressure in the extended position even though the patellofemoral force is low because the contact area is small.

Activities using more pronounced knee flexion angles usually involve large patellofemoral compressive forces. These include descending stairs (4,000 N), maximal isometric extension (6,100 N), kicking (6,800 N), the parallel squat (14,900 N), isokinetic knee extension (8,300 N), rising from a chair (3,800 N), and jogging (5,000 N) (66). In activities that use lesser amounts of knee flexion, the force is much less. Examples include ascending stairs (1,400 N), walking (840 to 850 N), and bicycling (880 N) (115). The activities with high patellofemoral forces should be limited or avoided by individuals with patellofemoral pain.

The patellofemoral compressive force and the quadriceps femoris force both increase at the same rate with knee flexion in weight bearing. If the leg extends against a resistance, such as in a leg extension machine or weight boot, the quadriceps femoris force increases, but the patellofemoral force decreases from flexion to extension. Because the function in a weight-lifting extension exercise is opposite that in daily activities that use flexion in the weight-bearing position, the use of a weight-bearing closed kinetic chain activity is preferable. At knee flexion angles greater than 60°, the patellar tendon force is only half to two-thirds that of the quadriceps tendon force (115).

Those with pain in the patellar region should avoid exercising at angles greater than 30° to avoid large flexing moments and patellofemoral compression forces. However, in extension, when the patellofemoral force is low, the anterior shear force is high, making terminal extension activities contraindicated for any ACL injury (115). There is a reversal at 50° of flexion, when the shear force is low and the patellofemoral compression force is high.

ANKLE AND FOOT

The ankle and foot are subjected to significant compressive and shear forces in both walking and running. In walking, a vertical force 0.8 to 1.1 times BW comes at heel strike. The magnitude of this force decreases to about 0.8 times BW in the midstance to 1.3 times BW at toe-off (33,139). This force, along with the contraction force of the plantarflexors, creates a compression force in the ankle.

In walking, the compression force in the ankle joint can be as high as 3 times BW at heel strike and 5 times BW at toe-off. A shear force of 0.45 to 0.8 times BW is also present, primarily as a result of the shear forces absorbed from the ground and the position of the foot relative to the body (27,33,55). In running, the peak ankle joint forces are predicted to range from 9 to 13.3 times BW. The peak Achilles tendon force can be in the range of 5.3 to 10 times BW (27). The ankle joint is subjected to forces similar to those in the hip and knee joints. Amazingly, the ankle joint has very little incidence of osteoarthritis. This may be partly attributable to the large weight-bearing surface in the ankle, which lowers the pressure on the joint.

The subtalar joint is subjected to forces equivalent to 2.4 times BW, with the anterior articulation between the talus, calcaneus, and navicular recording forces as high as 2.8 times BW (33,139). Large loads on the talus must be expected because it is the keystone of the foot. Loads travel into the foot from the talus to the calcaneus and then forward to the navicular and cuneiforms.

During locomotion, forces applied to the foot from the ground are usually applied to the lateral aspect of the heel, travel laterally to the cuboid, and then transfer to the second metatarsal and the hallux at toe-off. In Figure 6-45, the path of the forces across the plantar surface of the foot is shown. The greatest percentage of support time is spent in contact with the forefoot and the first and second metatarsals. If the contact time of the second metatarsal is longer than that of the first metatarsal, a condition known as **Morton toe** develops, and the pressure on the head of the second metatarsal is greatly increased (139). This pattern of foot strike and transfer of the forces across the foot depends on a variety of factors and can vary with speed, foot type, and the foot contact patterns of individuals.

Forces in running are 2 times greater than those seen in walking. At foot strike, the forces received from the ground create a vertical force of 2.2 times BW and 0.5 times BW shear force. A vertical force of 2.8 times BW and a shear force of 0.5 times BW are produced at toe-off (33,139). With the addition of the muscular forces, the compressive forces can be as high as 8 to 13 times BW in running. The anterior shear forces can be in the range of 3.3 to 5.5 times BW, the medial shear force in the range of 0.8 times BW, and lateral shear force in the range of 0.5 times BW (33). Forces are large because the foot must transmit them between the body and the foot as well as the ground and the body. Given the injury record for the ankle and the foot, the foot is resilient and adaptable to the forces it must control with each step in walking or running.

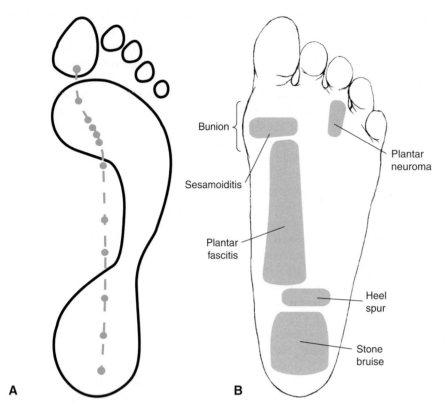

A

B

FIGURE 6-45 **(A)** Forces applied to the plantar surface of the foot during gait normally travel a path from the lateral heel to the cuboid and across to the first and second metatarsals. **(B)** High loading and extreme foot positions have been associated with a variety of injuries.

Summary

The lower extremity absorbs very high forces and supports the body's weight. The lower limbs are connected by the pelvic girdle, making every movement or posture of the lower extremity or trunk interrelated.

The pelvic girdle serves as a base for lower extremity movement and a site for muscular contraction, and it is important in the maintenance of balance and posture. The pelvic girdle consists of three coxal bones (ilium, ischium, and pubis) that are joined in the front at the pubic symphysis and connected to the sacrum in the back (sacroiliac joint). The pelvic and sacral movements of flexion, extension, posterior and anterior tilt, and rotation accompany movements of the thigh and the trunk.

The femur articulates at the acetabulum on the anterolateral surface of the pelvis. This ball-and-socket joint is well reinforced by strong ligaments that restrict all movements of the thigh except for flexion. The femoral neck is angled at approximately 125° in the frontal plane, and an increase (coxa valga) or decrease (coxa vara) in this angle influences leg length and lower extremity alignment and function. The angle of anteversion in the transverse plane also influences the rotation characteristics of the lower extremity.

The hip joint allows considerable movement in flexion (120° to 125°) produced by the hip flexors, iliopsoas, rectus femoris, sartorius, pectineus, and tensor fascia latae. The range of hyperextension is 10° to 15°. Hip extension is produced by the hamstrings, semimembranosus, semitendinosus, biceps femoris, and gluteus maximus. Abduction range of motion is 30° and is produced by the gluteus medius, gluteus minimus, tensor fascia latae, and piriformis. Adduction (30°) is produced by the gracilis, adductor longus, adductor magnus, adductor brevis, and pectineus. Internal rotation through approximately 50° is produced by the gluteus minimus, gluteus medius, gracilis, adductor longus, adductor magnus, tensor fascia latae, semimembranosus, and semitendinosus. External rotation through 50° is produced by the gluteus maximums, obturator externus, quadratus femoris, obturator internus, piriformis, and inferior and superior gemellus.

Movements of the thigh are usually accompanied by a pelvic movement and vice versa. For example, hip flexion in an open chain is accompanied by a posterior tilt of the pelvis. This reverses in a closed-chain weight-bearing position in which hip flexion is accompanied by an anterior movement of the pelvis on the femur.

The hip muscles can produce greater strength in the extension because of the large muscle mass from the hamstrings and the gluteus maximus. Extension strength is maximized from a hip flexion position. Strength output in the other movements can also be maximized with accompanying knee flexion for hip flexion strength facilitation, accompanying thigh flexion for the abduction movement, slight abduction for adduction facilitation, and hip flexion for the internal rotators.

Conditioning exercises for the lower extremity are relatively easy to implement because they include common movements associated with daily living activities. A closed kinetic chain exercise is beneficial for the lower extremity because of the transfer to daily activities. Because of the many two-joint muscles surrounding the hip joint, the position of adjacent joints is important. The hip flexors are best exercised with the person supine or hanging. The extensors are maximally stretched using a hip flexion position with the knees extended. The abductors, adductors, and rotators require creative approaches to conditioning because they are not easy to isolate.

The hip joint is durable and accounts for only a very small percentage of injuries to the lower extremity. Common soft-tissue injuries to the region include tendinitis of the gluteus medius; strain to the rectus femoris, hamstrings, iliopsoas, and piriformis; bursitis; and iliotibial band friction syndrome. Stress fractures are also more prevalent at sites such as the anterior iliac spine, pubic rami, ischial tuberosity, greater and lesser trochanters, and femoral neck. Common childhood disorders to the hip joint include congenital hip dislocation and Legg–Calvé–Perthes disease. The hip joint is also a site where osteoarthritis is prevalent in later years.

The knee joint is very complex and is formed by the articulation between the tibia and the femur (tibiofemoral joint) and the patella and the femur (patellofemoral joint). In the tibiofemoral joint, the two condyles of the femur rest on the tibial plateau and rely on the collateral ligaments, cruciate ligaments, menisci, and joint capsule for support. The patellofemoral joint is supported by the quadriceps tendon and the patellar ligament. The patella fits into the trochlear groove of the femur, which also offers stabilization to the patella.

An important alignment feature at the knee joint is the Q-angle, the angle representing the position of the patella with respect to the femur. An increase in this angle increases the valgus stress on the knee joint. High Q-angles are most common in females because of their wider pelvic girdles.

Flexion at the knee joint occurs through approximately 120° to 145° and is produced by the hamstrings, biceps femoris, semimembranosus, and semitendinosus. Accompanying flexion is internal rotation of the tibia, which is produced by the sartorius, popliteus, gracilis, semimembranosus, and semitendinosus. As the knee joint flexes and internally rotates, the patella also moves down in the groove and then moves laterally.

Extension at the knee joint is produced by the powerful quadriceps femoris muscle group, which includes the vastus lateralis, vastus medialis, rectus femoris, and vastus intermedius. When the knee extends, the tibia externally rotates via action by the biceps femoris. At the end of extension, the knee joint locks into the terminal position by a screw-home movement in which the condyles rotate into their final positions. In extension, the patella moves up in the groove and terminates in a resting position that is high and lateral on the femur.

The strength of the muscles around the knee joint is substantial, with the extensors being one of the strongest muscle groups in the body. The extensors are stronger than the flexors in all joint positions but not necessarily at all joint speeds. The flexors should not be significantly weaker than the extensors, or the injury potential around the joint will increase.

Conditioning of the knee extensors is an easy task because these muscles control simple lowering and rising movements. Closed-chain exercises are also very beneficial for the extensors because of their relation to daily living activities. The flexors are also exercised during a squat movement because of their action at the hip joint but can best be isolated and exercised in a seated position.

The knee is the most frequently injured joint in the body. Traumatic injuries damage the ligaments or menisci, and numerous chronic injuries result in tendinitis, iliotibial band syndrome, and general knee pain. Muscle strains to the quadriceps femoris and hamstrings are also common. The patella is a site for injuries such as subluxation and dislocation and other patellar pain syndromes, such as chondromalacia patella.

The foot and ankle consist of 26 bones articulating at 30 synovial joints, supported by more than 100 ligaments and 30 muscles. The ankle, or talocrural joint, has two main articulations, the tibiotalar and tibiofibular joints. The tibia and fibula form a mortise over the talus defined on the medial and lateral sides by the malleoli. Both sides of the joint are strongly reinforced by ligaments, making the ankle very stable.

The foot moves at the tibiotalar joint in two directions, plantarflexion and dorsiflexion. Plantarflexion can occur through a range of motion of approximately 50° and is produced by the gastrocnemius and soleus with some assistance from the peroneal muscles and the toe flexors. Dorsiflexion range of motion is approximately 20°, and the movement is created by the tibialis anterior and the toe extensors.

Another important joint in the foot is the subtalar or talocalcaneal joint, in which pronation and supination occur. At this joint, the rotation of the lower extremity and forces of impact are absorbed. Pronation at the subtalar articulation is a triplane movement that consists of calcaneal eversion, abduction, and dorsiflexion with the foot off the ground and calcaneal eversion, talar adduction, and plantarflexion with the foot on the ground in a closed chain. Muscles responsible for creating eversion are the peroneals, consisting of the peroneus longus, peroneus brevis, and peroneus tertius. Supination, the reverse movement, is created in the open chain through calcaneal inversion, talar adduction, and plantarflexion and in the closed chain through calcaneal inversion, talar abduction, and dorsiflexion. Muscles responsible for producing inversion are the tibialis anterior, tibialis posterior, and hallux flexors and extensors. The range of motion for pronation and supination is 20° to 62°.

The midtarsal joint also contributes to pronation and supination of the foot. These two joints, the calcaneocuboid and the talonavicular, allow the foot great mobility if the axes of the two joints lie parallel to each other. This is beneficial in the early portion of support, when the body is absorbing forces of contact. When these axes are not parallel, the foot becomes rigid. This is beneficial in the later portion of support, when the foot is propelling the body up and forward. Numerous other articulations in the foot, such as the intertarsal, tarsometatarsal, metatarsophalangeal, and interphalangeal joints, influence both total foot and toe motion.

The foot has two longitudinal arches that provide both shock absorption and support. The medial arch is higher and more dynamic than the lateral arch. The longitudinal arches are supported by the plantar fascia running along the plantar surface of the foot. Transverse arches running across the foot depress and spread in weight bearing. The shape of the arches and the bony arrangement determine foot type, which can be normal, flat, or high arched and flexible or rigid. An extremely flat foot is termed pes planus, and a high-arched foot is called pes cavus. Other foot alignments include forefoot and rear foot varus and valgus, a plantarflexed first ray, and equinus positions that influence function of the foot.

Plantarflexion of the foot is a very strong joint action and is a major contributor to the development of a propulsion force. Dorsiflexion is weak and not capable of generating high muscle forces.

The muscles of the foot and ankle receive a considerable amount of conditioning in daily living activities such as walking. Specific muscles can be isolated through exercises. For example, the gastrocnemius can be strengthened in a standing heel raise and the soleus in a seated heel raise. The intrinsic muscles of the foot can be exercised by drawing the alphabet or drawing figure eights with the foot or by just going barefoot.

The foot and ankle are frequently injured in sports and physical activity. Common injuries are ankle sprains; Achilles tendinitis; posterior, lateral, or medial tibial syndrome; plantar fasciitis; bursitis; metatarsalgia; and stress fractures.

The muscles of the lower extremity are major contributors to a variety of movements and sport activities. In walking, the hip abductors control the pelvis, the hamstrings control the amount of hip flexion and provide some of the propulsive force, and the hip flexors are active in the swing phase. In running, the hip joint motions and the muscular activity increase, but the same muscles used in walking are also used. At the knee joint, the quadriceps femoris serves as a shock absorption mechanism and a power producer for walking, running, and stair climbing. In cycling, the quadriceps femoris is responsible for a significant amount of power production. Ankle joint muscles such as the gastrocnemius and soleus are also important contributors to walking, running, stair climbing, and cycling.

The lower extremity must handle high loads imposed by muscles, gravity, and forces coming up from the ground. Loads absorbed by the hip joint can range from 2 to 10 times BW in activities such as walking, running, and stair climbing.

The knee joint can handle high loads and commonly absorbs 1 to 10 times BW in activities such as walking, running, and weight lifting. A maximum flexion position should be evaluated for safety, given the high shear forces that are present in the position. Patellofemoral forces can also be high, in the range of 0.5 to 8 times BW, in daily living activities. The patellofemoral force is high in positions of maximum knee flexion. The foot and ankle can handle high loads, and the forces in the ankle joint range from 0.5 to 13 times BW in walking and running. The subtalar joint also handles forces in the magnitude of 2 to 3 times BW.

REVIEW QUESTIONS

True or False

1. _____ A meniscus is found in both the knee and ankle joints.

2. _____ Knee flexion, as during a sitting, tightens the ACL.

3. _____ The hamstrings are best exercised in a standing position.

4. _____ The vastus medialis is only active from 90° of knee flexion until full extension.

5. _____ The male pelvis is narrower than the female pelvis.

6. _____ Strong contractions of the hip adductors are capable of causing an avulsion fracture on the greater trochanter.

7. _____ The lateral ankle is more prone to sprains than the medial side.

8. _____ Forefoot valgus, or pronation, occurs when the lateral side of the forefoot lifts.

9. _____ A deep squat position is a vulnerable position because it results in shear force at the knee.

10. _____ The patellofemoral force is greater going downstairs than going upstairs.

11. _____ Weight-bearing pronation consists of dorsiflexion, abduction, and eversion.

12. _____ In an open-chain position, anterior tilt of the pelvis accompanies hip extension.

13. _____ Coxa vara, which is more common in athletic females than males, increases the effectiveness of hip abductors.

14. _____ The angle of inclination of the femur is smallest at birth and continues to increase throughout life.

15. _____ The hip muscles can generate the greatest strength in flexion.

16. _____ Sartorius, gracilis, semitendinosus, semimembranosus, and popliteus internally rotate the tibia.

17. _____ During cycling, the quadriceps are activated eccentrically and concentrically while the hamstrings are silent.

18. ____ In single-leg stance, load at the hip joint is about three times BW because of abductor muscle contraction.

19. ____ External tibial rotation accompanies flexion at the knee joint.

20. ____ Hip hyperextension is always greater in running than walking because velocity is greater.

21. ____ A larger Q-angle increases the valgus stress on the knee joint.

22. ____ Hip flexors are best exercised in a supine position because of their biarticular nature.

23. ____ Bowleggedness is also termed genu varum.

24. ____ A hamstring-to-quadriceps strength ratio less than 0.75 may indicate a strength imbalance.

25. ____ The hip joint has good ligamentous support in all movement directions.

Multiple Choice

1. When trying to obtain maximum plantarflexion strength, one should ____.
   a. have flexed knees and slightly plantarflexed ankle
   b. have extended knees and slightly plantarflexed ankle
   c. have flexed knees and slightly dorsiflexed ankle
   d. have extended knees and slightly dorsiflexed ankle

2. In which foot type is there greater internal rotation of the foot relative to the tibia?
   a. High-arched foot
   b. Normal foot
   c. Flat foot
   d. Equinus foot

3. Prominent hip adductors include ____.
   a. gracilis, adductor longus, and adductor magnus
   b. adductor magnus, adductor longus, and semitendinosus
   c. sartorius, iliopsoas, and semimembranosus
   d. gracilis, adductor brevis, and popliteus

4. Which statement is false concerning the acetabulum?
   a. It faces anteriorly, laterally, and inferiorly
   b. It has thick cartilage on all surfaces
   c. The three pelvis bones articulate in the acetabulum cavity
   d. It contacts 20% to 25% of the head of the femur

5. The sacral movements are ____.
   a. flexion, extension, abduction, adduction, and rotation
   b. flexion, extension, nutation, and counternutation
   c. flexion, extension, and rotation
   d. flexion and extension

6. The gluteus medius ____.
   a. is less efficient when femoral anteversion is greater
   b. on the left side contracts during swing phase of the right leg
   c. is a powerful hip abductor and external rotator
   d. may result in pelvic drop if it is weak, also leading to increased femoral abduction and knee varus

7. Which of the following does not limit plantarflexion range of motion?
   a. Ligaments and joint capsule
   b. Talus and tibia
   c. Gastrocnemius and soleus muscles
   d. Calcaneus

8. Which of the following may lead to injury of the Achilles tendon?
   a. Repetitive vigorous contractions of gastrocnemius and soleus
   b. Abrupt forward propulsion after backward movement
   c. Excessive Q-angle
   d. All of the above
   e. Both a and b

9. The muscles that attach into the iliotibial band include the:
   a. gluteus medius
   b. tensor fascia latae
   c. sartorius
   d. All of the above
   e. Both a and b

10. The medial meniscus ____.
   a. is more likely to heal if tears occur on the interior versus periphery
   b. more mobile than the lateral meniscus
   c. is crescent shaped
   d. only attaches to the tibia on the anterior side

11. Which motion is not resisted by the multiple ligaments crossing the hip?
   a. Flexion
   b. Internal rotation
   c. Adduction
   d. Abduction
   e. None of the above

12. Which of the following help stabilize the knee joint?
   a. Medial and lateral collateral ligaments
   b. Joint capsule
   c. Semimembranosus
   d. All of the above
   e. Both a and b

13. During the support phase of running, ____.
   a. pronation accompanies external rotation of the knee
   b. pronation accompanies flexion of the knee
   c. supination accompanies external rotation of the knee
   d. supination accompanies extension of the knee
   e. Both a and d
   f. Both b and c

14. Which is not associated with a short Achilles tendon?
   a. Early heel rise during gait
   b. Toe walking
   c. Limited pronation
   d. Limited dorsiflexion motion during gait

15. The pelvis tilts ____.
   a. anteriorly during hip extension when the feet are planted
   b. anteriorly during hip flexion when the feet are planted

c. posteriorly during open-chain hip flexion
d. posteriorly during open-chain hip extension
e. Both a and d
f. Both b and c

16. Which is true of the patella?
a. It moves medially into the groove as the knee begins to flex
b. Its movement primarily depends on the strength of the quadriceps and menisci morphology
c. It moves more proximal as the knee flexes
d. All of the above
e. Both a and b

17. Which direction does the knee typically have the least range of motion?
a. Flexion
b. Internal/external rotation
c. Abduction/adduction
d. Hyperextension

18. Osteoarthritis is common in elderly and is characterized by ____.
a. growth of osteoblasts
b. joint space narrowing
c. degeneration of joint cartilage and subchondral bone
d. All of the above
e. Both b and c

19. Which is not true of the gluteus maximus?
a. It acts to extend both the hips and trunk
b. It contributes more to hip extension strength than do the hamstrings
c. It is more active during stair climbing than level walking
d. It is the most massive muscle
e. It also externally rotates the thigh

20. Patellofemoral compression force ____.
a. is greater during walking than squatting
b. increases as knee flexion increases
c. is high in terminal extension exercises
d. can be as high as 10 times BW in a squat exercise

21. Which is not true during cycling?
a. The hamstrings are very active during the middle of the cycle
b. The major power producers are the quadriceps
c. The plantarflexors are active while the dorsiflexors remain silent
d. Co-contraction of the quadriceps femoris and the hamstrings throughout the entire cycle

22. The knee can flex through a greater range of motion when the ____.
a. foot is pronating
b. thigh is hyperextended
c. foot is supinating
d. thigh is flexed
e. All of the above
f. Both a and b

23. The knee joint is considered to be a ____.
a. modified hinge joint
b. condyloid joint
c. double condyloid joint
d. All of the above

24. In regard to the plantar fascia, which is not true?
a. Sesamoid bones embedded within the fascia increase the mechanical advantage during the windlass mechanism
b. It originates at the calcaneus and inserts on the talus, cuneiforms, and navicular
c. Inflammation and pain associated with fasciitis usually occur at the origin not insertion
d. It is most prevalent in high-arched feet and individuals with Achilles tendons or leg length discrepancies

25. Which state is true when comparing walking and running?
a. Ankle compression forces are present in running but not walking
b. Vertical ground reaction forces are five times higher during running
c. There is a braking phase in both
d. Lower extremity joint motions are typically greater in walking because of increased stance time

References

1. Adelaar, R. (1986). The practical biomechanics of running. *American Journal of Sports Medicine*, 14:497-500.
2. Adkins, S. B., Figler, R. A. (2001). Hip pain in athletes. *American Family Physician*, 61:2109-2118.
3. Amendola, A., Wolcott, M. (2002). Bony injuries around the hip. *Sports Medicine and Arthroscopy Review*, 10:163-167.
4. Anderson, F. C., Pandy, M. G. (2001). Static and dynamic optimization solutions for gait are practically equivalent. *Journal of Biomechanics*, 34:53-161.
5. Anderson, K., et al. (2001). Hip and groin injuries in athletes. *The American Journal of Sports Medicine*, 29:521-533.
6. Andersson, E.A., et al. (1997). Abdominal and hip flexor muscle activation during various training exercises. *European Journal of Applied Physiology*, 75:115-123.
7. Apkarian, J., et al. (1989). Three-dimensional kinematic and dynamic model of the lower limb. *Journal of Biomechanics*, 22:143-155.
8. Areblad, M., et al. (1990). Three-dimensional measurement of rear foot motion during running. *Journal of Biomechanics*, 23:933-940.
9. Bates, B. (1983). Foot function in running: Researcher to coach. In J. Terauds (Ed.). *Biomechanics in Sports*. Del Mar, CA: Academic Publishers, 293-303.
10. Bauer, J. J., et al. (2001). Quantifying force magnitude and loading rate from drop landings that induce osteogenesis. *Journal of Applied Biomechanics*, 17:142-152.
11. Baum, B. S., Li, L. (2003). Lower extremity muscle activities during cycling are influenced by load and frequency. *Journal of Electromyography & Kinesiology*, 13:181-190.
12. Bazzoli, A., Pollina, F. (1989). Heel pain in recreational runners. *The Physician and Sportsmedicine*, 17:55-56.

13. Bergmann, G., et al. (2001). Hip contact forces and gait patterns from routine activities. *Journal of Biomechanics*, 34: 859-871.
14. Besier, T. F., et al. (2005). Patellofemoral joint contact area increases with knee flexion and weight bearing. *Journal of Orthopaedic Research*, 23:345-350.
15. Beutler, A. I., et al. (2002). Electromyographic analysis of single leg closed chain exercises: Implications for rehabilitation after anterior cruciate ligament reconstruction. *Journal of Athletic Training*, 37:13-18.
16. Beynnon B. D., et al. (1997). The strain behavior of the anterior cruciate Ligament during squatting and active extension: A comparison of an open- and a closed-kinetic chain exercise. *The American Journal of Sports Medicine*, 25: 823-829.
17. Beynnon, B. D., et al. (2001). Ankle ligament injury risk factors: A prospective study of college athletes. *Journal of Orthopaedic Research*, 19:213-220.
18. Blackburn, T. A., Craig, E. (1980). Knee anatomy: A brief review. *Physical Therapy*, 60:1556-1560.
19. Blackburn, T. A., et al. (1982). An introduction to the plica. *Journal of Orthopaedic and Sports Physical Therapy*, 3:171-177.
20. Blazevich, A. J. (2000). Optimizing hip musculature for greater sprint running speed. *NSCA Strength & Conditioning Journal*, 22:22-27.
21. Boyd, K. T., et al. (1997). Common hip injuries in sport. *Sports Medicine*, 24:273-288.
22. Brody, D. M. (1980). Running injuries. *Clinical Symposium*, 32:2-36.
23. Brown, D. A. (1996). Muscle activity patterns altered during pedaling at different body orientations. *Journal of Biomechanics*, 10:1349-1356.
24. Brown, L. P., Yavarsky, P. (1987). Locomotor biomechanics and pathomechanics: A review. *Journal of Orthopaedic and Sports Physical Therapy*, 9:3-10.
25. Browning, K. H. (2001). Hip and pelvis injuries in runners: Careful examination and tailored management. *The Physician and Sportsmedicine*, 29:23-34.
26. Buchbinder, M. R., et al. (1979). The relationship of abnormal pronation to chondromalacia of the patella in distance runners. *Podiatric Sports Medicine*, 69:159-162.
27. Burdett, R. G. (1982). Forces predicted at the ankle during running. *Medicine & Science in Sports & Exercise*, 14: 308-316.
28. Bynum, E. B., et al. (1996). Open versus closed chain kinetic exercises after anterior cruciate ligament reconstruction. A prospective randomized study. *The American Journal of Sports Medicine*, 23:401-406.
29. Cerny, K., et al. (1990). Effect of an unrestricted knee-ankle-foot orthosis on the stance phase gait in healthy persons. *Orthopedics*, 13:1121-1127.
30. Chesworth, B. M., et al. (1989). Validation of outcome measures in patients with patellofemoral syndrome. *Journal of Sports Physical Therapy*, 10(8):302-308.
31. Clark, T. E., et al. (1983). The effects of shoe design parameters on rearfoot control in running. *Medicine & Science in Sports & Exercise*, 5:376-381.
32. Colby, S. (2000). Electromyographic and kinematic analysis of cutting maneuvers. *The American Journal of Sports Medicine*, 28:234-240.
33. Czerniecki, J. M. (1988). Foot and ankle biomechanics in walking and running. *American Journal of Physical Medicine & Rehabilitation*, 67:246-252.
34. Davies, G. J., et al. (1980). Knee examination. *Physical Therapy*, 60:1565-1574.
35. Davies, G. J., et al. (1980). Mechanism of selected knee injuries. *Physical Therapy*, 60:1590-1595.
36. Dewberry, M. J., et al. (2003). Pelvic and femoral contributions to bilateral hip flexion by subjects suspended from a bar. *Clinical Biomechanics*, 18:494-499.
37. DiStefano, V. (1981). Anatomy and biomechanics of the ankle and foot. *Athletic Training*, 16:43-47.
38. Donatelli, R. (1987). Abnormal biomechanics of the foot and ankle. *Journal of Orthopaedic and Sports Physical Therapy*, 9:11-15.
39. DonTigny, R. L. (1985). Function and pathomechanics of the sacroiliac joint: A review. *Physical Therapy*, 65:35-43.
40. Draganich, L. F., et al. (1989). Coactivation of the hamstrings and quadriceps during extension of the knee. *The Journal of Bone & Joint Surgery*, 71:1075-1081.
41. Drez, D. Jr., et al. (1982). Nonoperative treatment of double lateral ligament tears of the ankle. *American Journal of Sports Medicine*, 10:197-200.
42. Drysdale, C. L., et al. (2004). Surface electromyographic activity of the abdominal muscles during pelvic-tilt and abdominal-hollowing exercises. *Athletic training*, 39: 32-36.
43. Earl, J. E. (2005). Gluteus medius activity during 3 variations of isometric single-leg stance. *Journal of Sport Rehabilitation*, 14:1-11.
44. Earl, J. E., et al. (2001). Activation of the VMO and VL during dynamic mini-squat exercises with and without isometric hip adduction. *Journal of Electromyography & Kinesiology*, 11:381-386.
45. Engsberg, J. R., Andrews, J. G. (1987). Kinematic analysis of the talocalcaneal/talocrural joint during running support. *Medicine & Science in Sports & Exercise*, 19:275-284.
46. Ericson, M. O., Nisell, R. (1986). Tibiofemoral joint forces during ergometer cycling. *The American Journal of Sports Medicine*, 14:285-290.
47. Erickson, M. O., et al. (1986). Power output and work in different muscle groups during ergometer cycling. *European Journal of Applied Physiology*, 55:229-235.
48. Escamilla, R. F., et al. (1998). Biomechanics of the knee during closed kinetic and open kinetic chain exercises. *Medicine & Science in Sports & Exercise*, 30:556-569.
49. Fiore, R. D., Leard, J. S. (1980). A functional approach in the rehabilitation of the ankle and rearfoot. *Athletic Training*, 15:231-235.
50. Fleming, B. C., et al.(2005). Open- or closed-kinetic chain exercises after anterior cruciate ligament reconstruction? 33:134-140.
51. Freedman, W., et al. (1976). EMG patterns and forces developed during step-down. *American Journal of Physical Medicine*, 55:275-290.
52. Fukubayashi, T., Kurosawa, H. (1980). The contact area and pressure distribution pattern of the knee: A study of normal and osteoarthritic knee joints. *Acta Orthopaedica Scandinavica*, 51:871-879.
53. Garrison, J. G., et al. (2005). Lower extremity EMG in male and female college soccer players during single-leg landing. *Journal of Sport Rehabilitation*, 14:48-57.
54. Gehlsen, G. M ., et al. (1989). Knee kinematics: The effects of running on cambers. *Medicine & Science in Sports & Exercise*, 21:463-466.

55. Giddings, V. L., et al.(2000). Calcaneal loading during walking and running. *Medicine & Science in Sports & Exercise*, 32, 627-634.

56. Godges, J. J., et al. (1989). The effects of two stretching procedures on hip range of motion and gait economy. *Journal of Orthopaedic and Sports Physical Therapy*, 10(9):350-357.

57. Grana, W. A., Coniglione, T. C. (1985). Knee disorders in runners. *The Physician and Sportsmedicine*, 13:127-133.

58. Grelsamer, R. P., et al. (2005). Men and women have similar Q angles : A clinical and trigonometric evaluation. *The Journal of Bone & Joint Surgery. British Volume*, 87:1498-1501.

59. Grieve, G. P. (1976). The sacroiliac joint. *Journal of Anatomy*, 58:384-399.

60. Hamilton, J. J., Ziemer, L. K. (1981). Functional anatomy of the human ankle and foot. In R. H. Kiene, K. A. Johnson (Eds.). *Proceedings of the AAOS Symposium on the Foot and Ankle*. St. Louis, MO: C.V. Mosby, 1-14.

61. Halbach, J. (1981). Pronated foot disorders. *Athletic Training*, 16:53-55.

62. Heller, M. O., et al. (2001). Musculoskeletal loading conditions at the hip during walking and stair climbing. *Journal of Biomechanics*, 34:863-893.

63. Hintermann, B. E. A. T. (1999). Biomechanics of the unstable ankle joint and clinical implications. *Medicine & science in sports & exercise*, 31(7 Suppl), S459-69.

64. Hodge, W. A., et al. (1987). The influence of hip arthroplasty on stair climbing and rising from a chair. In J. L. Stein (Ed.). *Biomechanics of Normal and Prosthetic Gait*. New York: American Society of Mechanical Engineers, 65-67.

65. Hole, J. W. (1990). *Human Anatomy and Physiology* (5th Ed.). Dubuque, IA: William C. Brown.

66. Hunt, G. C. (1985). Examination of lower extremity dysfunction. In J. Gould, G. J. Davies (Eds.). *Orthopaedic and Sports Physical Therapy*. St. Louis, MO: C.V. Mosby, 408-436.

67. Hurwitz, D.E., et al. (2003). A new parametric approach for modeling hip forces during gait. *Journal of Biomechanics*, 36:113-119.

68. Hutson, M. A., Jackson, J. P. (1982). Injuries to the lateral ligament of the ankle: Assessment and treatment. *British Journal of Sports Medicine*, 4:245-249.

69. Inman, V. T. (1959). The influence of the foot-ankle complex on the proximal skeletal structures. *Artificial Limbs*, 13:59-65.

70. Jacobs. C., et al. (2005). Strength and fatigability of the dominant and nondominant hip abductors. *Journal of Athletic Training*, 40:203-206.

71. James, S. L., et al. (1978). Injuries to runners. *American Journal of Sports Medicine*, 6:40-50.

72. Johnson, M. E., et al. (2004). Age-related changes in hip abductor and adductor joint torques. *Archives of Physical Medicine and Rehabilitation*, 85:593-597.

73. Jorge, M., Hull, M. L. (1986). Analysis of EMG measurements during bicycle pedalling. *Journal of Biomechanics*, 19:683-694.

74. Kapandji, I. A. (1970). *The Physiology of the Joints* (Vol. 2). Edinburgh: Churchill Livingstone.

75. Kernozek, T. W., et al. (2005). Gender differences in frontal and sagittal plane biomechanics during drop landings. *Medicine & Science in Sports & Exercise*, 37:1003-1012.

76. Kempson, G. E., et al. (1971). Patterns of cartilage stiffness on the normal and degenerative human femoral head. *Journal of Biomechanics*, 4:597-609.

77. Kettlecamp, D. H., et al. (1970). An electrogoniometric study of knee motion in normal gait. *The Journal of Bone & Joint Surgery*, 52(suppl A):775-790.

78. Kosmahl, E., Kosmahl, H. (1987). Painful plantar heel, plantar fasciitis, and calcaneal spur: Etiology and treatment. *Journal of Orthopaedic and Sports Physical Therapy*, 9:17-24.

79. Krebs, D.E., et al. (1998) Hip biomechanics during gait. *Journal of Sports Physical Therapy*, 28:51-59.

80. Kvist, J., Gillquist, J. (2001). Sagittal plane knee translation and electromyographic activity during closed and open kinetic chain exercises in anterior cruciate ligament-deficient patients and control subjects. *The American Journal of Sports Medicine*, 29:72-82.

81. Lafortune, M. A., Cavanagh, P. R. (1985). Three-dimensional kinematics of the patella during walking. In B. Jonsson (Ed.). *Biomechanics X-A*. Champaign, IL: Human Kinetics, 337-341.

82. Lafortune, M. A., et al. (1992). Three-dimensional kinematics of the human knee during walking. *Journal of Biomechanics*, 25:347-357.

83. Larson, R. L. (1973). Epiphyseal injuries in the adolescent athlete. *Orthopedic Clinics of North America*, 4:839-851.

84. Laubenthal, K. N., et al. (1972). A quantitative analysis of knee motion during activities of daily living. *Physical Therapy*, 52:34-42.

85. Leib, F. J., Perry, J. (1971). Quadriceps function: An electromyographic study under isometric conditions. *The Journal of Bone & Joint Surgery*, 53(suppl A):749-758.

86. Lieberman, D. E., et al. (2006). The human gluteus maximus and its role in running. *Journal of Experimental Biology*, 209:2143-55.

87. Lloyd-Smith, R., et al. (1985). A survey of overuse and traumatic hip and pelvic injuries in athletes. *The Physician and Sportsmedicine*, 13(10):131-141.

88. Locke, M., et al. (1984). Ankle and subtalar motion during gait in arthritic patients. *Physical Therapy*, 64:504-509.

89. Lovejoy, C. O. (1988). Evolution of human walking. *Scientific American*, 259(5):118-125.

90. Lutz, G. E., et al. (1993). Comparison of tibiofemoral joint forces during open-kinetic-chain and closed-kinetic-chain exercises. *The Journal of Bone & Joint Surgery*, 75:732-739.

91. Lyon, K. K., et al. (1988). Q-angle: A factor in peak torque occurrence in isokinetic knee extension. *Journal of Orthopaedic and Sports Physical Therapy*, 9:250-253.

92. MacKinnon, C. D., Winter, D. A. (1993). Control of whole body balance in the frontal plane during human walking. *Journal of Biomechanics*, 26:633-644.

93. Majewski, M., Klaus, S. H. (2006). Epidemiology of athletic knee injuries: A 10 year study. *Knee*, 13:184-188.

94. Mann, R. A., et al. (1986). Comparative electromyography of the lower extremity in jogging, running, and sprinting. *American Journal of Sports Medicine*, 14:501-510.

95. Markhede, G., Stener, G. (1981). Function after removal of various hip and thigh muscles for extirpation of tumors. *Acta Orthopaedica Scandinavica*, 52:373-395.

96. Markhede, G., Grimby, G. (1980). Measurement of strength of the hip joint muscles. *Scandinavian Journal of Rehabilitative Medicine*, 12:169-174.

97. Matheson G. O, et al. (1987). Stress fractures in athletes. A case study of 320 cases. *American Journal of Sports Medicine*, 15:46-58.

98. McFadyen, B. J., Winter, D. A. (1988). An integrated biomechanical analysis of normal stair ascent and descent. *Journal of Biomechanics*, 21:733-744.

99. McClusky, G. Blackburn, T. A. (1980). Classification of knee ligament instabilities. *Physical Therapy*, 60:1575-1577.

100. McLeod, W. D., Hunter, S. (1980). Biomechanical analysis of the knee: Primary functions as elucidated by anatomy. *Physical Therapy*, 60:1561-1564.

101. McPoil, T., Brocato, R. S. (1985). The foot and ankle: Biomechanical evaluation and treatment. In J. A. Gould, G. J. Davies (Eds.). *Orthopaedic and Sports Physical Therapy*. St. Louis, MO: C.V. Mosby, 313-341.

102. McPoil, T., Knecht, H. (1987). Biomechanics of the foot in walking: A functional approach. *Journal of Orthopedic and Sports Physical Therapy*, 7:69-72.

103. Metzmaker, J. N., Pappas, A. M. (1985). Avulsion fractures of the pelvis. *American Journal of Sports Medicine*, 13: 349-358.

104. Milgrom, C., et al. (1985). The normal range of subtalar inversion and eversion in young males as measured by three different techniques. *Foot & Ankle International*, 6: 143-145.

105. Mital, M. A., et al. (1980). The so-called unresolved Osgood-Schlatter lesion: A concept based on fifteen surgically treated lesions. *The Journal of Bone & Joint Surgery*, 62(suppl A): 732-739.

106. Montgomery, W. H., et al. (1994). Electromyographic analysis of hip and knee musculature during running. *The American Journal of Sports Medicine*, 22:272-278.

107. Morrison, J. B. (1968). Bioengineering analysis of force actions transmitted by the knee joint. *Journal of Biomedical Engineering*, 3:164-170.

108. Murray, M.P., et al (1964). Walking patterns of normal men. *The Journal of Bone & Joint Surgery*, 46A:335-360.

109. Murray, R., et al. (2002). Pelvifemoral rhythm during unilateral hip flexion in standing. *Clinical Biomechanics*, 17: 147-151.

110. Murray, S. M., et al. (1984). Torque-velocity relationships of the knee extensor and flexor muscles in individuals sustaining injuries of the anterior cruciate ligament. *American Journal of Sports Medicine*, 12:436-439.

111. Nadler, S. F., et al. (2002). Hip muscle imbalance and low back pain in athletes: Influence of core strengthening. *Medicine & Science in Sports & Exercise*, 34:9-16.

112. Nadzadi, M. E., et al. (2003). Kinematics, kinetics, and finite element analysis of commonplace maneuvers at risk for total hip dislocation. *Journal of Biomechanics*, 36:577-591.

113. Neptune, R. R., Kautz, S. A. (2000). Knee joint loading in forward versus backward pedaling: implications for rehabilitation strategies. *Clinical Biomechanics*, 15:528-535.

114. Neumann, D. A., et al. (1988). Comparison of maximal isometric hip abductor muscle torques between hip sides. *Physical Therapy*, 68:496-502.

115. Nisell, R. (1985). Mechanics of the knee: A study of joint and muscle load with clinical applications. *Acta Orthopaedica Scandinavica*, 56:1-42.

116. Nissell, R., et al. (1989). Tibiofemoral joint forces during isokinetic knee extension. *The American Journal of Sports Medicine*, 17:49-54.

117. Nissan, M. (1979). Review of some basic assumptions in knee biomechanics. *Journal of Biomechanics*, 13:375-381.

118. Nordin, M., Frankel, V. H. (1989). Biomechanics of the hip. In M. Nordin & V. H. Frankel (Eds.). *Basic Biomechanics of the Musculoskeletal System*. Philadelphia, PA: Lea & Febiger, 135-152.

119. Noyes, F. R., et al. (1980). Knee ligament tests: What do they really mean? *Physical Therapy*, 60:1578-1581.

120. Noyes, F. R., Sonstegard, D. A. (1973). Biomechanical function of the pes anserinus at the knee and the effect of its transplantation. *The Journal of Bone & Joint Surgery*, 35(suppl A):1225-1240.

121. Nyland, J., et al. (2004). Femoral anteversion influences vastus medialis and gluteus medius EMG amplitude: Composite hip abductor EMG amplitude ratios during isometric combined hip abduction-external rotation. *Journal of Electromyography & Kinesiology*, 14:255-261.

122. O'Brien, M., Delaney, M. (1997) The anatomy of the hip and groin. *Sports Medicine and Arthroscopy Review*, 5:252-267.

123. Ono, T., et al. (2005). The boundary of the vastus medialis oblique and the vastus medialis longus. *Journal of Physical Therapy Science*, 17:1-4.

124. Oshimo, T. A., et al. (1983). The effect of varied hip angles on the generation of internal tibial rotary torque. *Medicine & Science in Sports & Exercise*, 15:529-534.

125. Osternig, L. R., et al. (1979). Knee rotary torque patterns in healthy subjects. In J. Terauds (Ed.). *Science in Sports*. Del Mar, CA: Academic, 37-43.

126. Osternig, L. R., et al. (1981). Relationships between tibial rotary torque and knee flexion/extension after tendon transplant surgery. *Archives of Physical and Medical Rehabilitation*, 62:381-385.

127. Perry, J. (1992). *Gait Analysis: Normal and Pathological Function*. Thorofare, NJ: Slack.

128. Polisson, R. P. (1986). Sports medicine for the internist. *Medical Clinics of North America*, 70:469-474.

129. Porterfield, J. A. (1985). The sacroiliac joint. In J. A. Gould, G. J. Davies (Eds.). *Orthopedic and Sports Physical Therapy*. St. Louis, MO: C.V. Mosby, 550-579.

130. Pressel, T., Lengsfeld, M. (1998). Functions of hip joint muscles. *Medical Engineering & Physics*, 20:50-56.

131. Radakovich, M., Malone, T. (1980). The superior tibiofibular joint: The forgotten joint. *Journal of Orthopaedic and Sports Physical Therapy*, 3:129-132.

132. Radin, E. L. (1980). Biomechanics of the human hip. *Clinical Orthopaedics*, 152:28-34.

133. Rand, M. K., Ohtsuki, T. (2000). EMG analysis of lower limb muscles in humans during quick change in running direction. *Gait and Posture*, 12:169-183.

134. Raschke, U. Chaffin, D. B. (1996). Trunk and hip muscle recruitment in response to external anterior lumbosacral shear and moment loads. *Clinical Biomechanics*, 11:145-152.

135. Reid, D. C., et al. (1987). Lower extremity flexibility patterns in classical ballet dancers and their correlation to lateral hip and knee injuries. *American Journal of Sports Medicine*, 15(4):347-352.

136. Roach, K. E., Miles, T. P. (1991). Normal hip and knee active range of motion: The relationship to age. *Physical Therapy*, 71:656-665.

137. Robbins, S. E., Hanna, A. M. (1987). Running-related injury prevention through barefoot adaptations. *Medicine & Science in Sports & Exercise*, 19:148-156.

138. Robinovitch, S. N., et al. (2000). Prevention of falls and fall-related fractures through biomechanics. *Exercise and Sport Sciences Reviews*, 28:74-79.

139. Rodgers, M. (1988). Dynamic biomechanics of the normal foot and ankle during walking and running. *Physical Therapy*, 68:1822-1830.

140. Rubin, G. (1971). Tibial rotation. *Bulletin of Prosthetics Research*, 10(15):95-101.

141. Salathe, E. P. Jr., et al. (1990). The foot as a shock absorber. *Journal of Biomechanics*, 23:655-659.

142. Saudek, C. E. (1985). The hip. In J. Gould, G. J. Davies (Eds.). *Orthopaedic and Sports Physical Therapy*. St. Louis, MO: C.V. Mosby, 365-407.

143. Savelberg, H. H., Meijer, K. (2004). The effect of age and joint angle on the proportionality of extensor and flexor strength at the knee joint. *Journal of Gerontology*, 59(suppl A): 1120-1128.

144. Schache, A. G., et al. (2000). Relation of anterior pelvic tilt during running to clinical and kinematic measures of hip extension. *British Journal of Sports Medicine*, 34:279-283.

145. Scott, S. H., Winter, D. A. (1991). Talocrural and talocalcaneal joint kinematics and kinetics during the stance phase of walking. *Journal of Biomechanics*, 24:734-752.

146. Scott, S. H., Winter, D. A. (1993). Biomechanical model of the human foot: Kinematics and kinetics during the stance phase of walking. *Journal of Biomechanics*, 26:1091-1104.

147. Segal, P., Jacob, M. (1973). *The Knee*. Chicago, IL: Year Book Medical.

148. Shaw, J. A., et al. (1973). The longitudinal axis of the knee and the role of the cruciate ligaments in controlling transverse rotation. *The Journal of Bone & Joint Surgery*, 56(suppl A): 1603-1609.

149. Simpson, K. J., Kanter, L. (1997). Jump distance of dance landings influencing internal joint forces: I. axial forces. *Medicine & Science in Sports & Exercise*, 29:916-927.

150. Slocum, D. B., Larson, R. L. (1963). Pes anserinus transplantation: A surgical procedure for control of rotatory instability of the knee. *The Journal of Bone & Joint Surgery*, 50(suppl A):226-242.

151. Soderberg, G. L. (1986). *Kinesiology: Application to Pathological Motion*. Baltimore, MD: Lippincott Williams & Wilkins, 243-266.

152. Stacoff, A., et al. (2005). Ground reaction forces on stairs: effects of stair inclination and age. *Gait and Posture*, 21:24-38.

153. Stergiou, N., et al. (1999). Asynchrony between subtalar and knee joint function during running. *Medicine & Science in Sports & Exercise*, 31:1645-1655.

154. Stergiou, N., et al. (2003). Subtalar and knee joint interaction during running at various stride lengths. *The Journal of Sports Medicine and Physical Fitness*, 43:319-326.

155. Stormont, D. M., et al. (1985). Stability of the loaded ankle. Relation between articular restraint and primary and secondary static restraints. *American Journal of Sports Medicine*, 13:295-300.

156. Taunton, J. E., et al. (1985). A triplanar electrogoniometer investigation of running mechanics in runners with compensatory overpronation. *Canadian Journal of Applied Sports Science*, 10:104-115.

157. Taylor, W. R., et al. (2004). Tibio-femoral loading during human gait and stair climbing. *Journal of Orthopaedic Research*, 22:625-632.

158. Tehranzadeh, J., et al. (1982). Combined pelvic stress fracture and avulsion of the adductor longus in a middle distance runner. *American Journal of Sports Medicine*, 10:108-111.

159. Tropp, H. (2002). Commentary: Functional ankle instability revisited. *Journal of Athletic Training*, 37:512-515.

160. Valderrabano, V., et al. (2006). Ligamentous posttraumatic ankle osteoarthritis. *The American Journal of Sports Medicine*, 34:612-620.

161. VanDenBogert, A. J., et al. (1999). An analysis of hip joint loading during walking, running and skiing. *Medicine & Science in Sports & Exercise*, 31:131-142.

162. van Ingen Schenau, G. J., et al. (1995). The control of mono-articular muscles in multijoint leg extensions in man. *Journal of Physiology*, 484:247-254.

163. Visser, J. J., et al. (1990). Length and moment arm of human leg muscles as function of knee and hip-joint angles. *European Journal of Applied Physiology*, 61(5-6): 453-460.

164. Vleeming, A., et al. (1990). Relation between form and function in the sacroiliac joint: Part I. Clinical anatomical aspects. *Spine*, 15:130-132.

165. Wallace, L. A., et al. (1985). The knee. In J. Gould, G. J. Davies (Eds.). *Orthopaedic and Sports Physical Therapy*. St. Louis, MO: C.V. Mosby, 342-364.

166. Wang, C., et al. (1973). The effects of flexion and rotation on the length patterns of the ligaments of the knee. *Journal of Biomechanics*, 6:587-596.

167. Wright, D., et al. (1964). Action of the subtalar and ankle-joint complex during the stance phase of walking. *The Journal of Bone & Joint Surgery*, 46(suppl A):361-383.

168. Yack, H. J., et al. (1993). Comparison of closed and open kinetic chain exercise in the anterior cruciate ligament-deficient knee. *American Journal of Sports Medicine*, 21:49.

169. Yates, J. W., Jackson, D. W. (1984). Current status of meniscus surgery. *The Physician and Sportsmedicine*, 12:51-56.

170. Yu, B., et al. (2006). Lower extremity biomechanics during the landing of a stop-jump task. *Clinical Biomechanics*, 21:297-305.

171. Zajac, F. E. (2002). Understanding muscle coordination of the human leg with dynamical simulations. *Journal of Biomechanics*, 35:1011-1018.

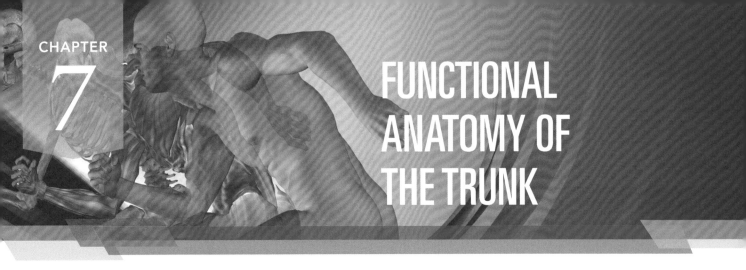

FUNCTIONAL ANATOMY OF THE TRUNK

OBJECTIVES

After reading this chapter, the student will be able to:

1. Identify the four curves of the spine, and discuss the factors that contribute to the formation of each curve.

2. Describe the structure and motion characteristics of the cervical, thoracic, and lumbar vertebrae.

3. Describe the movement relationship between the pelvis and the lumbar vertebrae for the full range of trunk movements.

4. Compare the differences in strength for the various trunk movements.

5. Describe specific strength and flexibility exercises for all of the movements of the trunk.

6. Describe some of the common injuries to the cervical, thoracic, and lumbar vertebrae.

7. Discuss the causes and sources of pain for the low back.

8. Discuss the influence of aging on trunk structure and function.

9. Identify the muscular contributions of the trunk to a variety of activities.

10. Explain how loads are absorbed by the vertebrae, and describe some of the typical loads imposed on the vertebrae for specific movements or activities.

OUTLINE

The Vertebral Column
    Motion Segment: Anterior Portion
    Motion Segment: Posterior Portion
    Structural and Movement Characteristics
      of Each Spinal Region
    Movement Characteristics of the
      Total Spine
    Combined Movements of the Pelvis
      and Trunk
Muscular Actions
    Trunk Extension
    Trunk Flexion
    Trunk Lateral Flexion
    Trunk Rotation
Strength of the Trunk Muscles
Posture and Spinal Stabilization
    Spinal Stabilization

Posture
Postural Deviations
Conditioning
    Trunk Flexors
    Trunk Extensors
    Trunk Rotators and Lateral Flexors
    Flexibility and the Trunk Muscles
    Core Training
Injury Potential of the Trunk
Effects of Aging on the Trunk
Contribution of the Trunk Musculature
    to Sports Skills or Movements
Forces Acting at Joints in
    the Trunk
Summary
Review Questions

The vertebral column acts as a modified elastic rod, providing rigid support and flexibility (48). The spine is a complex structure that provides a connection between the upper and lower extremities (64). There are 33 vertebrae in the vertebral column, 24 of which are movable and contribute to trunk movements. The vertebrae are arranged into four curves that facilitate support of the column by offering a springlike response to loading (37). These curves provide balance and strengthen the spine.

Seven cervical vertebrae form a convex curve to the anterior side of the body. This curve develops as an infant begins to lift his or her head; it supports the head and assumes its curvature in response to head position. The 12 thoracic vertebrae form a curve that is convex to the posterior side of the body. The curvature in the thoracic spine is present at birth. Five lumbar vertebrae form a curve convex to the anterior side, which develops in response to weight bearing and is influenced by pelvic and lower extremity positioning. The last curve is the sacrococcygeal curve, formed by five fused sacral vertebrae and the four or five fused vertebrae of the coccyx. Figure 7-1 presents the curvature of the whole spine as seen from the side and the rear.

The junction where one curve ends and the next one begins is usually a site of great mobility, which is also vulnerable to injury. These junctions are the cervicothoracic, thoracolumbar, and lumbosacral regions. Additionally, if the curves of the spine are exaggerated, the column will be more mobile, and if the curves are flat, the spine will be more rigid. The cervical and lumbar regions of the spinal column are the most mobile, and the thoracic and pelvic regions are more rigid (37).

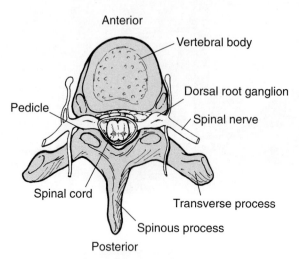

FIGURE 7-2 The vertebral column protects the spinal cord, which runs down the posterior aspect of the column through the vertebral foramen or canal. Spinal nerves exit at each vertebral level.

Besides offering support and flexibility to the trunk, the vertebral column has the main responsibility of protecting the spinal cord. As illustrated in Figure 7-2, the spinal cord runs down through the vertebrae in a canal formed by the body, pedicles, and pillars of the vertebrae, the disk, and a ligament (the ligamentum flavum). Peripheral nerves exit through the intervertebral foramen on the lateral side of the vertebrae, forming aggregates of nerve fibers and resulting in segmental innervations throughout the body.

The trunk, as the largest segment of the body, plays an integral role in both upper and lower extremity function because its position can significantly alter the function of the extremities. Trunk movement or position can be examined as a whole, or it can be examined by observing the movements or position of the different regions of the vertebral column or movement at the individual vertebral level. This chapter examines both the movement of the trunk as a whole and the movements and function within each region of the spine. The structural characteristics of the vertebral column are presented first, followed by an examination of the differences between the three regions of the spine: the cervical, thoracic, and lumbar.

The Vertebral Column

The functional unit of the vertebral column, the motion segment, is similar in structure throughout the spinal column, except for the first two cervical vertebrae, which have unique structure. The motion segment consists of two adjacent vertebrae and a disk that separates them (Fig. 7-3). The segment can be further broken down into anterior and posterior portions, each playing a different role in vertebral function.

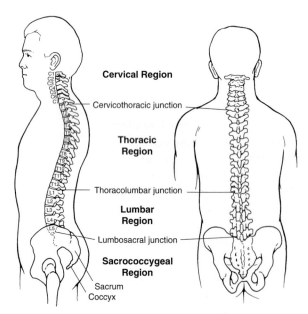

FIGURE 7-1 The vertebral column is both strong and flexible as a result of the four alternating curves. We are born with the thoracic and sacrococcygeal curves. The cervical and lumbar curves form in response to weight bearing and muscular stresses imposed on them during infancy.

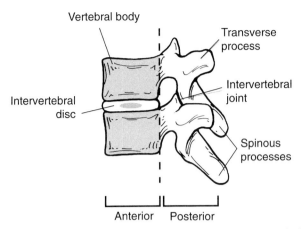

FIGURE 7-3 The vertebral motion segment can be divided into anterior and posterior portions. The anterior portion contains the vertebral bodies, intervertebral disk, and ligaments. The posterior portion contains the vertebral foramen, neural arches, intervertebral joints, transverse and spinous processes, and ligaments.

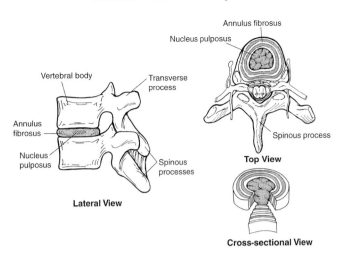

FIGURE 7-4 The intervertebral disk bears and distributes loads on the vertebral column. The disk consists of a gel-like central portion, the nucleus pulposus, which is surrounded by rings of fibrous tissue, the annulus fibrosus.

MOTION SEGMENT: ANTERIOR PORTION

The anterior portion of the motion segment contains the two bodies of the vertebrae, the intervertebral disk, and the anterior and posterior longitudinal ligaments. The two bodies and the disk separating them form a cartilaginous joint that is not found at any other site in the body.

Each vertebral body is tube shaped and thicker on the front side (15), where it absorbs large amounts of compressive forces. It consists of cancellous tissue surrounded by a hard cortical layer and has a raised rim that facilitates attachment of the disk, muscles, and ligaments. Also, the surface of the body is covered with hyaline cartilage, forming articular end plates into which the disk attaches.

Separating the two adjacent bodies is the intervertebral disk, a structure binding the vertebrae together while permitting movement between adjacent vertebrae. The disk is capable of withstanding compressive forces as well as torsional and bending forces applied to the column. The roles of the disk are to bear and distribute loads in the vertebral column and to restrain excessive motion in the vertebral segment. The load transmitted via the intervertebral disks distributes stress uniformly over the vertebral end plates and is also responsible for most of the mobility in the spine (32). Lateral, superior, and cross-sectional views of the disk are presented in Figure 7-4.

Each disk consists of the nucleus pulposus and the annulus fibrosus. The nucleus pulposus is a gel-like, spherical mass in the central portion of the cervical and thoracic disks and toward the posterior in the lumbar disks. The nucleus pulposus is 80% to 90% water and 15% to 20% collagen (12), creating a fluid mass that is always under pressure and exerting a preload to the disk. The nucleus pulposus is well suited for withstanding compressive forces applied to the motion segment.

During the day, the water content of the disk is reduced with compressive forces applied during daily activities, resulting in a shortening of the column by about 15 to 25 mm (1). The height and volume of the disks are reduced by about 20%, causing the disk to bulge radially outward and increase the axial loading on the posterior joints (1). At night, the nucleus pulposus imbibes water, restoring height to the disk. In elderly individuals, the total water content of the disk is lower (approximately 70%), and the ability to imbibe water is reduced, leaving a shorter vertebral column.

The nucleus pulposus is surrounded by rings of fibrous tissue and fibrocartilage, the annulus fibrosus. The fibers of the annulus fibrosus run parallel in concentric layers but are oriented diagonally at 45° to 65° to the vertebral bodies (39,95). Each alternate layer of fibers runs perpendicular to the previous layer, creating a crisscross pattern similar to that seen in a radial tire (35). When rotation is applied to the disk, half of the fibers tighten, and the fibers running in the other direction will be loosened.

The fibers that make up the annulus fibrosus consist of 50% to 60% collagen, providing the tensile strength in the disk (12). As a result of aging and maturation, the collagen is remodeled in the disk in response to changes in loading. This results in thicker annular fibers with higher concentrations of collagen fibers in the anterior disk area and thinner annular fibers in the lateral posterior portion of the disk because the fibers are less abundant. Fibers from the annulus fibrosus attach to the end plates of the adjacent vertebral bodies in the center of the segment and attach to the actual osseous material at the periphery of the disk (85). The fiber directions in the annulus fibrosus limit rotational and shearing motion between the vertebrae. The pressure on the tissue of the peripheral layer maintains the interspace between the end plates of the adjoining vertebrae (32). Tension is maintained in the annulus fibrosus by the end plates and by pressure exerted outward from the nucleus pulposus. The pressure tightens the outer layer and prevents radial bulging of the disk. Loss of disk tissue, such as occurs in aging, may impair spine function because of an increase in radial bulging, compression of the joints, or a reduction in space for the nerve tissue in the foramen (32).

The disk is both avascular and aneural, except for some sensory input from the outer layers of the annulus fibrosus. Because of this, healing of a damaged disk is unpredictable and not very promising.

The intervertebral disk functions hydrostatically when it is healthy, responding with flexibility under low loads and stiffly when subjected to high loads. When the disk is loaded in compression, the nucleus pulposus uniformly distributes pressure through the disk and acts as a cushion. The disk flattens and widens, and the nucleus pulposus bulges laterally as the disk loses fluid. This places tension on the annulus fibers and converts vertical compression force to tensile stress in the annulus fibers. The tensile stress absorbed by the annulus fibers is four to five times the applied axial load (60).

There are two weak points where disk injury is likely when subjected to high loads. First, the cartilage end plates, to which the disk is attached, are supported only by a thin layer of bone and thus are subject to fracture. Second, the posterior annulus is thinner and not attached as firmly as other portions of the disk, making it more vulnerable to injury (95).

The pressure in the disk increases linearly with increased compressive loads, with the pressure 30% to 50% greater than the applied load per unit area (15). The greatest change in disk pressure occurs with compression. During compression, the disk loses fluid, and the fiber angle increases (39). The disk is very resilient to the effects of a compressive force and rarely fails under compression. The cancellous bone of the vertebral body yields and fractures before the disk is damaged (39).

Movements such as flexion, extension, and lateral flexion generate a bending force that causes both compression and tension. With this asymmetrical loading, the vertebral body translates toward the loaded side, where compression develops, and the fibers are stretched on the other side, resulting in tension force.

In flexion, the vertebrae tilt anteriorly, forcing the nucleus pulposus posteriorly, creating a compression load on the anterior portion of the disk and a tension load on the posterior annulus. In extension the opposite occurs, as the upper vertebrae tilt posteriorly, driving the nucleus pulposus anteriorly and placing tensile pressure on the anterior fibers of the annulus.

In lateral flexion, the upper vertebrae tilt to the side of flexion, generating compression on that side and tension on the opposite side. Figure 7-5 illustrates disk behavior in flexion, extension, and lateral flexion.

As the trunk rotates, both tension and shear develop in the annulus fibrosus of the disk (Fig. 7-6). The half of the annulus fibers that are oriented in the direction of the rotation become taut, and the rest, which are oriented in

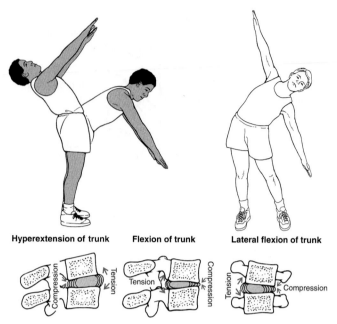

Hyperextension of trunk   **Flexion of trunk**   **Lateral flexion of trunk**

FIGURE 7-5 When the trunk flexes, extends, or laterally flexes, compressive force develops to the side of the bend and tension force develops on the opposite side.

the opposite direction, slacken. This increases the intradiscal pressure, narrows the joint space, and creates a shear force in the horizontal plane of rotation and tension in fibers oriented in the direction of the rotation. The peripheral fibers of the annulus fibrosus are subjected to the greatest stress during rotation (85).

The final structures of the anterior portion of the vertebral segment are the longitudinal ligaments running along the spine from the base of the occiput to the sacrum. The ligaments that act on the vertebral column are illustrated in Figure 7-7. The anterior longitudinal ligament is a very dense, powerful ligament that attaches to both the anterior disk and the vertebral bodies of the motion segment. This ligament limits hyperextension of the spine and restrains forward movement of one vertebra over another. It also maintains a constant load on the vertebral column and supports the anterior portion of the disk in lifting (35).

The posterior longitudinal ligament runs down the posterior surface of the vertebral bodies inside the spinal canal and connects to the rim of the vertebral bodies and the center of the disk. The posterolateral aspect of the segment is not covered by this ligament, adding to the vulnerability of this site for disk protrusion. It is broad in the cervical region and narrow in the lumbar region. This ligament offers resistance in flexion of the spine.

MOTION SEGMENT: POSTERIOR PORTION

The posterior portion of the vertebral motion segment includes the neural arches, intervertebral joints, transverse and spinous processes, and ligaments (Figs. 7-7 and 7-8).

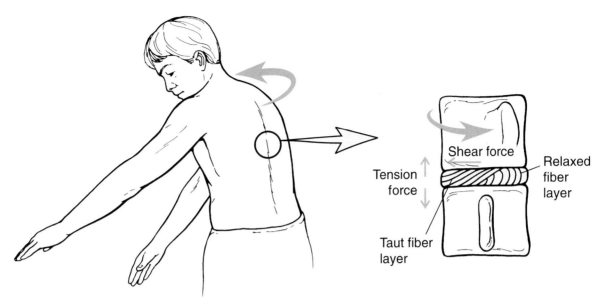

FIGURE 7-6 When the trunk rotates, half of the fibers of the annulus fibrosus become taut, and the rest relax. This creates tension force in the fibers running in the direction of the rotation and shear force across the plane of rotation.

The neural arch is formed by the two pedicles and two laminae, and together with the posterior side of the vertebral body, they form the vertebral foramen, in which the spinal cord is located. The bone in the pedicles and laminae is very hard, providing good resistance to the large tensile forces that must be accommodated. Notches above and below each pedicle form the intervertebral foramen, through which the spinal nerves leave the canal.

Projecting sideways at the union of the laminae and the pedicles are the transverse processes, and projecting posteriorly from the junction of the two laminae is the spinous process. The spinous and transverse processes serve as attachment sites for the spinal muscles running the length of the column.

The two synovial joints, termed the apophyseal joints, are formed by articulating facets on the upper and lower border of each lamina. The superior articulating facet is concave and fits into the convex inferior facet of the adjacent vertebra, forming a joint on each side of the vertebrae. The articulating facets are oriented at different angles in the cervical, thoracic, and lumbar regions of the spine, accounting for most of the functional differences between regions. These differences are discussed more specifically in a later section of this chapter.

The apophyseal joints are enclosed within a joint capsule and have all of the other characteristics of a typical synovial joint. Depending on the orientation of the facet joints, these joints can prevent the forward displacement of one vertebra over another and also participate in load bearing. In the hyperextended position, these joints bear 30% of the load (48). They also bear a significant portion of the load when the spine is flexed and rotated (30). Highest pressures in the facet joints occur with combined torsion, flexion, and compression of the vertebrae (10). The apophyseal joints protect the disks from excessive shear and rotation (1).

Five ligaments support the posterior portion of the vertebral segment (Fig. 7-7). The ligamentum flavum connects adjacent vertebral arches longitudinally, attaching laminae to laminae. This ligament has elastic qualities, allowing it to deform and return to its original length. It elongates with flexion of the trunk and contracts in extension. In the neutral position, it is under constant tension, imposing continual tension on the disk.

The supraspinous and the interspinous ligaments both run from one spinous process to another spinous process and resist both shear and forward bending of the spine. Finally, the intertransverse ligaments, connecting transverse process to transverse process, resist lateral bending of the trunk. The role of all of the intervertebral ligaments is to prevent excessive bending (1).

STRUCTURAL AND MOVEMENT CHARACTERISTICS OF EACH SPINAL REGION

Cervical Region

The cervical region has two vertebrae, the atlas (C1) and the axis (C2), that have structures unlike those of any other vertebra (Fig. 7-9). The atlas has no vertebral body and is shaped like a ring with an anterior and a posterior arch. The atlas has large transverse processes with transverse foramen through which blood supply travels. The atlas has no spinous process. Superiorly, it has a fovea, or dishlike depression, that holds the occiput of the skull.

The articulation of the atlas with the skull is called the atlantooccipital joint. At this joint, the head nods

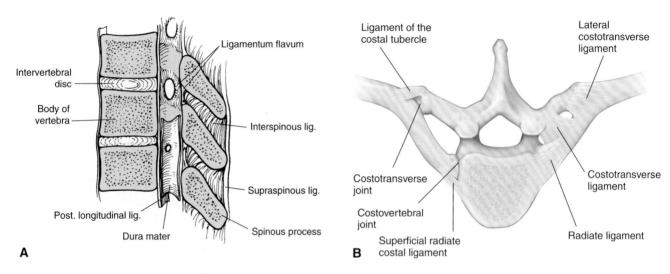

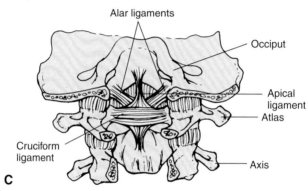

Ligment	Insertion	Action
Alar	Apex of dens TO medial occipital	Limits lateral flexion, rotation of head; holds dens in atlas
Apical	Apex of dens TO front foramen magnum	Holds dens in atlas and skull
Anterior longitudinal	Sacrum; anterior vertebral body and disc TO above anterior body and disc; atlas	Limits hyperextension of spine; limits forward sliding of vertebrae
Costotransverse	Tubercle of ribs TO transverse process of vertebrae	Supports rib attachment to thoracic vertebrae
Cruciform	Odontoid bone TO arch of atlas	Stabilizes odontoid, atlas; prevents posterior movement of dens in atlas
Iliolumbar	Transverse process TO spinous process	Limits lumbar motion in flexion, rotation
Interspinous	Spinous process TO transverse process	Limits flexion of trunk; limits shear forces acting on vertebrae
Intertransverse	Transverse process TO transverse process	Limits lateral flexion of trunk
Ligamentum flavum	Laminae TO laminae	Limits flexion of trunk; assists extension of trunk; maintains constant tension on disc
Ligamentum nuchae	Laminae TO laminae in cervical region; connects with supraspinous ligament	Limits cervical flexion; assists extension; maintains constant disc load
Posterior longitudinal	Posterior vertebral body and disc TO posterior body and disc of next vertebra	Limits flexion of trunk and lateral flexion
Radiate	Head of rib TO body of vertebra	Maintains rib to thoracic vertebra
Supraspinous	Spinous process TO spinous process of next vertebra	Limits flexion of trunk; resists forward shear force on spine

FIGURE 7-7 Ligaments of the spine. **(A)** There are a number of longitudinal ligaments that run the length of the spine. **(B)** The thoracic region of the spine has specialized ligaments that connect the ribs to the vertebrae. **(C)** In the cervical region, specialized ligaments connect the vertebrae to the occipital bone.

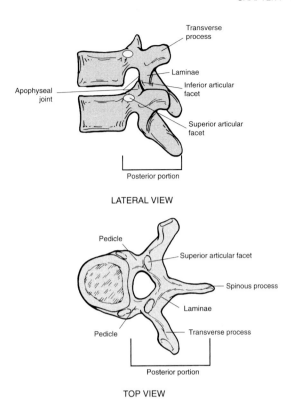

FIGURE 7-8   The posterior portion of the spinal motion segment is responsible for a significant amount of spinal support and restriction owing to its ligaments and structure. The posterior portion contains the only synovial joint in the spine, the apophyseal joint, which joins the superior and inferior facets of each vertebra.

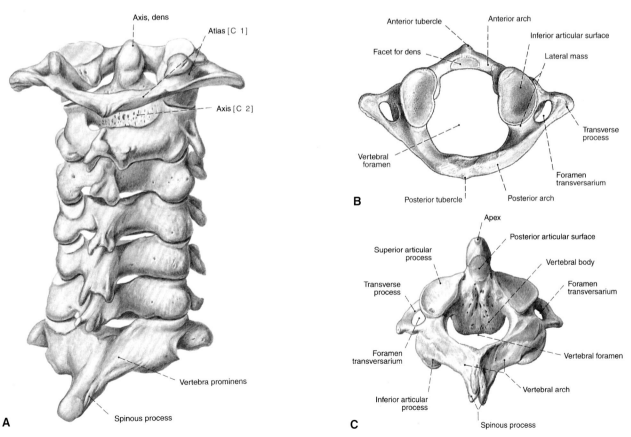

FIGURE 7-9   The cervical vertebrae **(A)** have two unique vertebrae, the atlas **(B)** and the axis **(C)**, that are very different from a typical vertebra **(D)** and have specialized functions of supporting the head. (Reprinted with permission from Sobotta J. 2001. In R. Putz, R. Pabst (Eds.). Atlas of Human Anatomy, Vol. 2, Trunk, Viscera, Lower Limb. Philadelphia, PA: Lippincott Williams & Wilkins, Figs. 720, 723, 725, 727.)

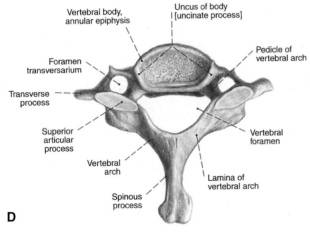

D

FIGURE 7-9 *(continued)*

on the spine because this joint allows free sagittal plane movements. This joint allows approximately 10° to 15° of flexion and extension (97) and no lateral flexion or rotation (87).

The weight of the head is transferred to the cervical spine via C2, the axis. The axis has a modified body with no articulating process on the superior aspect of the body and pedicles. Instead, the articulation with the atlas occurs via a pillar projecting from the superior surface of the axis that fits into the atlas and locks the atlas into a swivel or pivoting joint. The pillar is referred to as the odontoid process or dens.

The articulation between the atlas and the axis is known as the atlantoaxial joint and is the most mobile of the cervical joints, allowing approximately 10° of flexion and extension, 47° to 50° of rotation, and no lateral flexion (97). This joint allows us to turn our head and look from one side to the other. In fact, this articulation accounts for 50% of the rotation in the cervical vertebrae (97).

The remainder of the cervical vertebrae support the weight of the head, respond to muscle forces, and provide mobility. C3–C7 vertebrae have structures in the anterior and posterior compartments similar to those of the typical vertebrae. The bodies of the cervical vertebrae are small and about half as wide side to side as they are front to back. The cervical vertebrae also have short pedicles, bulky articulating processes, and short spinous processes. The transverse processes of the cervical vertebrae have a foramen where the arteries pass through. This is not found in other regions of the vertebral column. Figure 7-10 illustrates size, shape, and orientation differences across the regions of the spinal cord. A closer examination of structural differences between the cervical, thoracic, and lumbar vertebrae is presented in Figure 7-11.

The articulating facets in the cervical vertebrae face 45° to the transverse plane and lie parallel to the frontal plane (82), with the superior articulating process facing posterior and up and the inferior articulating processes facing anterior and down. In contrast to other regions of the vertebral column, the intervertebral disks are smaller laterally than the bodies of the vertebrae. The cervical disks are thicker ventrally than dorsally, producing a wedge shape and contributing to the lordotic curvature in the cervical region.

Because of the short spinous processes, the shape of the disks, and the backward and downward orientation of the articulating facets, movement in the cervical region is greater than in any other region of the vertebral column. The cervical vertebrae can rotate through approximately 90°, flex 20° to 45° to each side, flex through 80° to 90°, and extend through 70° (87). Maximum rotation in the cervical vertebrae occurs at C1–C2, maximum lateral flexion at C2–C4, and maximum flexion and extension at C1–C3 and C7–T1. Also, all cervical vertebrae move simultaneously in flexion.

In addition to the ligaments that support the whole vertebral column, some specialized ligaments are found in the cervical region. The locations and actions of these ligaments are presented in Figure 7-7.

Thoracic Region
One of the most restricted regions of the vertebral column is the thoracic vertebrae. Moving down the spinal column, the individual vertebrae increase in size; thus, the 12th thoracic vertebra is larger than the first one. The bodies become taller, and the thoracic vertebrae have longer pedicles than the cervical vertebrae (Fig. 7-11). The transverse processes on the thoracic vertebrae are long, and they angle backward, with the tips of the transverse processes posterior to the articulating facets. On the back of the thoracic vertebrae are long spinous processes that overlap the vertebrae and are directed downward rather than posteriorly, as in other regions of the spine.

The connection of the thoracic vertebrae to the ribs is illustrated in Figure 7-12. The thoracic vertebrae articulate with the ribs via articulating facets on the body of each vertebrae. Full facets are located on the bodies of T1 and T10–T12, and demifacets are located on T2–T9 to connect with the ribs. The thoracic vertebrae are supported by the ligaments presented earlier, along with four others that support the attachment between the ribs and the vertebral body and transverse processes (Fig. 7-7).

The apophyseal joints between adjacent thoracic vertebrae are angled at 60° to the transverse plane and 20° to the frontal plane, with the superior facets facing posterior and a little up and laterally and the inferior facets facing anteriorly, down, and medially (Fig. 7-11). Compared with the cervical vertebrae, the thoracic intervertebral joints are oriented more in the vertical plane.

The movements in the thoracic region are limited primarily by the connection with the ribs, the orientation of the facets, and the long spinous processes that overlap in the back. Range of motion in the thoracic region for flexion and extension combined is 3° to 12°, with very limited motion in the upper thoracic (2° to 4°) that increases in the lower thoracic to 20° at the thoracolumbar junction (10,97).

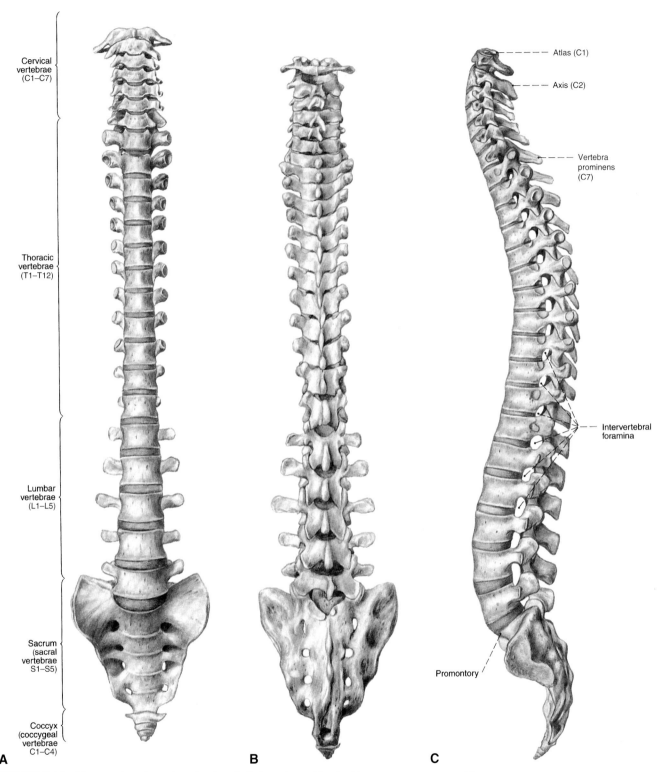

Cervical
vertebrae
(C1–C7)

Thoracic
vertebrae
(T1–T12)

Lumbar
vertebrae
(L1–L5)

Sacrum
(sacral
vertebrae
S1–S5)

Coccyx
(coccygeal
vertebrae
C1–C4)

Atlas (C1)

Axis (C2)

Vertebra
prominens
(C7)

Intervertebral
foramina

Promontory

A   **B**   **C**

FIGURE 7-10   The vertebrae in each region (cervical, thoracic, and lumbar) have common structural features with unique regional variations, as seen here in the anterior **(A)**, posterior **(B)**, and lateral **(C)** views. (Reprinted with permission from Sobotta J. [2001]. In R Putz, R. Pabst [Eds.]. *Atlas of Human Anatomy, Vol. 2, Trunk, Viscera, Lower Limb*. Philadelphia, PA: Lippincott Williams & Wilkins, Figs. 708–710.)

CERVICAL   THORACIC   LUMBAR

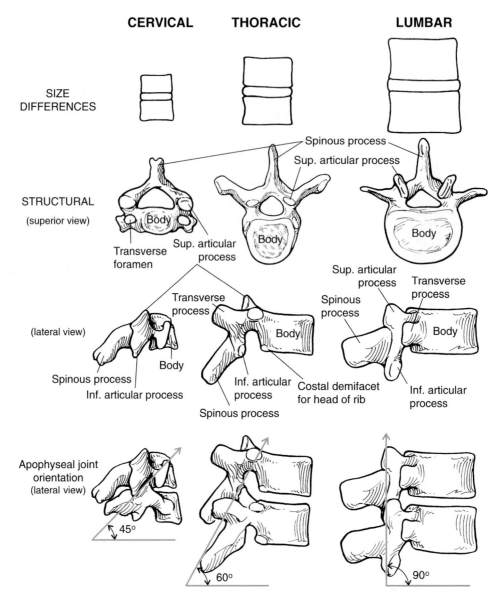

SIZE
DIFFERENCES

STRUCTURAL
(superior view)

Spinous process
Sup. articular process

Body

Body

Body

Transverse
foramen
Sup. articular
process

(lateral view)

Transverse
process

Sup. articular
process   Transverse
process
Spinous
process

Body   Body

Spinous process
Inf. articular process

Inf. articular
process

Spinous process

Costal demifacet
for head of rib

Inf. articular
process

Apophyseal joint
orientation
(lateral view)

45°

60°

90°

FIGURE 7-11   The cervical, thoracic, and lumbar vertebrae differ from each other. From the cervical to the lumbar region the bodies of the vertebrae become larger, and the transverse processes, spinous processes, and apophyseal joints all change their orientation.

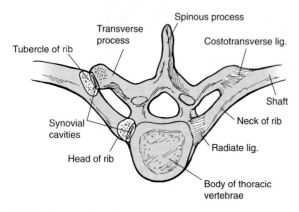

Spinous process
Transverse
process
Costotransverse lig.
Tubercle of rib

Shaft
Neck of rib
Synovial
cavities
Radiate lig.
Head of rib
Body of thoracic
vertebrae

FIGURE 7-12   The thoracic region is restricted in movement because of its connection to the ribs, which connect to a demifacet on the body of the thoracic vertebrae and a facet on the transverse process.

Lateral flexion is also limited in the thoracic vertebrae, ranging from 2° to 9° and again increasing as one progresses down through the thoracic vertebrae. Whereas in the upper thoracic vertebrae, lateral flexion is limited to 2° to 4°, in the lower thoracic vertebrae, it may be as high as 9° (10,97).

Rotation in the thoracic vertebrae ranges from 2° to 9°. Rotation range of motion is opposite to that of flexion and lateral flexion because it is maximum at the upper levels (9°) and is reduced at the lower levels (2°) (10,97).

The intervertebral disks in the thoracic region have a greater ratio of disk diameter to height of the disk than any other region of the spine. This reduces the tensile stress imposed on the vertebrae in compression by reducing the stress on the outside of the disk (60). Thus, disk injuries in the thoracic region are not as common as in other regions of the spinal column.

Lumbar Region

The large lumbar vertebra is the most highly loaded structure in the skeletal system. Figure 7-11 illustrates the characteristics of the lumbar vertebrae. The lumbar vertebrae are large, with wider bodies side to side than front to back. They are also wider vertically in the front than in the back. The pedicles of the lumbar vertebrae are short; the spinous processes are broad; and the small transverse processes project posteriorly, upward, and laterally. The disks in the lumbar region are thick; as in the cervical region, they are thicker ventrally than dorsally, contributing to an increase in the anterior concavity in the region. Frobin and colleagues (32) reported that the ventral disk height of the lumbar vertebrae remains fairly constant in the age range of 16 to 57 years, but there are gender differences and different disk heights at different levels of the vertebrae. The lumbar vertebrae are typically higher in males. Also, the highest disk height is found at L4–L5 and L5–S1.

The apophyseal joints in the lumbar region lie in the sagittal plane; the articulating facets are at right angles to the transverse plane and 45° to the frontal plane (97). The superior facets face medially and the inferior facets face laterally. This changes at the lumbosacral junction, where the apophyseal joint moves into the frontal plane and the inferior facet on L5 faces front. This change in orientation keeps the vertebral column from sliding forward on the sacrum.

The lumbar region is supported by the ligaments that run the full length of the spine and by one other, the iliolumbar ligament (Fig. 7-7). Another important support structure in the region is the thoracolumbar fascia, which runs from the sacrum and iliac crest up to the thoracic cage. This fascia offers resistance and support in full flexion of the trunk. The elastic tension in this fascia also assists with initiating trunk extension (35).

The range of motion in the lumbar region is large in flexion and extension, ranging from 8° to 20° at the various levels of the vertebrae (10,97). Lateral flexion at the various levels of the lumbar vertebrae is limited, ranging from 3° to 6°, and there is also very little rotation (1° to 2°) at each levels of the lumbar vertebrae (10,97). However, the collective range of motion in the lumbar region ranges from 52° to 59° for flexion, 15° to 37° for extension, 14° to 26° for lateral flexion, and 9° to 18° of rotation (93). A review of the range of motion at each level of the vertebral column is presented in Figure 7-13.

The lumbosacral joint is the most mobile of the lumbar joints, accounting for a large proportion of the flexion and extension in the region. Of the flexion and extension in the lumbar vertebrae, 75% may occur at this joint, with 20% of the remaining flexion at L4–L5 and 5% at the other lumbar levels (77).

MOVEMENT CHARACTERISTICS OF THE TOTAL SPINE

Motion in the spinal column is very small between each vertebra, but as a whole, the spine is capable of considerable

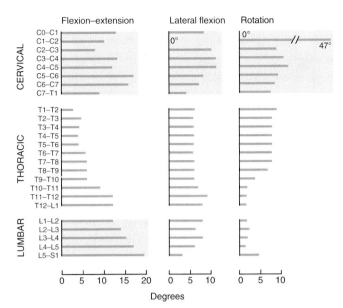

FIGURE 7-13 Range of motion at the individual motion segments of the spine is shown. The cervical vertebrae can produce the most range of motion at the individual motion segments. (Redrawn from White, A. A., Panjabi, M. M. [1978]. *Clinical Biomechanics of the Spine.* Philadelphia, PA: Lippincott Williams & Wilkins.)

range of motion. Motion is restricted by the disks and the arrangement of the facets, but motion can occur in three planes via active muscular initiation and control (90).

The movement characteristics of the total spine are presented in Figure 7-14. For the total spinal column, flexion and extension occur through approximately 110° to 140°, with free movement in the cervical and lumbar regions and limited flexion and extension in the thoracic region. The axis of rotation for flexion and extension lies in the disk unless there is considerable disk degeneration, which can move the axis of rotation out of the disk. Flexion of the whole trunk occurs primarily in the lumbar vertebrae through the first 50° to 60° and is then moved into more flexion by forward tilt of the pelvis (31). Extension occurs through a reverse movement in which first the pelvis tilts posteriorly and then the lumbar spine extends.

When flexion begins, the top vertebra slides forward on the bottom vertebra and the vertebra tilts, placing compressive force on the anterior portion of the disk. Both ligaments and the annulus fibers absorb the compressive forces.

On the back side, the superior portions of the apophyseal joints slide up on the lower facets, creating compression force between the facets and shear force across the face of the facets. These forces are controlled by the posterior ligaments, the capsules surrounding the apophyseal joints, posterior muscles, fascia, and posterior annulus fibers (85). The full flexion position is maintained and supported by the apophyseal capsular ligaments, intervertebral disks, supraspinous and interspinous ligaments, ligamentum flavum, and passive resistance from the back muscles, in that order (3).

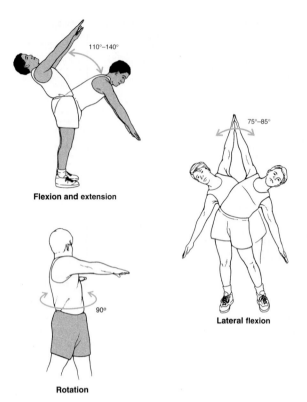

Flexion and extension

Rotation

Lateral flexion

FIGURE 7-14 The range of motion at the individual motion segment level is small, but in combination, the trunk is capable of moving through a significant motion range. Flexion and extension occur through approximately 110° to 140°, primarily in the cervical and lumbar region, with a very limited contribution from the thoracic region. The trunk rotates through 90°, with movement occurring freely in the cervical region, and with accompanying lateral flexion in the thoracic and lumbar regions. The trunk laterally flexes through 75° to 85°.

Lateral flexion range of motion is about 75° to 85°, mainly in the cervical and lumbar regions (Fig. 7-14). During lateral flexion, there is a slight movement of the vertebrae sideways, with disk compression to the side of the bend. Lateral flexion is often accompanied by rotation. In a relaxed stance, the accompanying rotation is to the opposite side of lateral flexion, that is, left rotation accompanying right lateral flexion.

If the vertebra is in full flexion, the accompanying rotation occurs to the same side, that is, right rotation accompanying right lateral flexion. This can vary by region of the spine. Also, an inflexible person usually performs some lateral flexion to obtain flexion in the trunk (2).

Rotation occurs through 90°, is free in the cervical region, and occurs in the thoracic and lumbar regions in combination with lateral flexion (Fig. 7-14). Generally, rotation is limited in the lumbar region. Right rotation in the thoracic or lumbar region is accompanied by some left lateral flexion.

The apophyseal joints are in a close-packed position in spinal extension, except for the top two cervical vertebrae, which are in a close-packed position in flexion. The total spine is in a close-packed position and is rigid during the military salute posture with the head up, shoulders back, and the trunk vertically aligned (35).

The flexibility of the regions of the trunk varies and is determined by the intervertebral disks and the angle of articulation of the facet joints. As pointed out earlier, mobility is highest at the junction of the regions. Mobility also increases in a region in response to restriction or rigidity elsewhere in the vertebral column.

COMBINED MOVEMENTS OF THE PELVIS AND TRUNK

The relationship of the movements of the pelvis to the movements of the trunk is discussed in Chapter 6. The movement synchronization between the pelvis and the trunk is referred to as the lumbopelvic rhythm. As shown in Figure 7-15, the lumbar curve reverses itself, flattens out (flexes), and curves in the opposite direction as trunk flexion progresses. This continues to a point at which the low back is rounded in full flexion of the trunk. Accompanying the movements in the lumbar vertebrae are flexion of the sacrum, anterior tilt of the pelvis, and extension of the sacrum. The pelvis also moves backward as weight is shifted over the hips.

Lumbar activity is maximum through the first 50° to 60° of flexion, after which anterior pelvic rotation becomes the predominant factor that increases trunk

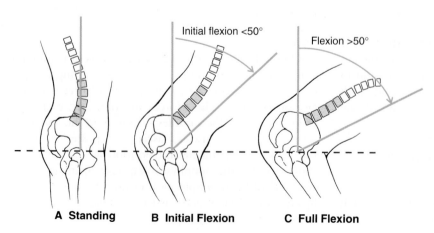

A Standing   **B Initial Flexion**   **C Full Flexion**

FIGURE 7-15 **(A)** In the normal standing posture, there is slight curvature in the lumbar region. **(B)** The first 50° of flexion takes place in the lumbar vertebrae as they flatten. **(C)** The continuation of flexion is a result of an anterior tilt of the pelvis.

flexion. On the return extension movement, pelvic posterior tilt dominates the initial stages of the extension, and lumbar activity reverses itself, dominating the later stages of trunk extension. The pelvis also moves forward as weight is shifted.

Movement relationships between the pelvis and the trunk during trunk rotation or lateral flexion are not as clear cut as in flexion and extension because of restrictions to the movement introduced by the lower extremity. The pelvis moves with the trunk in rotation and rotates right with trunk right rotation unless the lower extremity is forcing a rotation of the pelvis in the opposite direction. In this case, the pelvis may remain in the neutral position or rotate to the side exerting greater force.

Similarly, in lateral flexion of the trunk, the pelvis lowers to the side of the lateral flexion unless resistance is offered by the lower extremity, in which case the pelvis rotates to the opposite side (Fig. 7-16). The accompanying pelvic movements are determined by the trunk movement and the unilateral or bilateral positioning of the lower extremity.

The movement relationship between the pelvis and the trunk becomes complex when lower extremity movement, such as running, is performed in which the individual has different sequences of one limb moving on the ground and a limb moving off the ground. The lumbar spine flexes slightly and the pelvis tilts posteriorly during the loading phase with a quick reversal to lumbar extension and anterior pelvic tilt by midstance. Peak lumbar extension and anterior pelvic tilt occur right after toe-off. In the frontal plane, the spine laterally flexes to the right side, and the pelvis tilts to the left side during right foot contact and loading. This is followed by lumbar spine lateral flexion to the left side as the pelvis begins to elevate and tilt to the right until toe-off. Finally, the lumbar spine and the pelvis both rotate to the right with a right limb contact. The lumbar spine and pelvis rotate to the left during the support phase, but not at the same time (78).

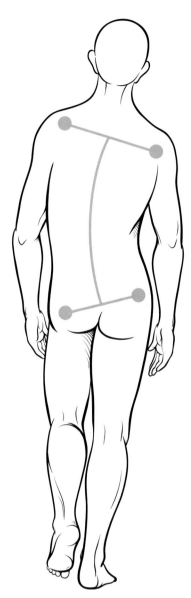

FIGURE 7-16 In walking and running, the trunk laterally flexes to the support side, but the pelvis lowers to the nonsupport side because of resistance offered by the lower extremity.

 ## Muscular Actions

Trunk extension is an important movement used to raise the trunk and to maintain an upright posture. The muscles typically get stronger as you come down the spine. The muscles actively used to extend the trunk also play dominant roles in trunk flexion; thus, it seems logical to first review the extensors.

TRUNK EXTENSION

The spinal extensors are graphically presented, and insertion, action, and nerve supply information is provided in Figure 7-17.

Numerous small muscles constitute the extensor muscle group. They can be classified into two groups, the erector spinae (iliocostalis, longissimus, and spinalis) and the deep posterior, or paravertebral, muscles (intertransversarii, interspinales, rotatores, and multifidus). These muscles run up and down the spinal column in pairs and create extension if activated as a pair or rotation or lateral flexion if activated unilaterally. Also, a superficial layer of muscle includes the trapezius and the latissimus dorsi. Although both the trapezius and the latissimus dorsi can influence trunk motion, they are not discussed in this chapter.

The three erector spinae muscles constitute the largest mass of muscles contributing to trunk extension. Extension is also produced by contributions from the deep vertebral muscles and other muscles specific to the region. These deep muscles contribute to trunk extension and other trunk movements, and they support the vertebral column, maintain rigidity in the column, and produce some of the finer movements in the motion segment (85).

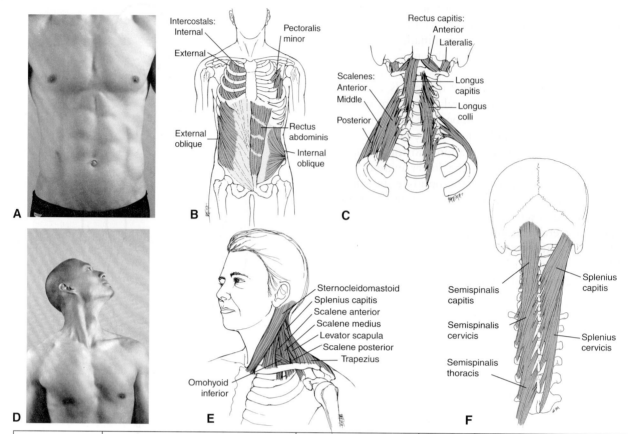

Muscle	Insertion	Nerve Supply	Flexion	Extension	Rotation	Lateral Flexion
External oblique (ABD)	Ninth to12th ribs alternating with l. dorsi, s. ant TO anterior superior spine; pubic tubercle, ant iliaccrest	Intercostal nerve; T7–T12	PM		PM: To opposite side	PM
Iliocostalis (ES) lumborum	Sacrum; spinous processes of L1–L5, T11, T12; iliac crest TO lower sixth or seventh ribs	Spinal nerves; dorsal rami		PM	PM: To same side	PM
Iliocostalis thoracis (ES)	Lower six ribs TO upper six ribs; transverse process of C7	Spinal nerves; dorsal rami		PM	PM: To same side	PM
Iliocostalis cervices (ES)	Third to sixth ribs TO transverse processes of C4–C6	Spinal nerves; dorsal rami		PM	PM: To same side	PM
Iliopsoas	Bodies of T12, L1–L5; transverse processes of L1–L5; inner surface of ilium, sacrum TO lesser trochanter	Femoral nerve; ventral rami; L1, L3	PM			
Internal oblique (ABD)	Iliac crest, lumbar fascia TO ribs 8–10; linea alba	Intercostal nerves; T7–T12, L1	PM		PM: To same side	PM
Interspinales (DP)	Spinous process TO spinous process	Spinal nerves; dorsal rami		PM		
Intertransversarii (DP)	Transverse process TO transverse process	Spinal nerves; ventral, dorsal rami		PM		PM
Longissimus (ES) thoracis	Posterior transverse process of L1–L5; thoraco-lumbar fascia TO transverse process of T1–T12	Spinal nerves; dorsal rami		PM	PM: To same side	PM
Longissimus cervices (ES)	Transverse process of T1–T5 TO transverse process of C4–C6	Spinal nerves; dorsal rami		PM	PM: To same side	PM
Longissimus capitis	Transverse process of T1–T5, C4–C7 TO mastoid process	Spinal nerves; dorsal rami		PM: head	PM: To same side	PM: Head
Longus capitis	Transverse process of C3–C6 TO occipital bone	Cervical nerves; C1–C3	PM: Cervical, head		PM: To same side	
Longus cervicis, colli	Transverse process of C3–C5; bodies of T1–T2; bodies of C5–C7, T1–T3 TO atlas; transverse process of C5–C6; bodies of C2–C4	Cervical nerves; C2–C7	PM: Cervical		PM: To same side	PM: Cervical
Multifundus (DP)	Sacrum; iliac spine; transverse processes L5–C4 TO spinous process of next vertebral side	Spinal nerves; dorsal rami		PM	PM: To opposite side	PM

FIGURE 7-17 Muscles acting on the spine, including the surface anatomy **(A)** and anterior muscles **(B)** of the trunk; deep anterior neck muscles **(C)**; surface anatomy **(D)** and muscles **(E)** of the lateral neck region; deep posterior muscles of the neck and upper back **(F)**, and surface anatomy **(G)** with corresponding superficial **(H)**, deep **(I)**, and pelvic **(J)** muscles of the posterior trunk.

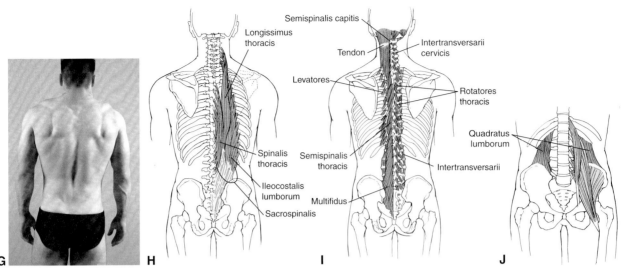

FIGURE 7-17   (continued)

Muscle	Insertion	Nerve Supply	Flexion	Extension	Rotation	Lateral Flexion
Quadratus lumborum	Iliac crest; transverse process of L2–L5 TO transverse process of L1–L2; last rib	Thoracic nerves; T12; lumbar nerves; ventral rami				PM
Rectus abdominis (ABD)	Fifth to seventh costal cartilage and xiphoid process TO pubic crest and syphysis	Intercostal nerve; T7–T12	PM			
Rotatores (DP)	Transverse process TO laminae of next vertebrae	Spinal nerves; dorsal rami		PM	PM: To opposite side	PM
Scaleni	Transverse process of cervical vertebrae TO ribs 1, 2	Cervical nerves	PM: Cervical		PM: Rotation to opposite side	PM: Cervical
Semispinalis capitis	C4–C6 facets; transverse process of C7 TO base of occipital flexion	Cervical nerves; dorsal rami		PM: Cervical	PM: Rotation to opposite side	PM: Cervical
Semispinalis cervicis	Transverse process of T1–T6 TO spinous process of C1–C5	Cervical nerves; dorsal rami		PM: Cervical	PM: Rotation to opposite side	PM: Cervical
Semispinalis thoracis	Transverse processes of T6–T10 TO spinous processes of T1–T4, C6, C7	Thoracic nerves; dorsal rami		PM	PM: Rotation to opposite side	PM
Spinalis thoracis (ES)	Spinous processes of L1–l2, T11–T12 TO spinous process of T1–T8 ligamentum nuchae	Spinal nerves; dorsal rami		PM	PM: To the same side	PM
Spinalis cervicis	Spinous process of C7 TO spinous process of axis	Spinal nerves; dorsal rami		PM		PM
Splenius capitis	Ligamentum nuchae; spinous process of C7, T1–T3 TO mastoid process, occipital bone	Cervical nerves; dorsal rami		PM: Cervical	PM: To the same side	PM: Cervical
Splenius cervicis	Spinous process of T3–T6 TO transverse process of C1–C3	Cervical nerves; dorsal rami		PM: Cervical	PM: To the same side	PM: Cervical
Sternocleidomastoid	Sternum, clavicle TO mastoid process	Accessory nerve; cranial nerve XI	PM: Head, cervical	PM: Rotation to opposite side	PM: Cervical	
Transverse abdominis (ABD)	Last 6 ribs; iliac crest; inguinal ligament; lumbo-dorsal fascia TO linea alba; pubic crest	Intercostal nerves; T7–T12, L1	*			

ABD, abdominals; DP, deep posterior muscles; ES, erector spinae.
* No specific action; increases internal abdominal pressure through compression.

There are some other muscles besides the erector spinae and the deep posterior muscle groups specific to each region. Figure 7-17 provides a full description of these muscles.

The erector spinae muscles are thickest in the cervical and lumbar regions, where most of the extension in the spine occurs. The multifidus is also thickest in the cervical and lumbar regions, adding to the muscle mass for generation of a trunk extension force.

The erector spinae and the multifidus muscles are 57% to 62% type I muscle fibers but also have types IIa and IIb fibers, making them functionally versatile so they can generate rapid, forceful movements while still resisting fatigue

for the maintenance of postures over long periods (89). In addition to providing the muscle force for extension of the trunk, these muscles provide posterior stability to the vertebral column, counteract gravity in the maintenance of an upright posture, and are important in the control of forward flexion (71).

TRUNK FLEXION

Flexion of the trunk is free in the cervical region, limited in the thoracic region, and free again in the lumbar region. Unlike the posterior extensor muscles, the anterior flexors do not run the length of the column. Flexion of the lumbar spine is created by the abdominals with assistance from the psoas major and minor. The flexion force of the abdominals also creates what little flexion there is in the thoracic vertebrae. The abdominals consist of four muscles: the rectus abdominis, internal oblique, external oblique, and transverse abdominis (see Fig. 7-17).

The internal and external oblique muscles and the transverse abdominis attach into the thoracolumbar fascia covering the posterior region of the trunk. When they contract, they place tension on the fascia, supporting the low back and reducing the strain on the posterior erector spinae muscles (9,71). The obliques are active in erect posture and in sitting, possibly stabilizing the base of the spine (83). The activity of the obliques drops off in a stooped standing posture as the load is transferred to other structures (83).

The transverse abdominis wraps around the trunk similar to a support belt and supports the trunk while assisting with breathing. The transverse abdominis applies tension to the linea alba, which is a fibrous connective tissue that runs vertically down the front that separates the rectus abdominis into right and left halves. If the linea alba is stabilized by transverse abdominis contraction, the obliques on the opposite side can act on the trunk. This muscle is also important for pressurizing the abdominal cavity (83) in activities such as coughing, laughing, defecation, and childbirth.

The abdominals consist of 55% to 58% type I fibers, 15% to 23% type IIa fibers, and 21% to 22% type IIb fibers (89). This fiber makeup, similar to that in the erector spinae muscles, allows for the same type of versatility in the production of short, rapid movements and prolonged movements of the trunk.

Two other muscles contribute to flexion in the lumbar region. First is the powerful flexor acting at the hip, the iliopsoas muscle, which attaches to the anterior bodies of the lumbar vertebrae and the inside of the ilium. The iliopsoas can initiate trunk flexion and pull the pelvis forward, creating lordotic posture in the lumbar vertebrae. Additionally, if this muscle is tight, an exaggerated anterior tilt of the pelvis may develop. If the tilt is not counteracted by the abdominals, lordosis increases, compressive stress on the facet joints develops, and the intervertebral disk is pushed posteriorly.

The second muscle found in the lumbar region is the quadratus lumborum, which forms the lateral wall of the abdomen and runs from the iliac crest to the last rib (Fig. 7-17). Although positioned to be more of a lateral flexor, the quadratus lumborum contributes to the flexion movement. It is also responsible for maintaining pelvic position on the swing side in gait (35).

When a person is standing or sitting upright, there is intermittent activity in both the erector spinae muscles and the internal and external obliques. The iliopsoas, on the other hand, is continuously active in the upright posture, but the rectus abdominis is inactive (81).

Flexion in the thoracic region, which is limited, is developed by the muscles of the lumbar and cervical regions. In the cervical region are five pairs of muscles that produce flexion if both muscles in the pair are contracting. If only one of the muscles in the pair contracts, the result is motion in all three directions, including flexion, rotation, and lateral flexion (85). The insertions, actions, and nerve supplies of these muscles are presented in Figure 7-17.

Flexion and Extension in the Spine

Have a partner stand straight and relaxed before slowly moving into a position of full flexion at the hip and back. Have them slowly return to the initial standing posture. All of the following structures play a role in this movement: posterior ligaments of the spine, abdominals and iliopsoas, extensors of the back, and extensors of the hip. Repeat the motion until you can determine the role of each of these structures and when they are being used.

Standing Toe-Touch

Movement into the fully flexed position from a standing posture is initiated by the abdominals and the iliopsoas muscles. After the movement begins, it is continued by the force of gravity acting on the trunk and controlled by the eccentric action of the erector spinae muscles. There is a gradual increase in the level of activity in the erector spinae muscles up to 50° to 60° of flexion as the trunk flexes at the lumbar vertebrae (6).

As the lumbar vertebrae discontinue their contribution to trunk flexion, the movement continues as a result of the contribution of anterior pelvic tilt. The posterior hip muscles, hamstrings, and gluteus maximus eccentrically work to control this forward tilt of the pelvis. As the trunk moves deeper into flexion, the activity in the erector spinae diminishes to total inactivity in the fully flexed position. In this position, the posterior ligaments and the passive resistance of the elongated erector spinae muscles control and resist the trunk flexion (48). The load on the ligaments in this fully flexed position is close to their failure strength (31), placing additional importance on loads sustained by the thoracolumbar fascia and the lumbar apophyseal joints.

As the trunk rises back to the standing position through extension, the movement is initiated by a contraction of the posterior hip muscles, gluteus maximus, and hamstrings, which flex and rotate the pelvis posteriorly. The erector spinae are active initially but are most active through the last 45° to 50° of the extension movement (71).

The erector spinae muscles are more active in the raising phase than in the lowering phase, being very active in the initial parts of the movements and again at the end of the extension movement, with some diminished activity in the middle of the movement. The abdominals can also be active in the return movement as they serve to control the extension movement (48).

TRUNK LATERAL FLEXION

Lateral flexion of the spine is created by contraction of muscles on both sides of the vertebral column, with most activity on the side to which the lateral flexion occurs. The most activity in lateral flexion of the trunk occurs in the lumbar erector spinae muscles and the deep intertransversarii and interspinales muscles on the contralateral side. The multifidus muscle is inactive during lateral flexion. If load is held in the arm during lateral flexion, there is also an increase in the thoracic erector spinae muscles on the opposite side.

The quadratus lumborum and the abdominals also contribute to lateral flexion. The quadratus lumborum on the side of the bend is in a position to make a significant contribution to lateral flexion. The abdominals also contract as the lateral flexion is initiated and remain active to modify the lateral flexion movement.

In the cervical spine, lateral flexion is further facilitated by unilateral contractions of the sternocleidomastoid, scalenes, and deep anterior muscles. Lateral flexion is quite free in the cervical region.

TRUNK ROTATION

The rotation of the trunk is more complicated in terms of muscle actions because it is produced by muscle actions on both sides of the vertebral column. In the lumbar region, the multifidus muscles on the side to which the rotation occurs are active, as are the longissimus and iliocostalis on the other side (8). The abdominals exhibit a similar pattern because the internal oblique on the side of the rotation is active, and the external oblique on the opposite side of the rotation is also active.

 ## Strength of the Trunk Muscles

The greatest strength output in the trunk can be developed in extension, averaging values of 210 Nm (newton-meters) for males (56). Reported trunk flexion strength is 150 Nm, or approximately 70% of the strength of the extensors. Lateral flexion is 145 Nm, or 69% of the extensor strength, and rotation strength is 90 Nm, or 43% of the extensor values (56). Female strength values are approximately 60% of the values recorded for males. In fact, other studies have shown women to be capable of generating only 50% of the lifting force of men for lifts low to the ground and 33% of the male lifting force for lifts high off the ground (99). In the cervical region, women have demonstrated as much as 20% to 70% less strength than men (19).

Taking into consideration all things such as forces generated by intra-abdominal pressure, ligaments, and other structures, the total extensor moment is slightly greater than the flexor moment (72). The abdominals contribute to one-third of the flexor moments, and the erector spinae contribute half of the extensor moments. In rotation, the abdominals dominate, with some contribution by the small posterior muscles (72).

Trunk position plays a significant role in the development of strength output in the various movements. Trunk flexion strength, measured isometrically, has been shown to improve by approximately 9% when measured from a position of 20° of hyperextension (85). Isometric trunk extension strength, measured from a position of 20° of trunk flexion, is 22% greater compared with a 20° trunk flexion position (85). Higher trunk flexion and extension strength values can also be achieved if the measurement is made with the person seated rather than supine or prone.

The strength of the trunk is significantly altered in a dynamic situation. There is a reported 15% to 70% increase in trunk moments during dynamic exertions accompanied by increases in antagonist and agonist muscle activity, an increase in intra-abdominal pressure, an increase in spinal load, and a reduction in the capacity of the muscles to respond to external loads (24). Because of higher levels of coactivity, there is greater loading on the spinal structures without contributing to the ability to offset external moments. It is suggested that trunk velocity and acceleration, especially in multiple directions, may be more accurate discriminators of low-back disorders than just range of motion because of the diminished strength and functional capacity that accompany the coactivation in faster dynamic movements (24).

Strength output while lifting an object using the trunk extensors also diminishes when there is greater horizontal distance between the feet and the hands placed on the object (33). In fact, the forces applied vertically to an object held away from the body are about half those of a lift completed with the object close to the body. Additionally, an increase in the width of a box decreases lifting capacity, and an increase in the length of a box has been shown to have no influence (33).

Lifting an object by pulling up at an angle reduces the load at the elbow, shoulder, and lumbar and hip regions but increases it at the knees and ankles. This type of lift decreases the compressive force on the lumbar vertebrae 9% to 15% (34). Also, 16% more weight can be lifted by a more freestyle lift than the traditional straight-back, bent-knee lift (34).

Posture and Spinal Stabilization

Efficiency of motion and stresses imposed on the spine are very much determined by the posture maintained in the trunk as well as trunk stability. Positioning of the vertebral segments is so important that a special section on posture and spinal stabilization is warranted.

SPINAL STABILIZATION

The spine is stabilized by three systems, including a passive musculoskeletal system, an active musculoskeletal subsystem, and the neural feedback system (67). The passive subsystem includes the vertebrae, facet articulations, joint capsules, intervertebral disks, and spinal ligaments. The active system includes the muscles and tendons that stabilize the spine, and the neural subsystem provides control. Stability in the spine increases and decreases with the demands placed on the structure. Stability decreases during periods of decreased muscle activity and increases when joint compressive forces increase (20). The smaller deep muscles of the spine control posture and the relationship of each vertebra to each other (73), and the larger, superficial muscles move the spine and disperse loads from the thorax to the pelvis. For stability, the spine requires activity in the small postural muscles to be stable.

Muscles that play an important role in spinal stabilization include the transverse abdominis, multifidus, erector spinae, and internal oblique. The transverse abdominis circles the trunk like a belt and increases intra-abdominal pressure and spinal stiffening. It is one of the first muscles to be active in both unexpected and self-loaded conditions (22). The multifidus is organized to act at the level of each vertebra and is active continuously in upright positions (92) and can make subtle adjustments to the vertebrae in any posture (51). The erector spinae is better suited for control of spinal orientation by nature of its ability to produce extension (73). Finally, the internal oblique works with the transverse abdominis to increase intra-abdominal pressure.

POSTURE

Standing Posture

To maintain an upright posture in standing, the S-shaped spine acts as an elastic rod in supporting the weight. A continuous forward bending action is imposed on the trunk in standing because the center of gravity lies in front of the spine. As a result of the forward bending action on the trunk, the posterior muscles and ligaments must control and maintain the standing posture.

There is more erector spinae activity in an erect posture than in a slouched posture. In the slouched posture, most of the responsibility for maintaining the posture is passed onto the ligaments and capsules. Any disruption in the standing posture or any postural swaying is controlled and brought back into alignment by the erector spinae, abdominals, and psoas muscles (66). All of these muscles are slightly active in standing, with more activity in the thoracic region than the other two regions (8).

Sitting Posture

Posture in the sitting position requires less energy expenditure and imposes less load on the lower extremity than standing (Fig. 7-18). Prolonged sitting, however, can have deleterious effects on the lumbar spine (85). Unsupported sitting is similar to standing, such that there is high muscle activity in the thoracic regions of the trunk with accompanying low levels of activity in the abdominals and the psoas muscles (7).

The unsupported sitting position places more load on the lumbar spine than standing because it creates a backward tilt, a flattening of the low back, and a corresponding forward shift in the center of gravity (48). This places load

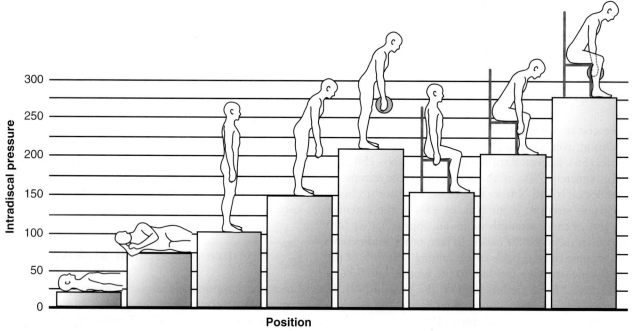

FIGURE 7-18   Compared with an erect standing posture, lower back loading is lowest in the supine position. Loading increases in a seated posture and increases more when leaning forward or slouching.

on the disks and the posterior structures of the vertebral segment. Sitting for a long time in the flexed position may increase the resting length of the erector spinae muscles (71) and overstretch the posterior ligamentous structures. A slouched sitting posture generates the largest disk pressures. Higher seating height can decrease the compressive force on the disks because of a more vertical posture, but increased loads are put on the lower extremity.

Working Postures

Biomechanical factors related to static work postures; seated and standing work postures; frequent bending and twisting; and lifting, pulling, and pushing are some of the risk factors for back injuries. Because of the high incidence of back-related injuries in the workplace, it is important to understand both causes and preventive measures.

Working postures can greatly influence the accumulation of strain on the low back (54), and both standing and sitting postures have appropriate uses in the workplace (28). A standing posture is preferred when the worker cannot put his or her legs under the work area and when more strength is required in the work task such as lifting and applying maximal grip forces. Strain can be reduced in a standing posture by using floor mats, using a foot rest, making sure the work station has adequate foot clearance, and wearing proper shoes (28). Muscle fatigue can also be reduced with several short breaks over the course of the work day. One of the most important factors for both standing and sitting is to avoid prolonged static postures.

Continuous flexion positions are a cause of both lumbar and cervical flexion injuries in the workplace. These postures can be eliminated by raising the height of the work station so that no more than 20° of flexion is present (85). The use of a footrest can also relieve strain.

Lifting tasks in the workplace can be the source of the low back pain, so guidelines should be established to reduce the risk. For example, the weights of objects being lifted should be lowered as lift frequency, lift distance, and object size increases. Weights should be lowered when lifting above shoulder height. Proper lifting technique, which involves maintaining a neutral spine, keeping the load close to the pelvis, avoiding trunk flexion and extension, and lifting with the lower extremity with a controlled velocity, is optimal for most tasks. In cases in which an object is awkwardly placed or when an object is in motion before the lift, it may be necessary to use a jerking motion. This places less torque on the lumbar extensor muscles (54).

Similar to the standing and working postures, the risk of injury from lifting can be minimized with regular breaks and by varying the work tasks (54). A fully flexed spine should be avoided in any lift because of the changes imposed on the major lumbar extensors. The fully flexed spine reduces the moment arms for the extensor muscles, decreases the tolerance to compressive loads, and transfers load from the muscles to the passive tissues (53). The workplace also has a high incidence of twisting, in which

the spine undergoes combined flexion and lateral bending. This posture maximally stretches the posterolateral structures, particularly the annulus (39). Twisting in the upright posture is limited by contact at the facet joints, but twisting in a flexed posture disengages the facet joints and shifts the resistance to the annulus fibrosus (50,70).

In seated work environments, a well-designed chair is important for providing optimal support because unsupported sitting results in disk pressures that are 40% higher than a standing posture (28). Prolonged sitting in a slouched flexion position maximally loads the iliolumbar ligaments because the loss of lumbar lordosis and the positioning of the upper body weight behind the ischial tuberosities (84). In supported sitting, the load on the lumbar vertebrae is lessened. A chair back reclined slightly backward and including lumbar support creates a seated posture that produces the least load on the lumbar region of the spine.

A lumbar backrest with free shoulder space is recommended for reducing some of the load. Higher backrests may not be effective because the ribcage is stiff. Backrests above the level just below the scapulae are not necessary (83). The work setting should be evaluated to determine high-risk lifting tasks such as repetitive bending, twisting, pushing, pulling, and lift and carry tasks.

POSTURAL DEVIATIONS

Postural deviations in the trunk are common in the general population (Fig. 7-19). In the cervical region, the curve is concave to the anterior side. This curve should be small and lie over the shoulder girdle. The head should be above the shoulder girdle. When the cervical curve is accentuated to the anterior side, lordosis is said to be

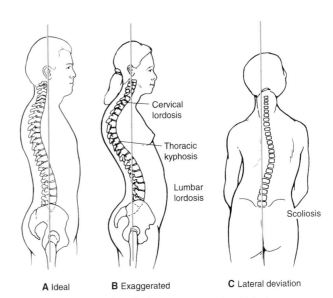

A Ideal   **B** Exaggerated   **C** Lateral deviation

FIGURE 7-19 (A) The ideal posture is one in which the curves are balanced but not exaggerated. **(B)** Curves can become exaggerated. **(C)** Lateral deviation of the spine, scoliosis, can create serious postural malalignment throughout the whole body.

present. Thus, cervical lordosis is an increase in the curve in the cervical region, often concomitant with exaggerated curves in other regions of the spine.

In the thoracic region, the curvature is concave to the posterior side. A rounded-shoulder posture may cause thoracic kyphosis, a common postural disorder in this region. The kyphotic thoracic region is also associated with osteoporosis and several other disorders.

The lumbar region, curving anteriorly, is subjected to forces that may be created by an exaggerated lumbar curve, termed lumbar lordosis or hyperlordosis. This accentuated swayback position is often created by anterior positioning of the pelvis or by weak abdominals. In the lumbar region, it is also not uncommon to have a flat back with decreased lumbar curve. This has been associated with a pelvis that is inclined upward at the front or with muscle tightness and rigidity in the spine.

The most serious of the postural disorders affecting the spine is scoliosis, a lateral deviation of the spine. The curve can be C shaped or S shaped depending on the direction and the beginning and ending segments. C-shaped scoliosis is designated when the deviation occurs in one region only. For example, a convex curve to the left in the cervical region is a left cervical C-shaped curve. In an S-shaped curve, the lateral deviation occurs in different regions and in opposite directions, as with a right thoracic, left lumbar convexity. Rotation can accompany the lateral deviation, creating a very complex postural malalignment. The cause of scoliosis is unknown, and it is more prevalent in females than in males.

Conditioning

The muscles around the trunk are active during most activities as they stabilize the trunk, move the trunk into an advantageous position for supplementing force production, or assist the limb movement. Because the low back is a common site of injury in sports and in the workplace, special attention should be given to exercises that strengthen and stretch this part of the trunk. Endurance training for the back muscles may be one of the better avenues for preventing back injury (54). Trunk exercises should also be evaluated for negative impact on trunk function and structure. A sample of trunk exercises is illustrated in Figure 7-20.

Trunk exercises should take place with the spine in the neutral position and use co-contraction of the abdominals. Co-contraction of the abdominals and the erector spinae increases spinal stiffness and stability, allowing for a better response to spinal loading (94). Spinal compression is increased with co-contraction, however, so the levels of co-contraction may need to be lessened for individuals with back pain who would be negatively affected by more compression.

Exercises creating excessive lordosis or hyperextension of the lumbar vertebrae should be avoided because they put excessive pressure on the posterior element of the spinal segment and can disrupt the facets or the posterior arch. Examples of such exercises are double-leg raises, double-leg raises with scissoring, thigh extension from the prone position, donkey kicks, back bends, and ballet arches. When selecting an exercise for the trunk, one should pay attention to its risks. The supine position produces the least amount of load on the lumbar vertebrae. The load in the supine position increases substantially, however, if the abdominals and the iliopsoas are activated.

TRUNK FLEXORS

The trunk flexors are usually exercised with some form of trunk or thigh flexion exercise from the supine position so that these muscles can work against gravity. Trunk flexion exercises should be evaluated for their effectiveness and safety. Three variations of trunk flexor exercises are shown in Figure 7-21. The abdominals are more active in trunk-raising activities than in double leg-raising activities (5). However, abdominal activity is important in leg-raising exercises to stabilize the pelvis and maintain rigidity in the trunk. Altering leg position between a bent or straight leg does not appear to significantly change the total muscle activity levels in the abdominals (5). The bent-knee positions do engage the hip flexors to a higher degree because of the reduced moment arm and decreased force capacity of the iliopsoas.

There is no one single best exercise for all abdominals at once. The best exercise for the rectus abdominis that maximizes activation and minimizes psoas activation is the curl, and the best exercise for the obliques is a side-supported position held either isometrically or dynamically (44).

TRUNK EXTENSORS

Extension of the trunk is usually developed through some type of lift using the legs and back. Figure 7-20 provides some examples of extensor exercises for stretching and strengthening these muscles.

Two basic types of lifts, the leg lift and the back lift, activate the erector spinae. The leg lift is the squat or deadlift exercise in which the back is maintained in an erect or slightly flexed posture and the knees are flexed. This lift has the least amount of erector spinae activity and imposes the lowest shear and compressive forces on the spine (74). The leg lift is begun with posterior tilt of the pelvis initiated by the gluteus maximus and the hamstrings. The erector spinae can be delayed and not involved until later in the leg lift, when the extension is increased. The delay is related to the magnitude of the weight being lifted, and the muscles usually do not become active until the initial acceleration is completed (6). Because there is considerable stress on the ligaments at the beginning of the lift, it is suggested that the erector spinae activity begin at the initial part of the lift to stabilize the back (48).

Muscle Group	Sample Stretching Exercises	Sample Strengthening Exercises	Other Exercises
Flexors	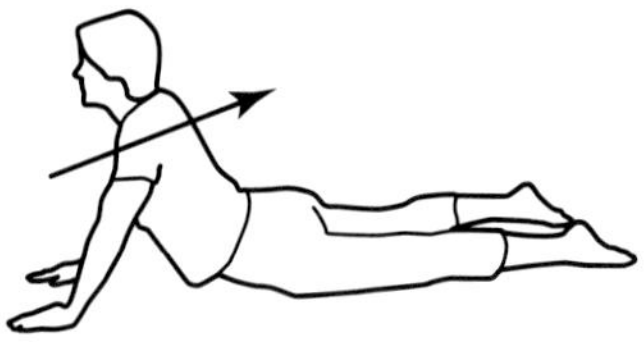 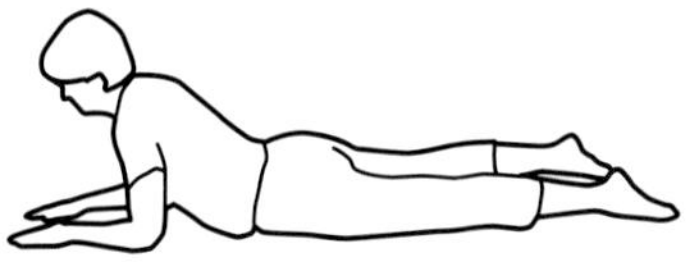  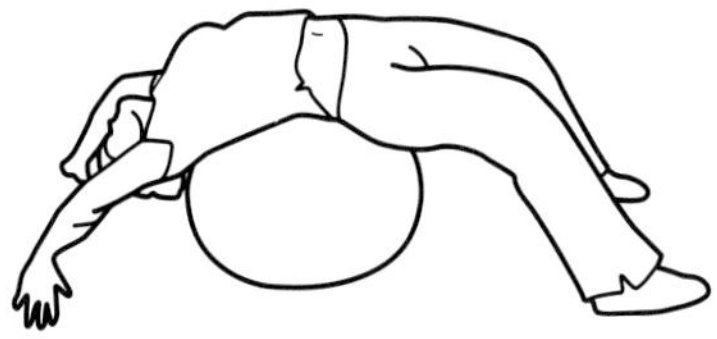	Captain's chair 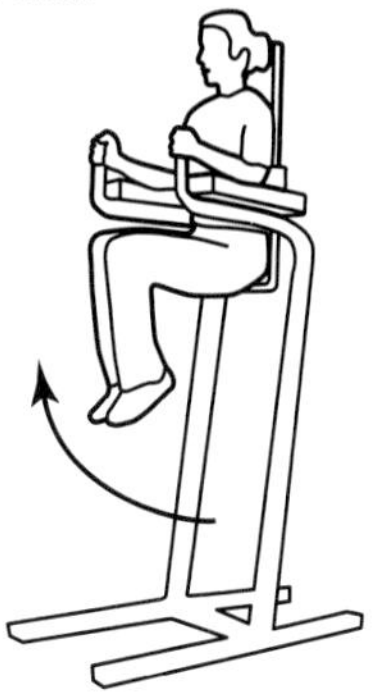 Abdominal machine  Weighted sit-ups 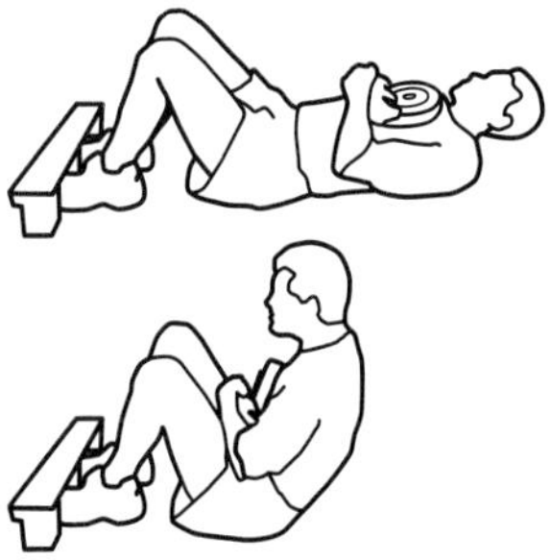	Crunch on stability ball Cable crunches Hanging leg raise Reverse sit-up Side bends with weight
Extensors	  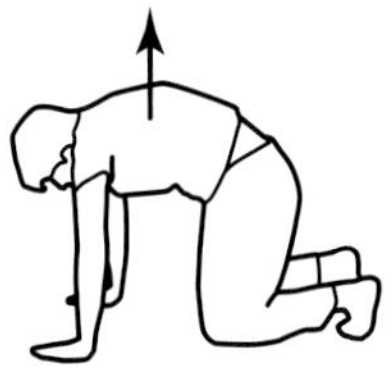	Roman chair 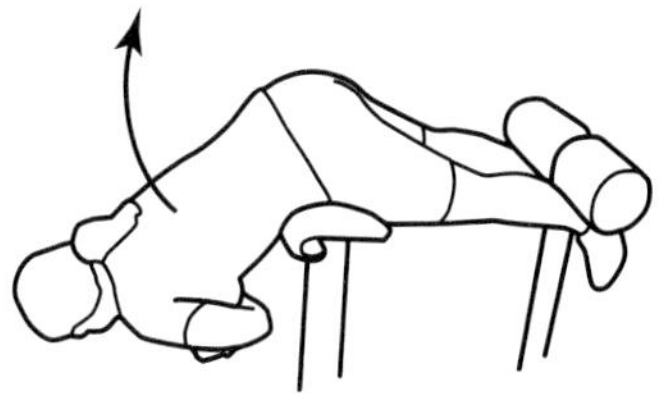 Back extension machine 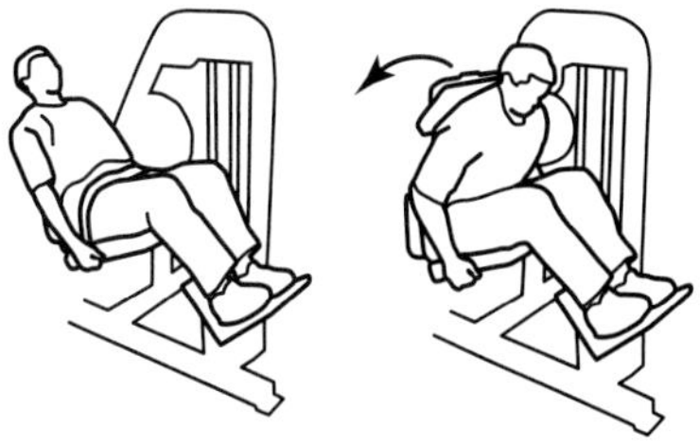	Seated row Hyperextension floor or bench Deadlift

FIGURE 7-20 Sample stretching and strengthening exercises for selected muscle groups. (*continued*)

Muscle Group	Sample Stretching Exercises	Sample Strengthening Exercises	Other Exercises
Extensors (cont.)	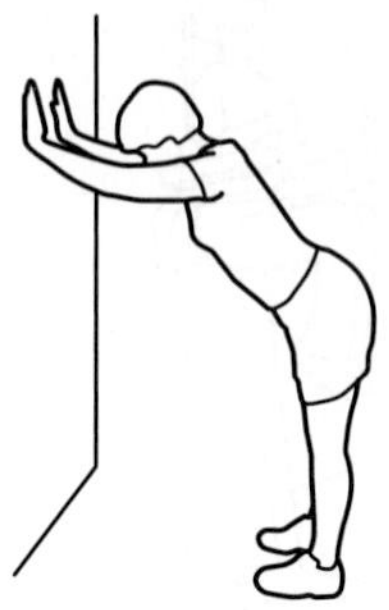	Good morning exercise 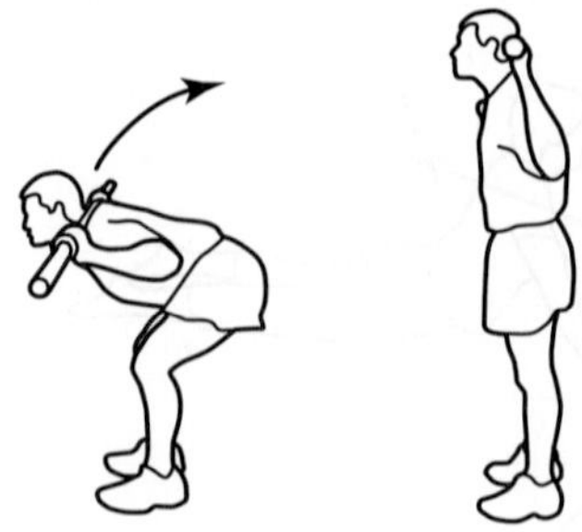	
Rotators	Hip roll 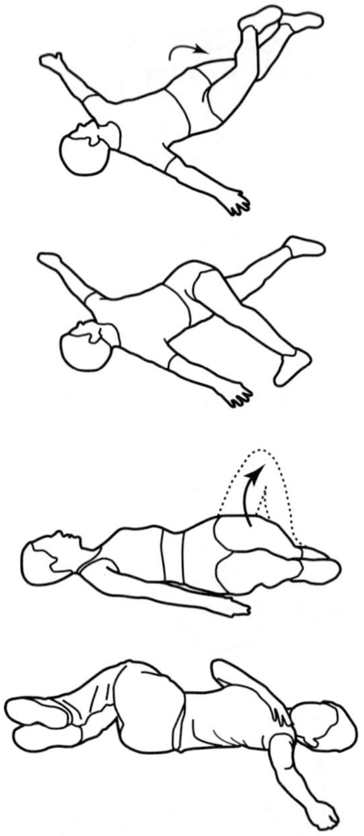	Bicycle 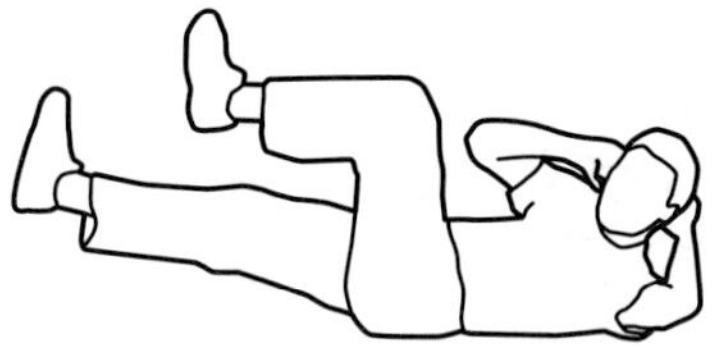 Standing rotations 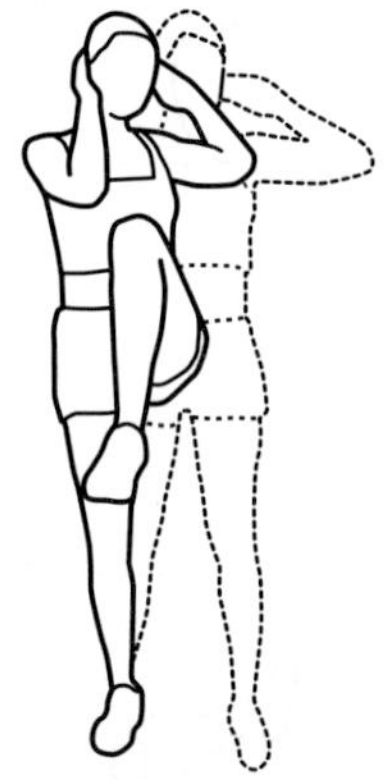	Reverse trunk twist
Lateral flexors		Dumbbell side bends 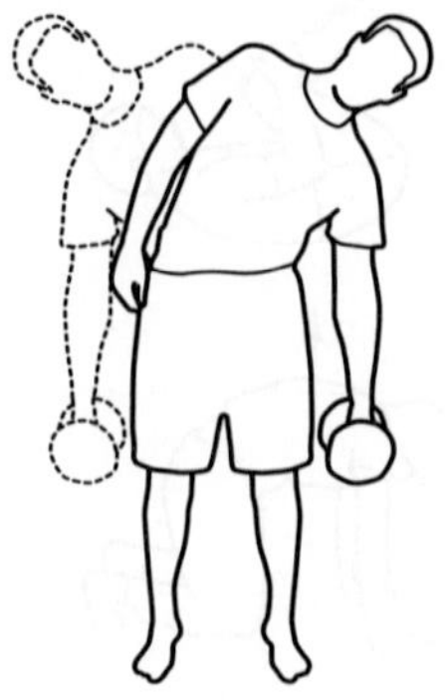	

FIGURE 7-20   (continued)

Exercise	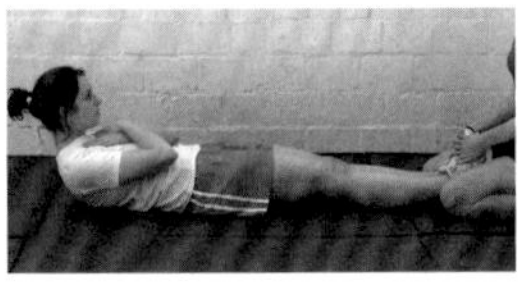			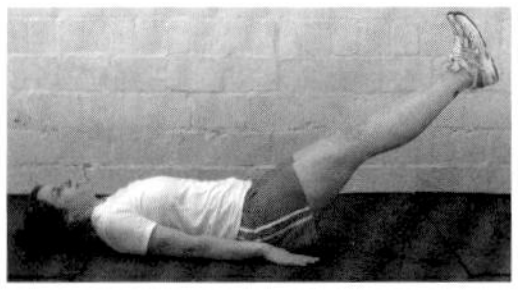
Characteristics	Curl-up: Lumbar in contact with the floor and no hip movement	Hip flexion sit-up: Lifting straight upper body with hip flexion	Sit-up: Combination of trunk and hip flexion; flex trunk first then flex hip	Leg-lift: One or two legs
Muscle Activity	• Moderate to high activity in the abdominals ○ Highest activity in rectus abdominis ○ Lowest activity in external oblique ○ Low activity in hip flexors ▪ Slightly higher with bent knees • No significant difference with holding feet or straight or bent knees	• Higher activity in abdominals than in curl-up • Even though external oblique activity is less than rectus abdominis and internal oblique, external oblique more active than in curl-up • High hip flexor activity • No change in abdominal activity with legs supported or unsupported	• Similar activity seen in hip flexion sit-up	• In bilateral leg lift, there is high activity in rectus femoris and external oblique • Little activity in abdominals if only one leg is lifted • High activity in hip flexors ○ Higher than in other activities with unsupported limbs ○ Rectus femoris activity similar to sit-up

FIGURE 7-21 The abdominals and hip flexors are used in different ways in exercises depending on the position of the trunk and hip joint.

In the back lift, the person bends over at the waist with the knees straight, as in the "good morning" exercise. This exercise creates the highest shear and compressive stress on the lumbar vertebrae, but the erector spinae are much more active in this type of exercise (74). In the back lift, the movement is initiated by the hamstrings and the gluteus maximus and then followed up by activity from the erector spinae. Extension of the spine begins approximately one-third of the way into the lift (6). In performing the back lift with no load, the erector spinae becomes active after the beginning of the lift, but if the lift is performed with weight, the erector spinae is active before the start of the lift (88).

In comparing leg lifts and back lifts, one must consider both the risks and the gains. The back lift imposes a greater risk of injury to the vertebrae because of the higher forces imposed on the system. Any stooping posture of the trunk imposes greater compression forces on the spine; consequently, a trunk flexion posture in a lift should be discouraged (23). The disk pressures are much higher in the back lift than in the leg lift, mainly because of the trunk position and distance (62). The erector spinae activity in the back lift is greater than that in the leg lift.

Trunk extensor activity increases with increases in trunk lean, and knee extension activity decreases as trunk lean decreases (55). Maximal erector spinae activity also occurs later in the back lift than the leg lift. Finally, abdominal activity is lower in the back lift than in the leg lift (88). Consequently, the back is not as well supported in the back lift as it is in the leg lift, creating additional potential for injury.

To work the extensors from the standing position by hyperextending the trunk requires an initial contribution from the erector spinae. This activity drops off and then picks up again later in the hyperextension movement (68). If resistance to the movement is offered, the activity in the lumbar erector spinae movement increases dramatically (43).

TRUNK ROTATORS AND LATERAL FLEXORS

The rotation and lateral flexion movements of the trunk are not usually emphasized in an exercise program. Some examples of rotation and lateral bending exercises are provided in Figure 7-20. There is some benefit to include some of these exercises in a training routine because rotation is an important component of many movement patterns. Likewise, lateral flexion is an important component of activities such as throwing, diving, and gymnastics.

Some individuals try to isolate the obliques by performing trunk rotation exercises against external resistance. The obliques are not isolated in this type of exercise because the erector spinae muscles are also actively involved. If a rotation exercise is added to an exercise set, caution should be used. No combined exercises should be performed in which the trunk is flexed or extended and then rotated. This loads the vertebrae excessively and unnecessarily. If rotation is to be included, it should be done in isolation. The same holds true for lateral flexion exercises that can be performed against a resistance from a standing or sidelying position.

Many strength routines for the trunk incorporate the use of an exercise ball. The advantage of exercises using this kind of ball is the improvement in posture because of the ongoing spine stabilization that is required while sitting or balancing on the ball. Exercises can progress from easy to difficult depending on the distance of the ball from the body.

FLEXIBILITY AND THE TRUNK MUSCLES

It is recommended that stretching exercises be functional range of motion activities that do not require extreme range of motions. In fact, with increased flexibility, there may be increased risk of injury (54). With this in mind, stretching the trunk muscles is easy and can be done from a standing or lying position. The lying positions offer stabilization of the lower extremity and the pelvis, which contribute to the movements if the exercise is performed from a stand. All stretching of the trunk muscles should be done through one plane only because movements through more than one plane at a time excessively load the vertebral segments.

Caution should be used in prescribing maximum trunk flexion exercises, such as touching the toes for the stretch of the extensors. Remember, the trunk is supported by the ligaments and the posterior elements of the segment in this position, and the loads on the disks are large, so an alternative exercise should be chosen.

Similarly, the sit–reach test is often used as a measure of both low-back and hamstring flexibility. It has been suggested that the sit–reach is primarily an assessment of hamstring, not low-back, flexibility (71). The sit–reach position has also been shown to increase the strain on the low back as a result of exaggerated posterior tilt of the pelvis. It is recommended that the sit–reach stretching exercise be done while maintaining a mild lordotic lumbar curve throughout to avoid the exaggerated curve.

Inflexibility in the trunk or posterior thigh influences the load and strains incurred during exercise. If the low back is inflexible, the reversal of the lumbar curve is restrained in forward flexion movements. This places an additional strain on the hamstrings. If the hamstrings are inflexible, rotation of the pelvis is restricted, placing additional strain on the low-back muscles and ligaments. Additionally, inhibition of forward rotation of the pelvis increases the overall compressive stress on the spine (71).

CORE TRAINING

The core is the area between the sternum and the knees, and exercises to this area emphasize the abdominals, low

back, and hips. The core muscles transmit forces between the upper body and lower body and provide spinal stability during lifting and everyday activity (86). Strengthening the core muscles can serve as a preventive measure for back injury or the reoccurrence of a back injury. Core exercises that focus specifically on the lumbar region of the trunk are illustrated in Figure 7-22. The curl-up targets the rectus abdominis; the side bridge targets the obliques, transverse abdominis, and quadratus lumborum; and the bird dog targets the back and hip extensors (54). Lateral flexion exercises also stimulate coactivation of the extensors and the flexors (46). Hollowing and bracing the low back against the ground as well as the cat camel exercise are known to increase activity in the transverse abdominis and internal oblique (11). The front bridge or cat camel is done standing or on all fours.

Injury Potential of the Trunk

Back pain has been shown to affect as high as 17% of U.S. workers, with higher injury rates among occupations such as carpentry, construction, nurses, and dentists (36). And 85% of the population of the Western world reports back pain at some point in their lives, with peak incidence of injury in the working years (13). Low back pain is a chronic problem for 1% to 5% of the general population and recurs in 30% to 70% of those with an initial low back problem (71). The sexes are affected equally. Low back pain is most common in the age range of 25 to 60 years, with the highest incidence of low back pain at age 40 years (71). Back pain is uncommon in children and athletes. Back sprain accounts for only 2% to 3% of the total sprains in the athletic population (25), but it is very debilitating. Back pain is a particular problem in sports that require high levels of bending and rotation, such as golf, gymnastics, and baseball. The major source of back pain is muscle or tendon strain, and only 1% to 5% of back pain is related to an injury of the intervertebral disk (14). Torn ligaments are not a common source of back pain, and most back injuries result from microtraumas to muscles and tendons from activities such as unbalanced lifting, prolonged static postures, chronic stress, and chronic sitting (14).

Back pain can be caused by compression on the spinal cord or nerve roots from an intervertebral disk protrusion or disk prolapse. Disk protrusions occur most frequently at the intervertebral junctions of C5–C6, C6–C7, L4–L5, and L5–S1 (45). Lumbar disk protrusions occur at a significantly higher rate than in any other region of the trunk. As shown in Figure 7-23, the disk protrusion may impinge on the nerve exiting from the cord, causing problems throughout the back and the lower extremity.

A disk injury commonly occurs to a motion segment that is compressed while being flexed slightly more than the normal limits of motion (3). Also, a significant amount of torsion, or rotation, of the trunk has been shown to tear fibers in the annulus fibrosus of the disk.

Bent knee sit-up

Front bridge

Bird dog

Camel (on all fours)

Curl-up

Supine bridge

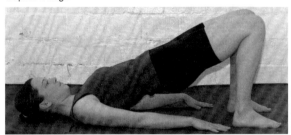

Camel (standing)

Side bridge

FIGURE 7-22 Exercises for the core focus on the abdominals and the trunk and hip extensors. Strengthening these muscles may prevent injury to the low back.

Anterior

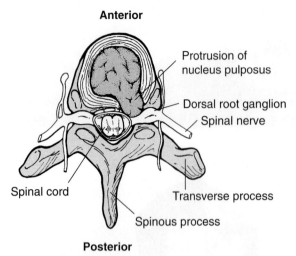

Posterior

FIGURE 7-23   Injury to a disk can be caused by extreme trunk flexion while the trunk is compressed or loaded. Rotation movements can also tear the disk. When a disk is ruptured, pressure can be put on the spinal nerves.

Progression

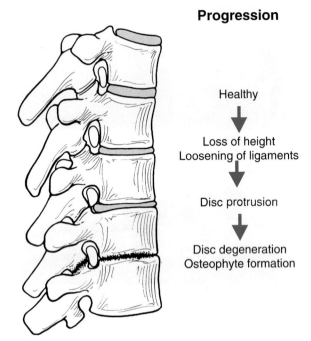

FIGURE 7-24   Disk degeneration narrows the joint space, causing a shortening of the ligaments, increased pressure on the disk, and stress on the apophyseal joints.

Pure compression to the spine usually injures the vertebral bodies and end plates rather than the disk. Likewise, maximal flexion of the trunk without compression may injure the posterior ligaments of the arch rather than the disk (3).

In the case of a disk prolapse, the nucleus pulposus extrudes into the annulus fibrosus either laterally or vertically. Vertical prolapse is more common than posterior prolapse, and the result is an anterolateral bulge of the annulus. This causes the bodies to tilt forward and pivot on the apophyseal joints, placing stress on the facets (95). A posterior or posterolateral prolapse of the disk into the spinal canal creates back pain and neurologic symptoms via nerve impingement. Schmorl's nodes is a condition in which a vertical prolapse of part of the nucleus pulposus protrudes into an end-plate lesion of the adjacent vertebrae (40). Damage to the disk is created through excessive load, failure of the inner posterior annulus fibers, and disk degeneration (10).

Disk degeneration in the early elderly years consists of a gradual process during which splits and tears develop in the disk tissue. The progression of disk degeneration is illustrated in Figure 7-24. Although the symptoms of disk degeneration may not appear until the early elderly years, the process may begin much earlier in life. It is common for disk degeneration to begin as the posterior muscles and ligaments relax, compressing the anterior portion and putting tension on the posterior portion of the disk. The tears in the disk are usually parallel to the end plates halfway between the end plate and the middle of the disk (95). As these tears get larger, there is potential for separation of the central portion of the disk. The splits and tears usually occur in the posterior and posterolateral portions of the disk along the posterior border of the marginal edges of the vertebral bodies (95). Eventually, the tears

may be filled with connective tissue and later with bone. Osteophytes develop on the periphery of the vertebral bodies, and cancellous bone is gradually laid down in the anterior portion of the disk where the pressure is great.

This condition can progress to a point of forming an osseous connection between two vertebral bodies that leads to further necrosis of the disk. Osteoarthritis of the apophyseal joints is also a byproduct of disk degeneration as added stress is placed on these joints. A disk that has undergone slight degeneration is also more susceptible to prolapse (3).

Fractures of the various osseous components of the vertebrae can occur. The fractures can be in the spinous processes, transverse processes, or laminae, or they can be compressive fractures of the vertebral body itself. Spondylolysis, shown in Figure 7-25, involves a fatigue fracture of the posterior neural arch at the pars interarticularis.

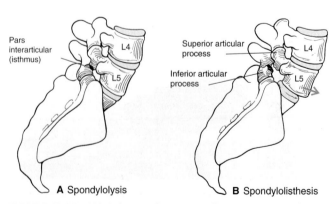

FIGURE 7-25   (A) A fatigue fracture to the pars interarticularis is called spondylolysis. (B) When the fracture occurs bilaterally, spondylolisthesis develops.

This injury is most common in sports requiring repeated flexion, extension, and rotation, such as gymnastics, weightlifting, football, dance, and wrestling (76). There is a 20.7% incidence of spondylolysis in athletes (41).

A typical example of an athlete who may fracture the neural arch is the football lineman. The lineman assumes a starting position in a three- or four-point stance with the trunk flexed. This flattens the low back, compresses and narrows the anterior portion of the disk, and stresses the transverse arch. When the lineman drives up with trunk extension and makes contact with an opponent, a large shear force across the apophyseal joint is created (76).

Another example of a spondylolysis-causing activity is pole vaulting. The vaulter extends the trunk at the plant and follows with rapid flexion of the trunk (75). The large range of motion occurring with rapid acceleration and deceleration is responsible for the development of the stress fracture. This condition is usually associated with repetitive activities, seldom with a single traumatic event (10).

With spondylolysis on both sides, spondylolisthesis can develop (Fig. 7-25). With a bilateral defect of the neural arch, the motion segment is unstable, and the anterior and posterior elements separate. The top vertebrae slip anteriorly over the bottom vertebrae. This condition is most common in the lumbar vertebrae, especially at the site of L5–S1, where shear forces are often high. The condition worsens with flexion of the spine, which adds to the anterior shear on the motion segment (96).

The cervical, thoracic, and lumbar regions of the trunk are subject to their specific injuries. In the cervical region of the spine, flexion and extension injuries, or whiplash injuries, are common. In whiplash, the head is rapidly flexed, straining the posterior ligaments or even dislocating the posterior apophyseal joints if the force is large (82). The rapid acceleration and deceleration of the head causes both sprain and muscular strain in the cervical region. During a rear-end impact, the body is thrown forward and the head is forced into hyperextension. This is followed by a rapid jerk of the head forward into flexion. This forceful whiplash can fracture the vertebral bodies through the wedging action in the flexion movement, which compresses the bodies together. The seventh cervical vertebra is a likely site of fracture in a flexion injury.

Flexion and compression injuries are also common in the cervical vertebrae and are seen in sports such as football and diving. The cervical spine straightens with flexion, creating a columnlike structure that lacks flexibility when contact is made. The disks, vertebral bodies, process, and ligaments resist this load, and when capacity is exceeded, vertebral dislocation and spinal cord impingement can result.

Injuries to the cervical vertebrae as a result of forceful extension include rupture of the anterior longitudinal ligament and actual separation of the annulus fibers of the disk from the vertebrae. Forceful hyperextension of the spine can be a part of whiplash injury, usually affecting the sixth cervical vertebra (85).

The cervical vertebrae are susceptible to injury in certain activities that subject the region to repeated forces. In diving, high jumping, and other activities with unusual landing techniques, individuals are subjected to repeated extension and flexion forces on the cervical vertebrae that may cause an injury (10). Positioning and posture of the cervical vertebrae are important in many of these activities that involve external contact in the region.

The thoracic region of the spine is not injured as frequently as the cervical and lumbar regions, probably because of its stabilization and limited motion as a result of interface with the ribs. The condition called Scheuermann disease is commonly found in the thoracic region. This disease is an increase in the kyphosis of the thoracic region from wedging of the vertebrae. The cause of Scheuermann disease is unknown, but it appears to be most prevalent in individuals who handle heavy objects. It is also common among competitive butterfly-stroke swimmers (10).

The lumbar region of the spine is the most injured, primarily because of the magnitude of the loads it carries. Low back pain can originate in any of a number of sites in the lumbar area. It is believed that in a sudden onset of pain, muscles are usually the problem, irritated through some rapid twisting or reaching movement. If the pain is of the low-grade chronic type, overuse is seen as the culprit (96).

Myofascial pain, common in the low back, involves muscle sheaths and tendons that have been strained as a result of some mechanical trauma or reflex spasm in the muscle (98). Muscle strain in the lumbar region is also related to the high tensions created while lifting from a stooped position.

Muscle spasms over time produce a dull, aching pain in the lumbar region. Likewise, a dull pain can be caused by distorted postures maintained for long periods. The muscles fatigue, the ligaments are stressed, and connective tissue can become inflamed as a result of poor posture.

Irritation of the joints in the lumbar region occurs most often in activities that involve frequent stooping, such as gardening and construction. Abnormal stress on the apophyseal joints is also common in activities such as gymnastics, ballet, and figure skating (98). Both spondylolysis and spondylolisthesis occur more frequently in the lumbar region than any other region of the trunk.

The intervertebral disks in the lumbar region have a greater incidence of disk prolapse than any other segment of the spinal column. A disk protrusion, as in any other area of the trunk, may impinge on nerve roots exiting the spinal cord, creating numbness, tingling, or pain in the adjacent body segments. Sciatica is such a condition. In it, the sciatic nerve is compressed, sending pain down the lateral aspect of the lower extremity.

The cause of low back pain is not clearly defined because of the multiple risk factors associated with the disorder. Some of these factors are repetitive work; bending and twisting; pushing and pulling; tripping, slipping, and falling; and sitting or static work posture (71). A low-back injury can be created through some uncoordinated or abnormal lift or through repetitive loading over time.

Low back pain associated with standing postures is related to positions maintaining hyperextension of the knees, hyperlordosis of the lumbar vertebrae, rounded shoulders, or hyperlordosis of the cervical vertebrae. In the seated posture, it is best to avoid crossing the legs at the knees because this position places stress on the low back. Likewise, positions that maintain the legs in an extended position with the hips flexed should be avoided because they accentuate lordosis in the low back.

Low-back injuries as a result of lifting are primarily a consequence of the weight of the load and its distance from the body. A correct lifting posture, as mentioned earlier in this chapter, is one with the back erect, knees bent, weight close to the body, and movement through one plane only (Fig. 7-26). This lifting technique minimizes the load imposed on the low back. A stooped lifting posture reduces the activity of the trunk extensors, and the forward moment is resisted by passive structures such as the disks, ligaments, and fascia. Lifting with flexed posture can place as much as 16% to 31% of the extensor moment on the passive structures (27), placing them at risk for injury.

A sudden maximal effort in response to an unexpected load is related to high incidence of back injury (49). The back extensors are slow postural muscles that may not generate force rapidly enough to prevent excessive spine bending or twisting with the application of a sudden load.

Unexpected loading can also increase the compressive force when the extensors contract to prevent a postural disturbance when a weight is unexpectedly placed in the hands (49). This could lead to a combination of high compression and bending stresses on the vertebrae.

Muscle strength and flexibility are also seen as predisposing factors for low back pain. Tight hamstrings and an inflexible iliotibial band have both been associated with low back pain (71). Weak abdominals are also related to low back pain. If the abdominals are weak, control over the pelvis is lacking, and hyperlordosis will prevail. The hyperlordotic position puts undue stress on the posterior apophyseal joints and the intervertebral disk. This is an important consideration in an activity such as a sit-up and curl-up.

Erector spinae muscle activity has also been shown to relate to the incidence of low back pain. Individuals with low back pain also have increased electrical activity and fatigue in the erector spinae muscle group (71). Even though there are inverse relationships between strength and flexibility and low back pain, the strength and flexibility of an individual may not predict whether that person will have low back pain. However, strength, flexibility, and fitness are predictive of the recurrence of low back pain (71).

Effects of Aging on the Trunk

The effects of aging on the spine may predispose someone to an injury or painful condition. During the process of aging, the flexibility of the spine decreases to as little as

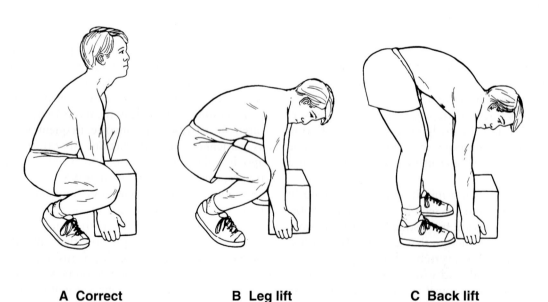

A Correct   **B Leg lift**   **C Back lift**

FIGURE 7-26   Low-back injury can be reduced if proper lifting techniques are used. The most important consideration is not whether the person uses the legs but where the weight is with respect to the body. Proper lifting technique has the weight close to the body with the head up and the back arched **(A)**. The leg lift technique **(B)** is no better than the back lift **(C)** if the weight is held far from the body. Both **(B)** and **(C)** should be avoided.

a 10th of that of younger individuals (61). There is also a corresponding loss of strength in the trunk muscles of approximately 1% per year (71). Between ages 30 and 80 years, the strength in cartilage, bone, and ligaments reduces by approximately 30%, 20%, and 18%, respectively (71).

The shape and length of the spine also change with aging. There is a smaller fluid region in the aging disk that places more stress on the annulus fibrosus (26). The disks may also lose height and create a shorter spine, although it has been reported that the ventral disk height is constant in both men and women in the age range of 16 to 57 years (32). There is also an increase in lateral bending of the trunk, an increase in thoracic kyphosis, and a decrease in lumbar lordosis (58). In the lumbar region specifically, there is a loss of mobility in the L5–S1 segment with an accompanying increase in the mobility of the other segments (40). It is not clear whether these age-related changes are a normal process of aging or are associated with abuse of the trunk, disuse of the trunk, or are disease related. It is clear that there is benefit in maintaining strength and flexibility in the trunk well into the elderly years.

 ## Contribution of the Trunk Musculature to Sports Skills or Movements

The contribution of the back muscles to lifting has been presented in an earlier discussion. Likewise, the contribution of the abdominals to a sit-up or curl-up exercise was evaluated. The trunk muscles also contribute to activities such as walking and running.

At touchdown, the trunk flexes toward the side of the limb making contact with the ground. It also moves back, and both of these movements are maximum at the end of the double support phase. After moving into single support, the trunk moves forward while still maintaining lateral flexion toward the support limb (91). As the speed of walking increases, there is a corresponding increase in lumbar range of motion accompanied by higher muscle activation levels (16).

For running, the movements in the support phase are much the same, with trunk flexion and lateral flexion to the support side. One difference is that whereas in walking, there is trunk extension at touchdown; in running, the trunk is flexed at touchdown only at fast speeds (90). At slower speeds, the trunk is extended at touchdown. For a full cycle in both running and walking, the trunk moves forward and backward twice per cycle.

Another difference between walking and running is the amount and duration of lateral flexion in the support phase. In running, the amount of lateral flexion is greater, but lateral flexion is held longer in the maximal position in walking than in running (91). There is one full oscillation of lateral flexion from one side to the other for every walking and running cycle.

As contact is made with the ground in both running and walking, there is a burst of activity in the longissimus and multifidus muscles. This activity can begin just before contact, usually as an ipsilateral contraction to control the lateral bending of the trunk. It is followed by a contraction of the contralateral erector spinae muscles, so that both sides contract (90).

There is a second burst of activity in these muscles in the middle of the cycle, occurring with contact of the other limb. Here, both the longissimus and the multifidus are again active. In the first burst of activity, the ipsilateral muscles are more active, but in this second burst, the contralateral muscles are more active (90). The activity of the erector spinae muscles coincides with extensor activity at the hip, knee, and ankle joints.

The lumbar muscles serve to restrict locomotion by controlling the lateral flexion and the forward flexion of the trunk (90). Cervical muscles serve to maintain the head in an erect position on the trunk and are not as active as the muscles in other regions of the spine.

A more thorough review of muscular activity is provided for a topspin tennis serve (Fig. 7-27). There is considerable activity in the abdominals and the erector spinae in the tennis serve. The most muscular activity is in the descending wind up and the acceleration phase (21). There is also considerable coactivation of the erector spinae and the abdominals to stabilize the trunk when it is brought back in a back arch in the descending windup and the subsequent acceleration. Both the internal and external oblique muscles are the most active of the trunk muscles. Because both the erector spinae and the abdominals are responsible for lateral flexion and rotation, there is unilateral activation of muscles to initiate left trunk lateral flexion and rotation to the right and left.

 ## Forces Acting at Joints in the Trunk

Loads applied to the vertebral column are produced by body weight, muscular force acting on each motion segment, prestress forces caused by disk and ligament forces, and external loads being handled or applied (48). The spine cannot support more than 20 N without muscular contraction (4,64). The muscleless lumbar spine can withstand a somewhat higher force (>100 N) before buckling (64).

Muscle tension is dependent on the position of the upper body segments and any external load being lifted. Any postural adjustments that move the load or body segments closer to the back will decrease how much muscle tension is needed to keep the system in balance. In Figure 7-28, the weights of the trunk and the load tend to rotate the trunk in a clockwise direction. The erector muscles in the lower back must contract to prevent this rotation. Reducing the muscle tension will decrease the lower back compressive force.

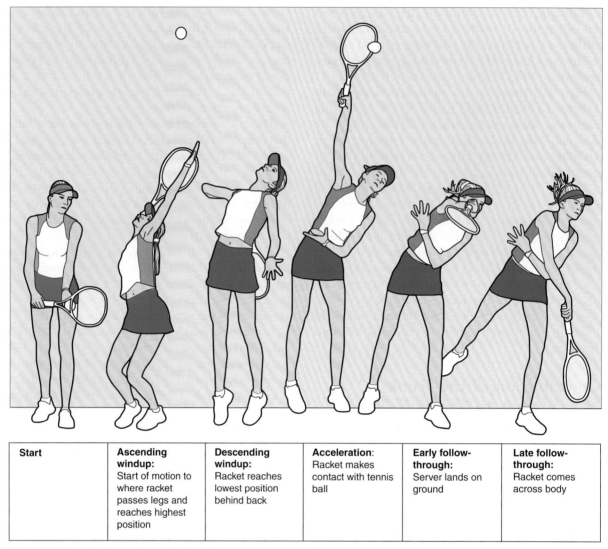

Start	Ascending windup: Start of motion to where racket passes legs and reaches highest position	Descending windup: Racket reaches lowest position behind back	Acceleration: Racket makes contact with tennis ball	Early follow-through: Server lands on ground	Late follow-through: Racket comes across body

FIGURE 7-27   Trunk muscles involved in the topspin tennis serve showing the level of muscle activity (low, moderate, and high) and the type of muscle action (concentric [CON] and eccentric [ECC]) with the associated purpose. (*continued*)

The disks, apophyseal joints, and intervertebral ligaments are the load-bearing structures. Compressive forces are applied perpendicular to the disk; thus, the line of action varies with the orientation of the disk. For example, in the lumbar vertebrae, only at the L3–L4 level is the compressive force vertical in upright standing (26). Compression forces are primarily resisted by the disk unless there is disk narrowing, where the resistance is offered by the apophyseal joints (26). For flexion bending moments, 70% of the moment is resisted by the intervertebral ligaments and 30% by the disks and in extension, and two-thirds of the moment is resisted by the apophyseal joints and the neural arch and a third by the disks (26). Lateral bending moments are resisted by the disks, and rotation is resisted by the disks and bony contact at the apophyseal joints (26).

The lumbar vertebrae handle the largest load, primarily because of their positioning, the position of the center of mass relative to the lumbar region, and greater body weight acting at the lumbar region than other regions of the spine. Of the compressional load carried by the lumbar vertebrae, 18% is a result of the weight of the head and trunk (57). The other source of substantial compression is muscle activity. Muscle forces protect the spine from excessive bending and torsion but subject the spine to high compressive forces. The compressive forces are increased with more lumbar flexion, and it is fairly common to see substantial increases in lumbar flexion with actions such as crossing legs (35% to 53%), squatting on the heels (70% to 75%), lifting weights from the ground (70% to 100%), and rapid lunging movements (100% to 110%) (26).

The axial load on the lumbar vertebrae in standing is 700 N. This can quickly increase to values greater than 3,000 N when a heavy load is lifted from the ground and can be reduced by almost half in the supine

Muscle	Ascending Windup: Start of Motion to Where Racquet Passes Legs and Reaches Highest Position			Descending Windup: Racquet Reaches Lowest Position behind Back			Acceleration: Racquet Makes Contact with Tennis Ball			Early Follow-through: Server Lands on Ground			Late Follow-through: Racquet Comes across the Body		
	Level	Action	Purpose	Level	Action	Purpose	Level	Action	Purpose	Level	Action	Purpose	Level	Action	Purpose
Erector spinae	Low	CON	Extension	Low-right Mod-left	CON CON CON	Rotation to the right Extension Left lateral flexion	Mod	Iso	Stabilize trunk	Mod	ECC	Control of trunk flexion	Mod-right Mod-left	ECC ECC	Control of trunk flexion Control of lateral flexion
External oblique	Low	CON	Rotation to the right	Low-right Mod-left	CON CON	Rotation to the right Left lateral flexion	Mod	CON	Trunk flexion	Low	Iso	Stabilize trunk	Low	ECC	Control of trunk rotation
Internal oblique	Low	CON	Rotation to the right	High	CON CON ECC	Rotation to the right Left lateral flexion Control of trunk extension	High	CON	Trunk flexion	Mod	Iso	Stabilize trunk	Mod	ECC	Control of trunk rotation
Rectus abdominis	Low	ECC	Control of trunk extension	Mod	ECC CON	Control of trunk extension Initiation of trunk flexion	Mod	CON	Trunk flexion	Low	CON	Trunk flexion	Low	ECC	Control of trunk rotation

FIGURE 7-27 (continued)

Source: Chow, J. W. (2003). Lower trunk muscle activity during the tennis serve. *Journal of Science and Medicine in Sport*, 6:512-518

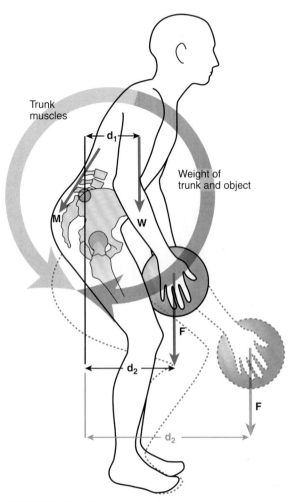

FIGURE 7-28 The weight of the trunk and the weight in the hands tend to rotate the trunk clockwise while the back muscles try to rotate the trunk in a counterclockwise direction. A balance of these torques ensure the trunk is in a static posture. Moving the weight further from the back will increase the clockwise tendency and require increased muscular contractions to maintain the posture. The total compressive force in the lower back is the sum of the trunk weight, the weight in the hands, and the muscle force.

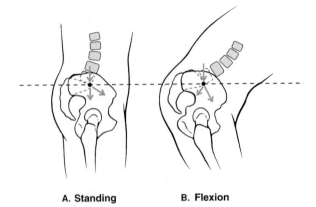

A. Standing   B. Flexion

FIGURE 7-29 **(A)** The shear force across the lumbosacral joint in standing is approximately 50% of the body weight. **(B)** If one flexes to where the sacral angle increases to 50°, the shear force can increase to as much as 75% of the body weight above the joint.

position (300 N) (15). Fortunately, the lumbar spine can resist approximately 9,800 N of vertical load before fracturing (61).

The load on the lumbar vertebrae is more affected by distance of the load from the body than the actual posture of lifting (61). For example, the magnitude of the compressive force acting on the lumbar vertebrae in a half squat is six to 10 times body weight (18). If the weight is taken farther as a result of flexion, compressive loads increase, even with postural adjustments such as flattening the lumbar curve (29).

Loads acting at the lumbar vertebrae can be as high as two to 2.5 times body weight in an activity such as walking (17). These loads are maximum at toe-off and increase with an increase in walking speed. Loads on the vertebrae in an activity such as walking are a result of muscle activity in the extensors and the amount of trunk lean in the walker (48). This is compared with loads of more than

four body weights in rowing, which is maximum in the drive phase as a result of muscle contraction and trunk position (59).

The direction of the force or load acting on the vertebrae is influenced by positioning. In a standing posture with the sacrum inclined 30° to the vertical, there is a shear force acting across the lumbosacral joint that is approximately 50% of the body weight above the joint (Fig. 7-29). If the sacral angle increases to 40°, the shear force increases to 65% of the body weight, and with a 50° sacral inclination, the force acting across the joint is 75% of the body weight above the joint (77).

Lumbosacral loads are also high in exercises such as the squat, in which maximum forces are generated at the so-called sticking point of the ascent. These loads are higher than loads recorded at either the knee or the hip for the same activity (65).

Loads are applied to the lumbar vertebrae even in a relaxed supine position of repose. The loads are significantly reduced because of the loss of the body weight forces but still present as a result of muscular and ligamentous forces. In fact, the straight-leg lying position imposes load on the lumbar vertebrae because of the pull of the psoas muscle. Flexing the thigh by placing a pillow under the knees can reduce this load.

Loads imposed on the vertebrae are carried by the various structural elements of the segment. The articulating facets carry large loads in the lumbar vertebrae during extension, torsion, and lateral bending but no loads in flexion (79). Facet loads in extension have been shown to be as high as 30% to 50% of the total spine load, and in arthritic joints, the percentage can be higher (38). The posterior and anterior ligaments carry loads in flexion and extension, respectively, but carry little load in lateral bending and torsion. The intervertebral disks absorb and distribute a great proportion of the load imposed on the vertebrae. The intradiscal pressure is 1.3 to 1.5 times the compressional load applied per unit area of disk (61,80), and the pressure increases linearly with loads up to 2,000 N (60). The load on the third lumbar vertebra in standing is approximately 60% of total body weight (62).

Pressures in sitting are 40% more than those in standing, but standing interdiscal pressures can be reduced by placing one foot in front of the other and elevating it (85). Whereas in standing, the natural curvature of the lumbar spine is increased, the curvature is reduced in sitting. The increased curvature reduces the pressure in the nucleus pulposus while at the same time increasing loading of the apophyseal joints and increasing compression on the posterior annulus fibrosus fibers (1). Pressures within the disk are large with flexion and lateral flexion movements of the trunk and small with extension and rotation (63). The pressure increases can be attributed to tension generated in the ligaments, which can increase intradiscal pressure

by 100% or more in full flexion (1). The lateral bend produces larger pressures than flexion and even more pressure if rotation added to the side bend causes asymmetrical bending and compression (8).

Intervertebral disks have been shown to withstand compressive loads in the range of 2,500 to 7,650 N (69). In older individuals, the range is much smaller, and in individuals younger than 40 years, the range is much larger (69).

The posterior elements of the spinal segment assist with load bearing. When the spine is under compression, the load is supported partially by the pedicles and pars interarticularis and somewhat by the apophyseal joints. When compression and bending loads are applied to the spine, 25% of the load is carried by the apophyseal joints. Only 16% of the loads imposed by compressive and shear

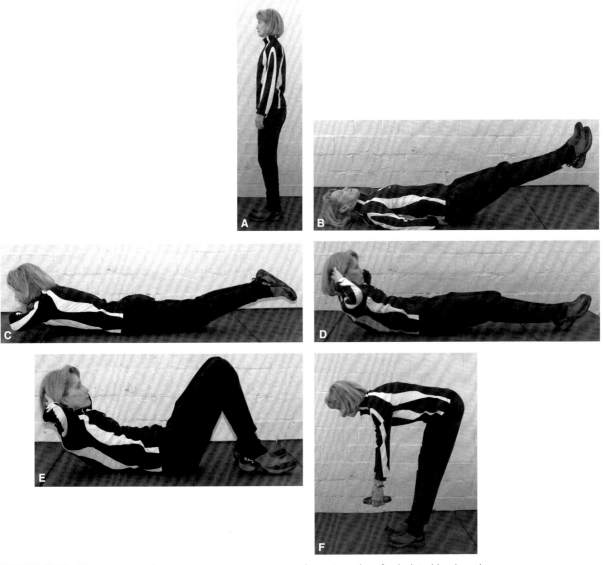

FIGURE 7-30 The representative postures or movements are shown in order of calculated load on the lumbar vertebrae using a miniaturized pressure transducer. The standing posture imposed the least amount of load (686 N) **(A)**, followed by the double straight-leg raise (1,176 N) **(B)**, back hyperextension (1,470 N) **(C)**, sit-ups with knees straight (1,715 N) **(D)**, sit-ups with knees bent (1,764 N) **(E)**, and bending forward with weight in the hands (1,813 N) **(F)**. (Adapted with permission from Nachemson, A. [1976]. Lumbar intradiscal pressure. In M. Jayson (Ed.). *The Lumbar Spine and Back Pain*. Kent: Pitman Medical Publishing Company Ltd, 407-433.)

forces are carried by the apophyseal joints (57). Any extension of the spine is accompanied by an increase in the compressive strain on the pedicles, an increase in both compressive and tensile strain in the pars articularis, and an increase in the compressive force acting at the apophyseal joints (42).

In full trunk flexion, the loads are maintained and absorbed by the apophyseal capsular ligaments, intervertebral disk, supraspinous and interspinous ligaments, and ligamentum flavum, in that order (3). The erector spinae muscles also offer some resistance passively.

In compression, most of the load is carried by the disk and the vertebral body. The vertebral body is susceptible to injury before the disk and will fail at compressive loads of only 3,700 N in the elderly and 13,000 in a young, healthy adult (47). In rotation, during which torsional forces are applied, the apophyseal joints are more susceptible to injury. During a forward bend movement, the disk and the apophyseal joints are at risk for injury because of compressive forces on the anterior motion segment and tensile forces on the posterior elements.

Loads in the cervical region of the spine are lower than in the thoracic or lumbar regions and vary with position of the head, becoming significant in extreme positions of flexion and extension (82). Loads on the lumbar disk have been calculated using a miniaturized pressure transducer (61). Approximate loads for various postures and exercises are presented in Figure 7-30, although the researchers recommend caution about the interpretation of absolute values and direct more attention to the relative values (61). Studies have demonstrated that the compressive load on the low back can be greater than 3,000 N in exercises such as sit-ups, and a sit-up with the feet fixed results in similar loads whether the bent-knee or straight-leg technique is used (52).

Summary

The vertebral column provides both flexibility and stability to the body. The four curves—cervical, thoracic, lumbar, and sacral—form a modified elastic rod. The cervical, thoracic, and lumbar curves are mobile, and the sacral curve is rigid.

Spinal column movement as a whole is created by small movements at each motion segment. Each motion segment consists of two adjacent vertebrae and the disk separating them. The anterior portion of the motion segment includes the vertebral body, intervertebral disk, and ligaments. Movement is allowed as the disk compresses. Within the disk itself, the gel-like mass in the center, the nucleus pulposus, absorbs the compression and creates tension force in the annulus fibrosus, the concentric layers of fibrous tissue surrounding the pulposus.

The posterior portion of the motion segment includes the neural arches, intervertebral joints, transverse and spinous processes, and ligaments. This portion of the motion segment must accommodate large tensile forces.

The range of motion in each motion segment is only a few degrees, but in combination, the trunk is capable of moving through considerable range of motion. Flexion occurs freely in the lumbar region through 50° to 60° and the total range of flexion motion is 110° to 140°. Lateral flexion range of motion is approximately 75° to 85°, mainly in the cervical and lumbar regions, with some contribution from the thoracic region. Rotation occurs through 90° and is free in the cervical region. Rotation takes place in combination with lateral flexion in the thoracic and lumbar regions.

Most lumbar spine movements are accompanied by pelvic movements, termed the lumbopelvic rhythm. In trunk flexion, the pelvis tilts anteriorly and moves backward. In trunk extension, the pelvis moves posteriorly and shifts forward. The pelvis moves with the trunk in rotation and lateral flexion.

The extension movement of the trunk is produced by the erector spinae and the deep posterior muscles running in pairs along the spinal column. The extensors are also very active, controlling flexion of the trunk through the first 50° to 60° of a lowering action with gravity. The abdominals produce flexion of the trunk against gravity or resistance. They also produce rotation and lateral flexion of the trunk with assistance from the extensors.

The trunk muscles can generate the greatest amount of strength in the extension movement, but the total extensor moment is only slightly greater than flexor moment. The strength output is also influenced by trunk position. In lifting, the extensor contribution diminishes the farther the object is horizontally from the body. The contribution of the various segments and muscles is also influenced by the angle of pull and the width of the object being lifted.

Posture and spinal stabilization is an important consideration in the maintenance of a healthy back. The spine is stabilized by three systems: a passive system, an active musculoskeletal system, and a neural feedback system. The transverse abdominis, erector spinae, and internal oblique play important roles in spinal stabilization. Both standing and sitting postures require some support from the trunk muscles. In the workplace, posture becomes an important factor, particularly if static positions are maintained for long periods of time. It is suggested that short breaks occur regularly over the course of the workday to minimize the accumulative strain in static postures. Postures that should be avoided include a slouched standing posture, prolonged sitting, unsupported sitting, and continuous flexion positions.v

Postural deviations are common in the general population. Some of the common postural deviations in the trunk are excessive lordosis, excessive kyphosis, and scoliosis. The most serious of these is scoliosis.

Conditioning of the trunk muscles should always include exercises for the low back. Additionally, trunk exercises should be evaluated in terms of safety and effectiveness. For example, the trunk flexors can be strengthened using a variety of trunk or hip flexion exercises, including the sit-up, curl-up, and the double-leg raise. Conditioning of the extensors can be done through the use of various lifts. Both the leg lift and the back lift are commonly used to strengthen the extensors. The back lift imposes more stress on the vertebrae and produces more disk pressure than the leg lift.

Stretching of the trunk muscles can be done either standing or lying, but it is recommended that stretching occurs through a functional range. The lying position offers more support for the trunk. Toe-touch exercises for flexibility should be avoided because of the strain to the posterior elements of the vertebral column.

The incidence of injury to the trunk is high, and it is predicted that 85% of the population will have back pain at some time in their lives. Back pain can be caused by disk protrusion or prolapse on a nerve but is more likely to be associated with soft-tissue sprain or strain. Disk degeneration occurs with aging and may eventually lead to reduction of the joint space and nerve compression. The spinal column can also undergo fractures in the vertebral body as a result of compressive loading or in the posterior neural arch associated with hyperlordosis (spondylolysis). When the defect occurs on both sides of the neural arch, spondylolisthesis develops: The vertebrae slip anteriorly over each other. Some injuries are specific to regions of the trunk, such as whiplash in the cervical region and Scheuermann disease in the thoracic vertebrae.

Changes in the spine associated with aging include decreased flexibility, loss of strength, loss of height in the spine, and an increase in lateral bending and thoracic kyphosis. It is not clear whether these changes are a normal consequence of aging or are related to disuse, misuse, or a specific disease process.

The contribution of the muscles of the trunk to sport skills and movements is important for balance and stability. The trunk muscles are active in both walking and running as the trunk laterally flexes, flexes and extends, and rotates. There is also considerable activity in the cervical region of the trunk as the head and upper body are maintained in an upright position. In a tennis serve, unilateral contraction of the abdominal and erector spinae occurs to initiate lateral flexion and rotation actions in the tennis serve. The obliques are the most active trunk muscle in the tennis serve.

The loads on the vertebrae are substantial in lifting and in different postures. The loads on the lumbar vertebrae can range from two to 10 times the body weight in activities such as walking and weight lifting. Loads on the actual intervertebral disks are influenced by a change in posture. For example, the pressures on the disk are 40% more in sitting than standing.

REVIEW QUESTIONS

True or False

1. ____ The transverse abdominis does not cause flexion, extension, or rotation of the trunk.

2. ____ Intervertebral disk height is minimal at the end of the day.

3. ____ Little flexion/extension motion occurs in the thoracic region.

4. ____ There are three prominent curves of the spine.

5. ____ Posterior tilt of the pelvis accompanies trunk extension.

6. ____ Compressive forces are perpendicular to the vertebral disk, even when a person is lying supine.

7. ____ The rectus abdominis is very active in a curl-up.

8. ____ Trunk flexion strength is two times greater than extension strength.

9. ____ The size of the vertebral body is indicative of the weight it bears.

10. ____ Sitting results in lower intervertebral disk pressures than standing.

11. ____ Crossing the legs at the knees while sitting places increased stress on the low back.

12. ____ Most low-back injuries occur to the ligaments.

13. ____ In full trunk flexion, most of the load is carried by the ligaments and disk.

14. ____ In walking and running, the trunk laterally flexes toward the swing limb.

15. ____ Aside from the annulus fibrosus, the disk is vascular and innervated by many nerves.

16. ____ The thoracic curvature begins to develop as an infant begins to lift his/her head.

17. ____ A well-designed chair can reduce seated lumbar pressures to less than standing pressures.

18. ____ Endurance training for the back muscles is better than strength or power training for preventing workplace injuries.

19. ____ Most back muscles are predominantly slow, type I muscle fibers.

20. ____ An exaggerated posterior curve in the lumbar region is termed lordosis or hyperlordosis.

21. ____ Neither the axis or atlas have a vertebral body.

22. ____ The erector spinae muscles respond first when a load is applied to the spine.

23. ____ In trunk extension, the disk moves posteriorly.

24. ____ Thoracic movement is limited primarily by bony articulations and morphology.

25. ____ Loads on the lumbar vertebrae can reach 10 times body weight during weight lifting.

Multiple Choice

1. Muscular stability of the spine is provided by ____.
   a. multifidus
   b. rectus abdominis
   c. erector spinae
   d. All of the above
   e. Both a and c

2. In a whiplash injury, ____.
   a. the car and head accelerate forward in synchrony
   b. the neck hyperextends
   c. strain develops on the anterior spine, while the posterior is compressed
   d. All of the above
   e. Both b and c

3. During walking and running ____.
   a. maximum lateral flexion is held longer when running
   b. longissimus and multifidus are active at contact
   c. trunk moves forward and backward once per cycle
   d. All of the above
   e. Both a and b

4. The atlas ____.
   a. articulates with the skull at the atlantooccipital joint
   b. has no spinous process
   c. has transverse foramen for blood vessels to pass through
   d. All of the above
   e. Both a and b

5. When bending forward at the hips ____.
   a. lumbar movement accounts for the first 50° to 60° of flexion, then anterior pelvic rotation occurs
   b. the spinal and hip extensors contract eccentrically
   c. the posterior fibers of the annulus fibrosus are in tension, while the anterior fibers are in compression
   d. All of the above
   e. Both a and c

6. Which provides the greatest resistance to trunk flexion?
   a. Apophyseal capsular ligaments
   b. Intervertebral disks
   c. Supraspinous and interspinous ligaments
   d. Ligamentum flavum
   e. Back extension muscles

7. If rotating to the right, ____.
   a. multifidus, longissimus, and iliocostalis on the right are active
   b. multifidus on the right is active and longissimus and iliocostalis on the left are active
   c. the right internal and external obliques are active
   d. the right internal oblique and left external oblique are active
   e. Both a and c
   f. Both b and d

8. The ligaments supporting the posterior portion of the vertebral segment include the ____.
   a. posterior longitudinal, ligamentum flavum, supraspinous, interspinous, and intertransverse
   b. ligamentum flavum, supraspinous, interspinous, and intertransverse
   c. anterior longitudinal, ligamentum flavum, supraspinous, and interspinous
   d. posterior longitudinal, supraspinous, and interspinous

9. The cause of scoliosis is ____.
   a. unknown
   b. lack of calcium
   c. poor posture in young girls
   d. unequal leg length
   e. muscle imbalances

10. Which is false in reference to spinal injuries?
    a. With a prolapsed disk, the nucleus pulposus extrudes into the annulus fibrosus
    b. Scheuermann disease results from wedging of the cervical vertebrae
    c. Tears in the intervertebral disks may be filled with connective tissue or bone
    d. Spondylolysis is a fracture of the pars interarticularis

11. Which is true of the abdominal muscles?
    a. Internal oblique causes rotation to the opposite side
    b. External oblique causes rotation to the same side
    c. Transverse abdominis contracts to increase abdominal pressure and spinal stabilization
    d. All of the above

12. Low back pain ____.
    a. may be caused by weak abdominals
    b. is common in athletes, middle-aged adults, and elderly
    c. affects men more than women
    d. All of the above
    e. Both a and c

13. One way to decrease compressive forces during lifting is to ____.
    a. move the load closer to the body
    b. flex more at the hips
    c. carry the object in one hand versus two
    d. All of the above

14. Contracted at the same time, the erector spinae muscles create _____, but if contracted unilaterally, they create _____.
    a. extension, rotation
    b. extension, lateral flexion
    c. flexion, lateral flexion
    d. Either a or b

15. The range of motion of the whole trunk segment is approximately ____.
    a. 70° of flexion, 45° of lateral flexion, and 50° of rotation
    b. 110° of flexion, 75° of lateral flexion, and 90° of rotation
    c. 110° of flexion, 50° of lateral flexion, and 100° of rotation
    d. 70° of flexion, 75° of lateral flexion, and 90° of rotation

16. Trunk flexion moment ____.
    a. is equal to the extension moment
    b. is greater than the extension moment
    c. is less than the extension moment
    d. may be larger or less than the extension moment, depending on the magnitude of lumbar curvature

17. A close-packed position exists for the apophyseal joints when ____.
    a. one is sitting or standing
    b. when the spine is fully flexed
    c. the atlas and axis are in flexion
    d. All of the above
    e. Both b and c

18. Low back pain is most common in the age range of ____.
    a. 25 to 60 years
    b. 40 to 50 years
    c. 40 to 65 years
    d. 60 to 75 years

19. Exercises that create excessive lordosis of the lumbar vertebrae should be avoided, including ____.
    a. double-leg raises
    b. donkey kicks
    c. back bends
    d. All of the above

20. Standing strain in the workplace can be reduced by ____.
    a. avoiding prolonged static postures
    b. standing on cushioned floor mats
    c. wearing work boots without midsoles
    d. All of the above
    e. Both a and b

21. Disk protrusions are most common in which region?
    a. Cervical
    b. Thoracic
    c. Lumbar
    d. Sacral

22. When the spine undergoes a combined flexion and lateral bending, stress is put on the:
    a. ligamentum flavum
    b. annulus fibrosus
    c. apophyseal joint
    d. Both a and b

23. A sit-up ____.
    a. with bent knees increases contribution of the abdominals relative to the hip flexors
    b. is an ideal exercise for all abdominals
    c. results in greater activation of the abdominals relative to double leg-raising exercises
    d. All of the above

24. A fully flexed spine should be avoided because it ____.
    a. transfers load from the muscles to the passive tissues
    b. decreases the tolerance to compressive loads
    c. reduces the moment arms for the extensor muscles
    d. All of the above

25. An aging spine does not result in ____.
    a. decreased spinal flexibility
    b. a greater fluid region in the disk
    c. less lumbar lordosis and increased thoracic kyphosis
    d. increased lateral bending

References

1. Adams, M. A., Dolan, P. (1995). Recent advances in lumbar spinal mechanics and their clinical significance. *Clinical Biomechanics*, 10(1):3-19.
2. Adams, M. A., Hutton, W. C. (1982). Prolapsed intervertebral disc: A hyperflexion study. *Spine*, 7:184-191.
3. Adams, M. A., et al. (1980). The resistance to flexion of the lumbar intervertebral joint. *Spine*, 5:245-253.
4. Andersen, T. B., et al. (2004). Movement of the upper body and muscle activity patterns following a rapidly applied load: The influence of pre-load alterations. *European Journal of Applied Physiology*, 91:488-492.
5. Andersson, E. A., et al. (1997). Abdominal and hip flexor muscle activation during various training exercises. *European Journal of Applied Physiology*, 75:115-123.
6. Andersson, G. B. J., et al. (1976). Myoelectric back muscle activity in standardized lifting postures. In P. V. Komi (Ed.). *Biomechanics 5–A*. Baltimore, MD: University Park Press, 520-529.
7. Andersson, G. B. J., et al. (1974). Myoelectric activity in individual lumbar erector spinae muscles in sitting. *Scandinavian Journal of Rehabilitative Medicine*, 3:91.
8. Andersson, G. B. J., et al. (1974). Quantitative electromyographic studies of back muscle activity related to posture and loading. *Orthopedic Clinics of North America*, 8:85-96.
9. Andersson, G. B. J., et al. (1977). Intradiscal pressure, intra-abdominal pressure and myoelectric back muscle activity related to posture and loading. *Clinical Orthopaedics*, 129:156-164.
10. Ashton-Miller, J. A., Schultz, A. B. (1988). Biomechanics of the human spine and trunk. In K. B. Pandolf (Ed.). *Exercise and Sport Sciences Reviews*. New York: Macmillan, 169-204.
11. Barnett, F., Gillard, W. (2005). The use of lumbar spinal stabilization techniques during the performance of abdominal strengthening exercises. *The Journal of Sports Medicine and Physical Fitness, 45*:38-44.
12. Beard, H. K., Stevens, R. L. (1976). Biochemical changes in the intervertebral disc. In M. Jayson (Ed.). *The Lumbar Spine and Back Pain*. Kent: Pitman Medical Publishing Company Ltd, 407-433.
13. Bigos, S. J., et al. (1986). Back injuries in industry: a retrospective study. III. Employee-related factors. *Spine*, 11(3): 252-256.
14. Bracko, M. R. (2004). Can we prevent back injuries? *ACSM's Health & Fitness Journal*, 8:5-11.

15. Broberg, K. B. (1983). On the mechanical behavior of inter-vertebral discs. *Spine*, 8:151-165.
16. Callaghan, J. P., et al. (1999). Low back three-dimensional joint forces, kinematics, and kinetics during walking. *Clinical Biomechanics*, 14:203-216.
17. Capazzo, A. (1984). Compressive loads in the lumbar vertebral column during normal level walking. *Journal of Orthopaedic Research*, 1:292.
18. Capazzo, A., et al. (1985). Lumbar spine loading during half squat exercises. *Medicine & Science in Sports & Exercise*, 17:613-620.
19. Chiu, T. T., et al.(2002). Maximal isometric muscle strength of the cervical spine in healthy volunteers. *Clinical Rehabilitation*, 16, 772-779.
20. Cholewicki, J., McGill, S. M. (1996). Mechanical stability of the in vivo lumbar spine: implications for injury and chronic low back pain. *Clinical Biomechanics*, 11:1-15.
21. Chow, J. W. (2003). Lower trunk muscle activity during the tennis serve. *Journal of Science and Medicine in Sport*, 6: 512-518.
22. Cresswell, A. G., et al. (1994). The influence of sudden perturbations on trunk muscle activity and intra-abdominal pressure while standing. *Experimental Brain Research*, 98:336-341.
23. Davies, P. R. (1981). The use of intra-abdominal pressure in evaluating stresses on the lumbar spine. *Spine*, 6:90-92.
24. Davis, K. G., Marras, W. S. (2000). The effects of motion on trunk biomechanics. *Clinical Biomechanics*, 15:703-717.
25. Dehaven, K. E., Lintner, D. M. (1986). Athletic injuries: Comparison by age, sport, and gender. *The American Journal of Sports Medicine*, 14:218-224.
26. Dolan, P., Adams, M. A. (2001). Recent advances in lumbar spinal mechanics and their significance for modeling. *Clinical Biomechanics*, 16:S8-S16.
27. Dolan, P., et al. (1994). Passive tissues help the back muscles to generate extensor moments during lifting. *Journal of Biomechanics*, 27:1077-1085.
28. Ebben, J. M. (2003). Improved ergonomics for standing work. *Occupational Health & Safety*, 72: 72-76.
29. Eklund, J. A. E., et al. (1983). A method for measuring the load imposed on the back of a sitting person. *Ergonomics*, 26:1063-1076.
30. El-Bohy, A. A., King, A. I. (1986). Intervertebral disc and facet contact pressure in axial torsion. In S. A. Lantz, A. I. King (Eds.). *Advances in Bioengineering*. New York: American Society of Mechanical Engineers, 26-27.
31. Farfan, H. F. (1975). Muscular mechanism of the lumbar spine and the position of power and efficiency. *Orthopedic Clinics of North America*, 6:135.
32. Frobin, W., Brinckmann, P., Biggemann, M., Tillotson, M., & Burton, K. (1997). Precision measurement of disc height, vertebral height and sagittal plane displacement from lateral radiographic views of the lumbar spine. *Clinical Biomechanics*, 12, (1 Suppl), pages S1-S63.
33. Garg, A. (1980). A comparison of isometric strength and dynamic lifting capability. *Ergonomics*, 23:13-27.
34. Garg, A., et al. (1983). Biomechanical stresses as related to motion trajectory of lifting. *Human Factors*, 25:527-539.
35. Gould, J. A. (1985). The spine. In J. A. Gould and G. J. Davies (Eds.). *Orthopaedic and Sports Physical Therapy*. St. Louis, MO: C.V. Mosby, 518-549.
36. Guo, H. R. (2002). Working hours spent on repeated activities and prevalence of back pain. *Occupational and Environmental Medicine*, 59:680-688.
37. Haher, T. R., et al. (1993). Biomechanics of the spine in sports. *Clinics in Sports Medicine*, 12:449-463.
38. Hedman, T. (1992). A new transducer for facet force measurement in the lumbar spine: Benchmark and in vitro test results. *Journal of Biomechanics*, 25:69-80.
39. Hickey, D. S., Hukins, D. W. L. (1980). Relation between the structure of the annulus fibrosus and the function and failure of the intervertebral disc. *Spine*, 5:106-116.
40. Hilton, R. C. (1976). Systematic studies of spinal mobility and Schmorl's nodes. In M. Jayson (Ed.). *The Lumbar Spine and Back Pain*. Kent: Pitman Medical Publishing Company Ltd, 115-131.
41. Hoshina, H. (1980). Spondylolysis in athletes. *The Physician and Sportsmedicine*, 3:75-78.
42. Jayson, M. I. V. (1983). Compression stresses in the posterior elements and pathologic consequences. *Spine*, 8:338-339.
43. Jonsson, B. (1970). The functions of individual muscles in the lumbar part of the erector spinae muscle. *Electromyography*, 10:5-21.
44. Juker, D., et al. (1998). Quantitative intramuscular myoelectric activity of lumbar portions of psoas and the abdominal wall during a wide variety of tasks. *Medicine & Science in Sports & Exercise*, 30:301-310.
45. Kelsey, J. L., et al. (1984). Acute prolapsed lumbar intervertebral disc: An epidemiological study with special reference to driving automobiles and cigarette smoking. *Spine*, 9:608-613.
46. Konrad, P., Schmitz, K., Denner, A. (2001). Neuromuscular evaluation of trunk training exercises. *Journal of Athletic Training*, 36:109-119.
47. Lander, J. E., et al. (1990). The effectiveness of weight-belts during the squat exercise. *Medicine & Science in Sports & Exercise*, 22:117-126.
48. Lindh, M. (1989). Biomechanics of the lumbar spine. In M. Nordin, V. H. Frankel (Eds.). *Basic Biomechanics of the Musculoskeletal System*. Philadelphia, PA: Lea & Febiger, 183-208.
49. Mannion, A. F., et al. (2000) Sudden and unexpected loading generates high forces on the lumbar spine. *Spine*, 842-852.
50. Marras, W. S., et al. (1998). Trunk muscle activities during asymmetric twisting motions. *Journal of Electromyography & Kinesiology*, 8:247-256.
51. McGill, S. M. (1991). Kinetic potential of the lumbar trunk musculature about three orthogonal orthopaedic axes in extreme postures. *Spine*, 16:809-815.
52. McGill, S. M. (1995). The mechanics of torso flexion: Sit-ups and standing dynamic flexion manoeuvres. *Clinical Biomechanics*, 10:184-192.
53. McGill, S. M., et al. (2000). Changes in lumbar lordosis modify role of the extensor muscles. *Clinical Biomechanics*, 15:777-780.
54. McGill, S. (2002). *Low Back Disorders. Evidence-Based Prevention and Rehabilitation*. Champaign, IL: Human Kinetics.
55. McLaughlin, T. M., et al. (1978). Kinetics of the parallel squat. *Research Quarterly*, 49:175-189.
56. McNeill, T., et al. (1980). Trunk strengths in attempted flexion, extension, and lateral bending in healthy subjects and patients with low-back disorders. *Spine*, 5:529-538.
57. Miller, J. A. A., et al. (1983). Posterior element loads in lumbar motion segments. *Spine*, 8:331-337.
58. Milne, J. S., Lauder, I. J. (1974). Age effects in kyphosis and lordosis in adults. *Annals of Human Biology*, 1:327-337.

59. Morris, F. L., et al. (2000). Compressive and shear force generated in the lumbar spine of female rowers. *International Journal of Sports Medicine*, 21:518-523.
60. Nachemson, A. (1960). Lumbar intradiscal pressure. Experimental studies of post-mortem material. *Acta Orthopaedica Scandinavica*, 43:10-104.
61. Nachemson, A. (1976). Lumbar intradiscal pressure. In M. Jayson (Ed.). *The Lumbar Spine and Back Pain*. Kent: Pitman Medical Publishing Company Ltd, 257-269.
62. Nachemson, A., Morris, J. M. (1964). In vivo measurements of intradiscal pressure: Discometry, a method for the determination of pressure in the lower lumbar discs. *The Journal of Bone & Joint Surgery*, 46(suppl A):1077-1092.
63. Nachemson, A., et al. (1979). Mechanical properties of human lumbar spine motion segments. Influence of age, sex, disc level and degeneration. *Spine*, 4:1-8.
64. Najarian, S., et al. (2005). Biomechanical effect of posterior elements and ligamentous tissues of lumbar spine on load sharing. *Bio-Medical Materials and Engineering*, 15:145-148.
65. Nisell, R., Ekholm, J. (1986). Joint load during the parallel squat in powerlifting and force analysis of in vivo bilateral quadriceps tendon rupture. *Scandinavian Journal of Sports Science*, 8:63-70.
66. Oddsson, L., Thorstensson, A. (1987). Fast voluntary trunk flexion movements in standing: Motor patterns. *Acta Physiologica Scandinavica*, 129:93-106.
67. Panjabe, M. M. (1992). The stabilizing system of the spine. Part 1. Function, dysfunction, adaptation, and enhancement. *Journal of Spinal Disorders*, 5:383-389.
68. Pauly, J. E. (1966). An electromyographic analysis of certain movements and exercises. I. Some deep muscles of the back. *Anatomical Record*, 155:223.
69. Perey, O. (1957). Fracture of the vertebral end plates in the lumbar spine. An experimental biomechanical investigation. *Acta Orthopaedica Scandinavica*, 25:1-101.
70. Plamondon, A., et al. (1995). Moments at the L5/S1 joint during asymmetrical lifting: effects of different load trajectories and initial load positions. *Clinical Biomechanics*, 10(3):128-136.
71. Plowman, S. A. (1992). Physical activity, physical fitness, and low-back pain. In J. O. Holloszy (Ed.). *Exercise and Sport Sciences Reviews*. Baltimore, MD: Lippincott Williams & Wilkins, 221-242.
72. Rab, G. T., et al. (1977). Muscle force analysis of the lumbar spine. *Orthopedic Clinics of North America*, 8:193.
73. Richardson, C. A., et al. (2002). *Therapeutic Exercise for Spinal Segmental Stabilization in Low Back Pain*. London: Churchill Livingstone.
74. Roozbazar, A. (1974). Biomechanics of lifting. In R. C. Nelson, C. A. Morehouse (Eds.). *Biomechanics IV*. Baltimore, MD: University Park Press, 37-43.
75. Rossi, F. (1978). Spondylolysis, spondylolisthesis and sports. *The Journal of Sports Medicine and Physical Fitness*, 18:317-340.
76. Saal, J. (1988). Rehabilitation of football players with lumbar spine injury: Part 1. *The Physician and Sportsmedicine*, 16:61-67.
77. Saunders, H. D. (1985). *Evaluation, Treatment and Prevention of Musculoskeletal Disorders*. Minneapolis, MN: Viking.
78. Schache, A. G. (2002). Three-dimensional and angular kinematics of the lumbar spine and pelvis during running. *Human Movement Science*, 21:273-293.
79. Schendel, M. J., et al. (1993). Experimental measurement of ligament force, facet force, and segment motion in the human lumbar spine. *Journal of Biomechanics*, 26:427-438.
80. Schultz, A. B., et al. (1982). Loads on the lumbar spine: Validation of a biomechanical analysis by measurements of intradiscal pressures and myoelectric signals. *The Journal of Bone & Joint Surgery*, 64-A:713-720.
81. Shah, J. S. (1976). Structure, morphology, and mechanics of the lumbar spine. In M. Jayson (Ed.). *The Lumbar Spine and Back Pain*. Kent: Pitman Medical Publishing Company Ltd, 339-405.
82. Shapiro, I., Frankel, V. H. (1989). Biomechanics of the cervical spine. In M. Nordin, V. H. Frankel (Eds.). *Basic Biomechanics of the Musculoskeletal System*. Philadelphia, PA: Lea & Febiger.
83. Snijders, C. J., et al. (1995). Oblique abdominal muscle activity in standing and in sitting on hard and soft surfaces. *Clinical Biomechanics*, 10:73-78.
84. Snijders, C. J., et al. (2004). The influence of slouching and lumbar support on iliolumbar ligaments, intervertebral discs and sacroiliac joints. *Clinical Biomechanics*, 19:323-329.
85. Soderberg, G. L. (1986). *Kinesiology: Application to Pathological Motion*. Baltimore, MD: Lippincott Williams & Wilkins.
86. Stephenson, J., Swank, A. M. (2004). Core training: Designing a program for anyone. *Strength & Conditioning Journal*, 26:34-38.
87. Swartz, E. E., et al. (2005). Cervical spine functional anatomy and the biomechanics of injury due to compressive loading. *Journal of Athletic Training*, 40:155-161.
88. Takala, E., et al. (1987). Electromyographic activity of hip extensor and trunk muscles during stooping and lifting. In B. Jonsson (Ed.). *Biomechanics X-A*. Champaign, IL: Human Kinetics.
89. Thorstensson, A., Carlson, H. (1987). Fiber types in human lumbar back muscles. *Acta Physiologica Scandinavica*, 131:195-202.
90. Thorstensson, A., et al. (1982). Lumbar back muscle activity in relation to trunk movements during locomotion in man. *Acta Physiologica Scandinavica*, 116:13-20.
91. Thorstensson, A., et al. (1984). Trunk movements in human locomotion. *Acta Physiologica Scandinavica*, 121:9-22.
92. Valencia, F. P., Munro, R. R. (1985). An electromyographic study of the lumbar multifidus in man. *Electromyography and Clinical Neurophysiology*, 25:205-221.
93. Van Herp, G., et al. (2000). Three-dimensional lumbar spinal kinematics: a study of range of movement in 100 healthy subjects aged 20 to 60+ years. *Rheumatology*, 39:1337-1340.
94. Vera-Garcia, F. J. (2006). Effects of different levels of torso activation on trunk muscular and kinematic responses to posteriorly applied sudden loads. *Clinical Biomechanics*, 21:443-455.
95. Vernon-Roberts, B. (1976). The pathology and interrelation of intervertebral disc lesions, osteoarthrosis of the apophyseal joints, lumbar spondylosis and low-back pain. In M. Jayson (Ed.). *The Lumbar Spine and Back Pain*. Kent: Pitman Medical Publishing Company Ltd, 83-113.
96. Weiker, G. G. (1989). Evaluation and treatment of common spine and trunk problems. *Clinics in Sports Medicine*, 8:399-417.
97. White, A. A., Panjabi, M. M. (1978). The basic kinematics of the spine. *Spine*, 3:12-20.
98. Wyke, B. (1976). The neurology of lower back pain. In M. Jayson (Ed.). *The Lumbar Spine and Back Pain*. Kent: Pitman Medical Publishing Company Ltd, 266-315.
99. Yates, J. W., et al. (1980). Static lifting strength and maximal isometric voluntary contractions of back, arm, and shoulder muscles. *Ergonomics*, 23:37-47.

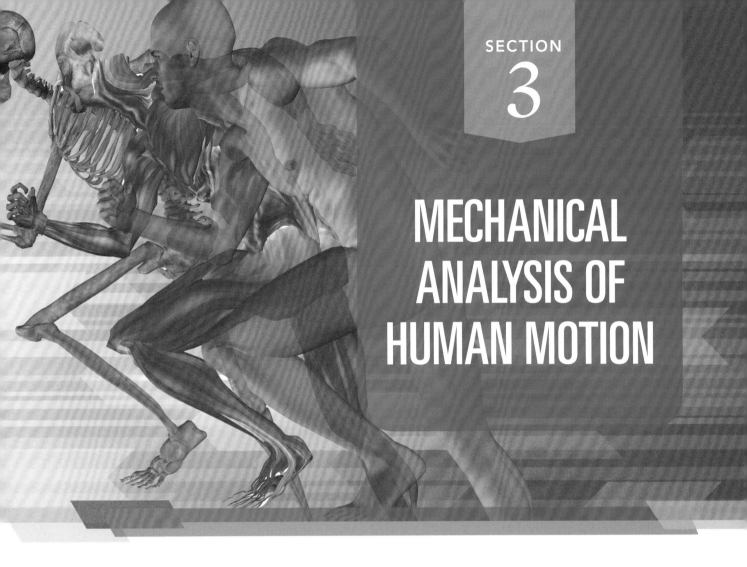

MECHANICAL ANALYSIS OF HUMAN MOTION

CHAPTER 8
Linear Kinematics

CHAPTER 9
Angular Kinematics

CHAPTER 10
Linear Kinetics

CHAPTER 11
Angular Kinetics

LINEAR KINEMATICS

OBJECTIVES

After reading this chapter, the student will be able to:

1. Describe how kinematic data are collected.

2. Distinguish between vectors and scalars.

3. Discuss the relationship among the kinematic parameters of position, displacement, velocity, and acceleration.

4. Conduct a numerical calculation of velocity and acceleration using the first central difference method.

5. Distinguish between average and instantaneous quantities.

6. Discuss various research studies that have used a linear kinematic approach.

7. Demonstrate knowledge of the three equations of constant acceleration.

8. Conduct a numerical calculation of the area under a parameter–time curve.

OUTLINE

Collection of Kinematic Data
   Reference Systems
   Movements Occur over Time
   Units of Measurement
   Vectors and Scalars
Position and Displacement
   Position
   Displacement and Distance
Velocity and Speed
   Slope
   First Central Difference Method
   Numerical Example
   Instantaneous Velocity
   Graphical Example
Acceleration
   Instantaneous Acceleration
   Acceleration and the Direction
     of Motion
   Numerical Example
   Graphical Example
Differentiation and Integration

Linear Kinematics of Walking
   and Running
   Stride Parameters
   Velocity Curve
   Variation of Velocity during Sports
Linear Kinematics of the Golf Swing
   Swing Characteristics
   Velocity and Acceleration of the Club
Linear Kinematics of Wheelchair
   Propulsion
   Cycle Parameters
   Propulsion Styles
Projectile Motion
   Gravity
   Trajectory of a Projectile
   Factors Influencing Projectiles
   Optimizing Projection Conditions
Equations of Constant Acceleration
   Numerical Example
Summary
Review Questions

The branch of mechanics that describes the spatial and temporal components of motion is called kinematics. The description of motion involves the position, velocity, and acceleration of a body with no consideration of the forces causing the motion. A kinematic analysis of motion may be either qualitative or quantitative. A qualitative kinematic analysis is a non-numerical description of a movement based on a direct observation. The description can range from a simple dichotomy of performance—good or bad—to a sophisticated identification of the joint actions. The key is that it is non-numerical and subjective. Examples include a coach's observation of an athlete's performance to correct a flaw in the skill, a clinician's visual observation of gait after application of a prosthetic limb, and a teacher's rating of performances in a skill test.

In biomechanics, the primary emphasis is on a quantitative analysis. The word quantitative implies a numerical result. In a quantitative analysis, the movement is analyzed numerically based on measurements from data collected during the performance of the movement. Movements may then be described with more precision and can also be compared mathematically with previous or subsequent performances. With the advent of affordable and sophisticated motion capture technology, quantitative systems are now readily available for use by coaches, teachers, and clinicians. Many of these professionals, who relied on qualitative analyses in the past, have joined researchers in the use of quantitative analyses. The advantages of a quantitative analysis are numerous. It provides a thorough, objective, and accurate representation of the movement. For example, podiatrists and physical therapists have at their disposal motion analysis tools that allow them to quantify the range of motion of the foot, movements almost impossible to track with the naked eye. These movements are important in the assessment of lower extremity function during locomotion.

A subset of kinematics that is particular to motion in a straight line is called linear kinematics. Translation or translational motion (straight-line motion) occurs when all points on a body or an object move the same distance over the same time. In Figure 8-1A, an object undergoes translation. The points A_1 and B_1 move to A_2 and B_2, respectively, in the same time following parallel paths. The distance from A_1 to A_2 and B_1 to B_2 is the same; thus, translation occurs. A skater gliding across the ice maintaining a pose is an example of translation. Although it appears that translation can occur only in a straight line, linear motion can occur along a curved path. This is known as curvilinear motion (Fig. 8-1B). While the object follows a curved path, the distance from A_1 to A_2 and B_1 to B_2 is the same and is accomplished in the same amount of time. For example, a sky diver falling from an airplane before opening the parachute undergoes curvilinear motion.

Collection of Kinematic Data

Kinematic data are collected for use in a quantitative analysis using several methods. Biomechanics laboratories, for example, may use accelerometers that measure the accelerations of body segments directly. The most common method of obtaining kinematic data, however, is high-speed video or optoelectric motion capture systems. The data obtained from high-speed video or optoelectric systems report the positions of body segments with respect to time. In the case of high-speed video, these data are acquired from the videotape by means of digitization. In optoelectric motion capture systems, markers on the body are tracked by a camera sensor that scans signals from infrared light-emitting diodes (active marker system), or the video capture unit serves as both the source and the recorder of infrared light that is reflected from a retroreflective marker (passive marker system). The location of the markers is sequentially fed into a computer, eliminating the digitization used in video systems. In all systems, the cameras are calibrated with a reference frame that allows for conversion between camera coordinates and a set of known actual coordinates of markers in the field of view.

REFERENCE SYSTEMS

Before any analysis, it is necessary to determine a spatial reference system in which the motion takes place. Biomechanists have many options in regard to a reference

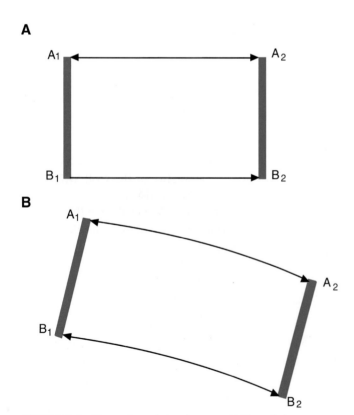

A

B

FIGURE 8-1 Types of translational motion. **(A)** Straight-line or rectilinear motion. **(B)** Curvilinear motion. In both A and B, the motion from A_1 to A_2 and B_1 to B_2 is the same and occurs in the same amount of time.

system. Most laboratories, however, use a Cartesian coordinate system. A Cartesian coordinate system is also referred to as a rectangular reference system. This system may either be two dimensional (2D) or three dimensional (3D).

> A computer program called MaxTRAQ is available for use to emphasize many of the concepts illustrated in this and later chapters. To obtain a copy of MaxTRAQ, go to the website below, and follow the instructions. After you have downloaded this program, it is strongly recommended that you use the tutorial to gain insight into how the program functions.
>
> Source: http://www.innovision-systems.com/lippincott/index.htm

A 2D reference system has two imaginary axes perpendicular to each other (Fig. 8-2A). The two axes (x, y) are positioned so that one is vertical (y) and the other is horizontal (x), although they may be oriented in any manner. It should be emphasized that the designations of these axes as x or y is arbitrary. The axes could easily be called a or b instead. What is important is to be consistent in naming the axes. These two axes (x and y) form a plane that is referred to as the x–y plane.

In certain circumstances, the axes may be reoriented such that one axis (y) runs along the long axis of a segment and the other axis (x) is perpendicular to the y-axis. As the

segment moves, the coordinate system also moves. Thus, the y-axis corresponding to the long axis of the segment moves with the result that the y-axis may not necessarily be vertical (Fig. 8-2B). This local reference system allows for the identification of a point on the body relative to an actual body segment rather than to an external reference point.

An ordered pair of numbers is used to designate any point with reference to the axes, with the intersection or origin of the axes designated as (0, 0). This pair of numbers is always designated in the order of the horizontal or x-value followed by the vertical or y-value. Thus, these are referred to as the ordinate (horizontal coordinate) and the abscissa (vertical coordinate), respectively. The ordinate (x-value) refers to the distance from the vertical axis, and the abscissa (y-value) refers to the distance from the horizontal axis. The coordinates are usually written as (horizontal; vertical; or x, y) and can be used to designate any point on the x–y plane.

A 2D reference system is used when the motion being described is planar. For example, if the object or body can be seen to move up or down (vertically) and to the right or to the left (horizontally) as viewed from one direction, the movement is planar. A 2D reference system results in four quadrants in which movements to the left of the origin result in negative x-values and movements below the origin result in negative y-values (Fig. 8-3). It is an advantage to place the reference system such that all of the points are within the first quadrant, where both x- and y-values are positive.

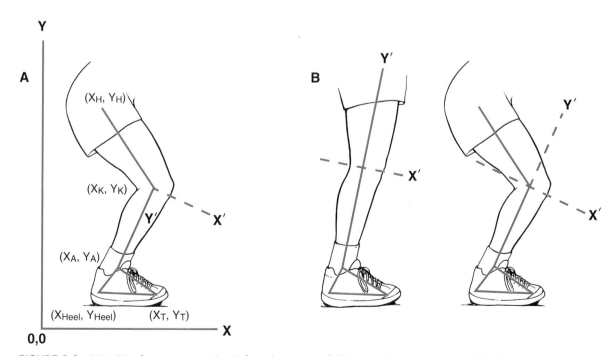

FIGURE 8-2 **(A)** A 2D reference system that defines the motion of all digitized points in a frame. **(B)** A 2D reference system placed at the knee joint center with the y-axis defining the long axis of the tibia.

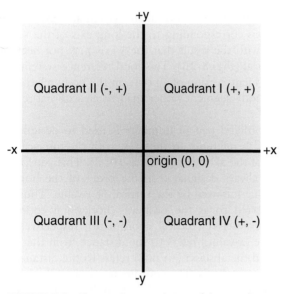

FIGURE 8-3   The quadrants and signs of the coordinates in a 2D coordinate system.

If an individual flexes and abducts the thigh while swinging it forward and out to the side, the movement would be not planar but 3D. A 3D coordinate system must be used to describe the movement in this instance. This reference system has three axes, each of which is perpendicular or orthogonal to the others, to describe a position relative to the horizontal or x-axis, to the vertical or y-axis, and to the mediolateral or z-axis. In any physical space, three pieces of information are required to accurately locate parts of the body or any point of interest because the concept of depth (z-axis; medial and lateral) must be added to the 2D components of height (y-axis; up and down) and width (x-axis; forward and backward). In a 3D system (Fig. 8-4), the coordinates are written as (horizontal; vertical; mediolateral; or x, y, z). The intersection

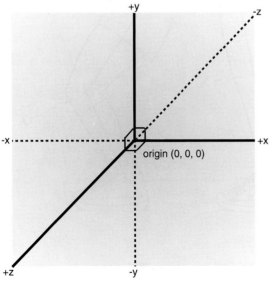

FIGURE 8-4   A 3D coordinate system.

of the axes or the origin is defined as (0, 0, 0) in 3D space. All coordinate values are positive in the first quadrant of the reference system, where the movements are horizontal and to the right (x), vertical and upward (y), and horizontal and forward (z). Correspondingly, negative movements are to the left (x), downward (y), and backward (z). In this system, the coordinates can designate any point on a surface, not just a plane, as in the 2D system. A 3D kinematic analysis of human motion is much more complicated than a 2D analysis and thus will not be addressed in this book.

Figure 8-5 shows a 2D coordinate system and how a point is referenced in this system. In this figure, point A is 5 units from the y-axis and 4 units from the x-axis. The designation of point A is (5,4). It is important to remember that the number designated as the x-coordinate determines the distance from the y-axis and the y-coordinate determines the distance from the x-axis. The distance from the origin to the point is called the resultant (r) and can be determined using the Pythagorean theorem as follows:

$$r = \sqrt{x^2 + y^2}$$

In the example from Figure 8-5:

$$r = \sqrt{5^2 + 4^2}$$
$$= 6.40$$

Before recording the movement, the biomechanist usually places markers on the end points of the body segments to be analyzed, allowing for later identification of the position and motion of that segment. For example, if the biomechanist is interested in a sagittal (2D) view of the lower extremity during walking or running, a typical placement of markers might be the toe, the fifth metatarsal, and the calcaneus of the foot; the lateral malleolus of the ankle; the lateral condyle of the knee; the greater trochanter of the hip; and the iliac crest. Figure 8-6 is a single frame of a recording illustrating a sagittal view of a runner using these specific markers. Appendix C presents 2D coordinates for one complete walking cycle using a whole-body set of markers.

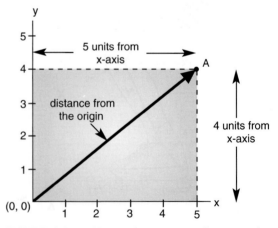

FIGURE 8-5   A 2D coordinate system illustrating the ordered pair of numbers defining a point relative to the origin.

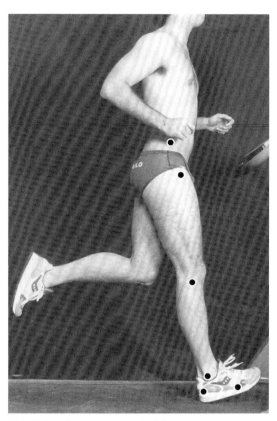

FIGURE 8-6 A runner marked for a sagittal kinematic analysis of the right leg.

For either a 2D or a 3D analysis, a global or stationary coordinate system is imposed on each frame of data, with the origin at the same location in each frame. In this way, each segment end point location can be referenced according to the same x–y (or x–y–z) axes and identified in each frame for the duration of the movement.

Refer to the walking data in Appendix C: Using the first frame, plot the ordered pairs of x, y coordinates for each of the segmental end points and draw lines connecting the segmental end points to create a stick figure.

If you have downloaded MaxTRAQ, you may use this program to digitize any of the video files and create a stick figure based on your digitized markers.

MOVEMENTS OCCUR OVER TIME

The analysis of the temporal or timing factors in human movement is an initial approach to a biomechanical analysis. In human locomotion, factors such as cadence, stride duration, duration of the stance or support phase (when the body is supported by a limb), duration of swing phase (when the limb is swinging through to prepare for the next ground contact), and the period of nonsupport may be investigated. The knowledge of the temporal patterns

of a movement is critical in a kinematic analysis because changes in position occur over time.

In a kinematic analysis, the time interval between each frame is determined by the sampling or frame rate of the camera or sensor. This forms the basis for timing the movement. Video cameras purchased in electronic stores generally operate at 24 to 30 fields or frames per second (fps). High-speed video cameras or motion capture units typically used in biomechanics can operate at 60, 120, 180, or 200 fps. At 60 fps, the time between each picture and frame is 1/60 seconds (0.01667 seconds); it is 1/200 seconds (0.005 seconds) at 200 fps. Usually, a key event at the start of the movement is designated as the beginning frame for digitization. For example, in a gait analysis, the first event may be considered to be the ground contact of the heel of the camera-side foot. With camera-side foot contact occurring at time zero, all subsequent events in the movement are timed from this event. The data collected for the walking trial in Appendix C are set up in this fashion, with data presented from time zero with the right-foot heel strike through to 1.15 seconds later, when the right-foot heel strike next occurs. The time of 1.15 seconds was computed from the sampling rate of 60 fps; the time between frames is 0.01667 seconds, and 69 frames were collected.

UNITS OF MEASUREMENT

If a quantitative analysis is conducted, it is necessary to report the findings in the correct units of measurement. In biomechanics, the metric system is used exclusively in scientific research literature. The metric system is based on the Système International d'Unités (SI). Every quantity of a measurement system has a dimension associated with it. The term dimension represents the nature of a quantity. In the SI system, the base dimensions are mass, length, time, and temperature. Each dimension has a unit associated with it. The base units of SI are the kilogram (mass), meter (length), second (time), and degrees Kelvin (temperature). All other units used in biomechanics are derived from these base units. The SI units and their abbreviations and conversion factors are presented in Appendix A. Because SI units are used most often in biomechanics, they are used exclusively in this text.

VECTORS AND SCALARS

Certain quantities, such as mass, distance, and volume, may be described fully by their amount or their magnitude. These are scalar quantities. For example, when one runs a race that is 5 km long, the distance or the magnitude of the race is 5 km. Additional scalar quantities that can be described with a single number include mass, volume, and speed. Other quantities, however, cannot be completely described by their magnitude. These quantities are called vectors and are described by both magnitude and direction. For example, when an object undergoes a displacement, the distance and the direction are important. Many

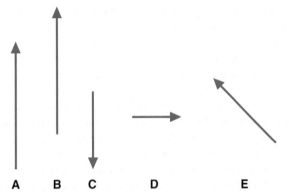

FIGURE 8-7 Vectors. Only vectors *A* and *B* are equal because they are equivalent in magnitude and direction.

of the quantities calculated in kinematic analysis are vectors, so a thorough understanding of vectors is necessary.

Vectors are represented by an arrow, with the magnitude represented by the length of the line and the arrow pointing in the appropriate direction (Fig. 8-7). Vectors are equal if their magnitudes are equal and they are pointed in the same direction.

Vectors can be added together. Graphically, vectors may be added by placing the tail of one vector at the head of the other vector (Fig. 8-8A). In Figure 8-8B, the vectors are not in the same direction, but the tail of *B* can still be placed at the head of *A*. Joining the tail of *B* to the head of *A* produces the vector *C*, which is the sum of *A* + *B*, or the resultant of the two vectors. Subtracting vectors is accomplished by adding the negative of one of the vectors. That is:

$$C = A - B$$

or

$$C = A + (-B)$$

This is illustrated in Figure 8-8C.

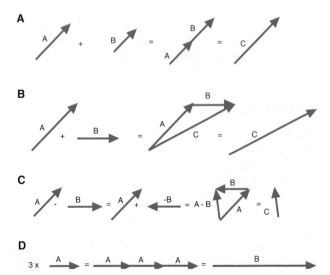

FIGURE 8-8 Vector operations illustrated graphically: **(A)** and **(B).** Addition. **(C)** Subtraction. **(D)** Multiplication by a scalar.

Vectors may also undergo forms of multiplication that are used mainly in a 3D analysis and, therefore, are not described in this book. Multiplication by a scalar, however, is discussed. Multiplying a vector by a scalar changes the magnitude of a vector but not its direction. Therefore, multiplying 3 (a scalar) times the vector *A* is the same as adding *A* + *A* + *A* (Fig. 8-8D).

A vector may also be resolved or broken down into its horizontal and vertical components. In Figure 8-9A, the vector *A* is illustrated with its horizontal and vertical components. The vector may be resolved into these components using the trigonometric functions sine and cosine (see Appendix B). A right triangle can consist of the two components and the vector itself. Consider a right triangle with sides *x*, *y*, and *a*, in which *a* is the hypotenuse of the right triangle (Fig. 8-9B). The sine of the angle theta (θ) is defined as:

$$\sin \theta = \frac{\text{length of side opposite } \theta}{\text{hypotenuse}}$$

or

$$\sin \theta = \frac{y}{r}$$

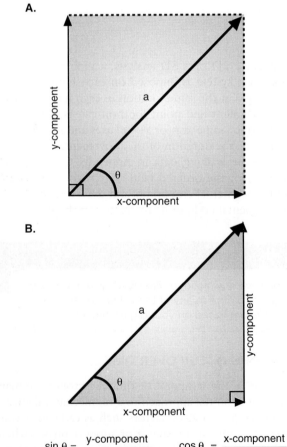

$$\sin \theta = \frac{y\text{-component}}{a} \qquad \cos \theta = \frac{x\text{-component}}{a}$$

FIGURE 8-9 Vector *a* resolved into its horizontal (*x*) and vertical (*y*) components using the trigonometric functions sine and cosine. **(A)** Components. **(B)** The components and vector form a right triangle.

The cosine of the angle θ is defined as:

$$\cos \theta = \frac{\text{length of side adjacent }\theta}{\text{hypotenuse}}$$

or

$$\cos \theta = \frac{x}{r}$$

If the vector components x and y and the resultant r form a right triangle and if the length of the resultant vector and the angle (θ) of the vector with the horizontal are known, the sine and cosine can be used to solve for the components.

For example, if the resultant vector has a length of 7 units and the vector is at an angle of 43°, the horizontal component is found using the definition of the cosine of the angle. That is:

$$\cos 43° = \frac{x}{a}$$

If the cos 43° is 0.7314 (see Appendix B), we can rearrange this equation to solve for the horizontal component:

$$x = a \cos 43°$$
$$= 7 \times 0.7314$$
$$= 5.12$$

The vertical component is found using the definition of the sine of the angle. That is:

$$\sin 43° = \frac{y}{a}$$

and if the sin 43° is 0.6820 (see Appendix B), we can rearrange this equation to solve for the vertical component y:

$$y = a \sin 43°$$
$$= 7 \times 0.6820$$
$$= 4.77$$

The lengths of the horizontal and vertical components are therefore 5.12 and 4.77, respectively. These two values identify the point relative to the origin of the coordinate system.

Often the vectors will be facing directions relative to the origin that are not in the first quadrant (Fig. 8-3). Take, for example, the vector illustrated in Figure 8-10. In this case, a vector of length 12 units lies at an angle of 155°, placing it in the second quadrant, where the x values are to the left and negative. Resolution of this vector into horizontal and vertical components can be computed in a number of ways, depending on which angle you choose to calculate. The vertical component of the vector can be computed using:

$$y = a \sin \theta_1$$
$$= a \sin 155°$$
$$y = 12 \times 0.4226$$
$$= 5.07$$

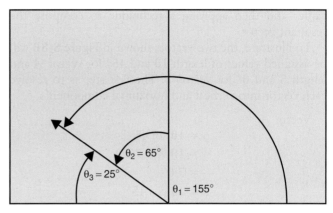

FIGURE 8-10 The orientation of a vector can be described relative to a variety of references, including the right horizontal (θ_1), the vertical (θ_2), and the left horizontal (θ_3).

or if you choose to use θ_2:

$$y = a \cos \theta_2$$
$$y = a \cos 65°$$
$$y = 12 \times 0.4226$$
$$= 5.07$$

or if you choose to use θ_3:

$$y = a \sin \theta_3$$
$$y = a \sin 25°$$
$$y = 12 \times 0.4226$$
$$= 5.07$$

Similarly, the horizontal component of the vector can be computed using the same angles:

$$x = a \cos \theta_1$$
$$x = a \cos 155°$$
$$x = 12 \times -0.9063$$
$$= -10.88$$

or if you choose to use θ_2:

$$x = a \sin \theta_2$$
$$x = a \sin 65°$$
$$x = 12 \times 0.9063$$
$$= -10.88$$

(x is negative in Quadrant II), or if you choose to use θ_3:

$$x = a \cos \theta_3$$
$$x = a \cos 25°$$
$$x = 12 \times 0.9063$$
$$= -10.88$$

It is common to work with multiple vectors that must be combined to evaluate the resultant vector. Vectors can be graphically combined by connecting the vectors head to tail and joining the tail of the first one with the head of the last one to obtain the resultant vector (Fig. 8-8). This can also be done by first resolving each vector into x and y components using the trigonometric technique described earlier and then applying a technique to compose the resultant vector.

To illustrate, the two vectors shown in Figure 8-8B will be assigned values of length 10 and 45° for vector A and length 5 and 0° for vector B. The first step is to resolve each vector into vertical and horizontal components.

Vector A:

$$y = 10 \sin 45°$$
$$y = 10 \times 0.7071$$
$$= 7.07$$
$$x = 10 \cos 45°$$
$$x = 10 \times 0.7071$$
$$= 7.07$$

Vector B:

$$y = 5 \sin 0°$$
$$y = 5 \times 0.0$$
$$= 0$$
$$x = 5 \cos 0°$$
$$x = 5 \times 1.000$$
$$= 5.00$$

To find the magnitude of the resultant vector, the horizontal and vertical components of each vector are added and resolved using the Pythagorean theorem:

	Horizontal Components	Vertical Components
Vector A	7.07	7.07
Vector B	5.00	0.00
Sum (S)	12.07	7.07

$$C = \sqrt{x^2 + y^2}$$
$$C = \sqrt{12.7^2 + 7.07^2}$$
$$= \sqrt{145.69 + 49.99}$$
$$= \sqrt{195.68}$$
$$= 13.99$$

To find the angle of resultant vector, the trigonometric functions the tangent and the arctangent or inverse tangent (see Appendix B) are used. In this example, these functions can be used to calculate the angle between the vectors:

$$\tan \theta = S \frac{y - \text{component}}{x - \text{component}}$$
$$\theta = \arctan \left(\frac{7.07}{12.07} \right)$$
$$\theta = \arctan (0.5857)$$
$$= 30.36°$$

The resultant vector C has a length of 13.99 and an angle of 30.36°. This composition of multiple vectors can be applied to any number of vectors.

 ## Position and Displacement

POSITION

The position of an object refers to its location in space relative to some reference. Units of length are used to measure the position of an object from a reference axis. Because the metric system is always used in biomechanics, the most commonly used unit of length is the meter. For example, a platform diver standing on a 10-m tower is 10 m from the surface of the water. The reference is the water surface, and the diver's position is 10 m above the reference. The position of the diver may be determined throughout the dive with a height measured from the water surface. As previously mentioned, the analysis of video or sensor frames determines the position of a body or segment end point relative to two references in a 2D reference system, the x-axis and the y-axis. The walking example in Appendix C has the 2D reference frame originating on the ground in the middle of the experimental area. This makes all y values positive because they are relative to the ground and all x values positive or negative depending on whether the body segment is behind (2) or in front (1) of the origin in the middle of the walking area.

DISPLACEMENT AND DISTANCE

When the diver leaves the platform, motion occurs, as it does whenever an object or body changes the position. Objects cannot instantaneously change the position, so time is a factor when considering motion. Motion, therefore, may be defined as a progressive change of position over time. In this example, the diver undergoes a 10-m displacement from the diving board to the water. Displacement is measured in a straight line from one position to the next. Displacement should not be confused with distance.

The distance an object travels may or may not be a straight line. In Figure 8-11, a runner starts the race, runs to point A, turns right to point B, turns left to point C, turns right to point D, and then turns left to the finish. The distance run is the actual length of the path traveled. Displacement, on the other hand, is a straight line between the start and the finish of the race.

Displacement is defined both by how far the object has moved from its starting position and by the direction it moved. Because displacement inherently describes the magnitude and direction of the change in position, it is a vector quantity. Distance, because it refers only to how far an object moved, is a scalar quantity.

The capitalized Greek letter delta (Δ) refers to a change in a parameter; thus, Δs means a change in s. If s represents

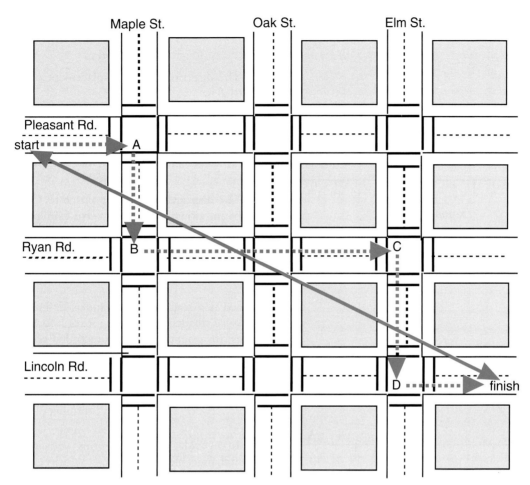

FIGURE 8-11 A runner moves along the path followed by the dotted line. The length of this path is the distance traveled. The length of the solid line is the displacement.

the position of a point, then Δs is the displacement of that point. Subscripts f and i refer to the final position and the initial position, respectively, with the implication that the final position occurred after the initial position. Mathematically, displacement (Δs) is for the general case:

$$\Delta s = s_f - s_i$$

where s_f is the final position and s_i is the initial position. Displacement for each component of position may also be calculated as follows:

$$\Delta x = x_f - x_i$$

for horizontal displacement and

$$\Delta y = y_f - y_i$$

for vertical displacement.

The resultant displacement may also be calculated using the Pythagorean relationship as follows:

$$r = \sqrt{\Delta x^2 + \Delta y^2}$$

For example, if an object is at position A (1, 2) at time 0.02 seconds and position B (7, 7) at time 0.04 seconds (Fig. 8-12A), the horizontal and vertical displacements are:

$$\Delta x = 7 \text{ m} - 1 \text{ m}$$
$$= 6 \text{ m}$$
$$\Delta y = 7 \text{ m} - 2 \text{ m}$$
$$= 5 \text{ m}$$

The object is displaced 6 m horizontally and 5 m vertically. The movement may also be described as to the right and upward relative to the origin of the reference system. The resultant displacement or the length of the vector from A to B may be calculated as:

$$r = \sqrt{6^2 \text{ m} + 5^2 \text{ m}}$$
$$= 7.81 \text{ m}$$

The direction of the displacement of the vector from A to B may be calculated as:

$$\theta = \arctan \left(\frac{5}{6} \right)$$
$$\theta = 0.833$$
$$= 39.8°$$

Therefore, the point is displaced 7.81 m up and to the right of the origin at 39.8°.

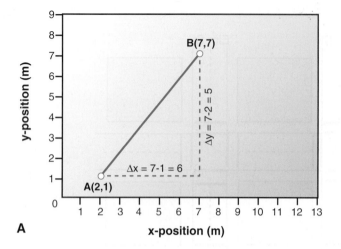

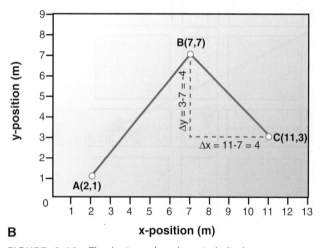

A

B

FIGURE 8-12 The horizontal and vertical displacements in a coordinate system of the path from **(A)** *A* to *B* and **(B)** *B* to *C*.

Using MaxTRAQ, import the video file of the woman walking. Find the frame at which the right foot first contacts the ground. What is the horizontal, vertical, and resultant displacement of the head between this frame and the subsequent frame when both feet are in contact with the ground?

Consider Figure 8-12B. In a successive position to *B*, the object moved to position *C* (11, 3). The displacement is:

$$\Delta x = 11 \text{ m} - 7 \text{ m}$$
$$= 4 \text{ m}$$
$$\Delta y = 3 \text{ m} - 7 \text{ m}$$
$$= 24 \text{ m}$$

The object would have been displaced 4 m horizontally and 4 m vertically, or 4 m to the right away from the *y*-axis

and 4 m down toward the *x*-axis. The resultant displacement between points *B* and *C* is:

$$r = \sqrt{4^2 \text{ m} + 4^2 \text{ m}}$$
$$= 5.66 \text{ m}$$

The direction of the displacement of the vector from *A* to *B* is:

$$\theta = \arctan\left(\frac{-4}{4}\right)$$
$$\theta = \arctan(-1)$$
$$= -45°$$

The displacement from point *B* to *C* is 5.66 m to the right and down toward the *x*-axis from point *B* at an angle of 45° below the horizontal.

Velocity and Speed

Speed is a scalar quantity and is defined as the distance traveled divided by the time it took to travel. In automobiles, for example, speed is recorded continuously by the speedometer as one travels from place to place. In the case of the automobile, speed is measured in miles per hour or kilometers per hour. Thus:

$$\text{Speed} = \frac{\text{distance}}{\text{time}}$$

In everyday use, the terms velocity and speed are interchangeable, but whereas velocity, a vector quantity, describes magnitude and direction, speed, a scalar quantity, describes only magnitude. In road races, the start is usually close to the finish, and the velocity over the whole race may be quite small. In this case, speed may be more important to the participant.

Velocity is a vector quantity defined as the time rate of change of position. In biomechanics, velocity is generally of more interest than speed. Velocity is usually designated by the lowercase letter *v* and time by the lower case letter *t*. Velocity can be determined by:

$$v = \frac{\text{displacement}}{\text{time}}$$

Specifically, velocity is:

$$v = \frac{\text{position}_f - \text{position}_i}{\text{time at final position} - \text{time at initial position}}$$
$$= \frac{\text{change in position}}{\text{change in time}}$$
$$= \frac{\Delta s}{\Delta t}$$

The most commonly used unit of velocity in biomechanics is meters per second (m/s or m · s⁻¹), although any unit of length divided by a unit of time is correct as long as it is appropriate to the situation. The units for velocity can be determined by using the formula for velocity and dividing the units of length by units of time.

$$\text{Velocity} = \frac{\text{displacement(m)}}{\text{time(seconds)}}$$
$$= \text{m/s or m} \cdot \text{s}^{-1}$$

Consider the position of an object that is at point A (2, 4) at time 1.5 seconds and moved to point B (4.5, 9) at time 5 seconds. The horizontal velocity (v_x) is:

$$v_x = \frac{4.5\,\text{m} - 2\,\text{m}}{5\,\text{seconds} - 1.5\,\text{seconds}}$$
$$= \frac{2.5\,\text{m}}{3.5\,\text{seconds}}$$
$$= 0.71\ \text{m/s}$$

The vertical velocity (v_y) could be similarly determined by:

$$v_y = \frac{9\,\text{m} - 4\,\text{m}}{5\,\text{seconds} - 1.5\,\text{seconds}}$$
$$= \frac{5\,\text{m}}{3.5\,\text{seconds}}$$
$$= 1.43\ \text{m/s}$$

The resultant magnitude or overall velocity can be calculated using the Pythagorean relationship as follows:

$$v = \sqrt{0.71^2 + 1.43^2}$$
$$= \sqrt{2.55}$$
$$= 1.60\ \text{m/s}$$

The resultant direction of the velocity is:

$$\tan\theta = \frac{y}{x}$$
$$\theta = \arctan\left(\frac{1.43}{0.71}\right)$$
$$\theta = \arctan(2.04)$$
$$= 63.92°$$

A sample of velocity measures are presented in Table 8-1. As you can see, there is a wide range of velocities in human movement, from the range of 0.7 to 1 m/s for a slow walk to the range of 43 to 50 m/s for a club head in the golf swing.

Using MaxTRAQ, import the video file of the woman walking. Find the frame at which the right foot first contacts the ground. Digitize the right ear in this frame and four frames later. The time between frames is 0.0313 seconds. What are the horizontal, vertical, and resultant velocity of the head between this frame and the subsequent frame?

TABLE 8-1	Sample Linear Velocity Examples	
Action		**Linear Velocity (m/s)**
Golf club head forward velocity at impact (23)		43
High jump approach velocity		7–8
High jump horizontal and vertical velocity at takeoff (8)		4.2, 4
Long jump approach velocity (22)		9.5–10
Pitching, fast ball velocity at release (10)		35.1
Pitching, curve ball velocity at release (10)		28.2
Vertical velocity at takeoff, squat jump, and countermovement jump (11)		3.43, 3.8
Hopping vertical velocity (11)		1.52
Walking forward velocity		0.7–3
Race walking		4
Running, sprinting		4–10
Wheelchair propulsion (30)		1.11–2.22

SLOPE

Figure 8-13 is an illustration of the change in horizontal position or position along the x-axis as a function of time. In this graph, the geometric expression describing the change in horizontal position (Δx) is called the rise. The expression that describes the change in time (Δt) is called the run. The slope of a line is:

$$\text{Slope} = \frac{\text{rise}}{\text{run}} = \frac{\Delta x}{\Delta t}$$

The steepness of the slope gives a clear picture regarding the velocity. If the slope is very steep, that is, a large number, the position is changing rapidly, and the velocity is great. If the slope is zero, the object has not changed the position, and the velocity is zero. Because velocity is a vector, it can have both positive and negative slopes. Figure 8-14 shows positive, negative, and zero slopes.

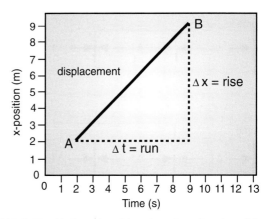

FIGURE 8-13   Horizontal position plotted as a function of time. The slope of the line from A to B is $\frac{\Delta x}{\Delta t}$.

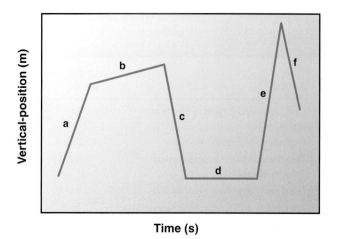

FIGURE 8-14 Different slopes on a vertical position versus time graph. Slopes a, b, and e are positive. Slopes c and f are negative; d has zero slope.

Lines *a* and *b* have positive slopes, implying that the object was displaced away from the origin of the reference system. Line *a* has a steeper slope than line *b*, however, indicating that the object was displaced a greater distance per unit time. Line *c* illustrates a negative slope, indicating that the object was moving toward the origin. Line *d* shows a zero slope, meaning that the object was not displaced either away from or toward the origin over that time. Lines *e* and *f* have identical slopes, but *e*'s slope is positive and *f*'s is negative.

FIRST CENTRAL DIFFERENCE METHOD

The kinematic data collected in certain biomechanical studies are based on positions of the segment end points generated from each frame of video with a time interval based on the frame rate of the camera. This presents the biomechanist with all of the information needed to calculate velocity. When velocity over a time interval is calculated, however, the velocity at either end of the time interval is not generated; that is, the calculated velocity cannot be assumed to occur at the time of the final position or at the time of the initial position. The position

of an object can change over a period less than the interval between video frames. Thus, the velocity calculated between two video frames represents an average of the velocities over the whole time interval between frames. An average velocity, therefore, is used to estimate the change in position over the time interval. This is not the velocity at the beginning or end of the time interval. If this is the case, there must be some point in the time interval between frames when the calculated velocity occurs. The best estimate for the occurrence of this velocity is at the midpoint of the time interval. For example, if the velocity is calculated using the data at frames 4 and 5, the calculated velocity would occur at the midpoint of the time interval between frames 4 and 5 (Fig. 8-15A).

If data are collected at 60 fps, the positions at video frames 1 to 5 occur at the times 0, 0.0167, 0.0334, 0.0501, and 0.0668 seconds. The velocities calculated using this method occur at the times 0.0084, 0.0251, 0.0418, and 0.0585 seconds. This means that after using the general formula for calculating velocity, the positions obtained from the video and velocities calculated are not exactly matched in time. Although this problem can be overcome, it may be inconvenient in certain calculations.

To overcome this problem, the most often used method for calculating velocity is the first central difference method. This method uses the difference in positions over two frames as the numerator. The denominator in the velocity calculation is the change in time over two time intervals. The formula for this method is:

$$v_{xi} = \frac{x_{i+1} - x_{i-1}}{2\Delta t}$$

for the horizontal component and

$$v_{yi} = \frac{y_{i+1} - y_{i-1}}{2\Delta t}$$

for the vertical component.

This infers that the velocity at frame i is calculated using the positions at frame i_{11} and frame i_{21}. Use of $2\Delta t$ renders the velocity at the same time as frame i because that is the midpoint of the time interval. For example, if

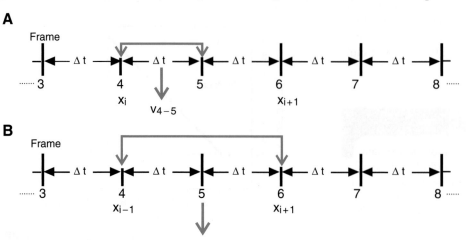

FIGURE 8-15 The location in time of velocity. **(A)** Using the traditional method over a single time interval. **(B)** Using the first central difference method.

the velocity at frame 5 is calculated, the data at frames 4 and 6 are used. If the time of frame 4 is 0.0501 seconds and frame 6 is 0.0835 seconds, the velocity calculated using this method would occur at time 0.0668 seconds, or at frame 5 (Fig. 8-15B).

Similarly, if the velocity at frame 3 is calculated, the positions at frames 2 and 4 are used. Because the time interval between the two frames is the same, the change in time would be 2 times Δt. If the horizontal velocity at the time of frame 13 is calculated, the following equation would be used:

$$v_{x_{13}} = \frac{x_{14} - x_{12}}{t_{14} - t_{12}}$$

The location of the calculated velocity would be at t_{13}, or the same point in time as frame 13. This method of computation exactly aligns in time the position and velocity data. It is assumed that the time intervals between frames of data are constant. As pointed out previously, this is usually the case in biomechanical studies.

The first central difference method uses the data point before and after the point where velocity is calculated. One problem is that data will be missing at the beginning and end of the video trial. This means that either the velocity at the beginning and end of the trial are estimated or some other means are used to evaluate the velocity at these points. A simple method is to collect and analyze several frames before and after the movement of interest. For example, if a walking stride was analyzed, the first contact of the right foot on the ground might be picked as the beginning event for the trial. In that case, at least one frame before that event would be analyzed to calculate the velocity at the instant of right-foot contact. Similarly, if the ending event in the trial is the subsequent right-foot contact, at least one frame beyond that event would be analyzed to calculate the velocity at the end event. In practice, biomechanists generally digitize several frames before and after the trial.

NUMERICAL EXAMPLE

The data in Table 8-2 represent the vertical movement of an object over 0.167 seconds. In this set of data, the rate of the camera was 60 fps, so that t was 0.0167 seconds. The object starts at rest, first moves up for 0.1002 seconds and then moves down beyond the starting position before returning to the starting position.

To illustrate, using the formula for the first central difference method, the computation of the velocity at the time for frame 3 is as follows:

$$v_{y3} = \frac{y_4 - y_2}{t_4 - t_2}$$

$$= \frac{0.27\,\text{m} - 0.15\,\text{m}}{0.0501\,\text{seconds} - 0.0167\,\text{seconds}}$$

$$= 3.59\,\text{m/s}$$

Table 8-2 shows the calculation of the velocity for each frame using the first difference method. Figure 8-16 shows the position and velocity profiles of this movement. Each of these calculated velocities represents the slope of the straight line, indicating the rate of position change within that time interval or the average velocity over that time interval. As the position changes rapidly, the slope of the velocity curve becomes steeper, and as the position changes less rapidly, the slope is less steep.

Refer to the walking data in Appendix C. Compute the horizontal and vertical velocity of the knee joint through the total walking cycle using the first difference method. Graph both the horizontal and vertical velocity and discuss the linear kinematic characteristics of the knee joint through the stance (frames 1 to 41) and swing (frames 41 to 69) phases.

TABLE 8-2	Calculation of Velocity from a Set of Position–Time Data		
Frame	Time (s)	Vertical Position (y) (m)	Vertical Velocity (v_y) (m/s)
1	0.0000	0.00	
2	0.0167	0.15	(0.22 − 0.00) / (0.0334 − 0.00) = 6.59
3	0.0334	0.22	(0.27 − 0.15) / (0.0501 − 0.0167) = 3.59
4	0.0501	0.27	(0.30 − 0.22) / (0.0668 − 0.0334) = 2.40
5	0.0668	0.30	(0.20 − 0.27) / (0.0835 − 0.0501) = − 2.10
6	0.0835	0.20	(0.00 − 0.30) / (0.1002 − 0.0668) = − 8.98
7	0.1002	0.00	(− 0.26 − 0.20) / (0.1169 − 0.0835) = −13.77
8	0.1169	− 0.26	(− 0.30 − 0.00) / (0.1336 − 0.1002) = − 8.98
9	0.1336	− 0.30	(− 0.22 − (− 0.26)) / (0.1503 − 0.1169) = 1.20
10	0.1503	− 0.22	(0.00 − (− 0.30)) / (0.1670 − 0.1336) = 8.98
11	0.1670	0.00	

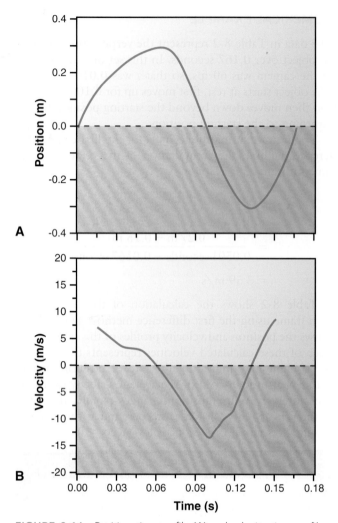

A

B

FIGURE 8-16   Position–time profile **(A)** and velocity–time profile **(B)** of the data in Table 8-2.

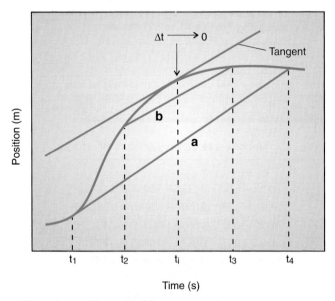

FIGURE 8-17   The slope of the secant *a* is the average velocity over the time interval t_1 to t_4. The slope of secant *b* is the average velocity over the time interval t_2 to t_3. The slope of the tangent is the instantaneous velocity at the time interval t_i when the time interval is so small that in effect it is zero.

INSTANTANEOUS VELOCITY

Even when using the first central difference method, an average velocity over a time interval is computed. In some instances, it may be necessary to calculate the velocity at a particular instant. This is called the instantaneous velocity. When the change in time, Δt, becomes smaller and smaller, the calculated velocity is the average over a much briefer time interval. The calculated value then approaches the velocity at a particular instant in time. In the process of making the time interval progressively smaller, the *t* will eventually approach zero. In the branch of mathematics called calculus, this is called a limit. A limit occurs when the change in time approaches zero. The concept of the limit is graphically illustrated in Figure 8-17. If the velocity is calculated over the interval from t_1 to t_2, as is done using the first central difference method, the slope of a line called a secant is calculated. A secant line intersects a curved line at two points on the curve. The slope of this secant is the average velocity over the time interval t_1 to t_2. When change in time approaches zero, however, the slope line actually touches the curve at only one point. This slope line is actually a line tangent to the curve, that is, a line that touches the curve at only one point. The slope of the tangent represents the instantaneous velocity because the time interval is so small that it may as well be zero.

Instantaneous velocity, therefore, is the slope of a line tangent to the position–time curve. In calculus, instantaneous velocity is expressed as a limit. The numerator in a limit is represented by d*x* or d*y*, meaning a very, very small change in position in the horizontal or vertical positions, respectively. The denominator is referred to as d*t*, meaning a very, very small change in time. For the horizontal and vertical cases, the formulae for instantaneous velocity expressed as limits:

$$\text{limit } v_x = \frac{dx}{dt}$$
$$dt \rightarrow 0$$
$$\text{limit } v_y = \frac{dy}{dt}$$
$$dt \rightarrow 0$$

For the instantaneous horizontal velocity, this is read as d*x*/d*t*, or the limit of v_x as d*t* approaches zero. It is also known as the derivative of *x* with respect to *t*. Similarly, the instantaneous vertical velocity, d*y*/d*t*, is the limit of v_y as d*t* approaches zero or the derivative of *y* with respect to *t*.

GRAPHICAL EXAMPLE

It is possible to graph an estimation of the shape of a velocity curve based on the shape of the position–time profile. The ability to do this is critical to demonstrate our understanding of the concepts previously discussed. Two such concepts will be used to construct the graph: (a) the slope and (b) the local extremum. A local extremum is the point at which a curve changes the direction (when

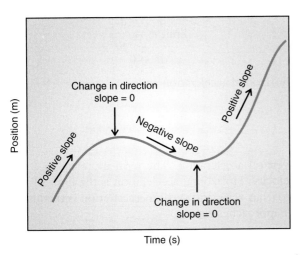

FIGURE 8-18 Local extrema (slope 0) on a position–time graph.

it reaches a maximum or a minimum). The slope at this point is zero, so the derivative of the curve at that point in time will be zero (Fig. 8-18). That is, when the position changes the direction, the velocity at the point of the change in direction will be instantaneously zero.

In Figure 8-19A, the horizontal position of an object is plotted as a function of time. The local extrema, the points at which the curve changes direction, are indicated as P_1, P_2, and P_3. At these points, by definition, the velocity will be zero. If the velocity curve is to be constructed on the same time line, these points can be projected to the velocity–time line, knowing that the velocity at these points will be zero. The slopes of each section of the

position–time curve are 1) positive, 2) negative, 3) positive, and 4) negative. From the beginning of the motion to the local extremum P_1, the object was moving in a positive direction, but at the local extremum P_1, the velocity was zero. The corresponding velocity curve in this section must increase positively and then become less positive, thereby returning to zero. In section 2 of the position–time curve, the slope is negative, indicating that the velocity must be negative. The local extrema, P_1 and P_2, however, indicate that the velocity at these points will be zero. Thus, in section 2, the corresponding velocity curve starts at zero, increases negatively, and then becomes less negative, returning to zero at P_2. Similarly, the shape of the velocity curve can be generated for sections 3 and 4 on the position curve (Fig. 8-19B).

Acceleration

In human motion, the velocity of a body or a body segment is rarely constant. The velocity often changes throughout a movement. Even when the velocity is constant, it may be so only when averaged over a large time interval. For example, in a distance race, the runner may run consecutive 400 m distances in 65 seconds, indicating a constant velocity over each distance. A detailed analysis, however, would reveal that the runner actually increased and decreased velocity, with the average over the 400 m being constant. In fact, it has been shown that runners decrease and then increase velocity during each ground contact with each foot (2). If velocity continually changes, it would appear that these variations in velocity should be noted. In addition, the rate at which velocity changes can be related to the forces that cause movement.

The rate of change of velocity with respect to time is called acceleration. In everyday usage, accelerating means speeding up. In a car, when the accelerator is depressed, the speed of the car increases. When the accelerator is released, the speed of the car decreases. In both instances, the direction of the car is not a concern because speed is a scalar. Acceleration, however, refers to both increasing and decreasing velocities. Because velocity is a vector, acceleration must also be a vector.

Acceleration, usually designated by the lowercase letter a, can be determined thus:

$$a = \frac{\text{change in velocity}}{\text{change in time}}$$

More generally,

$$a = \frac{\text{velocity}_f - \text{velocity}_i}{\text{time at final position} - \text{time at initial position}}$$

$$a = \frac{\text{change in velocity}}{\text{change in time}}$$

$$= \frac{\Delta v}{\Delta t}$$

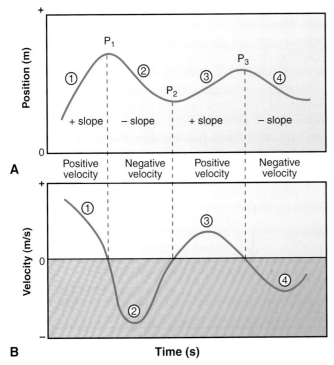

FIGURE 8-19 The position–time curve **(A)** and the respective velocity–time curve **(B)** drawn using the concepts of local extrema and slopes.

The units of acceleration are the unit of velocity (m/s) divided by the unit of time (second) resulting in meters per second per second (m/s/s) or m/s² or m·s⁻²:

$$\text{acceleration} = \frac{\text{velocity (m/s)}}{\text{time (second)}}$$

This is the most common unit of acceleration used in biomechanics.

The first central difference method is used to calculate acceleration in many biomechanical studies. The use of this method means that the calculated acceleration is associated with a time in the movement in which a calculated velocity and a digitized point are also associated. The first central difference formula for calculating acceleration is analogous to that for calculating velocity:

$$a_{xi} = \frac{vx_{i+1} - vx_{i-1}}{2\Delta t}$$

for the horizontal component and

$$a_{xi} = \frac{vy_{i+1} - vy_{i-1}}{2\Delta t}$$

for the vertical component. For example, to calculate the acceleration at frame 7, the velocity values at frames 8 and 6 and 2 times the time interval between individual frames would be used.

INSTANTANEOUS ACCELERATION

Because acceleration represents the rate of change of a velocity with respect to time, the concepts regarding velocity also apply to acceleration. Thus, acceleration may be represented as a slope indicating the relationship between velocity and time. On a velocity–time graph, the steepness and direction of the slope indicate whether the acceleration is positive, negative, or zero.

Instantaneous acceleration may be defined in an analogous fashion to instantaneous velocity. That is, instantaneous acceleration is the slope of a line tangent to a velocity–time graph or as a limit:

$$\text{limit } a_x = \frac{dv_x}{dt}$$
$$dt \to 0$$

for horizontal acceleration and

$$\text{limit } a_y = \frac{dv_y}{dt}$$
$$dt \to 0$$

for vertical acceleration. The term dv refers to a change in velocity. Horizontal acceleration is the limit of v_x as dt approaches zero, and vertical acceleration is the limit of v_y as dt approaches zero.

ACCELERATION AND THE DIRECTION OF MOTION

One complicating factor in understanding the meaning of acceleration relates to the direction of motion of an object. The term accelerate is often used to indicate an increase in velocity, and the term decelerate to describe a decrease in velocity. These terms are satisfactory when the object under consideration is moving continually in the same direction. Even if velocity and, therefore, acceleration change, the direction in which the object is traveling may not change. For example, a runner in a 100-m sprint race starts from rest or from a zero velocity. When the race begins, the runner increases velocity up to the 50-m point, and acceleration is positive. After the 50-m mark, the runner's velocity may not change for some of the race; there is zero acceleration. Having crossed the finish line, the runner reduces velocity; this is negative acceleration. Eventually, the runner comes to rest, at which point velocity equals zero. Throughout the race, the runner moved in the same direction but had positive, zero, and negative acceleration. Therefore, it is clear that the sign of the acceleration cannot be determined solely from the direction of motion.

Consider an athlete completing a shuttle run that consists of one 10-m run away from a starting position, followed by a 10-m run back to the starting position. The two sections of this run are illustrated in Figure 8-20.

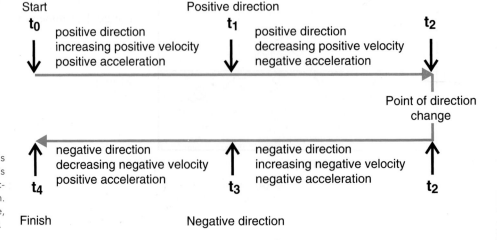

FIGURE 8-20 Motion to the right is regarded as positive and to the left is negative. Positive or negative velocity is based on the direction of motion. Acceleration may be positive, negative, or zero based on the change in velocity.

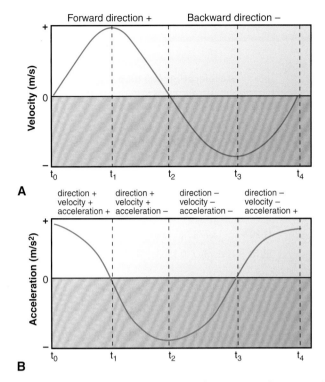

A

Forward direction +   Backward direction –

direction +
velocity +
acceleration +

direction +
velocity +
acceleration –

direction –
velocity –
acceleration –

direction –
velocity –
acceleration +

B

FIGURE 8-21 The graphical relationship between acceleration and direction of motion during a shuttle run (t_2 denotes when the runner changed direction).

The first 10-m section of the run may be considered a run in a positive direction. The runner increases velocity and then, approaching the turnaround point, must decrease the positive velocity. Thus, the runner must have a positive acceleration followed by negative acceleration. Figure 8-21 presents an idealized horizontal velocity profile and the corresponding horizontal acceleration for the shuttle run. The 10-m run in one direction from t_0 to t_2 illustrates that the positive velocity as change in position was constantly away from the y-axis. In addition, whereas the slope of the velocity curve from t_0 to t_1 is positive, indicating positive acceleration when the runner increases velocity, the slope of the velocity curve from t_1 to t_2 is negative, resulting in negative acceleration as the runner decreases velocity in anticipation of stopping and turning around.

At the turnaround point, the runner, now running in a negative direction, increases the negative velocity (Fig. 8-20), resulting in a negative acceleration. Approaching the finish line, the runner must decrease negative velocity to have positive acceleration. This is illustrated graphically in Figure 8-21; from t_2 to t_4, the velocity is negative because the object moved back toward the y-axis or the reference point. The slope of the velocity curve from t_2 to t_3 is negative, indicating negative acceleration.

Continuing toward the finish, the runner begins to decrease his or her velocity in the negative direction. This decrease in negative velocity is a positive acceleration and is illustrated in section t_3 to t_4 because the slope of the velocity curve is positive. Thus, the sign of the acceleration can only be determined from a combination of the direction of motion and whether the velocity is increasing or decreasing. Both positive and negative accelerations can result without the object changing direction. If the final velocity is greater than the initial velocity, the acceleration is positive. For example:

$$a = \frac{v_f - v_i}{t_f - t_i}$$

$$= \frac{10\,\text{m/s} - 3\,\text{m/s}}{3\,\text{seconds} - 1\,\text{second}}$$

$$= \frac{7\,\text{m/s}}{2\,\text{seconds}}$$

$$= 3.5\,\text{m/s}$$

If, however, the final velocity is less than the initial velocity, the acceleration is negative. For example:

$$a = \frac{v_f - v_i}{t_f - t_i}$$

$$= \frac{4\,\text{m/s} - 10\,\text{m/s}}{5\,\text{seconds} - 3\,\text{seconds}}$$

$$= \frac{-6\,\text{m/s}}{2\,\text{seconds}}$$

$$= -3\,\text{m/s}$$

In the first case, it is said that the object is accelerating, and in the latter, decelerating. These terms become confusing, however, when the object actually changes direction. For the sake of easing confusion, it is best that the terms acceleration and deceleration be avoided; the use of positive acceleration and negative acceleration is encouraged.

NUMERICAL EXAMPLE

The velocity data calculated from Table 8-2 representing the vertical (y) position of an object will be used to illustrate the first central difference method of calculating acceleration. Table 8-3 presents the time at each frame, the vertical position, the vertical velocity, and the calculated vertical acceleration for each frame.

For example, to calculate the acceleration at the time of frame 4:

$$a_{y4} = \frac{v_5 - v_3}{t_5 - t_3}$$

$$= \frac{-2.10\,\text{m/s} - 3.59\,\text{m/s}}{0.0668\,\text{seconds} - 0.0334\,\text{seconds}}$$

$$= -170.36\,\text{m/s}^2$$

Figure 8-22 represents graphs of the velocity and acceleration profiles of the complete movement. As the velocity increases rapidly, the slope of the acceleration curve becomes steeper, and as the velocity changes less rapidly, the slope is less steep.

SECTION III Mechanical Analysis of Human Motion

300

TABLE 8-3 Calculation of Acceleration from a Set of Velocity–Time Data

Frame	Time (s)	Vertical Position (y) (m)	Vertical Velocity (v_y) (m/s)	Acceleration (a_y) (m/s²)
1	0.0000	0.00		
2	0.0167	0.15	6.59	
3	0.0334	0.22	3.59	$(2.40 - 6.59) / (0.0501 - 0.0167) = -125.50$
4	0.0501	0.27	2.40	$(-2.10 - 3.59) / (0.0668 - 0.0334) = -170.32$
5	0.0668	0.30	-2.10	$(-8.98 - 2.40) / (0.0835 - 0.0501) = -340.64$
6	0.0835	0.20	-8.98	$(-13.77 - (-2.10)) / (0.1002 - 0.0668) = -349.60$
7	0.1002	0.00	-13.77	$(-8.98 - (-8.98)) / (0.1169 - 0.0835) = 0.00$
8	0.1169	-0.26	-8.98	$(1.20 - (-13.77)) / (0.1336 - 0.1002) = 448.21$
9	0.1336	-0.30	1.20	$(8.98 - (-8.98)) / (0.1503 - 0.1169) = 537.72$
10	0.1503	-0.22	8.98	
11	0.1670	0.00		

GRAPHICAL EXAMPLE

Previously, an estimation of the shape of the relationship between position and velocity was graphed using the concepts of slope and local extrema. It is also possible to graph an estimation of the shape of an acceleration curve based on the shape of the velocity–time profile. Again, the two concepts of the slope and the local extrema are used, this time on a velocity–time graph. Figure 8-23A represents the horizontal velocity of the data presented in Figure 8-19. The local extrema of the

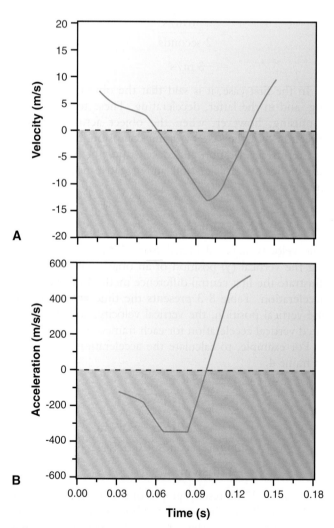

FIGURE 8-22 Velocity–time profile **(A)** and acceleration–time profile **(B)** for Table 8-3.

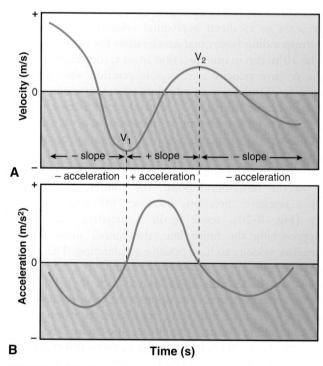

FIGURE 8-23 The relationship between the velocity–time curve and the acceleration–time curve drawn using the concepts of local extrema and slopes.

velocity curve, where the curve changes direction, are indicated as v_1 and v_2. At these points, the acceleration is zero. Constructing the acceleration curve on the same time line as the velocity curve allows projection of the occurrence of these local extrema from the velocity curve time line to the acceleration–time line.

The slopes of each section of the velocity–time curve are (a) to v_1, negative; (b) v_1 to v_2, positive; and (c) beyond v_2, negative. The velocity curve to v_1 has a negative slope, but the curve reaches the local extremum at v_1. The corresponding acceleration curve of this section (Fig. 8-23B) is negative, but it becomes zero at the local extremum v_1. Between v_1 and v_2, the velocity curve has a positive slope. The acceleration curve between these points in time will begin with a zero value at the time corresponding to v_1, become more positive, and eventually return to zero at a time corresponding to v_2. Similar logic can be used to describe the construction of the remainder of the acceleration curve.

Differentiation and Integration

Discussion thus far is of kinematic analysis based on a process whereby position data are accumulated first. Further calculations may then take place using the position and time data. When velocity is calculated from displacement and time or when acceleration is calculated from velocity and time, the mathematics is called differentiation. The solution of the process of differentiation is called a derivative. A derivative is simply the slope of a line, either a secant or tangent, as a function of time. Thus, when velocity is calculated from position and time, differentiation is the method used to calculate the derivative of position. Velocity is called the derivative of displacement and time. Similarly, acceleration is the derivative of velocity and time.

In certain situations, however, acceleration data may be collected. From these data, velocities and positions may be calculated based on a process that is opposite to that of differentiation. This mathematical process is known as integration. Integration is often referred to as antidifferentiation. The result of the integration calculation is called the integral. Velocity, then, is the time integral of acceleration. The following equation describes the above statement:

$$v = \int_{t_1}^{t_2} a\,\mathrm{d}t$$

where $\int_{t_1}^{t_2}$ r represents the integration sign. This expression reads that velocity is the integral of acceleration from time 1 to time 2. The terms t_1 and t_2 define the beginning and end points between which the velocity is evaluated. Likewise, position is the integral of velocity:

$$s = \int_{t_1}^{t_2} v\,\mathrm{d}t$$

The meaning of the integral, however, is not quite as obvious as that of the derivative. Integration requires calculating the area under a velocity–time curve to determine the average displacement or the area under an acceleration–time curve to determine the average velocity. The integration sign

$$\int_{t_1}^{t_2}$$

is an elongated s; it indicates summation of areas between time t_1 and time t_2.

The area under an acceleration–time curve represents the change in velocity over the time interval. This can be demonstrated by analysis of the units in calculating the area under the curve. For example, taking the area under an acceleration–time curve involves multiplying an acceleration value by a time value:

Area under the curve = acceleration × time

$$= \frac{m}{s^2} \times s$$
$$= \frac{m}{s \times s} \times s$$
$$= m/s$$

The area under the curve would have units of velocity. Thus, a measure of velocity is the area under an acceleration–time curve. This area represents the change in velocity over the time interval in question. Similarly, the change in displacement is the area under a velocity–time curve.

Figure 8-24 illustrates the concept of the area under the curve. Two rectangles represent a constant acceleration of 3 m/s² for 6 seconds in the first portion of the curve and constant acceleration of 7 m/s² for 2 seconds. The area of a rectangle is the product of the length and

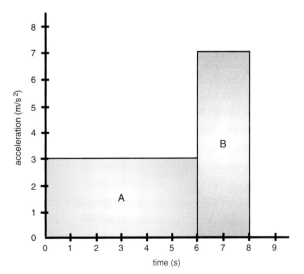

FIGURE 8-24 An idealized acceleration–time curve. Area A equals 3 m/s² × 6 seconds or 18 m/s. This represents the change in velocity over the time interval from 0 to 6 seconds. The change in velocity for area B is 14 m/s.

width of the rectangle. The area under the first rectangle, A, is 3 m/s² times 6 seconds, or 18 m/s. In rectangle B, the area is 7 m/s² times 2 seconds, or 14 m/s. The total area is 32 m/s. This value represents the average velocity over this time period.

Velocity–time and acceleration–time curves do not generally form rectangles as in the previous examples, so the computation of the integral is not quite so simple. The technique generally used is called a Riemann sum. It depends on the size of the time interval, dt. If dt is small enough, and it generally is in a kinematic study, the integral or area under the curve can be calculated by progressively summing the product of each data point along the curve and dt. For example, if the curve to be integrated is a horizontal velocity–time curve, the integral equals the change in position. If the horizontal velocity–time curve is made up of 30 data points, each 0.005 seconds apart, the integral would be:

$$\int_{t_i}^{t_{30}} v_{xi}\,\mathrm{d}t = \mathrm{d}s$$

and to find the area under the curve:

$$\mathrm{d}s = \sum_{i=1}^{30} (v_{xi} \times \mathrm{d}t)$$

The Riemann sum calculation generally gives an excellent estimation of the area under the curve.

Linear Kinematics of Walking and Running

A kinematic analysis describes the positions, velocities, and accelerations of bodies in motion. It is one of the most basic types of analyses that may be conducted because it is used only to describe the motion with no reference to the causes of motion. Kinematic data are usually collected, as previously described, using high-speed video cameras or sensors and positions of the body segments are generated through digitization or other marker recognition techniques. To illustrate kinematic analysis in biomechanics, the study of human gait is used here as an example. The most studied forms of human gait are walking and running.

STRIDE PARAMETERS

In both locomotor forms of movement, the body actions are cyclic, involving sequences in which the body is supported first by one leg and then the other. These sequences are defined by certain parameters. Typical parameters such as the stride and step are presented in Figure 8-25. A locomotor cycle or stride is defined by events in these sequences. A stride is defined as the interval from one event on one limb until the same event on the same limb in the following contact. Usually an event such as the first instant of foot contact defines the beginning of a stride. For example, a stride could be defined from heel contact of the right limb to subsequent heel contact on the right limb. The stride can be subdivided into steps. A step is a portion of the stride from an event occurring on one leg to the same event occurring on the opposite leg. For example, a step could be defined as foot contact on the right limb to foot contact on the left limb. Thus, two steps equal one stride, also called one gait cycle.

Stride length and stride rate are among the most commonly studied linear kinematic parameters. The distance covered by one stride is the stride length, and the number of strides per minute is the stride rate. Running and walking velocity is the result of the relationship between stride rate and stride length. That is:

Running speed = Stride length × Stride rate

Velocity can be increased by increasing stride length or stride rate or both. Examples of stride characteristics ranging from a slow walk-up through a sprint are presented in Table 8-4, which clearly shows adjustments in stride rate and stride length that contribute to the increase in the velocity. The stride can be lengthened only so much; in fact, from walking speeds of 0.75 m/s on, pelvic rotation begins to contribute to the stride lengthening (31). Many studies (9,19,22,28) have shown that in running,

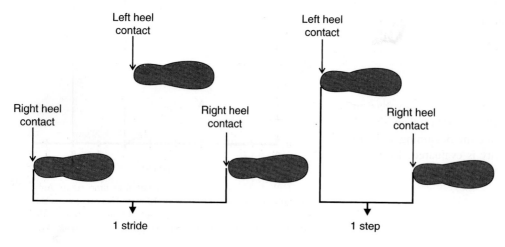

FIGURE 8-25 Stride parameters during gait.

TABLE 8-4	Stride Characteristic Comparison Between Walking and Running		
Variable	Walking (17,25)	Running (22,25)	Sprint
Speed (m/s)	0.67–1.32	1.65–4.00	8.00–9.00
Stride length (m)	1.03–1.35	1.51–3.00	4.60–4.50
Cadence (steps/min)		79.00–118.00	132.00–200.00
Stride rate (Hz)	0.65–0.98	1.10–1.38	1.75–2.00
Cycle time (s)	1.55–1.02	0.91–0.73	0.57–0.50
Stance (% of gait cycle)	66.00–60.00	59.00–30.00	25.00–20.00
Swing (% of gait cycle)	34.00–40.00	41.00–70.00	75.00–80.00

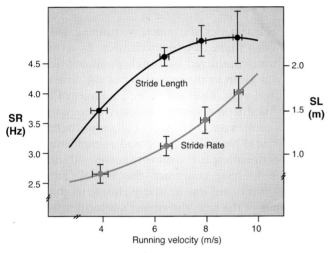

FIGURE 8-26 Changes in stride length and stride rate as a function of running velocity. (Adapted from Luhtanen, P., Komi, P. V. [1973]. Mechanical factors influencing running speed. In E. Asmussen, K. Jorgensen (Eds.). *Biomechanics VI-B*. Baltimore, MD: University Park Press.)

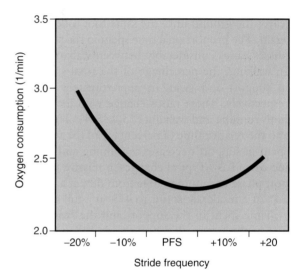

FIGURE 8-27 Oxygen consumption as a function of stride frequency.

both stride rate and stride length increase with increasing velocity, but the adjustment is not proportional at higher velocities. This is illustrated in Figure 8-26. For velocities up to 7 m/s, increases are linear, but at higher speeds, there is a smaller increment in stride length and a greater increment in stride rate. This indicates that when sprinting, runners increase their velocity by increasing their stride rate more than their stride length. A runner initially increases velocity by increasing stride length. However, there is a physical limit to how much an individual can increase stride length. To run faster, therefore, the runner must increase his or her stride rate.

It has been shown that individuals chose a walking or running speed (preferred locomotor speed) and a preferred stride length at that preferred walking speed (18). Deviation from the preferred stride length at the preferred speed has serious consequences for the individual. Researchers have shown that increasing or decreasing stride length while keeping locomotor velocity constant can increase the oxygen cost of locomotion (12,18). This is illustrated in Figure 8-27.

Refer to the walking data in Appendix C. Calculate the step length, step frequency, stride length, stride frequency, and cadence (steps per minute). Calculate the walking velocity.

Each individual has a preferred speed at which he or she opts to start running instead of walking faster. This speed is usually somewhere around 2 m/s. The walking velocity at which individuals switch to a run is higher than the running velocity at which they shift back to a walk (20).

Gait parameters are adjusted when physical or environmental conditions offer constraints to the gait cycle. For example, an individual with a limiting physical impairment usually walks with a slower velocity and cadence by increasing the support phase of the cycle, decreasing the swing phase, and shortening the step length (25). Many individuals with cerebral palsy have significant gait restrictions evidenced by slow velocities, short strides, slow cadence, and more time spent in double support. Environmental factors also influence gait; for example, when the walking surface becomes slippery, most individuals reduce their step length. This minimizes the chance of falling by increasing the heel strike angle with the ground and decreasing the potential for foot displacement on the slippery surface (3).

The running and walking stride can be further subdivided into support (or stance) and nonsupport (or swing) phases. The support or stance phase occurs when the foot is in contact with the ground, that is, from the point of foot contact until the foot leaves the ground. The support phase is often subdivided further into heel strike followed by foot flat, midstance, heel rise, and toe-off. The nonsupport or swing phase occurs from the point that the foot leaves the ground until the same foot touches the ground again. The proportional time spent in the stance and swing phases varies considerably between walking and running. In walking, the percentage of the total stride time spent in support and swing is approximately 60% and 40%, respectively. These ratios change with increased speed in both running and walking (Table 8-4). The absolute time and the relative time (a percentage of the total stride time) spent in support decreases as running and walking speeds increase (1,2). Typical changes in relative time of the support phase in running range from 68% at a jogging pace to 54% at a moderate sprint to 47% at a full sprint.

Time spent in the support and the swing phase is just one of the factors that distinguish walking from running. The other factor that determines if gait is a walk or a run is whether one foot is always on the ground or not. In walking, one foot is always on the ground, with a brief period when both feet are on the ground, creating a sequence of alternating single and double support (Fig. 8-28). In running, the person does not always have one foot on the ground; an airborne phase is followed by alternating single-support phases.

Using MaxTRAQ, import the video file of the woman walking. What is the length of time spent in single support and double support during one stride? (Note: The time between frames is 0.0313 seconds.)

VELOCITY CURVE

The linear kinematics of competitive running and walking has also been studied by biomechanists. In several cases, athletes were considered as a single point and no consideration was given to the movement of the arms and legs as individual units. Over the years, a number of researchers have tried to measure the velocity curve of a runner during a sprint race (14). A. V. Hill, who later won the Nobel Prize in physiology, proposed a simple mathematical model to represent the velocity curve, and subsequent investigations have confirmed this model (Fig. 8-29). Most sprinters conform relatively closely to this model. At the start of the race, the runner's velocity is zero. The velocity increases rapidly at first but then levels off to a constant value. This means that the runner accelerates rapidly at first, but the acceleration decreases toward

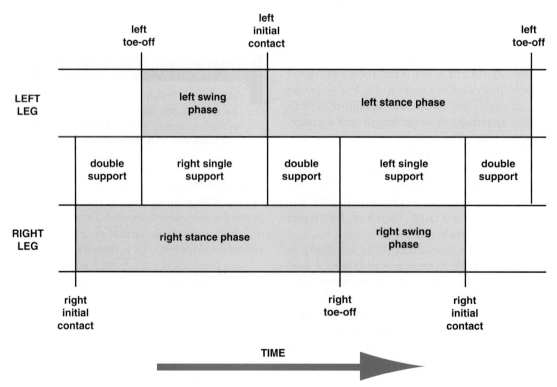

FIGURE 8-28 Support and double support during walking.

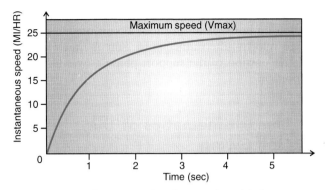

FIGURE 8-29 Hill's proposed mathematical model of a sprint race velocity curve. (Adapted from Brancazio, P. J. [1984]. *Sport Science*. New York: Simon & Schuster.)

the end of the run. The sprinter cannot increase velocity indefinitely throughout the race. In fact, the winner of a sprint race is usually the runner whose velocity decreases the least toward the end of the race. Figure 8-30 illustrates the displacement, velocity, and acceleration data for the women's 100-m final in the 2000 Olympics. The graphs demonstrate similar characteristics for Marion Jones (Fig. 8-28A) and Savatheda Fynes (Fig. 8-28B), even though they finished first and seventh, respectively. In a study of female sprinters (6), it was reported that the sprinters reached their maximum velocity between 23 and 37 m in a 100-m race. It was also reported that these sprinters lost an average of 7.3% from their maximum velocities in the final 10 m of the race. These two trends

(A) *Marion Jones*

(B) *Savatheda Fynes*

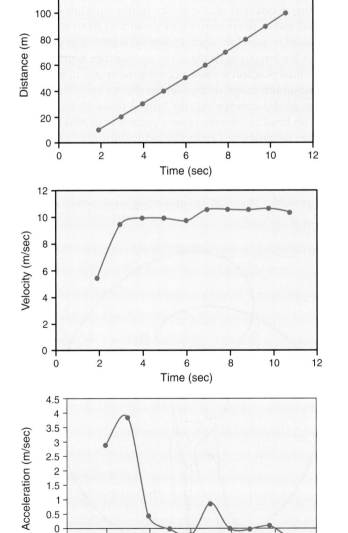

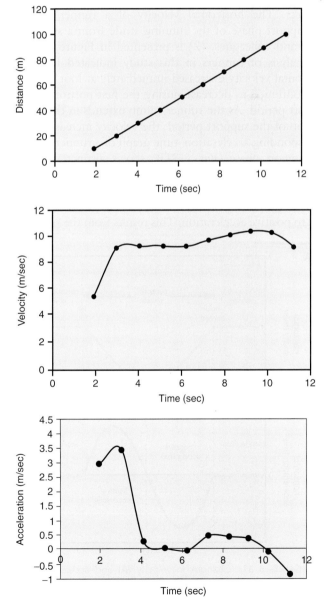

FIGURE 8-30 Distance (*top*), velocity (*middle*), and acceleration (*bottom*) curves for the 2000 Olympics women's 100-m final performance of Marion Jones **(A)** and Savatheda Fynes **(B)**. (Source of split times: http://sydney2000.nbcolympics.com/)

were also present in the 100-m women's final shown in Figure 8-30.

The fastest instantaneous velocity of a runner during a race has not yet been measured during competition. Average speed can be readily calculated, however. Marion Jones and Maurice Greene, in their gold medal performances at the 2000 Olympic Games, covered 100 m in 10.75 and 9.87 seconds, respectively, for an average speed of 9.30 and 10.13 m/s, or speeds equivalent to 20.8 and 22.7 mph, respectively.

VARIATION OF VELOCITY DURING SPORTS

When average velocity was calculated over the race, note that this was not the velocity of the runner at every instant during the race. During a race, a runner contacts the ground numerous times, and it is important to note what occurs to the horizontal velocity during these ground contacts. The horizontal velocity of a runner during the support phase of the running stride from a study by Bates and colleagues. (2) is presented in Figure 8-31A. An analysis of runners in this study indicated that the horizontal velocity decreased immediately at foot contact and continued to decrease during the first portion of the support period. As the runner's limb extends in the latter portion of the support period, the velocity increases. The corresponding acceleration–time graph of a runner during the support phase (Fig. 8-31B) shows distinct negative and positive accelerations. It can be seen that the runner instantaneously has zero acceleration during the support phase, representing the transition from negative acceleration to positive acceleration. This results from the runner's

slowing down during the first portion of support and speeding up in the latter portion. To maintain a constant average velocity, the runner must gain as much speed in the latter portion of the support phase as was lost in the first portion.

Linear Kinematics of the Golf Swing

SWING CHARACTERISTICS

The purpose of the golf swing is to generate speed in the club head and to control the club head so that it is directed optimally for contact with the ball. Although many of the important biomechanical characteristics of the swing are angular, the linear kinematics of the club head ultimately determine the success of the golf swing. Figure 8-32 shows the path of the club head in the swing. Starting at position *A*, the golfer brings the club back up behind and in front of the lead shoulder (*B*) to allow the club head to travel through a longer distance. The purpose of this backswing is to place the appropriate segments in an optimal position for force development and to establish the maximum range of motion for the subsequent downswing. In the downswing, the critical phase in the swing, the club head accelerates at rates greater than 800 m/s^2 to prepare for contact. Contact is made with the ball at the original starting position (*A*), where the club head is still accelerating. Peak velocity is obtained shortly after impact. Club head velocities at impact in the range of 40 m/s are very possible; they can be much higher in some golfers.

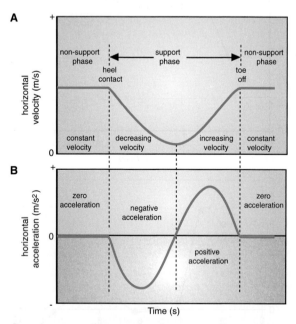

FIGURE 8-31 Changes in velocity **(A)** and acceleration **(B)** during the support phase of a running stride. (Adapted from Bates, B. T., et al. [1979]. Variations of velocity within the support phase of running. In J. Terauds and G. Dales (Eds.). *Science in Athletics.* Del Mar, CA: Academic.)

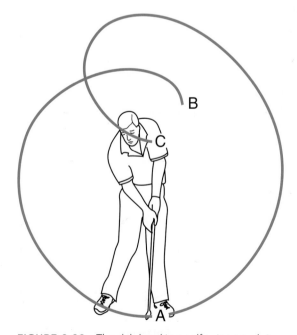

FIGURE 8-32 The club head in a golf swing travels in a curved path through a considerable distance (from A to B to C), allowing time to develop velocity in the club head.

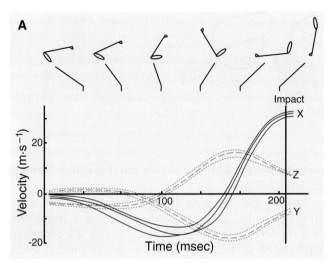
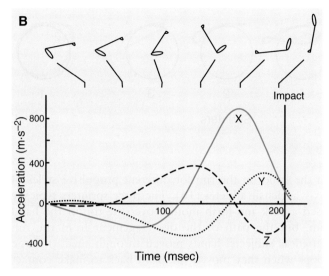

FIGURE 8-33   Velocity–time **(A)** and acceleration–time **(B)** graphs of a driver segment center of gravity toward or away from the ball (X), vertically (Y), and in toward or away from the body (Z). (Adapted from Neal, R. J., Wilson, B. D. [1985]. 3D Kinematics and kinetics of the golf swing. *International Journal of Sport Biomechanics*, 1:221–232.)

The time to complete the total swing may be in the range of 1,000 ms, with the downswing phase accounting for 210 ms, or a little over 20% of the time. After impact is complete, the follow-through phase decelerates the club until the swing is terminated at C.

VELOCITY AND ACCELERATION OF THE CLUB

Figure 8-33A illustrates the 3D velocity of the club (center of gravity) during the downswing phase (24). The motion of the club was recorded in three dimensions to determine linear kinematic characteristics toward the ball forward or backward (x), up or down (y), and away from or toward the golfer (z). In the initial portion of the downswing, with the club still up behind the head, the velocity in the x direction is backward, away from the ball, in the negative direction as the club is brought from the top of backswing around the body and toward the ball. In the last half of the downswing, the x velocity climbs sharply and becomes positive as the club is delivered toward the ball. The peak forward velocity is achieved after impact.

The z or mediolateral velocity starts out negative, indicating that from the top of the backswing to approaching the halfway point in the downswing, the club is moving toward the golfer and then shifts positive as the club is swinging away from the golfer. This trend is opposite to that in the vertical direction, where the velocity in the y direction starts out with small movements upward and reverses to a downward movement as the club is brought down to the ball.

The corresponding acceleration curves in Figure 8-33B identify critical phases in the swing at which maximal accelerations are obtained. Maximum acceleration in the direction of the ball (x) occurs 40 ms before impact, reaching a value of 870 m/s² (24). Acceleration continues on through impact, even though it is small. There is also vertical acceleration, reaching maximum just before impact and still accelerating through the impact. These profiles or trends would look very much the same for all clubs, but there would be a reduction in the values such as club head velocity which decreases from the driver to the nine-iron because of differences in club parameters.

Linear Kinematics of Wheelchair Propulsion

CYCLE PARAMETERS

Many individuals with spinal cord injury or other serious musculoskeletal impairments use a wheelchair for locomotion. Propelling a manual wheelchair involves cyclic body actions, using sequences in which both hands are in contact with the rim or not. The typical wheelchair cycle includes a propulsive phase with the hand pushing on the hand rim of the wheelchair followed by a nonpropulsive phase when the hand is brought back to the start of another propulsion phase. In the nonpropulsive phase, three actions describe the phase, starting with disengagement as the hand releases the hand rim at the end of propulsion followed by recovery as the hand is brought back up to the top of the hand rim to start the propulsion again and finally, contact when the hand touches the rim. The amount of time spent in contact and the range of displacement of the hand forward and backward varies with individual preferences and wheelchair configurations (e.g., seat position). Even in world-class athletes, the cycle patterns vary. Figure 8-34 shows displacement

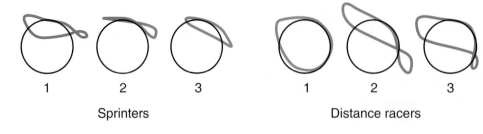

FIGURE 8-34 Hand displacement patterns for six Olympic wheelchair athletes illustrating differences in propulsion styles between sprinters (200 m) and distance racers (1,500 m). (Adapted from Higgs, C. [1984]. Propulsion of racing wheelchairs. In C. Scherrell (Ed.). *Sport and Disabled Athletes.* Champaign, IL: Human Kinetics, 165–172.)

of the hand on the rim for different propulsive styles of six wheelchair athletes (16). The three sprinters typically used a back-and-forth motion over the top of the hand rim, but the path of the disengagement from the rim varied as subjects' hands moved through small or large loops when they moved the hand back to make contact with the rim for propulsion. In the three distance racers, the pattern of the hand motion was more circular but still varied quite significantly between individuals.

PROPULSION STYLES

Two wheelchair propulsion styles have been identified (27). One is the pumping technique seen in the sprinters in Figure 8-35A, in which the hand moves back and forth horizontally with relatively large displacements away from the hand rim. The other popular technique is the circular technique (Fig. 8-35B), in which the hand moves in a circular path along the hand rim. The push phase in the circular pattern accounts for a larger percentage of the total propulsion cycle (43.0%) than the same phase in the

pumping pattern (34.7%), suggesting that it may be more efficient.

Vertical and horizontal displacements of the wrist, elbow, shoulder, and neck during wheelchair propulsion for a moderately active individual with T3–T4 paraplegia is presented in Figure 8-33A. In the propulsion cycle, the wrist travels forward and backward in a straight path, indicating a push pattern of propulsion. The neck moved forward and backward through a range of approximately 5.9 cm and at an average of about 1 cm for every 6.7 cm of wrist movement. Peak acceleration of the hand on the rim occurred close to the end of the push phase and is in the range of 32 m/s².

To increase the velocity of wheelchair propulsion, the cycle time is reduced by increasing the cycle frequency. This occurs as a result of shifting the start and end angles to the front of the hand rim without changing the angular components of the push angle (30). The pattern of action in the forearm changes from a push–pull pattern in the slower velocity to more of a push pattern with support from a continuous trunk flexion.

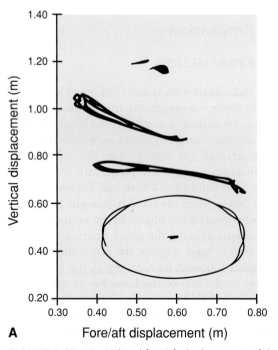

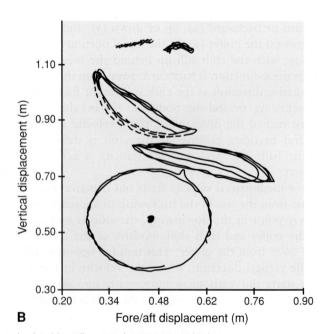

FIGURE 8-35 Vertical and fore/aft displacements of the neck, shoulder, elbow, and wrist over multiple wheelchair propulsion cycles show differences between two subjects incorporating the pumping action style **(A)** and a circular action style **(B)**. (Courtesy of Joe Bolewicz, RPT.)

Using MaxTRAQ, import the video file of the wheel-chair athlete. Digitize the axle of the wheelchair in the frame at which the individual initiates the propulsion phase and then in the frame when the propulsion ends. Calculate the distance the wheelchair has traveled.

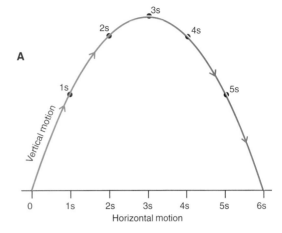

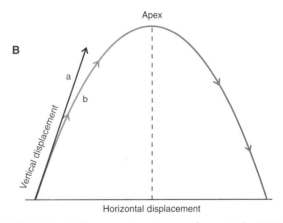

FIGURE 8-36 **(A)** The parabolic trajectory of a projectile. **(B)** Path a represents the trajectory of a projectile without the influence of gravity. Path b is a trajectory with gravity acting. Path b forms a parabola.

 ## Projectile Motion

Projectile motion refers to motion of bodies projected into the air. This type of motion implies that the projectile has no external forces acting on it except for gravity and air resistance. Projectile motion occurs in many activities, such as baseball, diving, figure skating, basketball, golf, and volleyball. The motion of a projectile is a special case of linear kinematics in which we know what changes in velocity and acceleration are going to occur after the object leaves the ground.

For the following discussion, air resistance will be considered negligible because it is relatively small compared with gravity. Depending on the projectile, different kinematic questions may be asked. For example, in the long jump or the shot put, the horizontal displacement is critical. In the high jump and pole vaulting, however, vertical displacement must be maximized. In biomechanics, understanding the nature of projectile motion is critical.

GRAVITY

The force of gravity acting on a projectile results in constant vertical acceleration of the projectile. The acceleration due to gravity is approximately 9.81 m/s^2 at sea level and results from the attraction of two masses, the earth and the object. Gravity uniformly accelerates a projectile toward the earth's surface. However, not all objects that travel through the air are projectiles. Objects that are propelled, such as airplanes and objects that are aerodynamic such as a boomerang, are not generally classified as projectiles.

TRAJECTORY OF A PROJECTILE

The flight path of a projectile is called its trajectory (Fig. 8-36A). The point in time at which an object becomes a projectile is referred to as the instant of release. Gravity continuously acts to change the vertical motion of the object after it has been released. The flight path followed by a projectile in the absence of air resistance is a parabola (Fig. 8-36A). A parabola is a curve that is symmetrical about an axis through its highest point. The highest point of a parabola is its apex.

If gravity did not act on the projectile, it would continue to travel indefinitely with the same velocity as when it was released (Fig. 8-36B). In space, beyond the earth's

gravitational pull, a short firing burst of a vehicle's rocket will result in a change in velocity. When the rocket ceases to fire, the velocity at that instant remains constant, resulting in zero acceleration. Because there is no gravity and no air resistance, the vehicle will continue on this path until the engine fires again.

FACTORS INFLUENCING PROJECTILES

Three primary factors influencing the trajectory of a projectile are the projection angle, projection velocity, and projection height (Fig. 8-37).

Projection Angle
The angle at which the object is released determines the shape of its trajectory. Projection angles generally vary from 0° (parallel to the ground) to 90° (perpendicular to the ground), although in some sporting activities, such as ski jumping, the projection angle is negative. If the projection angle is 0° (parallel to the horizontal), the trajectory is essentially the latter half of a parabola because it has zero vertical velocity and is immediately acted upon by gravity to pull it to the earth's surface. On the other hand, if the

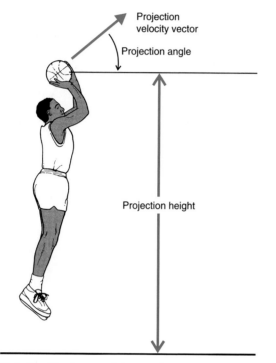

FIGURE 8-37 The factors influencing the trajectory of a projectile are projection velocity, projection angle, and projection height.

TABLE 8-5	Projection Angles Used in Selected Activities	
Activity	Angle (°)	Reference
Racing dive	5–22	5
Ski jumping	–4	21
Tennis serve	3–15	6
Discus	15–35	9
High jump (flop)	40–48	7

jumpers have a projection angle of 40° to 48° using the flop high jump technique (7). On the other hand, if one tried to jump for maximal horizontal distance such as in a long jump, the projection angle would be much smaller. This has also proved to be the case: Long jumpers have projection angles of 18° to 27° (13). Table 8-5 illustrates the projection angles reported in the research literature for several activities. Positive angles of projection (i.e., angles greater than zero) indicate that the object is projected above the horizontal, and negative angles of projection refer to those less than zero or below the horizontal. For example, in a tennis serve, the serve is projected downward from the point of impact.

Projection Velocity

The velocity of the projectile at the instant of release determines the height and distance of the trajectory as long as all other factors are held constant. The resultant velocity of projection is usually calculated and given when discussing the factors that influence the flight of a projectile. The resultant velocity of projection is the vector sum of the horizontal and vertical velocities. It is necessary, however, to focus on the components of the velocity vector because they dictate the height of the trajectory and the distance the projectile will travel. Similar to other vectors, the velocity of projection has a vertical component (v_y) and a horizontal component (v_x).

The magnitude of the vertical velocity is reduced by the effect of gravity (−9.81 m/s for every second of upward flight). Gravity reduces the vertical velocity of the projectile until the velocity equals zero at the apex of the projectile's trajectory. The vertical velocity component, therefore, determines the height of the apex of the trajectory. The vertical velocity also affects the time the projectile takes to reach that height and consequently the time to fall to earth.

The horizontal component of the projection velocity is constant throughout the flight of the projectile. The range or the distance the projectile travels is determined by the product of the horizontal velocity and the flight time to the final position. The magnitude of the distance that the projectile travels is called the range of the projectile. For example, if a projectile is released at a horizontal velocity

projection angle is 90°, the object is projected straight up into the air with zero horizontal velocity. In this case, the parabola is so narrow as to form a straight line.

If the projection angle is between 0° and 90°, the trajectory is parabolic. Figure 8-38 displays theoretical trajectories for an object projected at various angles with the same speed and height of projection.

The optimal angle of projection for a given activity is based on the purpose of the activity. Intuitively, it would appear that jumping over a relatively high object such as a high jump bar would require quite a steep projection angle. This has proved to be the case: High

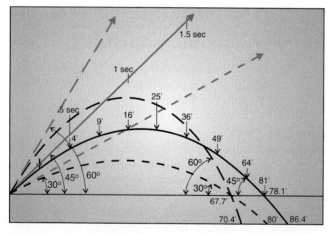

FIGURE 8-38 Theoretical trajectories of a projectile projected at different angles keeping velocity (15.2 m/s) and height (2.4 m) constant. (Adapted from Broer, M. R., Zernike, R. F. [1979]. *Efficiency of Human Movement*, 4th ed. Philadelphia, PA: WB Saunders.)

of 13.7 m/s, the projectile will have traveled 13.7 m in the first second, 27.4 m after 2 seconds, 40.1 m after 3 seconds, and so on.

The angle of projection affects the relative magnitude of the horizontal and vertical velocities. If the angle of projection is 40° and the projection velocity is 13.7 m/s, the horizontal component of the projection velocity is the product of the velocity and the cosine of the projection angle or 13.7 m/s and cosine 40° or 10.49 m/s. The vertical component is the product of the projection velocity and the sine of the projection angle (or 13.7 m/s) and sine 40° (or 8.81 m/s).

To understand in general how the angle of projection affects the velocity components, consider that the cosine of 0° is 1 and decreases to zero as the angle increases. If the cosine of the angle is used to represent the horizontal velocity, the horizontal velocity decreases as the angle of projection increases from 0° to 90° (Fig. 8-39). Also, the sine of 0° is zero and increases to 1 as the angle increases. Consequently, if the sine of the angle is used to represent vertical velocity, the vertical velocity increases as the angle increases from 0° to 90° (Fig. 8-39). It can readily be seen that as the angle gets closer to 90°, the horizontal velocity becomes smaller and the vertical velocity becomes greater. As the angle gets closer to 0°, the horizontal velocity becomes greater and the vertical velocity gets smaller.

At 45°, however, the sine and cosine of the angle are equal. For any given velocity, therefore, horizontal velocity equals vertical velocity. It would appear that 45° would be the optimum angle of projection because for any velocity, the horizontal and vertical velocities are equal. This is true under certain circumstances to be discussed in relation to projection height. Generally, if the maximum range of the projectile is critical, an angle to optimize the horizontal velocity, or an angle less than 45°, would be appropriate. Thus, in activities such as the long jump and

shot putting, the optimal angle of projection is less than 45°. If the height of the projectile is important, an angle greater than 45° should be chosen. This is the case in activities such as high jumping.

Projection Height

The height of projection of a projectile is the difference in height between the vertical takeoff position and the vertical landing position. Three situations greatly affect the shape of the trajectory. In each case, the trajectory is parabolic, but the shape of the parabola may not be completely symmetrical; that is, the first half of the parabola may not have the same shape as the second half.

In the first case, the projectile is released and lands at the same height (Fig. 8-40A). The shape of the trajectory is symmetrical, so the time for the projectile to reach the apex from the point of release equals the time for the projectile to reach the ground from the apex. If a ball is

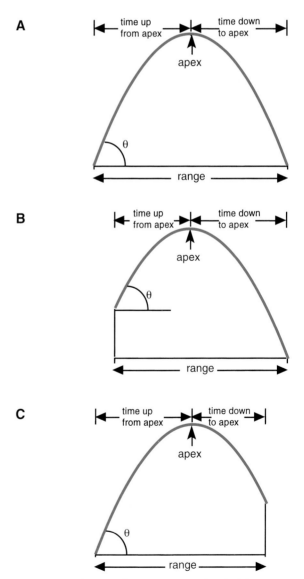

FIGURE 8-40   Influence of a zero projection height (A), positive projection height (B) and negative projection height (C) on the shape of the trajectory of a projectile.

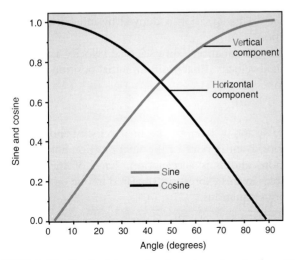

FIGURE 8-39   Graph of sine and cosine values at angles from 0° to 90°. Sin 45° cosine 45°.

kicked from the surface of a field and lands on the field's surface, the relative projection height is zero, so the time up to the apex is equal to the time down from the apex.

In the second situation, the projectile is released from a point higher than the surface on which it lands (Fig. 8-40B). The parabola is asymmetrical, with the initial portion to the apex less than the latter portion. In this case, the time for the projectile to reach the apex is less than the time to reach the ground from the apex. For example, if a shot-putter releases the shot from 2.2 m above the ground and the shot lands on the ground, the height of projection is 2.2 m.

In the third situation, the projectile is released from a point below the surface on which it lands (Fig. 8-40C). Again, the trajectory is asymmetrical, but now the initial portion to the apex of the trajectory is greater than the latter portion. Thus, the time for the projectile to reach the apex is greater than the time for the projectile to reach the ground from the apex. For example, if a ball is thrown from a height of 2.2 m and lands in a tree at a height of 4 m, the height of projection is –1.8 m.

Generally, when the projection velocity and angle of projection are held constant, the higher the point of release, the longer the flight time. If the flight time is longer, the range is greater. Also, for maximum range, when the relative height of projection is zero, the optimum angle is 45°; when the projection height is above the landing height, the optimum angle is less than 45°; and when the projection height is below the landing height, the optimum angle is greater than 45°. The effect of landings that are lower than takeoffs is shown in Figure 8-38.

OPTIMIZING PROJECTION CONDITIONS

To optimize the conditions for the release of a projectile, the purpose of the projectile must be considered. As discussed previously, the three primary factors that affect the flight of a projectile are interrelated and affect both the height of the trajectory and the distance traveled. Although it may seem intuitive that because the height of the apex and the length of the trajectory of the projectile are both affected by the projection velocity, increasing the projection velocity increases both of these parameters, this common perception is incorrect. The choice of an appropriate projection angle dictates whether the vertical or the horizontal velocity is increased with increasing projection velocity. In addition, the angle of projection can be affected by the height of projection.

The relative importance of these factors is illustrated in the following example. If an athlete puts the shot with a velocity of 14 m/s at an angle of 40° from a height of 2.2 m, the distance of the throw is 22 m. If each factor is increased by a given percentage (10% in this case) as the other two factors are held constant, the relative importance of each factor may be calculated. Increasing

the velocity to 15.4 m/s results in a throw of 26.2 m, increasing the angle to 44° results in a throw of 22 m, and increasing the height of projection to 2.4 m results in a throw of 22.2 m. It is readily evident that increasing the velocity of projection increases the range of the throw more substantially than increasing either the angle or the height of projection. The three factors are interrelated, however, and any change in one results in a change in the others.

 Equations of Constant Acceleration

When a projectile is traveling through the air, only gravity and air resistance act upon it. If air resistance is ignored, only gravity is considered to act on the projectile. The acceleration due to gravity is constant, so the projectile undergoes constant acceleration. Using the concepts from the previous section, equations of constant acceleration, or projectile motion, can be determined based on the definitions of velocity and acceleration. Three such expressions involve the interrelationships of the kinematic parameters time, position, velocity, and acceleration. These expressions are often referred to as the equations of constant acceleration. The first equation expresses final velocity as a function of the initial velocity, acceleration, and time:

$$v_f = v_i + at$$

where v_f and v_i refer to the final velocity and the initial velocity, a is the acceleration, and t is the time.

In the second equation, final position is expressed as a function of initial position, initial velocity, acceleration, and time:

$$s_f = s_i + v_i t + \tfrac{1}{2}at^2$$

where v_i is the initial velocity, t is the time, and a is the acceleration. The variable in this expression may refer to the horizontal or vertical case and is the change in position or the distance that the object travels from one position to another. This equation is derived by integrating the first equation.

The last equation expresses final velocity as a function of initial velocity, final position, initial position, and acceleration.

$$v_f^2 = v_i^2 + 2a(s_f - s_i)$$

where v_f and v_i refer to the final velocity and the initial velocity, s_f and s_i refer to the final position and the initial position, and a is acceleration. Each of the kinematic variables in this expression appeared in one or both of the previous equations.

NUMERICAL EXAMPLE

The equations of constant acceleration use parameters that are basic to linear kinematics. The three equations of

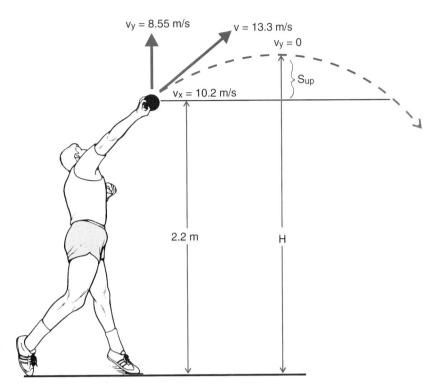

FIGURE 8-41 Conditions during the flight of the shot. Initial conditions are v = 13.3 m/s, projection angle 40°, and projection height 2.2 m.

constant acceleration thus provide a useful method of analyzing projectile motion. If calculating the range of a projectile, for example, the following expression can be used:

$$\text{Range} = \frac{v^2 \times \sin\theta \times \cos\theta + v_x \times \sqrt{(v_y)^2 + 2gh}}{g}$$

where v is the velocity of projection, θ is the angle of projection, h is the height of release of the projection, and g is the acceleration due to gravity. Suppose a shot-putter releases the shot at an angle of 40° from a height of 2.2 m with a velocity of 13.3 m/s. Figure 8-41 illustrates what is known about the conditions of the projectile at the instant of projection and the shape of the trajectory based on our previous discussion. Using the previous equation, the range can be calculated as follows:

$$\text{Range} = \frac{\begin{array}{c}13.3\,\text{m/s}^2 \times \sin 40 \times \cos 40 + 10.19\,\text{m/s} \\ \times \sqrt{8.55\,\text{m/s}^2 + 2(9.81\,\text{m/s}^2)(2.2\,\text{m})}\end{array}}{9.81\,\text{m/s}^2}$$

$$= \frac{\begin{array}{c}176.89 \times 0.6428 \times 0.766 \\ + 10.19 \times \sqrt{73.10 + 43.16}\end{array}}{9.81}$$

$$= \frac{87.09 + 10.19 \times 10.78}{9.81}$$

$$= \frac{87.09 + 109.84}{9.81}$$

$$= 20.07\ \text{m}$$

The same problem can be solved using the table method with the equations of constant acceleration (see Appendix D).

Summary

Biomechanics is a quantitative discipline. One type of quantitative analysis involves linear kinematics or the study of linear motion with respect to time. Linear kinematics involves the vector quantities, position, velocity, and acceleration and the scalar quantities displacement and speed. Displacement is defined as the change in position. Velocity is defined as the time rate of change of position and is calculated using the first central difference method as follows:

$$v = \frac{s_{i+1} - s_{i-1}}{2\Delta t}$$

Acceleration is defined as the time rate of change of velocity and is also calculated using the first central difference method as follows:

$$a = \frac{v_{i+1} - v_{i-1}}{2\Delta t}$$

The process of calculating velocity from position and time or acceleration from velocity and time is called differentiation. Calculating the derivative via differentiation entails finding the slope of a line tangent to the parameter–time curve. The opposite process to differentiation is called integration. Velocity may be calculated as the integral of acceleration and position as the integral of velocity. Integration implies calculating the area under the parameter–time curve. The method of calculating the area under a parameter–time curve is called the Riemann sum.

Projectile motion involves an object that undergoes constant acceleration because it is uniformly accelerated by gravity. The flight of a projectile, its height and distance, is affected by conditions at the point of release: the angle, velocity, and relative height of projection. Three equations govern constant acceleration. The first expresses final velocity, v_f, as a function of initial velocity, v_i; acceleration, a; and time, t. That is:

$$v_f = v_i + at$$

The second equation expresses position, s, as a function of initial velocity, v_i; acceleration, a; and time, t. That is:

$$s = v_i t + \tfrac{1}{2}at^2$$

The third equation expresses final velocity, v_f, as a function of initial velocity, v_i; initial position, s_i; final position, s_f; and acceleration, a:

$$v_f^2 = v_i^2 + 2a(s_f - s_i)$$

These equations may be used to calculate the range of a projectile. However, a general equation for the range of a projectile is:

$$\text{Range} = \frac{v^2 \times \sin\theta \times \cos\theta + v_x \times \sqrt{(v_y)^2 + 2gh}}{g}$$

Equation Review for Linear Kinematics

Purpose	Given	Formula
Vector composition, magnitude	Horizontal and vertical components	$r^2 = x^2 + y^2$
Vector composition, angle	Horizontal and vertical components	$\tan\theta = y/x$
Vector resolution, vertical	Magnitude and direction of vector	$y = r\sin\theta$
Vector resolution, horizontal	Magnitude and direction of vector	$x = r\cos\theta$
Time between video frames	Camera frame, sampling rate	Time (s) = 1/frame rate
Calculate position	Starting position relative to origin, constant velocity (zero acceleration), time	$s = s_i + v_i t$
Calculate position	Starting position at origin, constant velocity (zero acceleration), time	$s = v_i t$
Calculate position	Initial velocity, time, constant acceleration	$s = v_i t + \tfrac{1}{2}at^2$
Calculate position	Initial velocity zero, time, constant acceleration	$s = \tfrac{1}{2}at^2$
Calculate average velocity	Displacement and time	$v = (x_2 - x_1)/(t_2 - t_1)$
Calculate average velocity	Initial and final velocity	$v = (v_i + v_f)/2$
Calculate final velocity	Initial velocity, constant acceleration, and time	$v_f = v_i + at$
Calculate final velocity	Starting velocity zero, constant acceleration, time	$v = at$
Calculate final velocity	Velocity at time = zero, constant acceleration, initial position relative to origin, final position	$v_f = \sqrt{(v_i^2 + 2a(x_f - x_i))}$
Calculate final velocity	Initial velocity zero, constant acceleration, initial and final position	$v_f^2 = 2as$
		$v = \sqrt{(2a(x_f - x_i))}$
Calculate acceleration	Final velocity and displacement	$a = v_f^2/2d$
Calculate average acceleration	Velocity and time	$a = (v_2 - v_1)/(t_2 - t_1)$
Calculate time	Displacement, constant acceleration	$t = \sqrt{(2d/a)}$
Calculate time in air for projectile beginning and landing at same height	Vertical velocity, constant acceleration	$t = 2v_y/a$
Calculate distance of projectile	Resultant velocity, initial angle of release, constant acceleration	$s = r^2 \sin 2\theta/a$

REVIEW QUESTIONS

True or False

1. ____ Using digitization, calculating shoulder abduction and adduction during a jumping jack is a qualitative analysis.

2. ____ During a long jump, only translational motion occurs.

3. ____ The x- and y-axes in a 2D rectangular reference system are always oriented horizontally and vertically, respectively.

4. ____ With a coordinate reference system originating at the ankle joint, the axes of the coordinate system will always be horizontal and vertical during forward movements, such as walking.

5. ____ A 2D reference frame has two axes and two planes.

6. ____ Velocity is a vector.

7. ____ Acceleration is defined as the time rate of change of velocity.

8. ____ Displacement and distance are scalar quantities.

9. ____ The slope of a line tangent to a velocity by time curve indicates instantaneous acceleration.

10. ____ Vectors can be subtracted graphically by taking the opposite of one and placing its tail at the head of the other vector.

11. ____ The area under a velocity by time graph represents the displacement.

12. ____ Negative acceleration indicates that the direction of motion is away from the origin in the negative direction.

13. ____ The maximum point on a position by time graph indicates the point where velocity is also maximum.

14. ____ Change in velocity is equivalent to the area under an acceleration by time curve.

15. ____ A positive value on a velocity by time curve indicates that the corresponding position may be positive or negative.

16. ____ Stride length and step length are synonyms.

17. ____ During sprinting, a greater percentage of the gait cycle is spent in stance relative to walking because higher forces need to be generated.

18. ____ At faster walking speeds, one first increases his/her step frequency and then stride length.

19. ____ Running velocity is a product of stride length and stride rate; therefore, increasing one increases velocity.

20. ____ People are more efficient if they increase their stride length 10% more than preferred for a given running velocity.

21. ____ In a golf swing, the club head is near its maximum horizontal velocity when contacting the ball.

22. ____ Horizontal acceleration of a golf club head is less than its maximum at the moment of impact in a golf swing.

23. ____ Cycle frequency decreases as wheelchair propulsion velocity increases.

24. ____ Range of a projectile is maximized using a projection angle less than 45° if landing height is lower than release height.

25. ____ The horizontal velocity of a shot put is slowed down by gravity 9.81 m/s every second it is in the air.

Multiple Choice

1. Convert the rectangular coordinates of (111, 222) to polar coordinates.
   a. (192.3, 26.6°)
   b. (192.3, 63.4°)
   c. (248.2, 26.6°)
   d. (248.2, 63.4°)

2. Convert the polar coordinates of (202, 202°) to rectangular coordinates.
   a. (−187.3, −75.7)
   b. (−75.7, −187.3)
   c. (81.6, 81.6)
   d. (−187.3, 75.7)

3. Figure 8-11 depicts the path of a runner. If the runner starts and ends at the end of each block, and if each block is a square with lengths of 180 m, what is the magnitude of the resultant displacement of the runner?
   a. 3,600 m
   b. 1,080 m
   c. 894 m
   d. 805 m

4. A swimmer completes 10 laps in a 50-m swimming pool, finishing where she started. What were the linear distance and the linear displacement?
   a. Distance = 500 m; displacement = 500 m
   b. Distance = 500 m; displacement = 0 m
   c. Distance = 0 m; displacement = 0 m
   d. Distance = 0 m; displacement = 500 m

5. During a volleyball serve, the ball leaves the hand with an initial velocity of 10 m/s angled 41° from the horizontal. What are the horizontal and vertical velocities of the ball?
   a. v_x = 7.5 m/s; v_y = 6.6 m/s
   b. v_x = 6.6 m/s; v_y = 7.5 m/s
   c. v_x = 5.2 m/s; v_y = 8.6 m/s
   d. v_x = 8.6 m/s; v_y = 5.2 m/s

6. At takeoff, the horizontal and vertical velocities of a high jumper are 2.0 and 3.9 m/s, respectively. What are the resultant velocity and angle of takeoff?
   a. v = 19.2 m/s; θ = 62.9°
   b. v = 5.9 m/s; θ = 27.1°
   c. v = 4.4 m/s; θ = 62.9°
   d. v = 4.4 m/s; θ = 27.1°

7. Given a right triangle in quadrant I with hypotenuse = 28.8 cm, side X = 1.0 cm, find the length of side Y and the size of the other two angles.
   a. Side Y = 28.9 cm, $\theta 1$ = 86.2°, $\theta 2$ = 3.9°
   b. Side Y = 26.9 cm, $\theta 1$ = 86.0°, $\theta 2$ = 4.0°
   c. Side Y = 28.7 cm, $\theta 1$ = 86.2°, $\theta 2$ = 3.8°
   d. Side Y = 30.7 cm, $\theta 1$ = 86.5°, $\theta 2$ = 3.5°

8. Suppose an individual moves from point s_1 (1, 0) to point s_2 (1, 11) to point s_3 (−4, 4) and ends at point s_4 (2, −7). What are the horizontal, vertical, and resultant displacements?
   a. Horizontal = 11 units; vertical = 14 units; resultant = 7.1 units
   b. Horizontal = 1 unit; vertical = −7 units; resultant = 7.1 units
   c. Horizontal = 0 unit; vertical = 8 units; resultant = 8 units
   d. Horizontal = 1 units; vertical = 7 units; resultant = 25 units

9. Combine the following two velocity vectors to find the resultant vector. Vector A = 5.5 m/s at 210° and vector B = 10.7 units at 82°.
   a. Resultant = 8.51 m/s; θ = 67.4°
   b. Resultant = 8.51 m/s; θ = 112.6°
   c. Resultant = 72.3 m/s; θ = 22.6°
   d. Resultant = 72.3 m/s; θ = 112.6°

10. An individual runs 20 km in 89 minutes. What was the average speed in meters per second?
    a. 0.27
    b. 0.90
    c. 3.75
    d. 225

11. A bobsled accelerates from rest at a constant rate of 3 m/s². How fast is it going after 3.5 seconds?
    a. 6.5 m/s
    b. 7 m/s
    c. 9 m/s
    d. 10.5 m/s

12. The initial velocity of a baseball is 35 m/s at 10°. It was released at a height of 2.0 m. How high above the ground and how far horizontally is the object when it is 1.12 seconds into the flight?
    a. Vertical = 2.66 m; horizontal = 38.61 m
    b. Vertical = 0 m; horizontal = 38.61 m
    c. Vertical = 0.66 m; horizontal = 6.81 m
    d. Vertical = 2.66 m; horizontal = 6.81 m

13. A triple jumper needs a velocity of 9.2 m/s to make a good jump. If he is accelerating at 1.65 m/s², how much time does he need to reach the velocity?
    a. 4.58 seconds
    b. 5.71 seconds
    c. 3.50 seconds
    d. 5.58 seconds

14. A vaulter is trying to reach a velocity of 7 m/s at the end of a 14-m runway. How quickly must he accelerate?
    a. 0.50 m/s²
    b. 1.75 m/s²
    c. 2.00 m/s²
    d. 3.50 m/s²

15. If a gymnast leaves the ground with a vertical velocity of 9.81 m/s when doing a back flip, where in her trajectory will her center of mass be 1.0 seconds later?
    a. Approaching the apex
    b. At the apex
    c. Descending from the apex
    d. On the ground
    e. Not enough information

16. A child's vertical takeoff velocity when jumping on a trampoline is 5.3 m/s. Assuming takeoff and landing heights are the same, how long is the child airborne?
    a. 0.29 seconds
    b. 0.54 seconds
    c. 1.08 seconds
    d. 1.85 seconds

17. Two marbles have the same size and mass. Marble A is dropped from a height of 9.81 m and Marble B is given an initial horizontal velocity of 9.81 m/s and vertical velocity of 0 m/s. Which marble will contact the ground first?
    a. Marble A
    b. Marble B
    c. Both marbles will contact at the same time

Questions 18 to 24: A 1-kg discus is thrown with a velocity of 19 m/s at an angle of 35° from a height of 1.94 m. Neglect air resistance.

18. Calculate the vertical and horizontal velocity components.
    a. v_x = 15.56 m/s, v_y = 10.90 m/s
    b. v_x = 10.90 m/s, v_y = 15.56 m/s
    c. v_x = 30.19 m/s, v_y = 21.14 m/s
    d. v_x = 21.14 m/s, v_y = 30.19 m/s

19. Calculate the time to peak height.
    a. 1.11 seconds
    b. 1.59 seconds
    c. 2.15 seconds
    d. 2.22 seconds

20. Calculate the height of the trajectory from the point of release.
    a. 6.06 m
    b. 6.66 m
    c. 17.5 m
    d. 18.1 m

21. Calculate the total height of the parabola.
    a. 20.1 m
    b. 19.5 m
    c. 8.60 m
    d. 8.00 m

22. Calculate the time from the apex to the ground.
    a. 1.63 seconds
    b. 1.28 seconds
    c. 1.99 seconds
    d. 2.02 seconds

23. Calculate the total flight time.
   a. 3.13 seconds
   b. 3.10 seconds
   c. 2.39 seconds
   d. 2.74 seconds

24. Calculate the range of the throw.
   a. 29.9 m
   b. 42.6 m
   c. 26.1 m
   d. 37.2 m

25. A baseball leaving the bat at 46° at a height of 1.2 m from the ground clears a 3-m high wall 125 m from home plate. What is the initial velocity of the ball (ignore air resistance)?
   a. 34.9 m/s
   b. 35.3 m/s
   c. 35.0 m/s
   d. 36.9 m/s

References

1. Bates, B. T., Haven, B. H. (1974). Effects of fatigue on the mechanical characteristics of highly skilled female runners. In R. C. Nelson, C. A. Morehouse (Eds.). *Biomechanics IV.* Baltimore, MD: University Park Press, 119-125.

2. Bates, B. T., et al. (1979). Variations of velocity within the support phase of running. In J. Terauds, G. Dales (Eds.). *Science in Athletics.* Del Mar, CA: Academic, 51-59.

3. Brady, R. A., et al. (2000). Foot displacement but not velocity predicts the outcome of a slip induced in young subjects while walking. *Journal of Biomechanics*, 33:803-808.

4. Brancazio, P. J. (1984). *Sport Science*. New York: Simon & Schuster.

5. Broer, M. R., Zernike, R. F. *Efficiency of Human Movement*, 4th ed. Philadelphia, PA: Saunders College, 1979.

6. Chow, J. W. (1987). Maximum speed of female high school runners. *International Journal of Sports Biomechanics*, 3:110-127.

7. Dapena, J. (1980). Mechanics of translation in the Fosbury flop. *Medicine & Science in Sports & Exercise*, 12:37-44.

8. Dapena, J., Chung, C. S. (1988). Vertical and radial motions of the body during the take-off phase of high jumping. *Medicine & Science in Sports & Exercise*, 20:290-302.

9. Elliott, B. C., Blanksby, B. A. (1979). A biomechanical analysis of the male jogging action. *Journal of Human Movement Studies*, 5:42-51.

10. Elliott, B., et al. (1986). A three-dimensional cinematographic analysis of the fastball and curveball pitches in baseball. *International Journal of Sport Biomechanics*, 2:20-28.

11. Fukashiro, S., Komi, P. V. (1987). Joint moment and mechanical power flow of the lower limb during vertical jump. *International Journal of Sports Medicine*, 8:15-21.

12. Hamill, J., et al. (1995). Shock attenuation and stride frequency during running. *Human Movement Science*, 14:45-60.

13. Hay, J. G. (1986). The biomechanics of the long jump. *Exercise and Sport Science Review*, 14:401-446.

14. Henry, F. M., Trafton, I. (1951). The velocity curve of sprint running. *Research Quarterly*, 23:409-422.

15. Heusner, W. W. (1959). Theoretical specifications for the racing dive: Optimum angle for take-off. *Research Quarterly*, 30:25-37.

16. Higgs, C. (1984). Propulsion of racing wheelchairs. In C. Sherrel (Ed.). *Sport and Disabled Athletes.* Champaign, IL: Human Kinetics, 165-172.

17. Holden, J. P., et al. (1997). Changes in knee joint function over a wide range of walking speeds. *Clinical Biomechanics*, 12:375-382.

18. Holt, K. G., et al. (1991). Predicting the minimal energy cost of human walking. *Medicine & Science in Sports & Exercise*, 23(4):491-498.

19. Hoshikawa, T., et al. (1973). Analysis of running patterns in relation to speed. In *Medicine and Sport Vol. 8: Biomechanics III.* Basel: Karger, 342-348.

20. Hreljac, A. (1993). Preferred and energetically optimal gait transition speeds in human locomotion. *Medicine & Science in Sports & Exercise*, 25:1158-1162.

21. Komi, P. V., et al. (1974). *Biomechanics of Ski-Jumping.* Jyvaskyla, Finland: University of Jyvaskyla, 25-29.

22. Luhtanen, P., Komi, P. V. (1973). Mechanical factors influencing running speed. In E. Asmussen, K. Jorgensen (Eds.). *Biomechanics VI-B.* Baltimore, MD: University Park Press, 23-29.

23. Mason, B. R., et al. (1996). Biomechanical golf swing analysis. In R. Bauer (Ed.). *Proceedings of the XIII International Symposium for Biomechanics in Sports.* Thunder Bay, ON: International Society for Biomechanics in Sport, 67-70.

24. Neal, R. J., Wilson, B. D. (1985). 3D Kinematics and kinetics of the golf swing. *International Journal of Sport Biomechanics*, 1:221-232.

25. Ounpuu, S. (1994). The biomechanics of walking and running. *Clinics in Sports Medicine*, 13:843-863.

26. Owens, M. S., Lee, H. Y. (1969). A determination of velocities and angles of projection for the tennis serve. *Research Quarterly*, 40:750-754.

27. Sanderson, D. J., Sommer, H. J. (1985). Kinematic features of wheelchair propulsion. *Journal of Biomechanics*, 18:423-429.

28. Sinning, W. E., Forsyth, H. L.(1970). Lower limb actions while running at different velocities. *Medicine & Science in Sports & Exercise*, 2:28-34.

29. Terauds, J. (1975). Some release characteristics of international discus throwing. *Track and Field Review*, 75:54-57.

30. Vanlandewijck, Y. C., et al. (1994). Wheelchair propulsion efficiency: Movement pattern adaptations to speed changes. *Medicine & Science in Sports & Exercise*, 26:1373-1381.

31. Wagenaar, R. C., Emmerik, R. E. A. (2000). Resonant frequencies of arms and legs identify different walking patterns. *Journal of Biomechanics*, 33:853-861.

ANGULAR KINEMATICS

OBJECTIVES

After reading this chapter, the student will be able to:

1. Distinguish between linear, angular, and general motions.

2. Determine relative and absolute angles.

3. Discuss the conventions for the calculation of lower extremity angles.

4. Discuss the relationship among the kinematic quantities of angular distance and displacement, angular velocity, and angular acceleration.

5. Discuss the relationship between angular and linear motions, particularly displacement, velocity, and acceleration.

6. Discuss selected research studies that have used an angular kinematic approach.

7. Solve quantitative problems that employ angular kinematic principles.

OUTLINE

Angular Motion
Measurement of Angles
   Angle
   Units of Measurement
Types of Angles
   Absolute Angle
   Relative Angle
Lower Extremity Joint Angles
   Hip Angle
   Knee Angle
   Ankle Angle
   Rearfoot Angle
Representation of Angular Motion
   Vectors
Angular Motion Relationships
   Angular Position and Displacement
   Angular Velocity
   Angular Acceleration

Relationship between Angular and
   Linear Motions
   Linear and Angular Displacements
   Linear and Angular Velocities
   Linear and Angular Accelerations
Angle–Angle Diagrams
Angular Kinematics of Walking
   and Running
Lower Extremity Angles
   Rearfoot Motion
   Clinical Angular Changes
Angular Kinematics of the
   Golf Swing
Angular Kinematics of Wheelchair
   Propulsion
Summary
Review Questions

Angular Motion

Angular motion occurs when all parts of a body move through the same angle but do not undergo the same linear displacement. The subset of kinematics that deals with angular motion is **angular kinematics,** which describes angular motion without regard to the causes of the motion. Consider a bicycle wheel (Fig. 9-1). Pick any point close to the center of the wheel and any point close to the edge of the wheel. The point close to the edge travels farther than the point close to the center as the wheel rotates. The motion of the wheel is angular motion.

Angular motion occurs about an axis of rotation, that is, a line perpendicular to the plane in which the rotation occurs. For example, the bicycle wheel spins about its axle which is its axis of rotation. The axle is perpendicular to the plane of rotation described by the rim of the wheel (Fig. 9-1).

An understanding of angular motion is critical to comprehend how one moves. Nearly, all human movement involves rotation of body segments. The segments rotate about the joint centers that form their axes of rotation. For example, the forearm segment rotates about the elbow joint during flexion and extension of the elbow. When an individual moves, the segments generally undergo both rotation and translation. Sequential combinations of angular motion of multiple segments can result in linear motion of the segment end point seen in throwing and many other movements in which end point velocities are important. When the combination of rotation and translation occurs, it is described as general motion. Figure 9-2 illustrates the combination of linear and rotational motions. The gymnast undergoes translation as she moves across the ground. At the same time, she is rotating. The combination of rotation and translation is common in most human movements.

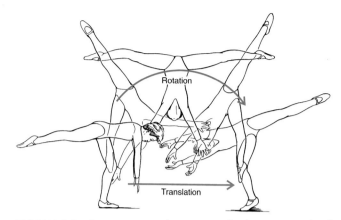

FIGURE 9-2 A gymnast completing a cartwheel as an example of general motion. The gymnast simultaneously undergoes both translation and rotation.

Measurement of Angles

ANGLE

An **angle** is composed of two lines, two planes, or a combination that intersects at a point called the **vertex** (Fig. 9-3). In a biomechanical analysis, the intersecting lines are generally body segments. If the longitudinal axis of the leg segment is one side of an angle and the longitudinal axis of the thigh segment is the other side, the vertex is the joint center of the knee. Angles can be determined from the coordinate points described in Chapter 8. Coordinate points describing the joint centers determine the sides and vertex of the angle. For example, an angle at the knee can be constructed using the thigh and leg segments. The coordinate

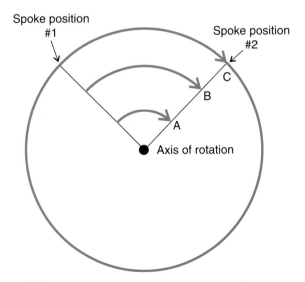

FIGURE 9-1 A bicycle wheel as an example of rotational motion. Points *A*, *B*, and *C* undergo the same amount of rotation but different linear displacements, with *C* undergoing the greatest linear displacement.

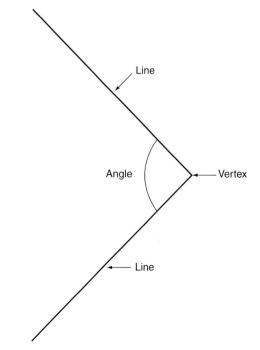

FIGURE 9-3 Components of an angle. Note that the lines are usually segments and the vertex of the angle is the joint center.

points describing the ankle and knee joint centers define the leg segment; the coordinate points describing the hip and knee joint centers define the thigh segment. The vertex of the angle is the knee joint center.

Definition of a segment by placing markers on the subject at the joint centers makes a technically incorrect assumption that the joint center at the vertex of the angle does not change throughout the movement. Because of the asymmetries in the shape of the articulating surfaces in most joints, one or both bones constituting the joint may displace relative to each other. For example, although the knee is often considered a hinge joint, it is not. At the knee joint, the medial and lateral femoral condyles are asymmetrical. Therefore, as the knee flexes and extends, the tibia rotates along its long axis and rotates about an axis through the knee from front to back. The location of the joint center, therefore, changes throughout any motion of the knee. The center of rotation of a joint at an instant in time is called the instantaneous joint center (Fig. 9-4). It is difficult to locate this moving axis of rotation without special techniques such as x-ray measurements. These measurements are not practical in most situations; thus, the assumption of a static instantaneous joint center must be made.

> Using MaxTRAQ, import the first video file of the golfer. What joints would you use to calculate the angle at the right shoulder? What is the vertex of the angle?

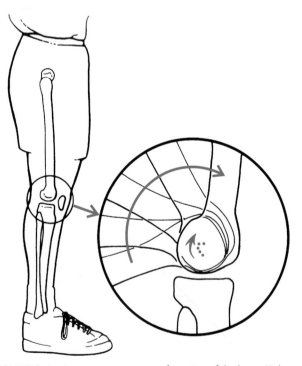

FIGURE 9-4 Instantaneous center of rotation of the knee. (Adapted from Nordin, M., Frankel, V. H. (Eds.) [1979]. *Biomechanics of the Musculoskeletal System*, 2nd ed.. Philadelphia, PA: Lea & Febiger.)

UNITS OF MEASUREMENT

In angular motion, three units are used to measure angles. It is important to use the correct units to communicate the results of this work clearly and to compare values from study to study. It is also essential to use the correct units because angle measurements may be used in further calculations. The first and most commonly used is the **degree**. A circle, which describes one complete rotation, transcribes an arc of 360° (Fig. 9-5A). An angle of 90° has sides that are perpendicular to each other. A straight line has a 180° angle (Fig. 9-5B).

The second unit of measurement describes the number of rotations or revolutions about a circle (Fig. 9-5A). One **revolution** is a single 360° rotation. For example, a triple jump in skating requires the skater to complete 3.5 revolutions in the air. The skater completes a rotation of 1,260°. This unit of measurement is useful in qualitative descriptions of movements in figure skating, gymnastics, and diving but is not useful in a quantitative analysis.

Although the degree is most commonly understood and the revolution is often used, the most appropriate unit for angular measurement in biomechanics is the **radian**. A radian is defined as the measure of an angle at the center

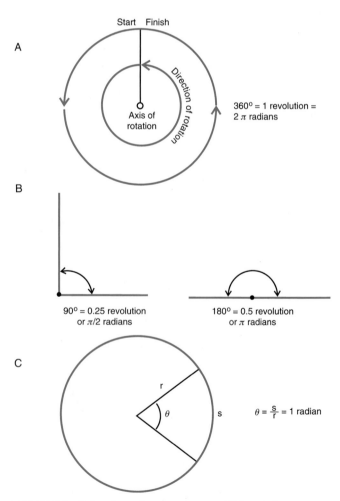

FIGURE 9-5 Units of angular measurement. **(A)** Revolution. **(B)** Perpendicular and straight lines. **(C)** Radian.

of a circle described by an arc equal to the length of the radius of the circle (Fig. 9-5C). That is:

$$\theta = s/r = 1 \text{ rad}$$

where $\theta = 1$ rad, s = arc of length r along the diameter, and r = radius of the circle. Because both s and r have units of length (m), the units in the numerator and denominator cancel each other out with the result that the radian is dimensionless.

In further calculations, the radian is not considered in determining the units of the result of the calculation. Degrees have a dimension and must be included in the unit of the product of any calculation. It is necessary, therefore, to use the radian as a unit of angular measurement instead of the degree in any calculation involving linear motion because the radian is dimensionless.

One radian is the equivalent of 57.3°. To convert an angle in degrees to radians, divide the angle in degrees by 57.3. For example:

$$\frac{72°}{57.3°} = 1.26 \text{ rad}$$

To convert radians to degrees, multiply the angle in radians by 57.3. For example:

$$0.67 \text{ rad} \times 57.3° = 38.4°$$

Angular measurement in radians is often determined in multiples of pi ($\pi = 3.1416$). Because 2π radians are in a complete circle, 180° may be represented as π radians, 90° as $\pi/2$ radians, and so on.

Although the unit of angular measurement in the Systéme International d'Unités (SI) is the radian and this unit must be used in further calculations, the angular motion concepts presented in the remainder of this chapter will use the degree for ease of understanding.

 **Types of Angles**

ABSOLUTE ANGLE

In biomechanics, two types of angles are generally calculated. The first is the **absolute angle**, which is the angle of inclination of a body segment relative to some fixed reference in the environment. This type of angle describes the orientation of a segment in space. Two primary conventions are used for calculating absolute angles. One involves placing a coordinate system at the proximal end point of the segment. The angle is then measured counterclockwise from the right horizontal. The most frequently used convention for calculating absolute angles, however, places a coordinate system at the distal end point of the segment (Fig. 9-6). The angle using this convention is also measured counterclockwise from the right horizontal. The absolute angles calculated using these two conventions are related and give comparable information. When calculating absolute angles, however, the convention used must

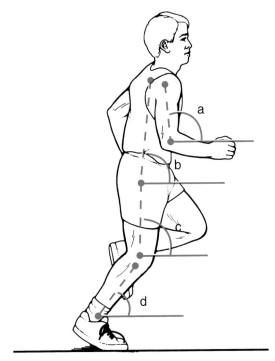

FIGURE 9-6 Absolute angles: The arm "a," trunk "b," thigh "c," and leg "d" of a runner.

be stated clearly. The absolute angle of a segment relative to the right horizontal is also called the segment angle.

Absolute angles are calculated using the trigonometric relationship of the **tangent**. The tangent is defined based on the sides of a right triangle. It is the ratio of the side opposite the angle in question and the side adjacent to the angle. The angle in question is not the right angle in the triangle. If the leg and thigh segment coordinate positions are considered, the absolute angles of both the thigh and leg segments can be calculated (Fig. 9-7).

To calculate the absolute leg angle, the distal end coordinate values are subtracted from the proximal end coordinate values and the ratio of y to x defines the tangent of the angle:

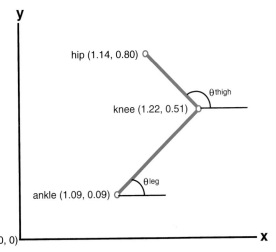

FIGURE 9-7 Absolute angles of the thigh and leg as defined in a coordinate system.

$$\tan \theta_{\text{leg}} = (y_{\text{proximal}} - y_{\text{distal}})/(x_{\text{proximal}} - x_{\text{distal}})$$
$$= (y_{\text{knee}} - y_{\text{ankle}})/(x_{\text{knee}} - x_{\text{ankle}})$$
$$= (0.51 - 0.09)/(1.22 - 1.09)$$
$$= 0.42/0.13$$
$$= 3.23$$

Next, the angle whose tangent is 3.23 is again determined using either the trigonometric tables (see Appendix B) or a calculator. This is called finding the inverse tangent and is written as follows:

$$\theta_{\text{leg}} = \tan^{-1} 3.23$$
$$= 72.8°$$

The absolute angle of the leg, therefore, is 72.8° from the right horizontal. This orientation indicates that the leg is positioned so that the knee is farther from the vertical (y) axis of the coordinate system than the ankle. That is, the knee joint is to the right of the ankle joint (see Fig. 9-7).

Similarly, to calculate the thigh angle, the coordinate values are substituted:

$$\tan \theta_{\text{thigh}} = (y_{\text{hip}} - y_{\text{knee}})/(x_{\text{hip}} - x_{\text{knee}})$$
$$= (0.80 - 0.51)/(1.14 - 1.22)$$
$$= 0.29/-0.08$$
$$= -3.625$$

Again, the angle whose tangent is –3.625 is determined as follows:

$$\theta_{\text{thigh}} = \tan^{-1} - 3.625$$
$$= -74.58°$$

This angle is clockwise from the left horizontal because we have moved into the second quadrant with the negative x-value. To convert the angle so that it is relative to the right horizontal and counterclockwise, it must be added to 180°, resulting in an absolute angle of 105.4° relative to the right horizontal (Fig. 9-8).

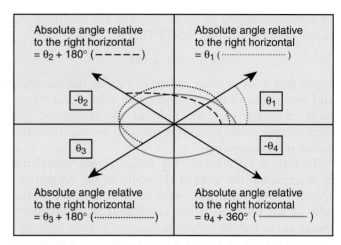

FIGURE 9-8 To calculate absolute angles relative to the right horizontal requires adjustments when the orientation is such that the differences between the proximal and distal end points indicate that the segment is not in the first quadrant.

An absolute thigh angle of 105.4° in the second quadrant indicates that the thigh is oriented such that the hip joint is closer to the vertical (y) axis and above the horizontal (x) axis of the coordinate system. In this case, the thigh would be oriented with the knee to the right of the hip in this reference system. When both x and y are negative, the value is in the third quadrant, and the angle is computed counterclockwise and relative to the left horizontal, so 180° is still added to adjust the absolute angle so that it is relative to the right horizontal. Finally, if there is only a negative y-value, the angle is in the fourth quadrant and taken clockwise and relative to the right horizontal, so 360° should be added to convert the absolute angle so that it is relative to the right horizontal in the counterclockwise direction.

Trunk, thigh, leg, and foot segmental end points for both the touchdown and the toe-off in walking are graphically illustrated in Figure 9-9. The corresponding calculations of the absolute angles shown in Table 9-1

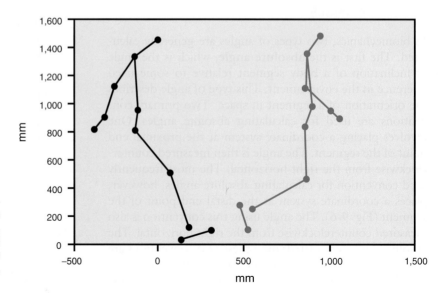

FIGURE 9-9 By plotting the segmental endpoints and creating a stick figure, the similarities or differences in position can be clearly observed. The differences in right foot touchdown (*black*) and right foot toe-off (*blue*) phases of a walking gait are apparent. See Appendix C Frame 1 and Frame 76, respectively.

TABLE 9-1	Absolute Angle Calculation for Touchdown and Toe-Off in Walking				
Frame	Trunk$_x$ = (Shoulder$_x$ to Greater Trochanter$_x$)	Trunk$_y$ = (Shoulder$_y$ to Great Trochanter$_y$)	Absolute Angle = Arctan (y/x)	Thigh$_x$ = (Greater Trochanter$_x$ to Knee$_x$)	Thigh$_y$ = (Greater Trochanter$_y$ to Knee$_y$)
1	−3.95	523.08	= −89.57° + 180° = 90.43°	−210.15	317.14
76	10.92	532.10	= 88.82°	−14.76	368.95

Absolute Angle = Arctan (y/x)	Leg$_x$ = (Knee$_x$ to Ankle$_x$)	Leg$_y$ = (Knee$_y$ to Ankle$_y$)	Absolute Angle = Arctan (y/x)	Foot$_x$ = (Heel$_x$ to Met$_x$)	Foot$_y$ = (Heel$_y$ to Met$_y$)	Absolute Angle = Arctan (y/x)
= −56.47° + 180° = 123.53°	−113.03	377.88	= −73.35° + 180° = 106.65°	−181.92	−67.95	= −0.48° + 180° = 200.48°
= −87.71° + 180° = 92.29°	313.49	218.66	= 34.90°	−51.20	169.39	= −73.18° + 180° = 106.82°

use the conventions discussed previously to convert all angles so they are taken counterclockwise with respect to the right horizontal. For example, the leg orientation in touchdown results in a negative x-position and positive y-position, so 180° is added to the final angle computation to make it relative to the right horizontal. In the case of toe-off, however, both x and y are positive, so there is no adjustment. Likewise, the foot orientation at touchdown results in both negative x and y, placing it in the third quadrant, where 180° is again added. These adjustments provide a consistent reference for the computation of the absolute angles.

Using MaxTRAQ, import the first video file of the woman walking (representing right heel contact). Digitize the right shoulder, right greater trochanter, right knee, and the right ankle. Calculate the absolute angles of the trunk, thigh, and leg.

RELATIVE ANGLE

The other type of angle calculated in biomechanics is the **relative angle** (Fig. 9-10A). This is the angle between longitudinal axes of two segments and is also referred to as the joint angle or the intersegmental angle. A relative angle (e.g., the elbow angle) can describe the amount of flexion or extension at the joint. Relative angles, however, do not describe the position of the segments or the sides of the angle in space. If an individual has a relative angle of 90° at the elbow and that angle is maintained, the arm may be in any of a number of positions (Fig. 9-10B).

Relative angles can be calculated using the **law of cosines**. This law, simply a more general case of the Pythagorean theorem, describes the relationship between the sides of a triangle that does not contain a

right angle. For our purposes, the triangle is made up of the two segments B and C and a line, A, joining the distal end of one segment to the proximal end of the other (Fig. 9-11).

In Figure 9-11, the coordinate points for two segments describing the thigh and the leg are given. To calculate the relative angle at the knee (θ), the lengths a, b, and c would be calculated using the Pythagorean relationship:

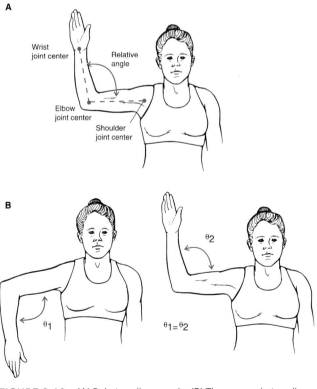

FIGURE 9-10 **(A)** Relative elbow angle. **(B)** The same relative elbow angle with the arm and forearm in different positions.

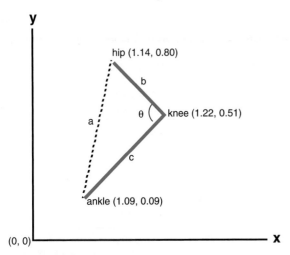

FIGURE 9-11 Coordinate points describing the hip, knee, and ankle joint centers and the relative angle of the knee (θ).

$$a = \sqrt{(x_b - x_a)^2 + (y_b - y_a)^2}$$
$$= \sqrt{(1.14 - 1.09)^2 + (0.80 - 0.09)^2}$$
$$= \sqrt{0.0025 + 0.5041}$$
$$= 0.71$$
$$b = \sqrt{(x_b - x_k)^2 + (y_b - y_k)^2}$$
$$= \sqrt{(1.14 - 1.22)^2 + (0.80 - 0.51)^2}$$
$$= \sqrt{0.0064 + 0.0841}$$
$$= 0.30$$
$$c = \sqrt{(x_k - x_a)^2 + (y_k - y_a)^2}$$
$$= \sqrt{(1.22 - 1.09)^2 + (0.51 - 0.09)^2}$$
$$= \sqrt{0.0169 + 0.1764}$$
$$= 0.44$$

The next step is to substitute these values in the law of cosine equation and solve for the cosine of the angle θ.

$$a^2 = b^2 + c^2 - 2 \times b \times c \times \cos\theta$$
$$\cos\theta = b^2 + c^2 - a^2 / 2 \times b \times c$$
$$\cos\theta = 0.30^2 + 0.44^2 - 0.71^2 / 2 \times 0.30 \times 0.44$$
$$\cos\theta = -0.833$$

To find the angle θ, the angle whose cosine is −0.833 can be determined using either trigonometric tables (see Appendix B) or a calculator with trigonometric functions. This process, known as finding the inverse cosine or arcos, is written as follows:

$$\theta = \cos^{-1} -0.833$$
$$\theta = 146.4°$$

Therefore, the relative angle at the knee is 146.4°. In this case, the knee is slightly flexed (180° representing full extension).

Using MaxTRAQ, import the third video file of the woman walking (representing midstance). Digitize the right iliac crest, right greater trochanter, right knee, and the right ankle. Calculate the relative angles of the hip and knee.

A relative angle can be calculated from the absolute values to obtain a result similar to computations using the law of cosines. The relative angle between two segments can be calculated by subtracting the absolute angle of the distal segment from the proximal segment. In the example using the thigh and lower leg, the following calculation is another option:

$$\theta_{relative} = \theta_{absolute\ thigh} - \theta_{absolute\ leg}$$
$$\theta_{relative} = -74.58° - 72.8°$$
$$\theta_{relative} = 147.4°$$

In clinical situations, the relative angle is most often calculated because it provides a more practical indicator of function and joint position. In quantitative biomechanical analyses, however, absolute angles are calculated more often than relative angles because they are used in a number of subsequent calculations. Regardless of the type of angle calculated, however, a consistent frame of reference must be used.

Unfortunately, many coordinate systems and systems of defining angles have been used in biomechanics, resulting in difficulty comparing values from study to study. Several organizations, such as the Canadian Society of Biomechanics and the International Society of Biomechanics, have standardized the representation of angles to provide consistency in biomechanics research, especially in the area of joint kinematics.

Lower Extremity Joint Angles

In discussing the angle of a joint such as the knee and ankle, it is imperative that a meaningful representation of the action of the joint be made. A special use of absolute angles to compute joint angles is very useful for clinicians and others interested in joint function. Lower extremity joint angles can be calculated using the absolute angles similar to the procedure described previously. A system of lower extremity joint angle conventions was presented by Winter (36). These lower extremity angle definitions are for use in a two-dimensional (2D) sagittal plane analysis only. In Winter's system, digitized points describing the trunk, thigh, leg, and foot are used to calculate the absolute angles of each (Fig. 9-12). From these absolute angles, joint angles can be computed. In such a biomechanical analysis, it is assumed that a right-side sagittal view is being analyzed. That is, the right side of the

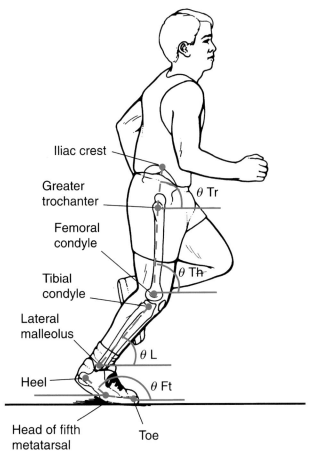

Iliac crest

Greater
trochanter

Femoral
condyle

Tibial
condyle

Lateral
malleolus

Heel

Head of fifth
metatarsal

Toe

θ Tr

θ Th

θ L

θ Ft

FIGURE 9-12 Definition of the sagittal view absolute angles of the trunk, thigh, leg, and foot. (After Winter, D. A. [1987]. *The Biomechanics and Motor Control of Gait.* Waterloo, ON: University of Waterloo Press.)

subject's body is closest to the camera and is considered to be in the *x–y* plane.

HIP ANGLE

Based on the absolute angles of the trunk and the thigh calculated, the hip angle is:

$$\theta_{hip} = \theta_{thigh\ absolute} - \theta_{trunk\ absolute}$$

In this scheme, if the hip angle is positive, the action at the hip is flexion; but if the hip angle is negative, the action is extension. If the angle is zero, the thigh and the trunk are aligned vertically in a neutral position. For example, the hip joint angle representing flexion and extension for the touchdown phase in walking (Table 9-1) would be:

$$\theta_{hip} = \theta_{thigh} - \theta_{trunk}$$
$$= 123.53° - 90.43°$$
$$= 33.1°$$

The joint angle of 33.1° at touchdown indicates that the thigh is flexing at the hip joint. In a human walking at a moderate pace, the hip angle oscillates ±35° about 0°; in running, the hip angle oscillates ±45°.

KNEE ANGLE

Using the absolute angle of the thigh and the leg, the knee joint angle is defined as:

$$\theta_{knee} = \theta_{thigh\ absolute} - \theta_{leg\ absolute}$$

In human locomotion, the knee angle is always positive (i.e., in some degree of flexion), and it usually varies from 0° to 50° throughout a walking stride and from 0° to 80° during a running stride. Because the knee angle is positive, the knee is always in some degree of flexion. If the knee angle gets progressively greater, the knee is flexing. If it gets progressively smaller, the knee is extending. A zero knee angle is a neutral position, and a negative angle indicates a hyperextension of the knee. The knee angle for the touchdown phase in the walking example (Table 9-1) is:

$$\theta_{knee} = \theta_{thigh} - \theta_{leg}$$
$$= 123.53° - 106.65°$$
$$= 16.88°$$

ANKLE ANGLE

The ankle angle is calculated using the absolute angles of the foot and the leg:

$$\theta_{ankle\ joint\ angle} = \theta_{leg} - \theta_{foot} + 90°$$

This may seem more complicated than the other lower extremity joint angle calculations. Without adding 90°, the ankle angle would oscillate about 90°, making interpretation of it difficult. Adding 90° makes the ankle angle oscillate about 0°. Thus, a positive angle represents dorsiflexion, and a negative angle represents plantarflexion.

The ankle angle for the touchdown phase in the walking example (Table 9-1) is:

$$\theta_{ankle} = \theta_{leg} - \theta_{foot} + 90°$$
$$= 106.65° - 200.48° + 90°$$
$$= -3.83°$$

This value indicates that the ankle is in plantarflexion at touchdown. The ankle angle generally oscillates ±20° during a natural walking stride and ±35° during a running stride. Lower extremity angles calculated for a walking stride using Winter's convention are presented in Figure 9-13.

Using MaxTRAQ, import the first video file of the woman walking (representing right heel contact). Digitize the right shoulder, right greater trochanter, right knee, and the right ankle. Calculate the absolute angles of the trunk, thigh, and leg (this was done in a previous assignment). Using these absolute angles, calculate the hip and knee angles according to Winter (36).

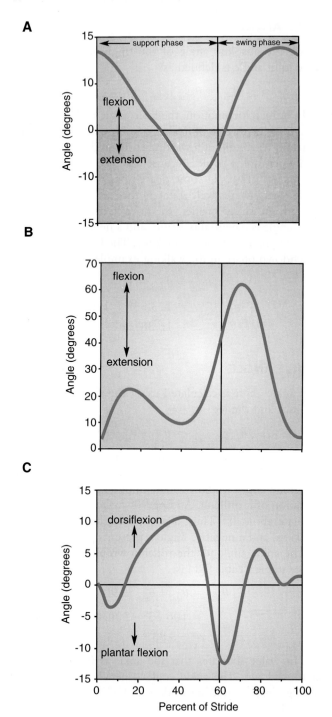

FIGURE 9-13 Graphs of the hip (**A**), knee (**B**), and ankle (**C**) angles during walking.

Joint angles calculated using the relative angle approach (law of cosines) and the same angles calculated from the absolute angles using Winter's (36) convention have exactly the same clinical meaning. In the relative angle approach, the joint angle that is calculated is the included angle between the two segments. Using the absolute angle approach, the joint angle that is calculated is the difference between the two segment angles. The interpretation of these angles is exactly the same. Both types of angles are presented in Figure 9-14.

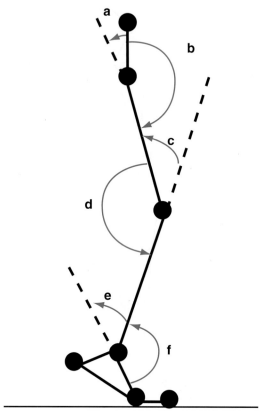

FIGURE 9-14 Joint angle representations using the relative angle and absolute angle calculations. Angles a, c, and e are calculated from the absolute angles, and angles b, d, and f use the relative angle calculations. Both representations are the same.

REARFOOT ANGLE

Another lower extremity angle that is often calculated in biomechanical analyses is the rearfoot angle. The motion of the subtalar joint in a 2D analysis is considered to be in the frontal plane. The rearfoot angle represents the motion of the subtalar joint. The rearfoot angle thus approximates calcaneal eversion and calcaneal inversion in the frontal plane. Calcaneal eversion and inversion are among the motions in the pronation and supination action of the subtalar joint. In the research literature, calcaneal eversion is often measured to evaluate pronation, and calcaneal inversion is measured to determine supination.

The rearfoot angle is calculated using the absolute angles of the leg and the calcaneus in the frontal plane. Two segment markers are placed on the rear of the leg to define the longitudinal axis of the leg. Two markers are also placed on the calcaneus (or the rear portion of the shoe) to define the longitudinal axis of the calcaneus (Fig. 9-15).

Researchers have reported that markers placed on the shoe rather than directly on the calcaneus do not give a true indication of calcaneal motion (30). In fact, it has been suggested that the rearfoot motion calculated when the markers are placed on the shoe is greater than when the markers are placed on the calcaneus for the same movement. Regardless of the positioning of the markers,

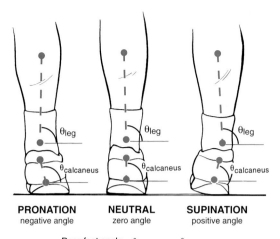

PRONATION
negative angle

NEUTRAL
zero angle

SUPINATION
positive angle

Rear foot angle = $\theta_{calcaneus} - \theta_{leg}$

FIGURE 9-15 Definition of the absolute angles of the leg and calcaneus in the frontal plane. These angles are used to constitute the rearfoot angle of the right foot.

they are used to calculate the absolute angles of the leg and heel; thus, the rear foot angle is:

$$\theta_{rearfoot} = \theta_{leg} - \theta_{calcaneus}$$

By this calculation, a positive angle represents calcaneal inversion, a negative angle represents calcaneal eversion, and a zero angle is the neutral position.

During the support phase of the gait cycle, the rearfoot, as defined by the rearfoot angle, is in an inverted position at the initial foot contact with the ground. At this instance, the rearfoot angle is positive. From that point onward until midstance, the rearfoot moves to an everted position; thus, the rearfoot angle is negative. At midstance position, the foot becomes less everted and moves to an inverted position at toe-off. The rearfoot angle becomes less negative and eventually positive at toe-off. Figure 9-16 is a representation of a typical rearfoot angle curve during the support phase of a running stride.

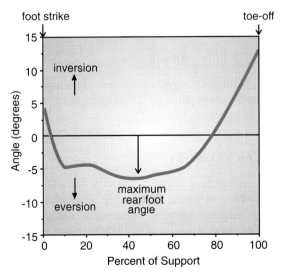

FIGURE 9-16 A typical rearfoot angle–time graph during running. Maximum rearfoot angle is indicated.

 # Representation of Angular Motion Vectors

Representing angular motion vectors graphically as lines with arrows, as is the case in linear kinematics, is difficult. It is essential, however, to determine the direction of rotation in terms of a positive or negative rotation. The direction of rotation of an angular motion vector is referred to as the **polarity** of the vector. The polarity of an angular motion vector is determined by the **right-hand rule**. The direction of an angular motion vector is determined using this rule by placing the curled fingers of the right hand in the direction of the rotation. The angular motion vector is defined by an arrow of the appropriate length that coincides with the direction of the extended thumb of the right hand (Fig. 9-17). The 2D convention generally used is that all segments rotating counterclockwise from the right horizontal have positive polarity and all segments rotating in clockwise have a negative polarity.

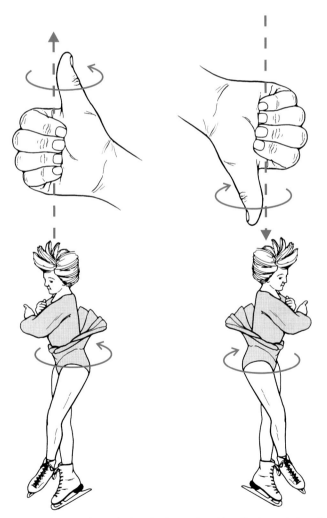

FIGURE 9-17 The right-hand rule used to identify the polarity of the angular velocity of a figure skater during a spin. The fingers of the right-hand point in the direction of the rotation, and the right thumb points in the direction of the angular velocity vector. The angular velocity vector is perpendicular to the plane of rotation.

Angular Motion Relationships

The relationships discussed in this chapter on angular kinematics are analogous to those in Chapter 8 on linear kinematics. The angular case is simply an analog of the linear case.

ANGULAR POSITION AND DISPLACEMENT

The angular position of an object refers to its location relative to a defined spatial reference system. In the case of a 2D system with the *y*-axis representing motion vertically up and down and the *x*-axis representing anterior to posterior motion, angular position is described in the *x–y* plane. A three-dimensional (3D) system adds a third axis, *z*, in the medial and lateral planes. Many clinicians use planes to describe angular positioning. For example, if the axes are placed with the origin at the shoulder joint, angular position of the arm in the *x–y* plane would be a flexion and extension position in the *y–z* plane abduction and adduction, and in the *x–z* plane, rotation. This system works well for describing joint angles but lacks precision for describing complex movements. Absolute angles can be computed relative to a fixed reference system placed at a joint or at another fixed point in the environment. As discussed earlier, angular position can also be computed relative to a line or plane that is allowed to move. It is common to present joint angles such as those shown in Figure 9-13 to document the joint actions in a movement such as walking.

The concepts of distance and displacement in the angular case must be discerned. Consider a simple pendulum swinging in the *x–y* plane through an arc of 70° (Fig. 9-18). If the pendulum swings though a single arc, the angular distance is 70°, but if it swings through 1.5 arcs, the angular distance is 105°. Angular distance is the total of all angular changes measured following its exact path. As in the linear case, however, angular distance is not the same as angular displacement.

Angular displacement is the difference between the initial and final positions of the rotating object

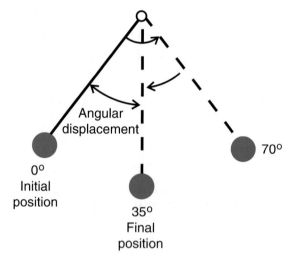

Angular displacement = 35° – 0° = 35°

FIGURE 9-19 Angular displacement is the difference between the initial position and the final position.

(Fig. 9-19). In the example of the pendulum, if the pendulum swings through two complete arcs, the angular displacement is zero because its final position is the same as the starting position. Angular displacement never exceeds 360° or 2π rad of rotation, but angular distance can be of any value. In discussing angular displacement, it is necessary to designate the direction of the rotation. Counterclockwise rotation is considered to be positive, and clockwise rotation is negative. With a 3D reference system placed at the shoulder joint, the positive *y*-axis would be upward, the positive *x*-axis would be posterior to anterior, and the positive *z*-axis would be medial to lateral. The corresponding positive joint movements about these axes would be flexion/extension (about the *x*-axis), internal/external rotation (about the *y*-axis), and abduction/adduction (about the *z*-axis).

If the absolute angle of a segment, theta (θ), is calculated for successive positions in time, the angular displacement (Δθ) is:

$$\Delta\theta = \theta_{final} - \theta_{initial}$$

The polarity, or sign, of the angular displacement is determined by the sign of Δθ as calculated and may be confirmed by the right-hand rule.

ANGULAR VELOCITY

Angular speed and **angular velocity** are analogous with linear speed and linear velocity in both definition and meaning. Angular speed is the angular distance traveled per unit of time. Angular speed is a scalar quantity and is generally not critically important in biomechanical analysis because it is not used in any further calculations.

Angular velocity, characterized by the Greek letter omega (ω), is a vector quantity that describes the time rate

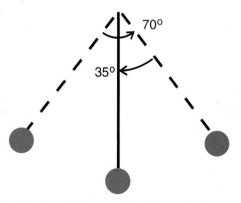

FIGURE 9-18 A swinging pendulum illustrating the angular distance over 1.5 arcs of swing.

of change of angular position. If the measured angle is θ, then the angular velocity is:

ω = Change in angular position/change in time

$= (\theta_{final} - \theta_{initial})/(time_{final} - time_{initial})$

$= \Delta\theta/\Delta t$

If the initial angle of a segment is 34° at time 1.25 seconds and the segment moves to an angle of 62° at time 1.30 seconds, the angular velocity would be:

$\omega = \Delta\theta/\Delta t$

$= (62° - 34°)/(1.30 \text{ seconds} - 1.25 \text{ seconds})$

$= 28°/0.05 \text{ seconds}$

$= 560°/\text{second}$

Angular speed and angular velocity are generally presented in degrees per second (°/s). If, however, as noted previously, any further computation is to be done using angular velocity, the units must be radians per second (rad/s).

In the previous example, the average angular velocity was calculated over the interval from 1.25 to 1.30 seconds. According to the discussion in the previous chapter, angular velocity represents the slope of a secant on an angular position–time graph over this interval. The instantaneous

angular velocity represents the slope of a tangent to an angular position–time graph and is calculated as a limit:

$$\text{Limit } \omega = d\theta/dt$$
$$dt \to 0$$

Angular velocity is thus the first derivative of angular position.

As in the linear case, the direction of the slope on an angle–time profile determines whether the angular velocity is positive or negative, and the steepness of the slope indicates the rate of change of angular position. If θ_{final} is greater than $\theta_{initial}$, ω is positive (i.e., the slope is positive), but if θ_{final} is less than $\theta_{initial}$, ω is negative (i.e., slope is negative). Both situations can be confirmed using the right-hand rule. If there is no change in the angle, the slope is zero and ω is zero.

The method used to calculate angular velocity over a series of frames of a kinematic analysis is the first central difference method. This method calculates the angular velocity at the same instant at which the data for angular position are available. For angular velocity, this formula is:

$$\omega_i = (\theta_{i+1} - \theta_{i-1})/(t_{i+1} - t_{i-1})$$

where θ_i is the angle at time t_i. Table 9-2 represents the thigh absolute angle data collected for one support phase

| TABLE 9-2 | Calculation of Angular Position, Velocity, and Acceleration of the Thigh During the Support Phase of Walking |||||||

Frame Number	Time (seconds)	GT x	GT y	Knee (K) x	Knee (K) y	Thigh Angle (°)	Thigh Velocity (°/s)	Thigh Acceleration (°/s/s)
1	0.0000	−127.4	817.4	81.5	499.7	123.33		
2	0.0083	−114.7	817.6	92.8	499.2	123.09	−29.52	
3	0.0166	−101.3	818.0	104.7	498.8	122.84	−31.33	−326.51
4	0.0249	−87.5	818.6	116.9	498.6	122.57	−34.94	−471.69
5	0.0332	−73.4	819.5	129.2	498.5	122.26	−39.16	−435.54
					$\omega_6 = (\theta_7 - \theta_5)/(t_7 - t_5) =$			
6	0.0415	−59.1	820.6	141.5	498.6	121.92	−42.17	−507.83
7	0.0498	−44.9	822.0	153.6	498.8	121.56	−47.59	−689.16
8	0.0581	−30.7	823.7	165.4	499.0	121.13	−53.61	−762.05
					$\alpha_9 = (\omega_{10} - \omega_8)/(t_{10} - t_8) =$			
9	0.0664	−16.7	825.6	176.8	499.3	120.67	−60.24	−907.23
10	0.0747	−2.8	827.7	187.6	499.6	120.13	−68.67	−1052.41
11	0.0830	11.0	829.9	197.9	499.9	119.53	−77.71	−1161.45
					$\theta_{12} = 180 + \text{atan}([GT_y - K_y]/[GT_x - K_x]) =$			
12	0.0913	24.6	832.3	207.5	500.1	118.84	−87.95	−1088.55
13	0.0996	38.0	834.8	216.4	500.2	118.07	−95.78	−980.12
14	0.1079	51.2	837.4	224.8	500.3	117.25	−104.22	−943.98
15	0.1162	64.3	840.1	232.5	500.4	116.34	−111.45	
16	0.1245	77.1	842.7	239.7	500.3	115.40		

in walking (see Appendix C). The rate of the camera was 120 frames per second, and every third frame is presented from touchdown (frame 0) to toe-off (frame 76). The time between each frame is $1/120 = 0.0083$ seconds; thus, the time between three frames is 0.0249 seconds. Using the first central difference method, the angular velocity is calculated from the absolute angular position for each frame. After it has been calculated, the values are typically graphed to observe the pattern of motion (Fig. 9-20). The results of the calculation and graphing of angular kinematics of the thigh indicate that for most of the support phase, the thigh is moving clockwise with respect to the knee joint. At heel strike, the thigh is in extreme hip flexion that is reduced as the trunk is brought over the support limb, moving the thigh clockwise (negative angular velocity). The thigh is vertically aligned in frame 39, and the trunk continues moving over the limb, forcing the thigh to continue in its clockwise rotation about the knee joint. At the end of the support phase (frame 63), the motion of the thigh reverses, and a counterclockwise movement of the thigh begins in preparation for toe-off (positive angular velocity).

Refer to the walking data in Appendix C. Using the first central difference method, calculate the angular velocity of the lower leg using the absolute segment position angles and graph the angular velocity.

ANGULAR ACCELERATION

Angular acceleration is the rate of change of angular velocity with respect to time and is symbolized by the Greek letter alpha (α).

Angular acceleration = Change in angular velocity/change in time

$$\alpha = (\omega_{final} - \omega_{initial})/(time_{final} - time_{initial})$$

$$\alpha = \Delta\omega/\Delta t$$

For ease of understanding, biomechanists generally present their results in degrees per second squared (deg/s²), but the most appropriate unit for angular acceleration is radians per second squared (rad/s²).

As with the linear case, angular acceleration is the derivative of angular velocity and represents the slope of a line (either a secant for average angular acceleration or a tangent for instantaneous angular acceleration). If α is the slope of a secant to an angular velocity–time profile, it represents an average acceleration over a time interval. If α is the slope of a tangent, the instantaneous angular acceleration has been calculated. This also implies that the slope may be positive ($\omega_{final} > \omega_{initial}$), negative ($\omega_{final} < \omega_{initial}$), or zero ($\omega_{final} = \omega_{initial}$). The direction of the angular acceleration vector may be confirmed using the right-hand rule. The instantaneous angular acceleration is calculated by:

$$Limit\ \alpha = d\omega/dt$$

$$dt \to 0$$

Again, in a kinematic analysis, the usual method of calculating angular acceleration is the first central difference method. The formula for angular acceleration for this method is:

$$\alpha_i = (\omega_{i+1} - \omega_{i-1})/(t_{i+1} - t_{i-1})$$

where θ_i is the angular velocity at time t_i. Table 9-2 presents the calculated angular acceleration of the thigh for selected frames in the support phase of walking.

As in the case for linear acceleration, the sign or polarity of angular acceleration does not indicate the direction of rotation. For example, positive angular acceleration may mean increasing angular velocity in the positive direction or decreasing angular velocity in the negative direction. Also, negative angular acceleration may indicate decreasing angular velocity in the positive direction or increasing angular velocity in the negative direction. The angular position, velocity, and acceleration of the thigh are presented in Table 9-2, with a corresponding graph in Figure 9-20. The

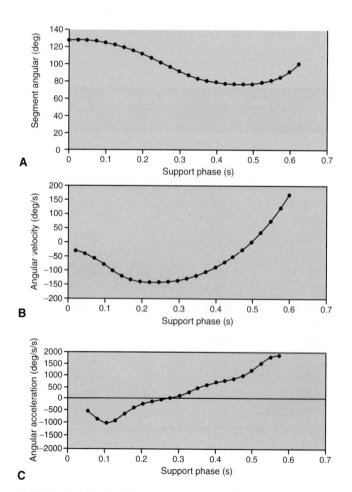

FIGURE 9-20 Graphic representations of the thigh's absolute angle **(A)**, angular velocity **(B)**, and angular acceleration **(C)** as a function of time for the support phase of walking (data from Appendix C).

angular acceleration of the thigh is negative (increasing angular velocity in the negative direction) and then positive (decreasing angular velocity in the negative direction) during the portion of the support phase where the thigh angular velocity is negative. The acceleration remains positive (increasing angular velocity in the positive direction) in the later stages of support when the thigh angular velocity changes from negative to positive (direction change).

> Refer to the walking data in Appendix C. Using the first central difference method, calculate and graph the angular acceleration of the leg.

 ## Relationship between Angular and Linear Motions

In many human movements, the motions of the segments constituting the movement are angular, whereas the result of the movement is linear. For example, a pitcher throws a baseball that travels linearly. However, the motions of the pitcher's segments resulting in the throw are rotational. For example, when it is necessary to know the linear motion of the hand, you must know that it depends on the angular motion of the segments of the upper extremity. This example suggests a mechanical relationship between linear and angular motions.

LINEAR AND ANGULAR DISPLACEMENTS

When the angular measure of an angle, the radian, is defined, it is noted that:

$$\theta = s/r$$

where θ is the angle subtended by an arc of length s that is equal to the radius of the circle. By rearranging this equation, the length of the arc can be presented as:

$$s = r\theta$$

Suppose the forearm, with length r_1, rotates about the elbow joint (Fig. 9-21). The arc described by the rotation—the distance that the wrist moves—is Δs_1 and the angle is $\Delta\theta$. The linear distance that the wrist travels is described as:

$$\Delta s_1 = r_1 \Delta\theta$$

Therefore, the linear distance that any point on the segment moves can be described if the distance of that point to the axis of rotation and the angle through which the segment rotates are known. Suppose another point on the arm is marked as s_2 with a distance of r_2 to the axis of rotation. The distance this point travels during the same angular motion is:

$$\Delta s_2 = r_2 \Delta\theta$$

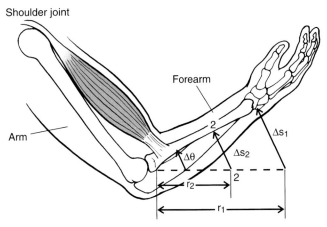

FIGURE 9-21 The relationship between linear and angular displacements.

Because r_1 is longer than r_2, the distance traveled by s_1 must be greater than s_2. Thus, the most distal points on a segment travel a greater distance than points closer to the axis of rotation. The value for the expression r is called the **radius of rotation** and refers to the distance of a point from the axis of rotation.

Consider that the change in the angle, $\Delta\theta$, is very small; then the length of the arc, Δs, can be approximated as a straight line. Therefore, a relationship between angular and linear displacement can be formulated. That is, when r is the radius of rotation:

Linear displacement = Radius of rotation × angular displacement

or

$$\Delta s = r \Delta\theta$$

or using calculus (i.e., when $d\theta$ is very small):

$$ds = r \, d\theta$$

For example, if the arm segment of length 0.13 m rotates about the elbow with an angular distance of 0.23 rad, the linear distance that the wrist traveled is:

$$\Delta s = r \Delta\theta$$

$$\Delta s = 0.23 \text{ rad} \times 0.13 \text{ m}$$

$$\Delta s = 0.03 \text{ m}$$

Δs has a unit of length m, which is the correct unit because it is a linear distance. Note that radians are dimensionless, so that the product radians times meters results in units of meters.

LINEAR AND ANGULAR VELOCITIES

The relationship between linear and angular velocities is similar to the relationship between linear and angular displacements. In the example in the last section, the arm, with length r, rotates about the elbow. The linear displacement of the wrist is the product of the distance r, the radius of rotation, and the angular displacement of the segment. Differentiating this equation with respect to time:

$$ds = r \, d\theta$$
$$ds/dt = r \, d\theta/dt$$
$$v = r\omega$$

Thus, the linear velocity of a point on a rotating body is the product of the distance of that point from the axis of rotation and the angular velocity of the body. The linear velocity vector in this expression is instantaneously tangent to the path of the object and is referred to as the **tangential velocity** or v_T (Fig. 9-22). That is, the linear velocity vector behaves as a tangent, touching the curved path at only one point. The linear velocity vector, therefore, is always perpendicular to the rotating segment.

For example, if the arm segment of length $r = 0.13$ m rotated with an angular velocity of 2.4 rad/s, the velocity of the wrist is:

$$v_T = r\omega$$
$$v_T = 0.13 \text{ m} \times 2.4 \text{ rad/s}$$
$$v_T = 0.31 \text{ m/s}$$

Linear velocity is expressed in meters per second, which results in this instance from the product of meters times radians per second because radians are dimensionless.

The relationship between linear velocity and angular velocity is a critical piece of information in a number of human movements, particularly those in which the performer throws or strikes an object. To increase the linear velocity of the ball, for example, a soccer player can either increase the angular velocity of the lower extremity segments or increase the length of the extremity by extending at the joints, or both, to gain the maximum range of a kick. For an individual, the major alternative is to increase the angular velocities of these segments. For example, Plagenhoef (23) reported foot velocities before the impact

of 16.33 to 24.14 m/s for several types of soccer kicks for the same individual. Because the segment lengths did not change substantially, if the foot velocity changed, the angular velocity certainly must have varied for each type of kick.

In some activities, however, the length of the radius r can change. For example, in golf, the clubs have varying lengths and club head lofts according to the desired distance for the ball to travel (Fig. 9-23). For example, the two-iron is longer than the nine-iron and has a different club head loft, with the nine-iron having a greater club head loft than the two-iron. If both clubs had the same loft, a two-iron shot would go farther than a nine-iron shot, given the same angular velocity of the golf swing, as it does for most expert golfers. Golfers often use the same club but vary the length, r, by choking up on the handle, that is, gripping the club closer to the middle of the shaft. Using this technique, a golfer may swing with the same angular velocity but vary the length, thereby varying the linear velocity of the club head.

LINEAR AND ANGULAR ACCELERATIONS

Note that the linear velocity vector calculated from the product of the radius and the angular velocity is tangent to the curved path and can be referred to as the tangential velocity. As previously stated:

$$v_T = r\omega$$

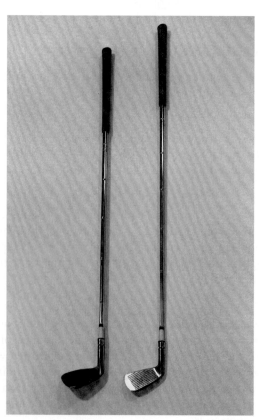

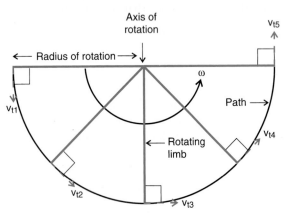

FIGURE 9-22 Tangential velocity of a rotating segment at different instants in time. The tangential velocity is perpendicular to the radius of rotation.

FIGURE 9-23 Comparison of the lengths of the shafts of golf clubs, a two-iron (left) and a nine-iron.

If the time derivative of this expression is determined, the relationship expresses the **tangential acceleration** in terms of the radius of rotation and the angular acceleration. The expression of the derivative is:

$$a_{\mathrm{T}} = \alpha r$$

where a_{T} is the tangential acceleration, r is the radius of rotation, and α is the angular acceleration. The tangential acceleration, similar to the tangential velocity vector, is a vector tangent to the curve and perpendicular to the rotating segment (Fig. 9-24). In any activity in which the performer spins to propel the implement (e.g., the discus throw), the purpose is to throw the object as far as possible. Therefore, to understand this activity, an understanding of tangential velocity and tangential acceleration is necessary. The time rate of change in the tangential velocity of the object along its curved path is the tangential acceleration. The peak tangential velocity is ideally reached just before the release of the object, at which time the tangential acceleration must be zero.

Consider a softball pitcher using an underhand pitch; further insight into another component of linear acceleration acting during rotational movement can be gained. As the pitcher moves the arm to the point of release of the pitch, the ball follows a curved path. Because the pitcher's arm is attached to the shoulder, the ball must follow the curved path produced by the rotation of the arm. Therefore, to continue on this path, the ball moves slightly inward and slightly downward at each instant in time until the ball is released (Fig. 9-25). That is, the ball is incrementally accelerated downward and inward toward the shoulder, or the axis of rotation.

Two components of acceleration produced by the rotation of a segment have been discussed: one tangential to the path of the segment and one along the segment toward the axis of rotation. These two accelerations are necessary for the ball in the pitcher's hand to continue on its curved path. The forward movement is the result of the tangential acceleration that has been previously discussed. Acceleration toward the axis or center of rotation, however, is called **centripetal acceleration** (Fig. 9-26).

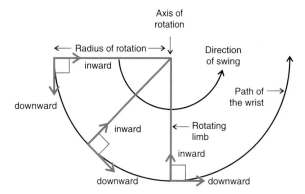

FIGURE 9-25 The directions of the acceleration components of the wrist of a softball or baseball pitcher during the downswing of the arm to the release of the ball. The wrist is accelerated in toward the shoulder and down tangential to the path of the wrist. These two vectors are perpendicular to each other.

The adjective centripetal means center seeking. Centripetal acceleration is also known as **radial acceleration**. Either name is correct, although for the remainder of this discussion, the term *centripetal acceleration* will be used.

To derive the formula for centripetal acceleration, the resultant linear acceleration of a segment end point, such as the wrist, of a rotating segment is:

$$a = \frac{\mathrm{d}v}{\mathrm{d}t}$$

Because the segment is rotating, the linear velocity is:

$$v_{\mathrm{T}} = \omega r$$

Substituting this into the acceleration equation, the acceleration becomes:

$$a = \mathrm{d}(\omega r)/\mathrm{d}t$$

If certain computational rules from calculus are applied, this equation becomes:

$$a = \omega \times \mathrm{d}r/\mathrm{d}t + \mathrm{d}\omega/\mathrm{d}t \times r$$

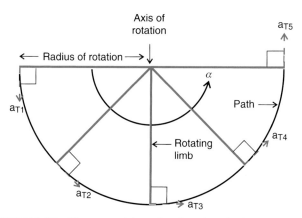

FIGURE 9-24 The tangential acceleration of a swinging segment. It is perpendicular to the swinging limb.

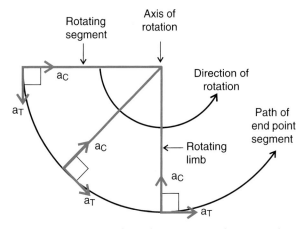

FIGURE 9-26 Tangential acceleration (a_{T}) and centripetal acceleration (a_{C}), which are perpendicular to each other. The tangential acceleration accelerates the segment end point downward, and the centripetal acceleration accelerates the end point toward the center of rotation. The result is motion along a curved path.

Because dr/dt is the linear velocity and $d\omega/dt$ is the angular acceleration of the segments, this expression is:

$$a = \omega v + \alpha r$$

The linear velocity, v, is equal to ωr, so the expression for the linear acceleration of the segment end point is:

$$a = \omega\omega r + \alpha r$$

or

$$a = \omega^2 r + \alpha r$$

Remember that the resultant acceleration has two components that are perpendicular to each other. This expression illustrates these two components. This explanation requires the use of vector calculus and is much more complicated in derivation than presented. The addition ($+$) sign in this expression means vector addition. It was previously determined that αr was the tangential acceleration so $\omega^2 r$ is centripetal acceleration. The expression for centripetal acceleration is:

$$a_C = \omega^2 r$$

Centripetal acceleration may also be expressed in the following form as a function of the tangential velocity and the radius of rotation. That is, if $v = \omega r$ is substituted into the centripetal acceleration equation, the equation becomes:

$$a_C = \frac{v^2}{r}$$

From this expression, it can be seen that the centripetal acceleration will increase if the tangential velocity increases or if the radius of rotation decreases. For example, the usual difference between an indoor running track and an outdoor track is that the indoor track is smaller and thus has a smaller radius. If a runner attempted to maintain the same velocity around the indoor turn as on an outdoor track, the centripetal acceleration would have to be greater for the runner to accomplish the turn. Generally, the runner cannot accomplish the turn at the same velocity as outdoors, so race times on indoor tracks are somewhat slower than on outdoor tracks.

Because centripetal and tangential accelerations are components of linear acceleration, they must be perpendicular to each other. The acceleration vector of these components may then be constructed. The resultant acceleration (Fig. 9-27) is computed using the Pythagorean relationship:

$$a = \sqrt{a_T^2 + a_C^2}$$

In computing either the tangential or the centripetal acceleration, the units of angular velocity and angular acceleration are radians per second and radians per second squared, respectively. The units of linear acceleration (meters per second squared) can result only when a radian-based unit is used in the computation.

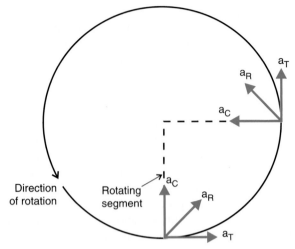

FIGURE 9-27   The resultant linear acceleration vector (a_R) comprised the centripetal and tangential acceleration components.

Angle–Angle Diagrams

In most graphical presentations of human movement, usually some parameter (e.g., position, angle, and velocity) is graphed as a function of time. In certain activities, such as locomotion, the motions of the segments are cyclic; that is, they are repetitive, with the end of one cycle at the beginning of the next. In these instances, an **angle–angle diagram** may be useful to represent the relationship between two angles during the movement. An angle–angle diagram is the plot of one angle as a function of another angle. That is, one angle is used for the x-axis and one for the y-axis. In an angle–angle diagram, one angle is usually a relative angle (angle between two segments) and the other is an absolute angle (angle relative to reference frame). For the angle–angle graph to be meaningful, a functional relationship between the angles should exist (Fig. 9-28). For example, whereas in studying an individual running, the relationship between the sagittal view ankle and knee angles may be meaningful, the relationship between the elbow angle and the ankle angle may not.

One problem with this type of diagram is that time cannot be easily represented on the graph. It can be presented, however, by placing marks on the angle–angle curve to represent each instant in time at which the data were calculated. These marks are placed at equal time intervals and give an indication of the angular distance through which each joint has moved in equal time intervals. Thus, angular velocity of the movement is represented, because the farther apart the marks are on the curve, the greater the velocity of the movement. Conversely, the closer together the marks, the less the velocity (Fig. 9-29).

Angle–angle diagrams have proved useful in the examination of the relationship between the rearfoot angle and the knee angle (1,34). This relationship is based on the

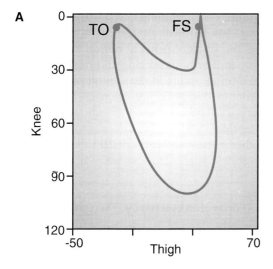

A

B

FIGURE 9-28 Angle–angle diagrams of the knee angle as a function of the thigh angle **(A)** and the knee angle plotted as a function of the ankle angle **(B)** for one complete running stride of an individual running at 3.6 m/s. TO, toe-off; FS, foot strike. (Adapted from Williams, K. R. [1985]. Biomechanics of running. *Exercise and Sports Sciences Reviews*, 14.)

related anatomical motions of the subtalar and knee joints. During the support phase of gait, the knee flexes at touchdown and continues to flex until midstance. At the same time, the foot lands in an inverted position and immediately begins to evert until midstance. Both knee flexion and subtalar eversion are associated with internal tibial rotation. After midstance, the knee extends and the subtalar joint inverts. Both of these joint actions result in external tibial rotation. These actions are presented in Figure 9-29 with the knee angle expressed as a relative angle and subtalar joint inversion and eversion expressed as an absolute angle.

Figure 9-30, an angle–angle diagram presented in a paper by van Woensel and Cavanagh (34), illustrates the relationship between the knee angle and the rearfoot angle in different shoe conditions. One of the three shoes used in this study was specifically designed to force the runner to pronate during support, another was designed to force the runner to supinate during support, and the third pair was a neutral shoe.

In many research studies, angle–angle diagrams are presented but not used in quantification of the movement. More recently, however, researchers have begun to use what is referred to as a modified vector coding technique to quantify angle–angle plots (12). This technique is used to determine the angle between each pair of contiguous points throughout a cycle (Fig. 9-31). The relative motion between these points has been used as a measure of coordination between the angles representing either segments or joints (12,24).

Using walking data in Appendix C, calculate the absolute angles of the thigh and leg for the frame denoting right foot contact and the next three frames. Using these absolute angles, calculate the relative motion between the thigh and the leg using the modified vector coding technique.

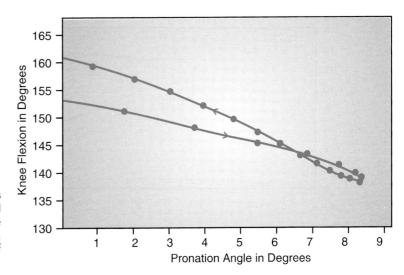

FIGURE 9-29 Angle–angle diagram of knee flexion as a function of subtalar pronation angle for an individual running at 6 min/mile on a treadmill. The dots on the curve indicate equal time intervals. (Adapted from Bates, B. T., et al. [1978, fall]. Foot function during the support phase of running. *Running*, 24:29.)

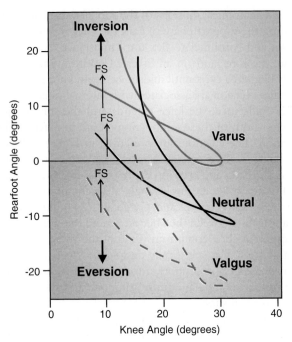

FIGURE 9-30   Knee–rearfoot angle–angle diagram of an individual in three types of running shoes. FS, foot strike. The varus shoe has a medial wedge, which mediates rearfoot pronation; the valgus shoe has a lateral wedge, which enhances rearfoot pronation; and the neutral shoe is a normal running shoe. (Adapted from van Woensel, W., Cavanagh, P. R. [1992]. A perturbation study of lower extremity motion during running. *International Journal of Sports Biomechanics*, 8:30–47.)

Angular Kinematics of Walking and Running

Many researchers have reported on how the lower extremity joint angles vary throughout the walking and running stride, particularly during the support portion of the stride. An angular kinematic analysis of walking and running typically includes a graphical presentation of joint actions as a function of time. Although some researchers have studied patterns of angular velocity and

acceleration in both running and walking, the major focus of investigation has been on the characteristics of angular positions and displacements at critical events in the locomotion cycle. For both walking and running, the greatest range of motion occurs in the sagittal plane, and segment movements in this plane are often used to describe gait characteristics. The calculation of sagittal plane angles can be accomplished with a 2D analysis. However, movement in the other planes can be as critical to successful gait, but obtaining these angles requires a 3D analysis.

Lower Extremity Angles

Sagittal, frontal, and transverse plane joint angular kinematic patterns for walking, running, and sprinting are shown in Figure 9-32. Although there are obvious magnitude differences where the angular displacements increase with speed of locomotion, the patterns are similar across the speeds of locomotion with some temporal phasing differences. The one exception is at the ankle joint, where there is less and less plantarflexion at heel strike as locomotion speed increases until a point in very fast running at which plantarflexion may be absent (6).

As contact is made with the ground in both walking and running, a loading response absorbs body weight. The angular kinematics that accompany this response are hip flexion, knee flexion, and ankle dorsiflexion. As the body continues over the foot in midstance, these movements continue until the terminal stages of stance, where there is a reversal into hip extension, knee extension, and plantarflexion.

The initial touchdown hip flexion angle for walking and running has been reported to be in the range of 35° to 40° and 45° to 50°, respectively (17,21,22). In the early phases of contact, the hip adducts in the reported range of 5° to 10° and 8° to 12° for walking and running, respectively. After touchdown, the amount of hip flexion reduces over the course of the support until toe-off, at which 0° to 3° of hip extension in walking and 3° to 5°

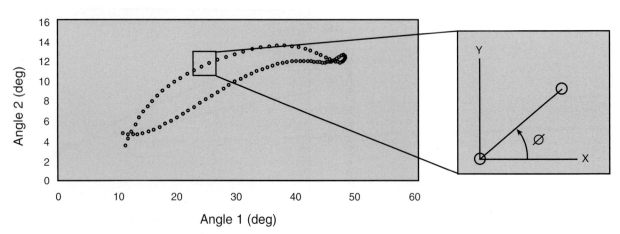

FIGURE 9-31   A representation of the modified vector coding technique. The angle between each pair of contiguous points is calculated relative to the right horizontal.

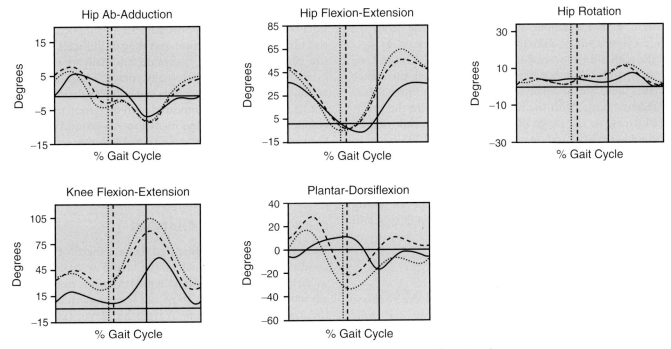

FIGURE 9-32 Gait analysis commonly includes a recording of angular kinematics across the gait cycle, including the support phase (percent of cycle up to the *vertical line*) and the swing phase (percent of cycle past the *vertical line*). The angular kinematic differences and similarities between walking (*solid line*), running (*dashed line*), and sprinting (*dotted line*) become apparent when graphed against one another. (Adapted from Novacheck, T. F. [1995]. *Instructional Course Lectures.* Park Ridge, IL, 44:497–506.)

of hip extension in running is reported (17,21,22). There is also hip movement into abduction at toe-off in the range of 2° to 5° for both walking and running. When the limb is off the ground in the swing phase, hip flexion maximum values are reported in the range of 35° to 50° for walking and 55° to 65° for running. Hip abduction in the initial portion of the swing phase is similar for walking and running, reported to be in the range of 3° to 8°. Hip adduction late in the swing phase varies more between walking and running, in the range of 0° to 5° and 5° to 15°, respectively (17,21,22).

The knee angle is flexed at touchdown and has been reported in the literature to be in the range of 10° to 15° for walking (17,21,22) and 21° to 40° (1,5,8,9,17,21,22) for running. After touchdown, the knee flexes to values ranging from 20° to 25° for walking and 38° to 60° for running, with the greater flexion occurring at faster speeds (1,2). The knee flexion movement helps to lower the body in stance. Maximal knee flexion occurs at midstance, after which the knee extends until toe-off. Full extension is not achieved at toe-off; values range from 10° to 40° in walking and 18° to 40° in running, depending on the speed (4,8,17,21,22). Greater extension values at toe-off are generally associated with faster speeds. In the swing phase, knee flexion is important to shorten the swing leg before the hip flexion brings the limb forward. The range of knee flexion in walking and running is reported to be 50° to 65° and 100° to 125°, respectively (17,21,22).

Again, although the magnitude of the knee angles at these specific instants in time during the support phase of running vary, the profile of the curve does not. The profile of the knee joint angle appears to be relatively stable and immune to distortion from influences such as running shoe construction (9,34) and delayed-onset muscle soreness (10).

In walking, the ankle is plantarflexed in a reported range of 5° to 6° at heel strike and moves into 10° to 12° of dorsiflexion before returning to 15° to 20° of plantarflexion at toe-off. During the swing phase of walking, the foot continues on through 18° to 20° of plantarflexion and then dorsiflexes in a reported range of 2° to 5° in preparation for the next heel strike. As the speed of locomotion increases, there is less plantarflexion at heel strike until dorsiflexion is the movement occurring at touchdown. Depending on the speed, the reported range of dorsiflexion at heel strike in running is 10° to 17°, increasing to 20° to 30° in midstance and then moving to plantarflexion at toe-off (range, 10° to 20°). Plantarflexion continues into the initial phases of the swing phase (range, 15° to 30°) and then moves, as in walking, into dorsiflexion in the reported range of 10° to 15° (17,21,22).

Alterations in the angular kinematics of the lower extremity joints during both walking and running occur in response to changes in the environment. For example, when running over a noncompliant surface, some individuals make a kinematic adjustment at impact by responding with more initial knee flexion at contact (7). Walking uphill brings about a number of adjustments in the lower extremity. For example, increasing the grade from 0% to 24%, the lower extremity adjustments start at heel strike with a 22% increase in dorsiflexion, 31% more knee flexion, and 23% more hip flexion (16). Over the stance phase, there are unequal adjustments at the three lower extremity joints, with the hip joint undergoing the greatest increase in the range of motion (+59%), followed by the ankle (+20%) and an actual decrease (−12%) range of motion at the knee (16). The primary adjustment to walking downhill during the stance phase occurs at the knee joint, where there is as much as 15° more knee flexion in early stance (15). Movement adjustments in the swing phase occur at the hip and ankle, with less hip flexion and less plantarflexion.

REARFOOT MOTION

A number of investigations have described the rearfoot angle during the support phase of walking and running. Excessive rearfoot motion during running has been hypothesized to cause a variety of lower extremity injuries, although little evidence directly relates excessive rearfoot motion and injury (5,20). In fact, a valid, clinical definition of excessive rearfoot motion has not yet been determined. From a functional standpoint, eversion of the calcaneus is necessary because it allows the foot to assume a flat position on the ground. Typically, maximum rearfoot angle values from −6° to −17° at midstance have been reported in the literature (5,11) for running and in the range of −9.2° to −12.9° for walking (14,26). This wide range in maximum values can be attributable to differences in the anatomical foot structure of individuals as well as the influence of footwear. It has been reported that more extreme rearfoot angles occur at midstance in running when the subjects wear racing shoes compared with training shoes (11). More extreme rearfoot eversion angles have also been reported for runners in a shoe with a very soft midsole than in a shoe with a firmer midsole (9). Although the rearfoot angle is related in motion to the knee angle by the action of tibial rotation, it is, unlike the knee angle, highly variable and certainly can be influenced by many factors.

The simultaneous action of these two lower extremity angles has been a topic of several investigations. Because internal tibial rotation accompanies knee flexion and subtalar joint eversion and both reach a maximum at midstance, the mistiming of these joint actions has been suggested as a possible mechanism for lower extremity injury (1). Hamill and colleagues (9) illustrated that the rearfoot angle could be changed by a running shoe with a very soft midsole, but the knee angle could not. These researchers reported that in a soft midsole running shoe, the maximum rearfoot angle occurs sooner in the support period than did maximum knee flexion. The subtalar joint also stays at this maximum while the knee begins to extend. Thus, they surmised that a twisting action may be applied to the tibia by the differential speeds at which the tibia rotated early in support and late in support. Because the tibia is a rigid structure and may be difficult to twist, the tibia may continue to rotate internally at the knee, even though it should externally rotate. This undesirable action at the knee may possibly cause knee pain in the runner. If these actions are repeated with each foot-to-ground contact and the runner has many foot-to-ground contacts, the runner may be subject to a knee injury that would prohibit training. This type of injury is often referred to as an **overuse injury**. It results from an accumulation of stresses rather than a single high-level, traumatic stress.

CLINICAL ANGULAR CHANGES

Locomotion is also influenced by a variety of medical conditions and functional limitations. For example, the walking gait of an individual with Parkinson disease usually exhibits small, quick steps and less range of motion in the lower extremity joints. A tight hip flexor (e.g., psoas muscle) in individuals with cerebral palsy can limit the hip extension during the stance phase (28). This causes an increase in pelvic tilt (2°). Specific adjustments in the joint kinematics of individuals with hemiplegia may show a reduction in the range of motion at the knee joint, with increases in range of motion at the ankle and excessive hip and knee motion in the swing phase. Finally, individuals with injury to one limb typically compensate for pain in one limb by altering range of motion in both limbs so they can increase the time spent on the limb with no pain.

Angular Kinematics of the Golf Swing

In Chapter 8, it was pointed out that the linear speed and path of the club head are important determinants of a successful golf shot. These two linear kinematic components of the swing are the result of a series of angular movements; thus, the angular kinematics of the golf swing is commonly the primary focus of golf instruction and evaluation. The golf swing can be accurately described using a double pendulum model, with one link being the arm rotating about the shoulder joint and the second link being the club and wrist, where the wrist acts as a hinge about which the club rotates (18). A third connected link has been suggested between the shoulder and what is referred to as the hub axis as the body rotates about a vertical axis (31). For purposes of this introduction to the golf swing, the primary focus will be on the double pendulum characteristics of the swing.

In a golf swing, the left arm (for a right-handed golfer) sets the plane of the swing (31). The swing plane is an elliptical plane around the body that brings the club head into contact with the ball from the inside, where the face of the club ends up perpendicular to the flight path of the ball. The swing is a pivot about an axis running through

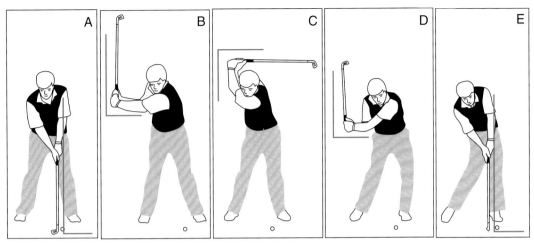

FIGURE 9-33 Critical angular positions in the phases representing the address (**A**), takeaway (**B**), top of backswing (**C**), downswing (**D**), and impact (**E**) determine the success of the golf swing.

the base of the neck with the head stationary (29). Many beginning golfers try to create an upright swing plane with the club brought straight back and straight forward. When the club head comes down to meet the ball with this swing plane, it is never square with the ball and puts spin on the ball at contact.

The angular positions of the club at various stages of the swing are good predictors of a successful swing. A frontal view of the golf swing provides a good perspective for evaluating club position (Fig. 9-33). In the first stage of the swing, the golfer addresses the ball. This position at the beginning of the swing should be the same as the position at impact. It establishes the arm and shoulder position that will bring the club into precise alignment for impact (29). The club and the left arm should form a straight line, and the club face should be aimed down a perpendicular line from the ball forward in a straight line (Fig. 9-34A). As the club is started in the backswing, there is an initial takeaway phase where the club head is taken back away from the ball. This is initiated with a weight shift to the rear that allows for greater range of motion at the hip and flattens the arc of the swing. A long takeaway is preferred: The club travels in a wide arc, and the wrist does not allow movement of the club until the hands are chest high. This increases the distance for the club head to travel as the shoulders are rotated farther from the target. At the end of the takeaway phase, the left arm should be horizontal to the ground, and the club should be vertical and perpendicular to the arm (Fig. 9-34B). Continuing to the top of the backswing, the upper body has rotated to allow the club to be positioned parallel to the ground again and parallel to the final target line for ball contact. The right elbow flexes at the end of the backswing to reduce the length and allow for more acceleration. The left arm continues to be straight and vertical. This position ensures that the club face will travel squarely to the ball at contact (Fig. 9-34C). From the top of the backswing, the downswing begins as the club shaft and the left arm drop in one piece to the position halfway down, where the left arm is again parallel to the ground and the club is

vertical (Fig. 9-34D). Hip rotation and the legs initiate this movement as they drive forward, dropping the right shoulder and the shaft into place. The impact position should duplicate the initial address position, with the left arm and club forming a straight vertical line and the club face traveling in a straight line through the ball (Fig. 9-34E). If these angular positions can be obtained within the context of a fluid swing, the ball will travel far and accurately.

The interaction of the arm and club links is shown in the displacement, velocity, and acceleration curves in the downswing phase illustrated in Figure 9-34. The displacement of the arm segment in the downswing is 100° to 270°, and the displacement of the club relative to the arm is 50° to 175°. As the shoulder displacement increases in the early phases, the wrist angle remains constant until it uncocked in the later stages of the downswing (18). This uncocking increases dramatically 80 to 100 ms before impact as the club is brought in line with the hands (19). The interaction between the arm and the club segments enhances the velocity and acceleration of the club at impact. This is illustrated in the angular velocity graph, where the arm velocity moves through a range of 250°/s, increasing to 800°/s and reducing velocity to 500°/s at impact. The resulting effect on the club segment is a build in velocity from zero initially to a culminating 2300 to 4000°/s at impact (18,19). Angular accelerations of the club are minimal in the beginning of the downswing and increase rapidly to values approaching 10,000°/s/s at a point where the angular acceleration of the arm is reduced to zero and begins the negative acceleration (18).

Using MaxTRAQ, import the first two video files of the golfer. Digitize the right shoulder, left shoulder, right elbow, left elbow, and the left wrist in each frame. Calculate the absolute angles of the upper arm and the lower arm.

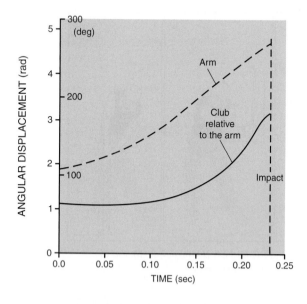

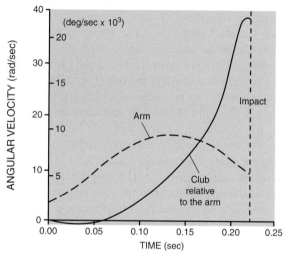

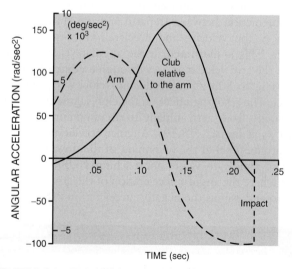

FIGURE 9-34 One model used to study the golf swing is the double pendulum. Displacement, velocity, and acceleration data for the arm (*dashed line*) and the club motion relative to the arm (*solid line*) illustrate the unique motion characteristics of each segment. (Adapted from Milburn, P. D. [1982]. Summation of segmental velocities in the golf swing. *Medicine & Science in Sports & Exercise*, 14:60–64.)

Angular Kinematics of Wheelchair Propulsion

Angular kinematic characteristics of the trunk and joint actions at the shoulder, elbow, and wrist are the focus of many investigations of wheelchair propulsion. Both linear and angular kinematics are constrained because the hand must follow the rim (32). Differences in hand position on the rim as well as different seat positions and other adjustments, however, can considerably alter the angular kinematics. A stick figure illustrating the sagittal angular positions of the arm, forearm, and hand segments during wheelchair propulsion is shown in Figure 9-35. The angular positions are shown for various stages in the event at a rim contact position that is −15° with respect to top dead center continuing on through +60° in 15° increments.

The range of motion in the elbow and shoulder joints has been reported to be an average of 55° to 62° of elbow flexion and extension, 60° to 65° of shoulder flexion and extension, 20° of shoulder abduction and adduction, 36°

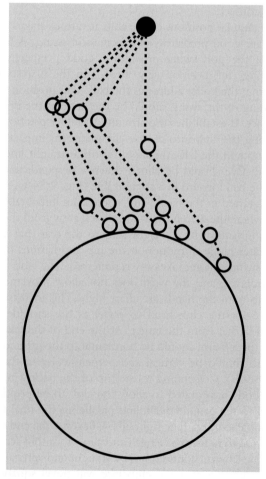

FIGURE 9-35 Angular positions of the upper extremity during wheelchair propulsion at 1.11 m/s. (Adapted from Van der Helm, F. C. T., Veeger, H. E. J. [1996]. Quasi-static analysis of muscle forces in the shoulder mechanism during wheelchair propulsion. *Journal of Biomechanics*, 29:39–52.)

of shoulder internal and external rotation, 35° of wrist flexion and extension, and 68° to 72° of wrist ulnar and radial flexion (13,25). An approximate value of 37° of pronation and supination has also been reported (3). The trunk contributes to wheelchair propulsion via flexion in the propulsive phase and extension in the recovery phase after hand release (33). Angular velocity and acceleration during wheelchair propulsion have not been studied extensively, but reported values approach 300°/s for elbow extension, even at slow speeds (1.11 m/s) (35).

If propulsion is made with a lever rather than on the hand rim, the angular kinematics change, requiring more elbow range of motion, less shoulder extension, more shoulder rotation, and more shoulder abduction (13). Likewise, speed-dependent changes are seen in angular displacement. It has been reported that with increased speed of propulsion, the trunk displacement increases and shoulder displacement decreases (27). Propelling up a slope also influences angular kinematics, resulting in greater trunk displacement and an increase in arm displacement at the shoulder. Finally, seat adjustments influence angular kinematics, depending on the direction and level of alteration (13).

Using MaxTRAQ, import the video files of the individual in the wheelchair. Digitize the right shoulder, right elbow, and right wrist for five frames after the initial propulsion phase. Calculate the absolute angles of the upper arm and the lower arm, the angular velocity, and the angular acceleration. Note that the time between frames is 0.0313 seconds.

Summary

Nearly all purposeful human movement involves the rotation of segments about axes passing through the joint centers; therefore, knowledge of angular kinematics is necessary to understand human movement. Angles may be measured in degrees, revolutions, or radians. If the angular measurement is to be used in further calculations, the radian must be used. A radian is equal to 57.3°.

Angles may be defined as relative and absolute, and both may be used in biomechanical investigations. A relative angle measures the angle between two segments but cannot determine the orientation of the segments in space. An absolute angle measures the orientation of a segment in space relative to the right horizontal axis placed at the distal end of the segment. How segment angles are defined must be clearly stated when presenting the results of any biomechanical analysis.

The kinematic quantities of angular position, displacement, velocity, and acceleration have the same relationship with each other as their linear analogs. Thus, angular

velocity is calculated using the first-order central difference method as follows:

$$\omega_i = (\theta_{i+1} - \theta_{i-1})/2\Delta t$$

Likewise, angular acceleration is defined as follows:

$$\omega_i = (\omega_{i+1} - \omega_{i-1})/2\Delta t$$

The techniques of differentiation and integration apply to angular quantities as well as linear quantities. Angular velocity is the first derivative of angular position with respect to time and angular acceleration is the second derivative. The concept of the slope of a secant and a tangent also applies in the angular case to distinguish between average and instantaneous quantities. Integration implies the area under the curve. Thus, the area under a velocity–time curve is the average angular displacement, and the area under an acceleration–time curve is the average angular velocity.

It is difficult to represent angular motion vectors in the manner in which linear motion vectors were represented. The right-hand rule is used to determine the direction of the angular motion vector. Generally, rotations that are counterclockwise are positive, and clockwise rotations are negative.

Sagittal view lower extremity angles were defined in this chapter using a system suggested by Winter (36). In this convention, ankle, knee, and hip angles were defined using the absolute angles of the foot, leg, thigh, and pelvis segments. The rearfoot angle measures the relative motion of the leg and the calcaneus in the frontal plane and is calculated from the absolute angles of the calcaneus and the leg.

There is a relationship between linear and angular motions. Comparable quantities of the two forms of motion may be related when the radius of rotation is considered. The linear velocity of the distal end of a rotating segment is called the tangential velocity and is calculated as follows:

$$v_T = \omega r$$

where ω is the angular velocity of the segment and r is the length of the segment. The derivative of the tangential velocity, the tangential acceleration, is:

$$a_T = \alpha r$$

where α is the angular acceleration of the rotating segment. The other component of the linear acceleration of the end point of the rotating segment is the centripetal or radial acceleration. This is expressed as:

$$a_C = \omega^2 r$$

The tangential and centripetal acceleration components are perpendicular to each other.

A useful tool in biomechanics is the presentation of angular motion in angle–angle diagrams. These diagrams generally present angles of joints that are anatomically functionally related. Time can be presented only indirectly on this type of graph, however. Recently, angle–angle plots have been quantified using a modified vector coding technique (12).

Equation Review for Angular Kinematics

Purpose	Given	Formula
Relative angle between two segments using the law of cosines	Length of segments a and b and distance between end of a and b (length c)	$\theta = \arccos(b^2 + c^2 - a^2)/(2 \times b \times c)$
Absolute angle	Endpoints – horizontal and vertical components	$\theta = \arctan([y_{proximal} - y_{distal}]/[x_{proximal} - x_{distal}])$
Calculate position	Starting position relative to origin, constant velocity (zero acceleration), and time	$\theta = \theta_{initial} + \omega_{initial}t$
Calculate position	Starting position at origin, constant velocity (zero acceleration), and time	$\theta = \omega_i t + \frac{1}{2}\alpha t^2$
Calculate position	Initial velocity, time, constant acceleration	$\theta = 1/2\alpha t^2$
Calculate position	Initial velocity = zero, time, constant acceleration	$\theta = \omega^2/2 \times \alpha$
Calculate final displacement	Final angular velocity; constant angular acceleration	$\omega = (\theta_2 - \theta_1)/(t_2 - t_1)$
Calculate average velocity	Displacement and time	$\omega = (\omega_{initial} + \omega_{final})/2$
Calculate average velocity	Initial and final velocity	$\omega_f = \omega_{initial} + \alpha t$
Calculate final velocity	Initial velocity, constant acceleration, and time	$\omega = \alpha t$
Calculate final velocity	Starting velocity = zero, constant acceleration, and time	$\omega = \sqrt{\omega_{initial}^2 + 2\alpha(\theta - \theta_{initial})}$
Calculate final velocity	Velocity at time = zero, constant acceleration, initial position relative to origin, final position	$\omega_f^2 = 2\alpha\theta$
Calculate final velocity	Initial velocity = zero, constant acceleration, initial and final position	$\omega = \sqrt{2\alpha(\theta - \theta_{initial})}$
Calculate acceleration	Final velocity and displacement	$\alpha = \omega_{final}^2/2\theta$
Calculate average acceleration	Velocity and time	$\alpha = (\omega_2 - \omega_1)/(t_2 - t_1)$
Calculate average acceleration	Displacement, time	$\alpha = 2\theta/t^2$
Calculate time	Displacement, constant acceleration	$t = \sqrt{2\theta/\alpha}$
Calculate linear distance	Radius, angular displacement	$s = r\theta$
Calculate linear velocity (tangential)	Radius, angular velocity	$v = r\omega$
Calculate linear acceleration (tangential)	Radius, angular acceleration	$a = r\alpha$
Centripetal acceleration	Radius, angular velocity	$a_c = \omega^2 r$
Centripetal acceleration	Radius, tangential linear velocity	$a_c = v^2/r$

REVIEW QUESTIONS

True or False

1. ____ An ankle joint with an angular velocity of −2.4 rad/s is plantarflexing.

2. ____ Choking up on a golf club increases the tangential and angular velocities of the club head.

3. ____ During running, hip flexion, knee extension, and ankle dorsiflexion occur during initial ground contact.

4. ____ When bowling, angular velocity of the arm about the shoulder is zero if it is swinging at a constant angular velocity.

5. ____ Points 1 and 40 cm from the center of a merry-go-round have the same angular and tangential velocities.

6. ____ Centripetal acceleration is proportional to the inverse of the distance of an object from the axis of rotation.

7. ____ The segment angle between the trunk and the arm abducted out to the side is 90°.

8. ____ A radius and a radian are the same thing.

9. ____ The absolute angle of the trunk when standing upright is 90°.

10. ____ In general, knee flexion angle is greater at foot contact when running versus walking.

11. ____ Tangential and centripetal accelerations are always parallel to each other in angular motion.

12. ____ Relative angles do not provide information on orientation of the body or segments in space.

13. ____ Tangential velocity of a point on a lever is a function of the length of the radius and the angular velocity.

14. ____ To calculate angular velocity when kicking a ball, your thigh length is the radius of rotation about the hip.

15. ____ Angular displacement can only range from 0 to 2π but angular velocity can range from negative to positive infinity.

16. ____ In the first half of stance, the rearfoot goes from everted to inverted in most runners.

17. ____ The law of cosines allows one to calculate segment angles not at right angles to each other.

18. ____ Except for the few individuals who can hyperextend their knees, the knee joint is always flexed and therefore a positive relative angle.

19. ____ If the right knee rotates counterclockwise about the foot, which is on the ground, angular velocity is positive.

20. ____ Relative joint angles are always calculated as proximal segment angle minus distal segment angle.

21. ____ Most movement in a triple axel in ice skating occurs in the transverse plane.

22. ____ A golf swing is best modeled as a single pendulum, with the arm, wrist, and club all rotating about the shoulder.

23. ____ For cyclical movements, time is plotted on the x-axis and a segment angle is plotted on the y-axis.

24. ____ Radians, degrees, and revolutions are appropriate units for angular distance and displacement.

25. ____ The right-hand rule is used to determine if a right or left horizontal is used to measure segment angles.

Multiple Choice

Questions 1–4: A hockey stick 1.1 m long completes a swing in 0.15 seconds through a range of 85°. Assume a uniform angular velocity.

1. What is the average angular velocity of the stick?
   a. 12.8 rad/s
   b. 567°/s
   c. 10.9 rad/s
   d. 623°/s

2. What is the linear distance moved by the end of the stick?
   a. 94 m
   b. 77 m
   c. 1.6 m
   d. 1.3 m

3. What is the tangential velocity of the end of the stick?
   a. 624 m/s
   b. 10.9 m/s
   c. 12.0 m/s
   d. 14.1 m/s

4. What is the tangential acceleration of the end of the stick?
   a. 145 m/s^2
   b. 132 m/s^2
   c. 36.3 m/s^2
   d. 72.6 m/s^2

5. During an elbow flexion exercise, the relative angle at the elbow is 103° at 0.67 seconds and 89° at 0.75 seconds. What is the average angular velocity of the elbow?
   a. −5.0°/s
   b. 5.0°/s
   c. −175°/s
   d. 175°/s

6. A softball player is only physically capable of swinging the bat at 600°/s. Can she increase the velocity of the ball she hits without further training or using a different bat?
   a. Yes, grasp the bat closer to the bottom so the radius of rotation increases.
   b. Yes, grasp the bat closer to the barrel so the radius of rotation decreases.
   c. No, she'll have to increase angular velocity or increase the mass of the bat.
   d. Angular velocity does not affect tangential velocity of the ball after impact.

7. If a skater is rotating during a spin at a constant angular velocity of 3.6 rad/s for 3.9 seconds, what is his angular acceleration?
   a. 14 rad/s^2
   b. 0.92 rad/s^2
   c. 0 rad/s^2
   d. 1.1 rad/s^2

8. A top is spinning at 5.6 rev/s and starts decelerating at 0.66 rev/s^2. How long will it take the object to stop?
   a. 0.12 seconds
   b. 3.7 seconds
   c. 6.6 seconds
   d. 8.5 seconds

9. The kicking phase of a football punt takes place over 0.25 seconds and includes hip flexion (89°) and knee extension (100°). If the thigh is 0.70 m long and the leg 0.68 m long, what is the tangential velocity of the foot resulting from these movements?
   a. 4.7 m/s
   b. 9.1 m/s
   c. 4.3 m/s
   d. 18.2 m/s

10. A discus thrower rotates through the last 50° of a turn in 0.12 seconds. The distance from his shoulder to his hand is 1.00 m and the distance from the axis of rotation to the center of the discus is 1.17 m. Calculate the average linear velocity of the discus.

a. 488 m/s
b. 416 m/s
c. 7.27 m/s
d. 8.51 m/s

11. An object is spinning at a constant angular velocity of 1000°/s. If it spins for 2.0 seconds, what is the angular distance traveled?
    a. 2,000 rad
    b. 34.9 rad
    c. 500 rad
    d. 8.73 rad

12. A hammer thrower releases the hammer after reaching an angular velocity of 16.2 rad/s. If the hammer is 180 cm from the axis of rotation, what is the linear velocity of the hammer at release?
    a. 29.2 m/s
    b. 11.1 cm/s
    c. 2916 m/s
    d. 9.00 m/s

13. Calculate the relative angle at the knee and the absolute angles of the thigh given the following positions in m: hip (2.000, 1.905), knee (2.122, 1.642), and ankle (1.897, 1.210).
    a. Abs = −65.1°; rel = −52.4°
    b. Abs = 114.9°; rel = −127.6°
    c. Abs = 114.9°; rel = 52.4°
    d. Abs = 24.9°; rel = 37.6°

14. During the support phase of walking, the absolute angle of the thigh has the following angular velocities:

Frame	Time (s)	Angular Velocity (rad/s)
38	0.6167	1.033
39	0.6333	1.511
40	0.6500	1.882
41	0.6667	2.190

    Calculate the angular acceleration at frame 39.
    a. 25.49 rad/s²
    b. 18.44 rad/s²
    c. 22.22 rad/s²
    d. 20.33 rad/s²

15. An ice skater rotating around a vertical axis decreases in angular velocity from 450°/s to 378°/s in 9.6 seconds. Find the angular acceleration.
    a. −7.5 rad/s²
    b. −0.13 rad/s²
    c. 7.5°/s²
    d. −72°/s²

16. A gymnast wants to complete three revolutions when she is airborne for 1.77 seconds. How quickly must she rotate on average?
    a. 1.69 rad/s
    b. 10.65 rad/s
    c. 33.4 rad/s
    d. 18.8 rad/s

17. A bike wheel makes 1.5 revolutions with an average angular velocity of 18 rad/s. You place your hand on the wheel to slow it down at a rate of 500°/s². How much angular distance is traveled before the wheel comes to a stop?
    a. 1064°
    b. 2152°
    c. 9.43 rad
    d. 28.0 rad

18. Which is true of a cyclist's pedals?
    a. The angular velocity is zero if his angular acceleration is zero.
    b. The angular displacement is always equal to the angular distance traveled.
    c. The angular distance is always greater than or equal to the angular displacement.
    d. Centripetal acceleration always points away from the axis of rotation.

19. A merry-go-round has a radius of 3.0 m. Child A runs around, pushing it with a velocity of 4.0 m/s. It takes Child B 4.71 seconds to make one revolution when he pushes the merry-go-round. Which child pushes it faster?
    a. Child A
    b. Child B
    c. They are equal
    d. Not enough information

20. A diver accelerates through a somersault at 802°/s. What is the angular displacement over 0.16 seconds?
    a. 20.5°
    b. 64.2°
    c. 128°
    d. 10.3°

21. An individual is running around a turn with a 17 m radius at 6.64 m/s. What is the runner's centripetal acceleration?
    a. 2.59 m/s²
    b. 2.56 m/s²
    c. 0.39 m/s²
    d. 114 m/s²

22. The final angular velocity of a golf swing was 400°/s with a constant angular acceleration of 501°/s². How far did the club rotate?
    a. 101°
    b. 319°
    c. 160°
    d. 200°

23. A cyclist averages 99 rpm during a race. What was his angular velocity in radians per second?
    a. 5.18
    b. 622
    c. 99
    d. 10.36

24. A female softball pitcher releases the ball with an angular velocity of 35.28 rad/s. It travels 12.5 m toward the batter in 0.372 seconds. What is the distance from her shoulder to the ball?

a. 1.05 m
b. 0.95 m
c. 0.92 m
d. 1.02 m

25. In the release phase of bowling, the bowler's arm and forearm travel through 0.24 and 0.28 rad, respectively, in 0.03 seconds. The length of the arm is 0.82 m and the forearm 0.51 m. What percent contribution to the tangential velocity of the ball comes from the rotation of the forearm?
    a. 42%
    b. 58%
    c. 35%
    d. 65%

References

1. Bates, B. T., et al. (1978). Foot function during the support phase of running. *Running*, 24:29.

2. Bates, B. T., et al. (1979). Functional variability of the lower extremity during the support phase of running. *Medicine & Science in Sports & Exercise*, 11(4):328-331.

3. Boniger, M. L., et al. (1997). Wrist biomechanics during two speeds of wheelchair propulsion: An analysis using a local coordinate system. *Archives Physical Medicine Rehabilitation*, 78:364-372.

4. Cavanagh, P. R., et al. (1977). A biomechanical comparison of good and elite distance runners. In P. Milvy (Ed.). *The Marathon: Physiological, Medical, Epidemiological, and Psychological Studies*. New York: New York Academy of Science, 328-345.

5. Clarke, T. E., et al. (1983). The effects of shoe design parameters of rearfoot control in running. *Medicine & Science in Sports & Exercise*, 15(5):376-381.

6. Czerniecki, J. M. (1988). Foot and ankle biomechanics in walking and running. *American Journal of Physical Medicine and Rehabilitation*, 67:246-252.

7. Dixon, S. J., et al. (2000). Surface effects on ground reaction forces and lower extremity kinematics in running. *Medicine & Science in Sports & Exercise*, 32:1919-1926.

8. Elliott, B. R., Blanksby, B. A. (1979). A biomechanical analysis of the male jogging action. *Journal of Human Movement Studies*, 5:42-51.

9. Hamill, J., et al. (1992). Timing of lower extremity joint actions during treadmill running. *Medicine & Science in Sports & Exercise*, 24:807-813.

10. Hamill, J., et al. (1990). Muscle soreness during running: Biomechanical and physiological considerations. *International Journal of Sports Biomechanics*, 7:125-137.

11. Hamill, J., et al. (1987). Effects of shoe type on cardiorespiratory responses and rearfoot control during treadmill running. *Medicine & Science in Sports & Exercise*, 20:515-521.

12. Heiderscheit, B. C., Hamill, J., Van Emmerik, R. E. A. (2002). Locomotion variability and patellofemoral pain. *Journal of Applied Biomechanics*, 18(2):110-121.

13. Hughes, C. J., et al. (1992). Biomechanics of wheelchair propulsion as a function of seat position and user-to-chair interface. *Archives of Physical Medicine and Rehabilitation*, 73:263-269.

14. Isacson, J., et al. (1986). Three dimensional electrogoniometric gait recording. *Journal of Biomechanics*, 19:627-635.

15. Kuster, M., et al. (1995). Kinematic and kinetic comparison of downhill and level walking. *Clinical Biomechanics*, 10:79-84.

16. Lange, G. W., et al. (1996). Electromyographic and kinematic analysis of graded treadmill walking and the implications for knee rehabilitation. *Journal of Orthopedic and Sports Physical Therapy*, 23:294-301.

17. Mann, R. A., Hagy, J. L. (1980). Biomechanics of walking, running, and sprinting. *American Journal of Sports Medicine*, 8:345-350.

18. Milburn, P. D. (1982). Summation of segmental velocities in the golf swing. *Medicine & Science in Sports & Exercise*, 14:60-64.

19. Neal, R. J., Wilson, B. D. (1985). 3D kinematics and kinetics of the golf swing. *International Journal of Sports Biomechanics*, 1:221-231.

20. Nigg, B. M., et al. (1983). Methodological aspects of sport shoe and sport surface analysis. In H. Matsui, K. Kobayashi (Eds.). *Biomechanics VIII-B*. Champaign, IL: Human Kinetics, 1041-1052.

21. Novacheck, T. F. (1995). Walking, running, and sprinting: A three dimensional analysis of kinematics and kinetics. *Instructional Course Lectures*, 44:497-506.

22. Ounpuu, S. (1994). The biomechanics of walking and running. *Clinics in Sports Medicine*, 13:843-863.

23. Plagenhoef, S. (1971). *Patterns of Human Motion*. Englewood Cliffs, NJ: Prentice-Hall.

24. Pollard, C. D., et al. (2005). Gender differences during an unanticipated cutting maneuver. *Journal of Applied Biomechanics*, 21:143-152.

25. Rodgers, M. M., et al. (1994). Biomechanics of wheelchair propulsion during fatigue. *Archives of Physical Medicine and Rehabilitation*, 75:85-93.

26. Ronsky, J. L., et al. (1995). Correlation between physical activity and the gait characteristics and ankle joint flexibility of the elderly. *Clinical Biomechanics*, 10:41-49.

27. Sanderson, D. J., Sommer, H. J. (1989). Kinematic features of wheelchair propulsion. *Journal of Rehabilitation Research*, 26:31-50.

28. Schwartz, M. H., et al. (2000). A tool for quantifying hip flexor function during gait. *Gait and Posture*, 12:122-127.

29. Shoup, T. E., Fabian, D. (1986). Lengths and lies: choosing golf equipment scientifically. *SOMA*, 1:16-23.

30. Stacoff, A., Nigg, B. M., Reinschmidt, C., et al. (2000). Tibiocalcaneal kinematics of barefoot versus shod running. *Journal of Biomechanics*, 33(11): 1387-1396.

31. Turner, A. B., Hills, N. J. (1999). A three-link mathematical model of the golf swing. In M. R. Farrally, A. J. Cochran (Eds.). *Science and Golf III*. Champaign, IL: Human Kinetics, 3-12.

32. Van der Helm, F. C. T., Veeger, H. E. J. (1996). Quasi-static analysis of muscle forces in the shoulder mechanism during wheelchair propulsion. *Journal of Biomechanics*, 29:39-52.

33. Vanlandewijck, Y. C., et al. (1984). Wheelchair propulsion efficiency: movement pattern adaptations to speed changes. *Medicine & Science in Sports & Exercise*, 26:1372-1381.

34. van Woensel, W., Cavanagh, P. R. (1992). A perturbation study of lower extremity motion during running. *International Journal of Sports Biomechanics*, 8:30-47.

35. Veeger, H. E. J., et al. (1991). Load on the upper extremity in manual wheelchair propulsion. *Journal of Electromyography & Kinesiology*, 1:270-280.

36. Winter, D. A. (1987). *The Biomechanics and Motor Control of Gait*. Waterloo, ON: University of Waterloo.

OBJECTIVES

After reading this chapter, the student will be able to:

1. Define *force* and discuss the characteristics of a force.

2. Compose and resolve forces according to vector operations.

3. State Newton's three laws of motion and their relevance to human movement.

4. Differentiate between a contact and a noncontact force.

5. Discuss Newton's law of gravitation and how it affects human movement.

6. Discuss the seven types of contact forces and how each affects human movement.

7. Represent the external forces acting on the human body on a free body diagram.

8. Define the impulse–momentum relationship.

9. Define the work–energy relationship.

10. Discuss the concepts of internal and external work.

11. Discuss the forces acting on an object as it moves along a curved path.

12. Discuss the relationships between force, pressure, work, energy, and power.

13. Discuss selected research studies that used a linear kinetic approach.

OUTLINE

Force
    Characteristics of a Force
    Composition and Resolution of Forces
Laws of Motion
    Law I: Law of Inertia
    Law II: Law of Acceleration
    Law III: Law of Action–Reaction
Types of Forces
    Noncontact Forces
    Contact Forces
Representation of Forces Acting
   on a System

Analysis Using Newton's Laws
   of Motion
Special Force Applications
    Centripetal Force
    Pressure
Linear Kinetics of Locomotion
Linear Kinetics of the Golf Swing
Linear Kinetics of Wheelchair
   Propulsion
Summary
Review Questions

In Chapters 8 and 9, we discussed linear and angular kinematics. Kinematics was defined as the description of motion with no regard to the cause of the motion. The motion described was translatory (linear), rotational (angular), or a combination of both linear and rotational (general). In this chapter, the concern is with the causes of motion. For example, how do we propel ourselves forward in running? Why does a runner lean on the curve of a track? What keeps an airplane in the air? Why do golf balls slice or hook? How does a pitcher curve a baseball? The search for understanding the causes of motion date to antiquity, and answers to some of these questions were suggested by such notables as Aristotle and Galileo. The culmination of these explanations was provided by the great scientist Sir Isaac Newton, who ranks among the greatest thinkers in human history for his theories on gravity and motion. In fact, the laws of motion described by Newton in his famous book *Principia Mathematica* (1687) form the cornerstone of the mechanics of human movement (14). The branch of mechanics that deals with the causes of motion is called **kinetics**. Kinetics is concerned with the forces that act on a system. If the motion is translatory, then **linear kinetics** is of concern. The basis for the understanding of the kinetics of linear motion is the concept of **force**.

 ## Force

Force is a very difficult concept to define. In fact, we generally define the term *force* by describing what a force can do. According to Newton's principles, objects move when acted upon by a force greater than the resistance to movement provided by the object. A force involves the interaction of two objects and produces a change in the state of motion of an object by pushing or pulling it. The force may produce motion, stop motion, accelerate, or change the direction of the object. In each case, the acceleration of the object changes or is prevented from changing. A force, therefore, may be thought of as any interaction, a push or pull, between two objects that can cause an object to accelerate either positively or negatively. For example, a push on the ground generated by a forceful knee and hip extension may be sufficient to cause the body to accelerate upward and leave the ground—that is, jump.

CHARACTERISTICS OF A FORCE

Forces are vectors and as such have the characteristics of a vector, including magnitude and direction. Magnitude is the amount of force being applied. It is also necessary to state the direction of a force because the direction of a force may influence its effect, for example, on whether the force is pushing or pulling. Vectors, as described in Chapter 8, are usually represented by arrows, with the length of the arrow indicating the magnitude of the force and the arrowhead pointing in the direction in which the

force is being applied. In the International System (SI) of measurement, the unit for force is the newton (N), although for comparison value, forces are often represented in the literature as a ratio of force to body weight (BW) or force to body mass. Sample peak force values for a variety of movements, expressed as a function of BW, are presented in Table 10-1.

TABLE 10-1	Maximum Forces Acting on the Body	
Activity	**Relative Force (N/BW)**
Vertical jump, peak vertical	
Squat jump	1.4–8.3 (34)
Countermovement	2.2 (62)
Hopping	1.5–5.4 (26)
Landing on hard surface from 0.45 m	5–7 (61)
Dismount from horizontal bars	8.2–11.6 (61)
Single leg landing, double back somersault	9.3–10.6 (63)
Basketball rebound landing	1.3–6.0 (77)
Vertical jump, hard surface	>3 (48)
Vertical jump, soft surface	2 (48)
Triple jump, vertical forces	
Hop	7–10 (65)
Step	8–12 (65)
Jump	7.1–12.2 (65)
Triple jump, anteroposterior forces	
Hop	2.1–3.3 (65)
Step	1.7–3.2 (65)
Jump	1.7–3.9 (65)
Basketball jump shot, two-point range	
Vertical	2.6 (29)
Horizontal	0.5 (29)
Walking (vertical)	121.5
Compressive forces in the ankle joint	3–5.5 (66)
Reaction forces in the ankle joint	3.9–5.2 (66)
Reaction forces in the subtalar joint	2.4–2.8 (66)
Running (vertical)	2–3.5 (70)
Bone-on-bone force in the ankle joint	13 (70)
Patellar tendon force	4.7–6.9 (70)
Patellofemoral force	7.0–11.1 (70)
Plantar fascia force	1.3–2.9 (70)
Achilles tendon force	
Walking	3.9 (34)
Running	7.7 (34)
Peak forces acting at hip	
Walking	2.8–4.8 (7)
Jogging	5.5 (7)
Stumbling	7.2 (7)

Source in parenthesis.

Forces have two other equally important characteristics, the **point of application** and the **line of action**. The point of application of a force is the specific point at which the force is applied to an object. This is very important because the point of application most often determines whether the resulting motion is linear or angular or both. In many instances, a force is represented by a point of application at a specific point, although there may be many points of application. For example, the point of application of a muscular force is the center of the muscle's attachment to the bone, or the insertion of the muscle. In many cases, the muscle is not attached to a single point on the bone but is attached to many points, such as in the case of the fan-shaped deltoid muscle. In solving mechanical problems, however, it is considered to be attached to a single point. Other points of application are the contact point between the foot and the ground for activities such as jumping, walking, and running; hand contact with the ball for a baseball throw; and the contact point between the racquet and the ball in tennis.

The line of action of a force is a straight line of infinite length in the direction in which the force is acting. A force can be assumed to produce the same acceleration of the object if it acts anywhere along this line of action. Thus, if the force coming up from the ground in the last jump phase of a triple jumper has a line of action directed to 18° with respect to the horizontal, the jumper accelerates forward and upward in that direction. The orientation of the line of action is usually given with respect to an *x*, *y* coordinate system. The orientation of the line of action to this system is given as an angular position and is referred to as the angle of application. This angle is designated by the Greek letter theta (θ). The four characteristics of a force—magnitude, direction, point of application, and line of action—are illustrated in Figure 10-1A for a muscular force and in Figure 10-1B for a high-jump takeoff.

COMPOSITION AND RESOLUTION OF FORCES

Forces are vector quantities that have both magnitude and direction. As presented in the discussion of kinematic vectors in Chapter 8, a single force vector may be resolved into perpendicular components, or several forces can be resolved into one vector. That is, a single force vector can be calculated or composed to represent the net effect of all of the forces in the system. Similarly, given the resultant force, the resultant force can be resolved into its horizontal and vertical components. To do either, the trigonometric principles presented in Appendix B are applied.

Several types of force systems must be defined to compose or resolve systems of multiple forces. Any system of forces acting in a single plane is referred to as *coplanar*, and if they act at a single point, they are called *concurrent*. Any set of concurrent **coplanar forces** may be substituted by a single force, or the resultant, producing the same effect as the multiple forces. The process of finding this single force is called **composition** of force vectors.

When force vectors act along a single line, the system is said to be collinear. In this case, vector addition is used to compose the forces. Consider the force system in Figure 10-2A. The force vectors *a*, *b*, and *c* all act in the same direction and can be replaced by a single force, *d*, which is the sum of *a*, *b*, and *c*. Thus:

$$d = a + b + c$$
$$= 5\,\text{N} + 7\,\text{N} + 10\,\text{N}$$
$$= 22\,\text{N}$$

The force vector *d* would have the same effect as the other three force vectors. In Figure 10-2B, however, two of the force vectors, *a* and *b*, are acting in one direction, but the vector c is acting in the opposite direction. Thus, the force vector *d* is the algebraic sum of these three force vectors:

$$d = a + b - c$$
$$= 5\,\text{N} + 7\,\text{N} - 4\,\text{N}$$
$$= 8\,\text{N}$$

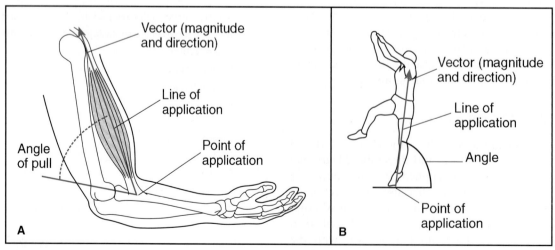

FIGURE 10-1   Characteristics of a force for an internal muscular force **(A)** and an external force generated on the ground in the high jump **(B)**.

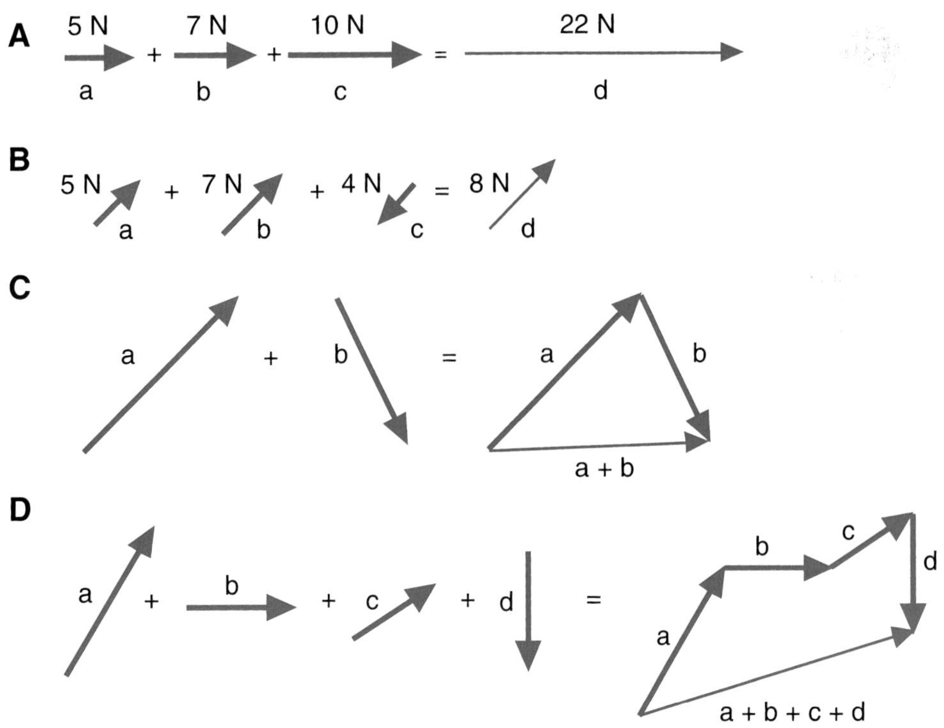

FIGURE 10-2 Force vector addition.

The force vector *d* still represents the net effect of these force vectors. In both of these examples, sets of **collinear force** vectors are present.

When the force vectors are not collinear but are coplanar, they may still be composed to determine the resultant force. Graphically, this can be done in exactly the same manner as described in Chapter 8 in the section on adding vectors. Consider Figure 10-2C. The force vectors *a* and *b* are not collinear, but they may be composed or added to determine their net effect. With the arrow of vector *a* placed at the tail of vector *b*, the resultant composed vector *c* is the distance between the tail of *a* and the arrow of *b*. This procedure is illustrated in Figure 10-2D with multiple vectors.

Multiple vectors can also be combined using trigonometric functions. First presented in Chapter 8, this involves first breaking each vector down into its components using **resolution**. After they are resolved into vertical and horizontal components, the orthogonal components for each vector are added, and the resultant vector is composed. To illustrate, the four vectors shown in Figure 10-2D will be assigned values of length 10 and $\theta = 45°$ for vector *A*, length 6 and $\theta = 0°$ for vector *B*, length 5 and $\theta = 30°$ for vector *C*, and length 7 and $\theta = 270°$ for vector *D*. The first step is to resolve each vector into vertical and horizontal components.

Vector *A*

$$y = 10 \text{ sin } 45°$$
$$= 10 \times 0.7071$$
$$= 7.07$$

$$x = 10 \text{ cos } 45°$$
$$= 10 \times 0.7071$$
$$= 7.07$$

Vector *B*

$$y = 6 \text{ sin } 0°$$
$$= 6 \times 0.0$$
$$= 0$$

$$x = 6 \text{ cos } 0°$$
$$= 6 \times 1.00$$
$$= 6.00$$

Vector *C*

$$y = 5 \text{ sin } 30°$$
$$= 5 \times 0.50$$
$$= 2.50$$

$$x = 5 \text{ cos } 30°$$
$$= 5 \times 0.866$$
$$= 4.33$$

Vector *D*

$$y = 7 \text{ sin } 270°$$
$$= 7 \times -1.00$$
$$= -7.00$$

$$x = 7 \text{ cos } 270°$$
$$= 7 \times 0.0$$
$$= 0$$

To find the magnitude of the resultant vector, the horizontal and vertical components of each vector are added and resolved using the Pythagorean theorem:

	Horizontal Components	Vertical Components
Vector A	7.07	7.07
Vector B	5.00	0.00
Vector C	4.33	2.50
Vector D	0.00	−7.00
Sum (Σ)	16.40	2.57

$$C = \sqrt{x^2 + y^2}$$
$$C = \sqrt{16.40^2 + 2.57^2}$$
$$= \sqrt{268.96 + 6.61}$$
$$= \sqrt{275.57}$$
$$= 16.60$$

To find the angle of resultant vector, the trigonometric function tangent is used:

$$\tan \theta = y\text{-component}/x\text{-component}$$
$$\theta = \arctan\left(\frac{2.57}{16.40}\right)$$
$$\theta = \arctan(0.1567)$$
$$= 8.91°$$

The characteristics of the resultant vector ($R = 16.60$, $\theta = 8.91°$) can be clearly confirmed by examining the resultant obtained by combining vectors graphically using the head-to-tail method.

Refer to the walking ground reaction force (GRF) data in Appendix C: Using the peak anteroposterior (F_y) and vertical force (F_z) for frame 21, calculate the resultant force and the angle of force application at this point.

Laws of Motion

The publication of the *Principia Mathematica* in 1687 by Sir Isaac Newton (1642 to 1727) astounded the scientific community of the day (14). In this book, he introduced his three laws of motion that we use to explain a number of phenomena. Although these laws have been superseded by Einstein's theory of relativity, we can still use Newton's basic principles as the basis for most analyses of human movement in biomechanics. Newton's three laws of motion have demonstrated how and when a force creates a movement and how it applies to all of the different types of forces previously identified. His work has provided the link between cause and effect. To fully understand the underlying nature of motion, it is necessary to understand the cause of the movement, not merely the description of the outcome. That is, the forces that cause motion must be fully understood. The statements of these laws are taken from a translation of Newton's *Principia Mathematica* (14).

LAW I: LAW OF INERTIA

Every body continues in its state of rest, or of uniform motion in a straight line, unless it is compelled to change that state by forces impressed on it (14).

The **inertia** of an object is used to describe its resistance to motion. Inertia is directly related to the mass of the object. Mass is a scalar and is the measure of the amount of matter that constitutes an object and is expressed in kilograms. An object's mass is constant, regardless of where it is measured, so that the mass is the same whether it is calculated on earth or on the moon. The greater the mass of an object, the greater its inertia and thus the greater the difficulty in moving it or changing its current motion.

For an object to move, the inertia of the object has to be overcome. Newton suggested that an object at rest—an object with zero velocity—would remain at rest. This seems obvious. A chair sitting in a room has zero velocity because it is not moving. Additionally, an object moving at a constant velocity would continue to do so in a straight line. This concept is not as obvious because the practical instances of individuals on the earth's surface rarely experience constant velocity motion. If it is noted that constant velocity results in zero acceleration just as zero velocity does, then it can be understood how this law holds for both cases. Therefore, the inertia of these objects would compel them to maintain their status at a constant velocity. Newton's first law of motion can be expressed as:

$$\text{If } \Sigma F = 0 \text{ then } \Delta v = 0$$

Note that ΣF refers to the net force and takes into account all forces acting on the object.

Overcoming the inertia of such objects requires a net external force greater than the inertia of the object. For example, if a barbell has a mass of 70 kg, a force greater than 686.7 N, or the product of the acceleration due to gravity (9.81 m/s²) and 70 kg, must be exerted to lift it. If an object is subjected to an external force that can overcome the inertia, the object will be accelerated. To get an object moving, the external force must positively accelerate the object. On the other hand, to stop the object from moving, the external force must negatively accelerate the object. Because body mass determines inertia, an individual with greater mass has to generate larger external forces to overcome inertia and accelerate.

LAW II: LAW OF ACCELERATION

The change of motion is proportional to the force impressed and is made in the direction of the straight line in which that force is impressed (14).

Newton's second law generates an equation that relates all forces acting on an object, the mass of the object, and the acceleration of the object. This relationship is expressed as:

$$\Sigma F = ma$$

This equation can also be used to define the unit of force, the newton. By substituting the units for mass and acceleration in the right-hand side, it can be seen that:

$$F = ma$$
$$\text{Newton} = \frac{\text{kg-m}}{\text{s}^2}$$

where kg-m is kilogram-meters. In this equation, the force is the net force acting on the object in question, that is, the sum of all of the forces involved.

In adding up all forces acting on an object, it is necessary to take the direction of the forces into account. If the forces exactly counteract each other, the net force is zero. If the sum of the forces is zero, the acceleration will also be zero. This case is also described by Newton's first law. If the net force produces acceleration, the accelerated object will travel in a straight line along the line of action of the net force.

Rearranging the equation described by Newton's second law allows another important concept in biomechanics to be defined. Acceleration was previously defined as the time rate of change of velocity, or dv/dt. Substituting this expression into the equation of the second law:

$$\Sigma F = m\frac{dv}{dt}$$

or

$$\Sigma F = \frac{m dv}{dt}$$

The product of mass and velocity in the numerator of the right-hand side of this equation is known as the momentum of an object. Momentum is the quantity of motion of an object. It is generally represented by the letter p and has units of kilogram-meters per second. For example, if a football player has a mass of 83 kg and is running at 4.5 m/s, his momentum is:

$$p = \text{mass} \times \text{velocity}$$
$$= 83 \text{ kg} \times 4.5 \text{ m/s}$$
$$= 373.5 \text{ kg-m/s}$$

Newton's second law can thus be restated:

$$\Sigma F = \frac{dp}{dt}$$

That is, force is equal to the time rate of change of momentum. To change the momentum of an object, an external force must be applied to the object. The momentum may increase or decrease, but in either case, an external force is required.

LAW III: LAW OF ACTION–REACTION

To every action there is always opposed an equal reaction; or, the mutual actions of two bodies upon each other are always equal and directed to contrary parts.

This law illustrates that forces never act in isolation but always in pairs. When two objects interact, the force exerted by object A on object B is counteracted by a force equal and opposite exerted by object B on object A. These forces are equal in magnitude but opposite in direction. That is:

$$\Sigma F_{\text{A on B}} = -\Sigma F_{\text{B on A}}$$

In addition, the force—the action—and the counterforce—the reaction—act on different objects. The result is that these two forces cannot cancel each other out because they act on and may have a different effect on the objects. For example, a person landing from a jump exerts a force on the earth, and the earth exerts an equal and opposite force on the person. Because the earth is more massive than the individual, the effect on the individual is greater than the effect on the earth. This example illustrates that although the force and the counterforce are equal, they may not necessarily have comparable results.

In human movements, an action force is generated on the ground or implement, and the reaction force generally produces the desired movement. As shown in Figure 10-3, the jumper makes contact with the ground and generates a large downward force because of the acceleration of the body combined with forces generated by body segments at contact, and a resulting reaction force upward controls the landing.

🧠⚙ Types of Forces

The forces that exist in nature and affect the way humans move may be classified in a number of ways. The most common classification scheme is to describe forces as contact or **noncontact force**s (11). A contact force involves

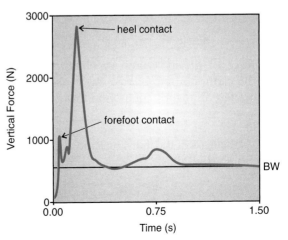

FIGURE 10-3 Vertical GRF during a landing from a jump.

the actions, pushes or pulls, exerted by one object in direct contact with another object. These are the forces involved, for example, when a bat hits a baseball or the foot hits the floor. In contrast to contact forces are those that act at a distance. These are called noncontact forces. As implied by the name, these are forces that are exerted by objects that are not in direct contact with one another and may actually be separated by a considerable distance.

NONCONTACT FORCES

In the investigation of human movement, the most familiar and important noncontact force is gravity. Any object released from a height will fall freely to the earth's surface, pulled by gravity. In Sir Isaac Newton's book, the *Principia Mathematica* (1687), he introduced his theory of gravity (14). With the law of gravitation, Newton identified gravity as the force that causes objects to fall to the earth, the moon to orbit the earth, and the planets to revolve about the sun. This law states: "The force of gravity is inversely proportional to the square of the distance between attracting objects and proportional to the product of their masses."

In algebraic terms, the law is described by the following equation:

$$F = \frac{Gm_1 m_2}{r^2}$$

where G is universal gravitational constant, m_1 is mass of one object, m_2 is mass of the other object, and r is the distance between the mass centers of the objects.

The constant value G was estimated by Newton and determined accurately by Cavendish in 1798. The value of G is 6.67×10^{-11} N-m²/kg².

The gravitational attraction of one object of a relatively small size to another object of similar size is extremely small and therefore can be neglected. In biomechanics, the objects of most concern are the earth, the human body, and projectiles. In these cases, the earth's mass is considerable, and gravity is a very important force. The attractive force of the earth on an object is called the **weight** of the object. This is stated as:

$$W = F_g = \frac{Gm_{object} - M_{earth}}{r^2}$$

The force of gravity causes an object to accelerate toward the earth at a rate of 9.81 m/s². Newton determined through his theories of motion that:

$$W = ma$$

where m is the mass of the individual and a is the acceleration due to gravity. Thus:

$$W = mg$$

where g is the acceleration due to gravity. BW is thus the product of the individual's mass and the acceleration due to gravity. It is apparent, therefore, that an individual's mass and BW are not the same. Whereas BW is a force and

the appropriate unit for BW is the newton, body mass is a scalar with units of kilograms. To determine a person's BW, simply multiply mass times the acceleration due to gravity (9.81 m/s²).

Because weight is a force, it has the attributes of a force. As a vector, it has a line of action and a point of application. An individual's total BW is considered to have a point of application at the center of mass and a line of action from the center of mass to the center of the earth. Because the earth is so large, this line of action is straight down toward the center of the earth.

The point of origin of the weight vector is called the center of gravity. This is a point about which all particles of the body are evenly distributed. Another term used interchangeably with *center of gravity* is *center of mass*, a point about which the mass of the segment or body is equally distributed. They differ in that the center of gravity refers only to the vertical direction because that is the direction in which gravity acts, but the center of mass does not depend on a vertical orientation. The computation of both the center of mass and the center of gravity is presented in Chapter 11.

The value for g, the acceleration due to gravity, depends on the square of the distance to the center of the earth. Because of the spin on its axis, the earth is not perfectly spherical. The earth is slightly flattened at the poles, resulting in shorter distances to the earth's center at the poles than at the equator. Thus, the points on the earth are not all equidistant from its center, and acceleration due to gravity, g, does not have the same value everywhere. The latitude—the position on the earth with respect to the equator—on which a long jump, for example, is performed can have a significant effect on the distance jumped. Another factor influencing the value of g is altitude. The higher the altitude, the lower the value for g. If one were weighing oneself and a minimal weight were desired, the optimum place for the weigh-in would be on the highest mountain at the earth's equator.

CONTACT FORCES

Because **contact forces** are those resulting from a direct interaction of two objects, the number of such forces is considerably greater than the single noncontact force discussed. The following contact forces are considered paramount in human movement: ground reaction force (GRF), joint reaction force, friction, fluid resistance, inertial force, muscle force, and elastic force.

Ground Reaction Force

In almost all terrestrial human movement, the individual is acted upon by the GRF at some time. This is the reaction force provided by the surface upon which one is moving. The surface may be a sandy beach, a gymnasium floor, a concrete sidewalk, or a grass lawn. If the individual is swinging from a high bar, the surface of the bar provides a reaction force. All surfaces on which an individual interacts provide a reaction force. The individual pushes against the ground with force, and the ground pushes back against the individual with equal force in the opposite direction (Newton's law of action–reaction). These forces affect both parties—the ground and the individual—and do not cancel out even though they are equal in magnitude but opposite in direction. Also, the GRF changes in magnitude, direction, and point of application during the period that the individual is in contact with the surface.

> Refer to the GRF data for walking in Appendix C. Calculate the resultant force using F_y (anteroposterior) and F_z (vertical) data for frames 18, 36, and 60. How does the direction of the force application change across the support phase?

As with all forces, the GRF is a vector and can be resolved into its components. For the purpose of analysis, it is commonly broken down into its components. These components are orthogonal to each other along a three-dimensional coordinate system (Fig. 10-4). The components are usually labeled F_z vertical (up–down), F_y anteroposterior (forward–backward), and F_x, mediolateral (side-to-side). In the reporting of three-dimensional data, however, some researchers label the axes as F_y (vertical), F_x (anteroposterior), and F_z (mediolateral). The former convention is used in this book because it is the most often used system. It is probably much clearer to represent the GRF components as $F_{vertical}$, $F_{anteroposterior}$, and $F_{mediolateral}$. Regardless of the convention used, the anteroposterior and mediolateral components are referred to as shear components because they act parallel to the surface of the ground.

Biomechanists record the GRF components using a **force platform**. A force platform is a sophisticated measuring scale usually imbedded in the ground, with its

FIGURE 10-5 A typical laboratory force platform setup.

surface flush with the surface of the ground, on which the individual performs. A typical force platform experimental setup for the data collected in Appendix C is shown in Figure 10-5. This device measures the forces of the foot on the performing surface or the force of an individual just standing on the platform. Force platforms have been used since the 1930s (28) but became more prominent in biomechanics research in the 1980s.

Although forces are measured in newtons, GRF data are generally scaled by dividing the force component by the individual's BW, resulting in units of times BW. In other instances, GRFs may be scaled by dividing the force by body mass, resulting in a unit of newtons per kilogram of body mass.

> Refer to the GRF walking data in Appendix C. Locate the maximum vertical (F_z), anteroposterior (F_y), and mediolateral (F_x) forces (N). Report peak forces scaled to BW and to body mass (N/kg).

GRF data have been used in many studies to investigate a variety of activities. Most studies, however, have dealt with the load or impact on the body during landings, either from jumps or during support phase of gait. For example, GRFs have been studied during the support phase of running (17,19), walking (36), and landings from jumps (27,53).

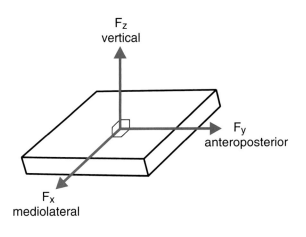

Direction of motion

FIGURE 10-4 GRF components. The origin of the force platform coordinate system is at the center of the platform.

A vertical component curve of a single foot contact of an individual landing from a jump is presented in Figure 10-3. Only the vertical component is presented because it is of much greater magnitude than the other components and because the major interest in landings has been the effect of impacts on the human body. In the landing curve, the first peak represents the initial ground contact with the forefoot. The second peak is the contact of the heel on the surface. Generally, the second peak is greater than the first peak. Some individuals, however, land flatfooted and have only one **impact peak**. When the individual comes to rest on the surface, the vertical force curve settles at the individual's BW. Although the magnitude of the vertical component at impact in running is three to five BW, the vertical component in landing can be as much as 11 times BW, depending on the height from which the person drops (27,53).

It should be noted that the GRF is the sum of the effects of all masses of the segments times the acceleration due to gravity. That is, the sum of the product of the masses and accelerations of each segment. This sum reflects the center of mass of the individual. Consequently, the GRF acts at the center of mass of the total body (Fig. 10-6). Dividing a force by the mass, the result would be acceleration. For example, the equation relating the vertical GRF to acceleration is given by Newton's second law as follows:

$$\Sigma F = ma$$

Dividing both sides of the equation by body mass (m):

$$a = \Sigma F/m$$

This value reflects the acceleration of the center of mass. Many researchers have related the vertical GRF component to the function of the foot during landings. Because the GRF acts on the center of mass, the relationship between a GRF component and the foot function is tenuous at best.

Joint Reaction Force

In many instances in biomechanical analyses, segments are examined, either singly or one at a time in a logical order. When a joint reaction force analysis is conducted, the segment is separated at the joints, and the forces acting across the joints must be considered. For example, if one is standing still, the thigh exerts a downward force on the leg across the knee joint. Similarly, the leg exerts an upward force of equal magnitude on the thigh (Fig. 10-7). This is the net force acting across the joint and is referred to as the **joint reaction force**. In most analyses, the magnitude of this force is unknown, but it can be calculated given the appropriate kinematic and kinetic data, in addition to anthropometric data describing the body dimensions.

Some confusion exists as to whether the joint reaction force is the force of the distal bony surface of one segment acting on the proximal bony surface of the contiguous segment. The joint reaction force does not, however, reflect this **bone-on-bone force** across a joint. The actual bone-on-bone force is the sum of the actively contracting muscle forces pulling the joint together and the joint reaction force. Because the force generated by the actively contracting muscles is not known, the bone-on-bone force

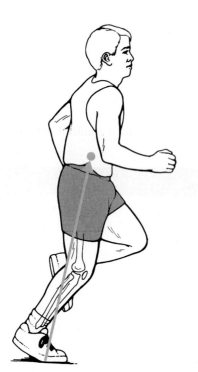

FIGURE 10-6 The GRF vector acts through the center of mass of the body.

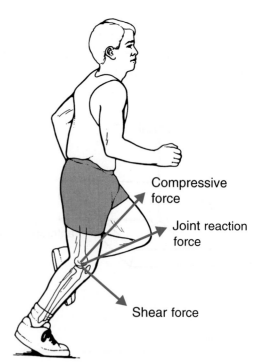

FIGURE 10-7 The joint reaction force of the knee with its shear and compressive components.

is difficult to calculate, although sophisticated calculations have been done to estimate bone-on-bone forces (91).

Friction

Friction is a force that acts parallel to the interface of two surfaces that are in contact during the motion or impending motion of one surface as it moves over the other. For example, the weight of a block resting on a horizontal table pulls the block downward, pressing it against the table. The table exerts an upward force on the block that is perpendicular or normal to the surface. To move the block horizontally, a horizontal force on the block of sufficient magnitude must be exerted. If this force is too small, the block will not move. In this case, the table evidently exerts a horizontal force equal and opposite to the force on the block. This interaction, the frictional force, is due to the bonding of the molecules of the block and the table at the places where the surfaces are in very close contact. Figure 10-8 illustrates this example.

Although it appears that the area of contact influences the force of friction, this is not the case. The force of friction is proportional to the normal force between the surfaces, that is:

$$F_f = \mu N$$

where μ is the **coefficient of friction** and N is the normal force or the force perpendicular to the surface. The coefficient of friction is calculated by:

$$\mu = \frac{F_f}{N}$$

The coefficient of friction is a dimensionless number. The magnitude of this coefficient depends on the nature of the interfacing surfaces. The greater the magnitude of the coefficient of friction, the greater the interaction between the molecules of the interfacing surfaces.

Continuing the block and table example, if a steadily increasing force is applied to the block, the table also applies an increasing opposite force resisting the movement. At the point where the pulling force is at a maximum and no movement results, the resisting force is called the maximum static friction force (F_{sMAX}). Before movement, it can be stated that:

$$F_{sMAX} \leq \mu_s N$$

where μ_s is the static coefficient of friction. At some point, however, the force is sufficiently large and the static friction force cannot prevent the movement of the block. This relationship simply means that if a block weighing 750 N is standing on a surface with μ_s of 0.5, it will take 50% of the 750 N normal force, or 375 N, of a horizontal force to cause motion between the block and the table. A μ_s of 0.1 would require a horizontal force of 75 N to cause motion and a μ_s of 0.8 would require a 600 N horizontal force. As can be seen, the smaller the coefficient of friction, the less horizontal force required to cause movement.

As the block slides along the surface of the table, molecular bonds are continually made and broken. Thus, after the two surfaces start moving relative to each other, it becomes somewhat easier to maintain the motion. The result is a force of sliding friction that opposes the motion. Sliding friction and rolling friction are types of **kinetic friction**. Kinetic friction is defined as:

$$F_k = \mu_k N$$

where μ_k is the dynamic coefficient of friction or the coefficient of friction during movement. It has been found experimentally that μ_k is less than μ_s and that μ_k depends on the relative speed of the object. At speeds of one centimeter per second to several meters per second, however, μ_k is relatively constant. Figure 10-9 illustrates the friction–external force relationship.

Although translational friction is important in human movement, **rotational friction** must also be considered. Rotational friction is the resistance to rotational or twisting movements. For example, the soles of the shoes of a basketball player accomplishing a pivot interact with the playing surface to resist the turning of the foot. Obviously, the player must be able to accomplish this movement during a game, so the rotational friction must allow this motion without influencing the other frictional characteristics of the shoe. A basketball player executing a 180°

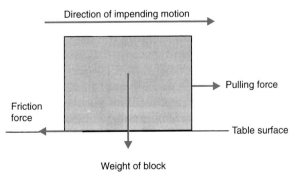

FIGURE 10-8 The forces acting on a block being pulled across a table.

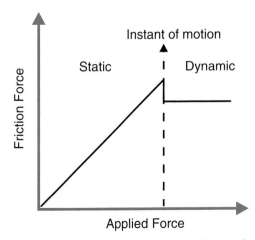

FIGURE 10-9 A theoretical representation of friction force as a function of the applied force. The applied force increases with the friction force until motion occurs.

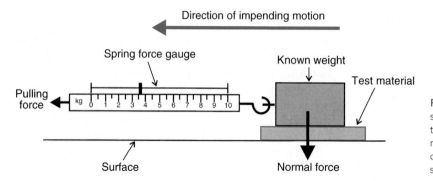

FIGURE 10-10 A towed sled device that measures applied force. The known weight constitutes the normal force, and the spring gauge measures the resistant horizontal force on the known weight. The coefficient of friction is computed by the ratio of the spring gauge force to the known weight.

pivot in a conventional basketball shoe on a wooden floor would have a rotational friction value 4.3 times greater than if the individual completed the same movement in gym socks (78). The measurement of rotational friction does not yield a coefficient of friction. The values used to compare rotational friction are based on the value of the resistance to rotation, usually measured on a force platform. For example, rotational friction values for a tennis shoe on artificial turf or artificial grass have been shown to range from 15.8 to 21.2 N-m (newton-meters) and 17.1 to 21.2 N-m, respectively (59). Translational friction and rotational friction are not independent of each other.

Friction is a complicated but important influence on human movement. Just to walk across a room requires an appropriate coefficient of friction between the shoe outsole and the surface of the floor. In everyday activities, one may try either to increase or to decrease the coefficient of friction, depending on the activity. For example, skaters prefer fresh ice because it has a low coefficient of friction. On the other hand, a golfer wears a glove to increase the coefficient of friction and get a better grip on the club.

Many types of athletes wear cleated shoes to increase the coefficient of friction and get better traction on the playing surface. Valiant (76) suggested that μ_s equaling 0.8 provides sufficient traction for athletic movements and any greater coefficient of friction may be unsafe. In certain situations, cleated shoes may, in fact, result in too much translational and/or rotational friction. This appears to be the case with artificial turf. Many injuries, such as turf toe and anterior cruciate ligament tears, have been related to too much friction force on artificial turf.

When the coefficient of friction is too small, a slip hazard results, but when the coefficient of friction is too great, a trip hazard occurs. In the workplace, slips and falls are numerous and often cause serious injury. Cohen and Compton (20) reported that 50% of 120,682 workers compensation cases in the state of New York from 1966 to 1970 were caused by slips. In England, Buck (12) cited 1982 statistics in which 14% of accidents in the manufacturing industry resulted from slips and trips. Thus, the coefficient of friction is a very important criterion in the design of any surface on which people perform, whether in the workplace or in athletics.

The static coefficient of friction of materials on different surfaces has been successfully measured using a towed sled device (2). This device, illustrated in Figure 10-10, involves an object of a known mass and a force-measuring gauge. The moving surface, usually some type of footwear, is placed beneath the mass. The gauge pulls the mass until it moves, and the force is measured by the gauge at the instant of movement. The coefficient of friction can then be calculated using the known mass and the force measured from the gauge. This type of measurement is known as a materials test measure; it does not involve human subjects.

It is difficult to measure the static or kinetic coefficient of friction accurately without sophisticated equipment. Both, however, may be measured with a force platform. The shear components F_y and F_x are, in fact, the frictional forces in the anteroposterior and mediolateral directions, respectively. If the normal force is known, the dynamic coefficient of friction may be estimated. Generally, this is done using the vertical (F_z) component as the normal force. Thus, the coefficient of friction can be determined by:

$$\mu = \frac{F_y}{F_z}$$

Several researchers have devised instruments to measure both translational and rotational friction. A device developed in the Nike Sports Science Laboratory is one such instrument (Fig. 10-11). It has been used to measure the friction characteristics of many types of athletic shoes.

In general, the magnitude of the coefficient of friction depends on the types of materials constituting the surfaces in contact and the nature of those surfaces. For example, a rubber-soled shoe would have a higher coefficient of friction on a wood gymnasium floor than a leather-soled shoe. Jogging shoes on an artificial track register static and dynamic frictional coefficients in the range of 0.7 to 1.1 and 0.7 to 1.0, respectively (59). This is compared with football shoes on artificial turf, for which the static and dynamic coefficients of friction range from 1.1 to 1.6 and 1.0 to 1.5, respectively (59). The relative roughness or smoothness of the contacting surfaces also affects the coefficient of friction. Intuitively, a rough surface has a higher coefficient of friction than a smooth surface. The addition of lubricants, moisture, or dust to a surface also greatly affects frictional characteristics. To determine the coefficient of friction, all of these factors must be considered.

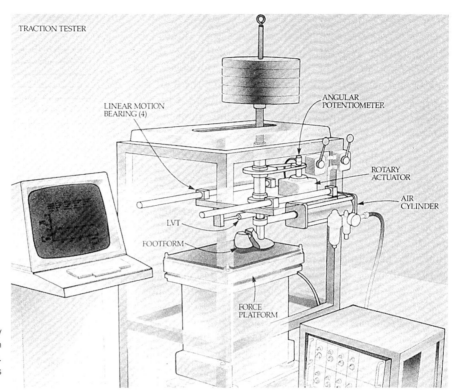

FIGURE 10-11 Device for mechanically evaluating translational and rotational friction characteristics of athletic footwear outsoles. This device was developed at the Nike Sports Research Laboratory.

Fluid Resistance

In many activities, human motion is affected by the fluid in which the activities are performed. Both air (a gas) and water (a liquid) are considered fluids. Thus, the motion of a runner is affected by the movement of air and that of a swimmer by the water or the air–water interface. Projectiles, whether humans or objects, are also affected by air. For example, anyone who has ever driven a golf ball into the wind will understand the effects of air on the golf ball.

Density and Viscosity

The two properties of a fluid that most affect objects as they pass through it are the fluid's density and viscosity. Density is defined as the mass per unit volume. Generally, the more dense the fluid, the greater the resistance it presents to the object. The density of air is particularly affected by humidity, temperature, and pressure. Viscosity is a measure of the fluid's resistance to flow. For example, water is more viscous than air, with the result that water resistance is greater than air resistance. Gases such as air become more viscous as the air temperature rises.

As an object passes through a fluid, it disturbs the fluid. This is true for both air and water. The degree of disturbance depends on the density and viscosity of the fluid. The greater the disturbance of the fluid, the greater the energy that is transmitted from the object to the fluid. This transfer of energy from the moving object to the fluid is called **fluid resistance**. The resultant fluid resistance force can be resolved into two components, lift and drag (Fig. 10-12).

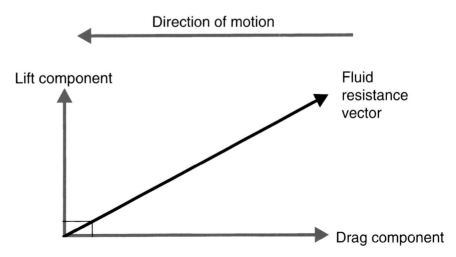

FIGURE 10-12 Fluid resistance vector with its lift and drag components.

Drag Force Component

Drag is a component of fluid resistance that always acts to oppose the motion. The direction of drag is always directly opposite to the direction of the velocity vector and acts to retard the motion of the object through the fluid. In most instances, *drag* is synonymous with *air resistance*. The magnitude of the drag component may be determined by:

$$F_{\text{drag}} = \frac{1}{2} C_d A \rho v^2$$

where C_d is a constant, the coefficient of drag; A is the projected frontal area of the object; the Greek letter rho (ρ) is the fluid viscosity; and v is the relative velocity of the object, that is, the velocity of the object relative to the fluid. The magnitude of the drag component is a function of the nature of the fluid, the nature and shape of the object, and the velocity of the object through the fluid.

Two types of drag must be considered. Drag as a result of friction between the object's surface and the fluid is referred to as **surface drag** or **viscous drag**. When an object moves through a fluid, the fluid interacts with the surface of the object, literally sticking to its surface. The resulting fluid layer is called the **boundary layer**. The fluid in the boundary layer is slowed down relative to the object as it passes it by. It results in the object pushing on the fluid and the fluid pushing on the object in the opposite direction. This interaction causes friction between the fluid in the boundary layer and the object's surface. This fluid friction opposes the motion of the object through the fluid. A fluid with high viscosity will generate a high drag component. In addition, the size of the object becomes more important if more surface is exposed to the fluid.

From the formula for drag force, it can be seen that drag force increases as a function of the velocity squared. The relative velocity of the fluid as it passes by the object actually determines how the object will interact with the fluid. At lower movement velocities of the object, the fluid passes the object in uniform layers of differing speed, with the slowest moving layers closest to the surface of the object. This is called **laminar flow** (Fig. 10-13A). Laminar flow occurs when the object is small and smooth and the velocity is small. The drag force consists almost entirely of surface or friction drag. On every object, however, there are points, called the points of separation, at which the fluid separates from the object. That is, the fluid does not completely follow the contours of the shape of the object. As the object moves through the fluid faster, the fluid moves by the object faster. When this occurs, the points of separation move forward on the object and the fluid separates from the contours of the object closer to the front of the object. Thus, at relatively high velocities, the fluid does not maintain a laminar flow; rather, **separated flow** occurs (Fig. 10-13B). Separated flow is also referred to as partially **turbulent flow**. In this instance, the fluid is unable to contour to the shape of the object,

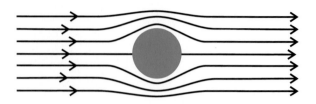

A Laminar Flow

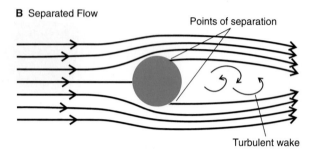

B Separated Flow

Points of separation

Turbulent wake

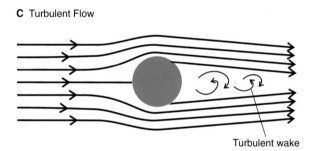

C Turbulent Flow

Turbulent wake

FIGURE 10-13   Flow about a sphere. **(A)** Laminar flow. **(B)** Separated flow. **(C)** Turbulent flow.

and the boundary layer separates from the surface. This produces turbulence behind the object. In power boating, for example, the turbulence behind the boat is called the wake. The wake of an object moving through a fluid is a low-pressure region. Separated flow occurs at a small relative velocity if the object is large and has a rough surface or under any conditions in which the fluid does not stick to the surface of the object.

When the fluid contacts the front of the object, an area of relatively high pressure is formed. The turbulent area behind the object is an area of pressure lower than the pressure at the front of the object. The greater the turbulence, the lower the pressure is behind the object. The net pressure differential between the front and back of the object retards the object's movement through the fluid. With increasing flow velocity, the point at which the boundary layer separates from the surface of the object moves farther to the front of the object, resulting in an even greater pressure differential and greater resistance. Drag resulting from this pressure differential is called **form drag**. In partially turbulent flow, both form and friction drag occur. As the wake increases, form drag dominates.

As the relative velocity of the object and the fluid increases, the whole boundary layer becomes turbulent. This type of fluid flow is called turbulent flow (Fig. 10-13C). Interestingly, the turbulence in the boundary layer actually moves the **point of separation** toward the back of the object, reducing the ability of the boundary layer to separate from the object. The net result is a reduction in the drag force.

Moving from partially turbulent flow to fully turbulent flow, thus decreasing the drag force, can be accomplished by streamlining and smoothing the surface of the object. Contrary to what one might expect, putting dimples on a golf ball or seams on a baseball, thereby roughening the surface, actually helps in this transition.

The shift from laminar to partially turbulent flow can be forestalled, hence drag minimized, with a streamlined shape or smooth surface or both. Athletes such as sprinters, cyclists, swimmers, and skiers generally wear smooth suits during their events (Fig. 10-14). A significant 10% reduction of drag occurs when a speed skater wears a smooth body suit (80). Wearing smooth clothing also prevents such things as long hair, laces, and loose-fitting clothing from increasing drag (47). Kyle (46) reported that loose clothing or thick long hair could raise the total drag 2% to 8%. He calculated that a 6% decrease in air resistance can increase the distance of a long jump by 3 to 5 cm.

Streamlining the shape of the object involves decreasing the projected frontal area. The projected frontal area of the object is the area of the surface that might come in contact with the fluid flow. Athletes in sports in which air resistance must be minimized manipulate this frontal area constantly. For example, a speed skater can assume any of a number of body positions during a race. A skater who has the arms hanging down in front presents a greater frontal area than one in an arms back racing position. The frontal areas in these speed skating positions are 42.21 and 38.71 m², respectively (84). Similarly, a ski racer assumes a tuck position to minimize the frontal area rather than the posture of a recreational skier. To decrease the drag component, a more streamlined position must be assumed. Streamlining helps minimize the pressure differential and thus the form drag on the object. For example, whereas the drag coefficient for a standing human figure is 0.92, it is 0.8 for a runner and 0.7 for a skier in a low crouch (46).

Much equipment has been designed to minimize fluid resistance. New bicycle designs; solid rear wheels on racing bicycles; clothing for skiers, swimmers, runners, and cyclists; bent poles for downhill skiers; new helmet designs; and so on have all contributed to help these athletes in their events. Research on streamlining body positions has also greatly aided athletes in many sports, such as cycling, speed skating, and sprint running (84).

Although it may seem counterintuitive, drag may also have a propulsive effect in some activities, particularly in swimming. This concept, called **propulsive drag**, was proposed by the famous Indiana University swimming coach,

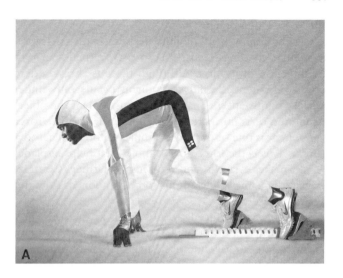

FIGURE 10-14   (**A** and **B**) Examples of clothing used by athletes to reduce drag (Photos provided by Nike, Inc.).

"Doc" Counsilman (21). This concept uses Newton's third law of action-reaction and states that, as the swimmer moves his or her hand through the water against the direction of motion, the reaction force of the water helps to propel the swimmer though the water. In addition, by changing the orientation of the hand as it moves through the water, a drag force is created in the direction opposite to the movement of the hand, thus further propelling the swimmer forward.

Lift Force Component

Lift is the component of fluid resistance that acts perpendicular to drag. Thus, it also acts perpendicular to the direction of motion. Although there is always a drag force component, the lift component occurs only under special circumstances. That is, lift occurs only if the object is spinning or is not perfectly symmetrical. The **lift force** component is one of the most significant forces in aerodynamics. This is the force, for example, that helps airplanes fly and makes a javelin and a discus go farther. Contrary to what the name suggests, this force component does not always oppose gravity.

Lift force is produced by any break in the symmetry of the airflow about an object. This can be shown in an object having an asymmetrical shape, a flat object being tilted to the airflow, or a spinning object. The effect makes the air flowing over one side of the object follow a different path than the air flowing over the other side. The result of this differential airflow is lower air pressure on one side of the object and higher air pressure on the other side. This pressure differential causes the object to move toward the side that has the lower pressure. The result is that the airfoil develops lift force in the direction of the lower pressure. The lift force concept is used, for example, on the wings of airplanes and on the spoilers on cars.

Lift force also contributes to the curved flight of a spinning ball that is critical in baseball and, most maddeningly, in golf. The spin on the ball results in the air flowing faster on one side of the ball and slower on the other side, creating a pressure differential. Consider the spinning ball in Figure 10-15. Side A of the ball is spinning against the airflow, causing the boundary layer to slow down on that side. On side B, however, because it is moving in the same direction as the airflow, the boundary layer speeds up. By Bernoulli's principle, this results in a pressure differential. This is comparable to the pressure differential about the airfoil. The ball, therefore, is deflected laterally toward the direction of the spin or the side on which there is a lower pressure area. This effect was first described by Gustav Magnus in 1852 and is known as the **Magnus effect**.

Baseball pitchers have mastered the art of putting just enough spin on the ball to curve its path successfully.

Soccer players spin their kicks to induce a curved path on the ball (Fig. 10-16). In this example, the defenders set up to prevent the ball from traveling in a straight line into the net. The player taking the free kick curves the ball around the defensive wall. Many golfers try not to put a sideways spin on the ball to avoid slicing or hooking the ball. They do, however, try to put backspin on the golf ball. The backspin, because of the Magnus effect, creates a pressure differential between the top and the bottom of the golf ball, with the lower pressure on the top. The golf ball gains lift and thus distance.

Just as there is a propulsive drag force in swimming, we can also discuss the concept of propulsive lift (21). A swimmer holds his or her hands so that they resemble airfoils. Pitching or changing the orientation of the hands puts the lift component in the desired direction of movement (5,69). Thus, lift can contribute to the forward motion of the swimmer.

Terminal Velocity

Drag forces help to explain why heavier objects seem to fall faster than light objects. In the absence of an atmosphere the only force acting on a falling object is weight and thus it accelerates at -9.81 m/s^2. In the presence of an atmosphere there is also a drag force acting upward. This drag force is small while moving slowly but increases as the velocity of the object increases. At some point during the acceleration of a falling object the drag force will be equal to the weight of the object. At this point, the net force is 0 and the object will no longer accelerate. The velocity at which this occurs is called the terminal velocity:

$$F_{drag} = F_{weight}$$

$$\frac{1}{2} C_d A \rho v^2 = mg$$

$$v = \sqrt{\frac{m \cdot g}{\frac{1}{2}(C_d \cdot A \cdot \rho)}}$$

where C_d is a constant, the coefficient of drag; A is the projected frontal area of the object; the Greek letter rho (ρ) is the fluid viscosity; m is the mass; g is the acceleration due to gravity; and v is the terminal velocity. In the stable arch position, a skydiver will reach a terminal velocity of

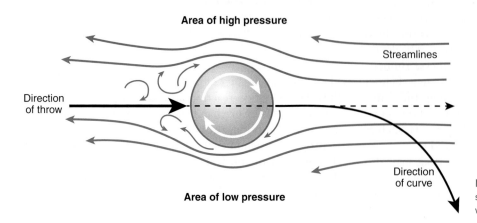

Area of high pressure

Streamlines

Direction of throw

Direction of curve

Area of low pressure

FIGURE 10-15 The Magnus effect on a spinning ball. Because of this effect, the ball will curve in the direction indicated.

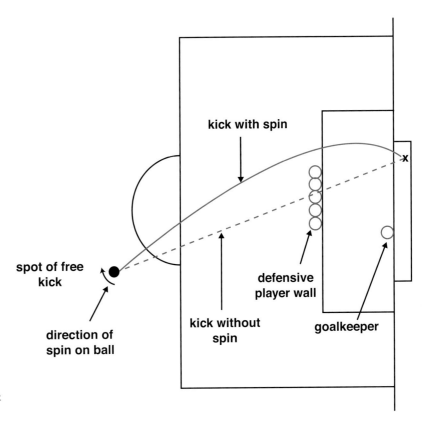

FIGURE 10-16 A soccer kick curves a free kick around a wall of defenders.

approximately 110 to 120 mph after 9 to 10 seconds of freefall (Fig. 10-17). Lighter objects of a similar shape will reach terminal velocity sooner and with a lesser velocity than heavier objects. Similarly weighted objects with a greater frontal area will generally have a lesser terminal velocity because of the increased drag force. A golf ball (0.45 N) will have a terminal velocity of approximately 40 m/s while a much heavier basketball (5.84 N) will have a lesser terminal velocity of around 20 m/s because of the increased frontal area.

FIGURE 10-17 A basic skydiving position is called the stable arch position. Frontal area is large to reduce terminal velocity and torques can be minimized to prevent rotation.

Inertial Force

In many instances in human movement, one segment can exert a force on another segment, causing a movement in that segment that is not due to muscle action. When this occurs, an inertial force has been generated. Generally, a more proximal segment exerts an inertial force on a more distal segment. For example, during the swing phase of running, the ankle is plantarflexed at takeoff and slightly dorsiflexed at touchdown. The ankle is relaxed during the swing phase, and in fact, the muscle activity at this joint is very limited. The leg also swings through, however, and exerts an inertial force on the foot segment, causing the foot to move to the dorsiflexed position. Similarly, the thigh segment exerts an inertial force on the leg.

Muscle Force

When a force was defined, it was noted that a force constituted a push or pull that results in a change in velocity. A muscle can generate only a pulling or tensile force and therefore has only unidirectional capability. The biceps brachii, for example, pulls on its insertion on the forearm to flex the elbow. To extend the elbow, the triceps brachii must pull on its insertion on the forearm. Thus, the movements at any joint must be accomplished by opposing pairs of muscles. Gravity also assists in the motion of segments.

In most biomechanical analyses, it is assumed that a muscle force acting across a joint is a net force. That is, the force of individual muscles acting across a joint cannot be taken into consideration. Generally, a number of muscles act across a joint. Each of these muscles constitutes an unknown value. Mathematically, the number of unknown values must have a comparable number of equations. Because there is no comparable number of equations, there cannot be a solution for each individual muscle force. If a solution were attempted, it would result mathematically in an indeterminate solution, that is, no solution. Thus, we can only calculate the net effect of all the muscles that cross the joint.

It is also assumed that the muscle force acts at a single point. This assumption is again not completely correct because the insertions of muscles are rarely, if ever, single points. Each muscle can be represented as a single force vector that is the resultant of all forces generated by the individual muscle fibers (Fig. 10-18). The force vector can be resolved into its components, one component (F_y) acting to cause a rotation at the joint and the other (F_x) acting toward the joint center. If the angle θ is considered, it can be seen that as θ gets large, such as when the joint is flexing, the rotational component increases while the component acting toward the joint center decreases. This can be assumed because:

$$F_y = F \sin \theta$$

and

$$F_x = F \cos \theta$$

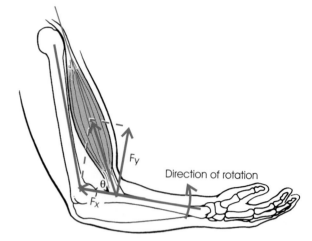

FIGURE 10-18   A muscle force vector, the angle of pull (θ), and its vertical and horizontal components.

At $\theta = 90°$, the rotational component is a maximum because the sin 90° = 1, but the component acting toward the joint center is zero because cos θ = 0.

The actual muscle force in vivo is very difficult to measure. To do so requires either a mathematical model or the placement of a measuring device called a force transducer on the tendon of a muscle. Many researchers have developed mathematical models to approximate individual muscle forces (41,71). To do this, however, it is necessary to make a number of assumptions, including the direction of the muscle force, point of application, and whether cocontraction occurs.

For example, a simple model to determine the peak Achilles tendon force during the support phase of a running stride would use the peak vertical GRF and the point of application of that force. It must be assumed that the line of this force is a specific distance anterior to the ankle joint and the Achilles tendon is located a specific distance posterior to the ankle joint. In addition, the center of mass of the foot and its location relative to the ankle joint center must be determined. In this rather simple example, the number of anatomical assumptions that must be made is evident.

The second technique, placing a force-measuring device on the tendon of a muscle, requires a surgical procedure. Komi and colleagues (45) and Komi (43) have placed force transducers on individuals' Achilles tendons. Komi (43) reported peak Achilles tendon forces corresponding to 12.5 BW while the subject ran at 6 m/s. Although the technique to measure in vivo muscle forces is not well developed for long-term studies, it can be used to analyze several important parameters in muscle mechanics.

Elastic Force

When a force is applied to a material, the material undergoes a change in its length. The algebraic statement that reflects this relationship is:

$$F = k\Delta s$$

where *k* is a constant of proportionality and Δs is the change in length. The constant *k* represents stiffness, or the ability of the material to be compressed or stretched. A stiffer material requires a greater force to compress or stretch it. This relationship is often applied to biological materials and represented in stress–strain relationships. The stress–strain relationship is explained in detailed in Chapter 2.

The effect of elastic force can be visualized in an example of a diver on a springboard. The diver uses BW as the force to deflect the springboard. The deflected springboard stores an elastic force that is returned as the springboard rebounds to its original state. The result is that the diver is flung upward. A considerable amount of work has been conducted to determine the elasticity of diving springboards used in competition (9,73).

In most situations, the biological tissue—muscles, tendons, and ligaments—do not exceed their elastic limit. Within this limit, these tissues can store force when they are stretched, much as a rubber band does. When the loading force is removed, the elastic force may be returned and, with the muscle force, contribute to the total force of the action. For example, using a prestretch before a movement increases the force output by inducing the elastic force potential of the surrounding tissues. There is, however, a time constraint on how long this elastic force can be stored. Attempts to measure the effect of stored elastic force have illustrated that using this force can affect oxygen consumption (3). Further estimates of stored elastic force in vertical jumping have been investigated by Komi and Bosco (44), who reported higher jumps using stored elastic force. It has been suggested by Alexander (1) that elastic force storage is important in the locomotion of humans and many animals, such as kangaroos and ostriches.

Representation of Forces Acting on a System

When one undertakes an analysis of any human movement, one must take into account a number of forces acting on the **system**. To simplify the problem for better understanding, a free body diagram is often used. A **free body diagram** is a stick figure drawing of the system showing the vector representations of the external forces acting on the system. In biomechanics, the system refers to the total human body or parts of the human body and any other objects that may be important in the analysis. It is critically important to define the system correctly; otherwise, extraneous variables may confound the analysis. External forces are those exerted outside the system rather than from inside the system. Thus, internal forces are not represented on a free body diagram.

After the system has been defined, the external forces acting upon the system must be identified and drawn. Figure 10-19 is a free body diagram of a total body sagittal

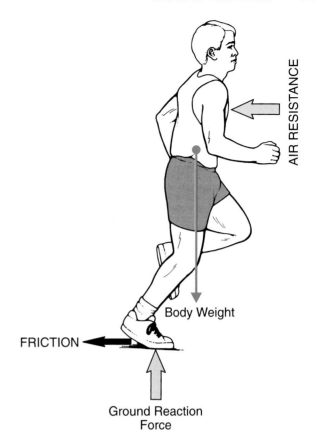

FIGURE 10-19 A free body diagram of a runner with the whole body defined as the system.

view of a runner. The external forces acting on the runner are the GRF, friction, fluid or air resistance, and gravity as reflected in the runner's BW. Vector representations of the external forces are drawn on the stick figure at the approximate point of application. If the runner is carrying an implement, such as a wrist weight, another force vector representing the weight of this implement must be added to the free body diagram (Fig. 10-20). For the most part, however, the four external forces noted earlier are the only ones identified on a total body diagram.

When a specific segment, not the total body, is defined as the system, the interpretation of what constitutes an external force must be clarified. In drawing a free body diagram of a particular segment, the segment must be isolated from the rest of the body. The segment is drawn disconnected from the rest of the body and all external forces acting on that segment are drawn. The muscle forces that cross the proximal or distal joints of that segment are external to the system and must be classified as external forces. As noted earlier, it is not possible to identify all of the muscles and their forces acting across a joint. An idealized net muscle force, that is, a single force vector, is used to represent the sum total of all muscle forces.

Figure 10-21 is a free body diagram of the forearm of an individual doing a biceps curl. The four external forces acting on this system that must be identified are the net biceps muscle force, the joint reaction force, the force

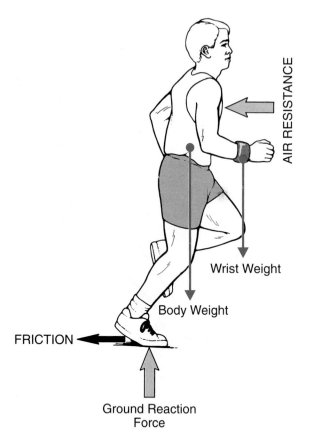

FIGURE 10-20 A free body diagram of a runner's wrist weight system.

of gravity on the arm represented by the weight of the forearm, and the force of gravity acting on the barbell or the weight of the barbell. These are drawn as they would act during this movement. In many instances, the joint reaction force and the net muscle force are not known but must be calculated. All other forces, such as the friction forces in the joints and the forces of the ligaments and tendons, are assumed to be negligible. Air resistance is also ignored.

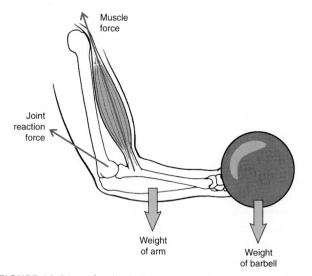

FIGURE 10-21 A free body diagram of the forearm during a biceps curl.

Free body diagrams are extremely useful tools in biomechanics. Drawing the system and identifying the forces acting upon the system define the problem and determine how to undertake the analysis.

ANALYSIS USING NEWTON'S LAWS OF MOTION

Multiple conceptual forms and variations of Newton's laws can describe the relationship between the kinematics and the kinetics of a movement. From Newton's law of acceleration ($F = ma$) arise three general approaches to exploring kinematic and kinetic interactions. These approaches can be categorized as the effect of a force at an instant in time, the effect of a force applied over a period of time, and the effect of a force applied over a distance (54). None of these methods can be considered better or worse than any other. The choice of which relationship to use simply depends upon which method will best answer the question that you are asking. Using the appropriate analytical technique, however, makes it possible to investigate the forces causing motion more effectively.

Effects of a Force at an Instant in Time
When considering the effects of a force and the resulting acceleration at an instant in time, Newton's second law of motion is considered:

$$\Sigma F = ma$$

Two situations based on the magnitude of the resulting acceleration can be defined. In the first situation, the resulting acceleration has a zero value. This is the branch of mechanics known as **statics**. In the second case, the resulting acceleration is a nonzero value. This area of study is known as **dynamics**.

Static Analysis
The static case is devoted to systems at rest or moving at a constant velocity. In both of these situations, the acceleration of the system is zero. When the acceleration of a system is zero, the system is said to be in equilibrium. A system is in equilibrium when, as stated in Newton's first law, it remains at rest or it is in motion at a constant velocity.

In translational motion, when a system is in equilibrium, all forces that are acting on the system cancel each other out, and the effect is zero. That is, the sum of all forces acting on the system must total zero. This is expressed algebraically as:

$$\Sigma F_{system} = 0$$

The forces in this equation can be further expressed in terms of the two-dimensional x- and y-components as:

$$\Sigma F_x = 0$$

and

$$\Sigma F_y = 0$$

Here the sum of the forces in the horizontal (x) direction must equal zero and the sum of the forces in the vertical (y) direction must equal zero.

The static case is simply a particular example of Newton's second law and can be described in terms of a cause-and-effect relationship. The left-hand side of these equations describes the cause of the motion, and the right-hand side describes the product or the result of the motion. Because all forces in the system are in balance, there is no acceleration. If the forces were not in balance, some acceleration would occur.

Figure 10-22 presents a free body diagram of a linear force system in which a 100-N box rests on a table. Gravity acts to pull the box downward on the table with a force of 100 N. Because the box does not move vertically, an equal and opposite force must act to support the box. In this coordinate system, up is positive and down is negative. The weight of the box, acting downward, thus has a negative sign. There are no horizontal forces acting in this example. Thus, the reaction force R_y is:

$$\Sigma F_y = 0$$
$$-100 \text{ N} + R_y = 0$$
$$R_y = 100 \text{ N}$$

R_y is the reaction force equal to the weight of the box. Because the weight of the box acts negatively, the reaction force must act positively or in the opposite direction to the weight of the box.

Consider a system with multiple forces acting. In Figure 10-23, a tug-of-war is presented as a linear force system. In this example, the two contestants on the right balance the three contestants on the left. The contestants on the left exert forces of 50, 150, and 300 N, respectively. These forces can be considered to act in a negative horizontal direction. Assuming a static situation, the reaction force (R_y) to produce equilibrium can be calculated. Thus:

$$\Sigma F_x = 0$$
$$-50 \text{ N} - 150 \text{ N} - 300 \text{ N} + R_x = 0$$
$$R_x = 50 \text{ N} + 150 \text{ N} + 300 \text{ N}$$
$$R_x = 500 \text{ N}$$

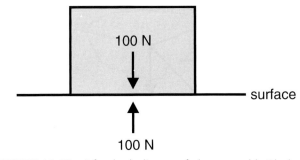

FIGURE 10-22 A free body diagram of a box on a table. The box is in equilibrium because there are no horizontal forces and the sum of the vertical forces is zero.

The two contestants on the right must exert a reaction force of 500 N in the positive direction to produce a state of equilibrium.

The linear force system previously presented is a relatively simple example of the static case, but in many instances in human motion, the forces are nonparallel. In Figure 10-24A, two nonparallel forces F_1 and F_2 act on a rigid body in addition to the weight of the rigid body. For this system to be in equilibrium, a third force (F_3) must act through the intersection of the two nonparallel forces. The free body diagram in Figure 10-24B illustrates that the horizontal component of F_3, acting in a positive direction, must counterbalance the sum of the horizontal components of the nonparallel forces F_1 and F_2. Also, the vertical components F_1 and F_2 must be counterbalanced by the weight of the rigid body and by the vertical component of F_3. If $F_1 = 100$ N, the components of F_1 are:

$$F_{1x} = F_1 \cos 30°$$
$$= F_1 \cos 150° \text{ (relative to right horizontal)}$$
$$= 100 \text{ N} \times \cos 150°$$
$$= -86.6 \text{ N}$$

$$F_{1y} = F_1 \sin 30°$$
$$= 100 \text{ N} \times \sin 150° \text{ (relative to right horizontal)}$$
$$= 100 \text{ N} \times \sin 150°$$
$$= 50.0 \text{ N}$$

FIGURE 10-23 Tug-of-war. The system is in equilibrium because the sum of the forces in the horizontal direction is zero. No movement to the left or the right can occur.

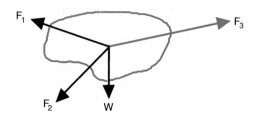

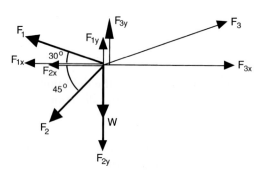

FIGURE 10-24   **(A)** A force system in which the sum of the forces is zero. **(B)** A free body diagram of the force system shows the horizontal and vertical components of all forces.

and if $F_2 = 212.13$ N, the components of F_2 are:

$F_{2x} = F_2 \cos 45°$
    $= 212.13$ N $\times \cos 225°$ (relative to right horizontal)
    $= -150$ N

$F_{2y} = F_2 \sin 45°$
    $= 212.13$ N $\times \sin 225°$ (relative to right horizontal)
    $= -150$ N

The weight of the rigid body, 50 N, also acts in a negative vertical direction. Thus:

$$\Sigma F_y = 0$$
$$F_{3y} + F_{1y} + F_{2y} + W = 0$$
$$F_{3y} + 50 \text{ N} - 150 \text{ N} - 50 \text{ N} = 0$$
$$F_{3y} = -50 \text{ N} + 150 \text{ N} + 50 \text{ N}$$
$$F_{3y} = 150 \text{ N}$$

F_{3y} must have a magnitude of 150 N to maintain the system in equilibrium in the vertical direction. In the horizontal direction:

$$\Sigma F_x = 0$$
$$F_{3x} + F_{1x} + F_{2x} = 0$$
$$F_{3x} - 86.6 \text{ N} - 150 \text{ N} = 0$$
$$F_{3x} = 86.6 \text{ N} + 150 \text{ N}$$
$$F_{3x} = 236.6 \text{ N}$$

To balance the two nonparallel forces in the horizontal direction, a force of 236.6 N is required. The resultant force, F_3, can be determined using the Pythagorean relationship:

$$F_3 = \sqrt{F_{3x}^2 + F_{3y}^2}$$
$$F_3 = \sqrt{236.6^2 + 150^2}$$
$$F_3 = 280 \text{ N}$$

The F_3 force orientation can be determined using trigonometric functions:

$$\theta_{F3} = \arctan(F_y/F_x)$$
$$= \arctan(150/236.6)$$
$$= \arctan(0.6340)$$
$$= 32.37°$$

The forces F_1 and F_2 and the weight of the rigid body are counteracted by the force F_3, thus keeping the system in equilibrium.

A second condition that determines whether a system is in equilibrium occurs when the forces in the system are not concurrent. **Concurrent forces** do not coincide at the same point, so they cause rotation about some axis. These rotations all sum to zero, however. Because this is a static case, no rotation occurs. This will be discussed in more detail in Chapter 11.

Static models have been developed to evaluate such tasks such as material handling and lifting. A free body diagram of the joint reaction forces and forces acting at the center of mass of the segment is created. Figure 10-25 is a static lifting model (18) that shows the linear forces acting on the body at the shoulder, elbow, wrist, hip, knee, ankle joints, and the ground contact. This model is not complete until the angular components are also included (see Chapter 11).

Dynamic Analysis

A static analysis may be used to evaluate the forces on the human body when acceleration is insignificant (4). When accelerations are significant, however, **dynamic analysis** must be undertaken. A dynamic analysis should be used, therefore, when accelerations are not zero. The equations for a dynamic analysis were derived from Newton's second law of motion and expanded by the famous Swiss mathematician Leonhard Euler (1707 to 1783). The equations of motion for a two-dimensional case are based on:

$$\Sigma F = ma$$

Linear acceleration may be broken down into horizontal (x) and vertical (y) components. As in the static two-dimensional analysis, independent equations are used in a dynamic two-dimensional linear kinetic analysis:

$$\Sigma F_x = ma_x$$
$$\Sigma F_y = ma_y$$

where x and y represent the horizontal and vertical coordinate directions, respectively, a is the acceleration of the center of mass, and m is the mass. The forces acting on a body may be any one of the previously discussed forces such as muscular, gravitational, contact, or inertial. The gravitational forces are the weights of each of the segments.

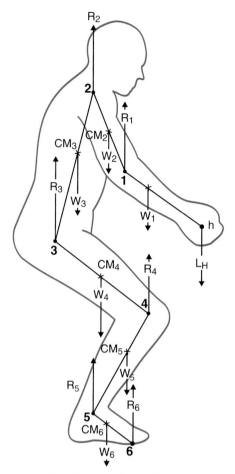

FIGURE 10-25 A free body diagram of a sagittal view static lifting model showing the linear forces at the joints and segments. (Adapted from Chaffin, D. B., Andersson, G. B. J. [1991]. *Occupational Biomechanics*, 2nd ed. New York: Wiley.)

The contact forces can be reactions—forces with another segment, the ground, or an external object—and the inertial forces are ma_x and ma_y. Using the equations of dynamic motion, the forces acting on a segment can be calculated.

In moving from static to dynamic analysis, the problem becomes more complicated. In the static case, no accelerations were present. In the dynamic case, linear accelerations and the inertial properties of the body segments resisting these accelerations must be considered. In addition, there is a substantial increase in the work done to collect the data necessary to conduct a dynamic analysis. Because the forces that cause the motion are determined by evaluating the resulting motion itself, a technique called an **inverse dynamics** approach will be used. This method is often referred to as a Newton–Euler inverse dynamics approach. This approach calculates the forces based on the accelerations of the object instead of measuring the forces directly.

In using the inverse dynamics approach, the system under consideration must be determined. The system is usually defined as a series of segments. The analysis on a series of segments is generally conducted beginning with the most distal segment, proceeding proximally up to the

next segment, and so on. Several assumptions must be made when using this approach. The body is considered to be a rigid linked system with frictionless pin joints. Each link, or segment, has a fixed mass and a center of mass at a fixed point. Finally, the moment of inertia about any axis of each segment remains constant. Moment of inertia is discussed in Chapter 11.

As stated previously, the dynamic case is more complicated than the static case. As a result, only a limited example of a single segment will be presented. In Figure 10-26 a free body diagram of the foot of an individual during the swing phase of the gait cycle is presented for the linear forces acting on the segment.

During the swing phase of gait, no external forces other than gravity act on the foot. It can be seen that the only linear forces acting on the foot are the horizontal and vertical components of the joint reaction force and the weight of the foot acting through the center of mass. The joint reaction force components can be computed with the two-dimensional linear kinetic equations defining the dynamic analysis. First, the horizontal joint reaction force may be defined:

$$\Sigma F_x = ma_x$$

Because there are no horizontal forces other than the horizontal joint reaction force, this equation becomes:

$$R_x = ma_x$$

If the mass of the foot is 1.16 kg and the horizontal acceleration of the center of mass of the foot is -1.35 m/s^2, the horizontal reaction force is:

$$R_x = 1.16 \text{ kg} \times -1.35 \text{ m/s}^2$$
$$R_x = -1.57 \text{ N}$$

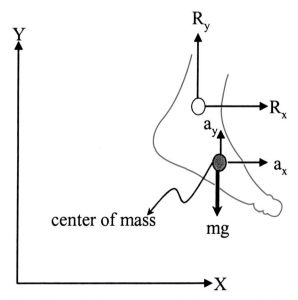

FIGURE 10-26 Free body diagram of the foot segment during the swing phase of a walking stride showing linear forces and accelerations.

Next, the vertical joint reaction force component may be defined from:

$$\Sigma F_y = ma_y$$

There is, however, a vertical force other than the vertical joint reaction force. This force is the weight of the foot itself, so the vertical forces are described as:

$$R_y - mg = ma_y$$

and solving for R_y, the equation becomes:

$$R_y = ma_y - mg$$

If the vertical acceleration of the center of mass of the foot is 7.65 m/s², then:

$$R_y = (1.16 \text{ kg} \times 7.65 \text{ m/s}^2) + (1.16 \text{ kg} \times 9.81 \text{ m/s}^2)$$
$$R_y = 20.3 \text{ N}$$

Refer to the walking data in Appendix C. Identify the maximum vertical peak force. Using the body mass of the participant (50 kg), compute the vertical acceleration at that point. Repeat for the maximum anteroposterior and mediolateral forces.

Effects of Force Applied over a Period of Time

For motion to occur, forces must be applied over time. Manipulating the equation describing Newton's second law of motion allows generation of an important physical relationship in human movement that describes the concept of forces acting over time. This relationship relates the momentum of an object to the force and the time over which the force acts. This relationship is derived from Newton's second law:

$$F = m \times a$$

Because $a = dv/dt$, this equation can be rewritten as:

$$F = m \times \frac{dv}{dt}$$

and further:

$$F = d\left(\frac{m \times v}{dt}\right)$$

If each side is multiplied by dt to remove the fraction on the right-hand side of the equation, the resulting equation is:

$$F \times dt = d(m \times v)$$

or

$$F \times dt = mv_{final} - mv_{initial}$$

The quantity mv (mass × velocity) refers to the momentum of the object. The right-hand side of this equation then refers to the change in momentum. The left-hand side of this equation, the product of $F \times dt$, is a quantity known as an **impulse** and has units of newton-seconds (N · s). Impulse is the measure that is required to change the momentum

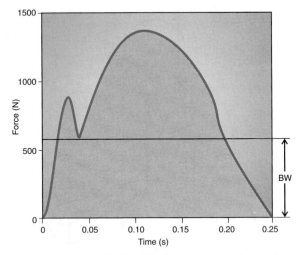

FIGURE 10-27 Vertical GRF of a subject running at 5 m/s. The vertical impulse is the shaded area under the force–time curve.

of an object. The derived equation describes the **impulse–momentum relationship**.

Figure 10-27 illustrates the vertical component of the GRF of a single footfall of a runner. This figure represents a force applied over time as the foot is in contact with the surface, generating a downward force and receiving an equal and opposite reaction force. Impulse may be expressed graphically as the area under a force–time curve.

Consider an individual with a mass of 65 kg jumping from a squat position into the air. The velocity of the person at the beginning of the jump is zero. Video analysis reveals that the velocity of the center of mass at takeoff was 3.4 m/s. The impulse can then be calculated as:

$$F \times dt = mv_{final} - mv_{initial}$$
$$F \times dt = (65 \text{ kg} \times 3.4 \text{ m/s}) - (65 \text{ kg} \times 0 \text{ m/s})$$
$$F \times dt = 221 \text{ kg-m/s}$$

If we assume that the force application took place over 0.2 seconds, the average force applied would be:

$$F \times 0.02 \text{ seconds} = 221 \text{ kg-m/s}$$
$$F = \frac{221 \text{ kg-m/s}}{0.2 \text{ seconds}}$$
$$F = 1,105 \text{ N}$$

The nature of the force applied and the time over which it is applied determine how the momentum of the object is changed. To change the momentum of an object, a force of a large magnitude can be applied over a short period of time, or a force of a smaller magnitude over a long period of time. Of course, the tactic used depends on the situation. For example, in landing from a jump, the performer must change momentum from some initial value to zero. The initial momentum value at impact is a function of body mass times the velocity created by the force of gravity that causes the jumper to accelerate toward the ground at a rate of 9.81 m/s². At floor impact, an impulse is generated to change the momentum and

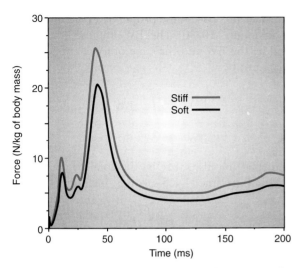

FIGURE 10-28 Vertical GRF for soft and stiff landings. Stiff landings have a 23% greater linear impulse than the soft landing. (Adapted from DeVita, P., Skelly, W. A. [1992]. Effect of landing stiffness on joint kinetics and energetics in the lower extremity. *Medicine & Science in Sports & Exercise*, 24:108–115.)

reduce it to zero as the jumper stops. If an individual lands with little knee flexion (locked knees), the impact force occurs over a very short period. If the individual lands and flexes the joints of the lower extremity, however, the impact force is smaller and occurs over an extended time. If in both cases, the jumper is landing with the same velocity and consequently the same momentum, the resulting impulses will also be the same, even though there are different force (large versus smaller force) and time (short versus extended) components.

An example of soft and rigid landings is presented in Figure 10-28. This example, taken from a study by DeVita and Skelly (23), shows a large peak force in the rigid landing compared with the smaller force in the soft landing. Because both appear to take place over about the same amount of time, there is a greater impulse in the rigid landings.

> Using the data in Appendix C, graph the vertical, anteroposterior, and mediolateral ground reaction curves for the support phase of walking and shade in the impulse for each curve.

One interesting research application for the use of the impulse–momentum relationship has been in vertical jumping. This area of research has used the force platform to determine the GRF during the vertical jump. Remember that the GRF reflects the force acting on the center of mass of the individual. Thus, researchers have used the impulse–momentum relationship on data collected from the force platform to determine the parameters necessary to investigate the height of the center of mass above its starting point during vertical jumping (44). Figure 10-29A

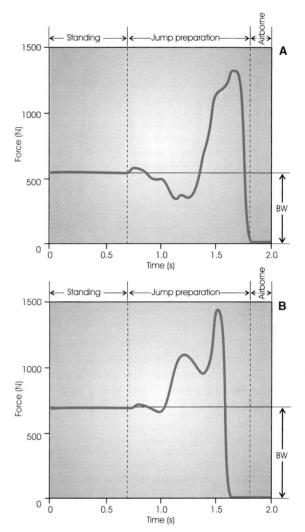

FIGURE 10-29 The vertical GRF of two types of vertical jumps: the countermovement jump (**A**) and the squat jump (**B**).

shows the vertical GRF profile of an individual starting at rest on the platform and jumping into the air. This is called a countermovement jump because the subject flexes at the knees and then swings the arms up as the knees are extended during the jump. Figure 10-29B illustrates a squat jump, in which the subject begins in a squat (knees flexed) and simply forcefully extends the knees to jump into the air.

In both cases represented in Figure 10-29, the impulse–momentum relationship can be used to calculate the peak height of the center of mass above the initial height during the jump. Consider the vertical GRF curve of a counter-jump in Figure 10-30. The constant vertical force in the initial portion of the curve is the individual's BW. If the BW line is extended to the instant of takeoff, the area beneath the curve describes the BW impulse (BW_{imp}). It may be calculated as an integral (i.e., determining the area under the curve):

$$BW_{imp} = \int_{ti}^{tf} BW \, dt$$

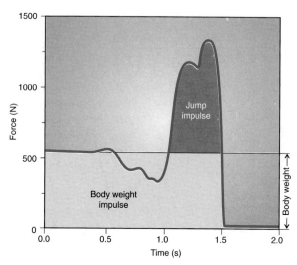

FIGURE 10-30 The vertical GRF of a countermovement jump illustrating the body weight impulse and the jump impulse.

where t_i to t_f represents the time interval when the subject is standing still on the force platform until the instant of takeoff. The total area under the force–time curve until the subject leaves the force platform may be designated as $total_{imp}$ and can be calculated as an integral:

$$total_{imp} = \int_{t_i}^{t_{jump}} F_x \, dt$$

where t_i to t_{jump} represents the time when the subject is on the platform before the jump. The impulse that propelled the subject into the air can then be determined by:

$$jump_{imp} = total_{imp} - BW_{imp}$$

The impulse–momentum relationship can then be formulated using the impulse that the subject generated to perform the jump. Thus:

$$jump_{imp} = m(v_f - v_i)$$

where v_i is the initial velocity of the center of mass and v_f is the takeoff velocity of the center of mass. Because $v_i = 0$:

$$jump_{imp} = m(v_f - 0)$$

By substituting the impulse of the jump ($jump_{imp}$) and the subject's body mass (m) in this equation, the velocity of the center of mass at takeoff for the vertical jump can be calculated. The height of the center of mass during the jump is then calculated based on the projectile equations elaborated upon in Chapter 8. Thus:

$$height_{cm} = \frac{v_{cm}^2}{2g}$$

This calculation has proved to be in good agreement with values calculated from high-speed film data. Komi and Bosco (44) reported an error of 2% from the computation on the force platform.

Consider the force–time profile of a countermovement jump in Figure 10-30. The BW impulse, the rectangle formed by the BW and the time of force application, was calculated to be:

$$BW_{imp} = \int_{t_0}^{t_{1.5}} BW \, dt$$

$$BW_{imp} = 620.80 \, \text{N} \cdot \text{s}$$

The total impulse, that is, the total force application from time 0 to time = 1.5 seconds, including the BW, is:

$$total_{imp} = \int_{t_i}^{t_f} F_x \, dt$$

$$total_{imp} = 772.81 \, \text{N} \cdot \text{s}$$

The jump impulse, therefore, is:

$$jump_{imp} = total_{imp} - BW_{imp}$$
$$jump_{imp} = 772.81 \, \text{N} \cdot \text{s} - 620.80 \, \text{N} \cdot \text{s}$$
$$jump_{imp} = 152.01 \, \text{N} \cdot \text{s}$$

Substituting this value in the impulse–momentum relationship, it is possible to solve for the velocity of the center of mass at takeoff. Thus, with the body mass of the jumper of 56.2 kg, the velocity of takeoff is:

$$jump_{imp} = m(v_f - v_i)$$
$$152.01 \, \text{N} \cdot \text{s} = 56.2 \, \text{kg}(v_f - 0)$$
$$v_f = \frac{152.01 \, \text{N} \cdot \text{s}}{56.2 \, \text{kg}}$$
$$v_f = 2.70 \, \text{m/s}$$

The height of the center of mass during the jump can be calculated by:

$$Height_{cm} = \frac{v_{cm}^2}{2g}$$

$$Height_{cm} = \frac{(2.70 \, \text{m/s})^2}{2 \times 9.81 \, \text{m/s}^2}$$

$$Height_{cm} = 0.373 \, \text{m}$$

Therefore, in this particular jump, the center of mass was elevated 37.3 cm above the initial height of the center of mass.

This calculation technique was used by Dowling and Vamos (25) to identify the kinetic and temporal factors related to vertical jump performance. They found a large variation in the patterns of force application between the subjects that made it difficult to identify the characteristics of a good performance. Interestingly, they reported that a high maximum force was necessary but not sufficient for a good performance. They concluded that the pattern of force application was the most important factor in vertical jump performance.

Refer to the walking data in Appendix C. Calculate the vertical, anteroposterior, and mediolateral impulse from contact (frame 0) to the first vertical peak (frame 18). Using the impulse values, calculate the velocities generated up to that point.

Effect of a Force Applied over a Distance

The term **work** is used to mean a variety of things. Generally, we consider work to be anything that demands mental or physical effort. In mechanics, however, work has a more specific and narrow meaning. Mechanical work is equal to the product of the magnitude of a force applied against an object and the distance the object moves in the direction of the force while the force is applied to the object. For example, in moving an object along the ground, an individual pushes the object with a force parallel to the ground. If the force necessary to move the object is 100 N and the object is moved 1 m, the work done is 100 N-m. The case cited, however, is a very specific one. More generally, work is:

$$W = F \times \cos \theta \times s$$

where F is the force applied, s is the displacement, and θ is the angle between the force vector and the line of displacement. The unit of mechanical work is derived from the product of force in Newtons and displacement in meters. The most commonly used units are the Newton-meter and the joule (J). These are equivalent units:

$$1 \text{ N-m} = 1 \text{ J}$$

In Figure 10-31A, the force is applied to a block parallel to the line of displacement, that is, at an angle of 0° to the displacement. Because $\cos 0° = 1$, the work

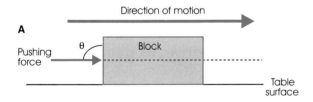

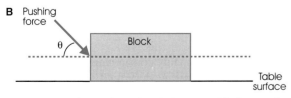

FIGURE 10-31 The mechanical work done on a block. **(A)** A force is applied parallel to the surface ($\theta = 0°$; thus $\cos \theta = 1$). **(B)** A force is applied at an angle to the direction of motion ($\theta = 30°$; thus, $\cos \theta = 0.866$).

done is simply the product of the force and the distance the block is displaced. Thus, if the force applied is 50 N and the block is displaced 0.1 m, the mechanical work done is:

$$\begin{aligned} W &= 50 \text{ N} \times \cos 0° \times 0.1 \text{ m} \\ &= 50 \text{ N} \times 1 \times 0.1 \text{ m} \\ &= 5 \text{ N-m} \end{aligned}$$

If the same force is applied at an angle of 30° over the same distance, d (Fig. 10-31B), the work done is:

$$\begin{aligned} W &= 50 \text{ N} \times \cos 30° \times 0.1 \text{ m} \\ W &= 50 \text{ N} \times 0.866 \times 0.1 \text{ m} \\ &= 4.33 \text{ N-m} \end{aligned}$$

Therefore, more work is done if the force is applied parallel to the direction of motion than if the force is applied at an angle.

As this discussion of work implies, work is done only when the object is moving and its motion is influenced by the applied force. If a force acts on an object and does not cause the object to move, no mechanical work is done because the distance moved is zero. During an isometric contraction, for example, no mechanical work is done because there is no movement. A weight lifter holding an 892-N (200-lb) barbell overhead does no mechanical work. In lifting the barbell overhead, however, mechanical work is done. If the barbell is lifted 1.85 m, the work done is:

$$\begin{aligned} W &= 892 \text{ N} \times 1.85 \text{ m} \\ &= 1650.2 \text{ J} \end{aligned}$$

This value assumes that the bar was lifted completely vertically.

Power

In evaluating the amount of work done by a force, the time over which the force is applied is not taken into account. For example, when the work done by the weight lifter to raise the barbell overhead was calculated, the time it took to raise the barbell was not taken into consideration. Regardless of how long it took to raise the barbell, the amount of work done was 1650.2 J. The concept of **power** takes into consideration the work done per unit of time. Power is defined as the rate at which a force does work:

$$P = \frac{\mathrm{d}w}{\mathrm{d}t}$$

where W is the work done and $\mathrm{d}t$ is the time period in which the work was done. Power has units of watts (W). The change in work is expressed in joules and the change in time in seconds. Thus:

$$1 \text{ W} = 1 \text{ J/s}$$

If power is plotted on a graph as a function of time, the area under the curve equals the work done.

If the weight lifter in the previous example raised the bar in 0.5 seconds, the power developed is:

$$P = 1,650.2 \text{ J}/0.5 \text{ seconds}$$
$$= 3,300.4 \text{ J/s}$$
$$= 3,300.4 \text{ W}$$

Decreasing the time over which the bar is lifted to 0.35 seconds increases the power developed by the weight lifter to 4,714.86 W. Although the work done remains constant, greater power must be developed to do the mechanical work more quickly.

Another definition for power can be developed by rearranging the formula. If the product of the force (F) and the distance over which it was applied (s) is substituted for the mechanical work done, the equation becomes:

$$P = \frac{d(F \times s)}{dt}$$

and rearranging this equation:

$$P = F \times \frac{ds}{dt}$$

Because ds/dt was defined in a previous chapter as the velocity in the s-direction, it can be readily seen that:

$$P = F \times v$$

where F is the applied force and v is the velocity of the force application.

Power is often confused with force, work, energy, or strength. Power, however, is a combination of force and velocity. In many athletic endeavors, power, or the ability to use the combination of force and velocity, is paramount. One such activity, weight lifting, has already been discussed, but there are many others, such as shot putting, batting in baseball, and boxing. Jumping also requires power. To generate a takeoff velocity of 2.61 m/s in a vertical jump, Harman and colleagues (40) reported a peak power generation of 3,896 W. In comparisons of jump techniques, researchers have noted differences in peak power output between countermovement jumps (men = 4,708 W; women = 3,069 W) and squat jumps (men = 4,620 W; women = 2,993 W) (68).

Energy

As with work, the mechanical term **energy** is often misused. Simply stated, energy is the capacity to do work. The many types of energy include light, heat, nuclear, electrical, and mechanical; the main concern of biomechanics is mechanical energy. The unit of mechanical energy in the metric system is the joule. The two forms of mechanical energy that will be discussed here are kinetic and potential energies.

Kinetic energy (KE) refers to the energy resulting from motion. An object possesses KE when it is in motion, that is, when it has some velocity. Linear KE is expressed algebraically as:

$$KE = \frac{1}{2} mv^2$$

where m is the mass of the object and v is the velocity. Because this expression includes the square of the velocity, any change in velocity greatly increases the amount of energy in the object. If the velocity is zero, the object has no KE. An approximate value for the KE of a 625 N runner would be 3,600 J; a swimmer of comparable BW would have a value of 125 J.

A moving body must have some energy because a force must be exerted to stop it. To start an object moving, a force must be applied over a distance. KE, therefore, is the ability of a moving object to do work resulting from its motion. The generation of a sufficient level of KE is especially important when projecting an object or body, such as in long jumping, throwing, and batting. For example, KE is developed in a baseball over the collision phase with the bat and projects the ball at velocities more than 100 mph. The KE before the collision has been demonstrated to be in the range of 320 and 115 J for the bat and ball, respectively (30). After the contact, the KE of the bat is reported to be reduced to 156 J, and the ball's KE increased to 157 J (28). With bat speeds in the range of 55 to 80 mph and incoming ball speeds of 85 to 100 mph, there is considerable interchange of KE.

Potential energy (PE) is the capacity to do work because of position or form. An object may contain stored energy, for example, simply because of its height or its deformation. In the first case, if a 30-kg barbell is lifted overhead to a height of 2.2 m, 647.5 J of work is done to lift the barbell. That is:

$$W = F \times s$$
$$= (30 \text{ kg} \times 9.81 \text{ m/s}^2) \times 2.2 \text{ m}$$
$$= 647.5 \text{ J}$$

As the barbell is held overhead, it has the PE of 647.5 J. The work done to lift it overhead is also the PE. PE gradually increases as the bar is lifted. If the bar is lowered, the PE decreases. PE is defined algebraically as:

$$PE = mgh$$

where m is the mass of the object, g is the acceleration due to gravity, and h is the height. The more work done to overcome gravity, therefore, the greater the PE.

An object that is deformed may also store PE. This type of PE has to do with elastic forces. When an object is deformed, the resistance to the deformation increases as the object is stretched. Thus, the force that deforms the object is stored and may be released as elastic energy. This type of energy, **strain energy** (SE), is defined as:

$$SE = \frac{1}{2} k \times \Delta x^2$$

where k is a proportionality constant and Δx is the distance over which the object is deformed. The proportionality constant, k, depends on the material deformed. It is often called the stiffness constant because it represents the object's ability to store energy.

It has already been discussed how certain tissues, such as muscles and tendons, and certain devices, such as springboards for diving, may store this SE and release it to aid in human movement. In athletics, numerous pieces of equipment achieve such an end. Examples are trampolines, bows in archery, and poles in pole vaulting. Perhaps the most sophisticated use of elastic energy storage is the design of the tuned running track at Harvard University. McMahon and Greene (52) analyzed the mechanics of running and the energy interactions between the runner and the track to develop an optimum design for the track surface. In the first season on this new track, an average speed advantage of nearly 3% was observed. Furthermore, it was determined that there was a 93% probability that any given individual will run faster on this new track (51).

Many instances in human motion can be understood in terms of the interchanges between KE and PE. The mathematical relationship between the different forms of energy was formulated by the German scientist von Helmholtz (1821 to 1894). In 1847, he defined what has come to be known as the **law of conservation of energy**. The main point of this law is that energy cannot be created or destroyed. No machine, including the human machine, can generate more energy than it takes in. It follows, therefore, that the total energy of a closed system is constant because energy does not enter or leave a closed system. A closed system is one that is physically isolated from its surroundings. This point can be phrased mathematically by stating:

$$TE = KE + PE$$

where TE is a constant representing the total energy of the system. In human movement, this occurs only when the object is a projectile, whereby the only external force acting upon it is gravity, because fluid resistance is neglected.

Consider the example of a projectile traveling up into the air. At the point of release, the projectile has zero PE and a large amount of KE. As the projectile ascends, the PE increases and the KE decreases because gravity is slowing the upward flight of the projectile. At the peak of the trajectory, the velocity of the projectile is zero and the KE is zero, but the PE is at its maximum. The total energy of the system does not change because increases in PE result in equal decreases in KE. On the downward flight, the reverse change in the forms of energy occurs. These changes in energy are illustrated in Figure 10-32.

When an object is moved, mechanical work is said to have been done on the object. Thus, if no movement occurs, no mechanical work is done. Because energy is the capacity to do work, intuitively, there should be some relationship between work and energy. This very useful relationship is called the **work–energy theorem**. This theorem states that the work done is equal to the change in energy:

$$W = \Delta E$$

where W is the work done and ΔE is the change in energy. That is, for mechanical work to be done, a change in the energy level must occur. The change in energy refers to all types of energy in the system, including kinetic, potential,

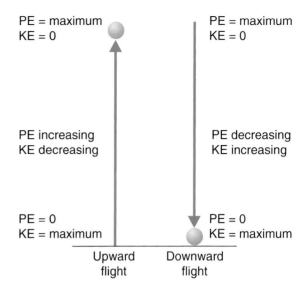

FIGURE 10-32 Changes in potential energy (PE) and kinetic energy (KE) as a ball is projected straight up and as it falls back to earth.

chemical, heat, and light. For example, if a trampolinist weighs 780 N and is at a peak height of 2 m above a trampoline bed, the PE based on the height above the trampoline bed is:

$$PE = 780 \text{ N} \times 2 \text{ m}$$
$$PE = 1,560 \text{ J}$$

Assuming no horizontal movement, at impact the KE is 1,560 J, while the PE is zero. The KE has this initial value, and as the trampoline bed deforms, the potential SE increases while the KE goes to zero. The work done on the trampoline bed is:

$$W = \Delta KE$$
$$W = 1,560 \text{ J} - 0 \text{ J}$$
$$W = 1,560 \text{ J}$$

This value, 1,560 J, is also the value of the potential SE of the trampoline bed. As the bed reforms, the potential SE changes from this value to zero and constitutes the work done by the trampoline bed on the trampolinist.

To evaluate the work done, the energy level of a system must be evaluated at different instants in time. This change in energy represents the work done on the system. For example, a system has an energy level of 26.3 J at position 2 and 13.1 J at position 1. The work done is:

$$W = \Delta E$$
$$W = 26.3 \text{ J} - 13.1 \text{ J}$$
$$W = 13.2 \text{ J}$$

Because this work has a positive value, it is considered positive work. On the other hand, if the energy level is 22.4 J at position 1 and 14.5 J at position 2, then:

$$W = \Delta E$$
$$W = 14.5 \text{ J} - 22.4 \text{ J}$$
$$W = 27.9 \text{ J}$$

Now the work done is negative and work is said to have been done on the system.

The work–energy relationship is useful in biomechanics to analyze human motion. Researchers have used this analytical method to determine the work done during a number of movements; however, the greatest use of this technique has been in the area of locomotion. The calculation of the total work done resulting from the motion of all of the body's segments is called **internal work**. This calculation has been used by many researchers, particularly those studying locomotion (60,64,85,87,90). For a single segment, the linear work done on the segment is:

$$W_s = \Delta KE + \Delta PE$$

$$W_s = \Delta\left(\frac{1}{2}\,mv^2\right) + \Delta(mgh)$$

where W_s is the work done on the segment, ΔKE is the change in linear KE of the center of mass of the segment, and ΔPE is the change in PE of the segment center of mass. Internal work of the total body can then be calculated as the sum of the work done on all segments:

$$W_b = \sum_{i-1}^{n} W_{si}$$

where W_b is the total body work and W_{si} is the work done on the ith segment. One major limitation of the calculation of internal work is that it does not account for all of the energy of a segment and so does not account for the energy of the total body. For example, the SE due to the deformation of tissue and the angular KE are not considered. Angular work is discussed in Chapter 11.

Internal work for the duration of an activity can be derived by summing the changes in the segment energies over time. That is, the change in energy at each instant in time is summed for the length of time that the movement lasts. Usually, this period in locomotion studies is the time for one stride. The analysis of the mechanical work done is extremely valuable as a global parameter of the body's behavior without a detailed knowledge of the motion. A variety of algorithms may be used to calculate internal work (15,60,64,85,87,90). These models have incorporated factors to quantify such parameters as positive and negative work, the effect of muscle elastic energy, and the amount of negative work attributable to muscular sources. The major difference among these algorithms is the way energy is transferred within a segment and between segments. Energy transfer within a segment refers to changes from one form of energy to another, as in the change from potential to kinetic. Energy transfer between segments refers to the exchange of the total energy of a segment from one segment to another. Presently, no consensus exists as to which model is most appropriate; it has been argued, in fact, that none of these methods is correct (13). Values of mechanical work for a single running stride may range from 532 to 1,775 W (86), although these values should be interpreted with caution.

If it is considered that all energy is transferred between segments, the point in the stride at which the transfer occurs can be illustrated. Figure 10-33 graphically illustrates the magnitude of between-segment energy transfer during the running stride. The magnitude of energy transfer decreases during support and increases from midstance to a maximum after toe-off (86).

When work is calculated over time, such as a locomotor stride, the result is often presented as power with units of watts. These power values have generally been scaled to body mass, resulting in units of watts per kilogram of body mass. Hintermeister and Hamill (42) investigated the relationship between mechanical power and energy expenditure. They reported that mechanical power significantly influenced energy expenditure independent of

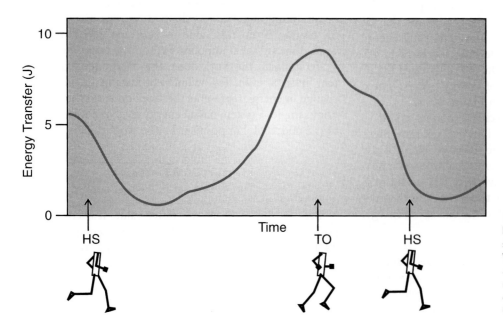

FIGURE 10-33 The magnitude of between-segment transfer during a single running stride. (Adapted from Williams, K. R. [1980]. A biomechanical and physiological evaluation of running efficiency. Unpublished doctoral dissertation, The Pennsylvania State University.)

the running speed. Using several algorithms representing different methods of energy transfer, mechanical power explained at best only 56% of the variance in energy expenditure. Note that these algorithms are often questioned as mechanically incorrect (13).

Two possible explanations have been offered for this rather weak relationship between mechanical work and energy expenditure. The first is that the methods of calculation are incomplete (13). The second involves the law of conservation of energy. This law states that energy is neither created nor destroyed but may be changed from one form to another, that is, PE may be changed to KE or heat energy. Not all of the energy can be or is used to perform mechanical work. In fact, most of the available energy of the muscle will go to maintain the metabolism of the muscle; only about 25% of the energy is used for mechanical work (74). Thus, the work–energy theorem, when used to determine the mechanical work of the body, does not account for all of the energy in the system.

External work may be defined as the work done by a body on an object. For example, the work done by the body to elevate the total body center of mass while walking up an incline is considered to be external work. External work is often calculated during an inclined treadmill walk or run as:

$$\text{External work} = \text{BW} \times \text{Treadmill speed} \times \text{Percent grade} \times \text{Duration}$$

where BW is the BW of the subject, percent grade is the incline of the treadmill, and duration is the length of time of the walk or run. The product of the treadmill speed, percent grade, and the duration is the total vertical distance traveled. If the percent grade is zero, that is, the walking surface is level, the vertical distance traveled is zero. Thus, walking on a level surface results in no external work being done.

⚙️ Special Force Applications

CENTRIPETAL FORCE

In Chapter 9, linear and angular kinematics were shown to be related using the situations in which an object moves along a curved path. It was demonstrated that centripetal acceleration acts toward the center of rotation when an object moves along a curved path. This is radial acceleration toward the center of the circle. The radial force occurring along a curved path that generates the acceleration is called the **centripetal force**. Using Newton's second law of motion, $F = ma$, a formula for the centripetal force can be generated. The force is not different from other forces and is generated by a push or pull. The force is called centripetal because of the effect: The force generates a change in the direction of the velocity. The magnitude of the centripetal or center-seeking force is calculated by:

$$F_C = m\omega^2 r$$

where F_C is the centripetal force, m is the mass of the object, ω is the angular velocity, and r is the radius of rotation. Centripetal force may also be defined as:

$$F_C = \frac{mv^2}{r}$$

where v is the tangential velocity of the segment.

Newton's third law states that for every action there is an equal and opposite reaction. For example, a runner moving along the curve of the running track applies a **shear force** to the ground, resulting in a shear GRF equal and opposite to the applied force. The shear reaction force constitutes the centripetal force. Figure 10-34 is a free body diagram of the runner moving along the curved path, showing the centripetal force, the vertical reaction force, and the resultant of these two force components. This centripetal force at the runner's foot tends to rotate the runner outward. To counteract this outward rotation, the runner leans toward the center of the curve. Hamill and colleagues (39) reported that this shear GRF increased as the radius of rotation decreased.

The resultant of the vertical reaction force and the centripetal force must pass through the center of mass of the runner. If the centripetal force increases, the runner leans more toward the center of rotation, and the resultant vector becomes less vertical. As mentioned in Chapter 9, banked curves on tracks reduce the shear force applied

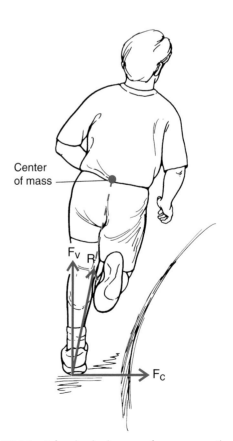

FIGURE 10-34 A free body diagram of a runner on the curve of a running track. F_C is the centripetal force, F_V is the vertical reaction force, and R is the resultant of F_C and F_V.

by the runner and thus reduce the effects of the centripetal force. As the bank increases, the runner's resultant force acts more longitudinally and the shear component decreases.

PRESSURE

Up to this point in the discussion of force, the way a force causes an object to accelerate to achieve a state of motion has been considered. It is also necessary to discuss how forces, particularly impact forces, are distributed. The concept of **pressure** is used to describe force distribution. Pressure is defined as the force per unit area. That is:

$$P = \frac{F}{A}$$

where F is a force and A is the area over which the force is applied. Pressure has units of N/m^2. Another unit of pressure often used is the pascal (Pa) or the kilopascal (kPa). One pascal is equal to 1 N/m^2. If an individual with a BW of 650 N is supported on the soles of the feet with an approximate area of 0.018 m^2, the pressure on the soles of the feet would be:

$$P = \frac{650\,N}{0.018\,m^2}$$
$$P = 36.11\,kPa$$

Another individual with a smaller BW, 500 N, and soles of the feet with an identical area would have a pressure of:

$$P = \frac{500\,N}{0.018\,m^2}$$
$$= 27.78\,kPa$$

If the heavier individual's foot soles had a larger area, 0.02 m^2, for example, the pressure would be 32.50 kPa. The pressure on the soles of the feet of the heavier individual is more than the pressure on the soles of the lighter individual, even though the BW is different. These pressures appear quite large, but imagine if these individuals were women wearing spike-heeled shoes, which have much less surface area than ordinary shoes. A more dramatic example would be if these individuals were wearing ice skates, which have a distinctly smaller area of contact with the surface than the sole of a normal shoe or a spike-heeled shoe. On the other hand, if these individuals were using skis or snow shoes to walk in deep snow, the pressure would be quite small because of the large area of the skis or snowshoes in contact with the snow. In this way, individuals walk on snow without sinking into it.

The concept of pressure is especially important in activities in which a collision results. Generally, when a force of impact is to be minimized, it should be received over as large an area as possible. For example, when landing from a fall, most athletes attempt a roll to spread the impact force over as large an area as possible. In the martial arts, considerable time is spent in learning how to fall correctly, specifically applying the pressure as force per unit area.

A number of sporting activities in which collisions abound have special protective equipment designed to reduce pressure. Examples are shoulder pads in football and ice hockey; shin pads in ice hockey, field hockey, soccer, and baseball (for the catcher); boxing gloves; and batting helmets in baseball. In all of these examples, the point of the design of the protective padding is to spread the impact force over as large an area as possible to reduce the pressure.

With the use of a force platform, it is possible to obtain a measure of the **center of pressure** (COP), a displacement measure indicating the path of the resultant GRF vector on the force platform. It is equal to the weighted average of the points of application of all of the downward-acting forces on the force platform. Because the COP is a general measurement, it may be nowhere near the maximal areas of pressure. However, it does provide a general pattern and has been extensively used in gait analysis. Cavanagh and Lafortune (17) showed different COP patterns for rear foot and midfoot strikers. Representative COP patterns are illustrated in Figure 10-35. Cavanagh and Lafortune suggested that COP information may be useful in shoe design, but these patterns have not been related successfully to foot function during locomotion. Miller (55) noted that COP data provide only restricted information on the overall pressure distribution on the sole of the foot.

Methods of measuring the local pressure patterns under the foot or shoe have been developed. An example of the type of data available on these systems is presented in Figure 10-36. Cavanagh and colleagues (16) developed such a measuring system and reported distinct local areas of high pressure on the foot throughout the ground contact phase. The greatest pressures were measured at the heel, on the metatarsal heads, and on the hallux.

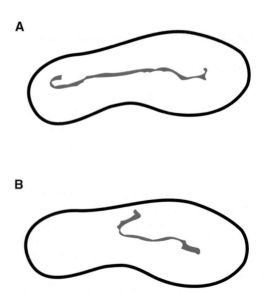

FIGURE 10-35 Center of pressure patterns for the left foot. **(A)** A heel–toe footfall pattern runner. **(B)** A midfoot foot strike pattern runner.

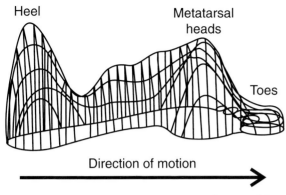

Heel

Metatarsal heads

Toes

Direction of motion

FIGURE 10-36 Pressure distribution pattern of a normal foot during walking. (Adapted from Cavanagh, P. R. [1989]. The biomechanics of running and running shoe problems. In B. Segesser, W. Pforringer (Eds.). *The Shoe in Sport.* London: Wolfe, 3–15.)

These researchers (16) also compared the pressure patterns during running barefoot and with various foam materials attached to the foot. They reported that peak pressures were reduced when wearing the foam materials, but the changes in the pressure pattern over the support period were similar. Foti and colleagues (32), using an in-shoe pressure measurement device, reported that softer midsole shoes distributed the foot-to-shoe pressure at heel contact during walking better than a hard midsole shoe. The implication is that softer midsole shoes provide a more cushioned feel to the wearer.

 Linear Kinetics of Locomotion

GRF profiles continually change with time and are generally presented as a function of time. The magnitude of the GRF components for running are much greater than for walking. The magnitude of the GRFs also varies as a function of locomotor speed (38,56), increasing with running speed. The vertical GRF component is much greater in magnitude than the other components and has received the most attention from biomechanists (Fig. 10-37). In walking, the vertical component generally has a maximum value of 1 to 1.2 BW, and in running, the maximum value can be 2 to 5 BW. The vertical force component in walking has a characteristic bimodal shape, that is, it has two maximum values. The first modal peak occurs during the first half of support and characterizes the portion of support when the total body is lowered after foot contact. The force rises above BW as full weight bearing takes place and the body mass is accelerated upward. The force then lowers as the knee flexes, partially unloading. The second peak represents the active push against the ground to move into the next step. Figure 10-38 presents a comparison of walking and running vertical GRF component profiles.

In running, the shape of the vertical GRF component depends on the footfall pattern of the runner (Fig. 10-39). These footfall patterns are generally referred to as the

A

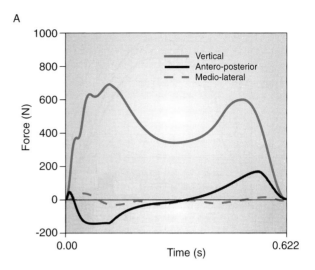

B

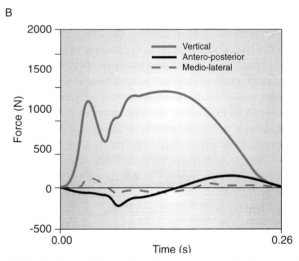

FIGURE 10-37 GRF for walking (**A**) and running (**B**). Note the difference in magnitude between the vertical component and the shear components.

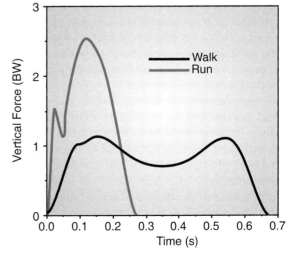

FIGURE 10-38 The vertical GRF for both walking and running.

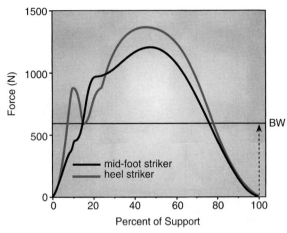

FIGURE 10-39 Vertical GRF profiles of a runner using a heel–toe footfall pattern and a runner who initially strikes the ground with the midfoot.

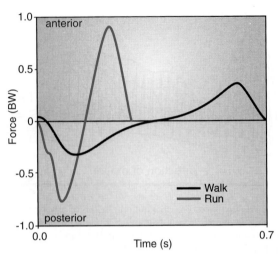

FIGURE 10-40 Anteroposterior (front-to-back) GRF for both walking and running.

heel–toe and midfoot patterns. The heel strike runner's curve has two discernible peaks. The first peak occurs very rapidly after the initial contact and is often referred to as the **passive peak**. The term *passive peak* refers to the fact that this phase is not considered to be under muscular control (58) and is influenced by impact velocity, contact area between the surface and the foot, the joint angles at impact, surface stiffness, and the motion of the segments (24). This peak is also referred to as the impact peak. The midfoot strike runner has little or no impact peak. The second peak in the vertical GRF component occurs during midsupport and generally has greater magnitude than the impact peak. Nigg (58) referred to the second peak as the **active peak**, indicating the role the muscles play in the force development to accelerate the body off the ground. Runners with either type of footfall pattern exhibit this peak.

The anteroposterior GRF component also exhibits a characteristic shape similar in both walking and running but of different magnitude (Fig. 10-40). The F_y component reaches magnitudes of 0.15 BW in walking and up to 0.5 BW in running. During locomotion, this component shows a negative phase during the first half of support as a result of a backward horizontal friction force between the shoe and the surface. This moves to positive near midstance, as force is generated by the muscles pushing back against the ground.

The mediolateral GRF component is extremely variable and has no consistent pattern from individual to individual. It is very difficult to interpret this force component without a video or film record of the foot contact. Figure 10-41 illustrates walking and running profiles for the same individual. The great variety in foot placement regarding toeing in (forefoot adduction) and toeing out (forefoot abduction) may be a reason for this lack of consistency in the mediolateral component. The range of foot placement was shown in one study to be from 12° of toeing in to 29° of toeing out, and toeing out at heel

strike has been shown to generate greater medial lateral forces and impulses (72). The magnitude of the mediolateral component ranges from 0.01 BW in walking to 0.1 BW in running.

Biomechanists have investigated GRFs to attempt to relate these forces to the kinematics of the lower extremity, particularly to foot function. Efforts have been made to relate these forces to the rear foot supination and pronation profiles of runners to identify possible injuries or aid in the design of athletic footwear (35,37). Because the GRF is representative of the acceleration of the total body center of mass, the use of GRF data for these purposes is probably extrapolating beyond the information provided by the GRFs.

To illustrate this point, a method of calculating the vertical GRF component proposed by Bobbert and colleagues (8) will be presented. In this method, the researchers used the kinematically derived values of the accelerations of the centers of mass of each of the body's segments.

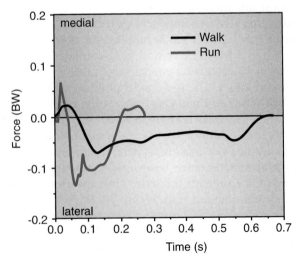

FIGURE 10-41 Mediolateral (side-to-side) GRF for both walking and running.

The vertical GRF component reflects the accelerations of the individual body segments resulting from the motion of the segments. The sum of the vertical forces of all body segments, including the effect of gravity, is the vertical GRF component. That is:

$$F_z = \sum_{i=1}^{n} m(a_{zi} - g)$$

where F_z is the vertical force component (forces directed upward are defined as positive), m_i is the mass of the ith segment, n is the number of segments, a_{zi} is the vertical acceleration of the ith segment (upward accelerations are defined as positive), and g is the acceleration due to gravity. The anteroposterior GRF component reflects the horizontal (i.e., in the direction of motion) accelerations of the individual body segments. Using similar methods, this force component can be computed as:

$$F_y = \sum_{i=1}^{n} (m_i a_{yi})$$

where a_{yi} is the horizontal acceleration of the ith segment. Similarly, the mediolateral GRF component reflects the side-to-side accelerations of the individual body segments:

$$F_x = \sum_{i=1}^{n} (m_i a_{xi})$$

where a_{xi} is the side-to-side acceleration of the ith segment. If the center of mass is a single point that represents the mass center of all the body's segments, the vertical component is:

$$F_z = m(a_z - g)$$

where m is the total body mass, a_z is the vertical acceleration of the center of mass, and g is the acceleration due to gravity. Similarly, the other components may be represented as the total body mass times the acceleration of the center of mass. That is:

$$F_y = ma_y$$
$$F_x = ma_x$$

Therefore, the GRF represents the force necessary to accelerate the total body center of gravity (55) and cannot be directly associated with lower extremity function. Caution, then, should be exercised in describing lower extremity function using GRF data.

Because the GRF relates to the motion of the total body center of mass, the anteroposterior force profile can be related to the acceleration profile of the center of mass during support. Chapter 8 discussed a study by Bates and colleagues (6), illustrating the horizontal velocity pattern of the center of mass during the support phase of the running stride (Fig. 10-42A). When this curve is differentiated, an acceleration curve is generated (Fig. 10-42B). This curve has the characteristic shape of the anteroposterior force component in that it has negative acceleration followed by positive acceleration. According to Newton's second law of motion, if each point along this curve was multiplied by the runner's

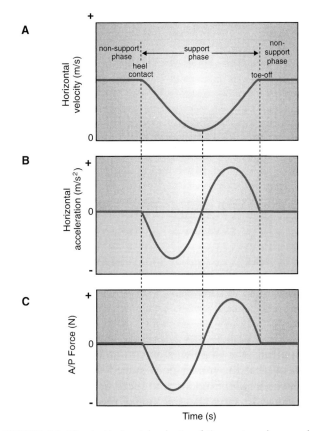

FIGURE 10-42 A. Horizontal velocity of the center of mass of a runner. **B.** If the velocity curve in (**A**) is differentiated, a horizontal acceleration curve of the center of mass is generated. **C.** Multiplying each point along this curve by the runner's body mass, the anteroposterior GRF is generated.

body mass (m), the anteroposterior GRF component would be generated as follows:

$$F_y = ma_y$$

Conversely, the acceleration curve could be calculated by dividing the F_y force component by the runner's body mass:

$$a_y = \frac{F_y}{m}$$

Either generating the curves using the kinematic procedure or collecting the anteroposterior GRF component leads to the same conclusion. The negative portion of the force component is often referred to as the *braking phase* and indicates a force against the runner serving to decrease velocity of the runner. The positive portion of the component is called the *propelling phase* and indicates a force in the direction of motion serving to increase velocity of the runner. If the running speed is constant, the negative and positive phases will be symmetrical, indicating no loss in velocity. If the negative portion of the curve is greater than the positive portion, the runner will slow down more than speed up. Conversely, if the positive portion is greater than the negative, the runner is speeding up.

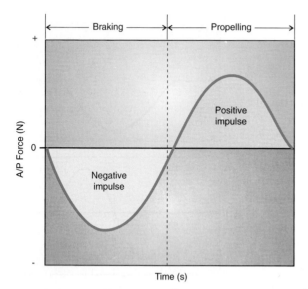

FIGURE 10-43 Anteroposterior GRF illustrating the braking and propulsion impulses.

Applying the impulse–momentum relationship again confirms that the runner does indeed slow down during the first portion of support and speed up in the latter portion (Fig. 10-43). The area under the negative portion of the force component or the negative impulse serves to slow the runner down, that is, to change incoming velocity to some lesser velocity value. The positive area or the positive impulse of the component serves to accelerate the runner, that is, to change velocity from some lesser value to some outgoing velocity. If the positive change in velocity equals the negative change in velocity, the individual is running at a constant speed. Figure 10-44 illustrates the changes in the braking and propelling impulses across a range of running speeds (38). In many instances in the

laboratory during collection of GRF data, the ratio of the negative impulse to the positive impulse is checked to determine if the runner is at a constant velocity, speeding up, or slowing down. Even if the individual is maintaining a constant running speed, the ratio of the positive to the negative impulse is rarely 1.0 for any given support period of a stride. The average ratio over a number of footfalls approaches the ratio of 1.0, however.

An exchange of mechanical energy occurs during both running and walking, although the energy fluctuations differ. External work in walking has two components, one caused by inertial forces as a result of speed changes in the forward direction and the other caused by the cyclic upward displacement of the center of gravity. The work done to accelerate in the lateral direction is only a small fraction of the total work, as is evidenced by the small forces and small displacements (75). The motion of the center of gravity in walking has been modeled as an inverted pendulum. In each step, the center of gravity is either behind or in front of the contact point between the foot and the ground (22). When the center of gravity is behind the point of contact, as during the heel strike phase of support, the GRFs cause a negative acceleration and KE decreases because of loss of forward speed. A concomitant occurrence with the loss of KE is an increase in the center of gravity as the body vaults over the support limb. This increases the gravitational PE, which reaches a maximum level in the middle of stance. As the center of gravity moves forward of the point of the contact, the KE increases because gravitational PE decreases with the reduction in the height of the center of gravity. This pendulum-like exchange between PE and KE allows savings as much as 65% of muscular work (75). This conservation of energy is not perfect, so the total energy of the center of gravity fluctuates (22). The

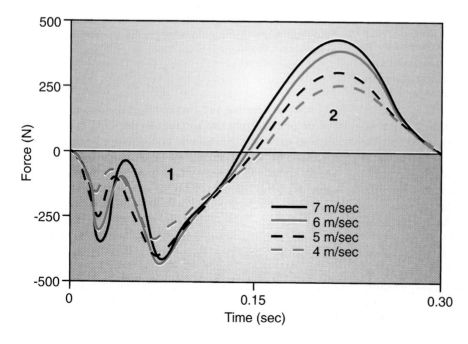

FIGURE 10-44 Changes in the anteroposterior GRF as a function of running speed. Area 1 is the braking impulse, and area 2 is the propulsion impulse. (Adapted from Hamill, J., et al. [1983]. Variations in GRF parameters at different running speeds. *Human Movement Science*, 2:47–56.)

net change in the overall mechanical energy in walking is actually small.

In running, the mechanical energy fluctuates more than in walking. KE in running is similar to walking, reaching minimum levels at midstance because of deceleration caused by the horizontal GRF and increasing in the latter half of the support phase. The PE is different than in walking because it is at a minimum at midstance because of compliance and flexion in the support limb (31). The overall vertical excursion of the center of mass is also less as the speed of running increases (31). There is not the pendulum-like exchange between PE and KE seen in running because the energies are in phase with each other compared with walking, in which they are 180° out of phase (31). The exchange of energy conserves less than 5% of the mechanical work required to lift and accelerate the center of mass (31). However, substantial mechanical energy is conserved through the storage and return of elastic energy in the tissues.

 ## Linear Kinetics of the Golf Swing

In the golf swing, substantial linear forces are generated on the ground in response to segmental accelerations. Other important force application sites are between the hand and the club and—most important—between the club head and the ball at contact. The GRFs vary between right and left limbs. High vertical GRFs are generated between the right foot and the ground in the backswing (right-handed golfer), and a rapid transfer of force to the left foot occurs before impact, resulting in a peak force that is more than 1 BW (88). In the mediolateral direction, a lateral GRF develops in the right foot up through the backswing and from the top of the backswing to just before impact and a medial force propels the body toward the direction of the ball. At impact, this reverses to slow the movement of the body from right to left (88). In the anteroposterior direction, a force is generated as the body rotates around a vertical axis, resulting in a backward force on the left foot and a forward force on the right foot in the backswing. This reverses in the downswing as the body rotates to impact (88). The pattern of GRF generation using the different clubs is essentially the same. However, there are changes in force magnitude with different clubs. A sample of the GRFs and the COP patterns for both feet are illustrated in Figure 10-45. Maximum vertical GRF values for a representative subject wearing regular spikes were 1,096 N at the front foot and 729 N at the rear foot (89). The maximum anteroposterior force in the anterior was 166 N generated in the front foot and was 143 N in the posterior direction, generated in the back foot. The maximum lateral force was generated in the front foot and was in the range of 161 N (89).

The forces acting on the body as a result of the swing have been shown to range between 40% and 50% of BW (49). These forces must be controlled by the golfer to produce an effective swing. A good golfer starts the downswing slowly, and the forces acting on the golfer consequently produce a smooth acceleration. An inexperienced golfer initiates the downswing with greater acceleration. The centripetal force produced by this rapid acceleration rotates the club, reduces the acceleration, and can actually cause negative acceleration later in the swing. Forces acting on the body as a result of the swing reach values around 40% of BW in the final third of the downswing (Fig. 10-46). These forces are directed backward and must be resisted by knee flexion and a wide base of support. The force produced by the swing is maximum at contact and is vertically directed and easily resisted by the body (49).

As a result of forces acting on the body, significant forces are generated at the knee joint. Compressive forces in the front and back knees reach values approximating

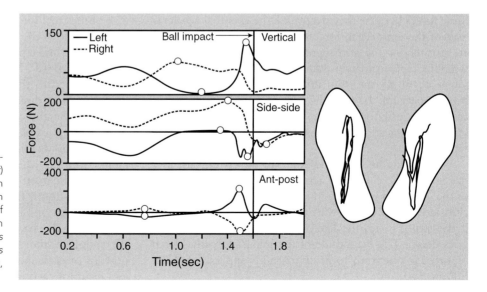

FIGURE 10-45 GRF and center of pressure for the front (*left*) and back (*right*) foot during the golf swing. (Adapted from Williams, K. R., Sih, B. L. [1999]. GRFs in regular-spike and alternative-spike golf shoes. In M. R. Farrally, A. J. Cochran (Eds.). *Science and Golf III: Proceedings of the 1998 World Scientific Congress of Golf.* Champaign, IL: Human Kinetics, 568–575.)

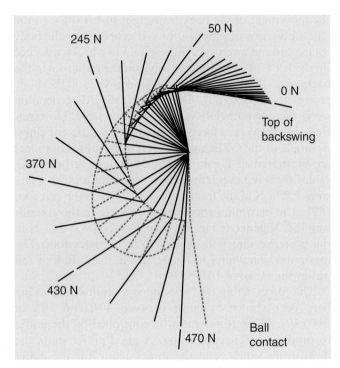

FIGURE 10-46 Forces generated as a result of the swing are shown for the golfer Bobby Jones. (Adapted from Mather, J. S. B. [2000]. Innovative golf clubs designed for the amateur. In A. J. Subic, S. J. Haake (Eds.). *The Engineering of Sport: Research, Development and Innovation.* Malden, MA: Blackwell Science, 61–68.)

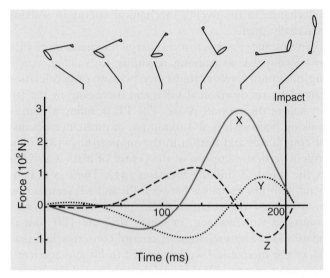

FIGURE 10-47 Forces generated at the wrist joint in all three directions. (Adapted from Neal, R. J., Wilson, B. D. [1985]. 3D kinematics and kinetics of the golf swing. *International Journal of Sports Biomechanics,* 1:221–232.)

100% and 72% of BW for the front and back knee joints, respectively (33). Shear forces are also developed. For a right-handed golfer, an anterior shear force is developed in the right knee (10% BW) and posterior shear forces are present in both the left and right knee joints with approximately 39% and 20% of BW in the left and right knee joints, respectively (33). For a left-handed golfer, the values for the right and left knee would be reversed.

Linear forces have also been measured at the wrist and shoulder. Resultant forces acting at the wrist and shoulder are shown in Figure 10-47. It is important to measure these forces because they determine the eventual acceleration of the club. Peak forces in the direction of the ball are greater for the arm segment (650 N) than the club segment (approximately 300 N), and the shoulder peak forces occur 85 ms from impact compared with 60 ms from impact for the club (57). Peak forces in the vertical and anteroposterior directions also occur earlier in the arm segment, suggesting some timing interaction between the segments.

When using an iron, the golf ball travels upward as a reaction to the club head action downward with the face of the club held in place and as a result of the angle of the club face. The contact with the ball is not upward, and if the swing is up, the likely result is a topped ball that goes down. The magnitude of the impact force has been reported in one study to be up to 15 kN applied for about 500 ms (50).

Linear Kinetics of Wheelchair Propulsion

To propel a wheelchair, the hand grasps the rim of the wheel and generates a pushing force. After the push phase, the hands return to the initial position before contact is again made with the rim. In this passive recovery phase, the inertial forces from the upper body movements can continue to influence the motion of the wheelchair (81) so that the backward swing of the trunk causes a reaction force that can propel the wheelchair forward.

The hand pushes on the rim at an angle, but only the force component tangential to the rim contributes to the propulsion (67). The propulsion force vector tangential to the rim is directed upward at the hand position of −15° to top dead center and directed downward at +60° from top dead center (79). This force has been shown to be only 67% of the total force applied to the hand rim. A representative sample of the hand rim forces produced at a velocity of 1.39 m/s and a power output of 0.5 W/kg is illustrated in Figure 10-48 (82). In this example, the downward forces applied to the hand rim are nearly twice the horizontal forward-directed forces. The outward force is the lowest of the three forces and increases only in the last third of the push phase. The actual propelling force can be calculated by dividing the torque at the hand rim by the radius from the wheel axle to the hand rim (83).

One of the main factors that determines the force application direction is the cost associated with each particular force application. If a force is applied perpendicular to a line from the hand to the elbow or from the hand to the shoulder, the cost increases at each joint (67). The posture of each individual influences the cost and effect as a result of each individual sitting in the wheelchair and

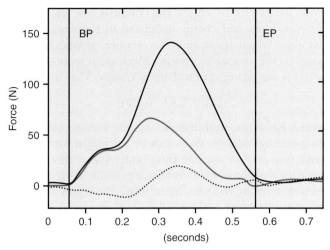

FIGURE 10-48 Vertical forces (*black*), anteroposterior forces (*blue*), and mediolateral forces (*dotted*) applied to the rim of the wheelchair by the hand. (Adapted from Veeger, H. E., et al. [1989]. Wheelchair propulsion technique at different speeds. *Scandinavian Journal of Rehabilitation Medicine*, 21:197–203.)

holding the rim at a certain point. Large joint reaction forces generated in the shoulder joint change with hand position. For example, the shoulder joint mean forces at top dead center and at 15° relative to top dead center have been computed to be 1,900 and 1,750 N, respectively, and are approximately 10 times the mean net forces at the joint (79).

The propulsion technique also influences the output. A circular propulsion technique has been shown to generate less mean power (37.1 W) than the pumping technique (44.4 W) (84). Individual differences in shoulder and elbow positions can also influence propulsion effectiveness. For example, if an individual lowers the shoulder during the push phase, a more vertical and less effective propulsive force might result.

The construction of the wheelchair can also influence propulsion. The axle position relative to the shoulder changes the push rim biomechanics significantly. If the vertical distance between the axle and the shoulder is increased, the push angle is decreased and the force available for propulsion is diminished (10). Improvements in wheelchair propulsion have been shown to be associated with a wheelchair with a more forward axle position (10). Also, cambered rear wheels whose top distance between the wheels is smaller than the bottom orient the hand rim to more closely resemble the force application. This facilitates a more effective use of elbow extension (83).

Summary

Linear kinetics is the branch of mechanics that deals with the causes of linear motion, or forces. All forces have magnitude, direction, point of application, and line of action. The laws governing the motion of objects were developed by Sir Isaac Newton and form the basis for the mechanical analysis of human motion:

1. **Law of inertia:** Every body continues in its state of rest or uniform motion in a straight line unless acted upon by an external force.
2. **Law of acceleration:** The rate of change of momentum of a body is proportional to the force causing it, and the change takes place in the direction of the force.
3. **Law of action–reaction:** For every action there is an equal and opposite reaction.

Forces may be categorized as noncontact or contact. The most important noncontact force acting during human movement is gravity. The contact forces include the GRF, joint reaction force, friction, fluid resistance, inertial force, muscle force, and elastic force.

The GRF, a direct application of Newton's third law, has three components: a vertical component and two shear components acting parallel to the surface of the ground. The joint reaction force, the net force acting across a joint, has compressive and shear components. Friction results from interaction between two surfaces and is a force that acts parallel to the interface of the two surfaces and in a direction opposite to the motion. The coefficient of friction is the quantification of the interaction of the two surfaces. Fluid resistance refers to the transfer of energy from an object to the fluid through which the object is moving. The fluid resistance vector has two components, lift and drag. Drag acts in a direction opposite to the direction of motion, and lift is perpendicular to the drag component. Inertia results from the force applied by one segment on another that is not caused by muscle actions. A muscle force is the pull of the muscle on its insertion, resulting in motion at a joint. Muscle forces are generally calculated as net forces, not individual muscle forces, although intricate mathematical procedures can evaluate individual muscle forces. An elastic force results from the rebound of a material to its original length after it has been deformed.

A free body diagram is a schematic illustration of a system with all external forces represented by vector arrows at their points of application. Internal forces are not presented on free body diagrams. Muscle forces are generally not represented on these diagrams unless the system involves a single segment.

Analyses using Newton's law are usually conducted using one of three calculations: the effect of a force at an instant in time ($F = ma$), the effect of a force applied over time (impulse–momentum relationship), and the effect of a force applied over a distance (work–energy theorem). In the first technique, the analysis may be a static case (when $a = 0$) or a dynamic case (when $a \neq 0$). The static two-dimensional linear case is determined using the following equations:

$$\Sigma F_x = 0 \text{ for the horizontal component}$$
$$\Sigma F_y = 0 \text{ for the vertical component}$$

The two-dimensional dynamic case uses the following equations:

$$\Sigma F_x = ma_x \text{ for the horizontal component}$$
$$\Sigma F_y = ma_y \text{ for the vertical component}$$

The impulse–momentum relationship relates the force applied over time to the change in momentum:

$$F \times dt = mv_{final} - mv_{initial}$$

The left-hand side of the equation ($F \times dt$) is the impulse, and the right-hand side ($mv_{final} - mv_{initial}$) describes the change in momentum. Impulse is defined as the area under the force–time curve and is thus equal to the change in momentum. This type of analysis has been used in research to evaluate the jump height of the center of mass in vertical jumping in association with the equations of constant acceleration.

Work is the product of the force applied and the distance over which the force is applied. Energy, the capacity to do work, has two forms, kinetic and potential. The relationship between work and energy is defined in the work–energy theorem, which states that the amount of work done is equal to the change in energy. Mechanical work is calculated via the change in mechanical energy. That is:

$$W = \Sigma KE + \Sigma PE$$

where KE is the translational kinetic energy and PE is the potential energy. Work can be calculated for either a single segment or for the total body. When this is done segment by segment, internal work or the work done on the segments by the muscles to move the segments is calculated. When the amount of work done is related to the time over which the work is done, the power developed is being evaluated.

Special force applications include definitions for centripetal force and pressure. Centripetal force is applied toward the center of rotation. Pressure is the force per unit area.

Equation Review for Linear Kinetics

Calculation	Given	Formula
Magnitude of resultant force	Horizontal, vertical forces	$R = \sqrt{F_x^2 + F_y^2}$
Angle of force application	Horizontal, vertical forces	$\tan \theta = \dfrac{F_y}{F_x}$
Vertical force	Resultant force, angle of application	$F_y = R \sin \theta$
Horizontal force	Resultant force, angle of application	$F_x = R \cos \theta$
Acceleration (a)	Force (F), mass (m)	$a = \dfrac{f}{m}$
Vertical acceleration (a)	Vertical force (F_y), mass (m)	$a_y = \dfrac{(F_y - mg)}{m}$
Force	Acceleration, mass	$F = ma$
Weight (W)	Mass	$W = ma = mg$
Momentum (p)	Mass, velocity	$\rho = mv$
Impulse	Momentum ($m \times v$)	$\text{Impulse} = mv_{final} - mv_{initial}$
Velocity	Force, time, mass	$V = \dfrac{F \times t}{m}$
Force	Momentum, time of force application	$F = \dfrac{mv}{t}$
Friction force (F_f)	Normal force (N), coefficient of friction μ	$F_f = \mu N$
Coefficient of friction	Frictional force and normal force	$\mu = \dfrac{F_f}{N}$
Coefficient of friction	GRF data	$\mu = \dfrac{F_y}{F_z}$
Fluid resistance drag force	Coefficient of drag (C_d), frontal area of object (A), fluid viscosity (r) Velocity of object relative to the fluid (v)	$F_{drag} = \dfrac{1}{2} C_d A \rho v^2$
Centripetal force	Mass of object (m), tangential velocity (v), radius of rotation (r)	$F_c = \dfrac{mv^2}{r}$
Centripetal force	Mass of object (m), angular velocity (ω), radius of rotation (r)	$F_c = m\omega^2 r$

Calculation	Given	Formula
Pressure	Force, area	$P = \dfrac{F}{A}$
Work	Force, displacement (*s*)	$W = Fs$
Work	Changes in kinetic energy (KE)	$W = \dfrac{1}{2}mv_2^2 - mv_1^2$
Horizontal work	Force, angle of force application, displacement	$W = F\cos\theta\, s$
Potential energy (PE)	Mass, vertical height (*h*)	$PE = mgh$
Kinetic energy (KE)	Mass, velocity	$KE = \dfrac{1}{2}mv^2$
Strain energy (SE)	Proportionality constant (*k*) distance object deformed (Δx)	$SE = \dfrac{1}{2}k\Delta x^2$
Power	Work (*W*), time (*t*)	$P = \dfrac{W}{t}$
Power	Force, velocity	$P = Fv$

REVIEW QUESTIONS

True or False

1. _____ An object's mass is a good estimate of its inertia.

2. _____ While you can pull or push an object, your muscles can technically only pull, or create tensile forces.

3. _____ When a force vector is resolved into its orthogonal components, the magnitude of the resultant vector is always larger than any of the components.

4. _____ As long as the distance between your center of mass and the earth's center of mass is constant, so is gravity.

5. _____ The amount of energy transmitted from you to the water when swimming is proportional to the disturbance of the water.

6. _____ During walking, the magnitude of GRF from highest to lowest is vertical, mediolateral, and anteroposterior.

7. _____ An object in motion will not stop unless some external force acts on it.

8. _____ Potential energy is proportional to an object's mass and inversely proportional to its height from the ground.

9. _____ Force is a vector.

10. _____ Two pushing and pulling forces acting on the same point at the same time can be added together and represented as one resultant force.

11. _____ Backspin imparted on a golf ball is an example of the Magnus effect, in which low pressure on the bottom and high pressure on the top results in the ball lifting upward.

12. _____ Only the force component tangential to the rim of a wheelchair causes propulsion.

13. _____ A force has to be applied over some time interval for an impulse to be generated.

14. _____ Frictional force depends on the contact area, and the coefficient of friction between the two objects of interest.

15. _____ If a 100 N force is applied to lift an object with a mass of 10 kg, that object will be accelerated upward.

16. _____ You performed vertical jumps on a force platform two times. The time you generated greater vertical impulse was the time you jumped higher.

17. _____ Gravity and friction are examples of external forces.

18. _____ The normal force is always directed vertically.

19. _____ Laminar flow is most likely to occur when the object passing through the fluid is small, has a smooth surface, and is moving slowly.

20. _____ Assuming your mass is constant when you sleep, you can decrease pressure by lying on your back versus your side.

21. _____ The center of pressure under your left foot when standing may be in a location of zero pressure.

22. _____ A pole vaulter in flight has only kinetic energy.

23. _____ When you are standing still, the sum of all of the external forces acting on your body always equals zero.

24. _____ A shuffleboard puck slows down as it slides on the floor. Its change in energy is equivalent to the change in power.

25. _____ Person A, who has half the mass of Person B, will be able to cause the same impulse on an object as Person B if she moves with twice the velocity of Person B.

Multiple Choice

1. A 100 N force is applied to a box at an angle of 60° to the horizontal via a rope. How much force "lifts" the box and how much "pulls" it along the surface?
   a. 86.6 N; 50 N
   b. 100 N; 100 N
   c. 50 N; 50 N
   d. 50 N; 86.6 N

2. The horizontal and vertical components of a force are 108 and 22 N, respectively. What is the magnitude of the resultant vector?
   a. 130 N
   b. 110 N
   c. 106 N
   d. 12,148 N

3. A 4.12 N ball needs to be accelerating in the vertical direction at 20 m/s^2 to reach its target height. How much force must be exerted in the vertical direction to accomplish this?
   a. 0.206 N
   b. 4.85 N
   c. 8.4 N
   d. 4.12 N

4. The force parallel to a surface is 380 N and the force perpendicular to the surface is 555 N. The coefficient of friction is 0.66. What is the frictional force?
   a. 841 N
   b. 617 N
   c. 251 N
   d. 366 N

5. In question 4, if the 380 N force is the anterior force applied by a foot on the ground when walking, what will the person do?
   a. Fall, because the frictional force is less than the applied force
   b. Fall, because the frictional force causes the foot to "stick" to the ground because it is greater than the anterior force
   c. Continue walking as normal since the frictional force is sufficient to prevent slipping
   d. Not enough information to answer

6. If an individual's leg exerts a force of 9 N as it moves 0.4 m within 0.2 seconds, what is the power generated by the leg?
   a. 18 W
   b. 45 W
   c. 0.72 W
   d. 4.5 W

7. If the static coefficient of friction of a basketball shoe on a particular playing surface is 0.58 and the normal force is 911 N, what horizontal force is necessary to cause the shoe to slide?
   a. <1,570 N
   b. >306 N
   c. >528 N
   d. <528 N

8. An individual lifts an 80 kg weight to a height of 2.02 m. When the weight is held overhead, what is the potential energy? What is the kinetic energy?
   a. PE = 0 N-m, KE = 162 N-m
   b. PE = 162 N-m, KE = 0 N-m
   c. PE = 0 N-m, KE = 1585 N-m
   d. PE = 1585 N-m, KE = 0 N-m

9. Calculate the impulse of a force that increases at a constant rate from 0.0 to 3.0 N over 3.0 seconds and then decreases at a different constant rate from 3.0 to 0.0 N in 4.0 seconds.
   a. 27.0 Ns
   b. 10.5 Ns
   c. 21 Ns
   d. 3.0 Ns

10. Consider the following free body diagram. Using static analysis, solve for the horizontal and vertical forces of C that will maintain this system in equilibrium if A = 331 N, B = 79 N, and W = 50 N.
    a. F_y = 342.5 N; F_x = 59.6 N
    b. F_y = –59.6 N; F_x = –342.5 N
    c. F_y = 59.6 N; F_x = 342.5 N
    d. F_y = –9.6 N; F_x = –342.5 N

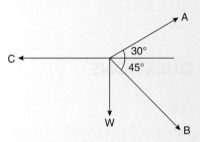

11. Calculate the height of the center of mass above its starting height during a squat jump based on the following information: BW = 777 N, total vertical force = 899 N, and time of force application = 0.93 seconds.
    a. 0.21 m
    b. 0.12 m
    c. 0.10 m
    d. 0.073 m

12. Calculate the takeoff velocity for a jump based on the following hypothetical graph (jumper's weight = 700 N).
    a. 1.40 m/s
    b. 0.71 m/s
    c. 16.1 m/s
    d. 1.96 m/s

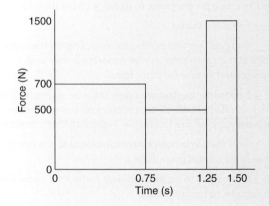

13. You stand on a 2% incline and later on a 10% incline. In which situation is frictional force lower?
    a. The 2% incline, because the normal force is less
    b. The 10% incline, because the normal force is less
    c. It remains constant because your mass and gravity remain constant
    d. Incline does not affect the frictional force

14. An object is being pushed by a stick horizontally across a table at a constant velocity. Which is(are) true?
    a. Linear acceleration equals zero.
    b. The applied force is equal to the kinetic frictional force.
    c. The applied force equals zero.
    d. All of the above
    e. Both a and b

15. A 1,300-kg car starts to roll down a road with a 30° incline. People rush to stop it. How much force must they apply to stop it?
    a. 1,300 N
    b. 11,044 N
    c. 12,753 N
    d. 6,377 N

16. A 150 kg football player is running toward another player at 11.1 m/s. How much force needs to be applied over 2.0 seconds to bring him to a stop?
    a. 1,665 N
    b. 13.5 N
    c. 833 N
    d. Need to know the other player's mass in order to solve.

17. You slide down the same hill on a sled when (a) there is snow at the bottom and (b) there is just grass at the bottom when the hill levels off. Assuming you reached the bottom of the hill with the same velocity, why do you not travel as far before you come to a stop when on the grass?
    a. It takes more work to slow you down in the snow
    b. You have less kinetic energy on the grass
    c. You do travel the same distance, it just takes you longer on the grass
    d. The work done by friction to slow you down happens faster on the grass

18. If the braking impulse in an anterior–posterior GRF by time curve for running is larger than the propulsion impulse, what is the runner doing?
    a. Slowing down
    b. Speeding up
    c. Running with a constant velocity
    d. Not moving

19. A 608 N woman dives from a 10-m platform. What is her potential and kinetic energy 7 m into the dive?
    a. KE = 1,824 J; PE = 4,255 J
    b. KE = 4,255 J; PE = 1,825 J
    c. KE = 41,744 J; PE = 17,903 J
    d. KE = 17,903 J; PE = 41,744 J

20. A constant force of 160 N acts on an object in the horizontal direction. The force moves the object forward 75 m in 2.3 seconds. What is the object's mass?
    a. 22.6 kg
    b. 5.64 kg
    c. 11.28 kg
    d. 60 kg

21. A 14,000-kg truck is traveling at 20 m/s. What would be the velocity of a 10,000-kg truck with the same momentum?
    a. 28 m/s
    b. 20 m/s
    c. 32 m/s
    d. 40 m/s

22. A running back is tackled with a force of 4,100 N by a linebacker weighing 1,000 N. What was the acceleration of the linebacker?
    a. 9.81 m/s^2
    b. 102 m/s^2
    c. 40.2 m/s^2
    d. 4.1 m/s^2

23. What is the pressure on the bottom of the foot for an 85-kg person on the balls of one foot making contact over an area of approximately 100 cm^2?
    a. 8,500 N/cm^2
    b. 0.85 N/cm^2
    c. 8.34 N/cm^2
    d. 83.4 N/cm^2

24. The work calculated at time 1 and time 2 was 925 and 998 N-m, respectively. Calculate the power if the time interval was 0.049 seconds.
    a. 811 W
    b. 18,878 W
    c. 39,245 W
    d. 1,923 W

25. How much power is generated in the horizontal direction by a force of 950 N applied to an object at an angle of 40°, causing the object to move horizontally 4 m in 1.6 seconds?
    a. 4,658 W
    b. 1,527 W
    c. 2,375 W
    d. 1,819 W

References

1. Alexander, R. M. (1984). Elastic energy stores in running vertebrates. *American Zoologist*, 24:85-94.
2. Andres, R. O., Chaffin, D. B. (1985). Ergonomic analysis of slip-resistance measurement devices. *Ergonomics*, 28:1065-1079.
3. Asmussen, E., Bonde-Peterson, F. (1974). Apparent efficiency and storage of elastic energy in human muscles during exercise. *Acta Physiologica Scandinavica*, 92:537-545.

4. Ayoub, M. M., Mital, A. (1989). *Manual Materials Handling*. London: Taylor & Francis.

5. Barthels, K. M., Adrian, M. J. (1975). Three-dimensional spatial hand patterns of skilled butterfly swimmers. In L. Lewillie, J. P. Clarys (Eds.). *Swimming II*. Baltimore, MD: University Park, 154-160.

6. Bates, B. T., et al. (1979). Variations of velocity within the support phase of running. In J. Terauds, G. Dales (Eds.). *Science in Athletics*. Del Mar, CA: Academic, 51-59.

7. Bergmann, G., et al. (1995). Is staircase walking a risk for the fixation of hip implants? *Journal of Biomechanics*, 28:535-553.

8. Bobbert, M. F., et al. (1991). Calculation of vertical ground reaction force estimates during running from positional data. *Journal of Biomechanics*, 24:1095-1105.

9. Boda, W. L., Hamill, J. (1990). A mechanical model of the Maxiflex "B" springboard. *Proceedings of the VI biannual meeting of the Canadian Society of Biomechanics*. Quebec City, QC: 109-110.

10. Boniger, M. L., et al. (1997). Wrist biomechanics during two speeds of wheelchair propulsion: An analysis using a local coordinate system. *Archives of Physical Medicine and Rehabilitation*. Published by the organizing committee of the sixth Canadian Society of Biomechanics meeting, 78:364-371.

11. Brancazio, P. J. (1984). *Sports Science*. New York: Simon & Schuster.

12. Buck, P. C. (1985). Slipping, tripping and falling accidents at work: A national picture. *Ergonomics*, 28:949-958.

13. Caldwell, G. E., Forrester, L. W. (1992). Estimates of mechanical work and energy transfers: demonstration of a rigid body power model of the recovery leg in gait. *Medicine & Science in Sports & Exercise*, 24(12):1396-1412.

14. Cajori, F. (1934). *Sir Isaac Newton's Mathematical Principles* (translated by Andrew Motte in 1729). Berkeley, CA: University of California.

15. Cavagna, G. A., et al. (1976). The sources of external work in level walking and running. *Journal of Physiology*, 262:639-657.

16. Cavanagh, P. R. (1989). The biomechanics of running and running shoe problems. In B. Segesser, W. Pforringer (Eds.). *The Shoe in Sport*. London: Wolfe, 3-15.

17. Cavanagh, P. R., Lafortune, M. A. (1980). Ground reaction forces in distance running. *Journal of Biomechanics*, 15:397-406.

18. Chaffin, D. B., Andersson, G. B. J. (1991). *Occupational Biomechanics* (2nd Ed.). New York: Wiley.

19. Clarke, T. E., et al. (1983). The effects of shoe cushioning upon ground reaction forces in running. *International Journal of Sports Medicine*, 4:376-381.

20. Cohen, H. H., Compton, D. M. J. (1982, June). Fall accident patterns. *Professional Safety*, 16-22.

21. Councilman, J. E. (1968). *Science of Swimming*. Englewood Cliffs, NJ: Prentice Hall, Inc.

22. Detrembleur, C., et al. (2000). Motion of the body center of gravity as a summary indicator of the mechanics of human pathological gait. *Gait and Posture*, 12:243-250.

23. DeVita, P., Skelly, W. (1992). Effect of landing stiffness on joint kinetics and energetics in the lower extremity. *Medicine & Science in Sports & Exercise*, 24:108-115.

24. Dixon, S. J., et al. (2000). Surface effects on ground reaction forces and lower extremity kinematics in running. *Medicine & Science in Sports & Exercise*, 32:1919-1926.

25. Dowling, J. J., Vamos, L. (1993). Identification of kinetic and temporal factors related to vertical jump performance. *Journal of Applied Biomechanics*, 9:95-110.

26. Dufek, J. S., Bates, B. T. (1990). The evaluation and prediction of impact forces during landings. *Medicine & Science in Sports & Exercise*, 22:370-377.

27. Dufek, J. S., Bates, B. T. (1991). Dynamic performance assessment of selected sport shoes on impact forces. *Medicine & Science in Sports & Exercise*, 23:1062-1067.

28. Elftman, H. (1939). Forces and energy changes in the leg during walking. *American Journal of Physiology*, 125:339-356.

29. Elliott, B. C., White, E. (1989). A kinematic and kinetic analysis of the female two point and three point jump shots in basketball. *Australian Journal of Science and Medicine in Sport*, 21:7-11.

30. Fallon, L. P., et al. (2000). Determining baseball bat performance using a conservation equations model with field test validation. In A. J. Subic, S. J. Haake (Eds.). *The Engineering of Sport: Research, Development and Innovation*. Oxford: Blackwell Science, 201-211.

31. Farley, C. T., Ferris, D. P. (1998). Biomechanics of walking and running: Center of mass movements to muscle action. In J. Holloszy (Ed.). *Exercise and Sport Science Reviews*. Baltimore, MD: Lippincott Williams & Wilkins 253-281.

32. Foti, T., et al. (1992). Influence of footwear on weight-acceptance plantar pressure distribution during walking. In R. Rodano (Ed.). *Biomechanics in Sports X*. Milan: Edi-Ermes, 243-246.

33. Gatt, C. J., et al. (1999). A kinetic analysis of the knees during a golf swing. In M. R. Farrally, A. J. Cochran (Eds.). *Science and Golf III: Proceedings of the 1998 World Scientific Congress of Golf*. Champaign, IL: Human Kinetics, 20-28.

34. Giddings, V. L., et al. (2000). Calcaneal loading during walking and running. *Medicine & Science in Sports & Exercise*, 32:627-634.

35. Hamill, J., Bates, B. T. (1988). A kinetic evaluation of the effects of in vivo loading on running shoes. *Journal of Orthopaedic & Sports Physical Therapy*, 10:47-53.

36. Hamill, J., et al. (1984). Ground reaction force symmetry during walking and running. *Research Quarterly for Exercise and Sport*, 55:289-293.

37. Hamill, J., et al. (1989). Relationship between selected static and dynamic lower extremity measures. *Clinical Biomechanics*, 4:217-225.

38. Hamill, J., et al. (1983). Variations in ground reaction force parameters at different running speeds. *Human Movement Science*, 2:47-56.

39. Hamill, J., et al. (1987). The effect of track turns on lower extremity function. *International Journal of Sports Biomechanics*, 3:276-286.

40. Harman, E. A., et al. (1990). The effects of arms and counter-movement on vertical jumping. *Medicine & Science in Sports & Exercise*, 22:825-833.

41. Hawkins, D., Hull, M. L. (1993). Muscle force as affected by fatigue: Mathematical model and experimental verification. *Journal of Biomechanics*, 26:1117-1128.

42. Hintermeister, R. A., Hamill, J. (1992, August). Mechanical power and energy in level treadmill running. *Proceedings of the North American Congress Of Biomechanics II*. Published by the organizing committee of the Fourth North American Congress on Biomechnics, Chicago, IL: 213-214.

43. Komi, P. V. (1990). Relevance of in vivo force measurements to human biomechanics. *Journal of Biomechanics*, 23:23-34.

44. Komi, P. V., Bosco, C. (1978). Utilization of stored elastic energy in leg extensor muscles by men and women. *Medicine & Science in Sports & Exercise*, 10:261-265.

45. Komi, P. V., et al. (1987). In vivo registration of Achilles tendon forces in man. I. Methodological development. *International Journal of Sports Medicine,* 8(S 1): S3-S8.

46. Kyle, C. R., & Wapert, R. A. (1989). The wind resistance of the human figure in sports. *In Proceedings First IOC World Congress on Sport Sciences,* US Olympic Committee, Colorado Springs, 287:287-288.

47. Kyle, C. R., Caizzo, V. L. (1986). The effect of athletic clothing aerodynamics upon running speed. *Medicine & Science in Sports & Exercise,* 18:509-513.

48. Lees, A. (1981). Methods of impact absorption when landing from a jump. *Engineering Medicine,* 10:207-211.

49. Mather, J. S. B. (2000). Innovative golf clubs designed for the amateur. In A. J. Subic, S. J. Haake (Eds.). *The Engineering of Sport: Research, Development and Innovation.* Malden, MA: Blackwell Science, 61-68.

50. Mather, J. S. B., Jowett, S. (2000). Three-dimensional shape of the golf club during the swing. In A. J. Subic, S. J. Haake (Eds.). *The Engineering of Sport: Research, Development and Innovation.* Malden, MA: Blackwell Science, 77-85.

51. McMahon, T. A., Greene, P. R. (1978). Fast running tracks. *Scientific American,* 239(6):148-163.

52. McMahon, T. A., Greene, P. R. (1979). The influence of track compliance on running. *Journal of Biomechanics,* 12:893-904.

53. McNitt-Gray, J. L. (1991). Kinematics and impulse characteristics of drop landings from three heights. *International Journal of Sports Biomechanics,* 7:201-224.

54. Miller, D. I., Nelson, R. C. (1973). *Biomechanics of Sport.* Philadelphia, PA: Lea & Febiger.

55. Miller, D. I. (1990). Ground reaction forces in distance running. In P. R. Cavanagh (Ed.). *Biomechanics of Distance Running.* Champaign, IL: Human Kinetics, 203-224.

56. Munro, C. F., et al. (1987). Ground reaction forces in running: a re-examination. *Journal of Biomechanics,* 20:147-155.

57. Neal, R. J., Wilson, B. D. (1985). 3D kinematics and kinetics of the golf swing. *International Journal of Sports Biomechanics,* 1:221-232.

58. Nigg, B. M. (1983). External force measurements with sports shoes and playing surfaces. In B. M. Nigg, B. Kerr (Eds.). *Biomechanical Aspects of Sports Shoes and Playing Surfaces.* Calgary, AB: University of Calgary, 11-23.

59. Nigg, B. M., et al. (2000). *Biomechanics and Biology of Movement.* Champaign, IL: Human Kinetics.

60. Norman, R. W., et al. (1976). Re-examination of the mechanical efficiency of horizontal treadmill running. In P. V. Komi (Ed.). *Biomechanics V-B.* Baltimore, MD: University Park, 87-93.

61. Ozguven, H. N., Berme, N. (1988). An experimental and analytical study of impact forces during human jumping. *Journal of Biomechanics,* 21:1061-1066.

62. Pandy, M. G., Zajac, F. E. (1991). Optimal muscular coordination strategies for jumping. *Journal of Biomechanics,* 24:1-10.

63. Panzer, V. P, et al. (1988). Lower extremity loads in landings of elite gymnasts. In G. deGroot, et al. (Eds.). *Biomechanics XI.* Amsterdam: Free University Press, 727-735.

64. Pierrynowski, M. R., et al. (1980). Mechanical energy transfer in treadmill walking. *Ergonomics,* 24:1-14.

65. Ramey, M. R., Williams, K. R. (1985). Ground reaction forces in the triple jump. *International Journal of Sports Biomechanics,* 1:233-239.

66. Rogers, M. M. (1988). Dynamic biomechanics of the normal foot and ankle during walking and running. *Physical Therapy,* 68:1822-1830.

67. Rozendaal, L. A., Veeger, H. E. J. (2000). Force direction in manual wheel chair propulsion: balance between effect and cost. *Clinical Biomechanics,* 15:S39-S41.

68. Sayers, S. P., et al. (1999). Cross-validation of three jump power equations. *Medicine & Science in Sports & Exercise,* 31:572-577.

69. Schleihauf, R. E. (1979). A hydrodynamic analysis of swimming propulsion. In J. Terauds, E. W. Bedington (Eds.). *Swimming III.* Baltimore, MD: University Park Press, 70-109.

70. Scott, S. H., Winter, D. A. (1990). Internal forces at chronic running injury sites. *Medicine & Science in Sports & Exercise,* 22:357-369.

71. Scott, S. H., Winter, D. A. (1991). Talocrural and talocalcaneal joint kinematics and kinetics during the stance phase of walking. *Journal of Biomechanics,* 24:743-752.

72. Simpson, K. J., Jiang, P. (1999). Foot landing position during gait influences ground reaction forces. *Clinical Biomechanics,* 14:396-402.

73. Sprigings, E., et al. (1989). Development of a model to represent an aluminum springboard in diving. *International Journal of Sports Biomechanics,* 5:297-307.

74. Stainsby W. N., et al. (1980). Exercise efficiency: validity of baseline subtractions. *Journal of Applied Physiology,* 48(3): 518-522.

75. Tesio, L., et al. (1998). The 3-D motion of the center of gravity of the human body during level walking. I. Normal subjects at low and intermediate walking speeds. *Clinical Biomechanics,* 13:77-82.

76. Valiant, G. A. (1987). Ground reaction forces developed on artificial turf. In H. Reilly and A. Lees (Eds.). *Proceedings of the First World Congress of Science and Medicine in Football.* London: E&FN, 143-158.

77. Valiant, G. A., Cavanagh, P. R. (1985). A study of landing from a jump: Implications for the design of a basketball shoe. In D. A. Winter, et al. (Eds). *Biomechanics IX-B.* Champaign, IL: Human Kinetics, 117-122.

78. Valiant, G. A., et al. (1986). Measurements of the rotational friction of court shoes on an oak hardwood playing surface. *Proceedings of the North American Congress on Biomechanics.* 295-296.

79. Van der Helm, F. C. T., Veeger, H. E. J. (1996). Quasi-static analysis of muscle forces in the shoulder mechanism during wheelchair propulsion. *Journal of Biomechanics,* 29:39-52.

80. van Ingen Schenau, G. J. (1982). The influence of air friction in speed skating. *Journal of Biomechanics,* 16:449-453.

81. Vanlandewijck, Y. C., et al. (1994). Wheelchair propulsion efficiency: Movement pattern adaptations to speed changes. *Medicine & Science in Sports & Exercise,* 26:1373-1381.

82. Veeger, H. E., et al. (1989). Wheelchair propulsion technique at different speeds. *Scandinavian Journal of Rehabilitation Medicine,* 21:197-203.

83. Veeger, H. E. J., et al. (1991). Load on the upper extremity in manual wheelchair propulsion. *Journal of Electromyography & Kinesiology,* 1:270-280.

84. Walpert, R. A., Kyle, C. J. (1989). Aerodynamics of the human body in sports. *Proceedings of the XII International Congress of Biomechanics,* 346-347.

85. White, S. C., Winter, D. A. (1985). Mechanical power analysis of the lower limb musculature in race walking. *International Journal of Sports Biomechanics,* 1:15-24.

86. Williams, K. R. (1980). A biomechanical and physiological evaluation of running efficiency. Unpublished doctoral dissertation, The Pennsylvania State University.

87. Williams, K. R., Cavanagh, P. R. (1983). A model for the calculation of mechanical power during distance running. *Journal of Biomechanics*, 16:115-128.

88. Williams, K. R., Cavanagh, P. R. (1983). The mechanics of foot action during the golf swing and implications for shoe design. *Medicine & Science in Sports & Exercise*, 15:247-255.

89. Williams, K. R., Sih, B. L. (1999). Ground reaction forces in regular-spike and alternative-spike golf shoes. In M. R. Farrally, A. J. Cochran (Eds.). *Science and Golf III: Proceedings of the 1998 World Scientific Congress of Golf.* Champaign, IL: Human Kinetics, 568-575.

90. Winter, D. A. (1979). A new definition of mechanical work done in human movement. *Journal of Applied Physiology*, 46: 79-83.

91. Winter, D. A. (1990). *Biomechanics and Motor Control of Human Movement,* 2nd ed. New York: Wiley.

CHAPTER 11

ANGULAR KINETICS

OBJECTIVES

After reading this chapter, the student will be able to:

1. Define *torque,* and discuss the characteristics of a torque.
2. State the angular analogs of Newton's three laws of motion and their impacts on human movement.
3. Discuss the concept of moment of inertia.
4. Understand the impact of angular momentum on human motion.
5. Define the concept of center of mass.
6. Calculate the segment center of mass and the total body center of mass.
7. Differentiate between the three classes of levers.
8. Define and conduct a static analysis on a single joint motion.
9. Define *stability,* and discuss its effect on human movement.
10. Define and conduct a dynamic analysis on a single joint motion.
11. Define the impulse–momentum relationship.
12. Discuss the relationships between torque, angular work, rotational kinetic energy, and angular power.
13. Define the work–energy relationship.

OUTLINE

Torque or Moment of Force
   Characteristics of a Torque
   Force Couple
Newton's Laws of Motion:
  Angular Analogs
   First Law: Law of Inertia
   Second Law: Law of Angular
    Acceleration
   Third Law: Law of Action–Reaction
Center of Mass
   Center of Mass Calculation:
    Segmental Method
Rotation and Leverage
   Definitions
   Classes of Levers

Second-Class Lever
Third-Class Lever
Types of Torque
Representation of Torques Acting
  on a System
Analysis Using Newton's Laws
  of Motion
   Effects of a Torque at an Instant in Time
   Effects of Torque Applied over
    a Period of Time
   Angular Work: Effects of Torque
    Applied over a Distance
Special Torque Applications
   Angular Power
   Energy

Angular Kinetics of
   Locomotion
Angular Kinetics of the
   Golf Swing

Angular Kinetics of Wheelchair
   Propulsion
Summary
Review Questions

Chapter 10 discussed the notion that movement does not occur unless an external force is applied. Also discussed were the characteristics of a force, two of which were the line of action and the point of application. If the line of action and the point of application of a force are critical, it would appear that the type of motion produced may depend on these characteristics. For example, a nurse pushing a wheelchair exerts two equal forces, one on each handle. The result is that the lines of action and points of application of the two forces cause the wheelchair to move in a straight line. What happens, though, when the nurse pushes the chair with only one arm, applying a force to only one of the handles? A force is still applied, but the motion is totally different. In fact, the wheelchair will translate and rotate (Fig. 11-1). The situation that has

just been described actually represents most of the types of motion that occur when humans move. It is rare that a force or a system of forces cause pure translation. In fact, the majority of force applications in human movement cause simultaneous translation and rotation.

The branch of mechanics that deals with the causes of motion is called kinetics. The branch of mechanics that deals with the causes of angular motion is called **angular kinetics**.

Torque or Moment of Force

When a force causes a rotation, the rotation occurs about a pivot point, and the line of action of the force must act at a distance from the pivot point. When a force is applied such that it causes a rotation, the product of that force and the perpendicular distance to its line of action is referred to as a **torque** or a **moment of force**. These terms are synonymous and are used interchangeably in the literature and will be used interchangeably in this text. A torque is not a force but merely the effect of a force in causing a rotation. A torque is defined, therefore, as the tendency of a force to cause a rotation about a specific axis. In a two-dimensional (2D) analysis, the axis about which the torque acts is neither the horizontal nor vertical axis. The torque acts about an axis that is perpendicular to the x–y plane. This axis is called the z-axis. Thus, the torques referred to in this chapter always act about the z-axis (Fig. 11-2).

CHARACTERISTICS OF A TORQUE

The two important components of a torque are magnitude of the force and the shortest or perpendicular distance from the pivot point to the line of action of the force. Also, any discussion of a torque must be with reference to a specific axis serving as the pivot point. Mathematically, torque is:

$$T = F \times r$$

where T is the torque, F is the applied force in newtons, and r is the perpendicular distance (usually in meters) from the line of action of the force to the pivot point. Because torque is the product of a force, with units of newtons, and a distance, with units of meters, torque has units of newton-meters (N-m). The distance, r, is referred to as the **torque arm** or the **moment arm** of the force (these terms may also be used interchangeably). This concept is illustrated in Figure 11-3. If the force acts directly

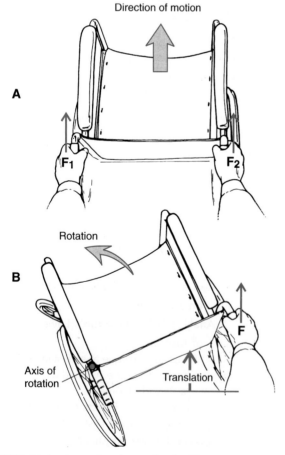

FIGURE 11-1   An overhead view of a wheelchair. **(A)** The wheelchair is translated forward by forces F_1 and F_2. **(B)** The wheelchair rotates and translates if only one force, F_1, is applied.

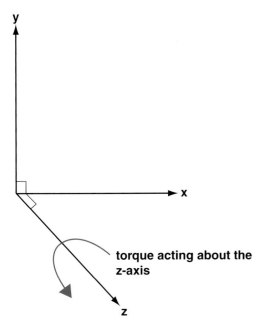

FIGURE 11-2 The z-axis is perpendicular to the x–y plane. In 2D, torques act about the z-axis. A positive torque would appear to come out of the page while a negative torque would go into the page.

through the pivot point or the axis of rotation, the torque is zero because the moment arm would be zero. Thus, regardless of the magnitude of the force, the torque would be zero (Fig. 11-3A).

In Figure 11-3B, a force of 20 N is applied perpendicular to a lever at a point 1.1 m from the axis. Because the force is not applied through the axis of rotation, the shortest distance to the axis is the perpendicular distance from the point of contact of the force to the axis. The torque generated by this force can be computed as:

$$T = F \times r$$
$$T = 20 \text{ N} \times 1.1 \text{ m}$$
$$T = 22 \text{ N-m}$$

Another method of calculating a torque uses trigonometric functions. An example of a 20-N force applied 1.44 m from the axis or rotation at an angle of 50° to the lever is shown in Figure 11-3C. This is a more common situation than the former case. To compute the torque, we can compute the moment arm by taking the product of the sine of the angle and the distance the force acts from the axis of rotation. This product would in essence

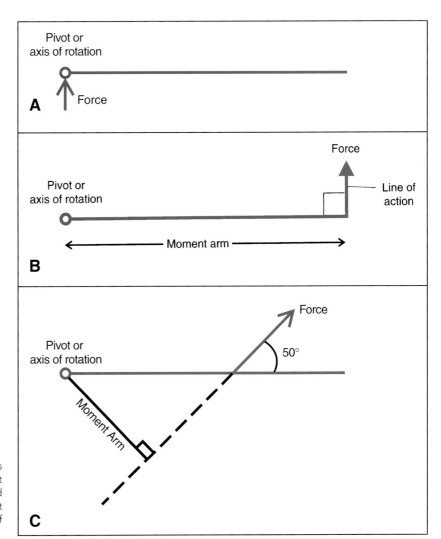

FIGURE 11-3 When a force is applied, it causes no rotation if it is applied through the pivot point (**A**), or it generates a torque if the force is applied a distance from the axis (**B** and **C**). The moment arm is the perpendicular distance from the line of action of the force to the axis of rotation.

give us the perpendicular distance of the line of application of the force to the axis of rotation. This computation would be:

$$T = F \sin \theta \times r$$
$$T = 20 \times \sin 50° \times 1.44 \text{ m}$$
$$T = 20 \text{ N} \times 0.766 \times 1.44 \text{ m}$$
$$T = 22 \text{ N-m}$$

In this case, the perpendicular distance of the moment arm is 1.1 m as we had in the previous example and the torque is the same. The second formula using the sine of the angle for the computation of torque is actually a more general case of the original formula. If the angle between the moment arm and the line of application of the force was 90°, then the sine of 90° is 1, and we simply assume we have a perpendicular distance. In any other case in which the angle is not 90°, taking product of the sine of the angle and the distance from the axis of rotation produces the perpendicular distance.

When the force is not applied through the pivot point, as in Figure 11-3B and C, a torque is said to result from an **eccentric force**, literally an off-center force. Although an eccentric force primarily causes rotation, it also causes translation. Examples of torque applications are illustrated in Figure 11-4. Force can be generated by muscles pulling a distance away from the joint (Fig. 11-4A), by the weight of a body segment acting downward away from a joint (Fig. 11-4B), or by a force coming up from the ground acting a distance away from the center of gravity (Fig. 11-4C).

Torque is a vector quantity and thus has magnitude and direction. The magnitude is represented by the quantity of the magnitude of the force times the magnitude of the moment arm perpendicular to the line of force application. The direction is determined by the convention of the right-hand rule. In this convention, the fingers of the

right hand are placed along line of the moment arm and curled in the direction of the rotation. The direction in which the thumb points indicates the positive or negative sign of the torque. As noted with angular measurement, a counterclockwise direction is considered positive (23), and a clockwise direction is negative. Thus, a counterclockwise torque is positive, and a clockwise torque is negative. Torques in a system may be evaluated through vector operations as has been described for forces. That is, torques may be composed into a resultant torque.

The concept of torque is prevalent in everyday life, for example, in the use of a wrench to loosen the nut on a bolt. Applying a force to the wrench produces rotation, causing the nut to loosen. Intuitively, the wrench is grasped at the end as in Figure 11-5A. Grasping it in this way maximizes the moment arm and hence the torque. If the wrench were grasped at its midpoint, the torque would be halved, even though the same force magnitude is applied (Fig. 11-5B). To achieve the same torque as in the first case, the force magnitude must be doubled. Thus, increasing the force magnitude, or increasing the moment arm, or both, can increase a torque.

The concept of torque is often used in rehabilitation evaluation. For example, if an individual has an injured elbow, the therapist may use a manual resistance technique to evaluate the joint. The therapist would resist the individual's elbow flexion by exerting a force at the mid-forearm position. This creates torque that the patient must overcome. As the individual progresses, the therapist may exert approximately the same force level at the wrist instead of the mid-forearm. By increasing the moment arm while keeping the force constant, the therapist has increased the torque that the patient must overcome. This rather simple technique can be helpful to the therapist in developing a program for the individual's rehabilitation.

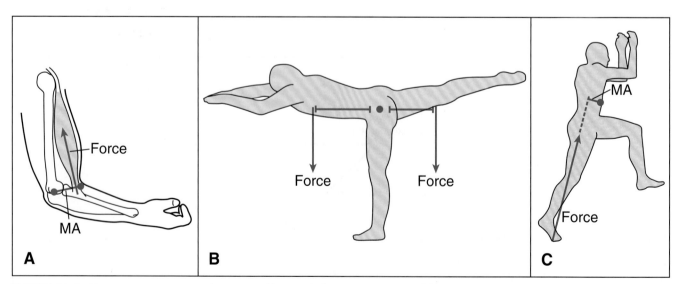

FIGURE 11-4 How torques are commonly generated by muscle force **(A)**, gravitational force **(B)**, and a ground reaction force **(C)**. MA, movement arm.

A

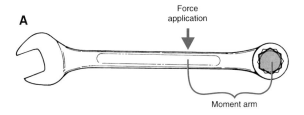

Force
application

Moment arm

B

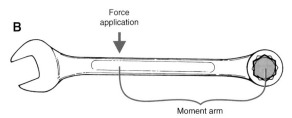

Force
application

Moment arm

FIGURE 11-5 A wrench with two points of force application. Grasping the wrench at the end **(A)** generates more torque than a grasp near the point of rotation **(B)** because the moment arm is greater at **(A)** than at **(B)**.

FORCE COUPLE

A gymnast who wishes to execute a twist about a longitudinal axis applies not one but two parallel forces acting in opposite directions. By applying a backward force with one foot and a forward force with the other, the gymnast creates two torques that produce a rotation about their longitudinal axis (Fig. 11-6). Such a pair of

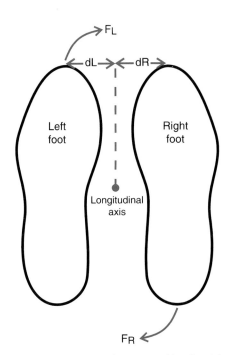

FIGURE 11-6 A torque $F_R \times d_R$ is created by the right foot while another torque, $F_L \times d_L$, is created by the left foot. Because these two torques are equal and in the same angular direction, the force couple will result in a rotation about the longitudinal axis through the center of mass.

forces is called a **force couple**. A force couple is two parallel forces that are equal in magnitude and that act in opposite directions. These two forces act at a distance from an axis of rotation and produce rotation about that axis. A force couple can be thought of as two torques or moments of force, each creating a rotation about the longitudinal axis of the gymnast. Torques, however, also cause a translation, but because the translation caused by each torque is in the opposite direction, the translation is canceled out. Thus, a force couple causes a pure rotation about an axis with no translation. By placing the feet slightly farther apart, the gymnast in Figure 11-6 can increase the moment arm and thus cause a great deal more rotation.

A force couple is calculated by:

$$\text{Force couple} = F \times d$$

where F is one of the equal and opposite forces and d is the distance between the lines of action of the two forces. Although no true force couples exist in human anatomy, the human body often uses force couples. For example, a force couple is created when one uses the thumb and forefinger to screw open the top on a jar.

Newton's Laws of Motion: Angular Analogs

The linear case of Newton's laws of motion was presented in Chapter 10. These laws can be restated to represent the angular analogs. Each linear quantity has a corresponding angular analog. For example, the angular analog of force is torque, of mass is moment of inertia, and of acceleration is angular acceleration. These analogs can be directly substituted into the linear laws to create the angular analogs.

FIRST LAW: LAW OF INERTIA

A rotating body will continue in a state of uniform angular motion unless acted on by an external torque. Stated mathematically as in the linear case:

$$\text{If } \Sigma T = 0 \text{ then } \Delta \omega = 0$$

That is, if the sum of the torques is zero, then the object is either in a state of rest or rotating at a constant angular velocity.

To completely understand this equation, however, we must first discuss the concept of inertia in the angular case. In the linear case, according to Newton's first law of motion, inertia is an object's tendency to resist a change in linear velocity. The measure of an object's inertia is its mass. The angular counterpart to mass is the **moment of inertia**. It is a quantity that indicates the resistance of an object to a change in angular motion. Unlike its linear counterpart, mass, the moment of inertia of a body is dependent not only on the mass of the object but also on the distribution of mass with respect to an axis of rotation.

The moment of inertia also has different values because an object may rotate about many axes. That is, the moment of inertia is not fixed but changeable.

If a gymnast rotates in the air in a layout body position, the way in which the moment of inertia changes can be illustrated. Suppose the gymnast rotates about the longitudinal axis passing through the center of mass of the total body. The center of mass is the point at which all of the mass appears to be concentrated; the calculations are presented later in this chapter. The mass of the gymnast is distributed along and relatively close to this axis that passes through the center of mass (Fig. 11-7A). If, however, the gymnast rotates about a transverse axis through the center of mass of the total body, the same mass is distributed much farther from the axis of rotation (Fig. 11-7B). Because there is a greater mass distribution rotating about the transverse axis than about the longitudinal axis, the moment of inertia is greater in the latter case. That is, there is a greater resistance to rotation about the transverse axis than about the longitudinal axis. The gymnast may also alter mass distribution about an axis by changing body position, as with assuming a tuck position, bringing more body mass closer to the transverse axis, thus decreasing the moment of inertia. In multiple aerial somersaults, gymnasts assume an extreme tuck position by almost placing the head between the knees in an attempt to reduce the moment of inertia. They do this to provide less resistance to angular acceleration and thus complete the multiple somersaults.

The concept of reducing the moment of inertia to enhance angular motion is also seen in running. During the swing phase, the foot is brought forward from behind the body to the point on the ground where the next foot–ground contact will be made. After the foot leaves the ground, however, the leg flexes considerably at the knee, and the foot is raised up close to the buttocks. The effect

of this action is to decrease the moment of inertia of the lower extremity relative to a transverse axis through the hip joint. This enables the limb to rotate forward more quickly than would be the case if the lower limb were not flexed. This action is a distinguishing feature of the lower extremity action of sprinters. Figure 11-8 illustrates the change in the moment of inertia of the lower extremity during the recovery action in running.

The calculation of moment of inertia is not trivial. If all objects are considered to be made up of a number of small particles, each with its own mass and its own distance from the axis of rotation, the moment of inertia can be represented in mathematical terms:

$$I = \sum_{i=1}^{n} m_i r_i^2$$

where I is the moment of inertia, n represents the number of particle masses, m_i represents the mass of the ith particle, and r_i is the distance of the ith particle from the axis of rotation. That is, the moment of inertia equals the sum of the products of the mass and the distance from the axis of rotation squared of all mass particles comprising the object. A dimensional analysis results in units of kilogram-meters squared (kg-m^2) for moment of inertia.

Consider the illustration in Figure 11-9. This hypothetical object is composed of five-point masses, each with a mass of 0.5 kg. The distances r_1 to r_5 represent the distance from the axis of rotation. The point masses are each 0.1 m apart, with the first mass being 0.1 m from the $y-y$ axis. Each point mass is 0.1 m from the $x-x$ axis. The moment of inertia about the $y-y$ axis is:

$$\begin{aligned}
I_{y-y} &= \sum_{i=1}^{n} m_i r_i^2 \\
&= m_1 r_1^2 + m_2 r_2^2 + m_3 r_3^2 + m_4 r_4^2 + m_5 r_5^2 \\
&= 0.5 \text{ kg} \times (0.1 \text{ m})^2 + 0.5 \text{ kg} \times (0.2 \text{ m})^2 + \\
&\quad 0.5 \text{ kg} \times (0.3 \text{ m})^2 + 0.5 \text{ kg} \times (0.4 \text{ m})^2 + \\
&\quad 0.5 \text{ kg} \times (0.5 \text{ m})^2 \\
&= 0.005 \text{ kg-m}^2 + 0.02 \text{ kg-m}^2 + 0.045 \text{ kg-m}^2 + \\
&\quad 0.08 \text{ kg-m}^2 + 0.125 \text{ kg-m}^2 \\
&= 0.275 \text{ kg-m}^2
\end{aligned}$$

If the axis of rotation is changed to the $x-x$ axis, the moment of inertia of the object about this axis would be:

$$\begin{aligned}
I_{x-x} &= \sum_{i=1}^{n} m_i r_i^2 \\
&= m_1 r_1^2 + m_2 r_2^2 + m_3 r_3^2 + m_4 r_4^2 + m_5 r_5^2 \\
&= 0.5 \text{ kg} \times (0.1 \text{ m})^2 + 0.5 \text{ kg} \times (0.1 \text{ m})^2 + \\
&\quad 0.5 \text{ kg} \times (0.1 \text{ m})^2 + 0.5 \text{ kg} \times (0.1 \text{ m})^2 + \\
&\quad 0.5 \text{ kg} \times (0.1 \text{ m})^2 \\
&= 0.005 \text{ kg-m}^2 + 0.005 \text{ kg-m}^2 + 0.005 \text{ kg-m}^2 + \\
&\quad 0.005 \text{ kg-m}^2 + 0.005 \text{ kg-m}^2 \\
&= 0.025 \text{ kg-m}^2
\end{aligned}$$

The change in the axis of rotation from the $y-y$ axis to the $x-x$ axis thus dramatically reduces the moment of inertia, resulting in less resistance to angular motion about

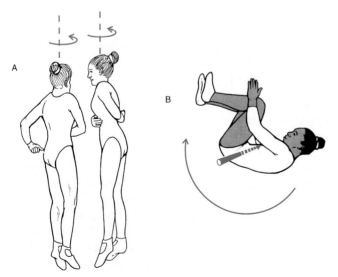

FIGURE 11-7 The mass distribution of an individual about the longitudinal axis through the total body center of mass (A) and about a transverse axis through the total body center of mass (B).

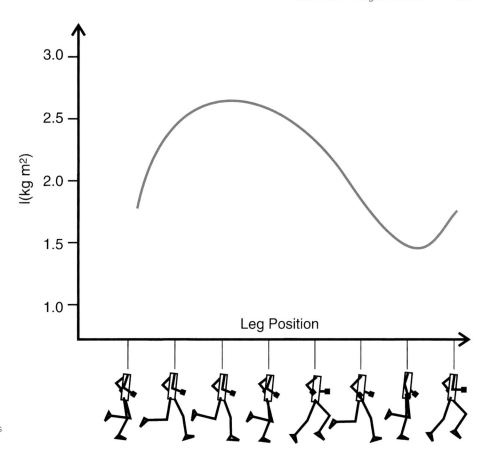

FIGURE 11-8   The changes in the leg's moment of inertia during the stride.

the $x-x$ axis than about the $y-y$ axis. If an axis that passes through the center of mass of the object is used, the mass of point 3 would not influence the moment of inertia because the axis passes directly through this point. Thus:

$$I_{CM} = \sum_{i-1}^{n} m_i r_i^2$$
$$= m_1 r_1^2 + m_2 r_2^2 + m_4 r_4^2 + m_5 r_5^2$$
$$= 0.5 \text{ kg} \times (0.2 \text{ m})^2 + 0.5 \text{ kg} \times (0.1 \text{ m})^2 +$$
$$0.5 \text{ kg} \times (0.1 \text{ m})^2 + 0.5 \text{ kg} \times (0.2 \text{ m})^2$$
$$= 0.02 \text{ kg-m}^2 + 0.005 \text{ kg-m}^2 + 0.005 \text{ kg-m}^2 +$$
$$0.02 \text{ kg-m}^2$$
$$= 0.05 \text{ kg-m}^2$$

From these examples, it should now be clear that the moment of inertia changes according to the axis of rotation.

In the human body, the segments are not as simply constructed as in the example. Each segment is made up of different tissue types, such as bone, muscle, and skin, which are not uniformly distributed. The body segments are also irregularly shaped. This means that a segment is not of uniform density, so it would be impractical to determine the moment of inertia of human body segments using the particle–mass method. Values for the moment of inertia of each body segment have been determined using a number of methods. Moment of inertia values have been

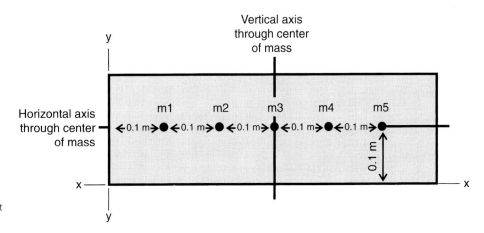

FIGURE 11-9   A hypothetical five-point mass system.

obtained experimentally. The values were generated from cadaver studies (10), mathematical modeling (20,21), and gamma-scanning techniques (63). Jensen (27) has developed prediction equations specifically for children based on the body mass and height of the child.

It is necessary for a high degree of accuracy to calculate a moment of inertia that is unique for a given individual. Most of the techniques that provide values for segment moments of inertia provide information on the segment radius of gyration; from this value, the moment of inertia may be calculated. The **radius of gyration** denotes the segment's mass distribution about the axis of rotation and is the distance from the axis of rotation to a point at which the mass can be assumed to be concentrated without changing the inertial characteristics of the segment. Thus, a segment's moment of inertia may be calculated by:

$$I = m(\rho l)^2$$

where I is the moment of inertia, m is the mass of the segment, l is the segment length, and the Greek letter rho (ρ) is the radius of gyration of the segment as a proportion of the segment length. For example, consider the leg of an individual with a mass and length of 3.6 kg and 0.4 m, respectively. The proportion of the radius of gyration to segment length is 0.302, based on the data of Dempster (10). This information is sufficient to calculate the moment of inertia of the leg about an axis through the center of mass of the leg. Thus, the moment of inertia is:

$$
\begin{aligned}
I_{CM} &= m(\rho l)^2 \\
&= 3.6 \text{ kg} \times (0.302 \times 0.4 \text{ m})^2 \\
&= 0.0525 \text{ kg-m}^2
\end{aligned}
$$

Table 11-1 illustrates the radius of gyration as a proportion of the segment length values from Dempster (10).

Using the radius of gyration technique, the moment of inertia about a transverse axis through the proximal and distal ends of the segment may also be calculated. The radius of gyration as a proportion of segment length about the proximal end of the leg in the previous example is 0.528. Therefore, the moment of inertia about the proximal end of the segment is calculated as:

$$
\begin{aligned}
I_{proximal} &= m(\rho_{proximal} l)^2 \\
&= 3.6 \text{ kg} \times (0.528 \times 0.4 \text{ m})^2 \\
&= 0.161 \text{ kg-m}^2
\end{aligned}
$$

At the distal end point of the leg segment, the moment of inertia is:

$$
\begin{aligned}
I_{distal} &= m(\rho_{distal} l)^2 \\
&= 3.6 \text{ kg} \times (0.643 \times 0.4 \text{ m})^2 \\
&= 0.238 \text{ kg-m}^2
\end{aligned}
$$

Using MaxTRAQ, import the midstance video file of the woman walking. Digitize the end points of the leg (i.e., knee and ankle), convert these values to real units (i.e., meters), and calculate the moment of inertia about transverse axes through the proximal and distal end points of the segment.

The moment of inertia value for any segment is usually given for an axis through the center of mass of the segment. The moment of inertia about an axis through the center of mass is the smallest possible value of any parallel axis through the segment. For example, in Figure 11-10, three parallel transverse axes are drawn through the leg

TABLE 11-1	Radii of Gyration as a Proportion of Segment Length about a Transverse Axis		
Segment	Center of Mass	Proximal	Distal
Head, neck, trunk	0.503	0.830	0.607
Upper arm	0.322	0.542	0.645
Arm	0.303	0.526	0.647
Hand	0.297	0.587	0.577
Thigh	0.323	0.540	0.653
Leg	0.302	0.528	0.643
Foot	0.475	0.690	0.690

Source: Dempster, W. T. (1955). Space requirements of the seated operator. *WADC Technical Report.* Wright-Patterson Air Force Base, 55–159.

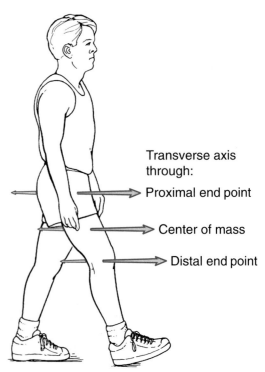

FIGURE 11-10 Parallel transverse axes through the proximal, distal, and center of mass points of the thigh.

Transverse axis through:
Proximal end point
Center of mass
Distal end point

segment. These axes are through the proximal end point, through the center of mass, and through the distal end point. Because the mass of the segment is distributed evenly about the center of mass, the moment of inertia about the center of mass axis is small, and the moments of inertia about the other axes are greater but not equal. This was illustrated in the previous moment of inertia calculations. Because the mass of most segments is distributed closer to the proximal end of the segment, the moment of inertia about the proximal axis is less than about a parallel axis through the distal end point.

The moment of inertia can be calculated about any parallel axis, given the moment of inertia about one axis, the mass of the segment, and the perpendicular distance between the parallel axes. This calculation is known as the **parallel axis theorem**. Assume that the moment of inertia about a transverse axis through the center of mass of a segment is known and it is necessary to calculate the moment of inertia about a parallel transverse axis through the proximal end point. This theorem would state that:

$$I_{prox} = I_{cm} + mr^2$$

where I_{prox} is the moment of inertia about the proximal axis, I_{cm} is the moment of inertia about the center of mass axis, m is the mass of the segment, and r is the perpendicular distance between the two parallel axes. From the calculation in the example of the leg, it was determined that the moment of inertia about an axis through the center of mass was 0.0525 kg-m². If the center of mass is 43.3% of the length of the segment from the proximal end, the moment of inertia about a parallel axis through the proximal end point of the segment can be calculated. If the length of the segment is 0.4 m, the distance between the proximal end of the segment and the center of mass is:

$$d = 0.433 \times 0.4 \text{ m}$$
$$= 0.173 \text{ m}$$

The moment of inertia about the proximal end point, then, is:

$$I_{prox} = I_{cm} + mr^2$$
$$= 0.0525 \text{ kg-m}^2 + 3.6 \text{ kg} \times (0.173 \text{ m})^2$$
$$= 0.161 \text{ kg-m}^2$$

This value is the same as was calculated using the radius of gyration and segment length proportion from Dempster's data (10). Thus, the moment of inertia about an axis through the center of mass is less than the moment of inertia about any other parallel axis through any other point on the segment.

SECOND LAW: LAW OF ANGULAR ACCELERATION

An external torque produces an angular acceleration of a body that is proportional to and in the direction of the torque and inversely proportional to the moment of inertia of the body.

This law may be stated algebraically as:

$$\Sigma T = I\alpha$$

where ΣT is the sum of the external torques acting on an object, I is the moment of inertia of the object, and α is the angular acceleration of the object about the z-axis (on an $X-Y$ plane). For example, if an individual abducts the arm from the body to a horizontal position, the torque at the shoulder results in an angular acceleration of the arm. The greater the moment of inertia of the arm about an axis through the shoulder, the less the angular acceleration of the segment.

The expression of the relationship in Newton's second law is again analogous to the linear case. That is, the sum of the external torques is equal to the time rate of change in angular momentum. That is:

$$T = I\alpha$$
$$T = I\frac{d\omega}{dt}$$
$$T = I\frac{dI\omega}{dt}$$

where $dI\omega$ is the change in angular momentum. Newton's second law can thus be restated:

$$T = \frac{dH}{dt}$$

where H is the angular momentum. That is, torque is equal to the time rate change of angular momentum. To change the angular momentum of an object, external torque must be applied to the object. The angular momentum may increase or decrease, but in either case, external torque is required. **Angular momentum** is the quantity of angular motion of an object and has units of kg-m²·s−1. Angular momentum is a vector and the right-hand rule determines the direction of the vector. Again, counterclockwise rotations are positive, and clockwise rotations are negative.

As with the linear case, by cross-multiplying the preceding equation, we get:

$$T \times dt = dI\omega$$

That is, the product of torque and the time over which it is applied is equal to the change in angular momentum. The quantity $T \times dt$ is referred to as the **angular impulse**.

When gravity is the only external force acting on an object, as in projectile motion, the angular momentum generated at takeoff remains constant for the duration of the flight. This principle is known as the **conservation of angular momentum** and is derived from Newton's first law that the angular momentum of a system will remain constant unless an external torque is applied to the system. Angular momentum is conserved during flight because the body weight vector, acting through the total body center of gravity, creates no torque (i.e., the moment arm is zero). No internal movements or torques generated at the segments can influence the angular momentum generated at takeoff. This principle enables divers and gymnasts to accomplish aerial maneuvers by manipulating their moments of inertia and angular velocities because their angular momentum is constant.

Consider the angular momentum of a gymnast performing an aerial somersault about a transverse axis through the total body center of mass (Fig. 11-11). The torque applied over time at the point of takeoff determines the quantity of angular momentum. During the flight phase, angular momentum does not change. The gymnast, however, may manipulate the moment of inertia to spin faster or slower about the transverse axis. At takeoff, the gymnast is in a layout position with a relatively large moment of inertia and a relatively small angular velocity or rate of spin. As the gymnast assumes a tuck position, the moment of inertia decreases and the angular velocity increases accordingly because the quantity of angular momentum is constant. Having completed the necessary rotation and in preparation to land, the gymnast opens up, assuming a layout position, increasing the moment of inertia and slowing the rate of spin. If these actions are done successfully, the gymnast will land on his or her feet.

To this point, only angular momentum about a single axis has been discussed. Angular momentum about one axis may be transferred to another axis. This occurs in many activities in which the body is a projectile. Although the total angular momentum is constant, it may be transferred, for example, from a transverse axis through the center of mass to a longitudinal axis through the center of mass. For example, a diver may twist about the longitudinal and initiate actions that produce a somersault about the transverse axis. This dive is known as a full twisting one-and-a-half somersault. Researchers have investigated the arm and hip movements to accomplish this change in

angular momentum (15,56,64,65). Other activities that use the principles of transferring angular momentum are freestyle skiing and gymnastics.

Rotations may be initiated in midair even when the total body angular momentum is zero. These are called zero momentum rotations. A prime example of this is the action of a cat when dropped from an upside-down position. The cat initiates a zero momentum rotation and lands on its feet. As the cat begins to fall, it arches its back, or pikes, to create two body sections, a front and a hind section, and two distinct axes of rotation (Fig. 11-12A). The cat's front legs are brought close to its head, decreasing the moment of inertia of the front section, and the upper trunk is rotated 180° (Fig. 11-12B). The cat extends its hind limbs and rotates the hind section in the opposite direction to counteract the rotation of the front segment. Because the moment of inertia of the hind section is greater than that of the front section, the angular distance that the hind section moves is relatively small. To complete the rotation, the cat brings the hind legs and tail into line with its trunk and rotates the back section about

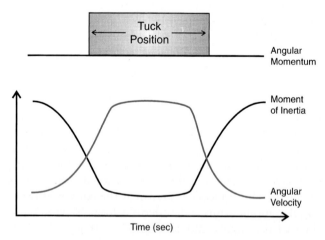

FIGURE 11-11 The angular momentum, moment of inertia, and angular velocity of a diver completing an aerial somersault. Throughout the aerial portion of the dive, the total body angular momentum of the diver is constant. When the diver is in a layout position, the moment of inertia decreases and the angular velocity increases in proportion. During the tuck portion of the dive, the angular velocity increases and the moment of inertia decreases in proportion.

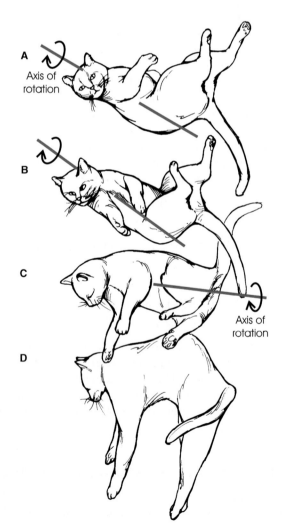

FIGURE 11-12 A cat initiating a rotation in the air in the absence of external torque.

an axis through the hind section (Fig. 11-12C). The reaction of the front portion of the cat to the hind section rotation is small because the cat creates a large moment of inertia by extending its front legs. Finally, the cat has rotated sufficiently to land upright on its four paws (Fig. 11-12D). The use of such actions in sports such as diving and track and field has received considerable attention (12,15).

During human movement, multiple segments rotate. When this happens, each individual segment has angular momentum about the segment's center of mass and also about the total body center of mass. The angular momentum of a segment about its own center of mass is referred to as the **local angular momentum** of the segment. The angular momentum of a segment about the total body center of mass is referred to as the **remote angular momentum** of the segment. A segment's total angular momentum is made up of both local and remote aspects (Fig. 11-13). Expressed algebraically:

$$H_{total} = H_{local} + H_{remote}$$

If the total angular momentum of an individual is calculated, the local aspects of each segment and the remote aspects of each segment must be included. Therefore,

$$H_{total} = \sum_{i-1}^{n} H_{local} + \sum_{i-1}^{n} H_{remote}$$

where i represents each segment and n is the total number of segments.

Local angular momentum is expressed as:

$$H_{local} = I_{cm}\omega$$

where H_{local} is the local angular momentum of the segment, I_{cm} is the moment of inertia about an axis through the segment center of mass, and ω is the angular velocity of the segment about an axis through the segment center of mass. The remote aspect of angular momentum is calculated as:

$$H_{remote} = md\omega'^2$$

where H_{remote} is the remote angular momentum, m is the mass of the segment, d is the distance from the segment

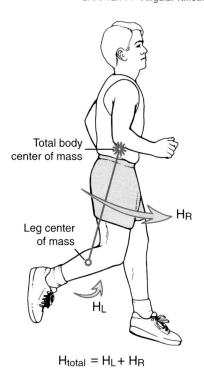

$$H_{total} = H_L + H_R$$

FIGURE 11-13 An illustration of the local (H_L) and remote (H_R) angular momenta of the leg segment.

center of mass to the total body center of mass, and ω' is the angular velocity of the segment about an axis through the total body center of mass (24). Figure 11-14 illustrates the proportion of local and remote angular momentum of the total angular momentum in a forward two-and-a-half rotation dive. In this instance, the remote angular momentum makes up a greater proportion of the total angular momentum than does the local angular momentum.

This technique for calculating total body angular momentum has been used in a number of biomechanics studies. Diving, for example, has been an area of study concerning the angular momentum requirements of many types of dives. Hamill and colleagues (19) reported that the angular momentum of tower divers increased as the number of rotations required in the dive increased (Fig. 11-15).

FIGURE 11-14 The relationship of total body angular momentum, total local angular momentum, and total remote angular momentum of a diver during a forward two-and-a-half rotation dive. The *hash mark* denotes the instant of takeoff. (Adapted from Hamill, J., et al. [1986]. Angular momentum in multiple rotation non-twisting dives. *International Journal of Sports Biomechanics*, 2:78–87.)

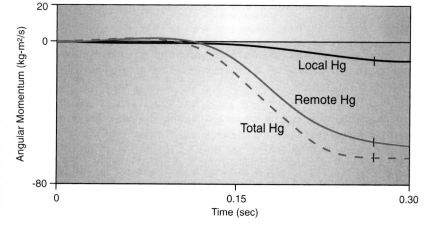

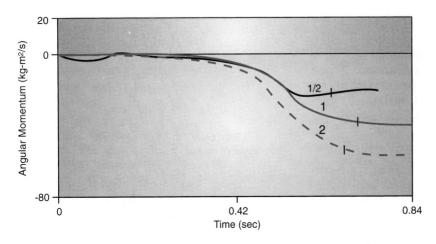

FIGURE 11-15 Profiles of back dives with half a rotation, one rotation, and two rotations depicting the buildup of angular momentum on the platform. The *hash marks* denote the instants of takeoff. (Adapted from Hamill, J., et al. [1986]. Angular momentum in multiple rotation non-twisting dives. *International Journal of Sports Biomechanics*, 2:78–87.)

Values as great as 70 kg-m²/s have been reported for springboard dives (37). The inclusion of a twisting movement with a multiple rotation dive further increases the angular momentum requirements (49).

Hinrichs (25) analyzed the motion of the upper extremities during running by considering angular momentum. He used a three-dimensional analysis to determine angular momentum about the three cardinal axes through the total body center of mass. Hinrichs reported that the arms made a meaningful contribution only about the vertical longitudinal axis. The arms generate alternating positive and negative angular momenta and tend to cancel out the opposite angular momentum pattern of the legs. These findings are illustrated in Figure 11-16. The upper portion of the trunk was found to rotate in conjunction with the arms, and the lower portion of the trunk rotates in conjunction with the legs.

THIRD LAW: LAW OF ACTION–REACTION

For every torque exerted by one body on another body, there is an equal and opposite torque exerted by the latter body on the former.

This law illustrates exactly the same principle as in the linear case. When two objects interact, the torque exerted by object A on object B is counteracted by a torque equal and opposite exerted by object B on object A. These torques are equal in magnitude but opposite in direction. That is:

$$\Sigma T_{A \text{ on } B} = -\Sigma T_{B \text{ on } A}$$

Generally, the torque generated by one body part to rotate that part results in a countertorque by another body part. This concept applies in activities such as long jumping. For example, the long jumper swings the legs forward and upward in preparation for landing. To counteract this lower body torque, the remainder of the body moves forward and downward, producing a torque equal and opposite to the lower body torque. Although the torque and countertorques are equal and opposite, the angular acceleration of these two body portions is different because the moments of inertia are different.

When equal torques are applied to different bodies, however, the resulting angular acceleration may not be the same because of the difference in the moments of inertia of the respective bodies. That is, the effect of one body on another may be greater than the latter body on the former. For example, when executing a pivot during a double play, the second baseman jumps into the air and throws the ball to first base. In throwing the ball, the muscles in the second baseman's throwing arm create a torque as the arm follows through. The rest of the body must counter this torque. These torques are equal and opposite but have a much different effect on the respective segments. While the arm undergoes a large angular acceleration, the body angularly accelerates much less because the moment of inertia of the body is greater than that of the arm.

CENTER OF MASS

An individual's body weight is a product of the mass and the acceleration due to gravity. The body weight vector originates at a point referred to as the **center of gravity**, or the point about which all particles of the body are evenly distributed. The point about which the body's mass is evenly distributed is referred to as the **center of mass**. The terms *center of mass* and *center of gravity* are often used synonymously. The center of gravity, however, refers only to the vertical direction because that is the direction in which gravity acts. The more general term is the center of mass.

If the center of mass is the point about which the mass is evenly distributed, it must also be the balancing point of the body. Thus, the center of mass can be further defined as the point about which the sum of the torques equal zero. That is:

$$\Sigma T_{cm} = 0$$

Figure 11-17 shows an illustration of two objects with different masses. Object A results in a counterclockwise torque about point C, and object B results in a clockwise torque, also about point C. If these two torques are equal, the objects are balanced, and point C may be considered

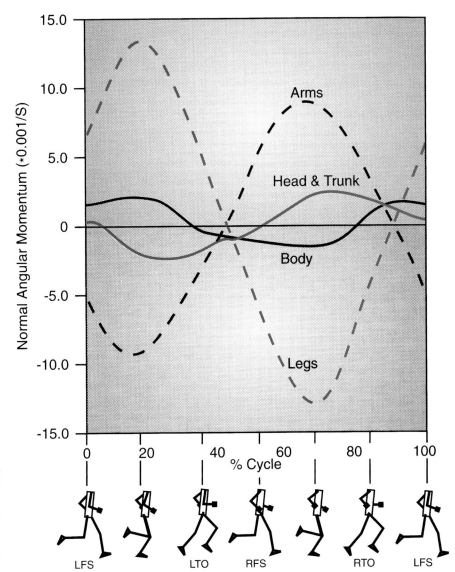

FIGURE 11-16 The vertical component of the angular momentum of the arms, head and trunk, legs, and total body of a runner at a medium running speed. (Adapted from Hinrichs, R. N. [1987]. Upper extremity function in running. II: Angular momentum considerations. *International Journal of Sports Biomechanics*, 3:242–263.)

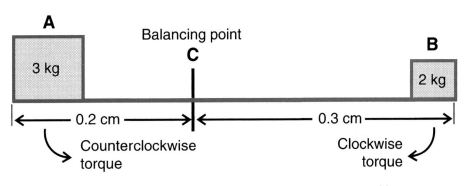

Torque A = (3 kg • 9.81 m/s²) • 0.2 m = 5.89 Nm

Torque B = (2 kg • 9.81 m/s²) • 0.3 m = 5.89 Nm

FIGURE 11-17 A two-point force system balanced at the center of mass.

the center of mass. This does not imply that the mass of these two object masses are the same but that the torques created by the masses are equal (Fig. 11-17).

The center of mass is a theoretical point whose location may change from instant to instant during a movement. The change in position of the center of mass results from the rapidly changing positions of the body segments during movement. In fact, the center of mass does not necessarily have to be inside the limits of the object. For example, the center of mass of a doughnut is within the inner hole but outside the physical mass of the doughnut. In the case of a human performer, the positions of the segments can also place the center of mass outside the body. In activities such as high jumping and pole vaulting, in which the body must curl around the bar, the center of mass is certainly outside the limits of the body (23).

CENTER OF MASS CALCULATION: SEGMENTAL METHOD

Segment Center of Mass Calculation

A number of methods can be used to compute the center of mass of an object using balancing techniques. The most common is to compute the center of mass of individual segments, which are then combined to provide the location of the center of mass of the total system. This approach, called the **segmental method**, involves knowledge of the masses and the location of the centers of mass of each of the body's segments. Two-dimensional coordinates from digitized data (x, y) and the previously mentioned properties of the segments are used to analyze one segment at a time and then calculate the total body center of mass. The estimation of the total body center of mass position is obtained by applying a model that assumes that the body is a set of rigid segments. The center of mass of the total body is calculated using the inertial parameters of each segment and its position.

Before presenting the computations for this method, the source of the information concerning the body segments must be approached. At least three methods for deriving this information have been utilized. They are measures based on cadaver studies, mathematical geometric modeling, and mass scanning.

Several researchers have presented formulae that estimate the mass and the location of the center of mass of the various segments based on cadaver studies (8,9,10,36). These researchers have generated regression or prediction equations that make it possible to estimate the mass and the location of the center of mass. Table 11-2 presents the prediction equations from Chandler and colleagues (8). These predicted parameters are based on known parameters, such as the total body weight and the length or circumference of the segment. An example of a regression equation based on Clauser and colleagues (9) for estimating the segment mass of the leg is as follows:

leg mass = 0.111 calf circumference + 0.047 tibial
height + 0.074 ankle circumference − 4.208

TABLE 11-2 Segment Weight Prediction Equations and Location of Center of Mass

Segment	Weight (N)	Center of Mass[a]
Head	0.032 BW − 18.70	66.3
Trunk	0.532 BW − 6.93	52.2
Upper arm	0.022 BW − 4.76	50.7
Arm	0.013 BW − 2.41	41.7
Hand	0.005 BW − 0.75	51.5
Thigh	0.127 BW − 14.82	39.8
Leg	0.044 BW − 1.75	41.3
Foot	0.009 BW − 2.48	40.0

[a]Location from proximal end as a percentage.
Source: Chandler, R. F. et al. (1975). *Investigation of Inertial Properties of the Human Body.* AMRL Technical Report. Wright-Patterson Air Force Base, 74–137.

where all dimensions of length are measured in centimeters. The location of the center of mass of the segment is usually presented as a percentage of the segment length from either the proximal or distal end of the segment.

Other researchers have used mathematical geometric models to predict the segment masses and locations for the center of mass calculation. This has been done by representing the individual body segments as regular geometric solids (1,20,21,22). Using this method, the body segments are represented as truncated cones (e.g., upper arm, forearm, thigh, leg, and foot), cylinders (e.g., trunk), or elliptical spheres (e.g., head and hand). One such model, the Hanavan model, is presented in Figure 11-18. Regression equations based on the geometry of these solids require the input of several measurements for each segment. For example, the thigh segment requires measurement of the circumference of the upper and lower thigh, the length of the thigh, and the total body mass to estimate the desired parameters for the segment.

The third method of determining the necessary segment characteristics is the gamma-scanning technique suggested by Zatsiorsky and Seluyanov (66). A measure of a gamma-radiation beam is made before and after it has passed through a segment. This allows calculation of the mass per unit surface area and thus generates prediction equations to determine the segment characteristics. The following example equation is based on these data and predicts the mass of the leg segment:

$$y = -1.592 + 0.0362 \times \text{body mass} + 0.0121 \times \text{body height}$$

where y is the leg mass in kilograms. The following equation will predict the location of the center of mass along the longitudinal axis for the same segment:

$$y = -6.05 - 0.039 \times \text{body mass} + 0.142 \times \text{body height}$$

where y is the location of the center of mass as a percentage of the segment length.

All of these methods have been used in the literature to determine the segment characteristics of the center of mass, and all give reasonable estimates of these parameters. In a typical biomechanical analysis, researchers collect height and weight measurements, collect position data on the segmental end points, and calculate the location of the center of mass and the proportionate weight of the segment using information provided by studies in the literature. Tables 11-3 and 11-4 present estimates of the center of mass location and proportionate weights generated from four studies (8,9,10,41). The thigh segment at toe-off (frame 76) from Appendix B will be used as an example to calculate these parameters. The subject in Appendix B is a female weighing 50 kg. Using Plagenhoef's data shown in Table 11-2 (41), the mass of the thigh of a female is 0.1175 times her total body mass. Thus, for this female subject, the mass of her thigh would be:

$$m_{thigh} = 0.1175 \times 50 \text{ kg}$$
$$m_{thigh} = 5.88 \text{ kg}$$

According to Plagenhoef's segmental data, the center of mass is at 42.8% of the length of the thigh measured from the proximal end along the long axis of the segment. Consider the segment end point coordinates at toe-off in Figure 11-19. The location of the center of mass in the x- or horizontal direction would be:

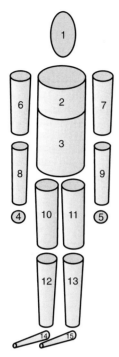

FIGURE 11-18 A representation of the human body using geometric solids. (Adapted from Miller, D. I., Morrison, W. E. [1975]. Prediction of segmental parameters using the Hanavan human body model. *Medicine & Science in Sports & Exercise,* 7(3):207-212.)

TABLE 11-3	Center of Mass Location: Percent of Segment Length from Proximal End			
Segment	Plagenhoef et al. (1983) (7 Men, 9 Women)	Clauser (1969) (13 Male Cadavers)	Dempster (1955) (8 Male Cadavers)	Chandler et al. (1975) (6 Male Cadavers)
Head, neck	M 55.0 F 55.0	M 46.4	M 50.0	M 66.3
Trunk	M 44.5 F 39.0	M 43.8	M 45.0	M 52.2
Whole trunk	M 63.0 F 56.9		M 60.4	
Upper arm	M 43.6 F 45.8	M 51.3	M 43.6	M 50.7
Forearm	M 43.0 F 43.4	M 39.0	M 43.0	M 41.7
Hand	M 46.8 F 46.8	M 48.0	M 49.4	M 51.5
Thigh	M 43.3 F 42.8	M 37.2	M 43.3	M 39.8
Lower leg	M 43.4 F 41.9	M 37.1	M 43.3	M 41.3
Foot	M 50.0 F 50.0	M 44.9	M 42.9	M 40.0

F, Female; M, Male.

TABLE 11-4	Segment Weight: Percent of Total Body Weight			
Segment	Plagenhoef et al. (1983) (37 Men; 100 Women)	Clauser (1969) (13 Male Cadavers)	Dempster (1955) (8 Male Cadavers)	Chandler et al. (1975) (6 Male Cadavers)
Head, neck	M 8.26 F 8.2	M 7.30	M 7.9	M 7.35
Trunk	M 46.8 F 45.22	M 50.7	M 51.1	M 51.66
Whole trunk	M 55.1 F 53.2			
Upper arm	M 3.25 F 2.90	M 2.60	M 2.70	M 3.26
Forearm	M 1.87 F 1.57	M 1.60	M 1.60	M 1.84
Hand	M 0.65 F 0.50	M 0.70	M 0.60	M 0.67
Thigh	M 10.50 F 11.75	M 10.3	M 9.70	M 9.4
Lower leg	M 4.75 F 5.35	M 4.30	M 4.50	M 4.01
Foot	M 1.43 F 1.33	M 1.50	M 1.40	M 1.45

F, Female; M, Male.

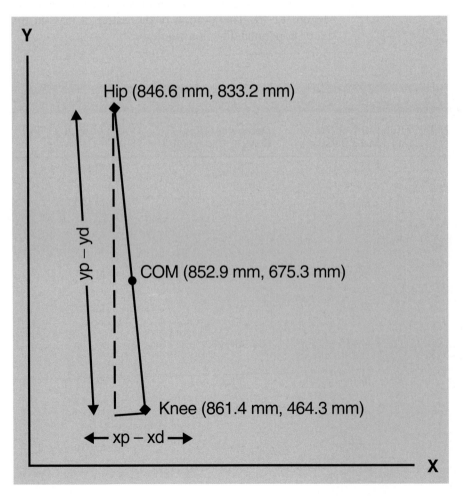

FIGURE 11-19 The thigh segment and the segment end point coordinates at an instant during toe-off in walking (see Appendix C). COM, center of mass.

$$x_{cm} = x_p - [\text{length of the segment in the} \atop x\text{-direction} \times 0.428]$$
$$x_{cm} = x_p - [(x_p - x_d) \times 0.428]$$
$$x_{cm} = 846.6 - [(846.6 - 861.4) \times 0.428]$$
$$x_{cm} = 852.9 \text{ mm}$$

where x_{cm} is the location of the center of mass, x_p is the location of the hip joint, and x_d is the location of the knee joint, all in the horizontal direction. Similarly, for the y- or vertical direction:

$$y_{cm} = y_p - [\text{length of the segment in the} \atop y\text{-direction} \times 0.428]$$
$$y_{cm} = y_p - [(y_p - y_d) \times 0.428]$$
$$y_{cm} = 833.2 - [(833.2 - 464.3) \times 0.428]$$
$$y_{cm} = 675.3 \text{ mm}$$

where y_{cm} is the location of the center of mass, y_p is the location of the hip joint, and y_d is the location of the knee joint, all in the vertical direction. Thus, the center of mass of the segment is (852.9, 675.3) in the reference frame. The center of mass must be between the values for the proximal and distal ends of the segments. This procedure must be carried out for each segment using the coordinates of the joint centers to define the segments.

Refer to the walking data in Appendix C. Compute the location of the center of gravity of the leg from touchdown (frame 0) to toe-off (frame 79), alternating every fourth frame. Plot the path of the center of gravity of the lower leg.

Using MaxTRAC, digitize the knee and ankle positions in the video of the woman in the midstance position of walking. Calculate the center of mass of the leg. Note: This woman has a body mass of 50 kg.

Total Body Center of Mass Calculation

After the segment center of mass locations have been determined, the total body center of mass can be calculated. Consider the illustration of a hypothetical three-segment model in Figure 11-20. The mass and the location of the center of mass of each segment have been previously determined. To determine the horizontal location of the center of mass of the three segments, the torque about the y-axis is calculated using the concept that the sum of the torques about the total system center of mass is zero. Four torques should be considered, three created by the segment centers of mass and one by the total system center of mass. Thus:

$$m_1 g x_1 + m_2 g x_2 + m_3 g x_3 = M g x_{cm}$$

where m is the mass of the respective segments, M is the total system mass, g is the acceleration due to gravity, x is the location of the segment centers of mass, and x_{cm} is the location of the system center of mass. Because the term g

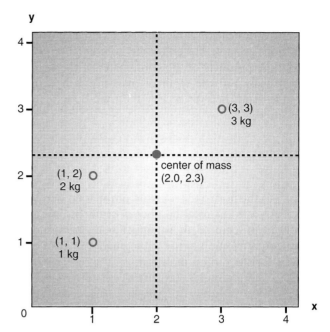

FIGURE 11-20 Location of the center of mass in a three-point mass system.

appears in every term in the equation, it can be removed from the equation, resulting in:

$$m_1 x_1 + m_2 x_2 + m_3 x_3 = M x_{cm}$$

After substitution of the values from Figure 11-20 into this equation, only one quantity is unknown, x_{cm}. Thus:

$$1(1) + 2(1) + 3(3) = 6x_{cm}$$
$$x_{cm} = \frac{12}{6}$$
$$x_{cm} = 2$$

Therefore, the center of mass is 2 units from the y-axis. Using the same procedure, the vertical location of the system center of mass can be located by determining the torques about the x-axis. Thus, the vertical location of the system center of mass can be calculated by:

$$m_1 y_1 + m_2 y_2 + m_3 y_3 = M y_{cm}$$
$$1(1) + 2(2) + 3(3) = 6y_{cm}$$
$$y_{cm} = \frac{14}{6}$$
$$y_{cm} = 2.3$$

The center of mass is 2.3 units from the x-axis. Thus, the center of mass of this total system is (2, 2.3). In the x-direction, the mass is evenly distributed between the left and right sides, and thus the location of the center of mass is basically in the center. This is not the case in the y-direction. The total mass is not proportioned evenly between the top and bottom because most of the mass is nearer the top of the system. As such, y_{cm} is closer to most of the mass of the system. The location of the system center of mass is represented in Figure 11-20 by the intersection of the *dashed lines*.

The previous example can be used to generalize the procedure for calculating the total body center of mass. In the previous example, for either the horizontal or vertical position, the products of the coordinates of the segment center of mass and the segment mass for each segment were added and then divided by the total body mass. In algebraic terms:

$$x_{cm} = \frac{\sum_{i=1}^{n} m_i x_i}{M}$$

where x_{cm} is the horizontal location of the total body center of mass, n is the total number of segments, m_i is the mass of the ith segment, M is the total body mass, and x_i is the horizontal location of the ith segment center of mass. Similarly, for the vertical direction:

$$y_{cm} = \frac{\sum_{i=1}^{n} m_i y_i}{M}$$

where y_{cm} is the vertical location of the total body center of mass, n is the total number of segments, m_i is the mass of the ith segment, M is the total body mass, and y_i is the vertical location of the ith segment center of mass.

This technique for calculating the total body center of mass is used in many studies in biomechanics. The computation is based on the segment characteristics and the coordinates determined from a kinematic analysis. For most situations, the body is thought of as a 14-segment model (head, trunk, and two each of upper arms, lower arms, hands, thighs, legs, and feet). In certain situations in which the actions of the limbs are symmetrical, an eight-segment model (head, trunk, upper arm, lower arm, hand, thigh, leg, and foot) will suffice, although the mass of the segments not digitized must be included.

As an example of this technique, consider the illustration of the jumper in Figure 11-21, which shows the coordinates of the segmental end points of the jumper. Using an eight-segment model (head and neck, trunk, upper arm, forearm, hand, thigh, lower leg, and foot), the coordinates of the location of each segment's center of mass is computed using one of the sets of available anthropometric data. For example, if it is assumed that the jumper is a male (body mass = 70 kg), segmental center of gravity locations can be calculated using Dempster's cadaver data (10). The location of the segmental center of mass measured from the proximal end is calculated in for each segment using estimated locations of the center of mass as demonstrated previously (Table 11-5):

$$COM_x = x_p + [(x_d - x_p) \times \text{estimated COM location} \text{ as a \% of segment length}]$$
$$COM_y = y_p + [(y_d - y_p) \times \text{estimated COM location} \text{ as a \% of segment length}]$$

In this example, the center of mass of each segment was calculated as a distance from the proximal end point. The x- and y-coordinate locations of the center of mass are next used to calculate each segmental torque by multiplying the center of mass location times the segment mass proportions. That is:

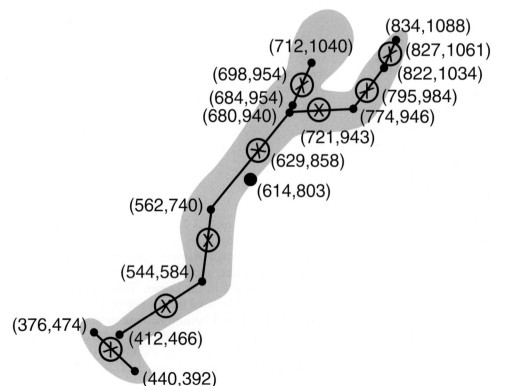

FIGURE 11-21 Segmental end points, segment centers of mass, and the total body center of mass for a jumper. The segment centers of mass are designated by an "X" and the total body center of mass by a black dot.

TABLE 11-5	Calculation of Total Body Center of Mass		
Segment	Segment COM Location Equation	Segment Mass × Segment COM 2x	Segment Mass × Segment COM 2y
Head, neck $x, y_{\text{top of head}} = (712, 1040)$ $x, y_{\text{mid-shoulder}} = (684, 954)$	$x = x_{\text{top of head}} + 0.50(x_{\text{mid-shoulder}} - x_{\text{top of head}})$ $= 712 + 0.5 \times (684 - 712) = 698$ $y = y_{\text{top of head}} + 0.50(y_{\text{mid-shoulder}} - y_{\text{top of head}})$ $= 1040 + 0.5 (954 - 1040) = 997$ $\text{COM}_{\text{head and neck}} = (698, 997)$	Weight ratio$_{\text{head and neck}}$ × COM $- x = 0.079 \times 698 = 55.14$	Weight ratio$_{\text{head and neck}}$ × COM $- y = 0.079 \times 997 = 78.76$
Trunk $x, y_{\text{mid-shoulder}} = (684, 954)$ $x, y_{\text{mid-hips}} = (562, 740)$	$x = x_{\text{mid-shoulder}} + 0.45(x_{\text{mid-hip}} - x_{\text{mid-shoulder}})$ $= 684 + 0.45(562 - 684) = 629.1$ $y = y_{\text{mid-shoulder}} - 0.45(y_{\text{mid-hip}} + y_{\text{mid-shoulder}})$ $= 954 + 0.45(740 - 954) = 857.7$ $\text{COM}_{\text{trunk}} = (629.1, 857.7)$	Weight ratio$_{\text{trunk}}$ × COM $- x = 0.511 \times 629.1 = 321.47$	Weight ratio$_{\text{trunk}}$ × COM $- y = 0.511 \times 857.7 = 438.29$
Right and left upper arm $x, y_{\text{elbow}} = (774, 946)$ $x, y_{\text{shoulder}} = (680, 940)$	$x = x_{\text{shoulder}} + 0.436(x_{\text{elbow}} - x_{\text{shoulder}})$ $= 680 + 0.436(774 - 680) = 721$ $y = y_{\text{shoulder}} + 0.436(y_{\text{elbow}} - y_{\text{shoulder}})$ $= 940 + 0.436(946 - 940) = 942.6$ $\text{COM}_{\text{upper arm}} = (721, 942.6)$	Weight ratio$_{\text{arm}}$ × COM $- x = 0.027 \times 721 = 19.47$	Weight ratio$_{\text{arm}}$ × COM $- y = 0.027 \times 942.6 = 25.45$
Right and left forearm $x, y_{\text{wrist}} = (822, 1034)$ $x, y_{\text{elbow}} = (774, 946)$	$x = x_{\text{elbow}} + 0.43(x_{\text{wrist}} - x_{\text{elbow}}) = 774 +$ $0.43(822 - 774) = 794.6$ $y = y_{\text{elbow}} + 0.43(y_{\text{wrist}} - y_{\text{elbow}}) = 946 +$ $0.43(1034 - 946) = 983.8$ $\text{COM}_{\text{forearm}} = (794.6, 983.8)$	Weight ratio$_{\text{forearm}}$ × COM $- x = 0.016 \times 794.6 = 12.71$	Weight ratio$_{\text{forearm}}$ × COM $- y = 0.016 \times 983.8 = 15.74$
Right and left hand $x, y_{\text{tip of finger}} = (834, 1088)$ $x, y_{\text{wrist}} = (822, 1034)$	$x = x_{\text{wrist}} + 0.494(x_{\text{tip of finger}} - x_{\text{wrist}})$ $= 822 + 0.494(834 - 822) = 827.9$ $y = y_{\text{wrist}} + 0.494(y_{\text{tip of finger}} - y_{\text{wrist}})$ $= 1034 + 0.494(1088 - 1034) = 1060.7$ $\text{COM}_{\text{hand}} = (827.9, 1060.7)$	Weight ratio$_{\text{hand}}$ × COM $- x = 0.006 \times 827.9 = 4.97$	Weight ratio$_{\text{hand}}$ × COM $- y = 0.006 \times 1060.7 = 6.36$
Right and left thigh $x, y_{\text{hip}} = (562, 740)$ $x, y_{\text{knee}} = (544, 584)$	$x = x_{\text{hip}} + 0.433(x_{\text{knee}} - x_{\text{hip}}) = 562 +$ $0.433(544 - 562) = 554.2$ $y = y_{\text{hip}} + 0.433(y_{\text{knee}} - y_{\text{hip}}) = 740 +$ $0.433(584 - 740) = 672.5$ $\text{COM}_{\text{thigh}} = (554.2, 672.5)$	Weight ratio$_{\text{thigh}}$ × COM $- x = 0.097 \times 554.2 = 53.76$	Weight ratio$_{\text{thigh}}$ × COM $- y = 0.097 \times 672.5 = 65.23$
Right and left lower leg $x, y_{\text{knee}} = (544, 584)$ $x, y_{\text{ankle}} = (412, 466)$	$x = x_{\text{knee}} + 0.433(x_{\text{ankle}} - x_{\text{knee}}) = 544 +$ $0.433(412 - 544) = 486.8$ $y = y_{\text{knee}} + 0.433(y_{\text{ankle}} - y_{\text{knee}}) = 584 +$ $0.433(466 - 584) = 532.9$ $\text{COM}_{\text{lower leg}} = (486.8, 532.9)$	Weight ratio$_{\text{lower leg}}$ × COM $- x = 0.045 \times 486.8 = 21.91$	Weight ratio$_{\text{lower leg}}$ × COM $- y = 0.045 \times 532.9 = 23.98$
Right and left foot $x, y_{\text{heel}} = (376, 474)$ $x, y_{\text{tip of toe}} = (440, 392)$	$x = x_{\text{heel}} + 0.429(x_{\text{tip of toe}} - x_{\text{heel}}) = 376 +$ $0.429(440 - 376) = 403.5$ $y = y_{\text{heel}} + 0.429(y_{\text{tip of toe}} - y_{\text{heel}}) = 474 +$ $0.429(392 - 474) = 438.8$ $\text{COM}_{\text{foot}} = (403.5, 438.8)$ TOTAL BODY COM (x, y)	Weight ratio$_{\text{foot}}$ × COM $- x$ $= 0.014 \times 403.5 = 5.65$ $\text{COM}_x = \Sigma\text{Segmental}$ $\text{COM}_x = 55.14 + 321.47 + 19.47$ $+ 19.47 + 12.71 + 12.71$ $+ 4.97 + 4.97 + 53.76$ $+ 53.76 + 21.91 + 21.91$ $+ 5.65 + 5.65 = 613.55$	Weight ratio$_{\text{foot}}$ × COM $- y = 0.014 \times 438.8 = 6.14$ $\text{COM}_y = \Sigma\text{Segmental}$ $\text{COM}_y = 78.76 + 438.29 + 25.45$ $+ 25.45 + 15.74 + 15.74$ $+ 6.36 + 6.36 + 65.23$ $+ 65.23 + 23.98 + 23.98$ $+ 6.14 + 6.14 = 802.76$

COM, center of mass.

Segment torque$_x$ = segment cm$_x$ × estimated segment mass proportion

Segment torque$_x$ = segment cm$_x$ × estimated segment mass proportion

The x_{cm} can then be calculated by summing the segmental

products and y_{cm} is similarly calculated using the equations presented previously in this chapter and expressed below.

$$x_{\text{cm}} = \frac{\sum\limits_{i=1}^{n} m_i x_i}{M}, \quad y_{\text{cm}} = \frac{\sum\limits_{i=1}^{n} m_i y_i}{M}$$

The total body center of mass coordinates in digitizing units, therefore, are (614, 803) and are indicated on Figure 11-21 by the larger solid circle.

> Refer to the data in Appendix C. Using Plagenhoef's data for a woman, calculate the location of the center of mass of a three-link system, including the thigh, lower leg, and foot, for frame 15.
>
> Using MaxTRAC, import the video file of the woman at midstance and digitize the right hip, right knee, right ankle, and right fifth metatarsal head. Using Plagenhoef's data for a woman, calculate the location of the center of mass of a three-link system, including the thigh, leg, and foot. Note: The woman's body mass is 58 kg.

 ## Rotation and Leverage

DEFINITIONS

The outcome of a torque is to produce a rotation about an axis. If rotations about a fixed point are considered, the concept of the lever can be discussed. A **lever** is a rigid rod that is rotated about a fixed point or axis called the **fulcrum**. A lever consists of a **resistance force**, an **effort force**, a barlike structure, and a fulcrum. In addition, two moment or lever arms are designated as the **effort arm** and the **resistance arm**. The effort arm is the perpendicular distance from the line of action of the effort force to the fulcrum. The resistance arm is the perpendicular distance from the line of action of the resistance force to the fulcrum. Because both the effort and resistance forces act at a distance from the fulcrum, they create torques about the fulcrum.

An anatomical example, such as the forearm segment, can be used to illustrate a lever (Fig. 11-22). The long bone of the forearm segment is the rigid barlike structure, and the elbow joint is the fulcrum. The resistance force may be the weight of the segment and possibly an added load carried in the hand or at the wrist. The effort force is produced by the tension developed in the muscles to flex the elbow. Figure 11-23 illustrates several examples of simple machines that are in effect different types of levers.

A lever may be evaluated for its mechanical effectiveness by computing its mechanical advantage (MA). **MA** is defined as the ratio of the effort arm to the resistance arm. That is:

$$MA = \frac{\text{effort arm}}{\text{resistance arm}}$$

In the construction of a lever, any of three situations may define the function of the lever. The simplest case is when MA = 1, that is, when the effort arm equals the resistance arm. In this case, the function of the lever is to alter the direction of motion or balance the lever but

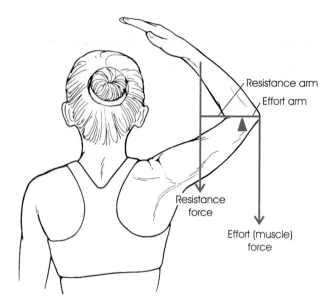

FIGURE 11-22 An anatomical lever showing the resistance arm, effort arm, and fulcrum (elbow joint).

not to magnify either the effort or resistance force. The second case is when the MA > 1, when the effort arm is greater than the resistance arm. In this case, the greater effort arm magnifies the torque created by the effort force. Thus, when MA > 1, the lever is said to magnify the effort force. In the third situation, MA < 1, and the effort arm is less than the resistance arm. In this case, a much greater effort force is required to overcome the resistance force. The effort force acts over a small distance, however, with the result that the resistance force is moved over a much greater distance in the same amount of time (Fig. 11-24). When MA < 1, therefore, velocity or speed of movement is said to be magnified.

CLASSES OF LEVERS

There are three classes of levers. In a **first-class lever**, the effort force and the resistance force are on opposite sides of the fulcrum. Everyday examples of this lever configuration are the seesaw, the balance scale, and the crowbar. A first-class lever may be configured many ways and may have an MA of 1, more than 1, or less than 1. First-class levers exist in the musculoskeletal system of the human body. The agonist and antagonist muscles simultaneously acting on opposite sides of a joint create a first-class lever. In most instances, however, the first-class lever in the human body acts with an MA of 1. That is, the lever acts to balance or change the direction of the effort force.

An example of the former is the action of the splenius muscles acting to balance the head on the atlanto-occipital joint (Fig. 11-25). The latter situation, in which the lever changes the direction of the effort force, is seen in the action of many bony prominences called processes. This type of first-class lever is a pulley. One such example is the action of the patella in knee extension, where the angle

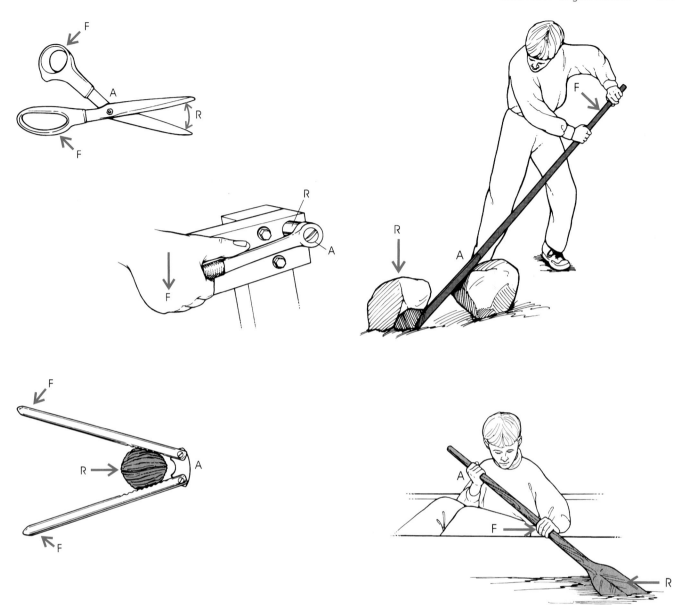

FIGURE 11-23   Levers.

of pull of the quadriceps muscles is altered by the riding action of the patella on the condylar groove of the femur.

SECOND-CLASS LEVER

In a **second-class lever**, the effort force and the resistance force act on the same side of the fulcrum. In this class of lever, the resistance force acts between the fulcrum and the effort force. That is, the resistance force arm is less than the effort arm and thus the MA is greater than 1. One example of a second-class lever in everyday situations is the wheelbarrow (Fig. 11-26). Using the wheelbarrow, effort forces can be applied to act against significant resistance forces provided by the load carried

FIGURE 11-24   A first-class lever in which the moment arm is less than 1, that is, the effort arm is less than the resistance arm. The linear distance moved by the effort force, however, is less than that moved by the resistance force in the same time.

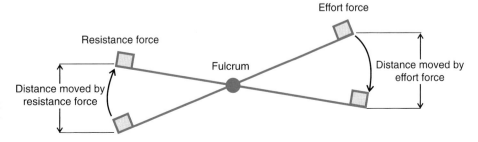

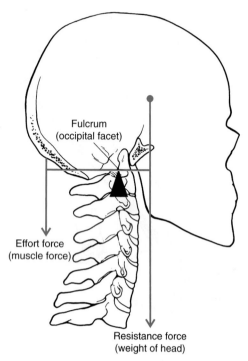

FIGURE 11-25 An anatomical first-class lever in which the weight of the head is the resistance force, the splenius muscles provide the effort force, and the fulcrum is the atlanto-occipital joint.

in the wheelbarrow. There are few examples of second-class levers in the human body, although the act of rising onto the toes is often proclaimed and also disputed as one such. This action is used in weight training and is known as a calf raise. Because there are so few examples of second-class levers in the human body, it is safe to say that humans are not designed to apply great forces via lever systems.

THIRD-CLASS LEVER

The effort force and the resistance force are also on the same side of the fulcrum in a **third-class lever**. In this arrangement, however, the effort force acts between the fulcrum and the line of action of the resistance force. As a result, the effort force arm is less than the resistance force arm and thus the MA is less than 1. An example of this type of lever is the shovel when the hand nearest the spade end applies the effort force (Fig. 11-27). Therefore, it would appear that a large effort force must be applied to overcome a moderate resistance force. In a third-class lever, a large effort force is applied to gain the advantage of increased speed of motion. This is the most prominent type of lever arrangement in the human body, with nearly all joints of the extremities acting as third-class levers. It is probably safe to conclude that from a design standpoint, greater speed of movement exemplified by third-class levers appears to be emphasized in the musculoskeletal system to the exclusion of greater effort force application ability of the second-class lever. Figure 11-28 illustrates a third-class lever arrangement in the human body.

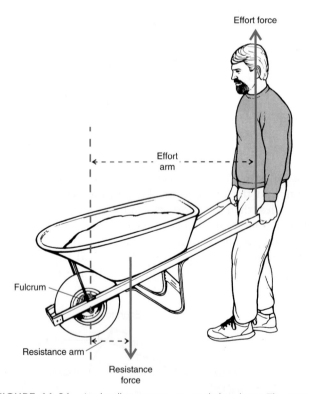

FIGURE 11-26 A wheelbarrow as a second-class lever. The resistance force is between the fulcrum and the effort force. Because the effort arm is greater than the resistance arm, MA > 1, and the effort force is magnified.

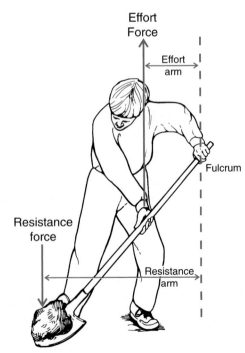

FIGURE 11-27 An individual using a shovel is a third-class lever.

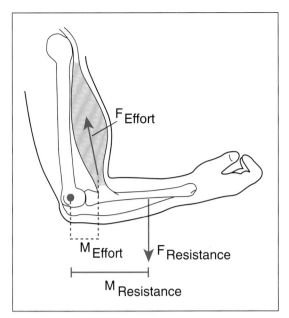

FIGURE 11-28 The arm held in flexion at the elbow is an anatomical third-class lever: The resistance force is the weight of the arm, the fulcrum is the elbow joint, and the effort force is provided by the elbow flexor muscles.

Types of Torque

A torque acting on a body is created by a force acting a distance away from the axis of rotation. Thus, any of the different types of forces discussed in Chapter 10 can produce a torque if applied in a direction that does not go through the axis or pivot point. Gravity, a noncontact force, generates a torque any time the line of gravity does not pass through the hip joint (pivot point). As illustrated in Figure 11-29, gravity acting on the trunk segment

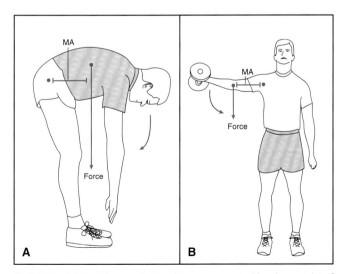

FIGURE 11-29 The gravitational torques created by the weight of body segments acting at a distance from the joint in movements such as trunk flexion **(A)** and an arm lateral raise **(B)** must be countered by muscular torques acting in the opposite direction. MA, moment arm.

produces a clockwise torque about the lumbar vertebrae (Fig. 11-29A), and the weight of the arm and dumbbell produces a clockwise torque about the shoulder joint (Fig. 11-29B). A muscular torque in the clockwise direction to hold the positions statically must counteract both of these gravitational torques.

Contact forces also produce torques if applied correctly. For example, torques are generated about the center of gravity in events such as diving and gymnastics by using the vertical ground reaction force in conjunction with body configurations that move the center of mass in front of or behind the force application. Consider Figure 11-30A, in which the ground reaction force generated in a backward somersault is applied a distance from the center of mass, causing a clockwise rotation about the center of mass. A muscle force also generates torques about the joint center, as shown in Figure 11-30B.

Representation of Torques Acting on a System

A free body diagram illustrating torques acting on a system is usually combined with linear forces to identify and analyze the causes of motion. Many biomechanical analyses start with a free body diagram for each body segment. Known as the rigid link segment model, it can take either a static or dynamic formulation. Consider the model of the dead lift shown in Figure 11-31 showing the lift (Fig. 11-31A) and the free body diagram for the leg, thigh, trunk, arm, and forearm segments (Fig. 11-31B). If a rigid link segment model is developed, forces acting at the joints (F_x, F_y) and the center of mass (W) can be indicated along with moments (M) acting at the joints.

Analysis Using Newton's Laws of Motion

Chapter 10 presented three variations of Newton's laws that describe the relationship between the kinematics and the kinetics of a movement. An angular analog can be generated for each of the three approaches. In most biomechanical analyses, both the linear and angular relationships are determined together to describe the cause-and-effect relationship in the movement. The linear analyses previously discussed presented three approaches categorized as the effect of a force at an instant in time, the effect of a force over a period of time, and the effect of a force applied over a distance. A thorough analysis also includes the angular counterparts and examines the effect of a torque at an instant in time, the effect of a torque over time, and the effect of a torque applied over a distance. Each approach provides different information and is useful in relation to the specific question asked about torque and angular motion.

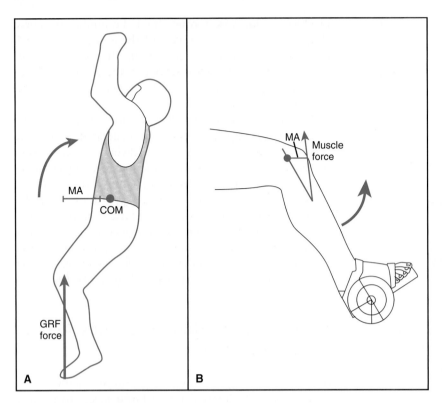

FIGURE 11-30 Contact forces such as ground reaction forces **(A)** and muscular forces **(B)** create torques because the line of action of the force does not go through the center of mass or joint axis, respectively. COM, center of mass; GRF, ground reaction force; MA, moment arm.

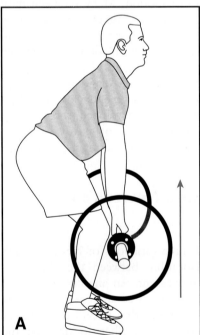

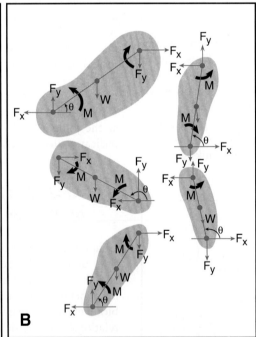

FIGURE 11-31 A squat **(A)** as free body diagram **(B)** using a linked segment model illustrating forces acting at the joints and the centers of mass, along with the moments acting at the joints.

EFFECTS OF A TORQUE AT AN INSTANT IN TIME

Newton's second law of motion is considered with the effects of a torque and the resulting angular acceleration. Thus:

$$\Sigma T = I\alpha$$

When the angular acceleration is zero, a static case is evaluated. A dynamic analysis results when the acceleration is not zero. Note that the torques act about the axis perpendicular to the $x-y$ plane (i.e., the z-axis) and the angular acceleration is also about this axis.

The application of Newton's second law for both angular and linear cause-and-effect relationships would consider both linear and angular equations. In linear motion, the effects of a force and the resulting accelerations at an instant in time are determined, but in angular motion, the effects of a torque and the resulting angular accelerations are determined.

Static Analysis

The static case involves systems at rest of moving at a constant velocity. A state of **equilibrium** exists when the acceleration of the system is zero. As illustrated in Chapter 10, linear equilibrium exists when the sum of the forces acting on the system equal zero. Equilibrium also depends on the balancing of torques acting on the system when forces are not concurrent. Concurrent forces do not coincide at the same point and thus cause rotation about some axis. These rotations all sum to zero, and the resulting system remains at rest or moves at a constant angular velocity. That is, the sum of the moments of force or torques in the system must sum to zero. Stated algebraically, therefore:

$$\Sigma T_{system} = 0$$

Previously, a convention was suggested in which moments or torques causing a counterclockwise rotation were considered to be positive and clockwise torques were considered to be negative. To satisfy this condition of equilibrium, therefore, the sum of the counterclockwise moments must equal the sum of the clockwise moments, and no angular acceleration can occur.

Consider the diagram in Figure 11-32, in which a first-class lever system is described. On the left side of the fulcrum, individual A weighs 670 N and is 2.3 m from the axis of rotation. This individual would cause a counterclockwise or positive moment. Individual B on the right side weighs 541 N, is 2.85 m from the fulcrum, and would cause a clockwise or negative moment. For this system to be in equilibrium, the clockwise torque must equal the counterclockwise torque. Thus:

$$T_A = 670 \text{ N} \times 2.3 \text{ m}$$
$$T_A = 1541 \text{ N-m}$$

and

$$T_B = 541 \text{ N} \times 2.85 \text{ m}$$
$$T_B = 1{,}541 \text{ N-m}$$

For this system to be in equilibrium:

$$\Sigma T = 0$$
$$T_A - T_B = 0$$
$$T_A = T_B$$

Because T_A is positive, T_B is negative, and their magnitudes are equal, these moments cancel each other, and no rotation occurs. With no further external forces, this system will come to rest in a balanced position.

Typically, multiple torques act on a system involving human movement. Figure 11-33A shows the forearm of an individual holding a barbell. To determine the action at the elbow joint, it would be helpful to know the moment caused by the muscles about the elbow joint. If the elbow joint is considered to be the axis of rotation, there are two negative or clockwise torques in this system. One negative torque is a result of the weight of the forearm and hand acting through the center of mass of the forearm–hand system, and the other is the result of the weight of the barbell. The counterclockwise or positive torque is a result of the muscle force acting across the elbow joint. The net moment of the muscles must be equal to the two negative moments for the system to be in equilibrium. Thus:

$$\Sigma T = 0$$
$$T_{muscle} - T_{arm\text{-}hand} - T_{barbell} = 0$$

Consider the free body diagram of this system illustrated in Figure 11-33B. The muscle moment that is necessary to keep the system in equilibrium can be calculated from the information in the free body diagram. The forearm–hand complex weighs 45 N, and the center of mass is 0.15 m from the elbow joint. The weight of the barbell is 420 N, and the center of mass of the barbell

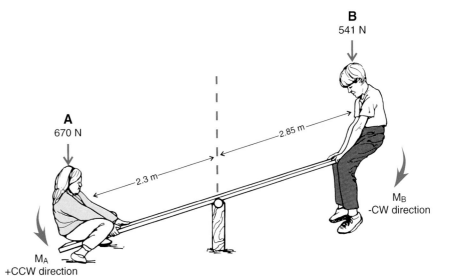

FIGURE 11-32   A first-class lever, a seesaw.

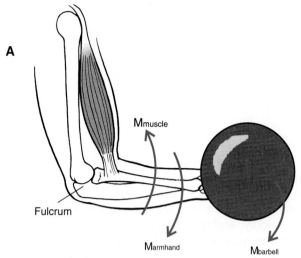

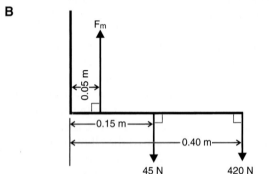

A

B

FIGURE 11-33   The forearm during an instant of a biceps curl **(A)** and a free body diagram of the system **(B)**.

is 0.4 m from the elbow joint. The moment due to the weight of the arm and hand is:

$$T_{arm\text{-}hand} = 45 \text{ N} \times 0.15 \text{ m} = 6.75 \text{ N-m}$$

and the moment due to the barbell is:

$$T_{barbell} = 420 \text{ N} \times 0.4 \text{ m} = 168 \text{ N-m}$$

The moment due to the muscle force can then be calculated as:

$$T_{muscle} - T_{arm\text{-}hand} - T_{barbell} = 0$$
$$T_{muscle} - 6.75 \text{ N-m} - 168 \text{ N-m} = 0$$
$$T_{muscle} = 6.75 \text{ N-m} + 168 \text{ N-m}$$
$$T_{muscle} = 174.75 \text{ N-m}$$

The muscle must create a torque of 174.75 N-m to counteract the weight of the forearm, hand, and barbell. This muscle torque cannot be directly attributed to any one muscle that crosses the joint. The muscle moment calculated is the net sum of all muscle actions involved. In this case, because the arm is being held in flexion, the net muscle moment is primarily due to the action of the elbow flexors, but it cannot be said exactly which elbow flexors are most involved. In fact, it might be surmised that because this is a static posture, there may be considerable co-contraction.

To determine the muscle action, the net torque about the joint must be considered. The net torque is the sum of all torques acting at the joint, in this case, the elbow. Thus:

$$T_{elbow} = T_{muscle} - T_{arm\text{-}hand} - T_{barbell}$$
$$T_{elbow} = 174.75 \text{ N-m} - 6.75 \text{ N-m} - 168 \text{ N-m}$$
$$T_{elbow} = 0$$

The net moment about the elbow joint is zero, indicating that the muscle action must be isometric. If the moment arm of the elbow flexors was estimated to be 0.05 m from the elbow joint, the muscle force would be:

$$F_{muscle} = \frac{T_{muscle}}{0.05 \text{ m}}$$
$$F_{muscle} = \frac{174.75 \text{ N-m}}{0.05 \text{ m}}$$
$$F_m = 3495 \text{ N}$$

It can be seen that the muscle force must be considerably greater than the other two forces because the moment arm for the muscle is very small compared with the moment arms for the forearm–hand or the barbell.

Consider Figure 11-34. In this free body diagram, the arm is placed in a posture similar to that in Figure 11-33, but this barbell weighs 100 N, and the measured muscle torque at the elbow is 180 N-m. All other measures in this situation are the same as in the previous example. Thus, evaluating the net moment at the elbow:

$$T_{elbow} = T_{muscle} - T_{arm\text{-}hand} - T_{barbell}$$
$$T_{elbow} = 180 \text{ N-m} - 6.75 \text{ N-m} - 40 \text{ N-m}$$
$$T_{elbow} = 133.25 \text{ N-m}$$

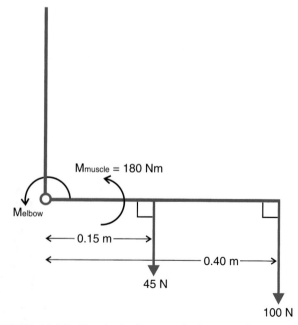

FIGURE 11-34   Free body diagram of a biceps curl at an instant when the forearm is horizontal.

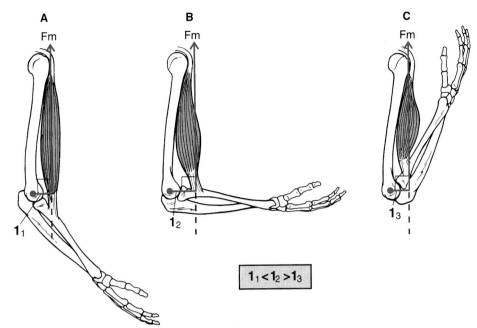

FIGURE 11-35 The change in magnitude of the moment arm of the biceps muscle throughout the range of motion. As the elbow flexes from an extended position, the moment arm becomes longer. As the arm continues to flex from the horizontal forearm position, the moment arm becomes shorter.

$$l_1 < l_2 > l_3$$

Because the net torque at the elbow is positive, the rotation is counterclockwise, and the muscle action therefore is interpreted as a flexor action.

In the previously described situations, the arm was parallel to the ground, and the torque arms from the axis of rotation were simply the distances measured along the segment to the lines of action of the forces. As the arm flexes or extends at the elbow joint, however, the torque arms change. In Figure 11-35, the moment arms for the muscle force are illustrated with the arm in three positions. With the elbow extended (Fig. 11-35A), the moment arm is rather small. As flexion occurs at the elbow (Fig. 11-35B), the moment arm increases until the arm is parallel to the ground. When flexion continues past this point (Fig. 11-35C), the moment arm again becomes smaller. The magnitude of the moment arm of the muscle force, therefore, depends on how much flexion or extension occurs at a joint. The change in moment arm during flexion and extension of the limb is also true for the moment arms for both the weight of the forearm and hand and for anything held in the hand.

In Figure 11-36, the arm is held at some angle θ below the horizontal. The value d describes the distance from the axis of rotation to the center of mass of the arm. The moment arm for the weight of the arm, however, is the distance a. The lines a and d form the sides of a right triangle with the line of action of the force. Therefore, with the angle θ, the cosine function can be used to calculate the length a. Thus:

$$\cos \theta = \frac{a}{d}$$
$$a = d \times \cos \theta$$

The arm is considered to have a 0° angle when it is parallel to the ground. As the arm extends, the angle increases until it is 90° at full extension. The cosine of 0° is 1, and as the angle becomes greater, the cosine gets smaller, until at 90° the cosine is 0. If the angle θ becomes greater as the arm is extended, the moment arm should become correspondingly smaller because the distance d does not change. When the angle θ is 90°, the arm is fully extended and the moment arm is zero because the line of action of the force due to the weight of the arm passes through the axis of rotation.

Consider an individual performing a biceps curl with a weight. The arm is positioned 25° below the horizontal,

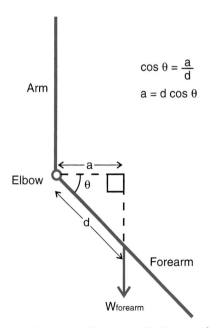

$$\cos \theta = \frac{a}{d}$$
$$a = d \cos \theta$$

FIGURE 11-36 The cosine of the angle of inclination of the forearm is used to calculate the moment arm when the forearm is not parallel to the horizontal.

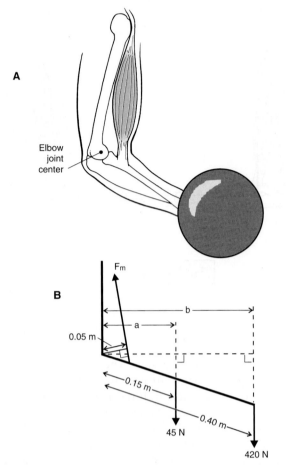

A

Elbow
joint
center

F_m

B

b

a

0.05 m

0.15 m

45 N

0.40 m

420 N

FIGURE 11-37 The forearm at an instant during a biceps curl **(A)** and the free body diagram of this position **(B)**. The arm is inclined below the horizontal.

so that the elbow is slightly extended (Fig. 11-37A). The corresponding free body diagram is presented in Figure 11-37B. The moment arm for the weight of the arm can be calculated from the distance from the axis of rotation to the center of mass and the angle at which the arm is positioned:

$$a = 0.15 \text{ m} \times \cos 25°$$
$$a = 0.15 \text{ m} \times 0.9063$$
$$a = 0.14 \text{ m}$$

Similarly, the moment arm for the weight of the barbell held in the hand can be calculated as:

$$b = 0.4 \text{ m} \times \cos 25°$$
$$b = 0.4 \text{ m} \times 0.9063$$
$$b = 0.36 \text{ m}$$

The moment arms are less than they would be if the arm were held parallel to the ground.

This problem may be solved for the muscle force using the same principles of a static analysis that were discussed previously. That is:

$$\Sigma T = 0$$

If, by convention, the moment due to the muscle is considered to be positive and the moments due to the weight of the arm and the weight of the barbell negative, then:

$$T_{\text{muscle}} - T_{\text{arm}} - T_{\text{weight}} = 0$$

Remember that the torque is the product of a force and its torque arm. Thus, we can substitute all known values into this equation. Rearranging it results in:

$$(F_{\text{muscle}} \times 0.05 \text{ m}) = (45 \text{ N} \times 0.14 \text{ m}) + (420 \text{ N} \times 0.36 \text{ m})$$

This equation can be solved to determine the muscle force necessary to maintain this position in a static posture. Thus:

$$F_{\text{muscle}} = \frac{6.3 \text{ N-m} + 151.2 \text{ N-m}}{0.05 \text{ m}}$$

$$F_{\text{muscle}} = 3,150 \text{ N}$$

The muscle must exert a force much greater than the weight of the arm and the barbell because the moment or torque arm for the muscle is relatively small. It appears that our body is at a great disadvantage when it comes to producing large moments of force about a joint. Most of our joints are arranged as third-class levers, however, indicating that range of motion is magnified. Muscles can, therefore, exert large forces, but only over very short periods of time.

Static Equilibrium: Stability and Balance

The concept of stability is closely related to that of equilibrium. **Stability** may be defined in much the same way as equilibrium, that is, as the resistance to both linear and angular acceleration. The ability of an individual to assume and maintain a stable position is referred to as **balance**. Even in a stable or balanced position, an individual may be subject to external forces.

If a body is in a state of static equilibrium and is slightly displaced by a force, the object may experience three conditions: return to its original position, continue to move away from its original position, or stop and assume a new position. If the object is displaced as a result of work done by a force and returns to its original position, it is said to be in a state of **stable equilibrium**. If the object is displaced and tends to increase its displacement, it is in a state of **unstable equilibrium**. A state of **neutral equilibrium** exists if the object is displaced by a force and does not return to the position from which it was displaced.

Figure 11-38 illustrates the various states of equilibrium. In Figure 11-38A, a ball on a concave surface exemplifies stable equilibrium. When it is displaced along one side of the surface by some force, it will return to its original position. In Figure 11-38B, an example of unstable equilibrium is presented: A force displaces a ball on a convex surface. The ball will come to rest in a new position, not its original position. Figure 11-38C illustrates

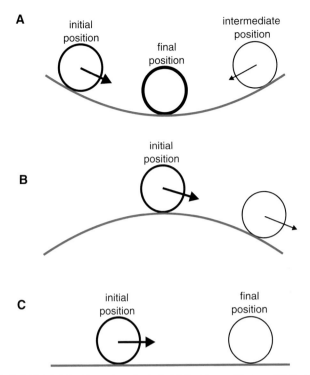

FIGURE 11-38 Examples of a ball. **(A)** Stable equilibrium. **(B)** Unstable equilibrium. **(C)** Neutral equilibrium.

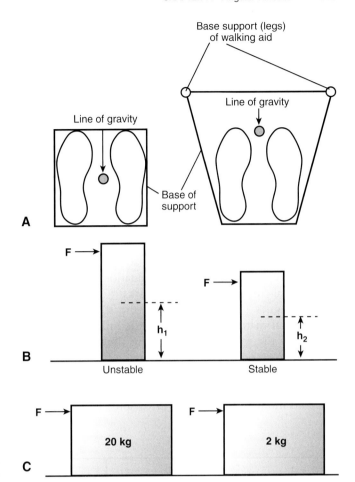

FIGURE 11-39 Factors that influence the stability of an object. **(A)** Increasing the base of support. **(B)** Lowering the center of mass. **(C)** Increasing the mass of the system.

neutral equilibrium. In this case, when a ball is placed on a flat surface and a force is applied, the ball will move to a new position.

With its multiple segments, the human body is much more complex than a ball, but it can assume the different states of equilibrium. An individual doing a headstand, for example, is in a position of unstable equilibrium, and a child sitting on a swing is in a state of stable equilibrium.

Several factors determine the stability of an object. The first factor is where the line of gravity falls with respect to the base of support. An object is most stable if the line of gravity is in the geometric center of the base of support. Increasing the area of the base of support generally increases the stability. A body may be stable in one direction, however, but not in another. For example, spreading one's feet apart increases the area of the base of support and makes the individual stable if pushed in mediolateral direction. It does not, however, help stability in the anteroposterior direction. Increasing the base of support to allow the line of gravity to fall within the base of support may be illustrated by the example of an individual using a walker, as in Figure 11-39A. The base of support is produced by the positions of the individual's legs and the legs of the walker. The walker increases the base of support, and the individual is positioned so that the line of action of the center of mass is in the geometric center of this base.

The stability of an object is also inversely proportional to the height of the center of mass. That is, an object

with a low center of mass tends to be more stable than an object with a high center of mass (Fig. 11-39B). If the two objects in Figure 11-39B undergo the same angular displacement as a result of the forces indicated, the line of gravity of the center of mass of the object on the left, the taller object, will move outside the limit of the base of support sooner than that of the object on the right (the shorter object). Therefore, the force dislodging the object on the left could be less than that for the object on the right. In football, for example, defensive linemen crouch in a three-point stance to keep their center of mass low. This enhances their stability so that they are less likely to be moved by the offensive linemen.

The final factor that influences stability is the mass of the object. According to the equations of motion, the greater the mass of an object, the greater its stability (Fig. 11-39C). Newton's second law states that the force applied to an object is proportional to the mass of the object and its acceleration. Thus, it takes more force to move an object with a greater mass. Moving a piano, for example, is extremely difficult because of its mass. Many sports such as wrestling and judo, in which stability is critical, take body mass into consideration by dividing the contestants into weight divisions because of the disproportionate stability of heavier individuals.

Applications of Statics

It appears that static analyses are limited in their usefulness because they describe situations in which no motion or motion at a constant velocity occurs. A **static analysis** of muscle forces and moments has been used extensively in ergonomics, however, even though the task may involve some movement. The evaluation of workplace tasks, such as lifting and manual materials handling, has been examined in considerable detail with static analyses. Use of static analysis to determine an individual's static or isometric strength is widely accepted in determining the ability of the person's lifting ability. Many researchers have suggested that this static evaluation should be used as a pre-employment screening evaluation for applicants for manual materials handling tasks (7,29,42). Lind and colleagues (32) reported that the static endurance in a manual materials handling task is influenced by the posture of the individual performing the task. A model to evaluate static strength evaluations of jobs has been developed by Garg and Chaffin (17). A free body diagram of one such lifting model is presented in Figure 11-40 (6). This model was presented in Chapter 10 to illustrate

a free body diagram for the linear forces acting on the system (Fig. 11-40A). Adding the moments of forces at each joint (Fig. 11-40B) allows the total lifting model to be evaluated (Fig. 11-40C).

Static analysis techniques have also been used in clinical rehabilitation. Quite often, no movement of a body or a body segment is desirable, so a static evaluation may be undertaken. For example, placing a patient in traction demands that a static force system be implemented. Many bracing systems for skeletal problems, such as scoliosis and genu valgum (knock-knee), use a static force system to counteract the forces causing the problem (Fig. 11-41). Static analyses have been used in the calculation of muscle forces and have been performed by multiple researchers on many joints (18,39,44).

Dynamic Analysis

As pointed out in Chapter 10, a dynamic analysis should be used when the accelerations are not zero. Newton's second law establishes the basis for the dynamic analysis by examining the force–acceleration relationship. In the linear case, the equations of motion for a 2D case are:

$$\Sigma F_x = ma_x$$
$$\Sigma F_y = ma_y$$

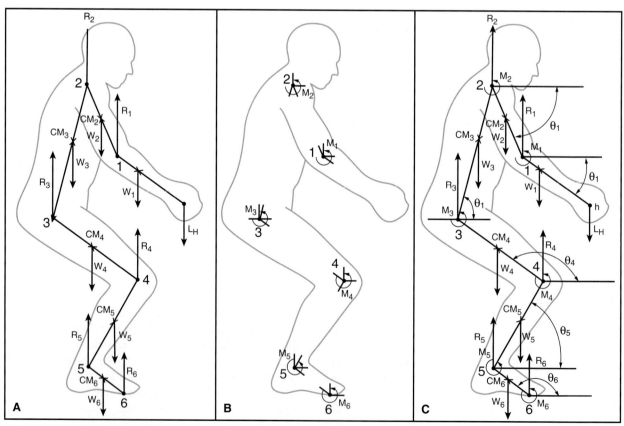

FIGURE 11-40 A free body diagram of a sagittal view static lifting model showing the linear reactive forces **(A)** and moments of forces **(B)** at the joints. These are combined to generate the total lifting model **(C)**. (Adapted from Chaffin, D. B., Andersson, G. B. J. [1991]. *Occupational Biomechanics*, 2nd ed. New York: Wiley.)

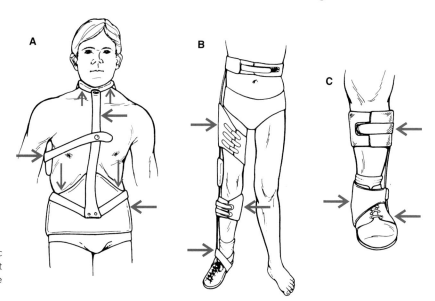

FIGURE 11-41 Bracing systems that illustrate static force systems. (**A**) A neck brace. (**B**) A three-point pressure brace to correct genu valgum. (**C**) A brace to correct a foot deformity.

where the linear acceleration is broken down into its horizontal (x) and vertical (y) components. The angular equivalent looking at the torque–angular acceleration relationship is:

$$\Sigma T_z = I\alpha_z$$

in which I is the moment of inertia and α is the angular acceleration. In a 2D system, angular acceleration occurs about the z-axis. If $\alpha_z = 0$, the motion is purely linear. If $a_x = 0$ and $a_y = 0$, the motion is purely rotational. If α_z, a_x, and $a_y = 0$, a static case exists. The torques acting on a body are created by a contact or gravitational force acting a distance from the axis or rotation. $I\alpha_z$ is the inertial torque similar to the linear case, ma_x and ma_y.

Angular accelerations and the inertial properties of the body segments resisting these accelerations must be considered in the dynamic case. As discussed in Chapter 10, in an **inverse dynamics** approach, each segment is evaluated from the most distal segment and works systematically up the segment links. Consider Figure 11-42, which illustrates a free body diagram of the foot during the swing phase of the gait cycle. In Chapter 10, a dynamic analysis of the linear forces was conducted by applying the 2D linear kinetic equations. Joint reaction forces at the ankle were determined to be -1.57 N for R_x and 20.3 N for R_y.

To determine the net moment acting at the ankle joint, all moments acting on the system must be evaluated. If the center of mass of the foot is considered as the axis of rotation, three moments are acting on this system, two as a result of the joint reaction forces and the net ankle moment itself. Because the joint reaction forces and their moment arms, the moment of inertia of the foot about an axis through the center of mass, and the angular acceleration of the foot are known, the net ankle moment can be calculated. Thus:

$$\Sigma M_{cm} = I_{cm}\alpha_z$$
$$M_{ankle} - M_{Rx} - M_{Ry} = I_{cm}\alpha_z$$

where M_{ankle} is the net moment at the ankle, M_{Rx} is the moment resulting from the horizontal reaction force at the ankle, M_{Ry} is the moment resulting from the vertical reaction force at the ankle, and $I_{cm}\alpha_z$ is the product of the moment of inertia of the foot and the angular acceleration of the foot. In the general equation of moments acting on the foot, the moments M_{Rx} and M_{Ry} will cause clockwise rotation of the foot about the center of mass of the foot. By convention, clockwise rotations are negative, so moments causing a clockwise rotation are negative. Substituting the appropriate values from Figure 11-42 and those calculated in the previous equations and rearranging the equation:

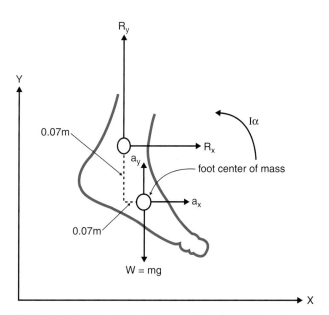

FIGURE 11-42 Free body diagram of the foot during the swing phase of a walking stride. Convention dictates that moments and forces at the proximal joint are positive, as indicated in the free body diagram.

$M_{\text{ankle}} = I_{\text{cm}}\alpha_z + M_{Rx} + M_{Ry}$

$M_{\text{ankle}} = (0.0096 \text{ kg-m}^2 \times -14.66 \text{ rad/s}^2) +$
$\qquad (0.07 \text{ m} \times -1.57 \text{ N}) + (0.07 \text{ m} \times 20.3 \text{ N})$

$M_{\text{ankle}} = -0.141 \text{ N-m} - 0.11 \text{ N-m} + 1.421 \text{ N-m}$

$M_{\text{ankle}} = 1.17 \text{ N-m}$

The net moment at this instant in time is positive, and we conclude that this is a counterclockwise rotation. A counterclockwise rotation of the foot indicates a dorsiflexion action. As with the static case, the exact muscles acting cannot be determined from this type of analysis. Thus, it cannot be stated whether the muscle activity is dorsiflexor concentric or dorsiflexor eccentric.

It was stated that a **dynamic analysis** usually proceeds from the more distal joints to the more proximal joints. The data from this calculation of the foot segment analysis are then used to calculate the net moment at the knee. The calculation continues to the thigh to calculate the hip moment. The analysis is conducted for each joint at each instant in time of the movement under consideration to create a profile of the net moments for the complete movement.

Applications of Dynamics

Dynamic analyses have been used in many biomechanical studies on a number of activities to determine the net moments at various joints. These activities include cycling (43), asymmetrical load carrying (11), weight lifting (31), jumping (57), and throwing (14). Figure 11-43 illustrates the net torques for the hip, knee, and ankle joints in a comparison study of one-legged and two-legged countermovement jumps by 10 volleyball players (57). The joint torques across the push-off phase demonstrated that the one-legged jump generated higher peak joint torques in all three joints and greater mean torques in the hip and ankle joints.

EFFECTS OF TORQUE APPLIED OVER A PERIOD OF TIME

For angular motion to occur, torques must be applied over a period of time. In the linear case, the product of a force applied over a period of time was referred to as the impulse. For the angular analog, the application of a torque over time is known as angular impulse. As in the linear case, we can derive this concept from Newton's second law of motion:

$$T = I\alpha$$
$$T = I \times \frac{d\omega}{dt}$$
$$T = \frac{d(I \times \omega)}{dt}$$
$$T \times dt = d(I \times \omega)$$

or

$$T \times dt = dI\omega_{\text{final}} - dI\omega_{\text{initial}}$$

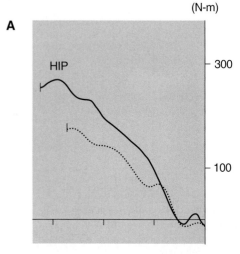

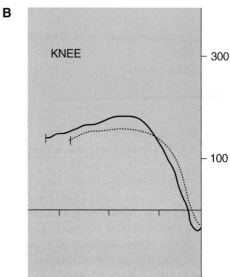

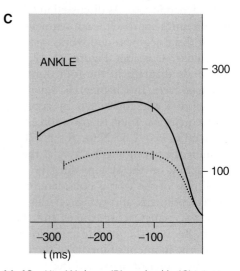

(N-m)

FIGURE 11-43 Hip **(A)**, knee **(B)**, and ankle **(C)** joint torques produced in the push-off phase of one-legged (*solid*) and two-legged (*dotted*) countermovement jumps. (Modified from Van Soest, A. J., et al. [1985]. A comparison of one-legged and two-legged countermovement jumps. *Medicine & Science in Sports & Exercise*, 17: 635–639.)

The left side of the equation represents the angular impulse, and the right side of the equation describes the change in angular momentum. This relationship is known as the **impulse–momentum relationship**. The equation demonstrates that when a torque is applied over a period of time, a change in the angular momentum occurs. Consider the handspring vault in Figure 11-44. The gymnast comes in contact with the horse after generating a large linear horizontal velocity during the approach. This velocity is converted to some vertical velocity and angular momentum. On contact with the horse, the contact with the horse (called a "blocking action") generates two torques. Because the gymnast contacts the horse at an angle, the contact with the horse generates both a vertical force (F_y) and a horizontal force (F_x). The vertical force on the horse increases the vertical velocity off the vault but creates a clockwise torque about the center of mass, which is the product of F_y and d_x. This angular impulse ($T \times t$) changes the angular momentum and decreases the counterclockwise rotation created by the angular momentum generated at takeoff from the board. The horizontal force on the horse acts in an opposite manner by decreasing horizontal velocity and increasing angular momentum generated on the board. A counterclockwise torque about the center of mass is generated via the impulse generated by F_x times d_y, resulting in an increase in the angular momentum from that generated at takeoff. Takei (53) indicated that the loss in angular momentum generated by the application of the vertical force was greater than the corresponding gain generated by the horizontal force, resulting in a loss of angular momentum while on the horse. In elite Olympic gymnasts, the angular momentum at contact averaged 95.3 kg-m²/s for the vaults with the highest scores and 92.4 kg-m²/s for the vaults with the lowest scores. In addition, the angular momentum was reduced by −33 and −27.4 kg-m²/s, respectively, during the horse contact phase as a result of these angular impulses (52).

ANGULAR WORK: EFFECTS OF TORQUE APPLIED OVER A DISTANCE

Mechanical **angular work** is defined as the product of the magnitude of the torque applied against an object and the angular distance that the object rotates in the direction of the torque while the torque is being applied. Expressed algebraically:

$$\text{Angular work} = T \times \Delta\theta$$

where T is the torque applied and $\Delta\theta$ is the angular distance. Because torque has units of N-m and angular

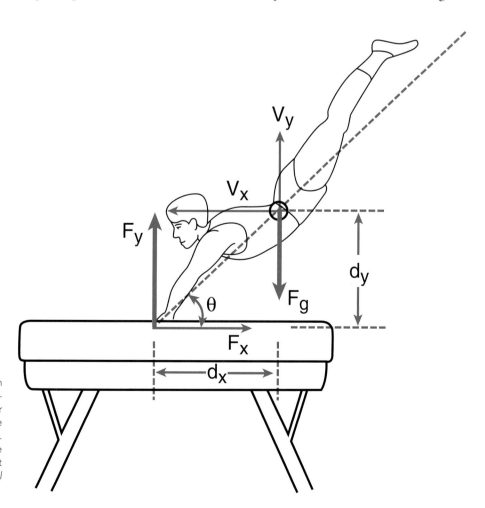

FIGURE 11-44 Torques generated on the horse by the vertical ($F_y \times d_x$) and antero-posterior force ($F_x \times d_y$) generate angular impulses about the center of mass of the vaulter. (Modified from Takei, Y. [1992]. Blocking and post flight techniques of male gymnasts performing the compulsory vault at the 1988 Olympics. *International Journal of Sports Biomechanics*, 8:87–110.)

distance has units of radians, the units for angular work are N-m and joules (J), the same units as in the linear case.

To illustrate angular work, if a 40.5 N-m torque is applied over a rotation of 0.79 rad, the work done in rotation is:

Angular work = $T \times \Delta\theta$

Angular work = 40.5 N-m $\times$ 0.79 rad

Angular work = 32.0 N-m or 32.0 J

Thus, 32 N-m of work is said to be done by the torque, T.

When a muscle contracts and produces tension to move a segment, a torque is produced at the joint, and the segment is moved through some angular displacement. The muscles that rotated the segment do mechanical angular work. To differentiate between the kinds of muscle actions, angular work done by muscles is characterized as either positive work or negative work. **Positive work** is associated with concentric muscle actions or actions in which the muscle is shortening as it creates tension. For example, if a weightlifter performs a biceps curl on a barbell, the phase in which the elbows flex to bring the barbell up is the concentric phase (Fig. 11-45A). During this motion, the flexor muscles of the weightlifter do work on the barbell. **Negative work**, on the other hand, is associated with eccentric muscle actions, or actions in which the muscle is lengthening as it creates tension. In a biceps curl, when the weightlifter is lowering the barbell, resisting the pull of gravity, the flexor muscles are doing negative work (Fig. 11-45B). In this instance, the barbell is doing work on the muscles. Although it has been found that positive work requires a greater metabolic expenditure than negative work, no direct relationship has been reported between mechanical work of muscles and physiologic work.

Special Torque Applications

Angular motion has special torque applications comparable to the force applications in the linear case. Many of the angular applications are direct analogs of the linear case and have similar definitions.

ANGULAR POWER

In Chapter 10, power was defined in the linear case as the work done per unit time or the product of force and velocity. **Angular power** may be similarly defined as:

$$\text{power} = \frac{dW}{dt}$$

where dW is the angular work done and dt is the time over which the work was done. Angular power may also be defined, as in the linear case, using the angular analogs of force and velocity, torque, and angular velocity. Angular power is the angular work done per unit time and is calculated as the product of torque and angular velocity:

$$\text{Angular power} = T \times \omega$$

where T is the torque applied in N-m and ω is the angular velocity in rad/s. Angular power thus has units of N-m/s, or W.

The concept of angular power is often used to describe mechanical muscle power. **Muscle power** is determined by calculating the net torque at the joint and the angular velocity of the joint. The net moment is assumed to describe the net muscle activity across a joint and does not represent any one particular muscle that crosses the joint but the net activity of all muscles. It also does not take into account the situation in which biarticulate muscles may be acting or the fact that there may be co-contraction of these muscles. This net muscle activity at a joint is simply described as flexor or extensor actions, but whether the muscle activity is concentric or eccentric cannot be ascertained directly from the joint moment. The net moment can, however, be used in conjunction with the angular velocity of the joint to determine the concentric or eccentric nature of the muscular action. As discussed previously, concentric actions of muscles are related to the positive

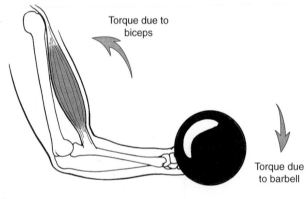

Torque due to biceps

Torque due to barbell

A   Torque due to biceps > Torque due to barbell

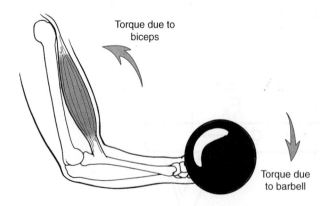

Torque due to biceps

Torque due to barbell

B   Torque due to biceps < Torque due to barbell

FIGURE 11-45 Positive muscular work **(A)** and negative muscular work **(B)** during a biceps curl of a barbell.

work of muscles and eccentric actions as the negative work of muscles. Because the work done by muscles is rarely constant with time, the concept of muscle power can be used. Muscle power is the time rate of change of work and is defined as the product of the net muscle moment and the joint angular velocity. It is expressed algebraically as:

$$P_{\text{muscle}} = M_j \times \omega_j$$

where P_{muscle} is the muscle power in units of W, M_j is the net muscle moment in N-m, and ω_j is the joint angular velocity in rad/s.

Muscle power may be either positive or negative. Because power is the time rate of change of work, the area under the power–time curve is the work done. For example, if M_j and ω_j are either both positive or both negative, muscle power will be positive. Positive muscle power indicates that the muscle is acting concentrically. If M_j is positive and ω_j is negative or M_j is negative and ω_j is positive, muscle power will be negative. Negative muscle power indicates a net eccentric action.

Figure 11-46 illustrates the positive and negative work possibilities of the elbow joint. In Figure 11-46A, M_j and v_j are positive, indicating a flexor moment with the arm moving in a flexor direction. The resulting muscle power is positive, indicating a concentric action of the elbow flexors. In Figure 11-46B, both M_j and v_j are negative, resulting in a positive muscle power or a concentric action. The arm is extending, indicating the muscle action is a concentric action of the elbow extensors. In Figure 11-46C, M_j is positive or a flexor moment, but the arm has a negative v_j, indicating extension. In this case, an external force is causing the arm to extend while the elbow flexors resist. This results in an eccentric action of the elbow flexors and is verified by the negative muscle power. Figure 11-46D illustrates the case in which an eccentric action of the elbow extensors occurs.

Figure 11-47 illustrates the angular velocity–time, net moment–time, and muscle power–time profiles during an elbow flexion followed by elbow extension. In the flexion phase of the movement, the net moment initially is positive, but it becomes negative as the arm becomes more flexed. The initial portion, therefore, results in a positive power or a concentric contraction of the flexor muscles. In the latter portion of the flexor phase, the

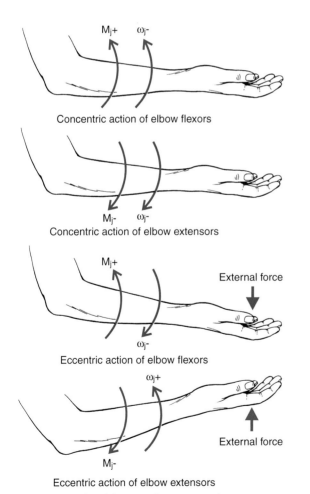

FIGURE 11-46 The definition of positive and negative power at the elbow joint: (A) and (B) result in positive power; (C) and (D) result in negative power.

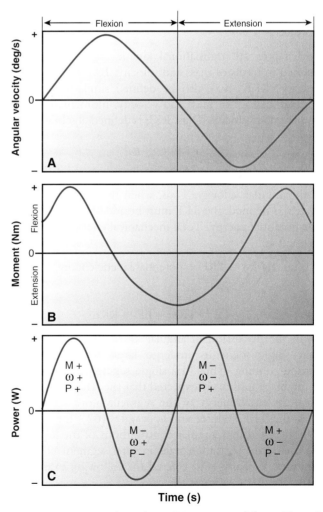

FIGURE 11-47 Angular velocity (A), moment of force (B), and muscle power (C) profiles during an elbow flexion–extension motion.

power is negative, indicating an eccentric muscle contraction. The eccentric contraction of the elbow extensor muscles occurs to decelerate the limb. Because the power is negative at this point, however, it does not mean that there is no flexor muscle activity. It simply means that the predominant activity is extensor. In the extension portion of the movement, the situation is reversed. In the initial portion of extension, the muscle activity is concentric extensor activity. In the latter portion, the muscle power profile again indicates flexor concentric activity to decelerate the limb.

The analysis of net moments of force and muscle power has been widely used in research in biomechanics. Winter and Robertson (63) and Robertson and Winter (46) have investigated the power requirements in walking. Robertson (45) described the power functions of the leg muscles during running to identify any common characteristics among a group of runners. Gage (16) reported on the use of the moment of force and the muscle power profiles as a preoperative and postoperative comparison. The analysis of muscle power in the lower extremity during locomotion, therefore, appears to be a powerful research and diagnostic tool.

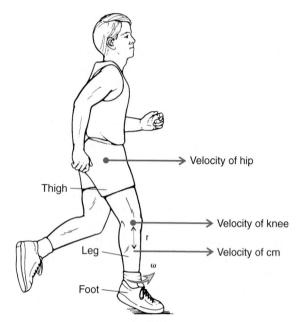

FIGURE 11-48 Theoretical representation of the translational and rotational velocities of the thigh and leg during running. (Adapted from Williams, K. R. [1980]. *A Biomechanical and Physiological Evaluation of Running Efficiency.* Unpublished doctoral dissertation, The Pennsylvania State University.)

ENERGY

In Chapter 10, translational kinetic energy (TKE) was defined in terms of mass and velocity. **Rotational kinetic energy** (RKE) may also be defined similarly using the angular analogs of mass and velocity, moment of inertia, and angular velocity. Thus, RKE is defined algebraically as:

$$RKE = \frac{1}{2} I\omega^2$$

where RKE is the RKE, I is the moment of inertia, and ω is the angular velocity. Thus, when the total energy of a system is defined, the RKE must be added to the TKE and the potential energy. Total mechanical energy is therefore defined as:

total energy = TKE + potential kinetic energy + RKE

or

$$TE = TKE + PE + RKE$$

In the discussion of angular kinematics, it was noted that single segments undergo large angular velocities during running. Thus, if a single segment is considered, it might be intuitively expected that the RKE might influence the total energy of the segment more significantly than the TKE. Winter and colleagues (62), for example, hypothesized that angular contributions of the leg would be important to the changes in total segment energy in running. Williams (59) offered the following example to illustrate that this is not the case.

Consider a theoretical model of a lower extremity segment undergoing both translational and rotational movements (Fig. 11-48). If the leg segment is considered, the linear velocity of the center of mass of the leg is:

$$v_{LEGcm} = \omega r$$

where ω is the angular velocity of the leg and r is the distance from the knee to the center of mass of the leg. If the body segment values of the leg are moment of inertia = 0.0393 kg-m², mass = 3.53 kg, and r = 0.146 m, the magnitude of the RKE is:

$$RKE = \frac{1}{2} I\omega^2$$
$$RKE = 0.5 \times 0.0393 \text{ kg-m}^2 \times \omega^2$$
$$RKE = 0.0197 \, \omega^2$$

The translational kinetic energy of the leg under the same circumstances is:

$$TKE = \frac{1}{2} mv^2$$

where m and v are the mass and the linear velocity of the leg. Substituting the expression ωr into this equation for the linear velocity, v, the equation becomes:

$$TKE = \frac{1}{2} m(\omega r)^2$$
$$TKE = 0.5 \times 3.53 \text{ kg} \times (0.146 \text{ m} \times \omega)^2$$
$$TKE = 0.0376 \, \omega^2$$

Because ω^2 has the same magnitude in both RKE and TKE, the two types of energy can be evaluated based on

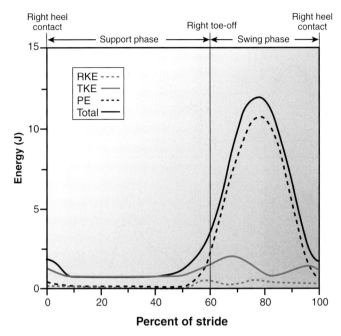

FIGURE 11-49 The relationship between total energy (*Total*), rotational kinetic energy (*RKE*), translational kinetic energy (*TKE*), and potential energy (*PE*) of the foot during a walking stride.

the values of 0.0197 times ω^2 for the rotational energy and 0.0376 times ω^2 for the translational energy. It can be seen that the TKE is almost double that of the RKE. In fact, Williams (59) stated that the RKE of most segments is much less than the TKE. Figure 11-49 illustrates the relationship between the energy components of the foot during a walking stride. In this figure, the magnitude of the total energy is mostly made up of the TKE during the support phase and the potential energy during the swing phase.

Work–Energy Relationship
Mechanical angular work was defined as the product of a torque applied to an object and the distance that the object moved during the torque application. Angular work is said to have been done on an object when a rotation occurs through some angular distance. Rotational energy has also been defined as the capacity to do angular work. Therefore, analogous to the linear case, the **work–energy theorem**, $W = \Delta E$, also applies. That is, for mechanical angular work to be done, a change in the rotational energy level must occur. The angular work done on an object is:

$$W_{\text{angular}} = \Delta \text{RE}$$
$$W_{\text{angular}} = \Delta \left(\frac{1}{2} \ I\omega^2 \right)$$

where W_{angular} is the angular work done on the object and ΔRE is the change in RKE about the center of mass. To calculate the total work done on the object, the other

forms of energy, such as potential and kinetic energy, must also be considered. With the inclusion of the additional form of energy, the work done on the object becomes:

$$W_{\text{object}} = \Delta \text{KE} + \Delta \text{PE} + \Delta \text{RE}$$
$$W_{\text{object}} = \Delta \left(\frac{1}{2} \ mv^2 \right) + \Delta (mgh) + \Delta \left(\frac{1}{2} \ I\omega^2 \right)$$

where W_{object} is the work done on the object, ΔKE is the change in the linear kinetic energy of the center of mass of the object $[\Delta(1/2 \ mv^2)]$, ΔPE is the change in potential energy of the object center of mass $[\Delta(mgh)]$, and RE is the change in rotational energy about the center of mass of the object $[\Delta(1/2 \ I\omega)^2]$. For example, in baseball batting, the goal is to generate maximum energy at contact so maximum work can be generated in the ball. Both the linear and rotational kinetic energies of the bat are important. Potential energy is also a factor because the bat stores potential energy in the handle that is later transferred as local kinetic energy at impact (13).

Angular Kinetics of Locomotion

The angular kinetics of locomotion, specifically the joint moments of force of the lower extremity, have been widely researched (3,60,61). Winter (60) stated that the resultant moment of force provided powerful diagnostic information when comparing injured gait with uninjured gait. Another common area of investigation for gait is muscle mechanical power. Muscle power is the product of the net joint moment and the joint angular velocity. Positive power results when there is a concentric muscle action, such as flexor moments accompanying segment movement in the flexion direction. Negative power is associated with eccentric muscle action when the net moment of force occurs in the opposite direction as the segment movement. For example, negative power would result when a net knee extension moment is generated as the knee is moving into flexion. It is common to see power fluctuate between negative and positive multiple times across the cycle in both walking and running.

Figure 11-50 illustrates the joint kinematics; the net muscle moments of force; and the corresponding powers at the hip, knee, and ankle during a walking stride. At the hip joint, there is a net hip extensor moment during the initial loading phase of support continuing through midsupport into late stance. In late stance, there is a power absorption as hip extension is decelerated via hip flexors (40). In preparation for toe-off, the hip flexors shorten to produce power for the initiation of the swing phase. Hip flexion continues into swing via power production from a hip flexor moment until it is terminated in late swing by a hip extensor moment.

At the knee joint, the loading response involves knee flexion controlled by a knee extensor moment from

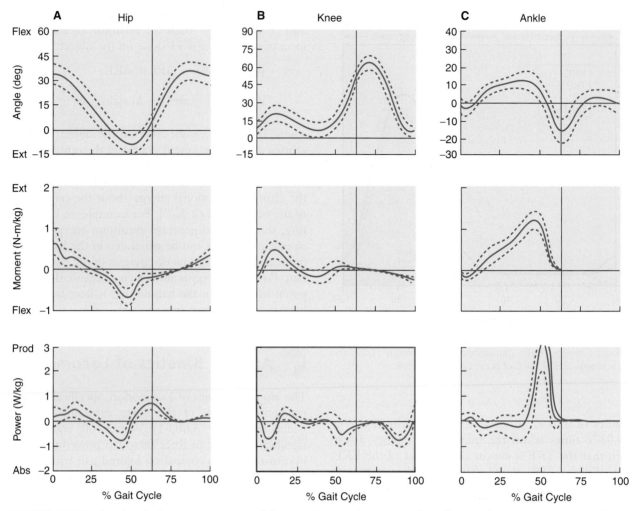

FIGURE 11-50 Angular displacement, moments of force, and power during a single walking stride. **(A)** Hip, **(B)** Knee. **(C)** Ankle. The transition from support to swing is indicated at the solid line. (Adapted from Ounpuu, S. [1994]. The biomechanics of walking and running. *Foot & Ankle Injuries*, 13:843–863.)

touchdown until midsupport. In late stance, there is again a knee flexor moment that moves to a small knee extensor moment. During swing, there is minimal power production until the terminal phase of swing, when the knee flexors act eccentrically to slow knee extension for contact.

The ankle joint exhibits a brief net dorsiflexor moment during the initial loading phase of stance as the foot is lowered to the ground. A transition to a net ankle plantarflexion moment first occurs through eccentric plantarflexion actions to control the leg's rotation over the foot. This is followed by a continuation of a net plantarflexion moment as the plantar flexors concentrically advance the limb into the swing phase. At the actual initiation of the swing phase, plantarflexion continues under the control of eccentric dorsiflexion activity. While in the swing phase, minimal power is produced at the ankle.

The joint moment and power patterns change with the speed of walking. For example, there are three basic moment patterns at the knee joint: a biphasic pattern, a flexor moment resisting an external extensor moment, and an extensor moment after heel strike (26). At slower speeds, there is more use of a flexor moment, with little knee flexion and small knee moments through midstance resulting in negative joint power (26). In a faster walk, there are greater knee flexion and extension moments and more energy generation early in the stance phase followed by energy absorption during early stance.

Figure 11-51 illustrates the joint kinematics; the net muscle moments of force; and the corresponding powers at the hip, knee, and ankle during a running stride (40). The moments of force and powers in running are greater in magnitude than those in walking. The lower extremity moments of force increase in magnitude with increases in locomotor speed. Cavanagh and colleagues (4) and Mann and Sprague (34) reported considerable variability in the magnitudes of the moments of force between subjects running at the same speed. For slow running, this variability is generally smallest at the ankle and greatest at the hip.

Similar to walking, the hip joint extends during both loading and propulsive stages of stance initially via concentric hip extension and later via eccentric hip flexion.

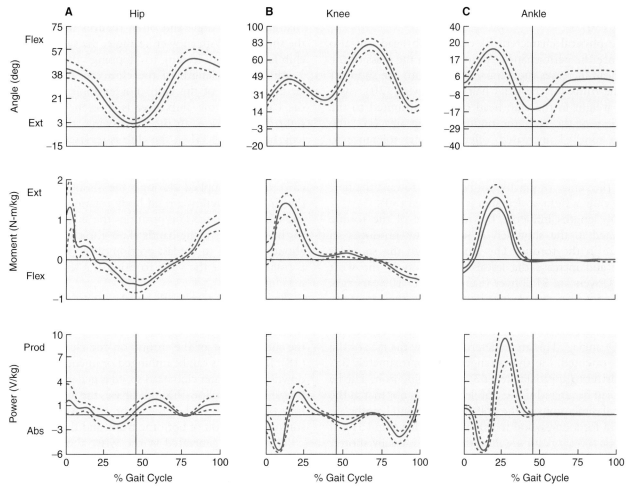

FIGURE 11-51 Angular displacement, moments of force, and power during a single running stride. **(A)** Hip, **(B)** Knee. **(C)** Ankle. The transition from support to swing is indicated at the solid line. (Adapted from Ounpuu, S. [1994]. The biomechanics of walking and running. *Foot & Ankle Injuries*, 13:843–863.)

Concentric hip flexion continues into the swing phase as the thigh is brought forward. This hip flexion continues into late swing when the hip extensors terminate the hip flexion movement and initiate a hip extension (40).

At the knee joint, the loading response is similar to that of walking, involving flexion controlled by the knee extensors to the point of midsupport. From that point, there is a net knee extensor moment. At the start of the swing phase, a small net knee extensor moment is associated with knee flexion. Later in the swing phase, a net knee flexor moment slows the rapidly extending knee.

Ankle joint net joint moments and powers are also similar to those of walking, depending on the style of running. For runners with a typical heel strike footfall pattern, the ankle joint exhibits a small net dorsiflexor moment during the loading phase followed by a net ankle plantarflexion moment for the remainder of the stance phase. At midstance, the net ankle plantarflexion moment controls rapid dorsiflexion, and in later stance, the plantarflexion moment produces rapid plantarflexion. Minimal ankle power is generated in the swing phase.

 Angular Kinetics of the Golf Swing

The actual physical dimensions and characteristics of the club influence the angular kinetics of a golf swing. The composition of the materials of both the shaft and the club head influences swing characteristics. Shafts made of graphite or composite materials are usually lighter and stronger, so using this type of club, a golfer can swing a lighter club faster while producing the same amount of angular work. Adding mass to a club increases the joint torque at the shoulder and the trunk in the latter portion of the swing (30). Use of stiffer shafts generally results in straighter shots. Use of flexible shafts generally results in longer shots, but these clubs are difficult to control, thus influencing accuracy (51). The reason the stiffer staff offers better control is because it has less bend and twist and it compresses the ball more, creating a flight that is more representative of the actual angle of the club (35). A steel club with good direction control may be better for a novice golfer than a graphite club that can achieve more distance but is harder to control. The most desirable physical characteristics of the club are a high moment of inertia and a low center of gravity (50).

The physical characteristics of the club head can also influence the performance. An increase in the mass of the club head increases the joint torque at both the shoulder and the trunk for the latter half of the swing (30). Also, if the weight in the club head can be distributed to the periphery of the face, the optimal hitting area (i.e., the "sweet spot") is increased, offering greater tolerance to off-center hits (50). If the center of gravity of the club head can be lowered, it can produce a higher flight path, and if the center of gravity can be moved toward the heel of the club, a right-to-left spin is promoted (50).

Club lengths influence the magnitude of the torque generated in the shoulder. A longer club produces an increase in the torque. This can open up the shoulder earlier and increase the linear acceleration of the wrist (30). Drivers are 5% longer than they were 10 years ago, and even if they can produce better results, they are more difficult to swing and control (35).

The angle that the shaft makes with the ground is called the "lie angle." This angle determines how the face of the club is oriented. Lie angles range from approximately 55° for a driver to approximately 63° for the nine-iron. The lie angle can be altered if the club length does not match the physical dimensions of the golfer. Sample club lengths are 110 cm for a one-wood to 90 cm for a nine-iron (51). If a club is too short for a golfer, the drive is usually shorter because of lower swing angular velocities. Likewise, if the club is too long, the golfer is required to stand up more and may choke up on the shaft. Standing more erect changes the lie angle and often reduces the control over the shot (51). With a lower lie angle, the sweet spot of the club face is raised, leading to "topping" the ball.

Altering the length of the club can affect the angular acceleration of the club. For example, at the top of the backswing, the club lever is shortened by increased right elbow flexion. In the downswing, the club lever is lengthened (33). When the club is reversed at the top of the backswing, the torque applied to the shaft actually causes it to bend where the head trails the line of the shaft (35). Torque applied through the shoulders accelerates the club, and there is a rigid body rotation with the wrist angle held constant. This positions the shaft perpendicular to the arms. Next, the hands allow the club to accelerate about the wrist. In the last portion of the swing, the shaft is rotated 90° at the wrist to bring the shaft face square into impact (35).

The distance a golf ball travels is related to the speed of the club head at contact. This speed is determined by the torque the golfer applies to the arm–club system and the management of the torque on the club by the wrists (28). If the wrists can be unhinged or uncocked later in the downswing, greater club head speed is generated for a given torque. In the follow-through phase, the angular momentum of the swing takes the right arm over the left and stimulates rotation of both the trunk and the head (33).

The torque generated at the wrist about the vertical (y), anteroposterior (z), and mediolateral (x) axes in the swing of a professional golfer is shown in Figure 11-52.

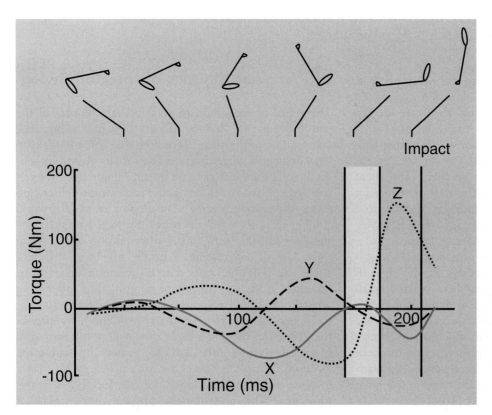

FIGURE 11-52   Joint torques acting at the wrist joint about the vertical (Y), anteroposterior (Z), and mediolateral (X) axes. The *solid line* indicates impact. (Adapted from Neal, R. J., Wilson, B. D. [1985]. 3D kinematics and kinetics of the golf swing. *International Journal of Sports Biomechanics*, 1:221–232.)

In the backswing, a positive torque is generated about the anteroposterior (z) axis as the angle between the club and the arm is maintained. At the initiation of the downswing, a negative torque is generated about the same axis to keep the club back and maintain the same angle between the club and the arm. At the bottom of the downswing, a large positive torque is developed as the wrist uncocks and accelerates the club to impact (38). Rotations about the mediolateral (x) axis are at a maximum as the club is taken back behind the head. Torques about the vertical (y) and anteroposterior (z) axes are zero at this point because the club is parallel to the vertical–anteroposterior plane. Torque about the vertical (y) axis is maximum at the initiation of the downswing as the body rotates around the support.

Angular Kinetics of Wheelchair Propulsion

Hand rim wheelchair locomotion is strenuous and involves a significant amount of mechanical work production by the upper extremity muscles. The magnitude of the joint moment at the shoulder and elbow joints is influenced by the direction of application of the linear force. A large elbow moment is present when the force application to the rim is perpendicular to a line from the hand to the elbow and is minimized when the force is applied along the line from the hand to the elbow (48). Likewise, propelling forces, which are perpendicular to the line between the hand and the shoulder, result in large shoulder moments.

The shoulder joint torques account for the majority of the external power in wheelchair propulsion (54). The peak torque in the shoulder joint is greater than those generated in the elbow joint (58). A very small torque is generated in the wrist joint, where a braking effect is generated (58). The torques generated at the elbow and wrist joints are one-third and one-fifth of the torques at the shoulder joint (5). Joint torques ranges for the joints approximate 5 to 9 N-m at the wrist joint, 9 to 25 N-m at the elbow joint, and 25 to 50 N-m at the shoulder joint (47,54). Figure 11-53 illustrates a typical net shoulder moment of a paraplegic subject during wheelchair propulsion (55).

At low speeds and resistances, wheelchair propulsion is generated primarily by upper extremity sagittal plane movement. Fifty percent of the shoulder joint work is lost as negative work at the wrist and elbow (5). As the workload and speed increase, a concomitant increase occurs in the direction and magnitudes of the joint torques. Greater extension and abduction at the shoulder coupled with more torque at the elbow joint contribute to the increased output (5). The increase in shoulder abduction is speculated to be a forced abduction because of rotation of the arm in a closed-chain system (54). A sample of the shoulder flexion and extension, shoulder adduction, elbow flexion, and wrist flexion and extension torque patterns for wheelchair propulsion are shown in Figure 11-54. The highest torques are generated via shoulder flexion and shoulder adduction. The shoulder flexion torque peaks just before the peak in elbow extension torque. The shoulder adduction torque controls the shoulder abduction that is created as a result of arm movement in

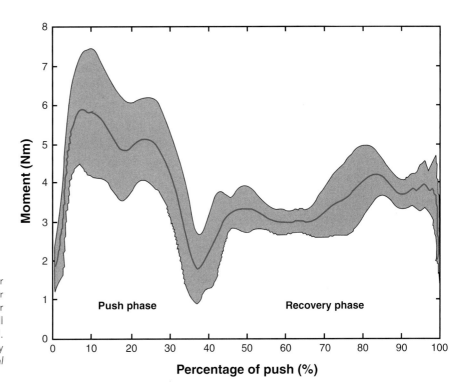

FIGURE 11-53 Example of the net shoulder moment (paraplegia subject) during wheelchair propulsion (mean and standard deviation over five trials; time normalized to 100% of a full cycle). (Adapted from van Drongelen, S., et al. [2005]. Mechanical load on the upper extremity during wheelchair activities. *Archives of Physical Medicine and Rehabilitation*, 86:1214–1220.)

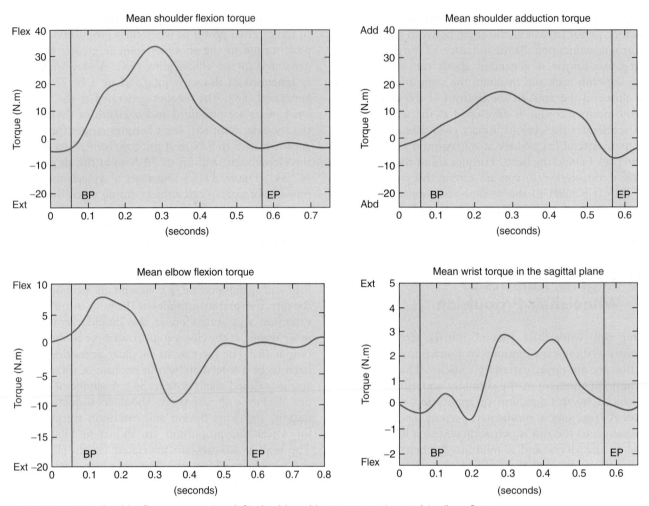

FIGURE 11-54 Shoulder flexion torque (*top left*), shoulder adduction torque (*top right*), elbow flexion torque (*bottom left*), and wrist torque in the sagittal plane (*bottom right*) for the propulsive stage of wheelchair propulsion. BP, beginning propulsion; EP, end propulsion. (Adapted from Veeger, H. E. J., et al. [1991]. Load on the upper extremity in manual wheelchair propulsion. *Journal of Electromyography & Kinesiology*, 1:270–280.)

a closed chain. There is a net flexor torque at the elbow at the beginning of the propulsion phase that shifts to an extensor torque, continuing on through the rest of propulsion. The wrist torque is primarily extensor throughout the majority of the propulsion phase. A pronation and a larger radial deviation moment act at the wrist over the majority of the propulsive phase (2).

Summary

Torque, *moment*, and the *moment of force* are terms that can be used synonymously. A torque results when the line of action of a force acts at a distance from an axis of rotation. The magnitude of a torque is the product of the force and the perpendicular distance from the axis of rotation to the line of action. The unit of torque is the N-m. Torques cause rotational motion of an object about an axis. A torque is a vector and thus must be considered in terms of magnitude and direction. The right-hand rule

defines whether the torque is positive (counterclockwise rotation) or negative (clockwise rotation).

Newton's laws of motion can be restated from the linear case to their angular analogs. The angular analogs of these laws are as follows:

1. A rotating body will continue in a state of uniform angular motion unless acted on by an external torque.
2. An external torque will produce an angular acceleration of a body that is proportional to and in the direction of the torque and inversely proportional to the moment of inertia of the body.
3. For every torque exerted by one body on another body, an equal and opposite torque is exerted by the latter body on the former.

Moment of inertia is the angular analog of mass. The magnitude of the moment of inertia of an object, or the resistance to angular motion, depends on the axis about which the object is rotating. The calculation of the segment moment of inertia can be calculated using the radius of gyration as:

$$I_{cm} = m(\rho_{cm}L)^2$$

where I_{cm} is the moment of inertia about a transverse axis through the center of mass, m is the mass of the segment, ρ is a proportion describing the ratio of the radius of gyration about the center of mass to the segment length, and L is the segment length.

Angular momentum is the angular analog of linear momentum and refers to the quantity of rotation of an object. The angular momentum of a multisegment body must be understood in terms of local and remote angular momenta. Local angular momentum is the angular momentum of a segment about its own center of mass. Remote angular momentum is the angular momentum of a segment about the total body center of mass. The total segment angular momentum is the sum of the segment's local and remote aspects. The total body angular momentum is the sum of the local and remote aspects of all segments. The angular analog of Newton's first law of motion is a statement of the law of conservation of angular momentum.

The concept of torque can be used to define the center of mass of an object. The sum of the torques about the center of mass of an object equals zero. That is:

$$\Sigma T = 0$$

This relationship defines the center of mass as the balancing point of the object.

The center of mass is commonly computed using the segmental method, which requires body segment parameters, such as the location of the center of mass and the segment proportion of the total body mass. The location of a segment center of mass requires the coordinates of both the proximal and distal ends of the segment. It is defined as:

$$s_{cm} = \text{proximal endpoint} - (\text{segment length} \times \% \text{ length of segment from segment's proximal end point})$$

$$s_{cm} = s_{proximal} - [(s_{proximal} - s_{distal}) \times \%L]$$

where s_{cm} is the location of the segment center of mass, $s_{proximal}$ is the coordinate point of the proximal end of the segment, s_{distal} is the coordinate point of the distal end of the segment, and $\%L$ is the location of the segment center of mass as a proportion of the segment length from the proximal end of the segment. Locating the total body center of mass with the segmental method also uses the concept that the sum of the torques about the center of mass is zero. This computation uses the formula:

$$s_{cm} = \sum_{i=1}^{n} \frac{m_i s_i}{M}$$

where s_{cm} is the location of the total body center of mass, n is the number of segments, m_i is the mass of the ith segment, s_i is the location of the segment center of mass, and M is the total body center of mass.

A lever is a simple machine with a balancing point called the fulcrum and two forces, an effort force and a resistance force. The MA of a lever is defined as the ratio of the effort moment arm to the resistance moment arm. Levers can magnify force (MA > 1), magnify speed of rotation (MA < 1), or change the direction of pull (MA = 1). The three classes of levers are based on the relationship of the effort and resistance forces to the fulcrum. In the human body, however, the third-class lever magnifying the speed of movement predominates. Most of the levers in the extremities are third-class levers.

Special torque applications include angular work, RKE, and angular power. These concepts may be developed by substituting the appropriate angular equivalent in the linear case. Angular work is defined as:

$$\text{Angular work} = T \times \Delta\theta$$

where T is the torque applied and $\Delta\theta$ is the angular distance over which the torque is applied. RKE is the capacity to do angular work and is defined as:

$$\text{RKE} = \frac{1}{2} I\omega^2$$

where I is the moment of inertia of the segment about its center of mass and ω is the angular velocity of the segment. Angular power is defined as either the rate of doing angular work:

$$\text{Angular power} = \frac{dW}{dt}$$

where W is the mechanical work done and dt is the time period over which the work is done or the product of torque and angular velocity:

$$\text{Angular power} = T \times \omega$$

where T is the torque and ω is the angular velocity.

Angular motion analysis can be conducted using one of three techniques: the effect of a torque at an instant in time ($T = I\alpha$), the effect of a torque applied over time (impulse–momentum relationship), or the effect of a torque applied over a distance (work–energy theorem).

In the first technique, the static 2D case is determined using the following equations:

$$\Sigma F_x = 0 \text{ for the horizontal component}$$
$$\Sigma F_y = 0 \text{ for the vertical component}$$
$$\Sigma T_z = 0 \text{ for rotation}$$

The dynamic 2D case uses the following equations:

$$\Sigma F_x = ma_x \text{ for the horizontal component}$$
$$\Sigma F_y = ma_y \text{ for the vertical component}$$
$$\Sigma T_z = I\alpha_z \text{ for rotation}$$

The purpose in both cases is to determine the net muscle moment about a joint. The impulse–momentum relationship relates the torque applied over time to the change in momentum:

$$T \times dt = I\omega_{final} - I\omega_{initial}$$

The left-hand side of the equation, $T \times dt$, is the angular impulse, and the right-hand side, $I\omega_{final} - I\omega_{initial}$,

describes the change in angular momentum. In the third type of analysis, mechanical angular work is calculated via the change in mechanical energy. That is:

$$W = \Delta RKE$$

where RKE is the RKE.

Equation Review for Angular Kinetics

Calculation	Given	Formula
Torque	Force (F), moment arm (r)	$T = F \times r$
Torque	Moment of inertia (I), angular acceleration (α)	$T = I\alpha$
Angular acceleration (a)	Torque (T), moment of inertia (I)	$\alpha = \dfrac{T}{I}$
Center of mass of system (x_{cm}, y_{cm})	Mass (m), location (x, y)	$x_{cm} = \dfrac{m_1 x_1 + m_2 x_2 + \cdots + m}{\text{Total mass}}$
		$y_{cm} = \dfrac{m_1 y_1 + m_2 y_2 + \cdots + m}{\text{Total mass}}$
Mechanical advantage (MA)	Length of effort arm (EA), resistance arm (RA)	$MA = \dfrac{EA}{RA}$
Moment of inertia (I)	Mass (m), distance from axis (r)	$I = mr^2$
Moment of inertia (I)	Angular acceleration (a), torque (T)	$I = \dfrac{T}{\alpha}$
Moment of inertia (I)	Mass (m), length (l), radius of gyration (r)	$I = m(\rho l)^2$
Angular momentum (ρ)	Moment of inertia (I), angular velocity (ω)	$H = I\omega$
Angular impulse	Angular momentum ($I \times \omega$)	Angular impulse $= I\omega_{final} - I\omega_{initial}$
Angular velocity	Torque, time, moment of inertia	$\omega = \dfrac{T \times t}{I}$
Angular work	Torque and displacement ($\Delta\theta$)	$W = T\Delta\theta$
Angular work	Changes in rotational kinetic energy (RKE)	$W = \Delta RKE$
Horizontal work	Force (F), angle of force application (θ), displacement (Δs)	$W = F \cos\theta\, \Delta s$
Potential energy (PE)	Mass (m), acceleration due to gravity (g), and vertical height (h)	$PE = mgh$
Rotational kinetic energy (RKE)	Moment of inertia (I), angular velocity (ω)	$RKE = \dfrac{1}{2}I\omega^2$
Angular power	Angular work (W), time (t)	$P = \dfrac{dW}{dt}$
Angular power	Torque (T), angular velocity (ω)	$P = T \times \omega$

REVIEW QUESTIONS

True or False

1. ____ In order for a force to cause a rotation, it must not pass through the pivot point.

2. ____ By convention, clockwise torques are considered to be positive.

3. ____ Third-class levers, where the effort and resisting forces are on the same side of the fulcrum, are most prominent in the body.

4. ____ Second-class levers require large effort forces relative to resistant forces and maximize the speed of movement.

5. ____ Decreasing moment of inertia by going from a laid out to tucked body position when doing a backflip allows one to rotate quicker.

6. ____ To work your deltoids more during a lateral arm raise, bend your elbows slightly, bringing the weight closer to your trunk.

7. ____ The moment arm for muscle force remains constant throughout the entire range of motion.

8. ____ Increasing an object's stability can be done by raising the height of the center of mass or increasing its mass.

9. ____ A positive net moment indicates that angular acceleration is occurring.

10. ____ The sum of torques about the center of mass of an object always equals zero.

11. ____ An MA > 1 indicates the effort arm exceeds the resistance arm, which magnifies the moment created by the effort force.

12. ____ Angular momentum increases if moment of inertia increases and decreases when angular velocity decreases.

13. ____ The only force creating a moment about the shoulder when an arm is held out to the side is the weight of the entire arm.

14. ____ Hip joint power is a proportional to hip joint moment and the velocity the thigh is rotating.

15. ____ The moment arm is always measured from the point of application of the force to the axis of rotation.

16. ____ Moment of inertia about an axis always passes through the center of mass of a segment.

17. ____ A force couple produces only rotation without translation because the two torques act in the same direction.

18. ____ Total body angular momentum can be increased during a dive by forcefully swinging the arms downward.

19. ____ The moment of inertia about an axis through the center of mass is less than the moment of inertia about any other perpendicular axis running through any other point on the segment.

20. ____ The final acceleration in the golf club during the downswing is attributed to the moment of inertia decreasing as the club center of mass gets closer to the axis of rotation.

21. ____ Once a gymnast leaves the vault, angular momentum remains constant while airborne if air resistance is neglected.

22. ____ Resistance of a thigh to rotate about its longitudinal axis is greater than its resistance to rotate about a mediolateral axis.

23. ____ During the stance phase of running, positive power at the hip, knee, and ankle joints indicates eccentric contractions.

24. ____ The direction of the torque vector changes the force vector rotates around the axis of rotation.

25. ____ Rotational kinetic energy of the foot is more influenced by the foot's angular velocity than its moment of inertia.

Multiple Choice

1. A force of 393 N is exerted 20 cm from the axis of rotation. What is the resulting moment of force?
   a. 786 N-m
   b. 78.6 N-m
   c. 19.7 N-m
   d. 197 N-m

2. An object has a moment of inertia of 184 kg-m^2. A torque of 84 N-m is applied to the object. What is the angular acceleration?
   a. 2.19 rad/s^2
   b. 15,460 rad/s^2
   c. 0.46 rad/s^2
   d. 125 rad/s^2

3. A 55-kg gymnast applies a vertical ground reaction force of 1,100 N at 0.17 m behind the center of mass during a forward somersault. Assume she is facing the right. What is the torque generated about the center of mass?
   a. 279 N-m
   b. −92 N-m
   c. 187 N-m
   d. −187 N-m

4. If a mass of 20.29 kg acts downward 0.29 m from the axis of rotation on one end of a board and another force of 85 N also acts downward, what is the moment arm of the second force to balance this system?
   a. 0.68 m
   b. 58 m
   c. 0.069 m
   d. The system is not balanced; angular rotation will occur.

5. What is the torque generated at the elbow by a 206-N force pulling on the forearm at an angle of 110° from the horizon at a point 16.5 cm from the elbow's axis of rotation? The forearm is positioned horizontal to the ground.
   a. 1,163 N-m
   b. 11.6 N-m
   c. 31.9 N-m
   d. 3,194 N-m

6. How much torque must be generated by the hip flexors to hold an 80-N ankle weight straight out at a 90° position? The ankle weight is 0.73 m from the hip joint. The thigh and leg weighs 130 N. The moment arm for the hip flexors muscle is 0.08 m.
   a. 58.4 N-m
   b. 106 N-m
   c. 1,323 N-m
   d. Not enough information

7. Calculate the rotational energy of a segment, given mass of the segment is 2.2 kg, moment of inertia is 0.57 kg-m^2, and angular velocity is 25 rad/s.
   a. 178 J
   b. 7.13 J
   c. 356 J
   d. 102 J

8. A force of 300 N is applied at a point 1.3 m from the axis of rotation, causing a revolving door to accelerate at 309.4 deg/s^2. What is the moment of inertia of the door from its axis of rotation?
   a. 72.2 kg-m^2
   b. 1.26 kg-m^2
   c. 2,106 kg-m^2
   d. 55.6 kg-m^2

9. If the net moment at a joint is 37.6 N-m and the angular velocity at the same instant in time is 4.40 rad/s, what is the angular power at that joint?
   a. 0.12 W
   b. 165 W
   c. 8.55 W
   d. 9,480 W

10. The center of mass of the following three-point system with masses of 6, 5, and 4 kg at the coordinates (6,0), (5,5), and (−4, −3), respectively, is ____.
    a. 5.1, 4.1
    b. 3.0, 0.87
    c. 5.1, 2.5
    d. 3.0, 1.4

11. What is the angular momentum if the force is 66 N, the lever arm is 7.7 m, and the time is 1.2 seconds?
    a. 424 kg-m²/s
    b. 79.2 kg-m²/s
    c. 508 kg-m²/s
    d. 610 kg-m²/s

12. Where is the center of mass of a segment in space if the proximal end is at (213, 400) and the distal end is at (378, 445) and the center of mass is 42.9% from the proximal end of the segment?
    a. 162, 191
    b. 142, 381
    c. 284, 419
    d. 381, 142

13. Calculate the moment of inertia of a baseball bat about its proximal end if its mass is 2.1 kg and has a radius of gyration of 0.58 and a length of 0.869 m.
    a. 0.614 kg-m²
    b. 0.920 kg-m²
    c. 0.533 kg-m²
    d. 1.12 kg-m²

14. A torque of 73 N-m results in an object rotating 52.72° in 0.86 seconds. How much angular work was done?
    a. 4,476 J
    b. 3,850 J
    c. 78.1 J
    d. 67.2 J

15. Why do the deltoids need to generate a greater torque when your arm–forearm is closer to your side versus straight out at 90°?
    a. Because the moment arm of the center or mass of the arm–forearm is larger.
    b. Because the component of the weight of the arm–forearm that causes a moment is larger.
    c. The torque is the same because the distance from the shoulder joint to the center of mass of the arm–forearm is constant.
    d. The torque is less because the moment arm of the center or mass of the arm–forearm is less.

16. Consider the following free body diagram. Using static analysis, solve for the muscle torque that will place this system in equilibrium, given mass of the leg and foot is 5.4 kg; distance from the knee joint to the center of mass of the leg–foot system 0.232 m; weight of the barbell 150 N; and distance from the knee joint to the center of mass of the barbell 0.514 m.
    a. 89.4 N-m
    b. 64.8 N-m
    c. 78.4 N-m
    d. Not enough information to solve N-m

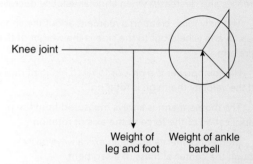

Knee joint

Weight of leg and foot   Weight of ankle barbell

17. If the leg is at an angle of 35° below the horizontal, calculate the moment arm of the torque caused by the weight of the leg, given that the distance to the center of mass of the leg 0.17 m from the knee joint.
    a. 0.10 m
    b. 0.14 m
    c. 0.15 m
    d. 0.17 m

18. Consider the following free body diagram. Using static analysis, solve for Achilles tendon force that will place this system in equilibrium if d1 is 0.043 m, d2 is 0.041 m, and d3 is 0.126 m.
    a. 239 N
    b. 5,578 N
    c. 5,557 N
    d. 10.3 N

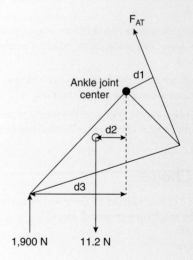

F_{AT}

Ankle joint center   d1

d2

d3

1,900 N   11.2 N

19. You are trying to make an abdominal crunch more difficult. Which of the following will accomplish?
    a. Hold your arms behind your head instead of across your chest.
    b. Swing your arms toward your legs as you crunch upward.
    c. Place a weight on your feet.
    d. You cannot make a crunch more difficult unless you hold additional weight.

20. Consider the following free body diagram. Using static analysis, solve for the moment at the elbow if $d1$ is 0.039 m, $d2$ is 0.142 m, and $d3$ is 0.450 m. What is the net muscle force?
    a. 2,957 N
    b. 2,393 N
    c. 93.3 N
    d. 1,084 N

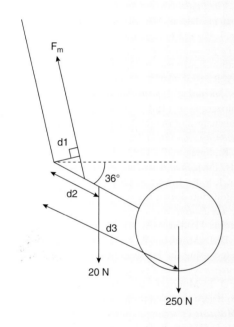

21. Consider the following diagram of the biceps brachii acting on the radius in two joint positions. The angle of pull of the biceps brachii force changes from 15° to 30°. If the muscle force is 1,600 N and the attachment

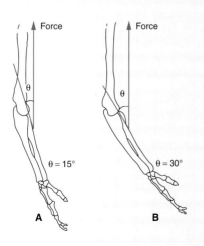

site of the muscle is 0.042 m from the joint axis, what is the change in joint torque applied by the biceps brachii from 15° to 30°?
    a. 16.2 N-m
    b. 17.4 N-m
    c. 33.6.0 N-m
    d. 67.2 N-m

22. What is total energy of the segment, given that mass is 3.9 kg; v_x is 1.45 m/s; v_y is 2.78 m/s; moment of inertia is 0.0726 kg-m^2; angular velocity is 9.0 rad/s; and the height of center of mass is 0.67 m?
    a. 396.1 J
    b. 66.91 J
    c. 65.04 J
    d. 47.74 J

23. What is the work done on the segment given the following information? The time between frames is 0.01 seconds.

	TKE	RKE	PE
Frame 2	48.8	2.5	37.8
Frame 130	6.0	2.0	3.9

    a. 60.3 J
    b. 8.4 J
    c. 77.2 J
    d. 42.8 J

24. During a knee extension exercise, the knee extensor muscle group is applying a torque of 250 N-m in an isometric contraction against the machine pad. If the knee joint angle is being held at an angle of 38° below the horizontal and the machine pad is 0.35 m from the knee joint, how much force is being applied at the pad?
    a. 250 N
    b. 1,160 N
    c. 714 N
    d. 906.5 N

25. During walking, the knee joint generates 60 N-m of extensor torque during the same interval of the stance phase when the knee joint moved from 8.02° of flexion to 14.9° of flexion in 0.02 seconds. Determine the power of the knee joint muscles.
    a. −360 W
    b. 360 W
    c. 20,640 W
    d. −20,640 W

References

1. Amar, J. (1920). *The Human Motor*. London: G. Routledge & Sons.
2. Boniger, M. L., et al. (1997). Wrist biomechanics during two speeds of wheelchair propulsion: An analysis using a local coordinate system. *Archives of Physical Medicine and Rehabilitation*, 78:364-371.

LIBRARY, UNIVERSITY OF CHESTER

3. Cavanagh, P. R., Gregor, R. J. (1976). Knee joint torque during the swing phase of normal treadmill walking. *Journal of Biomechanics*, 8:337-344.

4. Cavanagh, P. R., et al. (1977). A biomechanical comparison of elite and good distance runners. *Annals of New York Academy of Sciences*, 301-328.

5. Cerquiglini, S., et al. (1981). Biomechanics of wheelchair propulsion. In A. Moretti, et al. (Eds.). *Biomechanics VII-A.* Baltimore, MD: Park Press, 411-419.

6. Chaffin, D. B., Andersson, G. B. J. (1991). *Occupational Biomechanics,* 2nd ed. New York: Wiley.

7. Chaffin, D. B., et al. (1977). *Pre-employment Strength Testing in Selecting Workers for Materials Handling Jobs.* Cincinnati, OH: National Institute for Occupational Safety and Health. Pub. No. CDC-99–74-62.

8. Chandler, R. F., et al. (1975). Investigation of inertial properties of the human body. *AMRL Technical Report.* Wright-Patterson Air Force Base, 74-137.

9. Clauser, C. E. (1969). Weight, volume, and center of mass of segments of the human body. *AMRL Technical Report.* Wright-Patterson Air Force Base, 69-70.

10. Dempster, W. T. (1955). Space requirements of the seated operator. *WADC Technical Report.* Wright-Patterson Air Force Base, 55-159.

11. Devita, P., et al. (1991). Effects of asymmetric load carrying on the biomechanics of walking. *Journal of Biomechanics,* 24:119-1129.

12. Dyson, G. (1973). *The Mechanics of Athletics.* London: University of London.

13. Fallon, L. P., et al. (2000). Determining baseball bat performance using a conservation equations model with field test validation. In A. J. Subic, S. J. Haake (Eds.). *The Engineering of Sport: Research, Development, and Innovation.* Oxford: Blackwell Science, 201-211.

14. Feltner, M. E., Dapena, J. (1986). Dynamics of the shoulder and elbow joints of the throwing arm during the baseball pitch. *International Journal of Sports Biomechanics*, 2:235-259.

15. Frolich, C. (1979). Do springboard divers violate angular momentum conservation? *American Journal of Physics*, 47:583-592.

16. Gage, J. R. (1992). Millions of bits of data: How can we use it to treat cerebral palsy? *Proceedings of the Second North American Congress on Biomechanics*, 291-294.

17. Garg, A., Chaffin, D. B. (1975). A biomechanical computerized simulation of human strength. *Transactions of the American Institute of Industrial Engineers*, 7:1-15.

18. Ghista, D. N., Roaf, R. (1981). *Orthopaedic Mechanics: Procedures and Devices*, Vol. 2. New York: Academic Press.

19. Hamill, J., et al. (1986). Angular momentum in multiple rotation non-twisting platform dives. *International Journal of Sports Biomechanics*, 2:78-87.

20. Hanavan, E. P. (1964). A mathematical model of the human body. *AMRL Technical Report.* Wright-Patterson Air Force Base, 64-102.

21. Hatze, H. (1980). A mathematical model for the computational determination of parameter values of anthropomorphic segments. *Journal of Biomechanics*, 13:833-843.

22. Hatze, H. (1981). Estimation of myodynamic parameter values from observations on isometrically contracting muscle groups. *European Journal of Applied Physiology*, 46:325-338.

23. Hay, J. G. (1975). Straddle or flop. *Athletic Journal*, 55:83-85.

24. Hay, J. G., et al. (1977). A computational technique to determine the angular momentum of a human body. *Journal of Biomechanics*, 10:269-277.

25. Hinrichs, R. N. (1987). Upper extremity function in running. II: Angular momentum considerations. *International Journal of Sports Biomechanics*, 3:242-263.

26. Holden, J. P., et al. (1997). Changes in knee joint function over a wide range of walking speeds. *Clinical Biomechanics*, 12:375-382.

27. Jensen, R. K. (1987). The growth of children's moment of inertia. *Medicine & Science in Sports & Exercise*, 18:440-445.

28. Jorgensen, T. (1970). On the dynamics of the swing of a golf club. *American Journal of Physics*, 38:644-651.

29. Kamon, E., et al. (1982). Dynamic and static lifting capacity and muscular strength of steelmill workers. *American Industrial Hygiene Association Journal*, 43:853-857.

30. Kaneko, Y., Sato, F. (2000). The adaptation of golf swing to inertia property of golf club. In A. J. Subic, S. J. Haake (Eds.). *The Engineering of Sport: Research, Development and Innovation.* Malden, MA: Blackwell Science, 469-476.

31. Lander, J. E. (1984). *Effects of Center of Mass Manipulation on the Performance of the Squat Exercise.* Unpublished doctoral dissertation, University of Oregon.

32. Lind, A. R., et al. (1978). Influence of posture on isometric fatigue. *Journal of Applied Physiology*, 45:270-274.

33. Maddalozzo, G. F. (1987). An anatomical and biomechanical analysis of the full golf swing. *National Strength and Conditioning Association Journal*, 9:6-8, 77-79.

34. Mann, R., Sprague, P. (1982). Kinetics of sprinting. In J. Terauds (Ed.). *Biomechanics of Sports.* Del Mar, CA: Academic.

35. Mather, J. S. B., Jowett, S. (2000). Three-dimensional shape of the golf club during the swing. In A. J. Subic, S. J. Haake (Eds.). *The Engineering of Sport: Research, Development and Innovation.* Malden, MA: Blackwell Science, 77-85.

36. Miller, D. I., Morrison, W. E. (1975). Prediction of segmental parameters using the Hanavan human body model. *Medicine & Science in Sports & Exercise*, 7:207-212.

37. Miller, D. I., Munro, C. F. (1984). Body segment contributions to height achieved during the flight of a springboard dive. *Medicine & Science in Sports & Exercise*, 16:234-242.

38. Neal, R. J., Wilson, B. D. (1985). 3D kinematics and kinetics of the golf swing. *International Journal of Sports Biomechanics*, 1:221-232.

39. Olsen, V. L., et al. (1972). The maximum torque generated by the eccentric, isometric and concentric contractions of the hip abductor muscles. *Physical Therapy*, 52:149-158.

40. Ounpuu, S. (1994). The biomechanics of walking and running. *Foot & Ankle Injuries*, 13:843-863.

41. Plagenhoef, S., et al. (1983). Anatomical data for analyzing human motion. *Research Quarterly for Exercise and Sport*, 54:169-178.

42. Pytel, J. L., Kamon, E. (1981). Dynamic strength test as a predictor for maximal and acceptable lifting. *Ergonomics*, 24:663-672.

43. Redfield, R., Hull, M. L. (1986). On the relation between joint moments and pedalling rates at constant power in cycling. *Journal of Biomechanics*, 19:317-324.

44. Reilly, D. T., Martens, M. (1972). Experimental analysis of the quadriceps muscle force and the patello-femoral joint reaction force for various activities. *Acta Orthopaedica Scandinavica*, 43:126-137.

45. Robertson, D. G. E. (1987). Functions of the leg muscles during the stance phase of running. In B. Jonsson (Ed.). *Biomechanics X-B.* Champaign, IL: Human Kinetics, 1021-1027.

46. Robertson, D. G. E., Winter, D. A. (1980). Mechanical energy generation, absorption, and transfer amongst segments during walking. *Journal of Biomechanics*, 13:845-854.

47. Rogers, M. M. (1988). Dynamic biomechanics of the normal foot and ankle during walking and running. *Physical Therapy*, 68:1822-1830.

48. Rozendaal, L. A., Veeger, H. E. J. (2000). Force direction in manual wheel chair propulsion: Balance between effect and cost. *Clinical Biomechanics*, 15:S39-S41.

49. Sanders, R. H., Wilson, B. D. (1990). Angular momentum requirements of the twisting and non-twisting forward 1 1/2 somersault dive. *International Journal of Sports Biomechanics*, 3:47-53.

50. Shira, C. (2000). Advanced materials in golf clubs. In A. J. Subic, S. J. Haake (Eds.). *The Engineering of Sport: Research, Development and Innovation.* Malden, MA: Blackwell Science, 51-59.

51. Shoup, T. E., (1986). The anthropometric basis for fitting of golf clubs. In E. D. Rekow (Ed) *Medical Devices and Sporting Equipment.* New York: American Society of Mechanical Engineers, 13-16.

52. Takei, Y. (1991). A comparison of techniques used in performing the men's compulsory gymnastic vault at the 1988 Olympics. *International Journal of Sports Biomechanics*, 7:54-75.

53. Takei, Y. (1992). Blocking and postflight techniques of male gymnasts performing the compulsory vault at the 1988 Olympics. *International Journal of Sports Biomechanics*, 8:87-110.

54. Van der Helm, F. C. T., Veeger, H. E. J. (1996). Quasi-static analysis of muscle forces in the shoulder mechanism during wheelchair propulsion. *Journal of Biomechanics*, 29:39-52.

55. van Drongelen, S., et al. (2005). Mechanical load on the upper extremity during wheelchair activities. *Archives of Physical Medicine and Rehabilitation*, 86:1214-1220.

56. van Gheluwe, B. (1981). A biomechanical simulation model for airborne twists in backward somersaults. *Human Movement Studies*, 7:1-22.

57. Van Soest, A. J., et al. (1985). A comparison of one-legged and two-legged countermovement jumps. *Medicine & Science in Sports & Exercise*, 17:635-639.

58. Veeger, H. E. J., et al. (1991). Load on the upper extremity in manual wheelchair propulsion. *Journal of Electromyography & Kinesiology*, 1:270-280.

59. Williams, K. R. (1980). *A Biomechanical and Physiological Evaluation of Running Efficiency.* Unpublished doctoral dissertation, The Pennsylvania State University.

60. Winter, D. A. (1983). Moments of force and mechanical power in jogging. *Journal of Biomechanics*, 16:91-97.

61 Winter, D. A. (1984). Kinematic and kinetic patterns in human gait: variability and compensating effects. *Human Movement Science*, 3:51-76.

62. Winter, D. A., et al. (1976). Analysis of instantaneous energy of normal gait. *Journal of Biomechanics*, 9:253-257.

63. Winter, D. A., Robertson, D. G. E. (1978). Joint torque and energy patterns in normal gait. *Biological Cybernetics*, 142:137-142.

64. Yeadon, M. R., Atha, J. (1985). The production of a sustained aerial twist during a somersault without the use of asymmetrical arm action. In D. A. Winter et al. (Eds.). *Biomechanics IX-B.* Champaign, IL: Human Kinetics, 395-400.

65. Yeadon, M. R. (1989). Twisting techniques used in springboard diving. *Proceedings of the First IOC World Congress on Sport Sciences*, 307-308.

45. von Herzen, B. D. (1987). Encounters of the fore-muscle during the stance phase of running. In B. Johnson (Ed.), *Biomechanics X-B*. Champaign, IL: Human Kinetics, 1031-1037.

46. Aalderson, G. Ch. Ia., Winter, D. A. (1980). Mechanical energy generation, absorption, and transfer amongst segments during walking. *Journal of Biomechanics*, 13, 845-854.

47. Rogers, M. M. (1988). Dynamic biomechanics of the normal foot and ankle during walking and running. *Physical Therapy*, 68, 1822-1830.

48. Rorabeck, T. A., Vereer, Fr. F. (2000). Tissue reaction in anterior tibial-flexor production: Balance between electro-.... *Clinical Biomechanics*, 13, 538-541.

49. Sanders, S. H., Nilsson, B. D. (1980). Angular momentum requirements of the twisting and non-twisting forward 1 1/2 somersault dive. *International Journal of Sports Biomechanics*, 84, 58.

50. Shaw, C. (2000). Advanced materials in golf clubs. In A. J. Subic (Ed.), *The engineering of sport*. Oxford: Blackwell...

51. Shope, R. J. (1988). The anthropometric basis for jumping golf clubs. In P. D. Riddle (Ed.), *Medical Aspects and Sports Equipment*. New York: American Academy of Medicine...

52. Tatei, Y. (1991). A comparison of the torques used in popular ... Western consecutive gymnastics vault at the 1988 Olympic Games. *International Journal of Sports Biomechanics*, 7, 64-73.

53. Tatei, Y. (1991). Stickling and strength techniques of male gymnasts performing the compulsory vault at the 1988 Olympic Games. *International Journal of Sports Biomechanics*, 4, 62-70.

54. Vaughn, Hare, C. L., Vegeter, H. F. (1815-915). Cross-talk analytical muscle forces to the slipping and jumping... skeletal biomechanics. *Journal of Biomechanics*, 20, 50-53.

55. von Hoogstraten, S., et al. (2009). Medium at load on the appar extensility flange wheel mass reduction acceleration of ... Process Machine and Engineering, 66, 2, 54-1320.

56. Seegan, Okshaw, Th. (1981). A biomechanical simulation model for antenna swing in backward somersaults. *Human Movement Sciences*, 7, 122.

57. Van Supen, A. J., et al. (1885). A comparison of concurrent and motological constraint tensor temperature force. *Journal of Motor Behaviour*, 17, 635-656.

58. Vereer, Fr. Fr. A., et al. (1981). Load on the upper extremity in internal wheelchair propulsion. *Journal of Electromyography Kinesiology*, 1, 270-280.

59. Williams, K. R. (1990). A biomechanical view. *Biomechanics of Distance Running*. Champaign, IL: Human Kinetics.

60. Winter, D. A. (1983). Moments of force and mechanical power in jogging. *Journal of Biomechanics*, 16(1), 91-97.

61. Winter, D. A. (1988). Kinematic and kinetic patterns in human gait: variability and compensating effects. *Human Movement Sciences*, 3(1-2), 51-76.

62. Winter, D. A. (1979). Analysis of instantaneous energy of normal gait. *Journal of Biomechanics*, 9, 253-257.

63. Winter, D. A., Robertson, D. G. (1978). Joint torque and energy patterns in normal gait. *Biological Cybernetics*, 14, 129-142.

64. Yeadon, M. R. (1990). The production of a somersault rotation twist during somersault without the use of aerodynamical information. In J. M. A. Winter et al. (Eds.), *Biomechanics IX-B*. Champaign, IL: Human Kinetics, 395-400.

65. Yeadon, M. R. (1990). Twisting techniques used in spring-board diving. *Journal of Applied Biomechanics*, 8, and Sport Science, 301-308.

The Metric System and SI Units

All measurements in the biomechanics literature are expressed in terms of the metric system. This system of measurement uses units that are related to one another by some power of 10. Table A-1 presents the prefixes associated with these powers.

The standard length in the metric system is the meter. This standard measure was originally indicated by two scratches on a platinum–iridium alloy bar kept at the International Bureau of Weights and Measures in Sèvres, France. Table A-2 illustrates the use of these prefixes in common units of length.

The uniform system for the reporting of numerical values is known as the Système International d'Unites, or SI. This system was developed through international cooperation to standardize the report of scientific information. The base dimensions used in biomechanics are **mass, length, time, temperature, electric current, amount of substance,** and **luminous intensity**. The base units in SI corresponding to the base dimensions are the kilogram, the meter, the second, and the degree kelvin. Table A-3 presents the base units of SI.

Other units of measurement that are used in biomechanics are derived. These units are presented in Table A-4. Common conversions are shown in Table A-5.

TABLE A-1   Prefixes for the Powers of 10

Prefix	Multiplier	Symbol	Example
Giga	10^9	G	Gigabyte (Gb)
Mega	10^6	M	Megawatt (MW)
Kilo	10^3	k	Kilogram (kg)
Centi	10^{22}	c	Centimeter (cm)
Milli	10^{23}	m	Milligram (mg)
Micro	10^{26}	μ	Microsecond (μs)
Nano	10^{29}	v	Nanosecond (ns)

TABLE A-2   Units of Length

Metric Unit	Power	Symbol
Kilometer	10^{23}	km
Meter	—	m
Decimeter	10^{21}	dm
Centimeter	10^{22}	cm
Millimeter	10^{23}	mm
Micrometer	10^{26}	μm

TABLE A-3   Base Units of Measurement

Dimension	Unit	Symbol
Mass	Kilogram	kg
Length	Meter	m
Time	Second	s
Temperature	Degree kelvin	K
Electrical current	Ampere	A
Amount of substance	Mole	Mol
Luminous intensity	Candela	cd

TABLE A-4   Derived Units of Measurement

Dimension	Unit	Symbol
Acceleration	Meters per second squared	m/s^2
Angle	Radian	rad
Area	Meter squared	m^2
Capacitance	Farad	F
Concentration	Moles per meter cubed	mol/m^3
Density	Mass per unit volume	kg/m^3
Energy	Joule	J
Impulse	Force × time	Ns
Luminous flux	Lumen	lm
Moment of inertia	Kilogram-meters squared	kgm^2
Momentum	Kilogram-meters per second	kgm/s
Power	Watts	W
Pressure	Pascal	Pa
Resistance	Ohm	Ω
Speed	Meters per second	m/s
Torque	Newton meter	Nm
Voltage	Volt	V
Volume	Meter cubed	m^3
Work	Joule	J

TABLE A-5 Common Conversions

Length

1 inch = 25.40 mm	1 inch = 2.54 cm	1 foot = 0.3048 m	1 foot = 30.48 cm
1 yard = 0.9144 m	1 mile = 1609.34 m	1 cm = 0.3937 inches	1 cm = 0.0328 feet
1 m = 39.37 inches	1 m = 3.2808 feet	1 m = 1.0936 yard	1 m = 1.0936 yard
1 km = 0.6214 miles			

Area

1 in = 645.16 mm^2	1 ft^2 = 0.0929 m^2	1 yd^2 = 0.8361 m^2	1 cm^2 = 0.155 in^2
1 m^2 = 10.76 ft^2			

Volume

1 in^3 = 16.387 cm^3	1 ft^3 = 0.0283 m^3	1 yd^3 = 0.7645 m^3	1 yd^3 = 0.7645 m^3
1 cm^3 = 0.061 in^3	1 m^3 = 35.32 ft^3	1 m^3 = 1.308 yd^3	

Mass

1 oz = 28.349 g	1 lb = 0.4536 kg	1 slug = 14.5939 kg	1 slug = 32.2 lb
1 g = 0.0353 oz.	1 kg = 2.2046 lb	1 kg = 0.0685 slug	

Density

1 lb-ft^3 = 16.02 kg/m^3	1 slug-ft^3 = 515.38 kg/m^3

Moment of Inertia

1 slug-ft^2 = 1.36 kgm^2	1 lb-ft^2 = 0.042 kgm^2

Velocity

1 in/s = 24.4 mm/s	1 ft/s = 0.305 m/s	1 mph = 0.447 m/s	1 mph = 1.467 ft/s
1 m/s = 3.60 km/h	1 m/s = 3.28 ft/s	1 m/s = 2.237 m/h	1 cm/s = 0.0328 ft/s

Force

1 poundal = 0.1383 N	1 lb-force = 4.448 N	1 kg-force = 9.81 N	1 N = 0.102 kg
1 N = 0.2248 lb			

Pressure

1 poundal/ft^2 = 1.4881 Pa	1 pound-force /ft^2 = 47.889 Pa	1 pound-force/in^2 = 6.8947 Pa
1 mm mercury = 133.322 Pa		

Work and Energy

1 foot poundal = 0.0421 J	1 foot pound-force = 1.3558 J	1 foot pound-force = 0.1383 kgm	1 kgm = 7.2307 ft-lb
1 BTU = 1.0551 kJ	1 kilocalorie = 4.1868 kJ	1 J = 0.7376 ft-lb	1 J = 0.1020 kgm

Power

1 hp (British) = 745.700 W	1 hp (metric) = 735.499 W	1 hp (metric) = 735.499 W	1 ft-lb force/s = 1.3558 W

Torque

1 Nm = 0.74 lb-ft	1 lb-ft = 1.36 Nm	1 kgm = 7.23 lb-ft	1 kg-cm = 0.0723 lb-ft

BTU, British thermal unit.

Trigonometric Functions

Trigonometry is a branch of mathematics concerned with the measurements of the sides and angles of triangles and their relationships with each other. Many concepts in biomechanics require knowledge of trigonometry. A triangle is composed of three sides and three angles. The sum of the three angles of a triangle equals 180°. A right triangle is a triangle in which one of the angles is a right angle, that is, one of the angles equals 90°. The sum of the remaining angles, therefore, also equals 90°. Consider the triangle with vertices A, B, and C and sides of length AB, AC, and BC in Figure B-1.

The side opposite the right angle, AB, is always the longest side in the right triangle and is referred to as the hypotenuse. The other two sides are named according to which of the other angles is under consideration. If angle A is considered, the side AC is the adjacent side, and the side BC is the opposite side. If angle B is considered (as in the triangle on the right below), BC is the adjacent side and AC is the opposite side.

Based on the right triangle, trigonometric functions may be defined. A mathematical function is a quantity whose value varies and depends on some other quantity or quantities. Trigonometric functions vary with and depend on the values of the two acute (i.e., <90°) angles in a right triangle and the lengths of the sides of the triangle. The trigonometric functions are the ratios of the lengths of the sides of the triangle based on one of the two acute angles in the triangle. There are six such functions: sine (abbreviation is sin), cosine (abbreviation is cos), tangent (abbreviation is tan), cosecant, secant, and cotangent.

In biomechanics, only the first three of these functions are important. The trigonometric functions are thus defined for angle A (Fig. B-2A) as:

1. The sine of an angle is the ratio of the side opposite the angle to the hypotenuse.

$$\sin A = \frac{\text{opposite side}}{\text{hypotenuse}} = \frac{BC}{AB}$$

2. The cosine of an angle is the ratio of the side adjacent to the angle to the hypotenuse.

$$\cos A = \frac{\text{adjacent side}}{\text{hypotenuse}} = \frac{AC}{AB}$$

3. The tangent of an angle is the ratio of the side opposite to the angle to the side adjacent to the angle.

$$\tan A = \frac{\text{opposite side}}{\text{adjacent side}} = \frac{BC}{AC}$$

Similarly, the ratios for angle B in Figure B-2B may be defined as:

$$\sin A = \frac{\text{opposite side}}{\text{hypotenuse}} = \frac{BC}{AB}$$

$$\cos A = \frac{\text{adjacent side}}{\text{hypotenuse}} = \frac{AC}{AB}$$

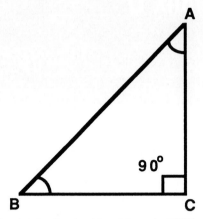

FIGURE B-1 A right triangle with a right angle C (90°) and two acute angles, A and B. A and B sum to 90°.

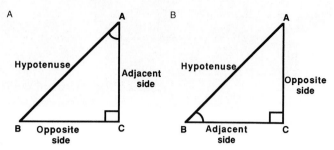

FIGURE B-2 Descriptions of the sides of a right triangle based on the acute angles A **(A)** and B **(B)**.

$$\tan A = \frac{\text{opposite side}}{\text{adjacent side}} = \frac{BC}{AC}$$

For any angle, the ratios formed by the sides of the right triangle will always be the same. For example, the sine of an angle of 32 will always equal 0.5299 regardless of the size of the sides of the triangle. The same is true for the cosine and tangent of the angle. The values for the sine, cosine, and tangent of angles can be presented in tables. Table B-1 presents these values for angles ranging from 0° to 90°.

The values in this table may also be used to determine the angle when the sides of the triangle are known. Consider the triangle in Figure B-3. The length of the hypotenuse, AC, is 0.05 m, and the length of the side opposite angle C is 0.03 m. The ratio of AB to AC is the sine of angle C. Thus:

$$\sin C = \frac{AB}{AC} = \frac{0.03 \text{ m}}{0.05 \text{ m}} = 0.6$$

If the sine values in Table B-1 are examined, it can be determined that the angle whose sine is 0.6 is approximately 37°. This is the arcsin of the angle. Thus:

$$C = \arcsin \frac{AB}{AC}$$

TABLE B-1	Trigonometric Functions								
Angle					Angle				
Degrees	Radian	Sine	Cosine	Tangent	Degrees	Radian	Sine	Cosine	Tangent
0	0.000	0.0000	0.0000	0.0000	19	0.332	0.3256	0.9455	0.3443
1	0.017	0.0175	0.9998	0.0175	20	0.349	0.3420	0.9397	0.3640
2	0.035	0.0349	0.9994	0.0349	21	0.367	0.3584	0.9336	0.3839
3	0.052	0.0523	0.9986	0.0524	22	0.384	0.3746	0.9272	0.4040
4	0.070	0.0698	0.9976	0.0699	23	0.401	0.3907	0.9205	0.4245
5	0.087	0.0872	0.9962	0.0875	24	0.419	0.4067	0.9135	0.4452
6	0.105	0.1045	0.9945	0.1051	25	0.436	0.4226	0.9063	0.4663
7	0.122	0.1219	0.9925	0.1228	26	0.454	0.4384	0.8988	0.4877
8	0.140	0.1392	0.9903	0.1405	27	0.471	0.4540	0.8910	0.5095
9	0.157	0.1564	0.9877	0.1584	28	0.489	0.4695	0.8829	0.5317
10	0.175	0.1736	0.9848	0.1763	29	0.506	0.4848	0.8746	0.5543
11	0.192	0.1908	0.9816	0.1944	30	0.524	0.5000	0.8660	0.5774
12	0.209	0.2079	0.9781	0.2126	31	0.541	0.5150	0.8572	0.6009
13	0.227	0.2250	0.9744	0.2309	32	0.559	0.5299	0.8480	0.6249
14	0.244	0.2419	0.9703	0.2493	33	0.576	0.5446	0.8387	0.6494
15	0.262	0.2588	0.9659	0.2679	34	0.593	0.5592	0.8290	0.6745
16	0.279	0.2756	0.9613	0.2867	35	0.611	0.5736	0.8192	0.7002
17	0.297	0.2924	0.9563	0.3057	36	0.628	0.5878	0.8090	0.7265
18	0.314	0.3090	0.9511	0.3249	37	0.646	0.6018	0.7986	0.7536

| TABLE B-1 | Continued |

	Angle					Angle			
Degrees	Radian	Sine	Cosine	Tangent	Degrees	Radian	Sine	Cosine	Tangent
38	0.663	0.6157	0.7880	0.7813	65	1.134	0.9063	0.4226	2.1445
39	0.681	0.6293	0.7771	0.8098	66	1.152	0.9135	0.4067	2.2460
40	0.698	0.6428	0.7660	0.8391	67	1.169	0.9205	0.3907	2.3559
41	0.716	0.6561	0.7547	0.8693	68	1.187	0.9272	0.3746	2.4751
42	0.733	0.6691	0.7431	0.9004	69	1.204	0.9336	0.3584	2.6051
43	0.750	0.6820	0.7314	0.9325	70	1.222	0.9397	0.3420	2.7475
44	0.768	0.6947	0.7193	0.9657	71	1.239	0.9455	0.3256	2.9042
45	0.785	0.7071	0.7071	1.0000	72	1.257	0.9511	0.3090	3.0777
46	0.803	0.7193	0.6947	1.0355	73	1.274	0.9563	0.2924	3.2709
47	0.820	0.7314	0.6820	1.0724	74	1.292	0.9613	0.2756	3.4874
48	0.838	0.7431	0.6691	1.1106	75	1.309	0.9659	0.2588	3.7321
49	0.855	0.7547	0.6561	1.1504	76	1.326	0.9703	0.2419	4.0108
50	0.873	0.7660	0.6428	1.1918	77	1.344	0.9744	0.2250	4.3315
51	0.890	0.7771	0.6293	1.2349	78	1.361	0.9781	0.2079	4.7046
52	0.908	0.7880	0.6157	1.2799	79	1.379	0.9816	0.1908	5.1446
53	0.925	0.7986	0.6018	1.3270	80	1.396	0.9848	0.1736	5.6713
54	0.942	0.8090	0.5878	1.3764	81	1.414	0.9877	0.1564	6.3138
55	0.960	0.8192	0.5736	1.4281	82	1.431	0.9903	0.1392	7.1154
56	0.977	0.8290	0.5592	1.4826	83	1.449	0.9925	0.1219	8.1443
57	0.995	0.8387	0.5446	1.5399	84	1.466	0.9945	0.1045	9.5144
58	1.012	0.8480	0.5299	1.6003	85	1.484	0.9962	0.0872	11.4300
59	1.030	0.8572	0.5150	1.6643	86	1.501	0.9976	0.0698	14.3010
60	1.047	0.8660	0.5000	1.7321	87	1.518	0.9986	0.0523	19.0810
61	1.065	0.8746	0.4848	1.8040	88	1.536	0.9994	0.0349	28.6360
62	1.082	0.8829	0.4695	1.8807	89	1.553	0.9998	0.0175	57.2900
63	1.100	0.8910	0.4540	1.9626	90	1.571	1.0000	0.0000	∞
64	1.117	0.8988	0.4384	2.0503					

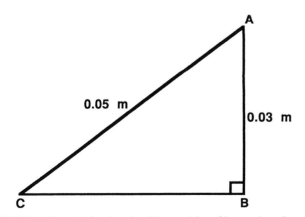

FIGURE B-3 A right triangle with two sides of known lengths and two unknown angles.

The ratio of the side AB to the hypotenuse AC is the cosine of angle A in Figure B-3. The ratio is still 0.6, and the angle whose cosine is 0.6 is approximately 53. This is referred to as the arccosine of an angle.

Two other useful trigonometric relationships are applicable to all triangles, not only right triangles. The first of these relationships is the law of sines, which states that the ratio of the length of any side to the sine of the angle opposite that side is equal to the ratio of any other side to the angle opposite that side. Consider the triangle in Figure B-4.

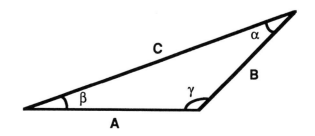

FIGURE B-4 A scalene triangle (no right angle; no two angles are the same).

For this triangle, the law of sines can be stated as follows:

$$\frac{A}{\sin\alpha} = \frac{B}{\sin\beta} = \frac{C}{\sin\gamma}$$

The other trigonometric relationship that is applicable to any triangle is the *law of cosines*. This relationship states that the square of the length of any side of a triangle is equal to the sum of the squares of the other two sides minus twice the product of the lengths of the other two sides and the cosine of the angle opposite the original side. Consider the triangle in Figure B-4. For side A of this triangle, the law of cosines can be stated as follows:

$$A^2 = B^2 + C^2 - 2BC\cos\alpha$$

Similarly, the law of cosines can be stated for sides B and C.

Sample Kinematic and Kinetic Data

Subject: 50-kg woman

Activity: Walking sampled at 120 Hz; one cycle from right foot touchdown to right foot touchdown

Markers: Head, R shoulder, R elbow, R wrist, R hand, R crest (IC), R greater trochanter (GT), R knee, R ankle, R heel, R fifth metatarsal (met), and R toe

Stride events: Right foot touchdown (RTD), left foot touchdown (LTD), and right foot toe-off (RTO)

TABLE C-1 Kinematic Data in Millimeters

Frame Number	Head$_x$	Head$_y$	Shoulder$_x$	Shoulder$_y$	Elbow$_x$	Elbow$_y$	Wrist$_x$	Wrist$_y$	Hand$_x$	Hand$_y$	IC$_x$	IC$_y$
−1	−28.0	1458.9	−155.1	1340.3	−272.4	1122.5	−337.1	891.2	−383.3	821.1	−129.7	959.4
0	−16.1	1460.1	−142.8	1340.4	−259.5	1122.3	−323.6	890.6	−369.1	820.3	−117.0	960.8
1	−5.8	1460.2	−132.4	1340.0	−249.1	1121.8	−313.2	889.9	−358.8	819.6	−105.7	961.0
2	5.5	1460.4	−120.8	1339.6	−237.4	1121.3	−301.5	889.2	−347.2	818.9	−93.3	961.2
3	17.6	1460.7	−108.5	1339.3	−225.0	1120.9	−288.9	888.6	−334.6	818.1	−80.1	961.6
4	30.1	1461.2	−95.7	1339.2	−211.9	1120.4	−275.4	887.9	−321.1	817.3	−66.5	962.1
5	42.9	1461.9	−82.6	1339.1	−198.3	1120.1	−261.2	887.3	−306.7	816.4	−52.6	962.9
6	55.9	1462.8	−69.2	1339.3	−184.3	1119.8	−246.2	886.7	−291.4	815.6	−38.6	963.9
7	69.1	1463.8	−55.7	1339.6	−169.9	1119.6	−230.6	886.2	−275.4	814.7	−24.5	965.1
8	82.2	1465.0	−42.1	1340.1	−155.3	1119.6	−214.3	885.7	−258.6	813.9	−10.5	966.6
9	95.4	1466.5	−28.4	1340.8	−140.4	1119.6	−197.4	885.4	−241.1	813.1	3.6	968.2
10	108.6	1468.0	−14.6	1341.7	−125.3	1119.7	−179.9	885.1	−222.8	812.3	17.6	970.1
11	121.7	1469.8	−0.9	1342.7	−110.0	1120.0	−161.8	884.9	−203.8	811.6	31.5	972.2
12	134.7	1471.6	12.8	1344.0	−94.5	1120.3	−143.1	884.8	−184.0	810.9	45.3	974.4
13	147.6	1473.6	26.6	1345.4	−78.9	1120.8	−123.9	884.9	−163.6	810.4	59.0	976.7
14	160.4	1475.8	40.2	1346.9	−63.1	1121.3	−104.2	885.0	−142.4	809.9	72.7	979.2
15	173.1	1478.0	53.9	1348.6	−47.2	1121.9	−84.0	885.2	−120.7	809.5	86.3	981.7
16	185.7	1480.3	67.5	1350.4	−31.1	1122.6	−63.4	885.6	−98.3	809.4	99.8	984.3
17	198.2	1482.7	81.1	1352.3	−14.9	1123.3	−42.4	886.1	−75.3	809.4	113.2	986.9
18	210.6	1485.2	94.7	1354.3	1.4	1124.2	−21.0	886.8	−51.8	809.6	126.5	989.4
19	222.8	1487.8	108.2	1356.4	17.9	1125.0	0.8	887.6	−27.8	810.0	139.7	991.9
20	235.0	1490.3	121.7	1358.5	34.5	1126.0	22.9	888.6	−3.3	810.7	152.7	994.4
21	247.1	1492.9	135.1	1360.7	51.2	1127.0	45.2	889.8	21.6	811.7	165.7	996.7
22	259.2	1495.5	148.5	1363.0	68.0	1128.0	67.9	891.1	46.8	812.9	178.6	998.9
23	271.2	1498.0	161.8	1365.3	84.8	1129.1	90.7	892.7	72.4	814.4	191.4	1001.0
24	283.1	1500.5	175.1	1367.6	101.8	1130.2	113.8	894.4	98.2	816.2	204.1	1002.9
25	295.1	1502.9	188.3	1369.9	118.8	1131.3	137.0	896.3	124.3	818.3	216.7	1004.7
26	307.0	1505.2	201.6	1372.1	135.8	1132.4	160.4	898.4	150.5	820.6	229.3	1006.3
27	318.9	1507.4	214.7	1374.2	152.9	1133.4	183.8	900.6	176.9	823.2	241.7	1007.7
28	330.7	1509.3	227.9	1376.2	169.9	1134.4	207.3	903.0	203.4	826.0	254.1	1008.9
29	342.6	1511.2	241.0	1378.1	187.0	1135.4	230.9	905.4	229.9	829.0	266.4	1009.9
30	354.5	1512.7	254.1	1379.7	204.0	1136.2	254.5	907.9	256.5	832.1	278.7	1010.7
31	366.3	1514.1	267.2	1381.1	221.0	1136.8	278.1	910.5	283.0	835.3	290.9	1011.3
32	378.2	1515.2	280.2	1382.3	238.0	1137.3	301.7	913.1	309.5	838.6	303.0	1011.6
33	390.1	1516.1	293.3	1383.3	254.9	1137.7	325.2	915.8	335.9	842.0	315.1	1011.8
34	402.0	1516.7	306.3	1383.9	271.8	1137.8	348.7	918.4	362.3	845.4	327.2	1011.7
35	413.9	1517.0	319.3	1384.2	288.7	1137.7	372.2	921.1	388.5	848.9	339.3	1011.4
36	425.9	1517.0	332.3	1384.2	305.5	1137.4	395.5	923.7	414.6	852.4	351.5	1010.8
37	437.8	1516.7	345.3	1383.9	322.3	1136.9	418.6	926.3	440.5	855.8	363.6	1010.1
38	449.8	1516.2	358.2	1383.3	339.1	1136.1	441.7	928.8	466.2	859.3	375.8	1009.1
39	461.8	1515.4	371.3	1382.4	355.9	1135.1	464.5	931.4	491.7	862.8	388.0	1007.8
40	473.8	1514.3	384.3	1381.2	372.6	1133.9	487.2	933.8	517.0	866.3	400.2	1006.4
41	485.9	1512.9	397.4	1379.7	389.3	1132.5	509.6	936.2	542.0	869.7	412.5	1004.8
42	498.0	1511.4	410.5	1378.0	406.0	1130.9	531.7	938.6	566.6	873.2	424.8	1002.9
43	510.1	1509.6	423.7	1376.1	422.6	1129.2	553.6	940.9	590.9	876.7	437.2	1000.9

Frame Number	GT$_x$	GT$_y$	Knee$_x$	Knee$_y$	Ankle$_x$	Ankle$_y$	R Heel$_x$	R Heel$_y$	Met$_x$	Met$_y$	R Toe$_x$	R Toe$_y$	Stride Events
−1	−151.9330	815.6667	60.0000	499.2000	179.6667	122.5333	137.0667	32.5333	318.8667	107.8000	389.6000	174.6000	
0	−138.8750	817.3138	71.2735	500.1740	184.3008	122.2974	139.1965	33.7988	321.1163	101.7447	393.5727	166.1098	RTD
1	−151.9	815.7	60.0	499.2	179.7	122.5	137.1	32.5	318.9	107.8	389.6	174.6	
2	−138.9	817.3	71.3	500.2	184.3	122.3	139.2	33.8	321.1	101.7	393.6	166.1	
3	−127.4	817.4	81.5	499.7	188.6	121.4	141.0	35.7	324.5	97.2	399.1	158.3	
4	−114.7	817.6	92.8	499.2	193.2	120.4	142.8	37.8	328.1	92.2	405.0	149.8	
5	−101.3	818.0	104.7	498.8	198.0	119.4	144.8	39.9	331.7	87.0	410.9	141.1	
6	−87.5	818.6	116.9	498.6	202.7	118.4	146.7	42.0	335.3	81.8	416.6	132.4	
7	−73.4	819.5	129.2	498.5	207.2	117.5	148.6	43.9	338.6	76.7	421.9	124.0	
8	−59.1	820.6	141.5	498.6	211.5	116.6	150.3	45.8	341.7	71.9	426.8	116.1	
9	−44.9	822.0	153.6	498.8	215.5	115.8	151.9	47.5	344.5	67.3	431.1	108.7	
10	−30.7	823.7	165.4	499.0	219.1	115.1	153.4	49.1	346.9	63.2	435.0	102.0	
11	−16.7	825.6	176.8	499.3	222.4	114.5	154.7	50.5	349.1	59.5	438.3	96.0	
12	−2.8	827.7	187.6	499.6	225.3	114.0	155.9	51.7	351.0	56.2	441.0	90.7	
13	11.0	829.9	197.9	499.9	227.8	113.6	156.9	52.8	352.5	53.4	443.4	86.1	
14	24.6	832.3	207.5	500.1	229.9	113.2	157.7	53.7	353.9	51.0	445.3	82.3	
15	38.0	834.8	216.4	500.2	231.8	113.0	158.4	54.5	355.0	49.0	446.9	79.1	
16	51.2	837.4	224.8	500.3	233.4	112.8	158.9	55.2	355.9	47.5	448.2	76.4	
17	64.3	840.1	232.5	500.4	234.7	112.6	159.3	55.7	356.6	46.2	449.2	74.3	
18	77.1	842.7	239.7	500.3	235.9	112.6	159.6	56.2	357.2	45.2	450.0	72.5	
19	89.8	845.3	246.3	500.3	236.9	112.6	159.8	56.6	357.7	44.5	450.6	71.2	
20	102.3	847.8	252.5	500.2	237.7	112.6	160.0	56.9	358.1	44.1	451.1	70.1	
21	114.7	850.3	258.3	500.1	238.5	112.8	160.1	57.2	358.5	43.7	451.5	69.2	
22	126.8	852.7	263.7	500.0	239.1	112.9	160.2	57.5	358.7	43.6	451.9	68.5	
23	138.9	855.1	268.8	499.8	239.7	113.1	160.2	57.7	359.0	43.5	452.1	67.8	
24	150.8	857.3	273.7	499.7	240.2	113.2	160.2	57.9	359.2	43.5	452.3	67.3	
25	162.6	859.4	278.3	499.5	240.7	113.4	160.3	58.1	359.3	43.6	452.5	66.8	
26	174.3	861.4	282.7	499.4	241.1	113.6	160.3	58.3	359.5	43.6	452.6	66.3	
27	186.0	863.2	286.9	499.2	241.5	113.9	160.4	58.4	359.7	43.7	452.8	65.8	
28	197.5	864.9	291.0	499.1	241.9	114.1	160.4	58.6	359.8	43.8	452.9	65.3	
29	209.0	866.5	294.9	498.9	242.2	114.3	160.5	58.8	359.9	43.9	453.0	64.8	
30	220.4	867.9	298.8	498.7	242.6	114.6	160.6	59.0	360.0	44.0	453.1	64.3	
31	231.8	869.1	302.5	498.6	242.9	114.8	160.7	59.2	360.1	44.1	453.3	63.8	
32	243.1	870.2	306.2	498.4	243.3	115.1	160.9	59.5	360.2	44.2	453.4	63.3	
33	254.4	871.0	309.8	498.2	243.6	115.3	161.0	59.7	360.3	44.2	453.5	62.8	
34	265.7	871.7	313.4	498.0	243.9	115.6	161.2	60.0	360.4	44.3	453.7	62.4	
35	276.9	872.2	317.0	497.8	244.2	115.9	161.4	60.3	360.5	44.3	453.8	61.9	
36	288.1	872.5	320.7	497.5	244.6	116.2	161.6	60.7	360.6	44.4	453.9	61.5	
37	299.4	872.6	324.4	497.2	244.9	116.5	161.8	61.1	360.6	44.4	454.1	61.1	
38	310.6	872.5	328.2	496.9	245.2	116.9	162.0	61.5	360.7	44.5	454.2	60.7	
39	321.9	872.2	332.1	496.6	245.6	117.2	162.3	62.0	360.7	44.6	454.4	60.3	
40	333.3	871.8	336.1	496.2	246.0	117.6	162.6	62.6	360.8	44.6	454.5	59.9	
41	344.7	871.1	340.4	495.7	246.4	118.0	162.9	63.2	360.8	44.7	454.7	59.6	
42	356.1	870.3	344.8	495.2	246.9	118.4	163.3	63.9	360.8	44.7	454.8	59.3	
43	367.6	869.4	349.5	494.6	247.4	118.9	163.6	64.6	360.8	44.8	455.0	58.9	

Frame Number	Head$_x$	Head$_y$	Shoulder$_x$	Shoulder$_y$	Elbow$_x$	Elbow$_y$	Wrist$_x$	Wrist$_y$	Hand$_x$	Hand$_y$	IC$_x$	IC$_y$
44	522.3	1507.7	437.0	1374.1	439.2	1127.3	575.1	943.1	614.7	880.1	449.7	998.7
45	534.5	1505.6	450.4	1371.9	455.7	1125.3	596.2	945.3	638.1	883.6	462.1	996.4
46	546.7	1503.3	463.9	1369.5	472.1	1123.3	616.9	947.4	661.0	887.0	474.7	994.0
47	559.0	1501.0	477.5	1367.2	488.4	1121.2	637.2	949.5	683.4	890.4	487.3	991.5
48	571.3	1498.7	491.2	1364.8	504.6	1119.0	657.1	951.5	705.2	893.8	500.0	988.9
49	583.8	1496.3	504.9	1362.5	520.6	1116.9	676.4	953.5	726.4	897.1	512.8	986.4
50	596.2	1494.0	518.8	1360.2	536.5	1114.9	695.3	955.3	746.9	900.3	525.7	983.8
51	608.8	1491.8	532.8	1358.0	552.1	1112.9	713.6	957.1	766.8	903.4	538.7	981.3
52	621.4	1489.6	546.8	1356.0	567.5	1111.0	731.4	958.7	786.0	906.4	551.9	978.9
53	634.0	1487.6	560.8	1354.1	582.7	1109.3	748.7	960.3	804.5	909.2	565.2	976.5
54	646.8	1485.8	574.9	1352.3	597.6	1107.6	765.3	961.7	822.3	911.9	578.7	974.3
55	659.6	1484.2	589.0	1350.8	612.2	1106.2	781.5	963.0	839.3	914.3	592.4	972.3
56	672.4	1482.7	603.1	1349.5	626.5	1104.9	797.0	964.1	855.7	916.4	606.4	970.4
57	685.3	1481.6	617.2	1348.4	640.5	1103.7	812.0	965.1	871.4	918.3	620.5	968.7
58	698.2	1480.6	631.2	1347.5	654.2	1102.9	826.4	965.9	886.4	919.8	634.9	967.3
59	711.1	1480.0	645.1	1346.9	667.6	1102.2	840.3	966.6	900.7	921.1	649.5	966.0
60	724.1	1479.6	658.8	1346.6	680.7	1101.7	853.6	967.1	914.3	922.0	664.3	965.1
61	737.0	1479.5	672.5	1346.5	693.5	1101.5	866.3	967.4	927.3	922.5	679.2	964.3
62	750.0	1479.6	686.0	1346.6	706.0	1101.4	878.6	967.5	939.6	922.7	694.3	963.8
63	762.9	1480.1	699.3	1347.0	718.1	1101.6	890.3	967.4	951.3	922.5	709.4	963.5
64	775.8	1480.8	712.5	1347.7	730.0	1102.0	901.4	967.2	962.3	922.0	724.5	963.4
65	788.8	1481.7	725.5	1348.5	741.6	1102.6	912.1	966.8	972.7	921.1	739.5	963.5
66	801.7	1482.8	738.3	1349.5	752.9	1103.3	922.3	966.2	982.6	919.8	754.5	963.8
67	814.6	1484.2	751.0	1350.7	763.9	1104.2	932.0	965.5	991.8	918.3	769.3	964.3
68	827.5	1485.7	763.5	1352.0	774.6	1105.3	941.2	964.6	1000.4	916.4	784.0	964.9
69	840.4	1487.4	775.8	1353.5	785.1	1106.4	950.0	963.6	1008.4	914.2	798.4	965.6
70	853.3	1489.2	787.9	1355.1	795.3	1107.6	958.3	962.4	1015.9	911.8	812.5	966.4
71	866.3	1491.1	799.9	1356.7	805.3	1109.0	966.2	961.1	1022.7	909.1	826.4	967.3
72	879.2	1493.1	811.8	1358.4	815.1	1110.4	973.6	959.6	1029.0	906.2	840.0	968.3
73	892.1	1495.2	823.4	1360.1	824.6	1111.8	980.6	958.0	1034.8	903.1	853.2	969.4
74	905.0	1497.3	834.9	1361.9	834.0	1113.3	987.2	956.3	1040.0	899.9	866.1	970.6
75	917.9	1499.5	846.3	1363.6	843.2	1114.8	993.3	954.5	1044.7	896.5	878.8	971.8
76	930.8	1501.7	857.5	1365.3	852.2	1116.3	999.1	952.6	1048.8	893.1	891.1	973.2
77	943.7	1503.9	868.7	1367.1	861.0	1117.8	1004.4	950.5	1052.4	889.6	903.3	974.6
78	956.6	1506.1	879.7	1368.8	869.7	1119.4	1009.4	948.5	1055.6	886.0	915.1	976.2
79	969.4	1508.3	890.6	1370.4	878.3	1120.9	1014.0	946.3	1058.2	882.5	926.9	977.8
80	982.3	1510.4	901.5	1372.0	886.7	1122.4	1018.2	944.1	1060.4	879.0	938.5	979.6
81	995.1	1512.4	912.4	1373.5	895.1	1124.0	1022.1	941.8	1062.2	875.5	950.0	981.4
82	1007.8	1514.4	923.3	1375.0	903.3	1125.5	1025.7	939.6	1063.6	872.0	961.6	983.3
83	1020.6	1516.2	934.2	1376.4	911.5	1126.9	1029.0	937.2	1064.6	868.6	973.1	985.3
84	1033.4	1518.0	945.2	1377.6	919.7	1128.4	1032.0	934.9	1065.2	865.3	984.8	987.2
85	1046.1	1519.5	956.3	1378.8	927.8	1129.8	1034.8	932.6	1065.5	862.1	996.5	989.2
86	1058.8	1520.9	967.4	1379.8	935.9	1131.2	1037.3	930.3	1065.6	858.9	1008.4	991.1
87	1071.5	1522.1	978.7	1380.7	944.1	1132.5	1039.7	928.0	1065.3	855.8	1020.5	992.9
88	1084.2	1523.1	990.1	1381.4	952.3	1133.7	1041.9	925.6	1064.8	852.8	1032.7	994.6
89	1096.9	1523.9	1001.7	1381.9	960.7	1134.8	1044.0	923.3	1064.2	850.0	1045.1	996.1
90	1109.5	1524.4	1013.4	1382.3	969.1	1135.7	1046.1	921.1	1063.3	847.2	1057.7	997.5

Frame Number	GT$_x$	GT$_y$	Knee$_x$	Knee$_y$	Ankle$_x$	Ankle$_y$	R Heel$_x$	R Heel$_y$	Met$_x$	Met$_y$	R Toe$_x$	R Toe$_y$	Stride Events
44	402.3	865.6	365.4	492.3	249.3	120.6	165.0	67.6	360.8	45.0	455.4	58.0	
45	414.0	864.1	371.4	491.4	250.1	121.3	165.5	68.8	360.9	45.1	455.6	57.7	
46	425.8	862.6	377.8	490.5	251.1	122.1	166.2	70.3	360.9	45.2	455.8	57.3	
47	437.6	860.9	384.6	489.5	252.2	123.0	166.9	72.0	361.0	45.2	455.9	57.0	
48	449.6	859.2	391.8	488.4	253.4	124.1	167.7	73.9	361.1	45.3	456.1	56.7	
49	461.7	857.4	399.5	487.2	254.9	125.3	168.7	76.1	361.3	45.4	456.3	56.3	
50	474.0	855.6	407.7	486.1	256.6	126.7	169.9	78.7	361.5	45.4	456.6	55.9	
51	486.4	853.9	416.4	484.9	258.5	128.4	171.2	81.7	361.8	45.5	456.8	55.5	
52	499.1	852.1	425.6	483.7	260.8	130.3	172.9	85.1	362.1	45.6	457.1	55.1	
53	511.9	850.4	435.5	482.5	263.4	132.4	174.9	88.9	362.5	45.6	457.4	54.6	
54	525.0	848.7	445.9	481.3	266.5	134.9	177.2	93.3	362.9	45.7	457.8	54.1	
55	538.3	847.1	456.9	480.1	270.0	137.6	180.1	98.2	363.5	45.9	458.2	53.5	
56	551.8	845.6	468.6	479.0	274.0	140.7	183.4	103.8	364.1	46.1	458.7	52.9	
57	565.6	844.2	481.0	477.8	278.6	144.1	187.4	109.9	364.9	46.4	459.3	52.2	
58	579.7	842.8	494.1	476.7	283.8	147.8	192.1	116.6	365.9	46.9	460.1	51.4	
59	594.0	841.6	507.9	475.6	289.7	151.9	197.6	124.0	367.1	47.5	461.0	50.4	LTD
60	608.6	840.4	522.5	474.5	296.4	156.3	204.0	131.9	368.6	48.4	462.1	49.4	
61	623.5	839.4	538.0	473.4	303.9	161.0	211.4	140.5	370.5	49.6	463.5	48.2	
62	638.5	838.4	554.3	472.4	312.3	166.1	220.0	149.7	372.8	51.1	465.1	46.9	
63	653.7	837.6	571.4	471.3	321.8	171.5	229.8	159.4	375.6	52.9	467.2	45.5	
64	669.1	836.8	589.4	470.2	332.2	177.1	240.9	169.5	379.1	55.2	469.8	44.0	
65	684.5	836.1	608.3	469.1	343.8	183.0	253.3	180.0	383.3	57.9	472.9	42.3	
66	700.0	835.4	628.0	468.1	356.6	189.1	267.1	190.7	388.5	61.1	476.7	40.6	
67	715.4	834.9	648.6	467.0	370.5	195.3	282.4	201.6	394.8	64.7	481.5	39.0	
68	730.8	834.4	670.0	466.0	385.6	201.6	299.0	212.4	402.3	68.8	487.2	37.4	
69	746.0	833.9	692.1	465.0	402.0	207.8	317.0	223.0	411.2	73.3	494.2	36.1	
70	761.1	833.5	714.9	464.2	419.6	214.1	336.3	233.2	421.6	78.1	502.6	35.1	
71	775.9	833.2	738.3	463.5	438.3	220.2	356.7	242.8	433.7	83.2	512.7	34.6	
72	790.6	832.9	762.3	463.0	458.2	226.0	378.3	251.7	447.7	88.4	524.7	34.5	
73	804.9	832.8	786.6	462.8	479.1	231.6	400.7	259.7	463.6	93.6	538.7	35.0	
74	819.1	832.8	811.3	462.9	501.1	236.8	423.9	266.8	481.5	98.7	555.0	35.9	
75	833.0	832.9	836.3	463.4	524.0	241.5	447.8	272.7	501.5	103.6	573.7	37.3	
76	846.6	833.2	861.4	464.3	547.9	245.6	472.3	277.5	523.5	108.1	594.7	39.1	RTO
77	860.1	833.7	886.5	465.7	572.6	249.1	497.3	281.0	547.5	112.1	618.1	41.0	
78	873.4	834.4	911.6	467.5	598.2	251.9	522.8	283.3	573.5	115.4	643.7	43.1	
79	886.5	835.3	936.6	469.8	624.5	254.0	548.8	284.4	601.1	118.0	671.5	45.1	
80	899.6	836.4	961.5	472.5	651.5	255.2	575.2	284.1	630.5	119.8	701.3	47.0	
81	912.5	837.7	986.0	475.7	679.2	255.6	602.0	282.7	661.3	120.8	732.9	48.7	
82	925.5	839.1	1010.3	479.2	707.5	255.2	629.2	280.0	693.4	121.0	766.2	50.2	
83	938.5	840.7	1034.2	482.9	736.4	253.9	656.9	276.1	726.8	120.4	800.9	51.5	
84	951.6	842.4	1057.8	486.9	765.8	251.8	685.0	271.1	761.3	119.0	837.0	52.6	
85	964.7	844.2	1080.9	491.0	795.7	249.0	713.7	265.0	796.8	117.1	874.3	53.5	
86	977.9	846.0	1103.5	495.1	826.2	245.4	742.9	257.8	833.3	114.6	912.5	54.3	
87	991.2	847.7	1125.7	499.2	857.2	241.1	772.6	249.8	870.7	111.6	951.7	55.0	
88	1004.6	849.5	1147.4	503.3	888.6	236.2	803.0	240.8	908.8	108.3	991.7	55.7	
89	1018.2	851.1	1168.6	507.1	920.6	230.7	834.0	231.1	947.7	104.8	1032.3	56.6	
90	1031.8	852.6	1189.3	510.8	953.0	224.8	865.6	220.8	987.2	101.2	1073.4	57.7	

Frame Number	Head$_x$	Head$_y$	Shoulder$_x$	Shoulder$_y$	Elbow$_x$	Elbow$_y$	Wrist$_x$	Wrist$_y$	Hand$_x$	Hand$_y$	IC$_x$	IC$_y$
91	1122.2	1524.6	1025.3	1382.4	977.7	1136.6	1048.1	918.8	1062.4	844.5	1070.3	998.7
92	1134.8	1524.5	1037.3	1382.3	986.4	1137.2	1050.1	916.6	1061.5	842.0	1083.1	999.7
93	1147.5	1524.2	1049.5	1382.0	995.4	1137.8	1052.1	914.4	1060.5	839.6	1096.0	1000.4
94	1160.1	1523.6	1061.9	1381.5	1004.5	1138.1	1054.3	912.3	1059.5	837.3	1108.9	1000.9
95	1172.7	1522.7	1074.5	1380.7	1013.8	1138.3	1056.6	910.2	1058.7	835.1	1122.0	1001.1
96	1185.4	1521.5	1087.2	1379.8	1023.4	1138.3	1059.1	908.2	1058.0	833.1	1135.1	1001.0
97	1198.1	1520.0	1100.1	1378.6	1033.2	1138.1	1061.8	906.3	1057.6	831.2	1148.2	1000.7
98	1210.7	1518.3	1113.1	1377.1	1043.2	1137.7	1064.8	904.5	1057.4	829.5	1161.4	1000.1
99	1223.5	1516.4	1126.3	1375.6	1053.5	1137.1	1068.1	902.8	1057.5	827.9	1174.7	999.2
100	1236.2	1514.3	1139.6	1373.8	1064.0	1136.4	1071.8	901.1	1058.0	826.5	1188.0	998.0
101	1249.1	1512.0	1153.0	1371.8	1074.6	1135.6	1075.9	899.6	1058.9	825.1	1201.4	996.7
102	1262.0	1509.6	1166.5	1369.8	1085.5	1134.6	1080.3	898.1	1060.3	823.9	1214.8	995.0
103	1274.9	1507.1	1180.1	1367.6	1096.5	1133.6	1085.2	896.8	1062.2	822.8	1228.3	993.2
104	1287.9	1504.5	1193.7	1365.3	1107.7	1132.4	1090.5	895.5	1064.6	821.8	1241.8	991.2
105	1300.8	1501.8	1207.3	1363.0	1119.0	1131.2	1096.3	894.3	1067.6	820.9	1255.3	989.0
106	1313.8	1499.2	1220.9	1360.6	1130.3	1130.0	1102.4	893.2	1071.2	820.0	1268.8	986.8
107	1326.7	1496.5	1234.5	1358.3	1141.6	1128.7	1109.0	892.2	1075.4	819.3	1282.3	984.4
108	1339.6	1493.8	1247.9	1355.9	1152.8	1127.5	1116.0	891.3	1080.1	818.5	1295.7	982.0
109	1352.3	1491.2	1261.1	1353.6	1164.0	1126.4	1123.3	890.5	1085.4	817.8	1308.9	979.5
110	1364.8	1488.8	1274.0	1351.4	1175.0	1125.3	1130.8	889.7	1091.1	817.2	1322.0	977.1
111	1377.0	1486.4	1286.5	1349.2	1185.7	1124.2	1138.5	889.0	1097.2	816.6	1334.7	974.8
112	1388.7	1484.2	1298.5	1347.2	1196.1	1123.3	1146.3	888.4	1103.5	816.0	1346.9	972.5
113	1399.9	1482.2	1309.9	1345.4	1206.0	1122.4	1153.9	887.8	1109.9	815.4	1358.5	970.4
114	1410.3	1480.4	1320.4	1343.7	1215.2	1121.7	1161.2	887.3	1116.2	814.9	1369.4	968.5
115	1419.7	1478.9	1330.0	1342.2	1223.6	1121.0	1168.1	886.9	1122.2	814.4	1379.3	966.8
116	1428.0	1477.6	1338.4	1341.0	1231.0	1120.5	1174.3	886.5	1127.8	814.0	1387.9	965.4
117	1435.0	1476.5	1345.3	1340.0	1237.2	1120.0	1179.6	886.3	1132.5	813.7	1395.2	964.2
118	1440.4	1475.7	1350.8	1339.2	1242.1	1119.7	1183.8	886.0	1136.4	813.4	1400.9	963.3
119	1444.3	1475.2	1354.7	1338.7	1245.6	1119.5	1186.8	885.9	1139.2	813.2	1404.9	962.7
120	1446.7	1474.9	1357.2	1338.4	1247.9	1119.3	1188.8	885.8	1141.0	813.1	1407.6	962.3
121	1448.5	1474.6	1358.9	1338.2	1249.5	1119.2	1190.2	885.7	1142.4	813.0	1409.4	962.1
122	1535.0	1468.6	1446.3	1331.3	1332.7	1115.0	1271.0	881.7	1222.5	807.7	1503.3	954.1

Frame Number	GT$_x$	GT$_y$	Knee$_x$	Knee$_y$	Ankle$_x$	Ankle$_y$	R Heel$_x$	R Heel$_y$	Met$_x$	Met$_y$	R Toe$_x$	R Toe$_y$	Stride Events
91	1045.5	853.9	1209.5	514.1	985.9	218.6	898.0	209.9	1027.2	97.7	1114.9	59.0	
92	1059.3	855.0	1229.1	517.2	1019.2	212.0	931.1	198.6	1067.7	94.2	1156.6	60.8	
93	1073.1	855.9	1248.3	519.9	1052.9	205.2	964.8	187.0	1108.6	91.1	1198.5	62.9	
94	1087.0	856.5	1266.9	522.2	1087.0	198.4	999.2	175.3	1149.8	88.3	1240.5	65.6	
95	1100.9	856.9	1285.0	524.0	1121.5	191.6	1034.4	163.5	1191.1	85.9	1282.3	68.9	
96	1114.9	857.0	1302.5	525.5	1156.1	184.8	1070.1	151.9	1232.5	84.0	1323.9	72.9	
97	1128.9	856.8	1319.5	526.5	1191.0	178.4	1106.4	140.4	1273.9	82.7	1365.1	77.5	
98	1142.9	856.3	1336.1	527.0	1226.0	172.2	1143.1	129.3	1315.1	82.1	1405.8	82.8	
99	1156.8	855.5	1352.1	527.1	1261.0	166.4	1180.1	118.7	1355.9	82.1	1445.9	88.9	
100	1170.8	854.4	1367.6	526.7	1295.9	161.1	1217.4	108.6	1396.2	82.9	1485.2	95.6	
101	1184.6	853.1	1382.6	526.0	1330.5	156.4	1254.6	99.3	1435.9	84.3	1523.6	102.9	
102	1198.4	851.5	1397.3	524.8	1364.6	152.2	1291.7	90.7	1474.7	86.4	1560.9	110.8	
103	1212.1	849.6	1411.5	523.3	1398.2	148.6	1328.3	82.9	1512.4	89.2	1597.1	119.1	
104	1225.7	847.5	1425.4	521.6	1430.9	145.6	1364.3	75.9	1548.8	92.4	1631.9	127.8	
105	1239.2	845.3	1439.0	519.6	1462.7	143.2	1399.4	69.8	1583.9	96.1	1665.2	136.6	
106	1252.5	842.9	1452.3	517.4	1493.3	141.3	1433.4	64.6	1617.3	100.1	1697.0	145.3	
107	1265.8	840.4	1465.4	515.2	1522.5	139.8	1465.9	60.2	1648.9	104.2	1727.0	153.9	
108	1278.8	837.8	1478.2	512.9	1550.1	138.7	1496.7	56.5	1678.6	108.3	1755.3	162.0	
109	1291.7	835.2	1490.8	510.6	1576.0	137.8	1525.6	53.4	1706.1	112.3	1781.7	169.6	
110	1304.2	832.7	1503.2	508.4	1600.1	137.1	1552.4	50.9	1731.5	115.9	1806.2	176.4	
111	1316.4	830.2	1515.2	506.3	1622.2	136.5	1576.9	48.8	1754.5	119.1	1828.7	182.4	
112	1328.2	827.9	1526.9	504.4	1642.2	135.9	1599.1	47.1	1775.1	121.7	1849.1	187.5	
113	1339.3	825.8	1538.1	502.7	1660.1	135.2	1618.6	45.6	1793.3	123.8	1867.4	191.6	
114	1349.7	823.8	1548.6	501.1	1675.7	134.5	1635.6	44.3	1809.1	125.2	1883.6	194.7	
115	1359.1	822.1	1558.2	499.8	1689.1	133.7	1650.0	43.1	1822.5	126.1	1897.5	197.1	
116	1367.3	820.6	1566.7	498.8	1700.2	132.9	1661.8	42.1	1833.4	126.6	1909.2	198.7	
117	1374.2	819.5	1573.9	497.9	1708.9	132.2	1671.0	41.3	1842.1	126.7	1918.5	199.7	
118	1379.6	818.6	1579.5	497.2	1715.5	131.5	1677.8	40.6	1848.5	126.6	1925.6	200.2	
119	1383.5	818.0	1583.5	496.8	1719.9	131.0	1682.4	40.1	1852.9	126.4	1930.4	200.5	
120	1386.0	817.6	1586.2	496.5	1722.7	130.7	1685.2	39.8	1855.6	126.2	1933.5	200.5	
121	1387.7	817.3	1588.0	496.3	1724.5	130.4	1687.0	39.6	1857.3	126.0	1935.5	200.5	RTD
122	1483.7	808.7	1679.0	487.3	1780.3	116.5	1731.3	38.7	1917.9	95.7	1937.5	200.4	

TABLE C-2   Kinetic Data in Newtons

Number	F_x	F_y	F_z	Events	Number	F_x	F_y	F_z	Events
−1	0.0	0.0	0.0		39	−10.7	−7.1	275.7	
0	1.6	−0.2	7.0	RTD	40	−12.3	−3.5	285.8	
1	−1.3	6.2	53.3		41	−12.1	−3.0	288.7	
2	1.7	9.5	111.1		42	−13.1	−0.5	304.0	
3	9.5	−4.7	184.4		43	−13.0	1.9	313.4	
4	21.1	−22.6	235.5		44	−14.3	5.1	337.0	
5	29.0	−41.4	289.7		45	−12.3	9.1	351.1	
6	30.0	−54.6	335.5		46	−13.1	13.0	379.4	
7	26.0	−66.6	385.8		47	−12.5	18.4	396.7	
8	18.2	−73.6	428.0		48	−14.5	22.6	428.9	
9	10.7	−82.6	459.7		49	−14.9	29.7	451.7	
10	6.9	−85.7	485.9		50	−16.0	34.6	482.5	
11	5.8	−89.9	499.8		51	−16.9	43.3	508.2	
12	1.4	−89.6	520.2		52	−18.2	49.0	533.8	
13	−5.7	−93.1	529.9		53	−20.2	57.4	558.3	
14	−14.6	−92.3	545.7		54	−21.5	63.2	574.1	
15	−19.1	−89.9	545.0		55	−23.8	71.7	592.9	
16	−25.3	−88.2	555.6		56	−24.5	78.3	600.9	
17	−26.9	−82.4	550.0		57	−26.5	85.8	610.6	
18	−28.1	−76.8	553.0		58	−24.9	93.0	604.6	
19	−25.9	−68.8	540.6		59	−24.9	99.0	604.4	
20	−24.8	−62.7	533.2		60	−22.2	104.5	583.8	
21	−22.7	−54.3	514.2		61	−20.4	107.4	570.0	
22	−21.1	−49.4	496.5		62	−17.0	110.5	535.1	
23	−19.6	−42.0	474.9		63	−14.3	110.7	503.3	
24	−17.8	−38.1	444.9		64	−9.7	109.1	452.5	
25	−17.3	−31.9	428.5		65	−6.3	101.1	404.7	
26	−16.1	−28.6	397.5		66	−3.8	91.2	345.5	
27	−17.3	−23.9	382.3		67	−3.1	77.4	285.3	
28	−15.6	−21.3	356.4		68	−3.6	66.5	228.9	
29	−16.3	−19.2	341.4		69	−3.1	51.3	171.8	
30	−14.5	−17.3	316.3		70	−4.3	39.7	128.9	
31	−14.2	−16.4	310.0		71	−4.4	26.8	88.2	
32	−11.9	−14.4	290.3		72	−4.6	18.2	61.7	
33	−11.2	−14.5	288.3		73	−2.4	10.2	34.1	
34	−9.9	−12.9	275.9		74	−2.0	5.4	25.2	
35	−9.6	−13.1	275.0		75	0.0	2.9	9.3	
36	−9.5	−10.4	269.7		76	−1.5	0.9	11.0	RTO
37	−9.5	−11.2	271.8						
38	−10.3	−7.6	273.8						

Appendix D

Numerical Example for Calculating Projectile

Projectile motion can be analyzed using the equations of constant acceleration because, while an object is in the air, the vertical acceleration is constant at -9.81 m/s^2 and the horizontal acceleration is constant at 0 m/s^2. The exact equations that you use and the order that you use them will depend on the information that you are given and the information that you are asked to solve for. There are too many possibilities to give an example of each so we have developed a method that will allow you to solve any projectile motion problem.

1) $v_f = v_i + at$
2) $s_f = s_i + v_i t + \frac{1}{2}at^2$
3) $v_f^2 = v_i^2 + 2a(s_f - s_i)$

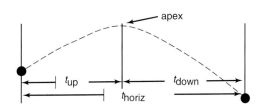

Step 1

Break up the vertical direction into an up and a down phase. The horizontal direction will have only one phase. There are six variables that the equations of constant acceleration use (two positions, two velocities, acceleration, and time). If we create a grid like the one below, we have a cell for each variable in each phase. Note that the final (f) variables for the VERT$_{up}$ phase are the same as the initial (i) variables for the VERT$_{down}$ phase. Also, horizontal velocity does not change so v_i and v_f are the same for HORIZ. Also, $t_{up} + t_{down} = t_{horiz}$.

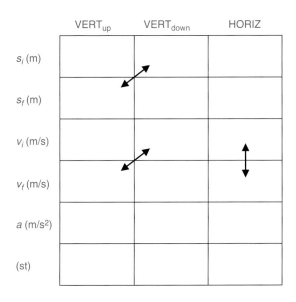

457

Step 2

Fill in what you know. You can always define your coordinate system so you can make the release position your origin or you could make the origin at the ground. v_f for the VERT$_{up}$ phase and v_i for the VERT$_{down}$ phase will always be zero because the object is changing the vertical direction at the apex of the flight. The grid will look like this if we choose the origin to be at the release position:

	VERT$_{up}$	VERT$_{down}$	HORIZ
s_i (m)	0		0
s_f (m)			
v_i (m/s)		0	
v_f (m/s)	0		
a (m/s²)	−9.81	−9.81	0
t (s)			

 ## Step 3

Read the problem and fill in any given values. For instance, if a shot putter releases a shot at an angle of 40° from a height of 2.2 m with a velocity of 13.3 m/s you can fill in the following additional cells by decomposing the velocity vector into vertical and horizontal components. You are not always given the same information to start with.

	VERT$_{up}$	VERT$_{down}$	HORIZ
s_i (m)	0		0
s_f (m)		−2.2	
v_i (m/s)	13.3 sin 40 = 8.55	0	13.3 cos 40 = 10.19
v_f (m/s)	0		10.19
a (m/s²)	−9.81	−9.81	0
t (s)			

 ## Step 4

If there is only one blank in a column, then you can use one of the equations of constant acceleration to find the value for that cell. If there is more than one blank choose a formula that has only one unknown and solve for it. Continue until all cells in the column have a value and then move to a different column. Note that if the release and landing are at the same height, you can use the fact that t_{up} and t_{down} equal ½ of t_{horiz} (the same amount of time is spent going up as going down).

	VERT$_{up}$	VERT$_{down}$	HORIZ
s_i (m)	0	3.73	0
s_f (m)	(eq. 2) s_f = 3.73	−2.2	(eq. 1) s_f = 20.08
v_i (m/s)	8.55	0	10.19
v_f (m/s)	0	(eq. 3) v_f = −10.78	10.19
a (m/s²)	−9.81	−9.81	0
t (s)	(eq. 1) t_{up} = 0.87	(eq. 1) t_{down} = 1.10	t_{up} + t_{down} t_{horiz} = 1.97

This procedure can be used to calculate any of the kinematic variables in the table as long as you are given enough information to start with. The distance that the shot traveled under these conditions was 20.08 m. It traveled 3.73 m higher than the release height and it was in the air for 1.97 seconds. In our initial discussion of projectiles and the equations of constant acceleration, air resistance was considered to be negligible. The mathematics of air resistance is well beyond the scope of this book because it requires solving differential equations. However, it may be of interest to know that if these initial conditions were used to solve this problem taking into account air resistance, a reduction of 1.46% in the range of the throw would be found. Considering air resistance, the throw would be 0.3 m less, or a total of 19.78 m.

Glossary

Abduction: Sideways movement of the segment away from the midline or sagittal plane.

Absolute Angle: Angle of a segment as measured from the right horizontal that describes the orientation of the segment in space.

Absolute Reference Frame: A reference frame in which the origin is at a joint center and does not move with the segment.

Acetabular Labrum: Rim of fibrocartilage that encircles the acetabulum, deepening the socket.

Acetabulum: The concave, cup-shaped cavity on the lateral, inferior, anterior surface of the pelvis.

Acromioclavicular Joint: Articulation between the acromion process of the scapula and the lateral end of the clavicle.

Actin: A protein of the myofibril, noticeable by its light banding. Along with myosin, it is responsible for the contraction and relaxation of muscle.

Action Potential: An electrical signal that travels through the nerve or muscle as the membrane potential changes because of the exchange of ions.

Active Insufficiency: The inability of a two-joint muscle to produce force when joint position shortens the muscle to the point where it cannot contract.

Active Peak: The peak of the running vertical ground reaction force curve during midstance. It is the result of active contraction of muscle.

Active Range of Motion: The range of motion achieved through some voluntary contraction of an agonist, creating the joint movement.

Adduction: Sideways movement of a segment toward the midline or sagittal plane.

Agonist: A muscle responsible for producing a specific movement through concentric muscle action.

All-or-None Principle: The stimulation of a muscle fiber that causes the action potential to travel over either the whole muscle fiber (activation threshold) or none of the muscle fiber.

Alpha Motoneuron: An afferent neuron with a large cell body in or near the spinal cord from which a long axon projects from the spinal cord to the muscle fibers that it innervates.

Amphiarthrodial Joint: A type of joint whose bones are connected by cartilage; some movement may be allowed at these joints. Also called cartilaginous joint.

Anatomical Cross Section: The cross section at a right angle to the direction of the fibers.

Anatomical Position: The standardized reference position used in the medical profession.

Anatomy: The science of the structure of the body.

Angle: A figure formed by two lines meeting at a point, the vertex.

Angle–Angle Diagram: A graph in which the angle of one segment is plotted as a function of the angle of another segment.

Angle of Inclination: Angle formed by the neck of the femur in the frontal plane.

Angular Acceleration: The change in angular velocity per unit time.

Angular Displacement: The difference between the final angular position and the initial angular position of a rotating body.

Angular Impulse: The torque multiplied by the duration of the torque. Also calculated as the integral of the torque with respect to time.

Angular Kinematics: The description of angular motion, including angular positions, angular velocities, and angular accelerations, without regard to the causes of the motion.

Angular Kinetics: An area of mechanics involving the causes of angular motion, mainly torques.

Angular Momentum: The angular quantity that changes when an angular impulse is applied to an object. Calculated as the product of the moment of inertia and the angular velocity.

Angular Motion: Motion about an axis of rotation in which different regions of the same object do not move through the same distance in the same time.

Angular Power: The product of angular work and angular velocity or the rate of doing angular work.

Angular Speed: The angular distance traveled divided by the time over which the angular motion occurred.

Angular Velocity: The time rate of change of angular displacement.

Angular Work: The torque multiplied by the angular distance that the object moves. Doing angular work on an object changes the angular energy.

Annular Ligament: Ligament inserting on the anterior and posterior margins of the radial notch; supports the head of the radius.

Antagonist: A muscle responsible for opposing the concentric muscle action of the agonist.

Anterior: A position in front of a designated reference point.

Anterior Compartment Syndrome: Nerve and vascular compression caused by hypertrophy of the anterior tibial muscles in a small muscular compartment.

Anteroposterior Axis: The axis through the center of mass of the body running from posterior to anterior.

Anteversion: The degree to which an anatomical structure is rotated forward. See retroversion for opposite.

Apex: The highest point of a parabola and the highest point a projectile reaches in its trajectory.

Aponeurosis: A flattened or ribbonlike tendinous expansion from the muscle that connects into the bone.

Apophysis: A bony outgrowth such as a process, tubercle, and tuberosity.

Apophysitis: Inflammation of the apophysis, or bony outgrowth.

Appendicular Skeleton: The bones of the extremities.

Articular Cartilage: Hyaline cartilage consisting of tough, fibrous connective tissue.

Articular End Plate: The region of bone in a joint that makes contact with another bone.

Asynchronous: Describing events that do not occur at the same time. In skeletal muscle contraction, the spacing of the activation of the motor unit.

Autogenic Facilitation: Internally generated excitation of the alpha motoneurons through stretch or some other input.

Avulsion Fracture: The tearing away of a part of bone when a tensile force is applied.

Axial Skeleton: The bones of the head, neck, and trunk.

Axis: The second cervical vertebra.

Axis of Rotation: The imaginary line about which an object rotates.

Axon: Neuron process carrying nerve impulses away from the cell body of the neuron. The pathway through which the nerve impulse travels.

Balance: The ability to resist linear and angular accelerations.

Ball-and-Socket Joint: A type of diarthrodial joint that allows motion through three planes.

Ballistic Stretching: Moving a limb to the terminal range of motion through rapid movements initiated by strong muscular contractions and continued by momentum.

Belly: The fleshy central portion of a muscle.

Bennett Fracture: Longitudinal fracture of the base of the first metacarpal.

Bicipital Tendinitis: Inflammation of the tendon of the biceps brachii.

Bilateral Deficit: The loss of both force and neural input to the muscles through bilateral activation of both limbs.

Biomechanics: The study of motion and the effect of forces on biological systems.

Bipennate: A feather-shaped fiber arrangement, in which the fibers run off of both sides of a tendon running through the muscle.

Bone Mineral Density: Amount of mineral measured per unit area or volume of bone tissue.

Bone-on-Bone Force: The force between two bones in the body.

Bone Strain: See Stress Fracture.

Boundary Layer: The layer of air molecules closest to an object. This layer tends to move with the object.

Boutonnière Deformity: A stiff proximal interphalangeal articulation caused by injury to the finger extensor mechanism.

Bursa: A fibrous fluid-filled sac between bones and tendons or other structures that reduces friction during movement.

Bursitis: Inflammation of a bursa.

Calcaneal Apophysitis: Inflammation at the epiphysis on the calcaneus.

Calcaneocuboid Joint: The articulation between the calcaneus and the cuboid bones; part of the midtarsal joint.

Calcaneofibular Ligament: Ligament inserting on the lateral malleolus and outer calcaneus; limits backward movement of the foot and restrains inversion.

Calcaneonavicular Ligament: Ligament inserting on the calcaneus and the navicular; supports the arch and limits abduction of the foot.

Cancellous Bone: A less dense form of bone tissue that consists of a sponge-like architecture. Usually found at the proximal and distal ends of long bones.

Capitulum: Eminence on the distal end of the lateral epicondyle of the humerus; articulates with the head of the radius at the elbow.

Capsular Ligament: Ligaments within the wall of the capsule; thickening in the capsule wall.

Capsule: A fibrous connective tissue that encloses the diarthrodial joint.

Cardinal Planes: The planes of the body that intersect at the total body center of mass.

Carpal Tunnel Syndrome: Pressure and constriction of the median nerve caused by repetitive actions at the wrist.

Carrying Angle: Angle between the ulna and the humerus with the elbow extended; 10° to 25°.

Cartilaginous Joint: A type of joint whose bones are connected by cartilage. Some movement may be allowed at these joints. Also called amphiarthrodial joint.

Cell Body: The portion of the neuron that contains the nucleus and a well-marked nucleolus. The cell body receives information through the dendrites and sends information through the axon. Also called the soma.

Center of Gravity: The point about which the gravitational force is balanced. Corresponds to the center of mass in most cases.

Center of Mass: The point about which the distribution of mass sums to zero.

Center of Pressure: The point about which the distribution of pressure sums to zero.

Central Nervous System: The brain and the spinal cord.

Centripetal Acceleration: The component of the linear acceleration directed toward the axis of rotation.

Centripetal Force: The force that keeps an object moving with a constant angular speed.

Chondrocytes: Cartilage cells.

Chondromalacia Patellae: Cartilage destruction on the underside of the patella; soft and fibrillated cartilage.

Circular Muscle: Concentrically arranged muscle around an opening or recess.

Circumduction: A movement that is a combination of flexion, adduction, extension, and abduction.

Clavicle: An S-shaped long bone articulating with the scapula and the sternum.

Close-Packed Position: The joint position with maximum contact between the two joint surfaces and in which the ligaments are taut, forcing the two bones to act as a single unit.

Closed-Chain Exercise: Exercises using eccentric and concentric muscle actions with the feet fixed on the floor. Movements begin with segments distal to the feet (trunk and thigh) and move toward the feet, as in the squat.

Coefficient of Friction: The ratio of the friction force to the normal force between two bodies.

Collagen: A connective tissue that is the main protein of skin, tendon, ligament, bone, and cartilage.

Collinear Force: Forces are collinear if they have the same line of action. They may be opposite in direction.

Complex Training: A workout technique that combines strength and plyometric training with sport specific exercises in an effort to train neural components, muscular strength and rate of force production.

Concentric: Muscle action in which tension causes visible shortening in the length of the muscle; positive work is performed.

Concurrent Forces: Multiple forces that pass through a common point.

Conduction Velocity: The speed at which an action potential is propagated.

Condylar Joint: A type of diarthrodial joint that is biaxial, with one plane of movement that dominates the movement in the joint.

Condyle: A rounded projection on a bone.

Congenital Hip Dislocation: A condition existing at birth in which the hip joint subluxates or dislocates for no apparent reason.

Conservation of Angular Momentum: The concept that the angular momentum of an object cannot change unless and external torque acts on it.

Contact Force: The forces between two objects that are making contact with each other.

Contractility: The ability of muscle tissue to shorten when the muscle tissue receives sufficient stimulation.

Contralateral: On the opposite side.

Convergent Muscle: Fan-shaped muscle with broad fibers that converge to a common insertion site.

Coplanar Forces: Two or more forces acting in the same plane.

Coracoacromial: The ligament connecting the coracoid process of the scapula to the acromion of the scapula.

Coracoclavicular: The joint or ligaments that connect the coracoid process of the scapula to the clavicle.

Coracohumeral: Ligaments that connect the coracoid process of the scapula to the humerus.

Coracoid Process: A curved process arising from the upper neck of the scapula; overhangs the shoulder joint.

Coronal Plane: Another name for the frontal plane. It divides the body in anterior and posterior portions.

Coronoid Fossa: Cavity in the humerus that receives the coronoid process of the ulna during elbow flexion.

Coronoid Process: Wide eminence on proximal end of ulna; forms the anterior portion of the trochlear fossa.

Cortical Bone: Dense, compact tissue on the exterior of bone that provides strength and stiffness to the skeletal system. Also called compact bone.

Coxa Plana: Degeneration and recalcification (osteochondritis) of the capitular epiphysis (head) of the femur; also called Legg-Calvé-Perthes disease.

Coxa Valga: An increase in the angle of inclination of the femoral neck (>125°).

Coxa Vara: A decrease in the angle of inclination of the femoral neck (<125°).

Cross-Sectional Area: The area of a slice of an object, usually a muscle or bone sliced perpendicular to the long axis.

Crossed Extensor Reflex: Reflex causing extension of a flexed limb when stimulated by rapid flexion or withdrawal by the contralateral limb.

Degeneration: Deterioration of tissue; a chemical change in the body tissue; change of tissue to a less functionally active form.

Degree: A unit of angular measurement: 1/360 of a revolution.

Degree of Freedom: The movement of a joint in a plane.

Deltoid Ligament: Ligament inserting on the medial malleolus, talus, navicular, and calcaneus; resists valgus forces and restrains plantarflexion, dorsiflexion, eversion, and abduction of the foot.

Dendrites: Processes on the neuron that receive information and transmit information to the cell body of the neuron.

Depolarization: A reduction in the potential of a membrane.

Depression: Movement of the segment downward (scapula and clavicle); return of the elevation movement.

Diaphysis: The shaft of a long bone.

Diarthrodial Joint: Freely movable joint; also called synovial joint.

Dislocation: Bone displacement; separation of the bony surfaces in a joint.

Distal: A position relatively far from a designated reference point.

Distal Femoral Epiphysitis: Inflammation of the epiphysis at the attachment of the collateral ligaments at the knee.

Distal Interphalangeal Joints: Joints separating the most distal portion of the phalanges and the next proximal bone.

Dorsal: See Posterior.

Dorsiflexion: Rotation of the foot up in the sagittal plane; movement toward the leg.

Downward Rotation: The action whereby the scapula swings toward the midline of the body.

Dynamic Analysis: Mechanical analysis of an object while it is accelerating.

Dynamics: The branch of mechanics in which the system being studied undergoes acceleration.

Eccentric: Muscle action in which tension is developed in the muscle and the muscle lengthens; negative work is performed.

Eccentric Force: A force that has a line of action that does not go through the axis of rotation.

Ectopic Bone: Bone formation that is displaced away from the normal site.

Effort Arm: In a lever system the effort arm is the perpendicular distance between the line of action of the effort force vector and the axis of rotation.

Elastic Material: A material that exhibits only elastic properties on a stress–strain curve.

Elastic Modulus: The linear portion of a stress–strain curve.

Elastic Region: The area of a stress–strain curve before the yield point. The area in which the material will return to its resting length when the applied force is removed.

Elasticity: The ability of muscle tissue to return to its resting length after a stretch is removed.

Electromyogram: The recorded signal of the electrical activity of the muscle.

Electromyography: The measurement of electrical activity of the muscle.

Elevation: Movement of a segment upward (e.g., of the scapula and clavicle).

Ellipsoid Joint: A type of diarthrodial joint with 2 degrees of freedom that resembles the ball-and-socket joint.

Endomysium: The sheath surrounding each muscle fiber.

Energy: The capacity to do work.

Epicondyle: Eminence on a bone above the condyle.

Epicondylitis: Inflammation of the epicondyle or tissues connecting to the epicondyle (e.g., medial and lateral epicondylitis).

Epimysium: A dense, fibrous sheath covering an entire muscle.

Epiphyseal Plate: The disk of cartilage between the metaphysis and the epiphysis of an immature long bone that permits growth in length.

Epiphysis: The ends of a long bone.

Equilibrium: Refers to 1) a state of rest in which multiple forces acting on an object are balanced or 2) a state of balance

Equinus: A limitation in dorsiflexion caused by a short Achilles tendon or tight gastrocnemius and soleus muscles.

Eversion: The movement in which the lateral border of the foot lifts so that the sole of the foot faces away from the midline of the body.

Excitation–Contraction Coupling: Electrochemical stimulation of the muscle fiber that initiates the release of calcium and the subsequent cross-bridging between actin and myosin filaments, which leads to contraction.

Extensibility: The ability of muscle tissue to lengthen beyond resting length.

Extension: Movement of a segment away from an adjacent segment so that the angle between the two segments is increased.

External Work: Work done by external forces.

Extracapsular Ligament: Ligament outside of the joint capsule.

Extrafusal Fiber: Fibers outside the muscle spindle; muscle fibers.

Facet: A small plane surface on a bone where it articulates with another structure.

Failure Point: The point on a stress–strain curve when the applied force causes a complete rupture of the material.

Fascia: Sheet or band of fibrous tissue.

Fascicle: A bundle or cluster of muscle fibers.

Fast-Twitch Fiber: Large skeletal muscle fiber innervated by the alpha-I motor neuron; has fast contraction times. The two subtypes of fast-twitch fibers are the low oxidative and high glycolytic (type IIb) and the medium oxidative and high glycolytic (type IIa).

Fatigue Fracture: A bone fracture due to abnormal stresses acting on normal bone.

Fibers: Elongated cylindrical structures containing cells that constitute the contractile elements of muscle tissue.

Fibrocartilage: A type of cartilage that has parallel thick, collagenous bundles.

First-Class Lever: A lever system in which the fulcrum is in between the effort force and the resistance force.

Flat Bone: A thin bone consisting of thin layers of cortical and cancellous bone.

Flat Muscle: Thin and broad-shaped muscle.

Flexion: Movement of a segment toward an adjacent segment so that the angle between the two is decreased.

Flexor Reflex: Reflex initiated by a painful stimulus that causes a withdrawal or flexion of the limb away from the stimulus.

Fluid Resistance Force: The force resisting the movement of on object through a fluid.

Force Couple: Two forces, equal in magnitude, acting in opposite directions, that produce rotation about an axis.

Force Platform: An instrument used to sense and record the dynamic ground reaction forces.

Forefoot: Region of the foot that includes the metatarsals and phalanges.

Forefoot Valgus: Eversion of the forefoot on the rear foot, with the subtalar joint in neutral position.

Forefoot Varus: Inversion of the forefoot on the rear foot with subtalar in the neutral position.

Form Drag: A type of fluid resistance resulting from a pressure difference in the fluid between different sides of the object.

Fracture: A break in a bone.

Free Body Diagram: A sketch of an object or objects and all of the external forces and moments acting on the object.

Frequency Coding: See Rate Coding.

Frequency Domain: An analysis technique whereby the power of the signal is plotted as a function of the frequency of the signal.

Friction: The force that resists motion when two objects slide against each other.

Frontal (Coronal) Plane: The plane that bisects the body into front and back halves.

Fulcrum: The object providing the pivot point in a lever system.

Functional Anatomy: The study of the body components needed to achieve a human movement or function.

Fundamental Position: A standardized reference position similar to the anatomical position.

Fusiform: Spindle-shaped fiber arrangement in a muscle.

Gamma Loop: A reflex arc that works with the stretch reflex, in which descending motor pathways synapse with both alpha and gamma motoneurons of the muscle fiber and the muscle spindle.

Gamma Motoneuron: A neuron that innervates the contractile ends of the muscle spindle.

Ganglia: Nerve cell bodies outside the central nervous system.

Genu Valgum: A condition in which the knees are abnormally close together with the space between the ankles increased; knock-knees.

Genu Varum: A condition in which the knees are abnormally far apart with the space between the ankles decreased; bowlegs.

Glenohumeral: The joint or ligaments connecting the glenoid fossa of the scapula to the head of the humerus.

Glenohumeral Joint: The articulation between the head of the humerus and the glenoid fossa on the scapula.

Glenoid Fossa: Depression in the lateral superior scapula that forms the socket for the shoulder joint.

Glenoid Labrum: Ring of fibrocartilage around the rim of the glenoid fossa that deepens the socket in the shoulder and hip joints.

Gliding Joint: A type of diarthrodial joint with flat surfaces that allows translation between the two bones; also called plane joint.

Golgi Tendon Organ: A sensory receptor located at the muscle–tendon junction that responds to tension generated during both stretch and contraction of the muscle. Initiates the inverse stretch reflex if the activation threshold is reached.

Hamstring: A group of muscles on the posterior thigh consisting of the semimembranosus, semitendinosus, and biceps femoris.

Haversian System: See Osteon.

Hinge Joint: A type of diarthrodial joint that allows 1 degree of freedom.

Horizontal Abduction: A combination of extension and abduction of the arm or thigh.

Horizontal Adduction: A combination of flexion and adduction of the arm or thigh.

Horizontal Extension (Abduction): Movement of an elevated segment (arm and leg) away from the body in the posterior direction.

Horizontal Flexion (Adduction): Movement of an elevated segment (arm and leg) toward the body in the anterior direction.

Horizontal Plane: A global plane that is parallel to the ground.

Hyaline Cartilage: See Articular Cartilage.

Hyperabduction: Abduction movement beyond the normal range of abduction.

Hyperadduction: Adduction movement beyond the normal range of adduction.

Hyperextension: Continuation of extension past the neutral position.

Hyperflexion: Flexion movement goes beyond the normal range of flexion.

Hyperpolarization: An increase in the potential of a membrane.

Hypertrophy: An enlargement or growth of tissue caused by an increase in the size of cells.

Hypothenar Eminence: The ridge on the palm on the ulnar side created by the presence of intrinsic muscles acting on the little finger.

Hysteresis: The mechanical energy lost by a material that has been deformed.

Iliac Apophysitis: Inflammation of the attachment sites of the gluteus medius and tensor fascia latae on the iliac crest.

Iliofemoral Ligament: Ligament inserting on the antero-superior spine of the ilium and intertrochanteric line of the femur; supports the anterior hip joint and offers restraint in extension and internal and external rotation.

Iliotibial Band: A fibrous band of fascia running from the ilium to the lateral condyle of the tibia.

Iliotibial Band Syndrome: Inflammation of the iliotibial band caused by rubbing on the lateral femoral epicondyle during knee flexion and extension.

Ilium: The superior bone of the pelvic girdle.

Impact Peak: The high frequency peak force that occurs when two objects collide quickly.

Impingement Syndrome: Irritation of structures above the shoulder joint due to repeated compression as the greater tuberosity is pushed up against the underside of the acromion process.

Impulse: Force multiplied by time or the area under a force by time curve.

Impulse-Momentum Relationship: The concept that it takes an impulse (force multiplied by time) to change the momentum (mass multiplied by velocity) of an object.

Inertia: The property of a mass that resists a change in velocity.

Inferior: A position below a designated reference point.

Infrapatellar Bursa: A bursa between the patellar ligament and the tibia.

Innervation Ratio: The average number of fibers controlled by each neuron in a muscle.

Insertion: The more distal attachment site of the muscle.

Intercarpal Joint: Articulation between the carpal bones.

Insufficiency Fracture: A bone fracture due to normal stresses acting on weak bone.

Intercondylar Eminence: Ridge of bone on the tibial plateau that separates the surface into medial and lateral compartments.

Intercondylar Notch: Convex surface on the distal posterior surface of the femur.

Internal Work: Work done by internal forces.

Interneuron: Small, connecting neuron in the spinal cord; can be excitatory or inhibitory.

Interosseous Ligament: Ligament connecting adjacent tarsals; supports the arch and the intertarsal joints.

Interosseous Membrane: A thin layer of tissue running between two bones (radius and ulna, tibia and fibula).

Interphalangeal Joint: Articulation between adjacent phalanges of the fingers and toes.

Intra-Articular Joint: Within the joint capsule.

Intrafusal Fiber: Fibers that are inside the muscle spindle.

Inverse Dynamics: The process of calculating forces and moments based on the kinematics and anthrompometrics of a body.

Inverse Stretch Reflex: Reflex initiated by high tension in the muscle, which inhibits contraction of the muscle through the Golgi tendon organ, causing relaxation of a vigorously contracting muscle.

Inversion: The movement in which the medial border of the foot lifts so that the sole of the foot faces away from the midline of the body.

Ipsilateral: On the same side.

Irritability: The capacity of muscle tissue to respond to a stimulus.

Ischiofemoral Ligament: Ligament inserting on the posterior acetabulum and iliofemoral ligament; restrains adduction and internal rotation of the thigh.

Ischium: The inferoposterior bone of the pelvic girdle.

Isokinetic Exercise: An exercise in which concentric muscle action is generated to move a limb against a device that is speed controlled. Individuals attempt to develop maximum tension through the full range of motion at the specified speed of movement.

Isometric: Muscle action in which tension develops but no visible or external change is seen in joint position; no external work is produced.

Isotonic Exercise: An exercise in which an eccentric or concentric muscle action (or both) is generated to move a specified weight through a range of motion.

Jersey Finger: Avulsion of a finger flexor tendon through forced hyperextension.

Joint Angle: Angle between two segments that is relative and does not change with body orientation.

Joint Reaction Force: The force at a joint that results from the weights and inertial forces of the segments above the joint. It does not include the muscle forces.

Kinematics: Area of study that examines the spatial and temporal components of motion (position, velocity, and acceleration).

Kinesiology: Study of human movement.

Kinetic Energy: The capacity to do work that an object possesses because of its velocity.

Kinetic Friction: The friction force between two objects that are moving relative to each other.

Kinetics: Study of the forces that act on a system.

Labyrinthine Righting Reflex: Reflex stimulated by tilting or spinning of the body, which alters the fluid in the inner ear. The body responds to restore balance by bringing the head to the neutral position or thrusting arms and legs out for balance.

Lamellae: The concentric tubes of collagen that encircle an osteon.

Laminar Flow: The smooth flow of a fluid over an object moving through the fluid. Laminar flow occurs at slower relative speeds and when the object is streamline. Pressure difference is minimized.

Lateral: A position relatively far from the midline of the body.

Lateral Collateral Ligament: Ligament inserting on the lateral epicondyle of the femur and head of the fibula; resists varus forces and is taut in extension.

Lateral Epicondyle: Projection from the lateral side of the distal end of the humerus giving attachment to the hand and finger extensors.

Lateral Flexion: A flexion movement of the head or trunk.

Lateral Tibial Syndrome: Pain on the lateral anterior leg caused by tendinitis of the tibialis anterior or irritation to the interosseous membrane.

Law of Conservation of Energy: The concept that indicates the total energy of a system remains constant unless acted upon by an external energy source.

Law of Cosines: The general case of the Pythagorean theorem.

Legg-Calvé-Perthes Disease: Degeneration and recalcification (osteochondritis) of the capitular epiphysis (head) of the femur; also called coxa plana.

Length–Tension Relationship: The relationship between the length of the muscle and the tension produced by the muscle; highest tensions are developed slightly past resting length.

Lesser Trochanter: A projection of bone on the posterior base of the femoral neck.

Lever: A simple machine that magnifies force or speed of movement.

Lift Force: A component of the air resistance force vector that is perpendicular to the direction of motion.

Ligament: A band of fibrous, collagenous tissue connecting the bone or cartilage to each other; supports the joint.

Line of Action: An infinite line extending along the direction of the force from the point where the force acts.

Linear Envelope: The process whereby a rectified electromyography signal has most of the high-frequency components removed via a low-pass filter.

Linear Kinetics: The study of the cause of translation (forces).

Linear Motion: Motion in a straight or curved line in which different regions of the same object move the same distance.

Lines of Action: An unbounded imaginary line going through a force vector.

Local Angular Momentum: The momentum about a local axis of rotation. This is usually the momentum about a segment center of mass.

Local Graded Potential: An excitatory or inhibitory signal in the nerve or muscle that is not propagated.

Longitudinal Arch: Two arches (medial and lateral) formed by the tarsals and metatarsals, which run the length of the foot and participate in both shock absorption and support while the foot is bearing weight.

Longitudinal Axis: The axis through the center of mass of the body running from top to bottom.

Loose-Packed Position: The joint position with less than maximum contact between the two joint surfaces and in which contact areas frequently change.

MA: The ratio of the effort arm to the resistance arm in a lever system. Indicates the advantage of the lever system.

Magnus Effect: The tendency of a spinning object to move in a curved path while moving through a fluid.

Mallet Finger: Avulsion injury to the finger extensor tendons at the distal phalanx; produced by a forced flexion.

Medial: A position relatively closer to the midline of the body.

Medial Collateral Ligament: Ligament inserting on the medial epicondyle of the femur, medial condyle of the tibia, and medial meniscus; resists valgus forces and restrains the knee joint in internal and external rotation; taut in extension.

Medial Epicondyle: Projection from the medial side of the distal end of the humerus giving attachment to the hand and finger flexors.

Medial Tension Syndrome: Also termed pitcher's elbow, medial pain brought on by excessive valgus forces that may cause ligament sprain, medial epicondylitis, tendinitis, or avulsion fractures to the medial epicondyle.

Medial Tibial Syndrome: Pain above the medial malleolus caused by tendinitis of the tibialis posterior or irritation of the interosseous membrane or periosteum; previously called shin splints.

Mediolateral Axis: The axis through the center of mass of the body running from right to left.

Meniscus: Disk-shaped or crescent-shaped fibrocartilage between two bones.

Metacarpophalangeal Joint: Articulation between the metacarpals and the phalanges in the hand.

Metaphysis: The wide shaft toward the end of a long bone.

Metatarsalgia: Strain of the ligaments supporting the metatarsals.

Metatarsophalangeal Joints: Articulations between the metatarsals and the phalanges in the foot.

Microtrauma: A disturbance or abnormal condition that is initially too small to be seen.

Midcarpal Joint: Articulation between the proximal and distal row of carpals in the hand.

Midfoot: Region of the foot that includes all of the tarsals except the talus and calcaneus.

Modeling: Bone resorption and deposit that forms bone at different sites and rates, resulting in altered size and shape.

Moment Arm: The perpendicular distance from the line of action of the force to the pivot point.

Moment of Force: The torque caused by an eccentric force.

Moment of Inertia: The resistance of an object to changes in angular motion. Determined by the mass and the distribution of mass.

Monosynaptic: A reflex arc consisting of one motor and one sensory neuron.

Monosynaptic Reflex Arc: The reflex arc whereby a sensory neuron is stimulated and facilitates the stimulation of a spinal motoneuron.

Morton Toe: A condition in which the second metatarsal is longer than the first metatarsal.

Motoneurons: Neurons that carry impulses from the brain and spinal cord to the muscle receptors.

Motor End Plates: A flattened expansion in the sarcolemma of the muscle that contains receptors to receive the expansions from the axonal terminals; also called the neuromuscular junction.

Motor Pool: Groups of neurons in the spinal cord that innervate a single muscle.

Motor Unit: The nerve and all of the muscle fibers that it innervates.

Multipennate: A feather-shaped fiber arrangement in which the muscle fibers run diagonally off one or both sides of a tendon running through the muscle.

Muscle Power: The quantity a muscle possesses because of the tension and the contraction velocity.

Muscle Spindle: An encapsulated sensory receptor that lies parallel to muscle fibers and responds to stretch of the muscle.

Muscle Volume: The amount of muscle space determined by the ratio of muscle mass divided by density.

Myelinated: Nerve fibers that have a myelin sheath composed of a fatty insulated lipid substance.

Myofibril: Rodlike strand contained within and running the length of the muscle fibers; contains the contractile elements of the muscle.

Myosin: A thick protein of the myofibril, noticeable by its dark banding. Along with actin, it is responsible for contraction and relaxation of the muscle.

Myotatic Reflex: Reflex initiated by stretching the muscle, which facilitates a contraction of the same muscle via muscle spindle stimulation; also called the stretch reflex.

Myotendinous Junction: The site where the muscle and tendon join, consisting of a layered interface as the myofibrils and the collagen fibers of the tendon meet.

Negative Work: Work that occurs when the force acting on an object opposes the motion.

Neuromuscular Junction: Region where the motoneuron comes into close contact with the skeletal muscle; also called the motor end plate.

Neuron: A conducting cell in the nervous system that specializes in generating and transmitting nerve impulses.

Neutral Equilibrium: Once an object is displaced the position remains constant.

Neutralizer: A muscle responsible for eliminating or canceling out an undesired movement.

Node of Ranvier: Gaps in the myelinated axon where the axon is enclosed only by processes of the Schwann cells.

Noncardial Planes: I think this refers to noncardinal planes. Cardinal plane is already defined.

Noncontact Forces: Contact forces are already defined Noncontact forces just means it is not a contact force.

Normal Stress: The amount of load per cross-sectional area applied perpendicular to the plane of a cross section of the loaded object.

Nuclear Bag Fiber: An intrafusal fiber within the muscle spindle that has a large clustering of nuclei in the center. The type Ia afferent neurons exit from the middle portion of this fiber.

Nuclear Chain Fiber: An intrafusal fiber within the muscle spindle with nuclei arranged in rows. Both type Ia and the type II sensory neurons exit from this fiber.

Nutation: See Sacral Flexion.

Olecranon Bursitis: Irritation of the olecranon bursae commonly caused by falling on the elbow.

Olecranon Fossa: A depression on the posterior distal humerus; creates a lodging space for the olecranon process of the ulna in forearm extension.

Olecranon Process: Projection on the proximal posterior ulna; fits into the olecranon fossa during forearm extension.

Open-Chain Exercise: Exercises in which the hand or foot is free to move.

Origin: The intersection of the axes of a reference system and the reference point from which measures are taken.

Origin: The more proximal attachment site of a muscle.

Osgood-Schlatter Disease: Irritation of the epiphysis at the tibial tuberosity caused by overuse of the quadriceps femoris muscle group.

Osseous: Having the nature or quality of bone.

Ossification: The formation of bone.

Osteoarthritis: Degenerative joint disease characterized by breakdown in the cartilage and underlying subchondral bone, narrowing of the joint space, and osteophyte formation.

Osteoblast: A type of bone cell responsible for bone deposition.

Osteochondral Fracture: Fracture at the bone and cartilage junction.

Osteochondritis Dissecans: Inflammation of the bone and cartilage resulting in splitting of pieces of cartilage into the joint (shoulder and hip).

Osteoclast: A type of bone cell responsible for bone resorption.

Osteocyte: A bone cell with processes that transport metabolites, communicate between cells, and help regulate mineral homeostasis.

Osteon: Long cylindrical structure in bone that serves as a weight-bearing pillar.

Overuse Injury: An injury caused by continual low-level stress on a body.

Pacinian Corpuscle: Sensory receptor in the skin that is stimulated by pressure.

Parallel: Two objects in which the distance between them is constant.

Parallel Axis Theorem: A method of calculating the moment of inertia about an axis of rotation of an object if you know the moment of inertia about a parallel axis.

Parallel Elastic Component: The passive component in a muscle model that behaviorally develops tension with elongation.

Passive Insufficiency: The inability of a two-joint muscle to be stretched sufficiently to allow a complete range of motion at all the joints it crosses because the antagonists cannot be further elongated.

Passive Peak: The impact peak of a vertical ground reaction force curve. Termed "passive" because it is not under muscular control.

Passive Range of Motion: The degree of motion that occurs between two adjacent segments through external manipulation, such as gravity and manual manipulation.

Patella Alta: Long patellar tendon.

Patella Baja: Short patellar tendon.

Patellar Tendon: The tendon connecting the patella to the tibial tuberocity.

Patellofemoral Joint: Articulation between the posterior surface of the patella and the patellar groove on the femur.

Patellofemoral Pain Syndrome: Pain around the patella.

Pelvifemoral Rhythm: The movement relationship between the pelvis and the femur during thigh movements at the hip.

Pennate: Muscle in which the fibers run diagonally to the tendon.

Pennation Angle: The angle made by the fascicles and the line of action (pull) of the muscle.

Perimysium: A dense connective tissue sheath covering the fascicles.

Periodization: An organized approach to training in which exercises are alternated for specific periods of time.

Periosteum: A white membrane of connective tissue that covers the outer surface of a bone except over articular cartilage.

Periostitis: Inflammation of the periosteum that is marked by tenderness and swelling on the bone.

Peripheral Nervous System: All nerve branches lying outside the brain and spinal cord.

Pes Anserinus: The combined insertion of the tendinous expansions from the sartorius, gracilis, and semitendinosus muscles.

Pes Cavus: High-arched foot.

Pes Planus: Flat foot.

Physiologic Cross Section: An area that is the sum total of all of the cross sections of fibers in the muscle, the area perpendicular to the direction of the fibers.

Pivot Joint: A type of diarthrodial joint that allows movement in one plane; pronation, supination, or rotation.

Plane: A two-dimensional space defined by three non-collinear points.

Plane Joint: A type of diarthrodial joint with flat surfaces that allows translation between the two bones; also called gliding joint.

Plantar Fascia: Fibrous band of fascia running along the plantar surface of the foot from the calcaneus to the metatarsophalangeal articulation.

Plantar Fasciitis: Inflammation of the plantar fascia.

Plantarflexed First Ray: Position of the first metatarsal below the plane of the adjacent metatarsal heads.

Plantarflexion: Movement of the foot downward in the sagittal plane; movement away from the leg.

Plastic Region: The region between the yield point and the failure point on a stress–strain curve; the region in which the material will not return to its initial length after it is deformed.

Plica: Ridge or fold in the synovial membrane.

Plyometric Training: Exercise that uses the stretch–contract sequence of muscle activity.

Plyometrics: A training technique that uses the stretch–shortening cycle to increase athletic power.

Point of Application: The point at which a force vector acts on an object.

Point of Separation: The point at which the boundary layer of fluid molecules separates from the object when the object is travelling through the fluid.

Polarity: The direction of rotation designated as positive or negative.

Positive Work: Work that occurs when the force acting on an object is in the same direction as the motion.

Porosity: The ratio of pore space to the total volume.

Posterior: A position behind a designated reference point.

Posterior Cruciate Ligament: Ligament inserting on the posterior spine of the tibia and the inner condyle of the femur; resists posterior movement of the tibia on the femur and restrains flexion and rotation of the knee.

Potential Energy: The capacity to do work that an object possesses because of its height.

Power: The product of force and velocity.

Power Grip: A powerful hand position produced by flexing the fingers maximally around the object at all three finger joints and the thumb adducted in the same plane as the hand.

Precision Grip: A fine-movement hand position produced by positioning the fingers in a minimal amount of flexion with the thumb perpendicular to the hand.

Pressure: The contact force per unit area of contact.

Progressive Overload: A gradual increase in the stress placed on the body during exercise by varying factors such as load, repetitions, speed, rest, and volume.

Pronation: Rotation in the forearm (radioulnar joints) or the foot (subtalar and midtarsal joints). Relative to anatomical position, radioulnar pronation causes the palm to face posteriorly. Subtalar pronation causes the plantar surface of the foot to face laterally.

Proprioceptive Neuromuscular Facilitation: Rehabilitation technique that enhances the response from a muscle through a series of contract–relax exercises.

Proprioceptor: A sensory receptor in the joint, muscle, or tendon that can detect stimuli.

Propriospinal Reflex: Reflex processed on both sides and at different levels of the spinal cord; an example is the crossed extensor reflex.

Propulsive Drag: The concept that drag forces can be used to propel the body forward such as in the freestyle stroke of swimming.

Protraction: The motion describing the separating action of the scapula.

Proximal: A position relatively closer to a designated reference point.

Pubic Ligament: Ligament inserting on the bodies of the right and left pubic bones; maintains the relationship between right and left pubic bones.

Pubic Symphysis: A cartilaginous joint connecting the pubic bones of the right and left coxal bones of the pelvis.

Pubis: The anterior inferior bone of the pelvic girdle.

Pubofemoral Ligament: Ligament inserting on the pubic part of the acetabulum, superior rami, and intertrochanteric line; restrains hip abduction and external rotation.

Q-Angle: The angle formed by the longitudinal axis of the femur and the line of pull of the patellar ligament.

Qualitative Analysis: A nonnumeric description or evaluation of movement that is based on direct observation.

Quantitative Analysis: A numeric description or evaluation of movement based on data collected during the performance of the movement.

Radial Acceleration: See Centripetal Acceleration.

Radial Flexion: The flexion movement of the hand toward the forearm on the thumb side of the hand.

Radian: The measure of an angle at the center of a circle described by an arc equal to the length of the radius of the circle (1 rad = 57.3°).

Radiate Ligament: Ligament inserting on the head of the ribs and body of the vertebrae; holds the ribs to the vertebrae.

Radiocarpal Joint: Articulation between the radius and the carpals (scaphoid and lunate).

Radiohumeral Joint: Articulation between the radius and the humerus.

Radioulnar Joint: Articulation between the radius and the ulna (superior and inferior).

Radius of Gyration: The distance that an object with a particular moment of inertia would have to be located from an axis of rotation if it were a point mass.

Radius of Rotation: The linear distance from the axis of rotation to a point on the rotating body.

Rate Coding: The frequency of the discharge of the action potentials. Also referred to as frequency coding.

Rear Foot: Region of foot that includes the talus and calcaneus; also called the hindfoot.

Rear Foot Varus: Inversion of the calcaneus with deviation of the tibia in the same direction.

Reciprocal Inhibition: Relaxation of the antagonistic muscle(s) while the agonist muscles produce a joint action.

Recruitment: A system of motor unit activation.

Rectification: The process whereby the negative portion of a raw electromyography signal is made positive so that the complete signal is positive.

Reference System: A system to locate a point in space.

Reflex: Involuntary response to stimuli.

Relative Angle (Joint Angle): The included angle between two adjacent segments.

Relative Reference Frame: A reference frame in which the origin is at the joint center and one of the axes is placed along one of the segments.

Remodeling: Sequential bone resorption and formation at the same site which does not change the size and shape of the bone.

Remote Angular Momentum: The momentum about a system axis of rotation. This is usually the momentum about the center of mass of the body.

Renshaw Cell: Interneuron that receives excitatory input from collateral branches of other neurons and then produces an inhibitory effect on other neurons.

Repolarization: A return to the resting potential of a membrane.

Residual Strain: The difference between the initial length of a material and the length when the material has gone beyond its yield point.

Resistance Arm: In a lever system the resistance arm is the perpendicular distance between the line of action of the resistance force vector and the axis of rotation.

Resistance Force: In a lever system this is the force that is resisting movement.

Resorption: A phase of bone remodeling in which bone is lost through osteoclastic activity.

Resting Potential: The voltage across the membrane at steady-state conditions.

Retraction: The motion describing the coming together action of the scapula.

Retrocalcaneal Bursitis: Inflammation of the bursa between the Achilles tendon and the calcaneus.

Retroversion: The degree to which an anatomical structure is rotated backward. See anteversion for opposite.

Revolution: A unit of measurement that describes one complete cycle of a rotating body.

Right-Hand Rule: The convention that designates the direction of an angular motion vector; the fingers of the right hand are curled in the direction of the rotation and the right thumb points in the direction of the vector.

Rotational Friction: The friction due to two objects in contact that are rotating relative to each other.

Rotational Kinetic Energy: The capacity to do work that an object possesses because of its angular velocity.

Rotation: A movement about an axis of rotation in which not every point of the segment or body covers the same distance in the same time.

Rotator Cuff: Four muscles surrounding the shoulder joint, the infraspinatus, supraspinatus, teres minor, and subscapularis.

Ruffini Ending: Sensory receptors in the joint capsule that respond to the change in joint position.

Rupture: An injury in which the tissue is torn or disrupted in a forcible manner.

Sacral Flexion: Anterior movement of the top of the sacrum.

Sacroiliac Joint: A strong synovial joint between the sacrum and the ilium.

Sacroiliitis: Inflammation at the sacroiliac joint.

Sacrum: A triangular bone below the lumbar vertebrae that consists of five fused vertebrae.

Saddle Joint: A type of diarthrodial joint that has two saddle-shaped surfaces, allowing 2 degrees of freedom.

Safety Factor: The ratio of the stress to reach the yield point to the stress of everyday activity.

Sagittal Plane: The plane that bisects the body into right and left sides.

Sarcolemma: A thin plasma membrane covering the muscle that branches into the muscle, carrying nerve impulses.

Sarcomere: One contractile unit of banding on the myofibril, running Z-band to Z-band.

Sarcopenia: Loss of muscle mass and decline in muscle quality with increased aging.

Sarcoplasm: The fluid enclosed within a muscle fiber by the sarcolemma.

Sarcoplasmic Reticulum: A membranous system within a muscle fiber that forms lateral sacs near the t-tubules.

Scapulohumeral Rhythm: The movement relationship between the humerus and the scapula during arm raising movements; the humerus moves 2 degrees for every 1 degree of scapular movement through 180 degrees of arm flexion or abduction.

Scapulothoracic Joint: A physiologic joint between the scapula and the thorax.

Schwann Cells: Cells that cover the axon and produce myelination, which are numerous concentric layers of the Schwann cell plasma membrane.

Screw-Home Mechanism: The locking action at the end of knee extension; external rotation of the tibia on the femur caused by incongruent joint surfaces.

Second-Class Lever: A lever system in which the resistance force is in between the effort force and the fulcrum.

Segment Angle: The angle of the segment with respect to the right horizontal that is absolute and that changes according to orientation of the body.

Segmental Method: The process of calculating the center of mass of a system of masses by taking a weighted average of the individual components.

Sensory Neuron: Neuron that carries impulses from the receptors in the body into the central nervous system.

Separated Flow: As the relative speed of an object moving through a fluid increases, the air molecules behind the object tend to separate leaving an air pocket or region of low pressure.

Series Elastic Component: The passive component in a muscle model that behaviorally develops tension in contraction and during elongation.

Shear Force: A force applied parallel to the surface, creating deformation internally in an angular direction.

Shear Stress: The amount of load per cross-sectional area applied parallel to the plane of a cross section of the loaded object.

Size Principle: The principle that describes the order of motor unit recruitment as a function of size.

Sliding Filament Theory: A theory describing muscle contraction whereby tension is developed in the myofibrils as the head of the myosin filament attaches to a site on the actin filament.

Slipped Capital Femoral Epiphysitis: Displacement of the capital femoral epiphysis of the femur caused by external forces that drive the femoral head back and medial to tilt the growth plate.

Slow-Twitch Fiber: Small skeletal muscle fiber innervated by the alpha-2 motor neuron, having a slow contraction time. This fiber is highly oxidative and poorly glycolytic.

Snapping Hip Syndrome: A clicking sound that accompanies thigh movements; caused by the hip capsule or iliopsoas tendon moving on a bony surface.

Soma: The portion of the nerve cell that contains the nucleus and well-marked nucleolus. The soma receives information from the dendrites and sends information through the axon; also called the cell body.

Spinal Nerves: The 31 pairs of nerves that arise from the various levels of the spinal cord.

Sprain: An injury to a ligament surrounding a joint; rupture of fibers of a ligament.

Stability: Refers to 1) a state of balance or 2) the ability of a joint to resist dislocation

Stabilizer: A muscle responsible for stabilizing an adjacent segment.

Stable Equilibrium: Exists when, after a force or torque is applied, an object returns to its original position.

Static Analysis: Analysis of the forces and torques acting on an object when they sum to zero. This occurs whenever the object is not moving or moving with constant velocity.

Static Stretching: Moving a limb to the terminal range of motion slowly and then holding the final position.

Statics: A branch of mechanics in which the system being studied undergoes no acceleration.

Sternoclavicular Joint: Articulation between the sternum and the clavicle.

Strain: Injury to the muscle, tendon, or muscle–tendon junction caused by overstretching or excessive tension applied to the muscle; tearing and rupture of the muscle or tendon fibers.

Strain Energy: The capacity to do work that an object possesses because of deformation of elastic materials.

Strap Muscle: A muscle shape lacking a central belly.

Strength: The maximum amount of force produced by a muscle or muscle group at a site of attachment on the skeleton; one maximal effort.

Stress: Force per unit area.

Stress Fracture: Microfracture of the bones developed through repetitive force application exceeding the structural strength of the bone or the rate of remodeling in the body tissue.

Stress–Strain Curve: A plot of the stress placed on a material against the strain imposed by the stress.

Stretch-Contract: See stretch-shortening cycle.

Stretch Reflex: Reflex initiated by stretching the muscle, which facilitates a contraction of the same muscle via muscle spindle stimulation; also called the myotatic reflex.

Stretch–Shortening Cycle: A common sequence of joint actions in which an eccentric muscle action, or prestretch, precedes a concentric muscle action.

Subacromial Bursae: The bursae between the acromion process and the insertion of the supraspinatus muscle.

Subacromial Bursitis: Inflammation of the subacromial bursae that is common to impingement syndrome.

Subluxation: An incomplete or partial dislocation between two joint surfaces.

Subtalar Joint: The articulation of the talus with the calcaneus; also called the talocalcaneal joint.

Superior: A position above a designated reference point.

Supination: Rotation in the forearm (radioulnar joints) or the foot (subtalar and midtarsal joints). Relative to anatomical position radioulnar supination causes the palm to face anteriorly. Subtalar supination causes the plantar surface of the foot to face medially.

Supraspinal Reflex: Reflex brought into the spinal cord but processed in the brain; an example is the labyrinthine righting reflex.

Surface Drag: A type of fluid resistance resulting from friction between the surface of the object and the fluid.

Synapse: The junction or point of close contact between two neurons or between a neuron and a target cell.

Synarthrodial Joint: A type of joint whose bones are connected by fibrous material; little or no movement is allowed at these joints; also called fibrous joint.

Synchronous: Describes events occurring at the same time. In muscular contraction, the concurrent activation of motor units.

Synergist: A muscle that performs the same motion as the agonist.

Synovial Fluid: Liquid secreted by the synovial membrane that reduces friction in the joint; the fluid changes viscosity in response to the speed of joint movement.

Synovial Joint: Freely movable joint; also called diarthrodial joint.

Synovial Membrane: Loose vascularized connective tissue that lines the joint capsule.

System: In a free-body diagram, the object or objects under consideration. A system could consist of a body segment, a body, or multiply bodies.

Talocrural Joint: The articulation of the tibia and fibula with the talus; the ankle joint.

Talofibular Ligament: Ligament inserting on the lateral malleolus and the posterior talus; limits plantarflexion and inversion; supports the lateral ankle.

Talonavicular Joint: Articulation between the talus and the navicular bones; part of the midtarsal joint.

Tangent: The ratio of the side opposite an angle to the adjacent angle in a right triangle.

Tangent: A line that touches a curve at only one place.

Tangential Acceleration: The change in linear velocity per unit time of a body moving along a curved path.

Tangential Velocity: The change in linear position per unit time of a body moving along a curved path.

Tarsometatarsal Joint: Articulation between the tarsals and metatarsals.

Tendinitis: Inflammation of a tendon.

Tendon: A fibrous cord, consisting primarily of collagen, by which muscles attach to the bone.

Tenosynovitis: Inflammation of the sheath surrounding a tendon.

Tetanus: The force response of muscle to a series of excitatory inputs, resulting in a summation of twitch responses.

Thenar Eminence: Ridge or mound on the radial side of the palm formed by the intrinsic muscles acting on the thumb.

Third-Class Lever: A lever system in which the effort force is in between the fulcrum and the resistance force.

Tibial Plateau: A level area on the proximal end of the tibia.

Tibiofemoral Joint: Articulation between the tibia and the femur; the knee joint.

Tibiofibular Joint (Inferior): Articulation between the distal end of the fibula and the distal end of the tibia.

Tibiofibular Joint (Superior): Articulation between the head of the fibula and the posterolateral inferior aspect of the tibial condyle.

Tibiotalar Joint: Articulation between the tibia and the talus.

Time Domain: A parameter that is presented as a function of time.

Tonic Neck Reflex: Reflex stimulated by head movements, which stimulates flexion and extension of the limbs. The arms flex with head flexion and extend with neck extension.

Torque: The product of the magnitude of a force and the perpendicular distance from the line of action of the force to the axis of rotation.

Torque Arm: See moment arm.

Trabeculae: Strands within cancellous bone that adapt to the direction of stress on the bone.

Traction Apophysitis: Inflammation of the apophysis (process, tuberosity) created by a pulling force of tendons.

Transverse Arch: An arch formed by the tarsals and metatarsals; runs across the foot, contributing to shock absorption in weight bearing.

Transverse Ligament: Ligament inserting on the medial and lateral meniscus; connects the menisci to each other.

Transverse Plane: A body plane that is perpendicular to the long axis of a segment or the body.

Transverse Tubules: An invagination of the sarcolemma that transmits the muscle action potential deep inside the muscle.

Traumatic Fracture: A break in a bone as a result of a single high-magnitude force application.

Trigger Finger: Snapping during flexion and extension of the fingers created by nodules on the tendons.

Trochlea: Medial portion of the distal end of the humerus; articulates with the trochlear notch of the ulna.

Trochlear Notch: A deep groove in the proximal end of the ulna; articulates with the trochlea of the humerus.

T-Tubule (Transverse Tubule): Structure in the sarcolemma that facilitates rapid communication between action potentials and myofilaments in the interior of the muscle.

Turbulent Flow: The chaotic flow of a fluid that occurs behind an object moving through a fluid. Turbulent flow occurs at faster relative speeds and when the object is not streamlined.

Twitch: The force response of a muscle to a single stimulation.

Type Ia Primary Afferent: A sensory fiber that measures velocity of stretch within the muscle spindle.

Type II Secondary Afferent: A sensory fiber that measures fiber length within the muscle spindle.

Ulnar Flexion: The flexion movement of the hand toward the forearm on the little finger side of the hand.

Ulnohumeral Joint: Articulation between the ulna and the humerus; commonly called the elbow.

Unipennate: A feather-shaped fiber arrangement in which the muscle fibers run diagonally off one side of the tendon.

Unstable Equilibrium: Exists when, after a force or torque is applied, an object continues to displace from the original position.

Upward Rotation: The action whereby the scapula swings out from the midline of the body.

Valgus: Segment angle bowed medially; medial force.

Varus: Segment angle bowed laterally; lateral force.

Vector Composition: The summation of two or more vectors to obtain a resultant vector.

Ventral: See Anterior.

Vertex: The intersection of two lines that form an angle.

Viscoelastic Material: A material that exhibits nonlinear properties on a stress–strain curve.

Viscous Drag: See surface drag.

Weight: The force acting on an object that results from the gravity between the object and the earth.

Windlass Effect: Tightening of the plantar fascia during hyperextension of the metatarsal–phalangeal joints resulting in a more rigid foot.

Work: The product of a force and the distance that the point of application travels in the direction of the force vector.

Work-Energy Theorem: The concept that work done on an object will change the energy of that object.

Yield Point: The point on a stress–strain curve at which the material reaches the plastic region.

Index

Note: Page numbers in *italics* denote figures; those followed by 't' denote tables

A

Abdominals, 256
Abduction, 13–14, *13*, 137
  horizontal, 15, *15*
Abscissa, 285
Absolute angle, *320*, 321–323, *322*, 323t
Absolute reference frame, 17, *17*
Acceleration
  angular, 330–331
  law of, 399–402, *400–402*
  of center of mass, 354
  centripetal (radial), 333, *334*
  constant, equations of, 312–313, *313*
  direction of motion and, *298*, 298–299, *299*
  first central distance method of calculating, 299, 300t, *300*
  graphical example, 300–301, *300*
  gravity-induced, 352
  instantaneous, 296
  law of, 350–351
  linear, 297–301, *298–300*, 300t
  numerical example, 300t, *300*
  positive versus negative, 299
  relationship between angular and linear, 331–334, *333, 334*
  tangential, 333, *333, 334*
  units, 298
Acceleration-time curve, *301*, 302
Accelerometer, 284
Acetabular labrum, 178
Acetabulum, 178, *179*
Achilles tendon, injuries, 224
Acromioclavicular joint
  anatomy and functional characteristics, *132*, 133–134
  injuries, 142
Acromioclavicular ligaments, 133
Actin, 66, *66*
Action potential, 67, 103, *103*, 105
Action-reaction, law of, 351, *351*, 402
Active insufficiency, 79
Active peak, 378
Active range of motion, 116
Adduction, 13, *13*, 137
  horizontal, 14–15, *15*
Afferent neuron, 110–112
  type Ia primary, 110, *111*, 111–112, *112*
  type II secondary, 110, *111*, 112
Agonist muscles, 74, *74*
Air resistance, 357

Airfoil, lift on, 360
All-or-none principle, 104
Alpha motoneuron, 104
Amphiarthrodial joint, 51, *53*
Amputated foot, stress-strain analysis, *27*
Anatomical cross-section area, 64
Anatomical movement descriptors, 10–16
  anatomical terms, *10*, 10–12, *11*
  basic movements, 12–14, *12–14*
  segment names, 10, *10*
  specialized movements, 14–16, *15*
  summary, 16t
Anatomical position or direction, *11*, 11–12
Anatomical starting position, 10, *10*
Anatomy, 5
Angle(s)
  absolute, *321*, 321–323, *322*, 323t
  of application, 348
  components, 319, *319*
  lower extremity joint, 324–327, *325–327*
  measurement, *319*, 319–321, *320*
  projection, *309*, 309–310, 310t
  relative, 10, *10*, *323*, 323–324, *324*
  segment, 321
  trigonometric functions, 445–448, *446–447*, 446t–447t
Angle-angle diagrams, 334–335, *335, 336*
Angular acceleration, 330–331
  law of, 399–402, *400–402*
Angular impulse, 399, 422–423
Angular kinematics, 318–342
  angle-angle diagrams, 334–335, *335, 336*
  angular motion, 319, *319*
  displacement, 328, *328*
  distance, 328, *328*
  equation review, 342
  of golf swing, 7, *8*, 338–340, *339, 340*
  lower extremity joint angles, 324–327, *325–327*
  measurement of angles, *319*, 319–321, *320*
  position, 328
  relationship to linear kinematics, 331–334, *331–334*
  review questions, 342–345
  speed, 328–330
  types of angles, 321–324
  units of measurement, *320*, 320–321
  vectors, 327, *327*
  velocity, 328–330
  of walking and running, 336, *337*
  of wheelchair propulsion, 340–341, *340*

Angular kinetics, 391–434
  center of mass, 402–404
  dynamic analysis, 420–422, *421, 422*
  equation review, 434
  of golf swing, 429–431, *430*
  Newton's laws of motion in, 395–410
  review questions, 434–437
  rotation and leverage, 410–413
  special torque applications, 424–427
  static analysis, 415–420, *415–421*
  torque. *See* Torque
  of walking and running, 428–429, 427–429
  of wheelchair propulsion, 431–432, *431–432*
Angular momentum, 399
  conservation of, 399–400
  local, 401
  remote, 401
Angular motion, *319*
Angular (rotational) motion, 6–7, *7*, 319, *319*
Angular power, 424–426, *425, 425*
Angular work, 423–424, *424*, 427
Ankle
  anatomy and function, 209–210, *210*
  axis of rotation, 211, *212*
  conditioning exercises, 222, *223*
  injury potential, 224–225
  inversion sprain, 42
  joint forces, *231*, 231
  knee and, combined movements, 217
  ligaments, 209–210, *211*
  movement characteristics, 217
  muscular actions, 218–221, *219–221*
  osteoarthritis, 225
  range of motion, 183–184, 217
  strength, 221–222
Ankle angle, 325–326
Annulus fibrosus, 243
Antagonistic muscles, 75, *75*
Anterior aspect, 11, *11*
Anterior compartment syndrome, 62, 225
Anterior cruciate ligament, 194, *197, 198*
  injuries, 207
  repair, rehabilitation after, 205
Anterior longitudinal ligament, 244
Anterior motion segment, 243–244, *243–244*
Anteroposterior axis, 18, *18*
Anteversion, femoral, 181–182, *181, 182*

Aponeurosis, muscle attachment via, 67, *69*
Apophyseal joints, 245
    loads on, 271
    osteoarthritis, 266
Apophysis, 41, *42*
Apophysitis
    calcaneal, 225
    iliac, 192
Appendicular skeleton, 10, *10*, 32
Application, point of, 348
Arches, foot, 214–217, *216*
Area under the curve concept, in linear
    kinematics, *301*, 302
Arm. *See also* Forearm; Shoulder girdle
    conditioning exercises, 152–154, *153*
    movements
        descriptors, 14, *15*, 16t
        relationship to shoulder girdle
            movements, 137, *138, 140,*
            140–140, *141*
    as segment, 9, 10, *10*
Arthritis, degenerative. *See* Osteoarthritis
Articular cartilage, 47–48
Articular end plate, 49
Atlantoaxial joint, 248
Atlantooccipital joint, *247–249*
Atlas (C1), *247*, 248
Attachment sites, bone, 29
Autogenic facilitation, 112
Avulsion fracture, 42, *42*
Axial skeleton, 10, *10*, 32
Axis (axes)
    reference frame, 17
    of reference systems, 17–21, *18–21*
    of rotation, 6, *7*, 18, *18*, 319, *319*
Axis (C2), *247, 248*
Axon, *102,* 102–103

B

Back lift, 263
Back pain
    etiology, 264, *266, 268,* 264–268
    prevention, 264, *265*
Balance, 418
Ball, exercise, 264
Ball-and-socket joint, 51, *52*
Ballistic stretching, 116
Barefoot locomotion, benefits, 222
Baseball. *See* Projectile motion; Throwing
Belly, muscle, 65, *66*
Bending forces, *40*, 40t, 43–44, *44*
Bending fracture, 43–44
Bennett's fracture, 160
Bernoulli's principle, 360
Biceps, parallel and rotatory components,
    73, *73*
Bicipital tendinitis, 146
Bicycle wheel, as example of rotational
    motion, *319, 319*
Bicycling, lower extremity in, 229–230, *229*
Bilateral deficit, training-induced, 115
Biomechanics
    definition, 4
    versus kinesiology, 4–5
    terminology, 3–22
Bipennate muscle,*63,* 64
Body segment names, 10, *10*
Body tissues, structural analysis, 26–29, *26–29*

Bone, 29–47. *See also* Bone tissue
    anatomical classification, 32–33, *33, 36*
    ectopic, forearm, 154
    failure, 37, *38*
    formation, 33–36
    physical activity and, 34–35
    injury threshold, 46, *46*
    loads on, 39–45
        bending forces, *40*, 40t, 43–44, *44*
        combined, 44–45, *45, 46*
        compression forces, 40–41, *41,* 40t
        shear forces, *40*, 40t, 42–43, *43*
        tension forces, *40*, 40t, 41–42, *42*
        torsional forces, *40*, 40t, 44, *45*
    loss, 36
    macroscopic structure, 30–36
    major, *31*
    mechanical properties, 36–47
    modeling, 34
    ossification, 34
    remodeling, 34
    resorption, 34
    strain, 45–47, *46*, 47t
    stress-strain curve, 37, *38, 70*
    transmission of muscle force to, 68–70,
        69, 70
Bone mineral density, physical activity and,
    34–35
Bone-on-bone force, 354
Bone tissue
    anisotropic characteristics, *38*, 38
    biomechanical characteristics, 29–36
    changes across lifespan, 34
    composition, 30
    functions, 29–30
    stiffness, 37–38, *38*
    strength, 37, *37*
        development, 43, *43*
    viscoelastic characteristics, 38, *39*
Bony articulations, 49–53, 53t. *See also*
    Joint(s)
Boundary layer, 358
Boutonniere deformity, 160
Brachial nerve plexus, trauma, 145
Bracing systems, 420, *421*
Brain, 100, *100*
Braking phase, 379
Brittle material, stress-strain curve, 29, *29,*
    38, *38*
Bursa, 135
Bursitis
    hip region, 193
    retrocalcaneal, 224
    scapular, 145
    subacromial, 146

C

Calcaneal apophysitis, 225
Calcaneocuboid joint, 214
Calcaneofibular ligament, sprain, 224
Calcaneus, 209
    avulsion fracture, 42
Cancellous bone, *32*, 32
Capsular ligaments, 48
Capsule, joint, 48, *49*
Cardinal plane, *18*, 18
Carpal tunnel syndrome, 160, 162, *162*
Carpometacarpal joints, *155,* 156–157

Carrying angle, elbow, 147, *149*
Cartesian coordinate system, 285–286
Cartilage, 47–49
    articular, 47–48
    fibro-, 48
Cell body, 102, *102*
Center of gravity, 352, 402
Center of mass, 352, 402–410, *403*
    acceleration of, 354
    definition, 402
    segmental calculation, *405*, 404–410,
        404t–406t, *406*
    total body calculation, *407*, 407–410,
        408, 409t
Center of pressure, 376, *376*
Center of rotation, 320, *320*
Central nervous system, 100, *100*
Centripetal acceleration, 333, *334*
Centripetal force, 375, *375*
Cervical lordosis, 260
Cervical spine
    anatomy and function, 245–251, *247,*
        249, 250
    compression fracture, 40
    injuries, 267
    loads on, 273
    movements, 248
    range of motion, *252*
Chair, rising from, 162
Chair design, seated work
    environment, 259
Children, hip conditions, 192
Chondrocytes, 47
Chondromalacia patellae, 209
Circular muscle, 63, *63*
Circular technique, wheelchair propulsion,
    308, *308*
Circumduction, *15*, 16
Clavicle
    anatomy and function, 132, *132*
    injuries, 142, 145
    movements, 132–133
Closed kinetic chain exercise, 90, *91*
Coefficient of friction, 355–356, *356*
Collagen, 30, 48
Collateral ligaments, 147, 149, *149,*
    196, 207
Colles fracture, 160
Collinear force, 349
Compartment syndrome, anterior, 62, 225
Compartments, muscle, 62, *63*
Complex training, 120
Compliant material, stress-strain curve,
    29, *29*
Compression forces, 40–41, *40–41,* 40t
Compression fracture, 40–41, *41*
Concentric muscle actions, 75, *75*
    comparison to eccentric and
        isometric actions, *76*, 76–77, *77*
    force-velocity relationship, 79, *80*
    prestretch for, *83*, 83, *84*
Concurrent force, 348, 366
Conditioning exercises
    for elbow and forearm, *153*
    for fingers and wrist, 159–160
    for hip joint muscles, *190–191*
    for knee joint, *196*
    for shoulder muscles, *143–144*

for trunk, *261–263*
  core training, *265*
  extensors, *261–262*
  flexors, *261, 263*
  lateral flexors, *262*
  rotators, *262*
Conduction velocity, 67
Condylar joint, 51, *52*
Condyles, lateral and medial, 196
Connective tissue
  conditioning exercises, 93
  as source of muscle soreness after
    exercise, 92
Contact force, 352–363
Contractile component, Hill muscle model,
  70, *71*
Contractility, muscle, 61
Contraction, muscle, 67, *68, 69*
Contralateral aspect, *11*, 12
Convergent muscle, 63, *63*
Coordinate system, Cartesian, 285–286
Coplanar force, 348
Coracoacromial ligament, 133
Coracoclavicular ligament, 133
Coracohumeral ligament, 135
Coracoid process, injuries, 145
Coronal plane, 18, *18*
Coronoid fossa, 147
Coronoid process, 147
Cortical bone, 30–32, *32*
Counternutation, 176, *176*
Coxa plana, 192
Coxa valga, 180, *181*
Coxa vara, 180, *181*
Crossed extensor reflex, 109, *110*
Cruciate ligaments, 196, *197, 198,* 207
Curvilinear motion, 284, *284*
Cycling, lower extremity in, 229–230, *229*

D

Deceleration, 299
Degenerative joint disease, *54*
Degree, in angular kinematics, 320, *320*
Degree of freedom, 16t, 20
Deltoid ligament, sprain, 224
Dendrites, 102, *102*
Dens, 248
Density, fluid, 357
Depolarization, 67, 105
Depression, 14, *15*, 132
Diaphysis, 32, *36*
Diarthrodial joint, 49–51
  characteristics, 49–50, *49*
  close-packed versus loose-packed
    positions, 50–51, *51*
  stability, 50–51
  types, 51, *52*
Differentiation, linear, 301
Direction, anatomical, *11*, 11–12
Disc. *See* Intervertebral disc
Displacement
  angular, 328, *328*
  linear, 293–295, *294, 295*
  relationship between angular and linear,
    331, 331–334
Distal aspect, 11, *11*

Distal femoral epiphysis, fracture, 43, *43*
Distal femoral epiphysitis, 208
Distal radioulnar joint, *155,* 155
Distance
  angular, 328, *328*
  linear, 293, *294*
Diving, angular momentum requirements,
  401, *402*
Dorsal aspect, *11*, 12
Dorsiflexion, *15*, 16, 210, *212,* 218, 221
Downhill walking, 338
Downward rotation, 14, *15*
Drag force, 358–359, *358, 359*
Ductile material, stress-strain curve, 37, *38*
Dynamics
  angular motion, 420–422, *421, 422*
  applications, 422, *422*
  inverse, 367, 421
  linear motion, 366–368, *367*
  versus statics, 9

E

Eccentric exercise, injury risk, 91, *92*
Eccentric force, 394
Eccentric muscle actions, 75–76, *75*
  comparison to concentric and isometric
    actions, *76,* 76–77, *77*
  force-velocity relationship, 81, *80*
Ectopic bone, forearm, 154
Effort arm, 410
Effort force, 410
Elastic force, 362–363
Elastic material, stress-strain curve, 27, *28*
Elastic modulus, 27, *27*
Elastic region, stress-strain curve, 27, *27*
Elasticity, muscle, 62
Elbow, 146–154
  carrying angle, 147, *149*
  conditioning exercises, 152–154, *153*
  injury potential, 154
  joint forces and moments, 165
  joints
    anatomy and functional characteristics,
      147–149, *148, 149*
    stability, 147, 149, *149*
  ligaments, 147, 149, *149*
  movement characteristics, 149
  muscular actions, 149–152, *150, 151*
  relative angle, 10, *11*
Elderly persons
  disc degeneration, *266,* 266
  hip conditions, 192
  resistance training, 88
  sarcopenia, 83–84
  spinal changes, 269
Electrodes, electromyographic, *120,* 120–121
Electromechanical delay, 115, *115,*
  122, *122*
Electromyogram, 120, *120*
Electromyography, 120–124
  applications, 122–123, *123*
  electrodes, *120,* 120–121
  factors affecting, 121, *121*
  limitations, 123
  records, 120, *120*
  signal amplification, 121
  signal analysis, 121–122, *122*
  training-induced changes, 115, *115*

Elevation, 14, *15*, 132
Ellipsoid joint, 51, *52*
EMG. *See* Electromyography
Endomysium, 65, *66*
Energy, 372–375
  conservation of, law of, 373
  potential, 272–373
  projectile motion and, 373, *373*
  relationship to work, 373–375, *374,* 427
  rotational, 426–427, *426*
  strain, 372
Epicondyles, 147, 194
Epicondylitis, 154, 160
Epimysium, 65, *66*
Epiphyseal fracture, 43, *43*
Epiphyseal plates, 34
Epiphysis, 32, *36*
Equations of constant acceleration, 312–313,
  313
Equilibrium, 415
  states of, 418–419, *419*
Equinus, 218
Erector spinae, 253, 256, 268
Ergonomics, electromyography applied to,
  123
Estrogen, osteoporosis and, 36
Eversion, foot, *15*, 16, 212–213, 217, 221
Excitation-contraction coupling, 67, *68*
Exercise
  conditioning
    for ankle/foot, 222–, *223*
    for connective tissue, 93
    for elbow and forearm, 152–154, *153*
    for fingers and wrist, 159–160, *161*
    for hip joint muscles, 189, *190–191*
    for knee joint, 196, 205–207
    for shoulder muscles, 142, *143–144*
    for trunk, 260–264, *261–263*
      core training, 264, *265*
      extensors, 260–263, *261–262*
      flexibility and, 264
      flexors, 260, *261, 263*
      lateral flexors, *262,* 263–264
      rotators, *262,* 263–264
    intensity, motor unit recruitment and,
      107, *107*
    plyometric, 83, *83,* 118–120, *119*
    strengthening. *See* Strength
      training; Weight lifting
    stretching. *See* Stretching exercises
Exercise ball, 264
Extensibility, muscle, 61–62
Extension, 12, *12*
  horizontal, 14, *15*
External oblique muscles, 256
External rotation, 13, *14, 15*
External work, 375
Extracapsular ligaments, 48
Extrafusal fibers, 113

F

Failure, bone, 37, *37*
Failure point, 29, *29*
Fascia, plantar, 215–216, *216*
Fascicle, 65, *66*
Fasciitis, plantar, 224
Fatigue, neural effects, 115
Fatigue fracture, 45, *46,* 47t

Female athletes, osteoporosis in, 36
Femoral anteversion, 181, *181, 182*
Femoral epiphysis
    distal, fracture, 43, *43*
    slipped capital, 192
Femoral epiphysitis, distal, 208
Femoral head, 178, *179*
Femoral neck, 180–181
    compression fracture, 40, *41*
    inclination angle, 180, *180, 181*
Femoral retroversion, *181,* 182
Femur, 42, 178, *179,* 193, *195*
Fibrocartilage, 48
Fibula, 209, *210*
Fingers
    anatomy, 155–157, *155–156*
    conditioning exercises, 159–160, *161*
    injury potential, 160, 162
    ligaments, *155*
    movement descriptors, 16t
    muscular actions, *155–156,* 157–159, *158*
    strength, 159, *159*
First central distance method
    of calculating acceleration, 299, 300t, *300*
    of calculating velocity, *294,* 294–295
First ray
    hypermobility, 214
    plantarflexed, 214, 218
Flat bone, 33, *36*
Flat muscle, 63, *63*
Flexibility, and trunk muscles, 264
Flexibility exercises. *See* Stretching exercises
Flexion, 12, *12*
    horizontal, 14–15, *15*
    lateral, 14, *15*
    plantar, 15, *15*
    radial, 15
    ulnar, 15
Flexor reflex, 109, *110*
Fluid resistance, 357–361, *357–361*
Foot. *See also* Ankle; Forefoot *entries;*
    Rearfoot *entries*
    amputated, stress-strain analysis, 26, *27*
    arches, 214–217, *216*
    conditioning exercises, 222, *223*
    eversion/inversion, *15,* 16, 212–213,
        217, 221
    function, alignment and, 217–218
    injury potential, 224–225
    intrinsic muscles, 221, *222*
    joint forces, *231,* 231
    joints, anatomy and functional
        characteristics, 209–214
    knee and, combined movements, 217
    ligaments, 209–210, *211*
    midfoot articulations, *210,* 214
    midtarsal joint, *210,* 214, *213*
    movement characteristics, 217
    movement descriptors, *15,* 15, 16t
    muscular actions, 218–221, *219–221*
    plantar fascia, 215–216, *216*
    pronation, 212, *212, 213,* 218, 221, *221*
    range of motion, 183–184
    as segment, 10, *10*
    strength, 221–222
    subtalar joint, *210,* 212–213, *212, 213*
    supination, 212–213, *213*
    talocrural joint, 209–212, *210*

Force(s)
    acting on body, 347t
    acting on system, representation, *363,*
        363–364, *364*
    bending, *40,* 40t, 43–44, *44*
    bone-on-bone, 354
    centripetal, *375, 375*
    characteristics, 347–348, *348*
    collinear, 348
    composition, 348, *349*
    compression, 40–41, *40–42,* 40t
    concurrent, 348, 366
    contact, 352–363
    coplanar, 348
    definition, 7, 347
    distribution, 376–377, *377*
    drag, 358–359, *358, 359*
    eccentric, 394
    effects
        at instant in time, 364–368
        over a distance, 371–375
        over a time period, 368–370
    effort, 410
    elastic, 362–363
    frictional, *354–357,* 355–356
    ground reaction
        description, 352–354, *353, 354*
        for golf swing, 381–382, *381–382*
        for walking and running, 377–380,
            377–381
        for wheelchair propulsion, *308,* 308
    inertial, 362
    joint. *See* Joint forces
    joint reaction, 354–355, *354*
    lift, 360–362, *360–362*
    muscle. *See* Muscle force
    noncontact, 351–352
    resistance, 410
    resolution, 348–350, *349*
    shear, *40,* 40t, 42–43, *43,* 375
    tension, *40,* 40t, 41–42, *42*
    torsional, *40,* 40t, 44, *45*
    types, 351–363
Force couple, 395
Force platform, 353, *353*
Force-time characteristics, tendon, 70
Force-velocity relationships
    factors influencing, 80–84, *81–84*
    muscle actions and, *79,* 79
Forearm
    anatomy, 146, *148,* 146
    conditioning exercises, 152–154, *153*
    injury potential, 154
    movement descriptors, 14–16, *15,* 16t
    muscle strength, 152
    muscular actions, 149–152, *150, 151*
    as segment, 10, *10*
Forefoot, 214
Forefoot valgus, 214, *215,* 218
Forefoot varus, 214, *214,* 214
Form drag, 358
Four-point bending load, *44,* 44
Fracture
    avulsion, 42, *42*
    bending, 43–44
    Bennett's, 160
    compression, 40–41, *41*
    epiphyseal, 43, *43*

    hurdler's, 192
    osteochondral, 225
    osteoporosis and, 35
    shear, 42–43
    ski boot, 43, *44*
    spiral, 44, *45*
    stress (fatigue), 45–47, *46,* 47t
    tensile, 41–42
    torsional, 44, *45*
    traumatic, 45
Free body diagram, *363,* 363–364, *364*
Frequency coding, motor units, 107–108,
    108
Frequency domain, electromyography, 121,
    122
Friction, *355–357,* 355–356
    coefficient of, 355–356, *357*
    kinetic (translational), 355
    rotational, 355
Frontal plane, 18, *19*
Frontal plane movements, 19, *20*
Fulcrum, 30, 410
Functional anatomy, 5
Functional resistance training, 90–91
Fundamental starting position, 10, *10*
Fusiform muscle, 63, *63*
Fynes, Savatheda, 305, *305*

G

Gait analysis. *See also* Running; Walking
    angular kinematics, 336, *337*
    angular kinetics, 427–429, *428–429*
    electromyography, 123, *123*
    linear kinematics, 302–306,
        302–307, 303t
    linear kinetics, *377–380,* 377–381
Gait cycle, *302,* 302–303, 303t, *304*
    phases, 226, 304
Gamma loop, 113, *113*
Gamma motoneuron, 109–110, *110,*
    112, *112*
Gamma scan, 404
Ganglia, 102, *102*
Gastrocnemius, 78, 218
Gender differences, pelvic girdle, 173, *173*
General motion, 319, *319*
Genu valgum, *199,* 199
Genu varum, *199,* 199
Glass, stress-strain curve, 38, *38*
Glenohumeral joint
    anatomy and functional characteristics,
        135–137, *136*
    impingement area, 135, *137*
    injuries, 145
    movement characteristics, 137, *138*
    stability, 135
Glenohumeral ligament, 135
Glenoid fossa, 135
Glenoid labrum, 135
Gliding movements, 20
Golf clubs, angular positions, *339*
Golf swing
    angular kinematics, 338–340, *339, 340*
    angular kinetics, 429–431, *430*
    double pendulum characteristics, 338, *340*
    length comparison of clubs, 332, *332*
    linear kinematics, 307–308, *308*
    linear kinetics, 381–382, *381–382*

swing characteristics, 307–308, *308*
  upper extremity muscular contribution, *164,* 164–165
  velocity and acceleration of the club, *307,* 307
Golgi tendon organ, 113, *114*
Gravity
  center of, 352, 402
  law of, 352
Gravity force, on projectile motion, 309
Greater trochanteric bursitis, 193
Greene, Maurice, 306
Grip strength, 159, *159*
Ground reaction force (GRF)
  description, 352–354, *353, 354*
  for golf swing, 381–382, *381–382*
  for walking and running, *377–380,* 377–381
  for wheelchair propulsion, 307–308, *308*
Gyration, radius of, 398, 398t

H

Hamstring group, 201
  injuries, 207
  as two-joint muscle, 78, *79*
Hamstring-to-quadriceps ratio, 205
Hanavan model, 404, *405*
Hand
  conditioning exercises, 159–160, *161*
  injury potential, 160, 162, *162*
  joints
    anatomy and functional characteristics, 155–157, *155–156*
    combined movements, 157
  ligaments, *155*
  movement descriptors, 16t
  muscular actions, *155–156,* 157–159, *158*
  as segment, 10, *10*
  strength, 159, *159*
Haversian system, 31
Head
  movement descriptors, 16t
  as segment, 10, *10*
Height, projection, 311–312, *311*
Hill, A. V., 304–305
Hill muscle model, 70, *71*
Hinge joint, 51, *52*
Hip angle, 325
Hip joint. *See also* Thigh
  anatomy, 178–183, *179–182*
  childhood disorders, 192
  compression forces, 40, *41*
  conditioning exercises, 189, *190–191*
  congenital dislocation, 192
  femoral anteversion, 181–182, *181, 182*
  femoral neck inclination angle, 180, *180, 181*
  femoral retroversion, *181,* 182
  forces, 230–232
  injuries, 192–193
  and knee joint, combined movements, 204
  ligaments, *175,* 180
  muscular actions, 184–188, *185–186*
  osteoarthritis, 193
  range of motion, 182, *182,* 183–184
  strength, 188–189
  and thigh movements, *182,* 182
Horizontal abduction, 15, *15*
Horizontal adduction, 15, *15*

Horizontal coordinates, 285, 286
Horizontal extension, 15, *15*
Horizontal flexion, 15, *15*
Horizontal plane, 18, *18*
Humerus
  head, 135
  spiral fracture, 44, *45*
Hurdler's fracture, 192
Hyperabduction, 13, *13*
Hyperadduction, 13, *13*
Hyperextension, 12, *12*
Hyperflexion, 12, *12*
Hyperpolarization, 67, 105
Hypertrophy, muscle, during resistance training, 85, *85*
Hypothenar eminence, 157
Hysteresis, 29, *29*

I

Iliac apophysitis, 192
Iliofemoral ligament, 180
Iliopectineal bursitis, 193
Iliopsoas, 256
Iliotibial band syndrome, 193, 207
Ilium, 174
Immobilization
  effects on muscle, 93
  osteoarthritis from, 53
Impact peak, 354, 378
Impingement area, glenohumeral joint, 135, *137*
Impingement syndrome, subacromial, 146
Impulse
  angular, 399, 422–423
  linear, 368
Impulse-momentum relationship, *368–369,* 368–370, 423
Inclination angle, femoral neck, 180, *180, 181*
Inertia
  law of, 350, 395–399, *396–398*
  moment of, *397*
Inertial force, 362
Inferior aspect, 11, *11*
Infrapatellar bursa, 198
Injury(ies). *See also specific anatomic location*
  bone
    examples, 47, 47t
    threshold, 47, *46*
  muscle
    cause and site, 91–92, *92*
    prevention, 92–93
  overuse. *See* Overuse injuries
  risk, eccentric exercise and, 91–92, *92*
Innervation ratio, 103
Instantaneous angular acceleration, 330
Instantaneous angular velocity, 329
Instantaneous joint center, 320, *320*
Instantaneous velocity, 296, *296*
Integration, 301–302, *301*
Intercarpal joint, *155,* 155–156
Intercondylar eminence, 194
Intercondylar notch, 194
Internal oblique muscles, 256, 258
Internal rotation, 14, 15, *14, 15*
Internal work, 374
Interneurons, 108, *108*
Interosseous ligaments, 214
Interosseous membrane, 147

Interphalangeal joints, 157, *210,* 214
Intersegmental (relative) angle, 10, *10, 323,* 323–324, *324*
Interspinous ligament, 245
Intertransverse ligament, 245
Intervertebral disc
  anatomy and function, *243,* 243–244
  compressive loads, 244, 273
  degeneration, *266,* 266
  injury, 244, 264, *266,* 268
  movements and forces on, *244–245,* 244
  water content, 243
Intra-articular ligaments, 48
Intrafusal fibers, muscle spindle, 109
Intrinsic muscles of foot, 221, 222
Inverse dynamics, 367, 421
Inverse stretch reflex, 113
Inversion, foot, 15, *16,* 213, 217, 221
Inversion sprain, 42
Ipsilateral aspect, *11,* 12
Irregular bone, 32, *33*
Irritability, muscle, 61
Ischial bursitis, 193
Ischiofemoral ligament, 180
Ischium, 174
Isokinetic exercise, 90, *90*
Isometric exercise, 89
Isometric muscle actions, 75, *75*
  comparison to concentric and eccentric actions, *76,* 76–77, *77*
Isotonic exercise, *89,* 89–90

J

Jersey finger, 160
Jogging. *See* Running
Joint(s)
  degenerative disease, 52–53, *54*
  instantaneous center of rotation, 320, *320*
  major, 53t
  types, 49–51
Joint (relative) angle, 10, *10, 323,* 323–324, *324*
Joint forces
  lower extremity, 230–232
  trunk, 269–271, *273–274, 273*
  upper extremity, 165
Joint movement
  degrees of freedom, 16t, 20
  gliding, 20
Joint reaction force, 354–355, *354*
Joint sensory receptors, 114, *114*
Jones, Marion, 305, *305*
Jumping
  prestretch for, 83, *83*
  vertical, 369–370, *369*

K

Kinematic chain, 21
Kinematics
  angular, 318–342. *See also* Angular kinematics
  data collection, 284–290
    methods for capturing data, 284
    reference systems, 284–287, *285–287*
    temporal factors, 287
    units of measurement, 287, *320,* 320–321
    vectors and scalars, 287–293, *291, 292*

Kinematic chain (*continued*)
  description, 7, *8*
  versus kinetics, 7–9, *8*
  linear, 283–314. *See also* Linear kinematics
  qualitative versus quantitative, 284
  relationship between angular and linear,
    331–334, *331–334*
  using Newton's laws of motion, 364–375
Kinesiology
  biomechanics versus, 4–5
  definition, 4
Kinetic chain exercise, 90, *91*
Kinetic energy, 372–375. *See also* Energy
Kinetic friction, 355
Kinetics, 7–9, *8*
  angular, 391–434. *See also* Angular kinetics
  linear, 346–385. *See also* Linear kinetics
  using Newton's laws of motion, 364–375
Knee angle
  definition, 325
  rearfoot angle and, 334–335, *335, 336*
Knee joint, 193–209
  bony landmarks, *195*
  conditioning exercises, *196,* 205–206
  and foot, combined movements, 217
  forces, 230–231
  and hip, combined movements, 204
  injury potential, 207–209
  instantaneous center of rotation, 320, *320*
  ligaments, *196, 197, 198,* 207
  menisci, *196,* 194–195, 208
  movement characteristics, 200–201, *200*
  muscular actions, 201–204, *202–203*
    extension, 201
    flexion, 204
    rotation, 204
  patellofemoral joint, 198–199, *198*
  planes and axes, *18*
  Q-angle, 198–199, *199*
  range of motion, 183–184, 200
  relative angle, 10, *10*
  strength, 204–205
  tibiofemoral joint, 194, *195,* 194–196,
    199, *199*
  tibiofibular joint, 199–200, *199*
Knee pain, 208
Kyphosis, thoracic, 260

L

Labyrinthine righting reflex, 109, *110*
Lamellae, 30–31
Laminar flow, 358, *358*
Lateral aspect, 11, *11*
Lateral collateral ligament, 147, 149,
    149, 196
  injuries, 207
Lateral condyle, 194
Lateral epicondyle, 147
Lateral epicondylitis, 154, 160
Lateral flexion, 14, *15*
Lateral rotation, 13, *14*
Lateral tibial syndrome, 224
Law of acceleration, 350–351
Law of action-reaction, 351, *351,* 402
Law of angular acceleration, 399–402,
    400–401
Law of conservation of energy, 373
Law of cosines, 323

Law of gravity, 352
Law of inertia, 350, 395–399, *396–398*
Laws of motion
  angular analogs, 395–410
  angular analysis using, 413–424
  description, 350–351, *351*
  linear analysis using, 364–375
Left lateral flexion, 14, *15*
Left rotation, 14, *14*
Leg
  movement descriptors, 16t
  as segment, 10, *10*
Leg lift, 263
Legg-Calvé-Perthes disease, 192
Length-tension relationship, muscle fiber,
    81, *81*
Lesser trochanter fracture, 193
Lever(s)
  definition, 410
  first-class, 410–411, *411, 412, 415*
  illustrations, 410, *410, 411*
  mechanical advantage, 410–411
  second-class, 411, *412*
  third-class, 412–413, *412–413*
Lever system, skeletal, 30
Lift force, 360–361, *360–361*
Lifting. *See also* Weight lifting
  free body diagram, 413, *414,* 420, *420*
  proper technique, 259, 268, *268*
Ligaments, 48
  avulsion, 42, *42*
  flexibility and, 116
  stress-strain curve, 49, *49*
Ligamentum flavum, 245
Limit concept, instantaneous velocity and,
    296, *296*
Line of action, 72, *73,* 348
Linear envelope, 121, *122*
Linear kinematics, 283–314
  acceleration, 297–301, *298–300,* 300t
  differentiation, 301
  displacement, 293–295, *294, 295*
  distance, 293, *294*
  equation review, 313–314
  of golf swing, 7, *8,* 306–307, *306*
  integration, 301–302, *301*
  position, 293
  of projectile motion, 309–312, *309–311,*
    310t
  relationship to angular kinematics,
    331–334, *331–334*
  review questions, 315–317
  speed, 292
  velocity, 292–297, *293–294, 296–287,* 293t
  of walking and running, 302–306,
    302–307, 303t
  of wheelchair propulsion, 307–308, *308*
Linear kinetics, 346–385
  equation review, 384–385
  forces. *See* Force(s)
  of golf swing, 381–382, *381–382*
  Newton's laws of motion in, 350–351
  review questions, 385–387
  of walking and running, *377–380,*
    377–381
  of wheelchair propulsion, 382–383, *383*
Linear (translatory) motion, 6–7, *6, 7,*
    284, *284*

with rotational motion, *319*
Local angular momentum, 401
Local extremum, on position-time curve,
    296, *296*
Local graded potential, 108, *108*
Locomotion
  angular kinematics, 336, *337*
  angular kinetics, 427–429, *428–429*
  barefoot, benefits, 222
  clinical angular changes, 338
  joint forces, *231,* 231
  linear kinematics, 302–306, *302–307,* 303t
  linear kinetics, *377–380,* 377–381
  lower extremity in, *226–227,* 226–229
  rearfoot motion, 338
  stride parameters, *302,* 302–303, 303t
  trunk muscles in, 269
  velocity curve, 304–306, *304, 305*
  velocity variations during, 306, *306*
Long bone, 32, *33, 34, 36*
Longitudinal arch, 214, 215, *216*
Longitudinal axis, 18, *18*
Lordosis
  cervical, 260
  lumbar, 259
Low back pain
  etiology, 264, *266,* 266–268, *268*
  prevention, 264, *265*
Lower extremity
  in cycling, 229–230, *229*
  functional anatomy, 172–234
    ankle and foot, 209–225
    knee joint, 193–209
    pelvis and hip complex, 173–193
  joint angles, 324–327, *325–327*
  joint forces, 230–232
    ankle and foot, *231,* 231
    hip, 230
    knee, 230–231
  in locomotion, *226–227,* 226–229
  nerves, *101*
  plyometric exercises, 119–120, *119*
  review questions, 234–236
  in stair ascent and descent, 225–226, *226*
Lumbar disc protrusion, 264
Lumbar lordosis, 259
Lumbar spine
  anatomy and function, *246,* 251
  compression fracture, 40, *41*
  core exercises, 264, *265*
  injuries, 267–268
  loads on, 269–271, *272–273,* 273
  range of motion, 248, *252*
  spondylolisthesis, *266,* 267
Lumbopelvic rhythm, *252,* 252–253, *253*
Lumbosacral joint, 268–269, 270, *273*

M

Magnus effect, 360
Mallet finger, 160
Mass, center of. *See* Center of mass
Materials test measure, 356, *356*
Measurement, units of, 287, 441, 442t–443t
Measurement units, 441, 442t–443t
Mechanical advantage, 410–411
Mechanical energy, 372–375. *See also* Energy
Mechanical properties of body tissues,
    measurement, 26–29, *26–31*

Medial arch, 215, *216*
Medial aspect, 11, *11*
Medial collateral ligament, 147, *149,* 196
    injuries, 207
Medial condyle, 194
Medial epicondyle, 147
Medial epicondylitis, 154, 160
Medial rotation, 13, *14*
Medial tension syndrome, 154
Medial tibial syndrome, 224
Median nerve compression, 160, 162, *162*
Mediolateral axis, 18, *18*
Mediolateral coordinates, 286
Meniscus
    anatomy and function, 48, *196,* 194–196
    injuries, 49, 208
Metacarpophalangeal joints, *155,* 157
Metal, stress-strain curve, 37, *38*
Metaphysis, 32, *36*
Metatarsal fracture, 42, 225
Metatarsalgia, 225
Metatarsophalangeal joint, *210,* 214
Metric system, 287
Microtrauma, 45
Midcarpal joint, *155,* 155–156
Midfoot, articulations, *210,* 214
Midtarsal joint, *210, 213,* 214
Minerals, bone, 30
Modified vector coding technique, 335, *336*
Moment arm, 72, *73, 392, 393*
Moment of force. *See* Torque
Moment of inertia, 395–399, *396*
Moments, upper extremity, 165
Momentum
    angular, 399–400, 401
    of object, 351
    relationship to impulse, *368–370,*
        368–370, 423
Monosynaptic connection, 110
Monosynaptic reflex arc, 112
Morton's neuroma, 225
Morton's toe, 231
Motion
    angular (rotational), 6–7, *7,* 319, *319*
    curvilinear, 284, *284*
    general, 319, *319*
    laws of. *See* Laws of motion
    linear (translatory), 6–7, *6, 7,* 284, *284*
    planar, 285
    projectile. *See* Projectile motion
    range of, 116–117
    straight-line, 284, *284*
Motion segment
    anterior, 243–244, *243–244*
    posterior, 244–245, *246–247*
Motoneurons, 102–108
    alpha, 103
    gamma, 109, *110,* 112–113, *113*
    structure, 102–103, *102*
Motor end plate, 67, *102,* 103
Motor pool, 106
Motor units
    all-or-none principle, 104
    fast twitch glycolytic (type IIb), 104–105,
        105
    fast twitch oxidative (type IIa), 104–105,
        105
    nervous system portion, 103–106, *104*

properties, 106
rate (frequency) coding, 107–108, *108*
recruitment, 106–107, *107*
slow twitch oxidative (type I), *105,* 105–106
structure, 67–68
types, 104–106, *104*
Movement analysis
    anatomical movement descriptors, 10–16,
        10–15, 16t
    core areas of study, 4–9, *5*
    reference systems, 17–21, *18–21*
    terminology, 3–22
Multifidus, 253, 257
Multipennate muscle, *63,* 64
Multivector resistance training, 91
Muscle
    actions. *See* Muscle actions
    anatomy, 65, *66*
    architecture, 63–65,*63*
    atrophy, 93
    attachments, 71–72, *71*
    angle of, 72–74, *73, 74*
    contractility, 61
    contraction, *68,* 68, *69*
    cross-section area, 64, 80
    development of torque, 72, *73*
    elasticity, 62
    extensibility, 61–62
    fatigue, electromyography, 123, *123*
    fibers. *See* Muscle fiber
    force generation. *See* Muscle force
    force-velocity relationships,
        factors influencing, 80–84, *81–84*
    muscle actions and, *79,* 79
    functions, 63
    groups, 63,*63*
    hypertrophy, during resistance training,
        85, *85*
    illustration, *61*
    injury
        cause and site, 91–92, *92*
        inactivity and, 93
        prevention, 92–93
    irritability, 61
    length, whole, 84
    mechanical model, 70–71, *71*
    neural activation, force output and, 82
    one-joint, 78
    origin versus insertion, 71–72, *72*
    pennation angle, 64–65, *64*
    power, 424–426, *426*
    preloading, 82–83, *83, 84*
    roles, 72–79
    sarcopenia, 83–84
    soreness, after exercise, 92
    strengthening. *See* Strength training
    stress-strain curve, *70*
    structure, 62–66, *63–66*
    two-joint, 78–79, *79*
    types, 61
    volume, 64
Muscle actions
    agonists and antagonists, 74, *74*
    comparison, *76,* 76–77, *77*
    concentric, 75, *75*
    eccentric, 75–76, *75*
    examples, 76
    force-velocity relationships and, *79,* 79

isometric, 75, *75*
neutralizers, 74–75, *74*
stabilizers, 74–75, *74*
Muscle contraction
    overview, 67, *68,* 69
    sliding filament theory, 67, *69*
Muscle fiber
    anatomy, 64, *66*
    fast twitch (type II), 65
    force-velocity relationship, 79
    intermediate fast twitch, 65
    length-tension relationship, 81, *81*
    parallel arrangements, 63,*63*
    penniform arrangements,*63,* 64
    slow-twitch (type I), 65
    types, 65, 82
Muscle force, 362, *362*
    electromyography, 122–123, *123*
    generation, 67–71, *68–71*
    neural control, 106–108, *107, 108*
    output, fiber type and, 85
    tendon influences, 70
    transmission to bone, 68–70, *70, 69*
Muscle spindle, 109–113, *110–113*
Musculotendinous unit, 70, *71*
Myelination, 103
Myofascial pain, low back, 267
Myofibril, 65, *66*
Myosin, 66, *66*
Myositis ossificans, forearm, 154
Myotatic reflex, 112, *112*
Myotendinous junction, 68–69

N

Neck, as segment, 10, *10*
Negative acceleration, 299
Negative work, 424, *424*
Nervous system
    control of muscle force, 106–108, *107, 108*
    general organization, *100,* 100–101, *101*
    motoneurons, 102–108
    motor units, 103–106, *104, 105*
    sensory receptors, 108–114
    training adaptations, 114–120
Neural arch, 245
Neuroma, Morton's, 225
Neuromuscular junction, 67, *102,* 103
Neurons. *See also* Motoneurons
    afferent, 110–112
        type Ia primary, 110, *111,* 111–112, *112*
        type II secondary, 110, *111,* 112
    sensory, 101
    structure, *102*
Neutral equilibrium, 419
Neutralizing muscles, 74, *74*
Newton, 350
Newton, Isaac
    law of gravity, 351
    laws of motion, 350–351, *351*
Newton-Euler inverse dynamics approach, 367
Nike Sports Science Laboratory, 356, *357*
Noncardinal plane, 18, *19*
Noncontact force, 352
Nonsupport (swing) phase of gait, 304, *304*
Nuclear bag fiber, 109, *110*
Nuclear chain fiber, 109, *110*
Nucleus pulposus, 243
Nutation, 176, *176*

O

Oblique muscles, 256, 258
Odontoid process, 248
Olecranon fossa, 147
Olecranon process, 147
  traction apophysitis, 154
One-joint muscle, 77–78
Open kinetic chain exercise, 90–91, *91*
Optoelectric systems, high-speed, 284
Ordinate, 285
Origin, of reference frame, 17–18
Osgood-Schlatter disease, 42, 209
Osseous tissue. *See* Bone tissue
Ossification, 34
Osteoarthritis, 52–53, *54*
  ankle joint, 225
  apophyseal joints, 266
  hip joint, 193
Osteoblast, 30
Osteochondral fracture, 225
Osteochondritis dissecans, 154, 225
Osteoclast, 30
Osteocyte, 30
Osteon, 31
Osteoporosis, 35–36
Overhand throwing. *See* Throwing
Overload, progressive, 87
Overuse injuries
  acromioclavicular joint, 142
  elbow, 154
  hand and fingers, 160
  running shoe and, 338
  wrist extensors, 154

P

Pacinian corpuscle, 114, *114*
Parallel axis theorem, 399
Parallel elastic component, Hill muscle
  model, 70, *71*
Parallel muscles, 63, *63*
Passive insufficiency, two-joint muscle,
  78–79
Passive peak, 378
Passive range of motion, 116
Patella, 196, *198*, 198–199
  compression fracture, 40
  injuries, 209
  movements, *200*, 200–201
Patella alta, 199, 209
Patella baja, 199, 208
Patellar tendon, 198
Patellofemoral compression force, 231
Patellofemoral joint, 198, *198*, 198–199
Patellofemoral pain
  extension exercises for, 205
  syndrome, 208
Pelvic complex, 173–193
  hip joint, 178–183, *178–182. See also* Hip
    joint
  injury potential, 192–193
  ligaments, *175*
  muscular actions, 184–188, *185–186*
  pelvic girdle, 173–178
Pelvic girdle, 173–178
  bones, 174
  gender differences, 173, *173*
  injuries, 42, *43*, 192–193
  movements, 176–177, *176–178*

sacroiliac joint, 174–175, *174*, 176–177
  and thigh, combined movement, 176,
    176, 182, 184
  and trunk, combined movement, 176,
    176, 252, 252–253, *253*
Pelvifemoral rhythm, 183
Pelvis, trunk and, combined movement,
  252, 253
Pennation angle, 64, *64*
Penniform muscle, *63*, 64
Perimysium, 65, *66*
Periodization, resistance training, 88
Periosteum, 32, *36*
Periostitis, 224
Peripheral nervous system, 100–101, *101*
Peroneal muscles, 221
Pes anserinus, 204
Pes cavus, 217
Pes planus, 217
Physical inactivity
  bone mineral density and, 35
  effects on muscle, 93
Physiological cross-section area, 64
Pinch, 159
Piriformis syndrome, 193
Pitcher's elbow, 154
Pivot joint, 51, *52*
Pivot point, *268*
Planar motion, 285
Plane(s), anatomical, 18–21, *18–21*
Plane joint, 51, *52*
Plantar fascia, 215–216, *216*
Plantar fasciitis, 224
Plantarflexed first ray, 214, 218
Plantarflexion, 16, *15, 212*, 217, 221
Plantaris, 218
Plastic region, stress-strain curve, 27, *27–28*, 38
Plica, 198
  injury, 209
Plyometric exercise, 83, *84*, 119–120, *119*
Point of application, 348
Point of separation, *358*, 359
Porosity, bone, 30
Position
  anatomical, *11*, 11–12
  angular, 328
  linear, 293
Position-time curve, *296*, 296–297, *297*
Positive acceleration, 299
Positive work, 424, *424*
Posterior aspect, *11, 12*
Posterior cruciate ligament, 196, *197, 198*
  injuries, 207
Posterior longitudinal ligament, 244
Posterior motion segment, 244–245,
  246–247
Posture
  deviations, 259–260, *259*
  sitting, 258–259
  standing, 258
  working, 259
Potential energy, 372–375
Power
  angular, 424–426, *425*
  muscle, 424–426, *425*
  and work, 80, 371–372
Power grip, 159, *159*
Precision grip, 159, *159*
Prehensile grip, 159

Preloading, 82–83, *83, 84*
Pressure, 376–377, *377*
Prestretch
  neural effects, 115
  technique, 82, *84*, 118
Progressive overload, 87
Projectile motion
  energy changes, 373, *373*
  equations of constant acceleration,
    312–313, *313*
  factors influencing, 309–312, *309–311*, 310t
  gravity force on, 309
  linear kinematics, 309–312, *309–311*, 310t
  optimizing, 312
  trajectory, 309, *309*
Projection angle, *309*, 309–310, 310t
Projection height, 311–312, *311*
Projection velocity, 310–311, *311*
Pronation, 15, *15*, 16, 147
  foot, 212, 213, *213*, 217, 221, *221*
  measurement, 326, *327*
Propelling phase, 379
Proprioceptive neuromuscular facilitation,
  117–118, *118*
Proprioceptors, 109
Propriospinal reflex, 109, *110*
Propulsive drag, 359
Propulsive lift, 360
Proteoglycan, 48
Protraction, 14, *15*, 133, 134
Proximal aspect, 11, *11*
Psoas, origin versus insertion, 72, *73*
Pubic ligament, 174
Pubic symphysis, 174
Pubis, 174
Pubofemoral ligament, 180
Pumping technique, wheelchair propulsion,
  308, *308*

Q

Q-angle, 198–199, *199*
Quadrants, 285, *286*
Quadratus lumborum, 256
Quadriceps femoris, 201, 204
  injuries, 207
Qualitative analysis, 4
Quantitative analysis, 4

R

Radial acceleration, 333, *333*
Radial collateral ligament, 147, 149, *149*
Radial flexion, 15
Radian, *320*, 320–321
Radiate muscle, 63, *63*
Radiocarpal joint, 155, *155*
Radiohumeral joint, 147, *148*
Radioulnar joint, 147, *148*
  distal, *155*, 155
  injuries, 154
Radius, injuries, 160
Radius of gyration, 398, 398t
Radius of rotation, 331
Range of motion, 116–117
Rate coding, motor units, 107–108, *108*
Rearfoot angle
  description, 326–327, *327*
  knee angle and, 334–335, *335, 336*
  during walking and running, 338

Rearfoot varus, 217
Reciprocal inhibition, 112
Rectangular reference system, 285
Rectification, 121, *122*
Rectilinear motion, 284, *284*
Rectus femoris, 78, *78*, 201
Reference frame, absolute versus relative, 17–18, *17*
Reference positions, starting, 10, *10*
Reference systems, 17–21, *18–21*
  2D, 285, *285*, 286, *286*
  3D, 286, *286*
  kinematic data collection, 284–287, *285–287*
  planes and axes, 18–21, *18–21*
  relative versus absolute, 17–18, *17*
Reflex
  crossed extensor, 109, *110*
  flexor, 109, *110*
  inverse stretch, 113
  labyrinthine righting, 109, *110*
  myotatic, 112, *112*
  propriospinal, 109, *110*
  simple, 109, *109*
  stretch, 109, 112, *112*
  supraspinal, 109, *110*
  tonic neck, 109, *110*
Reflex arc, monosynaptic, 112
Reflexes, *109*, 110
Relative angle, 10, *11, 323*, 323–324, *324*
Relative reference frame, 17, *17*
Remote angular momentum, 401
Renshaw cell, 108
Repolarization, 67, 105
Resistance
  air, 357
  fluid, 357–361, *357–362*
Resistance arm, 410
Resistance force, 410
Resistance training. *See* Strength training
Resting potential, 67
Resultant, 286
Retraction, 14, *15*, 133, 134
Retrocalcaneal bursitis, 224
Retroversion, femoral, *181*, 182
Review questions
  angular kinematics, 342–345
  angular kinetics, 434–437
  basic terminology, 22–23
  linear kinematics, 315–317
  linear kinetics, 385–387
  lower extremity, 230–232
  muscular system, 94–96
  nervous system, 124–126
  skeletal system, 55–56
  trunk, 275–277
  upper extremity, 167–169
Revolution (about a circle), 320, *320*
Riemann sum, 302
Right-hand rule, 327, *327*
Right lateral flexion, 14, *15*
Right rotation, 14, *14*
Rotation, 13–14, *14*
  axis (axes) of, 6, *7, 17, 18, 319, 319*
  downward, 14, *15*
  external, 13, 14, *14, 15*
  internal, 13, 14, *14, 15*
  radius of, 331

  upward, 14, *15*
  zero momentum, 400
Rotational friction, 355
Rotational kinetic energy, 426–427, *426*
Rotational motion, 6–7, *7*, 319, *319*
Rotator cuff muscles, 135, 139, *140*
  injuries, 145–146
Ruffini ending, 114, *114*
Running
  angular kinematics, 336, *337*
  angular kinetics, 427–429, *428–429*
  angular momentum requirements, 401, *401–402*
  energy transfer during, 374, *374*
  joint forces, *231*, 231
  linear kinematics, 302–306, *302–307*, 303t
  linear kinetics, *377–380*, 377–381
  lower extremity in, *226*, 226–229
  motor unit recruitment, 106–107
  multiple plane movements, 20
  pelvis and trunk movement relationship in, 252–253, *253*
  rearfoot angle-time graph, 327, *327*
  rearfoot motion, 338
  stress-strain relationship during, 28, *28*
  stride parameters, *302*, 302–303, 303t, *303*
  trunk muscles in, 269
  velocity curve, 304–305, *304, 305*
  velocity variations during, 306, *306*
Running shoes
  knee-rearfoot angle relationships, 334–335, *335*
  overuse injuries and, 338
Running speed, 302, 303

S

Sacral extension, 176, *176*
Sacral flexion, 176, *176*
Sacroiliac joint
  anatomy, 174, *174*
  injuries, 192
  movements, 176–177, *176*
Sacroiliitis, 192
Sacrum
  injuries, 192
  movements, 176–177, *176*
Saddle joint, 51, *52*
Safety factor, stress-strain relationship, 28
Sagittal plane, 18, *18*
Sagittal plane movements, 19, *19*
Sarcolemma, 65, *66*
Sarcomere, 66, *66*
Sarcopenia, 83–84
Sarcoplasm, 66, *66*
Sarcoplasmic reticulum, 66, *66*
Scalars, kinematic analysis, 287
Scaphoid, *155*, 155–156
Scapula
  anatomy and functional characteristics, 133, *134*
  bursitis, 145
  injuries, 145
  movements, 133–134, *134*
    descriptors, 14, *15*
    relationship to arm movements, 137, *138*
Scapulohumeral rhythm, 137, *138*

Scapulothoracic joint, *134,* 133–134
Scheuermann's disease, 267
Schmorl's nodes, 266
Schwann cell, *102,* 103
Sciatica, 267
Scoliosis, 260
Screw-home mechanism, 200–201
Secant line, 296, *296*
Segment angle, 17, *17,* 321
Segment names, 10, *10*
Sensory neurons, 101
Sensory receptors, 108–114
  Golgi tendon organ, 113, *114*
  muscle spindle, 109–113, *110–113*
  reflexes and, 108, *109*, 110
  tactile and joint, 114, *114*
Separated flow, 358, *358*
Separation, point of, 358, *358*
Series elastic component, Hill muscle model, 71, *72*
Sesamoid bone, *33*
Shear forces, 40, 40t, 42–43, *43*, 375
Shear fracture, 42–43
Shear strain, 39, *39*
Shear stress, 39, *39*
Shin splints, 42
Short bone, 32–33, *33*
Shoulder girdle, 132–146
  conditioning exercises, 142, *143–144*
  injury potential, 142, 145–146
  isometric force output, angle and, 74, *74*
  joint forces and moments, 165
  joints
    anatomy and functional characteristics, *132–137*, 132–137
    combined movement characteristics, 137–138, *138*
  ligaments, 132, *133,* 133–134, 135
  movement descriptors, 14, *15*, 16t
  muscle strength, 142
  muscular actions, 138–141, *139–141*
    abduction or flexion, 138, *139–140*
    adduction or extension, 140, *140*
    horizontal, 141
    internal and external rotation, 141, *141*
  range of motion, 137
  as segment, 10, *10*
Shoulder joint. *See* Glenohumeral joint
SI (Système International d'Unités) units, 287, 441, 442t–443t
Simple reflex, 109, *109*
Sit-reach test, 264
Sitting posture, 258–259
Size principle, motor unit recruitment, 106
Skeletal muscle. *See* Muscle *entries*
Skeletal system
  bones, 29–47. *See also* Bone
  bony articulations, 49–53, 53t. *See also* Joint(s)
  cartilage, 47–48
  injuries, 47t
  ligaments, 48–49
Ski boot fracture, 43, *44*
Sliding filament theory, 67, *69*
Slipped capital femoral epiphysis, 192
Slope, velocity and, 293, *294*
Snapping hip syndrome, 193
Soccer kick, 360, *361*

Soft tissue, flexibility and, 116
Soleus, 218
Soma, 102, *102*
Speed
    angular, 328–330
    linear, 292
Spinal cord, 100, *100*
Spinal nerves, 100–101, *101*
Spinal stabilization, 257
Spine. *See* Vertebral column
Spinous process, 244
Spiral fracture, 44, *45*
Spondylolisthesis, 42, *266*, 267
Spondylolysis, 41, *266*, 267
Squat, 205
Stability, factors influencing, 418–419, *419*
Stabilizing muscles, 74, *74*
Stable equilibrium, 418
Stair ascent and descent, lower extremity in,
    225–226, *226*
Standing posture, 258
Standing toe touch, 256
Starting position, anatomical, 10, *10*
Static stretching, 116
Statics
    angular motion, 415–421, *415–421*
    applications, 420, *420–421*
    versus dynamics, 9
    linear motion, 364–366, *365–367*
Step, definition, 302, *302*
Sternoclavicular joint
    anatomy and functional characteristics,
        132, 132–133
    injuries, 142
Stiff material, stress-strain curve, 29, *29*
Stiffness
    bone tissue, 38, *38*
    calculation, 27
Straight-line motion, 284, *284*
Strain. *See also* Stress-strain *entries*
    bone, 45–46, *46*, 47t
    measurement, 27
    normal, 39, *39*
    residual, 27
    shear, 39, *39*
Strain energy, *372*
Strap muscle, 63, *63*
Streamlining of shape, 359, *359*
Strength
    bone tissue, 36, *37*
    development, 43, *43*
    definition, 85
Strength training, 84–91. *See also* Weight
    lifting
    for ankle/foot, 221–222, *223*
    for elbow and forearm, 152, *153*
    for fingers and wrist, 159, *159*
    for hip joint muscles, 189, *190–191*
    intensity, 86
    for knee joint, *196*, 204–205
    modalities
        closed and open kinetic chain, 91, *91*
        functional training, 90–91
        isokinetic, 90, *90*
        isometric, 89
        isotonic, *89*, 89–90
        multivector training, 91
    neural adaptations, 114–115, *115*

for nonathlete, 88–89
principles, 85–89, *87*, 88t
program components, 85
rest intervals, 87
for shoulder muscles, 142, *143–144*
skeletal muscle hypertrophy during, 85, *85*
specificity, 85–86, *87*, 115
strength progression during, *86*
for trunk, 260–264, *261–263*
    core training, 264, *265*
    extensors, 260–263, *261–262*
    flexibility and, 264
    flexors, 260, *260, 262*
    lateral flexors, *262*, 263–264
    rotators, *262*, 263–264
    volume, 87–88
Stress
    measurement, 27–28
    normal, 39, *39*
    shear, 39, *39*
Stress fracture, 45–47, *46*, 47t
Stress-strain curve
    bone, 37, *37, 70*
    bone vertebral segments, 26, *27*
    compliant, stiff, and brittle materials, 29, *29*
    elastic material, 27, *28*
    elastic-plastic regions, 27, *27*, 38
    energy lost (hysteresis), 29, *29*
    energy stored, 28, *28*
    failure point, 29, *29*
    ligaments, 48, *49*
    muscle, *70*
    tendon, *70*
    viscoelastic material, 28–29, *29*
    yield point, 27–28, *27*
Stress-strain relationship
    during jogging, 28, *28*
    safety factor, 28
Stress-strain structural analysis, 26–29, *26–29*
Stretch-contract cycle, 82, *83, 84*, 118
Stretch reflex, 109, 112, *112*
    inverse, 113
Stretching, flexibility and, 116–118, *117*
Stretching exercises, 116–120
    for ankle/foot, 222, *223*
    ballistic, 116
    for elbow and forearm, 152, *153*
    for fingers and wrist, 159, *159*
    for hip joint muscles, 189, *190–191*
    for knee joint, *196*, 204–205
    overview, 116–118, *117*
    proprioceptive neuromuscular facilitation
        after, 117–118, *118*
    for shoulder muscles, 142, *143–144*
    static, 116
    for trunk, 260–264, *261–263*
        core training, 264, *265*
        extensors, 260–263, *261–262*
        flexibility and, 264
        flexors, 260, *261, 263*
        lateral flexors, *262*, 263–264
        rotators, *262*, 263–264
Striated muscle, 61
Stride
    definition, 302, *302*
    frequency, oxygen consumption and,
        302–303, *303*
    parameters, *302*, 302–303, 303t, *303*

Structural analysis
    stress-strain properties, 26–29, *26–29*
    types of materials, 28–29, *29*
Subacromial bursae, 135
Subacromial bursitis, 146
Subacromial impingement syndrome, 146
Subtalar joint
    anatomy and function, *210*, 212–213,
        212, 213
    forces, 231
    knee and, combined movements, 217
Superior aspect, 11–12, *11*
Supination, 15, *15*, 16, 147
    foot, 213, *213*
    measurement, 326, *327*
Support (stance) phase of gait, 304, *304*
Supraspinal reflex, 109, *110*
Supraspinous ligament, 251
Surface drag, 358
Swimming, 162
Synapse, *102*, 103
Synarthrodial (fibrous) joint, 51, *53*
Synchronous movements, 108
Synovial joint. *See* Diarthrodial joint
Synovial membrane, *49*, 50

T

T-tubule, 66, *66*
Tactile sensory receptors, 114, *114*
Talocalcaneal joint. *See* Subtalar joint
Talocrural joint. *See* Ankle
Talofibular ligament, sprain, 224
Talonavicular joint, 213
Talus, 209
    osteochondral fracture, 225
Tangent, 321–322
    inverse, 322
Tangential acceleration, 333, *333, 334*
Tangential velocity, 332, *332*
Tarsometatarsal joint, *210*, 214
Tendon
    characteristics, 68–70, *70*, , *71*
    force-time characteristics, 70
    muscle attachment via, 67, *69*
    stress-strain analysis, 26, *26*
    stress-strain curve, *70*
Tennis elbow, 154
Tennis serve, topspin, trunk muscles in, 269,
    271–272
Tenosynovitis, 160
Tensile fracture, 41–42
Tension forces, *40*, 40t, 41–42, *41*
Tension-length relationship, muscle fiber,
    81–82, *81*
Tetanus, 67, *69*
Thenar eminence, 157
Thigh
    angular kinematics, 329t, 330, *331*
    compartments, 62, *63*
    conditioning exercises, 189, *190–191*
    movement descriptors, 14, *15*, 16t
    movements, *182*, 182
    muscular actions, 184–188, *185–186*
        abduction, 187, *188*
        adduction, 187, *188*
        extension, 184, 186
        flexion, 184
        rotation, 188

pelvis and, combined movement, 176, *176,* 182, 183
as segment, 10, *10*
strength, 188–189
Thoracic kyphosis, 260
Thoracic spine
  anatomy and function, *246, 250,* 248–249, *250*
  connection to ribs, *250*
  injuries, 267
  movements, 248
  range of motion, *252*
Thoracolumbar fascia, 251
Three-point bending load, 43–44, *44*
Throwing
  elbow injuries, 154
  injuries, 146
  multiple plane movements, 20, *21*
  shoulder joint rotation, 141, *141*
  upper extremity muscular contribution, 162–165, *163–164*
Thumb
  carpometacarpal joint, 156–157
  interphalangeal joint, 157
  metacarpophalangeal joint, 157
  movement descriptors, 16t
  muscles, *155,* 159
Tibia, 194, *195,* 209, *210*
  strain rates, 45, *45, 46*
Tibial plateau, 194
Tibial tuberosity
  formation, 41
  tensile forces at, 41–42
Tibialis anterior, 221
Tibiofemoral compression force, 230
Tibiofemoral joint, 194, *195,* 194–196
Tibiofibular joint, 199–200, *199,* 209
Tibiotalar joint, 209
Time domain, electromyography, 121, *122*
Timing factors, kinematic analysis, 287
Toe(s)
  Morton's, 231
  movement descriptors, 16t
Toe touch, standing, 256–257
Toeing-in, 181, *182*
Tonic neck reflex, 109, *110*
Torque, 72, *73*
  acting on system, representation, 413, *414*
  calculation, 393–394
  characteristics, 392–394, *393–395*
  definition, 392
  effects
    at instant in time, 414–422
    over a distance, 423–424
    over a time period, 422–423
  special applications, 424–427
  types, 413, *413, 414*
  Z-axis, 392, *393*
Torque arm, 392, *393*
Torsional forces, *40,* 40t, 44, *45*
Torsional fracture, 44, *45*
Total body center of mass, *407,* 407–410, *408,* 409t
Towed sled device, 356, *356*
Trabeculae, *32,* 32
Traction apophysitis, olecranon process, 154

Training. *See also* Exercise
  neural adaptations, 114–120
  resistance. *See* Strength training
Trajectory, 309, *309*
Translation (translatory motion). *See* Linear (translatory) motion
Translational friction, 355
Transverse abdominis, 256
Transverse arch, 214, 216, *216*
Transverse plane, 18, *18*
Transverse plane movements, 19, *21*
Transverse process, 245
Transverse tubule, 66, *66*
Traumatic fracture, 45
Trigger finger, 160
Trigonometric functions, 445–448, *446–447,* 446t–447t
Trochlea, 147
Trochlear notch, 147
Trunk, 241–275
  aging effects, 268–269
  conditioning exercises, 260–264
    core training, 264, *265*
    extensors, 260–263, *261–262*
    flexibility and, 265
    flexors, 260, *261, 263*
    lateral flexors, *262,* 264–265
    rotators, *262,* 264–265
  contribution to sports skills or movements, 269, *271–272*
  injury potential, 264, *266,* 268, 266–268
  joint forces, 269–271, *273–274,* 273
  movement descriptors, 16t
  muscular actions, 253–257, *254–255*
    extension, 253
    flexion, 255–257
    lateral flexion, 257
    rotation, 257
  pelvis and, combined movement, 176, *176,* 252, 252–253, *253*
  range of motion, *252,* 251–252
  review questions, 275–277
  as segment, 10, *10*
  strength, 257
  vertebral column, 242–253. *See also* Vertebral column
Turbulent flow, 358, *359*
Twitch, 67, *69*
Two-joint muscle, 78–79, *79*
  injury risk, 91, *92*

U

Ulnar collateral ligament, 147, *149*
Ulnar flexion, 15
Ulnar nerve injury, 162
Ulnohumeral joint, 147, *148,* 149
Unipennate muscle, *63,* 65
Units of measurement, 287, 441, 442t–443t
Unstable equilibrium, 418
Uphill walking, 338
Upper extremity
  functional anatomy, 131–167
    elbow and radioulnar joints, 146–154
    shoulder complex, 132–146
    wrist and fingers, 154–162
  in golf swing, *164,* 164–165
  joint forces and moments, 165

  nerves, *101*
  in overhand throwing, 162–165, *163–164*
  plyometric exercises, 119–120, *119*
  review questions, 167–169
Upward rotation, 14, *15*

V

Valgus, 180, *181,* 196
Varus, 180, *181*
Vastus medialis, 201
Vectors. *See also* Force(s)
  addition, 291, *291*
  angular motion, 327, *327*
  combination, 292–293
  fluid resistance, 357–361, *357–362*
  force, composition and resolution, *349*
  kinematic analysis, 287–290, *291, 292*
  multiplication, 291, *291*
  polarity, 327, *327*
  resolution, 291, *291*–292, *292*
Velocity
  angular, 328–330
  first central distance method of calculating, *294,* 294–295
  graphical example, 296–297, *297*
  instantaneous, 296, *296*
  linear, 292–297, *293–294, 296–297,* 293t
  numerical example, 295, 295t, *296*
  projection, 310–311, *311*
  rate of change. *See* Acceleration
  relationship between angular and linear, 331–332, *332, 333*
  slope and, 293, *294*
  tangential, 332, *332*
Velocity curve, locomotion, 304–305, *304, 305*
Velocity-time curve, 296, 296–297, *297*
Ventral aspect, *11,* 12
Vertebral body, 243
Vertebral column, 242–253
  apophyseal joints, 245
  cervical region, *247,* 245–251, *247, 249, 250*
  conditioning exercises, 260–264, *261–263*
  curvature, *242*
  ligaments, 244–245, *246–247*
  loads on, 269–271, *273–274,* 273
  lumbar region, *249, 250,* 251
  motion segment
    anterior, 243–244, *243–244*
    posterior, 244, 245, *246–247*
  muscular actions, 253–257, *254–255*
  posture
    deviations, 259–260, *259*
    sitting, 258–259
    standing, 258
    working, 259
  range of motion, *252*
  regional variations, *249, 250*
  stabilization, 258
  strength, 257
  thoracic region, 248–249, *249, 250*
  total, movement characteristics, 251–252, *252*
Vertebral foramen, 245
Vertex, 319, *319*
Vertical coordinates, 285, 286

Vertical jumping, 369–370, *369–370*
Video, high-speed, 284
Viscoelastic material, stress-strain curve, 28–29, *29*
Viscosity, fluid, 357
Viscous drag, 358

W

Walking
  absolute angle calculations, *322*, 322–323, 323t
  absolute angle data for thigh during, 329–330, 329t, *330*
  angular kinematics, 336, *337*
  angular kinetics, 427–429, *428–429*
  energy components during, 427, *427*
  hip, knee, and ankle angles during, *326*
  joint forces, 231, *231*
  linear kinematics, 302–306, *302–307*, 303t
  linear kinetics, *377*, 377–381, *378*
  lower extremity in, *227*, 228
  motor unit recruitment, 106–107, *107*
  rearfoot motion, 338
  stride parameters, *302*, 302–303, 303t, *303*
  support and double support during, 304, *304*

trunk muscles in, 269
  two-joint muscle actions, 78, *78*
Warmup, neural effects, 115
Weight lifting. *See also* Strength training
  compression fracture, 41
  force-velocity relationship, 79
  free body diagram, 420, *420*
  proper technique, 259, 268, *268*
  training program, volume of work, 88
Weight of object, 352
Wheelchair propulsion
  angular kinematics, 340–341, *340*
  angular kinetics, 431–432, *431–432*
  cycle parameters, 307–308, *308*
  linear kinematics, 307–308, *308*
  linear kinetics, 382–383, *383*
  propulsion styles, 307–308, *308*
Whiplash, 267
Wolff's law, 34
Work
  angular, 423–424, *424*, 427
  external, 375
  internal, 374
  mechanical, 371
  power and, 371–372
  relationship to energy, 373–375, *374*, 427

Working posture, 259
Wrist
  conditioning exercises, 159–160, *161*
  injury potential, 154, 160, 162, *162*
  joints
    anatomy and functional characteristics, 155–157, *155–156*
    combined movements, 157
  ligaments, *155*
  movement descriptors, 15, *15*
  muscular actions, *155–156*, 157–159, *158*
  range of motion, 156

X

X, y, z space, 285
X-y plane, 285

Y

Yield point, 29, *29*

Z

Z-axis, 392, *393*
Zero momentum rotation, 400
Zero position, 10, *10*